The Human Brain and Spinal Cord

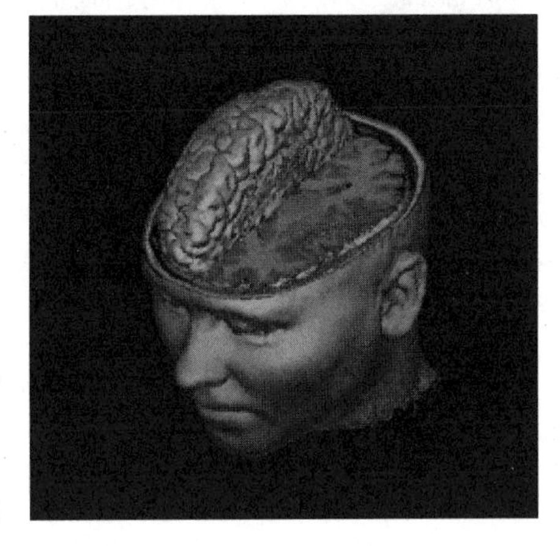

Functional Neuroanatomy and Dissection Guide

Second Edition

Lennart Heimer, MD
Departments of Otolaryngology—
Head and Neck Surgery, and
Neurological Surgery
University of Virginia

With 462 illustrations, mostly in color

Springer-Verlag
New York Berlin Heidelberg London Paris
Tokyo Hong Kong Barcelona Budapest

Lennart Heimer, MD
Departments of Otolaryngology—
Head and Neck Surgery, and
Neurological Surgery
University of Virginia
Charlottesville, VA 22908 USA

Cover illustration: A three-dimensional rendering of the brain in vivo for surgical planning.
From the Neurosurgical Visualization Laboratory, University of Virginia.

Library of Congress Cataloging-in-Publication Data
Heimer, Lennart.
The human brain and spinal cord: functional neuroanatomy and dissection guide / Lennart Heimer.—2nd ed.
p. cm.
Includes bibliographical references and index.
ISBN 0-387-94227-0.—ISBN 3-540-94227-0
1. Brain—Anatomy. 2. Spinal cord—Anatomy. 3. Human dissection.
I. Title.
[DNLM: 1. Central Nervous System—anatomy & histology.
2. Dissection. WL 101 H467m 1994]
QM455.H4 1994
611′.81—dc20
DNLM/DLC
for Library of Congress 94-241

Printed on acid-free paper.

Production managed by Karen Phillips; manufacturing supervised by Rhea Talbert.
Typeset by Best-set Typesetter Ltd., Hong Kong.
Printed and bound by Kingsport Press, Kingsport, TN.
Printed in the United States of America.

9 8 7 6 5 4 3 2 1

ISBN 0-387-94227-0 Springer-Verlag New York Berlin Heidelberg
ISBN 3-540-94227-0 Springer-Verlag Berlin Heidelberg New York

To Hanne-Björg

Preface to the Second Edition

The knowledge of the mammalian central nervous system has increased dramatically during the last decade, which has provided a major impetus for preparing the second edition of *The Human Brain and Spinal Cord*. For the medical profession this has been a revolutionary time, since modern imaging methods have provided unparalleled opportunities for anatomical and functional studies of the human body *in vivo*. It is now essential for the clinician to have an intimate knowledge of anatomy including the functional–anatomical systems in the brain and spinal cord. The new edition of this textbook reflects this progress in the sense that almost all of the chapters have been rewritten and several new figures have been included. Since the book was prepared primarily with the medical student in mind, the *Clinical Notes* and *Clinical Examples*, which are part of every chapter, have been updated and expanded.

As in the previous edition, a major part of the text is dedicated to a guide to the *Dissection of the Brain* and to individual chapters on the *Functional–Anatomical Systems* in the brain and spinal cord. These are complemented by short reviews on the *Development of the Nervous System*, *Neurohistology*, *Neuroanatomic Techniques* and *Neurotransmitters*, in addition to an *Atlas* based on myelin-stained sections with matching magnetic resonance images (MRI) in frontal, horizontal and sagittal planes. The myelin stained sections used in the *Atlas* are part of the Yakovlev Collection, and the photographs were taken by Mr. Paul Reimann, University of Iowa. The MRIs included in the atlas were prepared by Dr. John Snell, the Neurosurgical Visualization Laboratory, Department of Neurosurgery at the University of Virginia, and many of the MRIs showing central nervous system pathology were provided by Dr. Maurice Lipper, the Department of Radiology, also at the University of Virginia. Most of the new drawings were prepared by Ms. Anne Dunn. I am truly grateful for their contributions.

A caveat is in order for the first 5 figures in Chapter 10, which represent cross-sections through different levels of the brainstem. Considering the rapidly expanding reliance on in vivo imaging by the clinicians, figures 10–1 to 10–5 are presented with the posterior parts of the brainstem facing downwards, since this is the way the brainstem images appear in axial MRIs routinely used by neuroradiologists (see Chapter 5). This somewhat unconventional approach, suggested by Dr. Duane Haines, is directly relevant for the transfer of basic science information to clinical practice. All other brainstem sections in the book are presented in the conventional fashion with the posterior (dorsal) surface at the top.

During the preparation of this edition I have asked advice from many friends and colleagues, who have read part of the text, and I would especially like to thank the following people for their help and constructive criticism: Drs. George Alheid, Shirley Bayer, Fritz Dreifuss, Björn Folkow, Michael Forbes, Paul Francel, Charles Gilbert, Nedzad Gluhbegovic, John Gobel, Gunnar Grant, Sten Grillner, Irena Grofova, Patrice Guyenet, Clarke Haley, George Hanna, Glenn Hatton, John Jane, Paul Lambert, Ivan Login, Bruce Masterton, James Miller, Enrico Mugnaini, George Paxinos, Pasko Rakic, Chris Schaffrey, Oswald Steward, Gary Van Hoesen, Jan Voogd, Sarah Winans-Newman, László Záborszky, and Scott Zahm.

Finally, I like to thank the competent staff of Springer-Verlag for carrying out a most difficult assignment.

Lennart Heimer
June, 1994

Preface to the First Edition

This book was written to serve both as a guide for the dissection of the human brain and as an illustrated compendium of the functional anatomy of the brain and spinal cord. In this sense, the book represents an updated and expanded version of the book *The Human Brain and Spinal Cord* written by the author and published in Swedish by Scandinavian University Books in 1961.

The complicated anatomy of the brain can often be more easily appreciated and understood in relation to its development. Some insight about the coverings of the brain will also make the brain dissections more meaningful. Introductory chapters on these subjects constitute *Part I* of the book.

Part 2 is composed of the dissection guide, in which text and illustrations are juxtaposed as much as possible in order to facilitate the use of the book in the dissection room. The method of dissection is similar to dissection procedures used in many medical schools throughout the world, and variations of the technique have been published by several authors including Ivar Broman in the "Människohjärnan" (The Human Brain) published by Gleerups Förlag, Lund, 1926, and László Komáromy in "Dissection of the Brain," published by Akadémiai Kiadó, Budapest, 1947. The great popularity of the CT scanner justifies an extra laboratory session for the comparison of nearly horizontal brain sections with matching CT scans.

Since there is a tendency to rely heavily on expensive audiovisual aids in medical education, it seems especially important to promote the dissection of the human brain. No brain model or TV movie can match the efficiency of brain dissections in teaching the gross anatomy of the brain, and it is usually possible to secure a certain number of human brains at most medical schools. It seems important to emphasize, however, that a systematic description of the topographic anatomy of the brain is all the more important if brains cannot be obtained for dissection. With this in mind, the dissection guide has been richly illustrated and can be used as an introductory gross anatomy text without dissecting the brain.

Part 3 constitutes an illustrated account of the functional anatomy of the major parts and systems of the brain and spinal cord. Many of the schematic illustrations in part 3 have been modeled after the more elaborate drawings in Part 2, and they can be best appreciated on the basis of a reasonable understanding of the gross anatomy of the brain as outlined in the dissection guide. Considering the clinical importance of cerebrovascular diseases, the blood supply of the brain is discussed separately in the last chapter.

Physiologic, chemical and pharmacologic aspects have been briefly discussed whenever appropriate, but no effort has been made to introduce the basics of these subjects. The clinical relevance of neuroanatomy has been emphasized by including "Clinical Notes" at the end of each chapter, and some commonly seen disorders of the nervous system have been presented in the form of "Clinical Examples" in order to prepare the medical student for the practice of medicine. Many of the "Clinical Examples" have been illustrated with appropriate CT scans or angiograms. The reading lists at the end of the chapters do not pretend to be complete; they are presented as an encouragement for further studies, and they include some of the larger medical school texts, in which the various subjects are more fully discussed. Figures of the distribution of the peripheral nerve, finally, have been included in an appendix.

It is my hope that the book will serve a useful purpose in almost any type of neuroscience education. It should fortify and complement related studies in basic and clinical neuroscience.

Lennart Heimer

Acknowledgments to the First Edition

Many of my friends and colleagues have been kind enough to read one or several of the chapters, and it is with a great deal of gratitude that I acknowledge the help of Drs. Shirley Bayer, Theodore Blackstad, Anders Björklund, Irving Diamond, Björn Folkow, Ann Graybiel, Gunnar Grant, Gary Van Hoesen, Leonard Jarrard, Anders Lundberg, Eugene Millhouse, Enrico Mugnaini, Ulf Norsell, Alan Peters, Jim Petras, Gordon Shepherd, Jan Voogd, and Torsten Wiesel.

Colleagues at the University of Virginia, including Drs. George Alheid, Robert Cantrell, Fritz Dreifuss, John Jane, Ivan Login, Leon Morris, Edwin Rubel, and Richard Winn, have also been very helpful, and I would like to acknowledge specifically Dr. James Bennett, who was kind enough to prepare the "Clinical Examples" and Dr. James Q. Miller, who revised the "Clinical Notes."

The carbon dustings in the dissection guide were prepared by Mrs. Florence Kabir, who spared herself no effort in preparing for the final artwork. Some of the other drawings were prepared by Mrs. Jane Gordon and Mr. János Kálmánfi.

The photographs from various planes of myelin stained sections of the human brain, which appear in the dissection guide, were taken by Mr. Paul Reimann and Dr. Gary Van Hoesen from material in the Yakovlev Collection at the Armed Forces Institute of Pathology in Washington. I would like to thank Dr. Paul Yakovlev for collecting this magnificent material and Mr. Mohamad Haleem for making it easily accessible.

Drs. Victor Haughton and Michael Wolff of the Radiology Department of the Medical College of Wisconsin supplied most of the CT scans, whereas Dr. Leon Morris of the Radiology Department of the University of Virginia furnished the angiograms.

It has been a great pleasure to collaborate with the competent staff of Springer-Verlag in New York.

Contents

1

INTRODUCTION

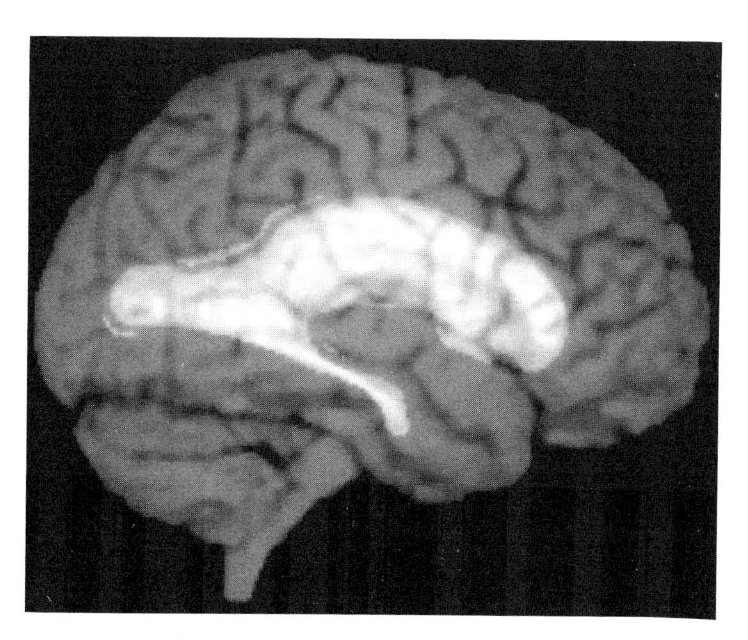

Three-dimensional magnetic resonance image (MRI) of the brain and lateral ventricle in a living person with no apparent neurologic symptoms. (Courtesy of Drs. Daniel R. Weinberger and Douglas W. Jones, National Institute of Mental Health.)

1

Basic Design and Terminology

In freshly cut sections of the brain and spinal cord some regions display a white glistening color, others a grayish color. The white matter represents collections of bundles containing long nerve fibers, which are surrounded by white glistening myelin sheaths. Nerve cell bodies, on the other hand, tend to aggregate in a superficial cortical sheath or in subcortical nuclei, referred to as gray matter.

If the differentiation between gray and white matter is accentuated by appropriate staining methods, it is easily appreciated that the central nervous system is composed of a large number of more or less distinct regions. Although it is important to have an understanding of these various subdivisions and their topography, such knowledge is not very useful by itself. A more meaningful picture will emerge from the knowledge of how the various subdivisions relate to each other and how groups of neurons are interconnected to form anatomic pathways and functional systems.

The Central and the Peripheral Nervous Systems

On the basis of gross anatomic features, the nervous system can be divided into two parts: the *central nervous system* (CNS), which consists of the brain and spinal cord (Fig. 1–1A), and the *peripheral nervous system* (PNS), which consists of the cranial and spinal nerves and their ganglia. The peripheral nerves connect the CNS with the sense organs and the effector organs (muscles and glands). Although it is useful to subdivide the nervous system in this manner, one should remember that functional–anatomic circuits and systems pay little attention to the boundaries between the central and peripheral nervous systems. For instance, the cranial and spinal nerves are described as part of the PNS, but their cells of origin (motor neurons) or their terminal branches (sensory neurons) are situated within the CNS.

Both the central and the peripheral nervous systems contain somatic as well as autonomic parts. The somatic parts of the nervous system control movements and innervate sensory organs, whereas the autonomic parts are concerned with the innervation of the visceral organs. Although the *autonomic nervous system* was originally conceived of as a peripheral system, albeit with cells of origin located in the spinal cord and brain stem, many other areas in the CNS, including the hypothalamus, amygdaloid body, and cerebral cortex, are also involved in the control of the visceral organs.

The Neuron

The nervous system contains about 100 billion nerve cells or neurons, and, in addition, an even larger number of neuroglial cells that subserve various supportive, metabolic, and phagocytic functions.

The neurons are specialized to receive, conduct, and transmit nervous impulses, and they have certain characteristics in common, even though they vary widely in shape and size. Two types of processes extend from the cell body (Fig. 1–2). The receiving processes, the *dendrites*, have a broad base and taper away from the cell body; they branch in the immediate neighborhood of the cell body. There usually are many dendrites per neuron. The *axon*, of which there is only one per cell, conducts the impulses away from the cell body. Axons vary greatly in length and diameter. The thicker axons, which conduct impulses more rapidly than the thinner ones, are insulated from the environment by a lipoprotein sheath, the *myelin sheath*. Thin axons are either unmyelinated or thinly myelinated. The axon loses its myelin sheath at its destination, and divides into several small branches, which typically end as swellings referred to as *axon terminals* or *boutons*. The boutons establish contacts with other neurons or with an effector organ. The site of interneuronal contact, i.e., where the impulses are transmitted from one cell to another, is called a *synapse* or *synaptic junction*. The neuron and its processes are discussed in more detail in Chapter 6.

Gray and White Matter

The distribution of gray and white matter can be easily appreciated in macroscopic sections through the brain and spinal cord (Figs. 1–1B and C). The basis for this distinction relates to the structure of the nerve cell. As already indicated, many axons are surrounded by a myelin sheath, which has a white glistening color; consequently, those parts that contain many myelinated axons are white, i.e., *white matter*. The parts that contain aggregations of nerve cell bodies embedded in a network of delicate nerve processes have a gray color, hence the term *gray matter*. There are two major territories of gray substance in the brain, the cerebral cortex on the surface and large subcortical nuclei, e.g., *caudate nucleus*, *putamen*, *globus pallidus*, and *thalamus*, in the interior. The gray substance in the spinal cord is located in the center and has a typical H-shaped appearance in transverse sections. It is surrounded by white matter, which contains long ascending and descending pathways.

Collections of Nerve Cell Bodies

The *cortex* is the superficial coat of gray matter in the cerebral hemispheres and in cerebellum. The nerve cell bodies in the cortex are arranged in more or less well-defined laminae or layers. Groups of nerve cell bodies in other parts of the brain and spinal cord are referred to as *nuclei*, or *columns* if they occur in long rows as they often do in the spinal cord. Note that the term column is also used for the white matter in the dorsal part of the spinal cord (*see below*). Accumulations of nerve cell bodies outside the CNS are called *ganglia*.

Collections of Nerve Fibers

A collection of nerve fibers with a common origin and destination constitutes a *tract*, e.g., the corticospinal (or pyramidal) tract. A tract does not necessarily form a well-defined bundle, since the fibers of a tract often intermingle with fibers of neighboring pathways.

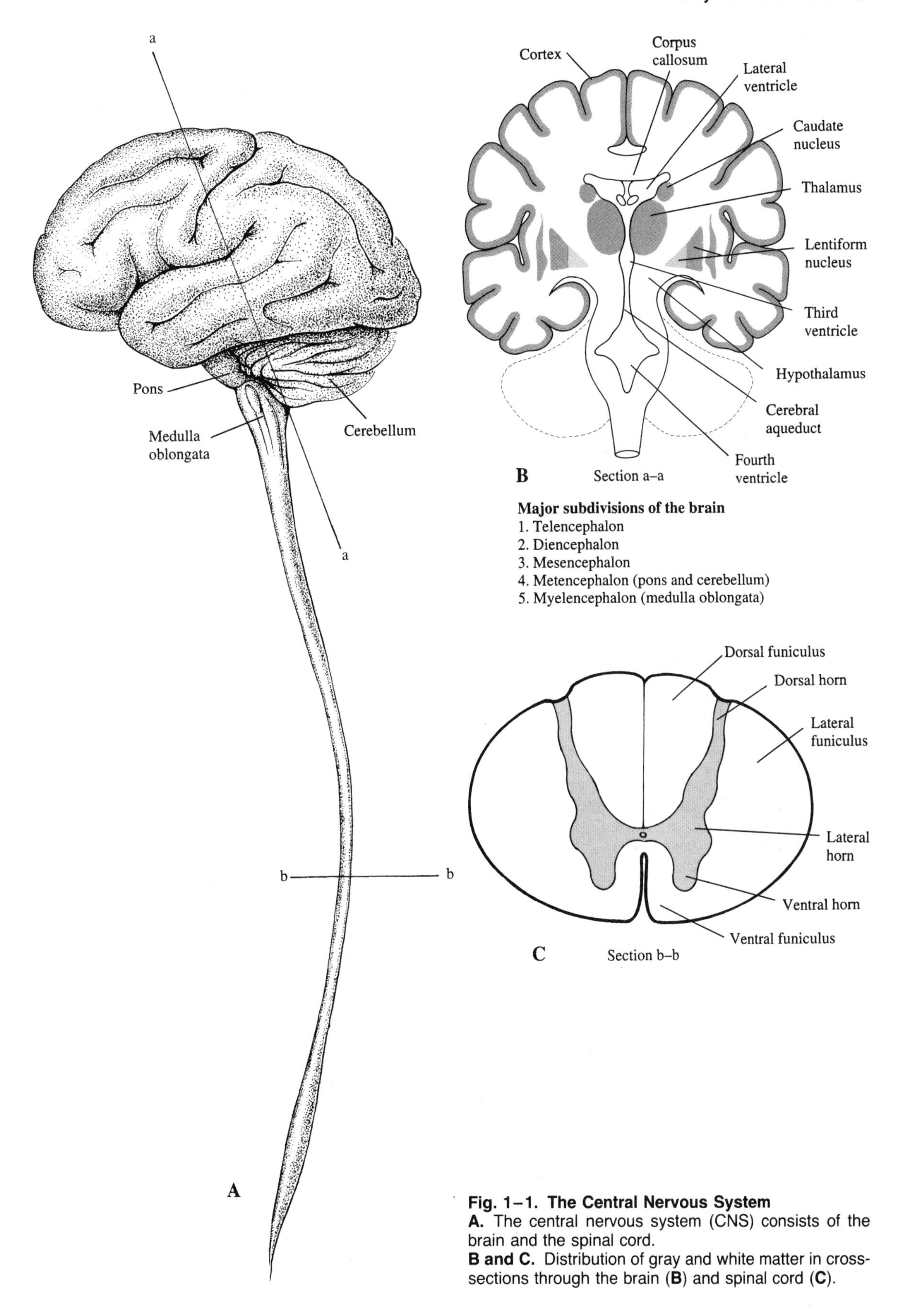

Fig. 1–1. The Central Nervous System
A. The central nervous system (CNS) consists of the brain and the spinal cord.
B and C. Distribution of gray and white matter in cross-sections through the brain (**B**) and spinal cord (**C**).

A distinct collection of nerve fibers is referred to as a *fasciculus*, *peduncle*, or *brachium*. Such collections of fibers often contain more than one tract. The term *lemniscus* is specifically used for ascending sensory fiber bundles in the brain stem. The white matter of the spinal cord, which contains the long ascending and descending tracts, is subdivided into *ventral* (*anterior*),[1] *lateral*, and *dorsal* (*posterior*) *funiculus* (Fig. 1–1C). The dorsal funiculus, however, is often referred to as the dorsal column, especially in the physiologic literature.

Fibers connecting similar areas on the two sides of the brain form well-defined bundles known as *commissures*. Many fiber bundles cross the midline in their course from one level of the nervous system to another. Such crossing fibers form a *decussation*.

Nerves, *nerve roots*, *nerve trunks*, and *rami* are examples of bundles of nerve fibers in the PNS.

Nervous Pathways

Efferent pathways project away from the region under consideration. *Afferent pathways* project to the region under consideration. *Extrinsic pathways* project into or out of the region under consideration. *Intrinsic pathways* are confined to the region under consideration.

Major Subdivisions of the Brain

The brain is subdivided into five major divisions: telencephalon, diencephalon, mesencephalon, metencephalon, and myelencephalon (Fig. 1–1B).

Telencephalon

The telencephalon (endbrain) consists of the two *cerebral hemispheres*, which form the largest part of the human brain. The hemispheres are connected by a massive bundle of commissural fibers, the *corpus callosum*. Each hemisphere is covered by a thin layer of gray substance, the *cerebral cortex*, which is heavily folded. The folds are called convolutions, or *gyri*, and the grooves are referred to as *sulci*, or *fissures*, if they are very deep. In the interior of each hemisphere is a centrally placed cavity, the *lateral ventricle*, and a large mass of white substance as well as several large nuclei referred to as the *basal ganglia*. Although the *thalamus* in the diencephalon (*see below*) is sometimes included in the basal ganglia, the term is most commonly used for the two telencephalic nuclei, the *caudate* and the *lentiform nuclei*, and some closely related brain stem nuclei, known to be of special importance for control of movements. The *claustrum*, whose functions are unknown, is sometimes included in the term basal ganglia.

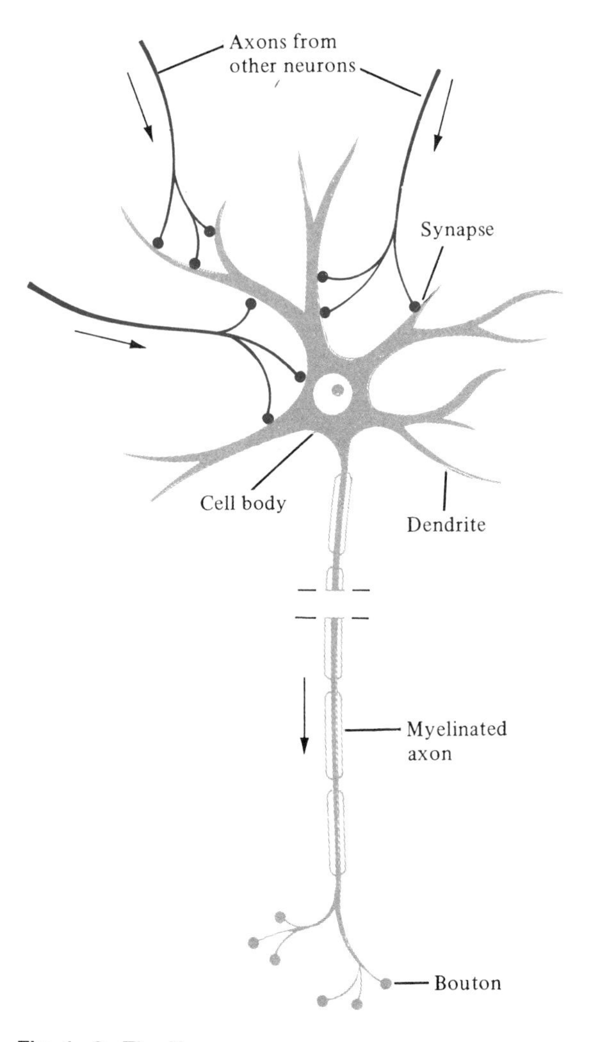

Fig. 1–2. The Neuron
Although neurons vary widely in size and shape, they have certain features in common. Most neurons are multipolar, i.e., they have several processes. The dendrites receive impulses from other neurons whereas the axon, of which there is only one to each cell, conducts the impulses away from the cell body.

Diencephalon

The diencephalon is divided into two halves by a slit-formed median cavity, the *third ventricle*, which is continuous via the small *interventricular foramen* (*of Monro*) with the lateral ventricle in each

1. The terms explaining position can be confusing. Depending on context, the term *anterior* can be synonymus with both *rostral* (from Latin, *rostrum* = beak) and *ventral* (from Latin, *venter* = belly), whereas the term *posterior* can be synonymus with *caudal* (from Latin, *cauda* = tail) and *dorsal* (from Latin, *dorsum* = back). The term *median* means a position in the midline. *Medial* signifies a position nearer to the median or the midsagittal plane, whereas *lateral* means a position farther from the median or the midsagittal plane.

hemisphere. The *thalamus* and *hypothalamus* are the two major divisions of the diencephalon. The thalamus is a collection of several nuclei with various functions and anatomic relations. The hypothalamus, likewise, contains a variety of different cell groups and pathways. It is well known for its regulation of autonomic and endocrine functions.

Mesencephalon

The mesencephalon (midbrain) is a short segment of the brain between the diencephalon and the metencephalon. Its internal structure is complicated, but its gross anatomic features are simple. It is traversed by a narrow canal, the *cerebral aqueduct*, which connects the third ventricle in the diencephalon with the fourth ventricle in the metencephalon.

Metencephalon

The metencephalon consists of the *cerebellum* and the *pons* (bridge). The cerebellum, which is concerned with coordination of movements, is located underneath the tentorium cerebelli in the posterior cranial fossa. It consists of the two *cerebellar hemispheres* joined in the midline by a narrow wormlike portion, the *vermis*. Like the cerebral hemispheres, the cerebellum is covered by a layer of gray substance, the *cerebellar cortex*, which is characterized by a large number of parallel folds, *folia*, separated by deep *fissures*.

Although the *pons* is usually said to be located ventral to the cerebellum, it is actually situated in front of the cerebellum when the head is in the normal upright position. When viewed from its ventral side, the pons appears as a bridge over a canal (the brain stem).

Myelencephalon

The myelencephalon, or *medulla oblongata*, is continuous with the spinal cord at the level of the foramen magnum. The cavity that is located between the cerebellum on the one hand and the pons and medulla oblongata on the other is referred to as the *fourth ventricle*.

The medulla oblongata, pons, and mesencephalon, which together form the *brain stem*, contain many important nuclear groups including the cranial nerve nuclei. They also contain important pathways connecting the spinal cord and cerebellum with the diencephalon and the telencephalon.

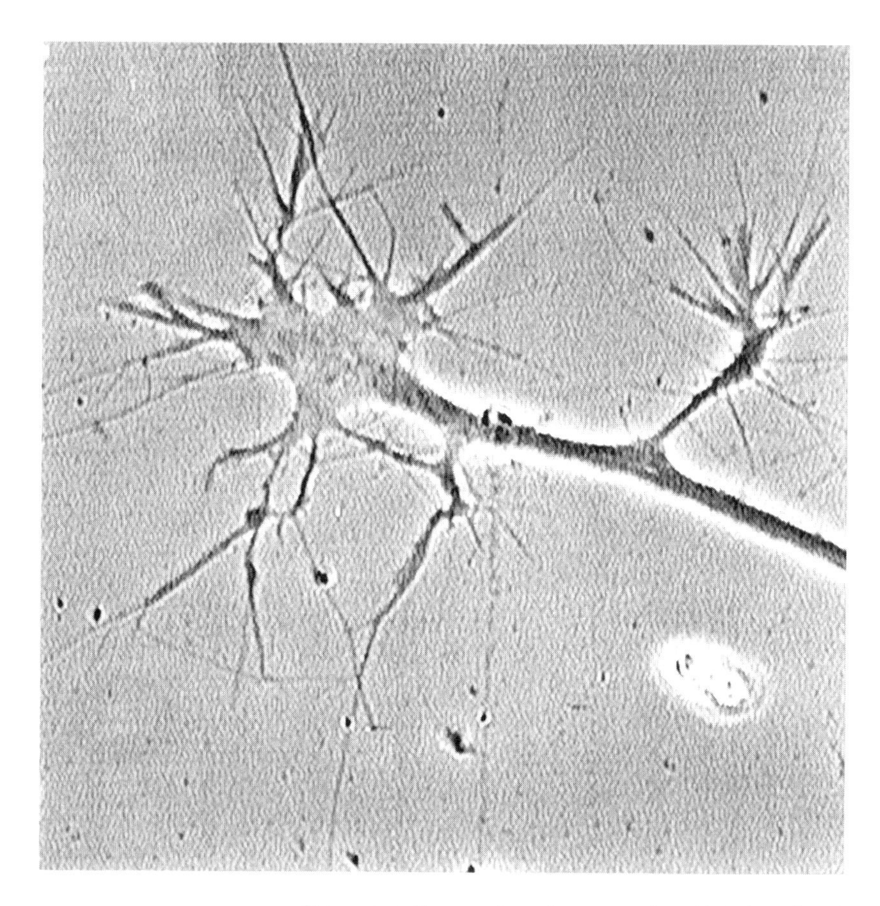

Nerve growth cone from the axon of a chick dorsal root ganglion neuron. (Courtesy of Drs. Paul Letourneau and Timothy Gomez, University of Minnesota, Minneapolis, Minnesota.)

2

Development of the Nervous System

Knowledge of the important developmental events will facilitate the understanding of the anatomy of the CNS. It will also make it easier to understand malformations of the nervous system that are met in clinical practice.

Proliferation and migration of cells in the nervous system start early in embryonic life and continue in some areas until after birth. In addition, neurons mature and neuronal processes develop their elaborate branching pattern and synaptic connections in both the pre- and postnatal periods. The precision with which synaptic connections are formed is as remarkable as the multitude of molecules and cellular mechanisms involved. Although the external shape of the brain and spinal cord is well developed at birth, the size of the CNS increases dramatically also in the neonatal period and during infancy. The protracted maturation of the nervous system makes it especially vulnerable to a variety of harmful environmental factors, including infections, toxic substances, malnutrition, etc.

Early Development of the Nervous System

Neural Tube

The first sign of a nervous system appears in the 3rd week, when the ectoderm on the dorsal side of the embryo thickens to form the *neural plate* (Fig. 2–1A). The mechanism whereby some of the ectodermal cells are transformed into specialized nervous tissue cells is known as induction. Peptides released from the underlying mesoderm apparently serve as inducers in this process.

The neural plate soon starts to fold and a groove is formed, the *neural groove* (Fig. 2–1B). The two side walls, the *neural folds*, approach each other in the midline and the groove is gradually transformed into the *neural tube*, which becomes separated from the rest of the ectoderm (Fig. 2–1C). The process by which the plate is transformed into a tube is called neurulation. The closure of the neural groove begins at the midcervical level and proceeds in both cranial and caudal directions. The last portions of the neural tube to close, located rostrally and caudally, are called the *anterior* and *posterior neuropores* (Fig. 2–1D). They are normally closed by the end of the 4th embryonic week, at which time the neural tube surrounds a fluid-filled central cavity.

The site of the closure of the rostral neuropore corresponds to the region of the lamina terminalis of the mature brain. It should be noted that the major divisions of the brain, i.e., the forebrain, midbrain, and hindbrain (*see below*), can be identified before the closure of the neural tube. A defective closure of the neural tube is a common cause of neurologic malformations (see Clinical Notes). The neuroepithelial cells that form the walls of the neural tube rapidly multiply and eventually give rise to all the neurons and macroglial cells (astrocytes, oligodendrocytes, and ependymal cells) of the brain and the spinal cord. Microglia are believed to be of mesodermal origin, and they apparently invade the CNS at the time the vasculature develops.

Neural Crest

As the neural plate folds to form the neural tube, the most lateral cells are pinched off. These cells, the *neural crest* cells (black dots in cross-sections Figs. 2–1B and C), develop in rostrocaudal sequence along the dorsolateral part of the tube. They become segmented into cell groups that give rise to the sensory ganglion cells of the spinal and cranial nerves. Many of the neural crest cells migrate to other parts of the body where they appear in a variety of cell types, including postganglionic autonomic neurons, chromaffin cells of the adrenal medulla, and supporting cells in the PNS such as satellite cells, Schwann cells, and neurolemma cells. All melanocytes, likewise, come from the neural crest.

Spinal Cord

Ventricular, Intermediate, and Marginal Zones

When the neural tube has closed, the neuroepithelial cells form a thick pseudostratified columnar epithelium (Fig. 2–2A). The cells multiply at a rapid rate in this zone, which is referred to as the *ventricular zone* (neuroepithelium). During the proliferative period the nuclei migrate in a characteristic fashion to and from the lumen at each division. In the interphase, when the undifferentiated neuroepithelial cells synthesize DNA, their form is wedge-shaped; the nuclei lie in the outer zone of the primitive ependymal zone and slender cytoplasmic processes extend toward the ventricular surface. During the mitotic cycle the nuclei migrate toward the lumen, where they divide within a few hours following the DNA synthesis. After the division, the nuclei pass away from the lumen and any given cell may continue in a lateral direction and participate in the formation of the *intermediate (mantle) zone* (Fig. 2–2A). In the intermediate zone the young neurons that have just been produced by their neuroepithelial precursors accumulate and differentiate. The axons of many of the neurons collect along the periphery of the intermediate zone, where they form the *marginal zone*. This description refers to the events in the caudal part of the neural tube, which develops into the spinal cord. The ventricular, intermediate, and marginal zones are represented by the ependymal layer, the gray substance, and the white substance in the mature spinal cord (Fig. 2–2B). The events in the rostral part of the neural tube, which gives rise to the brain, are more complicated (*see below*).

Basal and Alar Plates

A longitudinal groove, the *sulcus limitans*, appears on both sides of the central cavity and runs along the length of the spinal cord. This sulcus, which can be clearly recognized as far rostrally as the midbrain, divides the rapidly expanding side walls of the neural tube into a ventral thickening, the *basal plate*, and a dorsal thickening, the *alar plate* (Fig. 2–2A). The cell bodies in the basal plate form the *ventral* and *lateral gray columns*, which contain somatic and autonomic motor neurons, respectively. The cell bodies in the alar plate form the *dorsal gray column*. These columns are often referred to as the *ventral*, *lateral*, and *dorsal horns* when viewed in cross-section (Fig. 2–2B).

The marginal zone, which increases in thickness as more and more neurons send axons into it, forms the white matter of the spinal cord. As a result of continuous enlargement of the walls of the neural tube, the wide central cavity gradually narrows and eventually becomes the *central canal* of the spinal cord.

Neuronal Migration and Maturation

The development of neurons is an elaborate process, which begins with the early formation of the neuroepithelium and the subsequent proliferation and differentiation of cells into postmitotic immature neurons. These young neurons migrate from the ventricular zone to the intermediate zone where they mature by acquiring their unique pattern of neuronal processes. Whereas the distance of migration is in general rather short in the spinal cord, the migration routes in the developing brain can be long and sometimes rather complicated. During migration, many of the young neurons are guided by radially arranged glial fibers that span the wall of the neural tube from the ventricular surface to the pia[1]. The radial glial cells disappear later in development; some degenerate and others develop into astrocytes (see Neuroglia, Chapter 6).

How the neurons find their correct position or local address in the nervous system is not fully understood. Nonetheless, having settled in their final location, the young neurons mature by developing axonal and dendritic processes and establishing synaptic contacts. How this can be accomplished with such precision is a central question, which is attracting an ever-increasing number of young and curious scientists into the fields of developmental and molecular neurobiology. How can axons find their distant target, stop at the appropriate place, and establish the correct synaptic connections needed to produce the multitude of highly specific neuron circuits that characterize the nervous system? Much attention has focused on the tip of the growing axon, the growth cone[2]. The growth cone (see Vignette) is a highly specialized structure with great motility, and it is widely believed that the axons are guided to their appropriate targets by various mechanical and chemical interactions taking place between the growth cone and the environment.

Clearly, many different mechanisms are involved in the final sculpturing of the nervous system. Although the guidance provided by the axonal growth cones results in an orderly development of specific pathways, there is initially an overproduction of neurons and formation of too many and partly inappropriate connections. To adjust for these initial deficiencies, a number of corrective measures take place, including cell death and elimination of aberrant and superfluous processes and synapses. Finally, it is important to remember that nervous tissue remains remarkably plastic even after birth. Fine tuning of neuronal circuits depends on experience and activity; social or sensory deprivation of infants may have serious consequences. Indeed, mechanisms responsible for the initial wiring of the nervous system are also used to modify the adult nervous system in response to experience.

During development, many axons acquire a lipid-rich myelin sheath that greatly increases the efficiency and velocity of signal transmission. The myelination of axons, which is accomplished by Schwann cells in the peripheral nervous system and by oligodendrocytes in the CNS, starts in the second trimester and continues in some parts of the CNS into adult life. The myelinating process,

1. The intriguing hypothesis of glial-guided neuronal migration was first suggested by Enrico Mugnaini and Paul Fosträ̈nen (Zeitschrift für Zellforschung 1967, 77:115–143) in their studies on the cerebellar cortex and was developed further by the Yugoslavian-born scientist Pasco Rakic in his studies of cerebellar and cerebral cortices. The notion has recently been confirmed and extended to other parts of the brain.

2. The growth cone was discovered by Ramón y Cajal in 1880. Aided by the static images obtained in his, by our standards, primitive microscope, Cajal vividly captured the essence of the growth cone as a "sort of club or battering ram, endowed with exquisite chemical sensitivity, with rapid ameboid movements, and with certain impulsive force, thanks to which it is able to proceed forward and overcome obstacles met in its way, forcing cellular interstices until it arrives at its final destination.... What mysterious forces precede the appearance of these prolongations, promote their growth and ramification, and finally establish the protoplasmic kisses, the intercellular articulations that appear to constitute the final ecstasy of an epic love story." (citation from Ramón y Cajal's "Recollections of My Life" MIT Press, Cambridge, Mass.). When it became possible to study live growth cones in tissue culture, their unusual motility and "ameboid movements" were promptly confirmed, and "the mysterious forces" responsible for the movements and ramifications to the "final ecstasy" of synapse formation are gradually being revealed with the aid of cellular and molecular biological techniques.

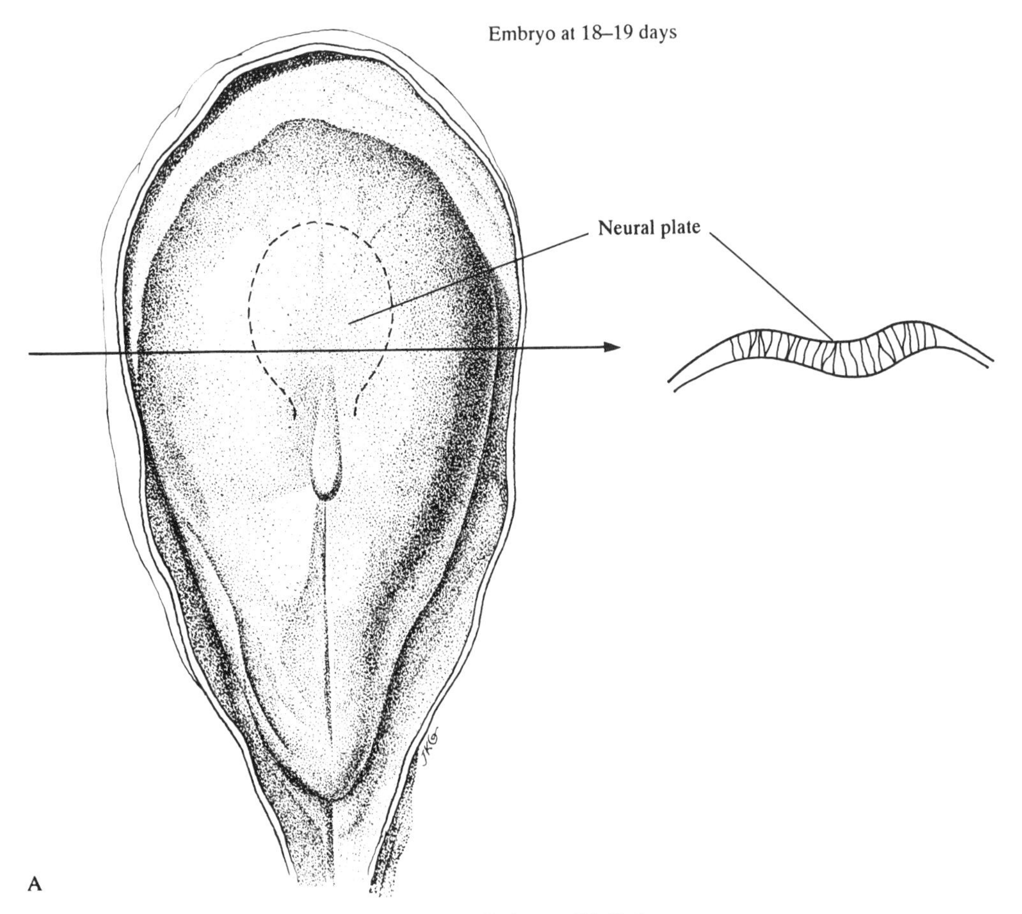
Embryo at 18–19 days
Neural plate
A
JKG

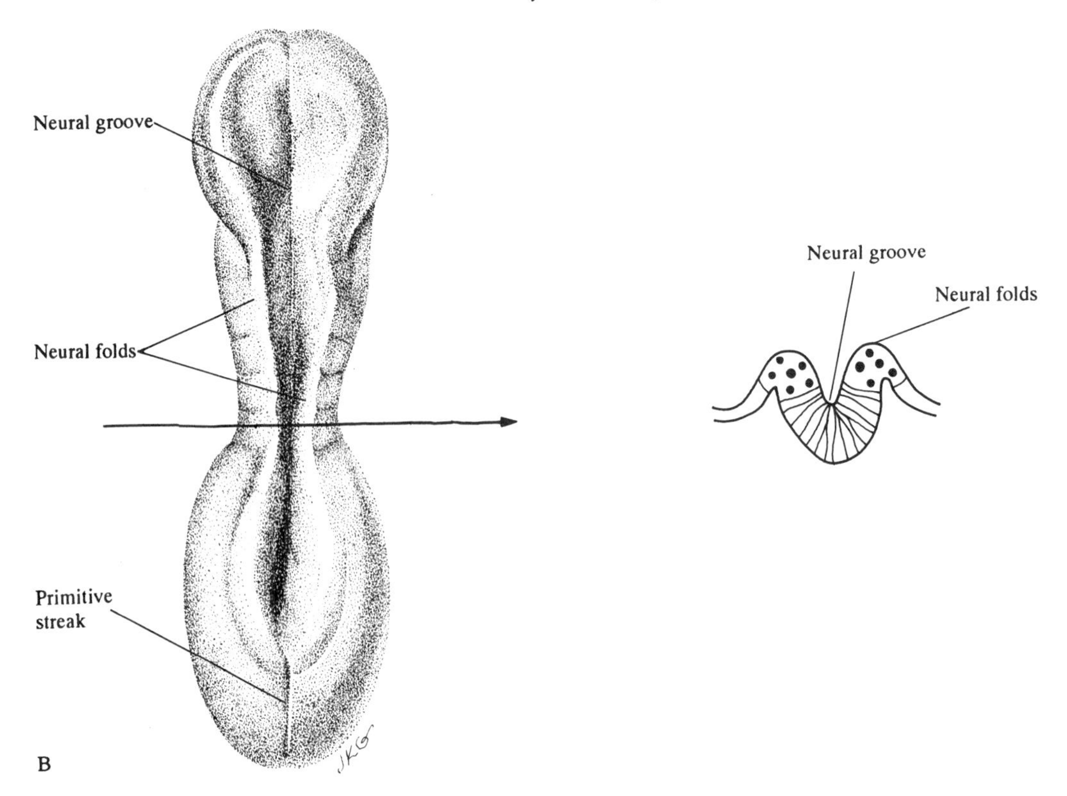
Embryo at 20–21 days
Neural groove
Neural folds
Primitive streak
Neural groove
Neural folds
B
JKG

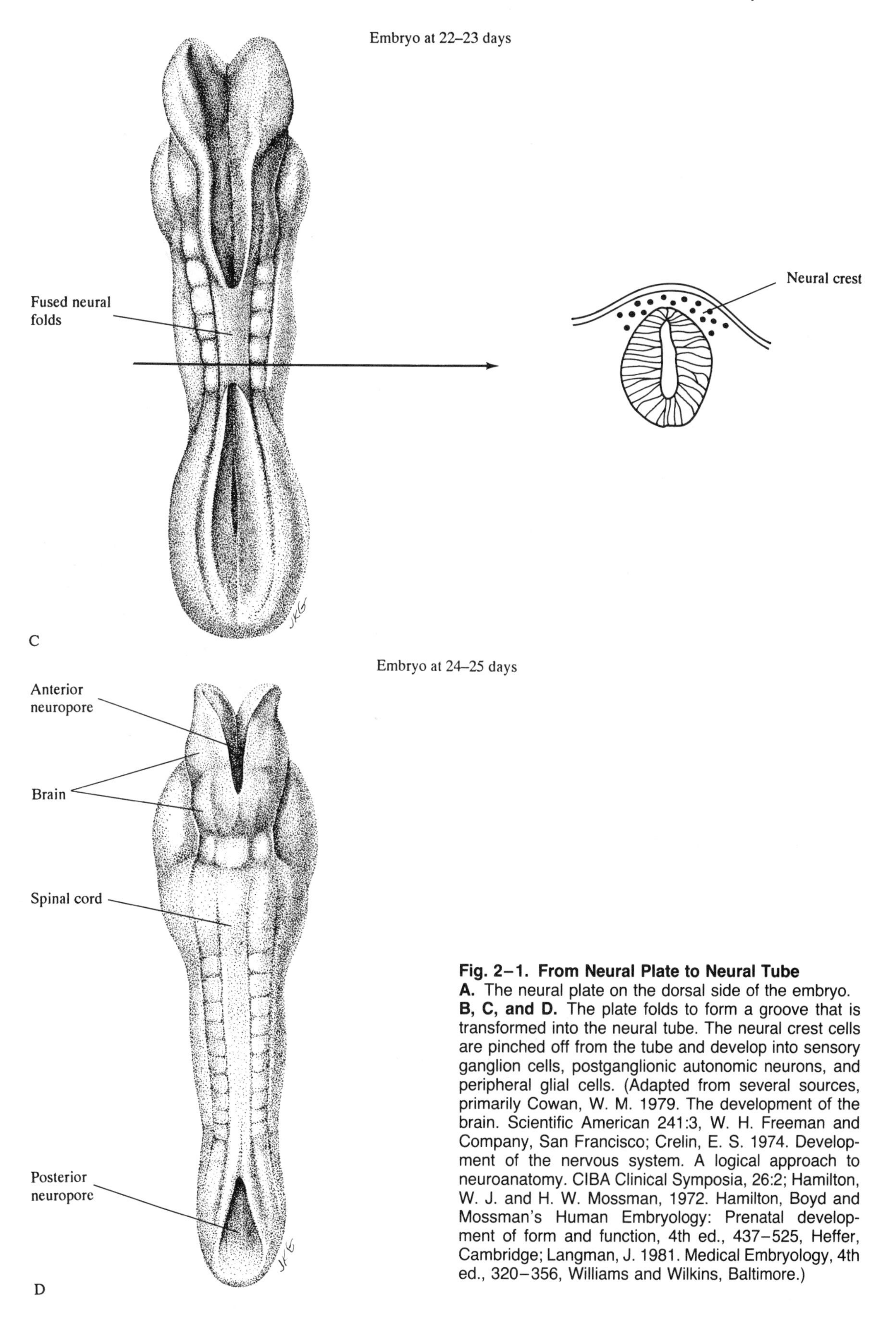

Fig. 2–1. From Neural Plate to Neural Tube
A. The neural plate on the dorsal side of the embryo.
B, C, and D. The plate folds to form a groove that is transformed into the neural tube. The neural crest cells are pinched off from the tube and develop into sensory ganglion cells, postganglionic autonomic neurons, and peripheral glial cells. (Adapted from several sources, primarily Cowan, W. M. 1979. The development of the brain. Scientific American 241:3, W. H. Freeman and Company, San Francisco; Crelin, E. S. 1974. Development of the nervous system. A logical approach to neuroanatomy. CIBA Clinical Symposia, 26:2; Hamilton, W. J. and H. W. Mossman, 1972. Hamilton, Boyd and Mossman's Human Embryology: Prenatal development of form and function, 4th ed., 437–525, Heffer, Cambridge; Langman, J. 1981. Medical Embryology, 4th ed., 320–356, Williams and Wilkins, Baltimore.)

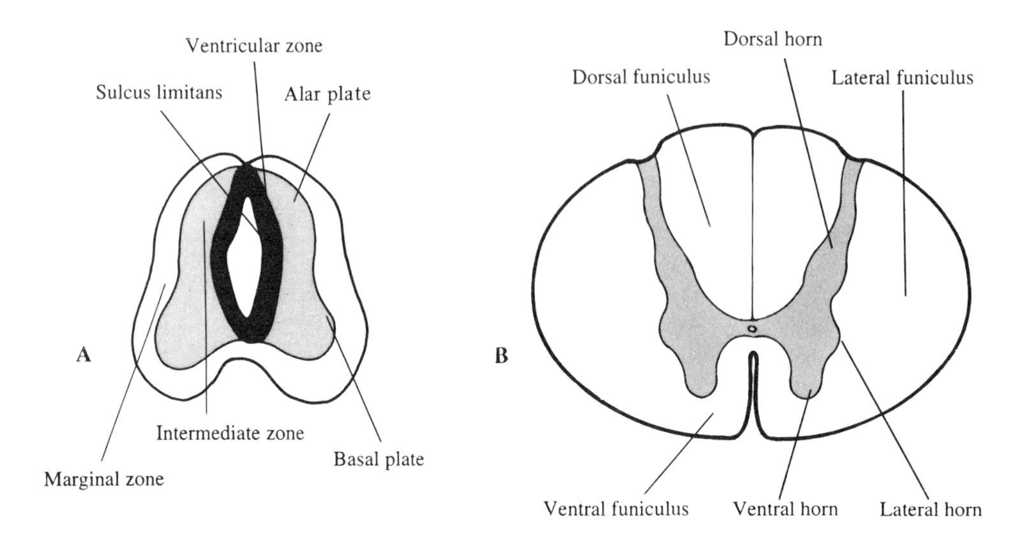

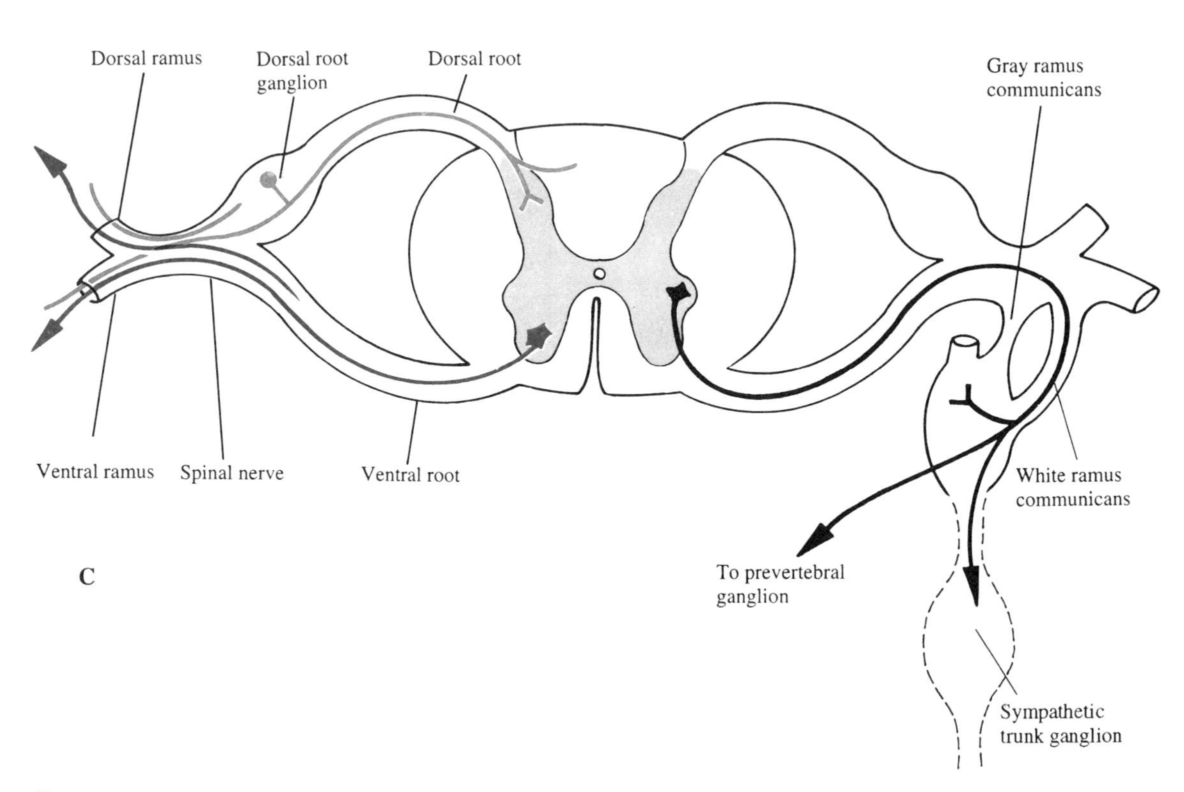

Fig. 2–2. Development of the Spinal Cord

A. The walls of the neural tube rapidly increase in thickness through cell proliferation. Sulcus limitans indicates the boundary between the basal and alar plates.

B. The basal plates develop into ventral and lateral horns, whereas the alar plates form the dorsal horns.

C. Schematic drawing showing the principal components of the spinal nerves. Somatic motor neuron, *red*; autonomic motor neuron, *black*; sensory neuron, *blue*.

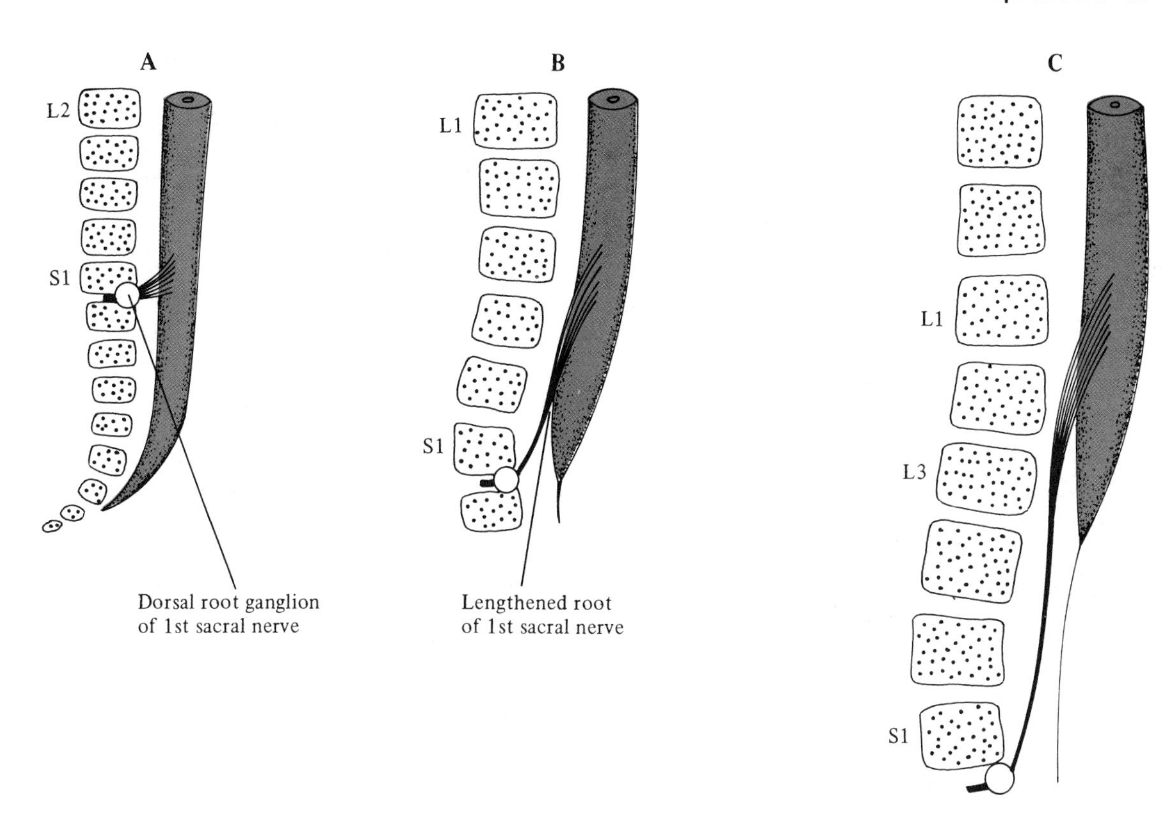

Fig. 2–3. Cauda Equina
Schematic drawings showing the relationship between the spinal cord and the vertebral column at various stages of development. **A,** approximately at 3rd month; **B,** end of 5th month; **C,** in the newborn (After Streeter, G. L., 1919. Factors involved in the formation of the filum terminale. American Journal of Anatomy 25:1–11 and after Langman, J. 1981. Medical Embryology, 4th ed., 320–356, Williams & Wilkins, Baltimore. Langman, 1981. With permission of Williams & Wilkins, Baltimore).

which is maximal during the first and second year after birth, is energy demanding, and various metabolic insults, including severe starvation, can significantly reduce the amount of myelin.

Dorsal Root Ganglia and Spinal Nerves

The pseudounipolar sensory cells of the dorsal root ganglia (spinal ganglia) develop from the neural crest. The axons of these sensory cells divide in a T-shaped pattern close to the cell bodies. The central processes grow into the spinal cord. These are the *dorsal roots* of the spinal nerves (Fig. 2–2C). The peripheral processes, which are related to different types of specialized receptors or which ramify as free nerve endings in various organs, join the *ventral root* fibers in the region of the intervertebral foramina to form the *spinal nerves*. Most of the axons in the ventral roots arise from somatic and autonomic motor neurons in the ventral and lateral gray columns of the spinal cord, and terminate in skeletal muscles or in autonomic ganglia.

Cauda Equina

In the embryo, the spinal cord extends the entire length of the vertebral canal, and the spinal nerves pass through the intervertebral foramina at the level of origin in the spinal cord (Fig. 2–3A). After the 3rd fetal month, however, the vertebral column and dura mater grow more rapidly than the spinal cord, and the various spinal cord segments will eventually lie somewhat higher than the corresponding vertebrae (Figs. 2–3B and C). The distance between the spinal cord segment and its corresponding vertebra becomes increasingly greater in the caudal segments. At birth, the spinal cord ends at the level of vertebra L3, but it reaches only vertebra L2 in the adult.

The differential growth of the spinal cord and the vertebral column explains the formation of the *cauda equina* (Fig. 2–4), which consists of the long lumbar and sacral roots that have been stretched during the development of the cord and the vertebral column. The caudal tip of the tube remains

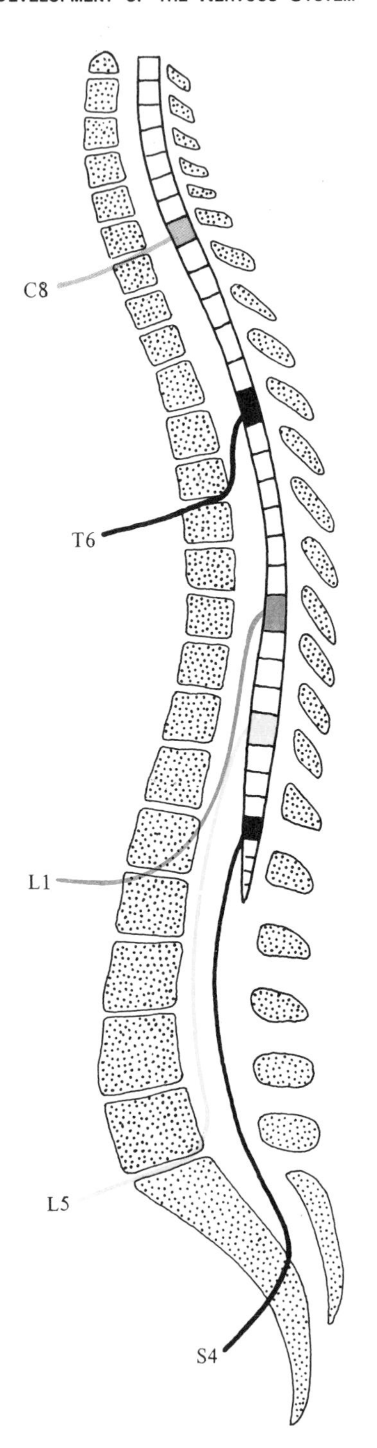

Fig. 2–4. Relationship between Spinal Cord Segments and Their Corresponding Vertebrae
The distance between the spinal cord segment and its corresponding vertebra becomes increasingly greater in the caudal direction. The cauda equina consists of the long lumbar and sacral roots that have been stretched during the development of the spinal cord and vertebral column.

attached to the coccygeal part of the vertebral column and becomes a thin filament, the *filum terminale*. The fact that the spinal cord of the adult reaches only to the second lumbar vertebra is of great clinical relevance (see Lumbar Puncture in Clinical Notes, Chapter 3).

Early Development of the Brain

In the 3rd embryonic week, a bend, the mesencephalic (midbrain) flexure, develops in the rostral parts of the neural folds before the closure of the neural tube is completed, and three divisions of the early brain can be recognized (Fig. 2–5A): *prosencephalon* = forebrain; *mesencephalon* = midbrain; and *rhombencephalon* = hindbrain. Subsequently, a cervical flexure develops at the junction between the hindbrain and spinal cord (Fig. 2–5A). Somewhat later, a compensatory flexure, the pontine flexure, develops between the other two flexures. (Fig. 2–5B). The prosencephalon can soon be divided into two parts, the *telencephalon*, which will become the cerebral hemispheres, and a median part, the *diencephalon*. At the same time the rhombencephalon divides into the *metencephalon* and the *myelencephalon*.

Myelencephalon

The myelencephalon develops into the *medulla oblongata*, which is continuous with the spinal cord. The medulla resembles the spinal cord macroscopically, but the typical spinal cord pattern with a centrally placed gray substance surrounded by white substance cannot be recognized. Further, the rostral part of the medulla becomes wider as the alar plates ((*a*) in Figs. 2–6A and B) move laterally to give room to the expanding fourth ventricle. This process results in the development of an extensive but thin roof plate covering the fourth ventricle. It also explains why the cranial sensory neurons as derivatives of the alar plates are located dorsolateral to the columns of cranial motor neurons (Fig. 2–6C), which derive from the basal plates ((*b*) in Figs. 2–6A and B). Young neurons from the alar plates migrate to several parts of the brain stem. Many of them aggregate in the ventrolateral part of the medulla oblongata where they form the *inferior olivary nucleus*, a prominent landmark closely related to the cerebellum. Another prominent structure in the ventral part of medulla is the *pyramid*, which contains corticospinal fibers. The dorsal part of the medulla contains cranial nerve nuclei, long ascending and descending tracts, and reticular nuclei (see Chapter 10).

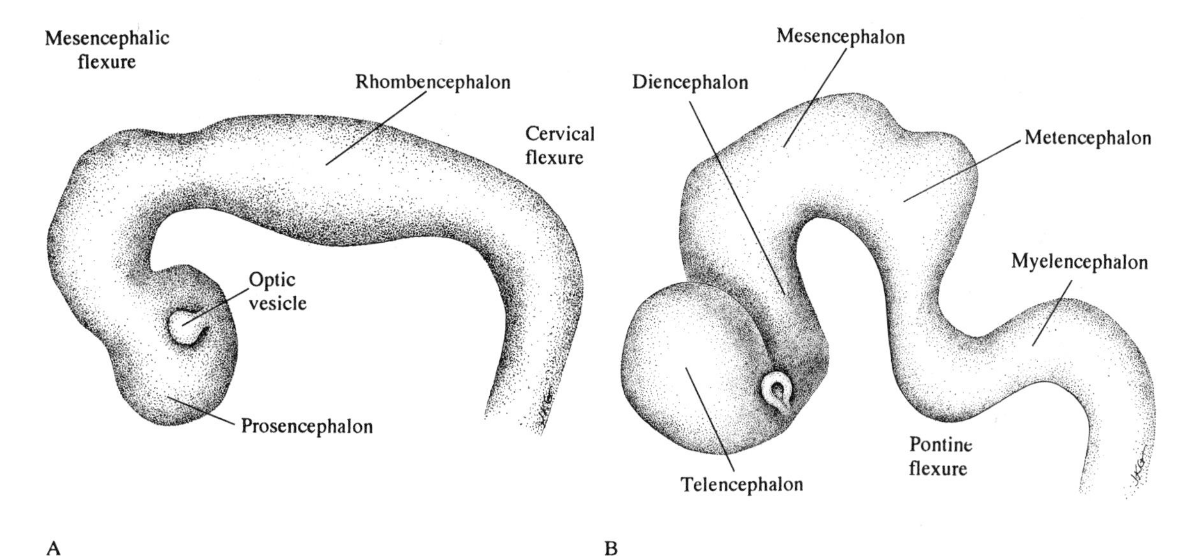

Fig. 2–5. Early Development of the Brain
A. A mesencephalic (midbrain) flexure develops in the region of the mesencephalon, and a slight bend, the cervical flexure, can be recognized at the junction of the rhombencephalon and the spinal cord.
B. A third flexure, the pontine flexure, develops between the metencephalon and myelencephalon and the brain now contains five parts (Modified after Hochstetter, F., 1919 and 1929. Beiträge zur Entwicklungsgeschichte des menschlichen Gehirns. Deuticke: Vienna and Leipzig.)

Metencephalon

Pronounced changes take place in the *metencephalon.* The early development of the *cerebellum* is obscured by the rather complicated three-dimensional form resulting from the appearance of the *pontine flexure.* Initially, the cerebellum develops bilaterally from the alar plates, and the cerebellar primordia bulge into the rhombencephalic cavity or fourth ventricle (Fig. 2–7A), whereas the extraventricular portions rapidly expand in a dorsal direction (Fig. 2–7B). Where the two cerebellar primordia (future hemispheres) meet in the midline, they form a wormlike portion referred to as the vermis. The surface area increases disproportionally to the rest of the cerebellum to accommodate the cerebellar cortex, which in the final stage is characterized by transversely oriented deep *fissures* separating narrow folds, called *folia.* Both the hemispheres and the vermis contain a heavily folded cortex as well as subcortical nuclear groups, the intracerebellar nuclei. The cerebellum is of special importance for the coordination of movements.

The ventral part of the metencephalon develops into the pons, the basal part of which is characterized by prominent collections of neurons, the pontine nuclei, which send their axons to the cerebellum. The pontine nuclei, like the inferior olivary nucleus, derive from the alar plates. The dorsal, tegmental part of the pons is occupied by cranial nerve nuclei, long fiber bundles, and the reticular formation (Fig. 2–7C). The cavity of the rhombencephalon becomes the fourth ventricle.

The Cerebellar Cortex

Initially, the cerebellar primordium, like the rest of the neural tube, consists of a ventricular zone with actively dividing cells. Immature neurons migrate in a radial direction to the superficial part of the intermediate layer (Fig. 2–8A) where they develop into the large *Purkinje cells* (Fig. 2–8B). During further development, another group of neuroepithelial cells, derived from the *rhombic lip* in the dorsolateral part of the alar plate, migrates along the pial surface to form a secondary germinal matrix, the *external germinal* (granular) *layer* (Fig. 2–8A). This layer retains its capacity to proliferate. Normally, migrating cells in the developing nervous system are postmitotic, i.e., they have lost their capacity to divide before they start to migrate.

The next stage in the development provides a graphic illustration of the finely tuned interactions taking place during development. The external granular cells, which have retained their mitotic activity, give rise to cells that start a second wave

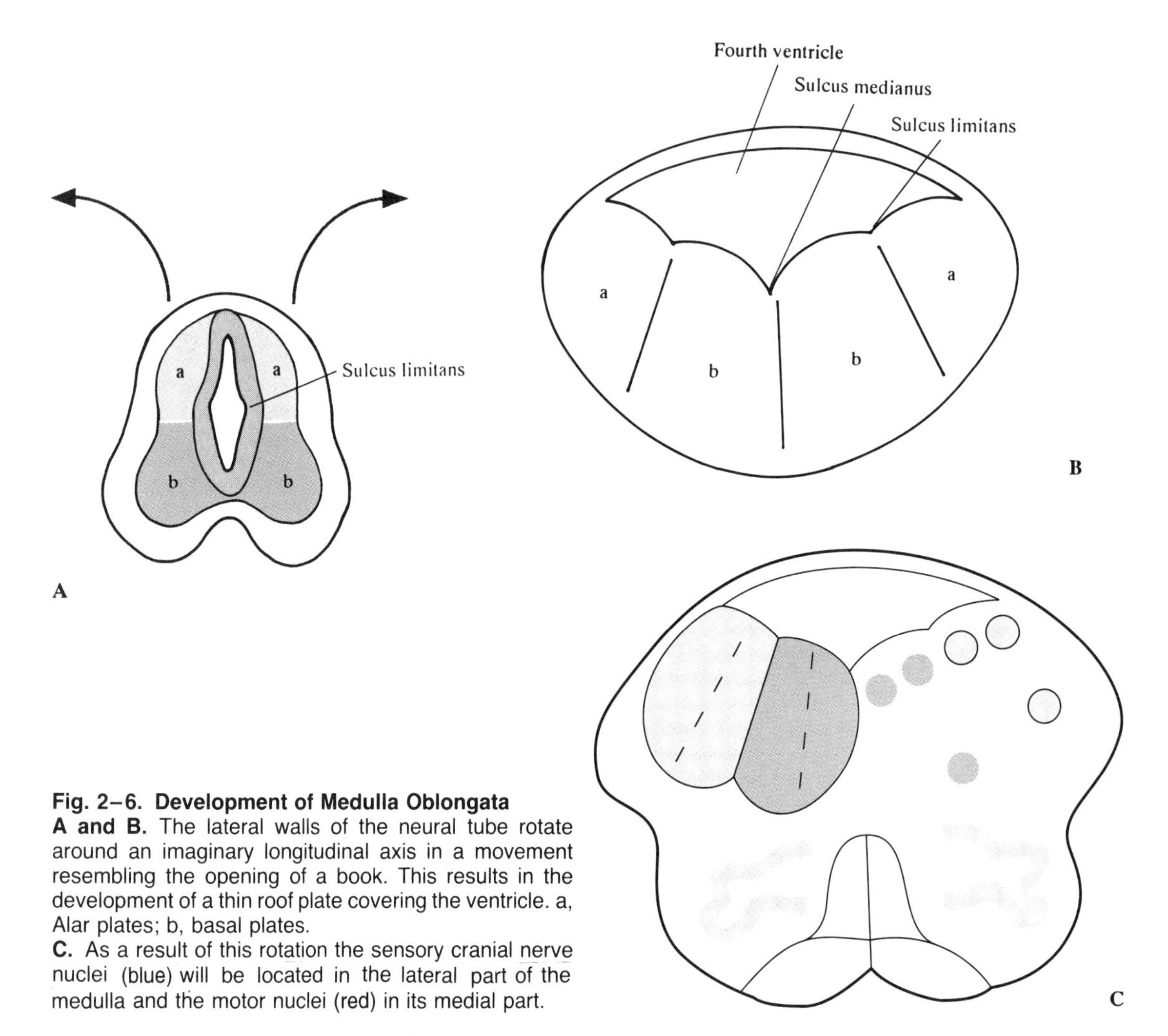

Fig. 2–6. Development of Medulla Oblongata
A and B. The lateral walls of the neural tube rotate around an imaginary longitudinal axis in a movement resembling the opening of a book. This results in the development of a thin roof plate covering the ventricle. a, Alar plates; b, basal plates.
C. As a result of this rotation the sensory cranial nerve nuclei (blue) will be located in the lateral part of the medulla and the motor nuclei (red) in its medial part.

Fig. 2–7. Development of Metencephalon
A. The rhombic lips expand rapidly in a dorsal direction and give rise to the cerebellum.
B. A drawing by Dr. Pasko Rakic showing the human brain at approximately 10 weeks of gestation. The arrow points to the attachment of the cut rhombencephalic membrane; IV = trochlear nerve. (Courtesy of Dr. Rakic. From Sidman, R. L. and P. Rakic, 1982. Development of the Human Central Nervous System. Charles C. Thomas: Springfield, Illinois.)
C. The ventral part of the metencephalon, the pons, can be subdivided into a small dorsal part and a large basal part, which contains the pontine nuclei.

→

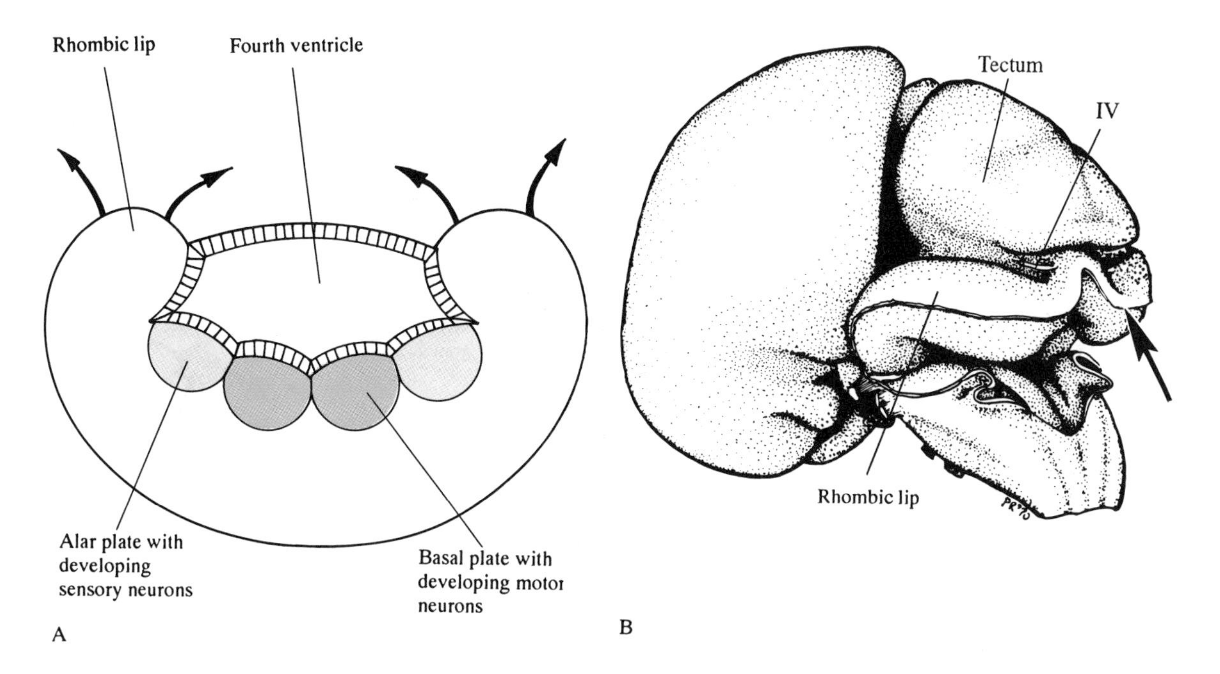
Rhombic lip
Fourth ventricle
Alar plate with developing sensory neurons
Basal plate with developing motor neurons
A
Tectum
IV
Rhombic lip
B

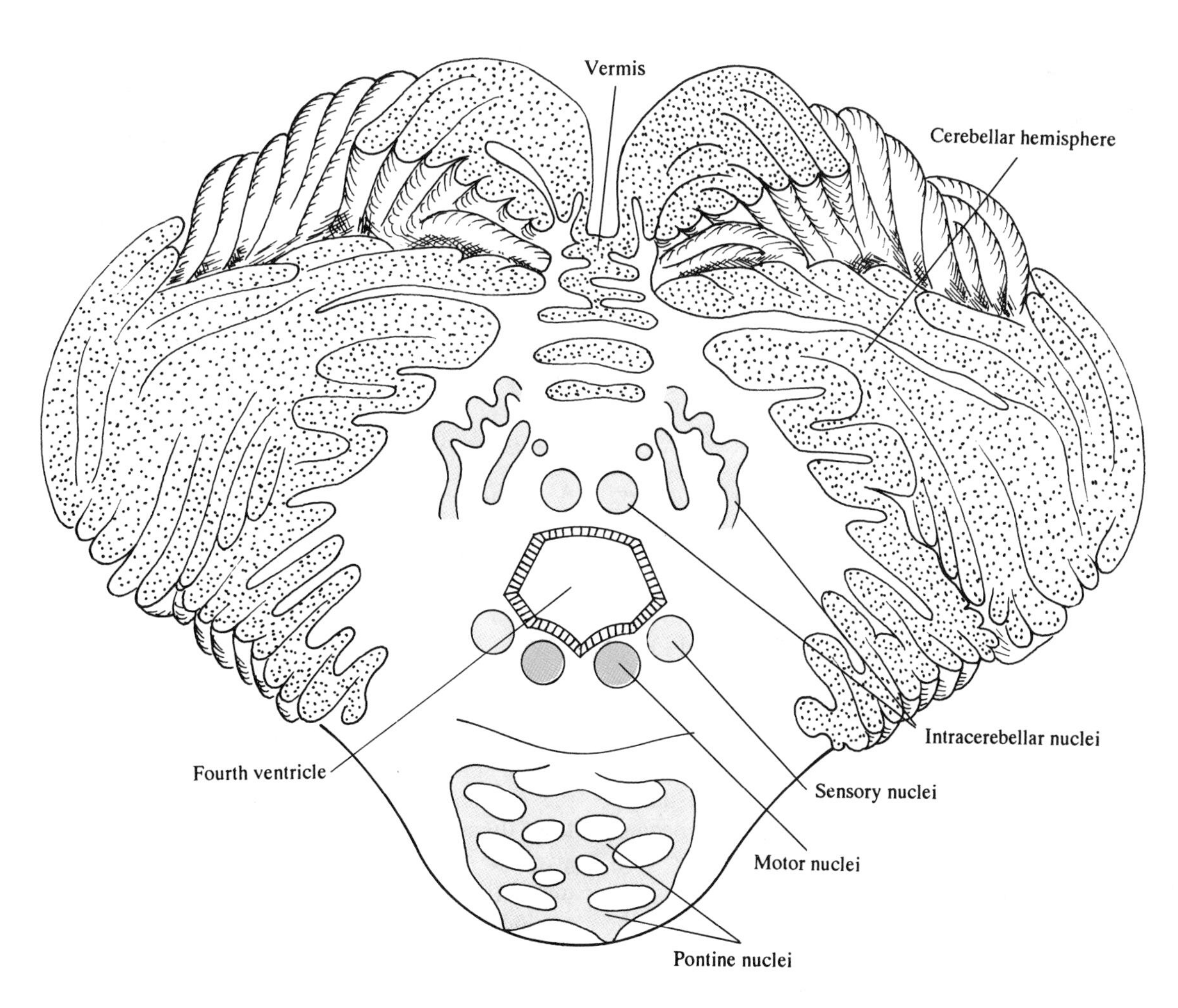
Vermis
Cerebellar hemisphere
Intracerebellar nuclei
Fourth ventricle
Sensory nuclei
Motor nuclei
Pontine nuclei
C

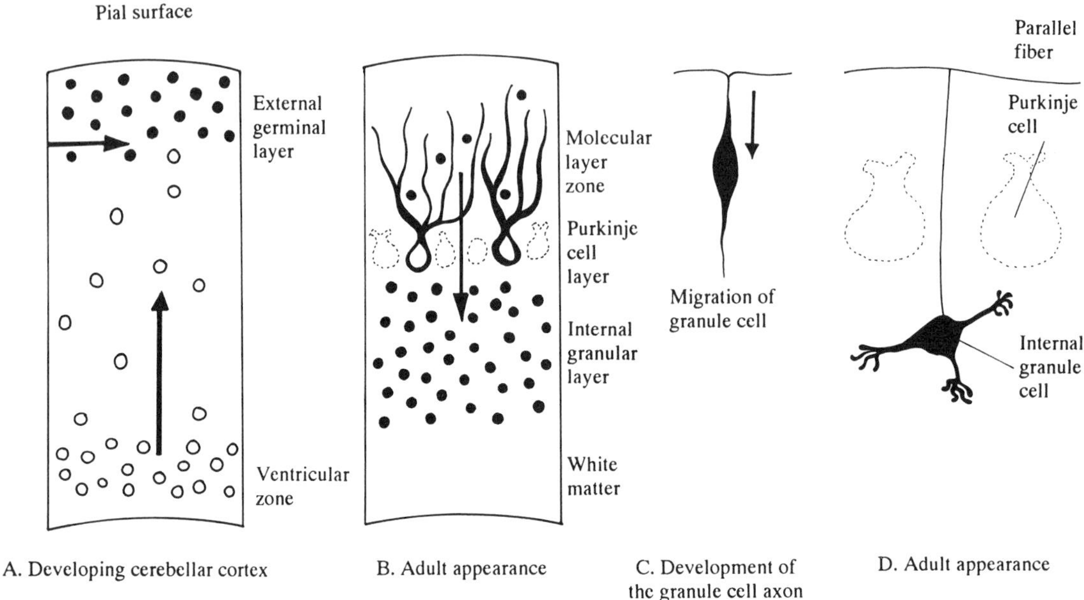

Fig. 2–8. Development of Cerebellar Cortex
A and B. Young neurons from the ventricular zone migrate in a radial direction to form the layer of the Purkinje cells. Another set of neuroepithelial cells migrate along the pial surface to form a secondary germinal matrix, the external germinal (granular) layer. The cells in this layer retain the capacity to divide and many of the daughter cells are destined to form the internal granular layer (see **C and D**).
C and D. Many of the external granular cells develop tangentially oriented axonal processes before they develop radial processes along which the cell bodies migrate inward to form the internal granular layer. During the migration, the cells leave behind a perpendicular process, giving the axon a typical T-shaped appearance. (Modified after Lund, R. D., 1978. Development and Plasticity of the Brain: An Introduction. Oxford University Press, New York).

of migration, but this time in an inward radial direction to form the *internal granular layer* (Fig. 2–8B). Whereas the majority of the cells in the nervous system generate their processes after they have reached their final destination, axonal growth may precede cell migration in some cases. This is exactly what happens in the cerebellar cortex, where the young neurons destined for the internal granular layer develop tangentially oriented axonal processes before they develop radial processes along which the cell bodies migrate inward to form the internal granular layer (Figs. 2–8C and D). In effect, during the migration the cell leaves behind a perpendicular process, giving the axon a characteristic T-shaped appearance. The tangential part of the axon is referred to as a *parallel fiber* (Fig. 17–1). The parallel fibers, and cells that do not take part in the second wave of migration, become important constituents of the *molecular layer* (e.g., basket and stellate cells; see Fig. 17–1).

The various cell types that are released from the external granular layer are closely related to radial glial processes (Bergmann astrocytes), which guide the migrating cells in their inward radial movement.

The intracerebellar nuclei, which are situated in the white medullary core of the cerebellum, are formed by cells from the ventricular layer.

Mesencephalon

A gradual thickening of the wall of the mesencephalon reduces the central cavity to a narrow passage, the *cerebral aqueduct* (Fig. 2–9). The alar plates develop into the *tectum* with the two paired superior colliculi at the level shown in Fig. 2–9, and the inferior colliculi in the candal part of mesencephalon. The superior colliculi are important centers for visual reflexes and the inferior colliculi serve as relay centers for auditory impulses. The *central* or *periaqueductal gray* around the cerebral aqueduct is also believed to come from the alar plate. Neuroblasts in the basal plate give rise to the oculomotor and trochlear nerve nuclei (IIIrd and IVth cranial nerve nuclei), which are located ventral to the central gray of the levels of the superior and inferior colliculi, respectively. It is not known if the *substantia nigra* and *red nucleus*, two prominent nuclei in the ventral part of the mesencephalon, arise from cells in the basal

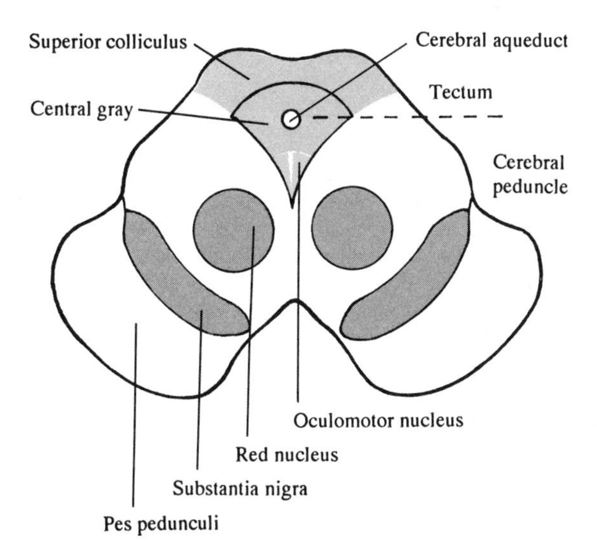

Fig. 2–9. Development of Mesencephalon
A gradual thickening of the wall of the neural tube reduces the central cavity to a narrow passage, the cerebral aqueduct. An imaginary horizontal line through the aqueduct divides the mesencephalon into a dorsal part, the tectum, and a massive ventral part consisting of the two cerebral peduncles.

or the alar plates. As more and more fibers connect the rapidly expanding forebrain with regions in the brain stem and spinal cord, the ventral part of the mesencephalon develops into the two prominent *cerebral peduncles*. Most of the descending projection fibers from forebrain regions accumulate in the *pes pedunculi*, which, together with the substantia nigra, forms the *base of the peduncle*. The dorsal part of the cerebral peduncle, the *tegmentum*, contains both ascending and descending pathways as well as reticular and cranial nerve nuclei.

Diencephalon

A sulcus limitans is not present rostral to the mesencephalon, and basal and alar laminae, therefore, cannot be identified in the forebrain. Three swellings in the lateral wall of the third ventricle give rise to the main divisions of the diencephalon: the *epithalamus*, *thalamus*, and *hypothalamus* (Fig. 2–10). The epithalamus, which is the smallest division, develops into the *habenular nuclei* and the *pineal gland* (epiphysis). The pineal gland, which seems to play a role in gonadal function, has a tendency to calcify, and since it is normally located in the median plane it serves as an important landmark on an X-ray film of the skull. An expansive lesion on one side of the brain tends to push the pineal gland away from the midline toward the opposite side.

The *sulcus hypothalamicus*, which appears as a longitudinal groove in the lateral wall of the third ventricle, separates the thalamus from the hypothalamus. The development of the thalamus, which is the largest part of the diencephalon, is closely related to the development of the cerebral cortex, and the two structures are closely interrelated in the mature brain. Many of the nuclear groups in the thalamus are well defined during the second half of the prenatal period. The hypothalamus, which constitutes a relatively small part of the diencephalon, contains many and often ill-defined nuclei that are closely associated with autonomic and endocrine functions.

As the thalamus and hypothalamus develop in the lateral walls of the diencephalon, the central cavity is gradually transformed into the slitlike *third ventricle*. Its roof is formed by a single layer of ependymal cells covered by a highly vascular mesenchyme. The invagination of these two structures into the third ventricle forms the choroid plexus (*see below*). The *lamina terminalis*, a derivative of the telencephalon, forms the rostral wall of the third ventricle. This is believed to be the area where the final closure of the anterior neuropore occurs. At an early stage, three swellings can be recognized in the floor of the third ventricle. These are, from rostral to caudal: the *optic chiasm*, *infundibulum*, and *mammillary body*.

The Eye

On both sides of the forebrain, an *optic vesicle* appears in the 3rd embryonic week (Figs. 2–11A and B). This vesicle, which remains attached to the brain by the *optic stalk*, apposes the ectoderm, in which a thickening is formed. This is the *lens placode*, which develops into the *lens* of the eye. The optic vesicle invaginates to form the retina with associated pigment layer, and as the nervous retina apposes the pigment layer, the cavity of the optic vesicle is obliterated (Fig. 2–11C). The optic stalk, whose lumen gradually becomes occluded by ingrowing axons from the retina, develops into the *optic nerve*. The nerves from each eye converge at the base of the brain to form the *optic chiasm*, in which the fibers from the medial halves of the two retinae cross the midline.

As the rim of the optic cup closes around the lens to become the *iris*, a groove, the *choroid fissure* remains on the ventral side of the optic cup (Fig. 2–11D). The hyaloid vessels, which reach the interior of the cup through this fetal cleft, develop into the *central artery and vein* of the retina (Fig. 2–11E). If the edges of the choroid fissure do not fuse properly, a malformation known as coloboma will result (see Clinical Notes).

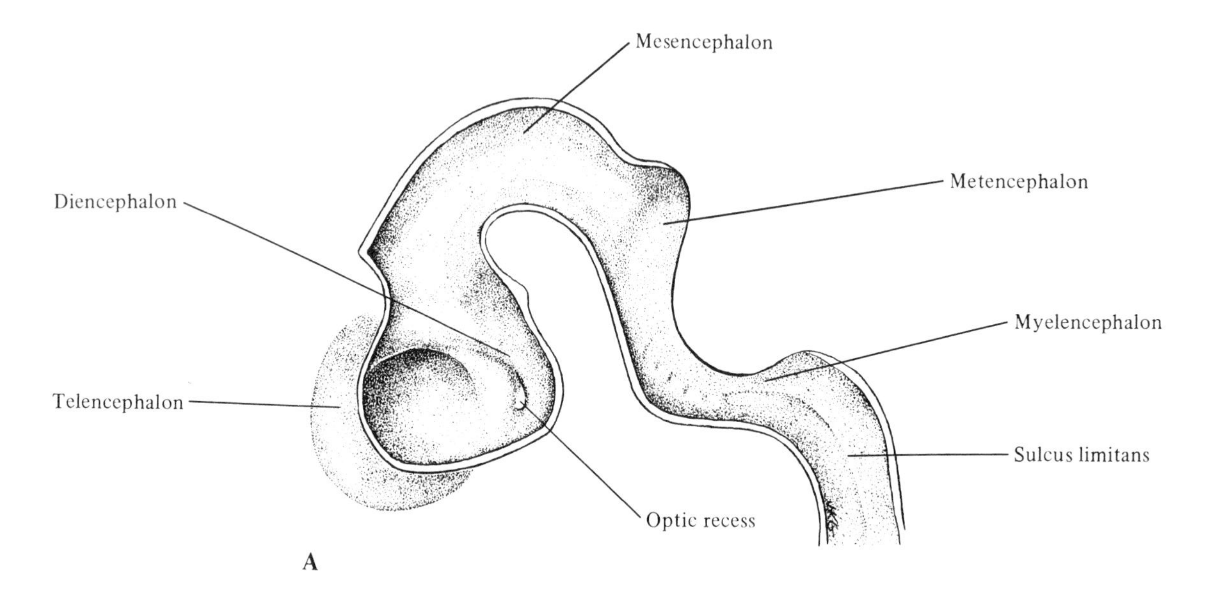

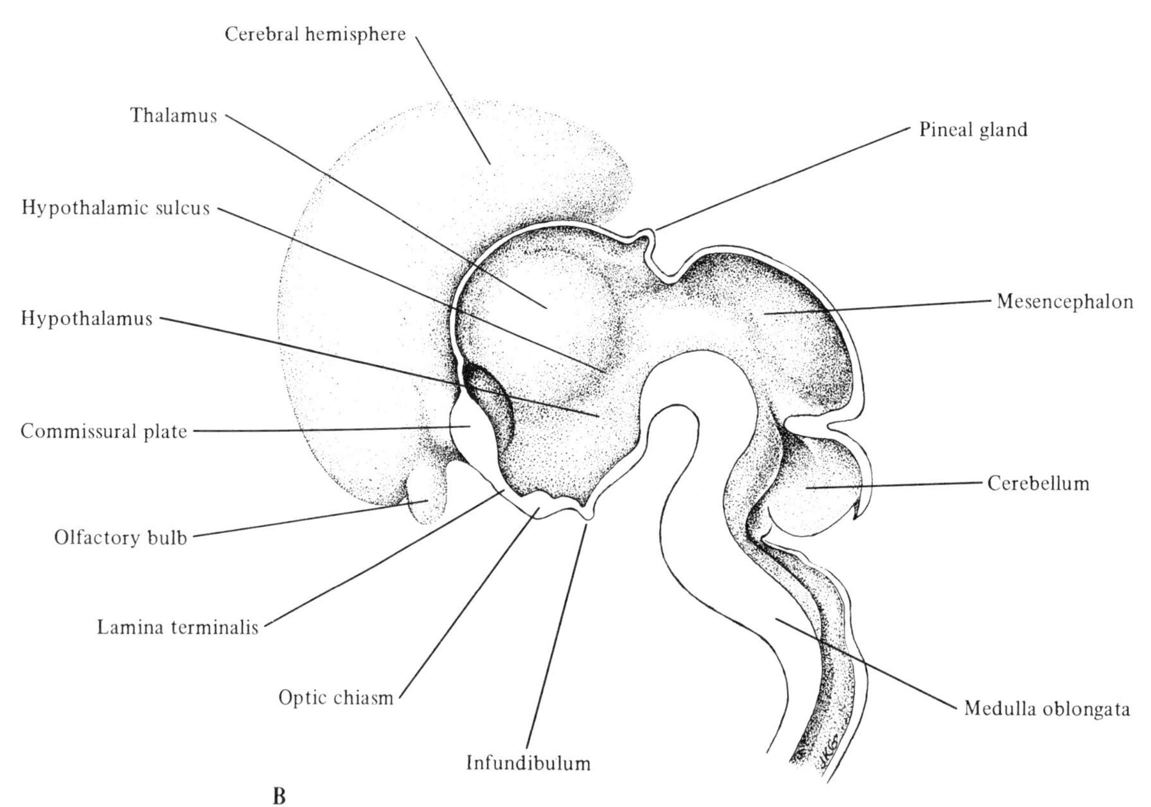

Fig. 2–10. Development of Diencephalon
A. Medial surface of the right half of the cranial part of the CNS in an 11-mm embryo showing the five brain vesicles.
B. The medial surface of the brain in a 43-mm human embryo. The sulcus hypothalamicus separates the thalamus from the hypothalamus. (Modified after Hines, M., 1922. Studies in the growth and differentiation of the telencephalon in man: The fissura hippocampi. J. Comp. Neurol. 34:73–171.)

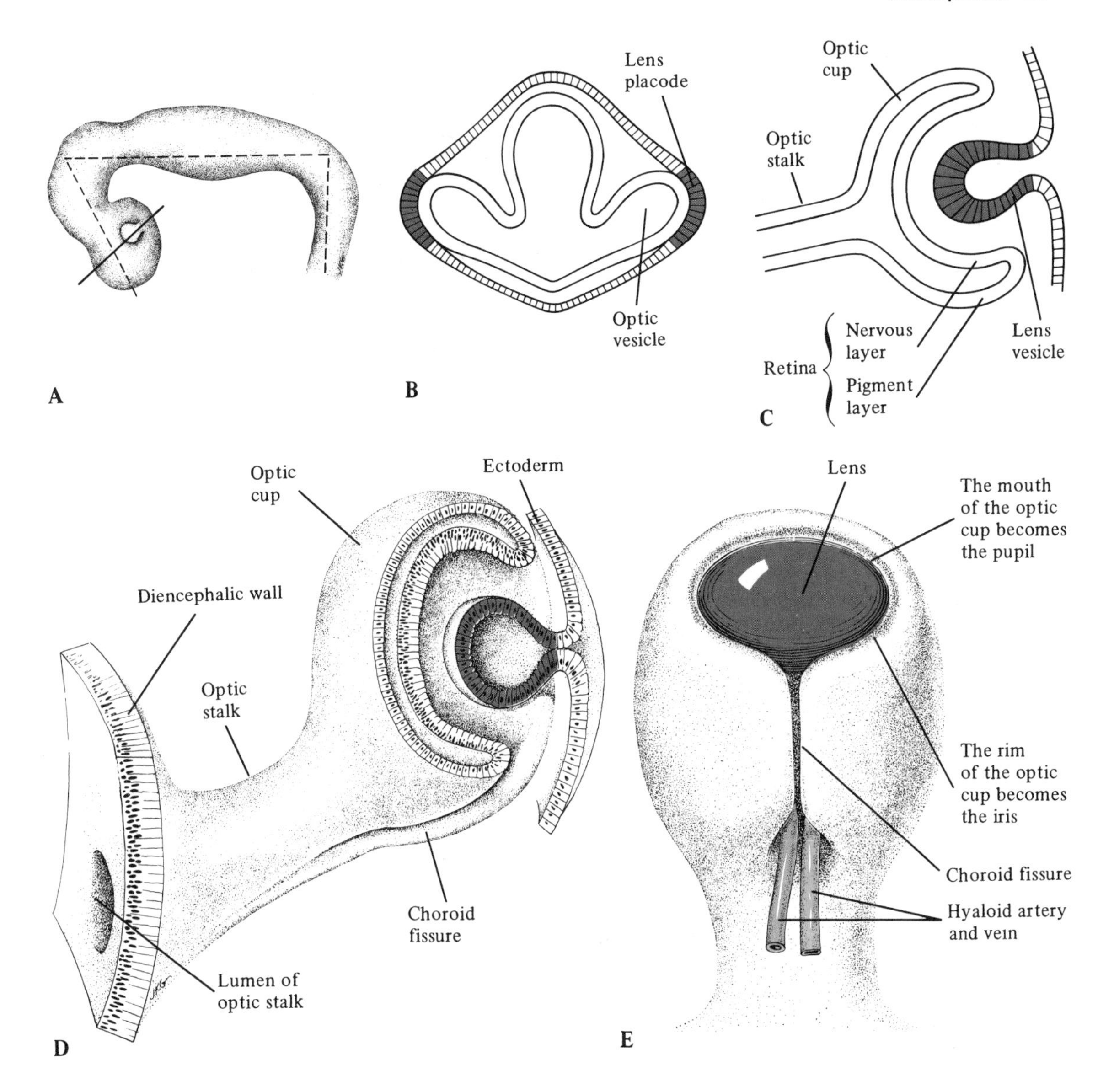

Fig. 2–11. Development of the Eye
A–C. The optic vesicles invaginate to form the retina, whereas the lens placode develops into the lens.
D and E. Three-dimensional drawings showing the optic cup and the choroid fissure with the hyaloid vessels, which develop into the central artery and vein of the retina. (Modified after Mann, J. C., 1950. The Development of the Human Eye. Cambridge University Press, London.)

The Hypophysis (Pituitary)

Each lobe of the hypophysis has a different origin. The *adenohypophysis* (anterior lobe, pars distalis, anterior hypophysis) develops from an ectodermal primordium that is situated external to the oropharyngeal membrane and in close apposition to the floor of the forebrain where the *neurohypophysis* develops from a downward evagination of the diencephalon, the infundibulum (Fig. 2–12A). The adenohypophysial primordium soon develops into an adenohypophysial pouch (Rathke's pouch), which keeps its close contact with the diencephalic evagination, developing into the neurohypophysis (Fig. 2–12B). The stem of the adenohypophysial pouch is connected to the roof of the mouth, and residues of the stem may persist as a craniopharyngeal canal through the sphenoid bone. Cells from the adenohypophysis extend along and around the pituitary stalk, forming the *pars tuberalis* (Fig. 2–12C). Remnants of the adenohypophysial

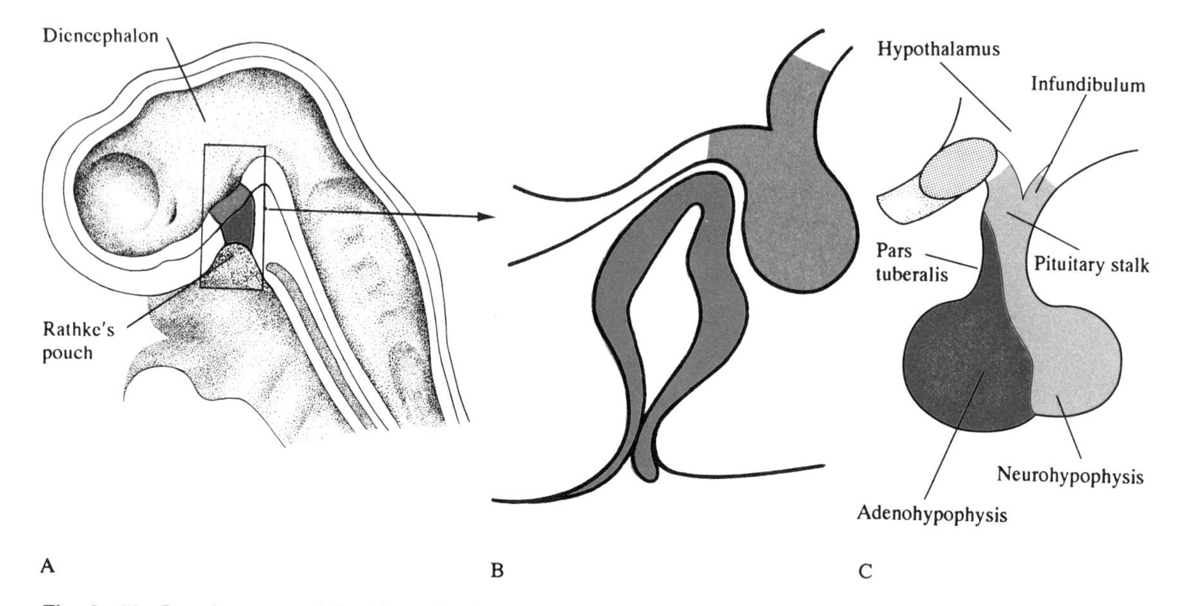

Fig. 2–12. Development of the Hypophysis
A–C. Successive stages in the development of the pituitary gland, which has two different origins. The neurohypophysis develops from the infundibulum in the floor of the diencephalon, whereas the adenohypophysis comes from Rathke's pouch in the primitive mouth cavity. (From Moore, K. L., 1977. The Developing Human: Clinically Oriented Embryology, 2nd ed. W. G. Saunders and Co., Philadelphia. With permission of the publisher.)

pouch sometimes give rise to a tumor called craniopharyngioma (see Clinical Notes).

Telencephalon

The Cerebral Hemispheres

The most dramatic developmental changes are seen in the forebrain where the two cerebral hemispheres rapidly expand to cover more and more of the rest of the brain (Figs. 2–13A and B). The expansion is not symmetric in all directions. The posteroventral part of the hemisphere, for instance, curves down and forward to form the temporal lobe, with the result that its anterior end, the temporal pole, gradually approaches the lower part of the frontal lobe (see arrows in Fig. 2–13C). The central lateral part of the hemisphere, the future insula, which is characterized by a more modest expansion, is therefore gradually covered by the rostral, dorsal, and posterior parts of the expanding hemisphere. These overlying parts are referred to as the *frontal*, *frontoparietal*, and *temporal opercula* (Fig. 2–13C). A small swelling, which appears on the anteroventral surface of the hemisphere at an early stage, enlarges and can soon be recognized as the olfactory bulb (Figs. 2–10B and 2–13B).

The large increase in the number of cortical cells and the local variations in the rate of proliferation result in an expanded surface area and the formation of grooves, i.e., sulci or fissures, that demarcate the different brain convolutions, gyri. Some of the more pronounced grooves, i.e., the *lateral sulcus*, *parieto-occipital sulcus*, and *preoccipital notch*, divide the hemisphere into four lobes: the frontal, parietal, occipital, and temporal lobes (Fig. 2–13C). The fifth lobe, the *insula*, is hidden in the depth of the lateral sulcus.

The Cerebral Cortex

At an early stage, the wall of the hemisphere, like that of the primordial spinal cord, consists of a ventricular zone with rapidly multiplying cells and a thin marginal zone whose cells derive from the ventricular layer and mature at an early stage (Fig. 2–14). The marginal zone becomes the most superficial layer, layer 1, in the mature cortex. Other young neurons migrate outward into an intermediate zone (arrow in Fig. 2–14C), where they form a cortical plate, which will give rise to cortical

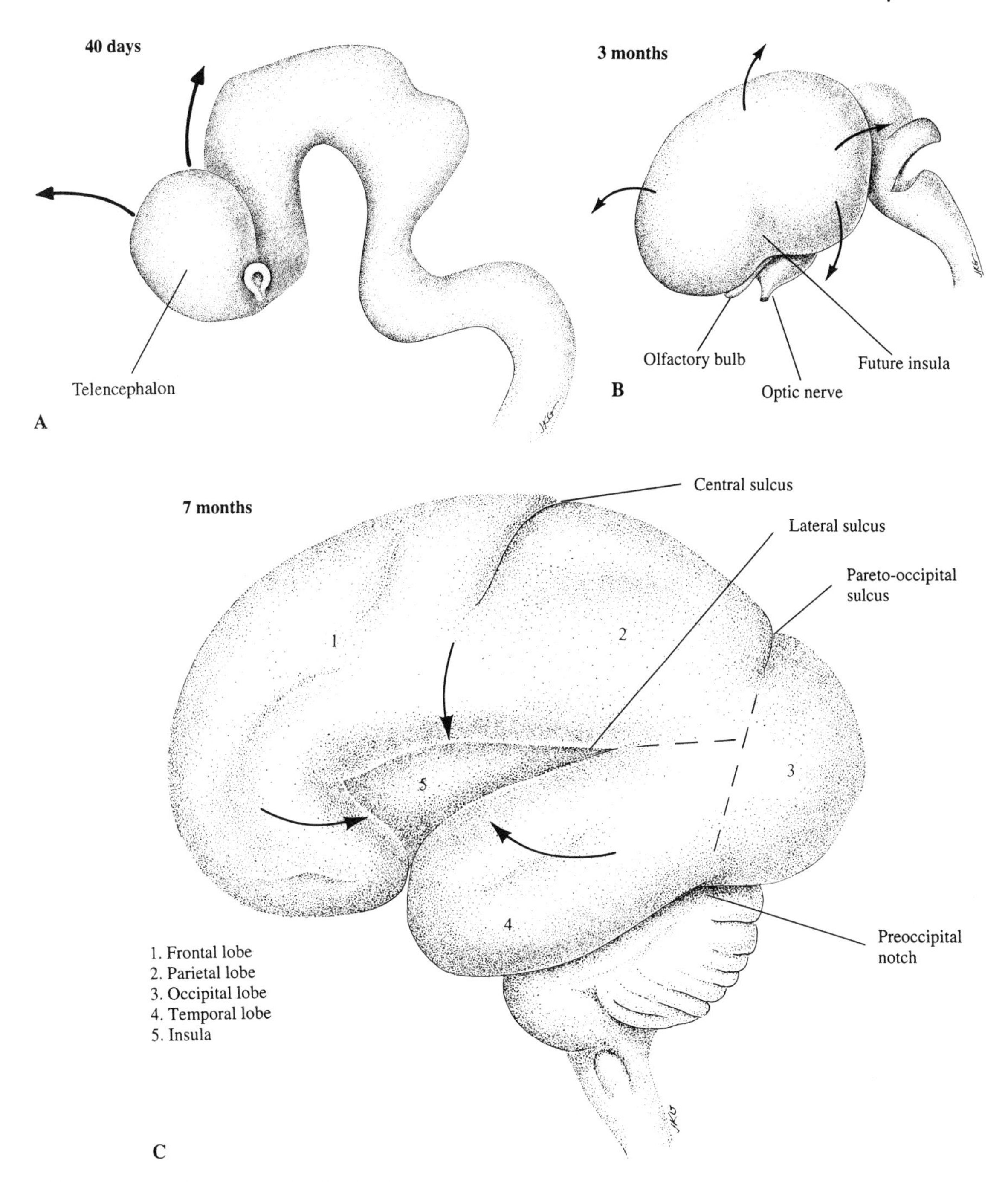

Fig. 2–13. Development of the Cerebral Hemisphere

A and B. The telencephalic vesicle expands rapidly to cover the rest of the brain, and the posteroventral part of the vesicle curves down and forward to form the temporal lobe.

C. The lateral surface of the brain of a 7-month-old foetus. The insular region is bounded by the frontal, frontoparietal, and temporal opercula. Some of the more pronounced grooves have appeared and the hemisphere can be divided into five lobes: the frontal, parietal, occipital, and temporal lobes, and the insula.

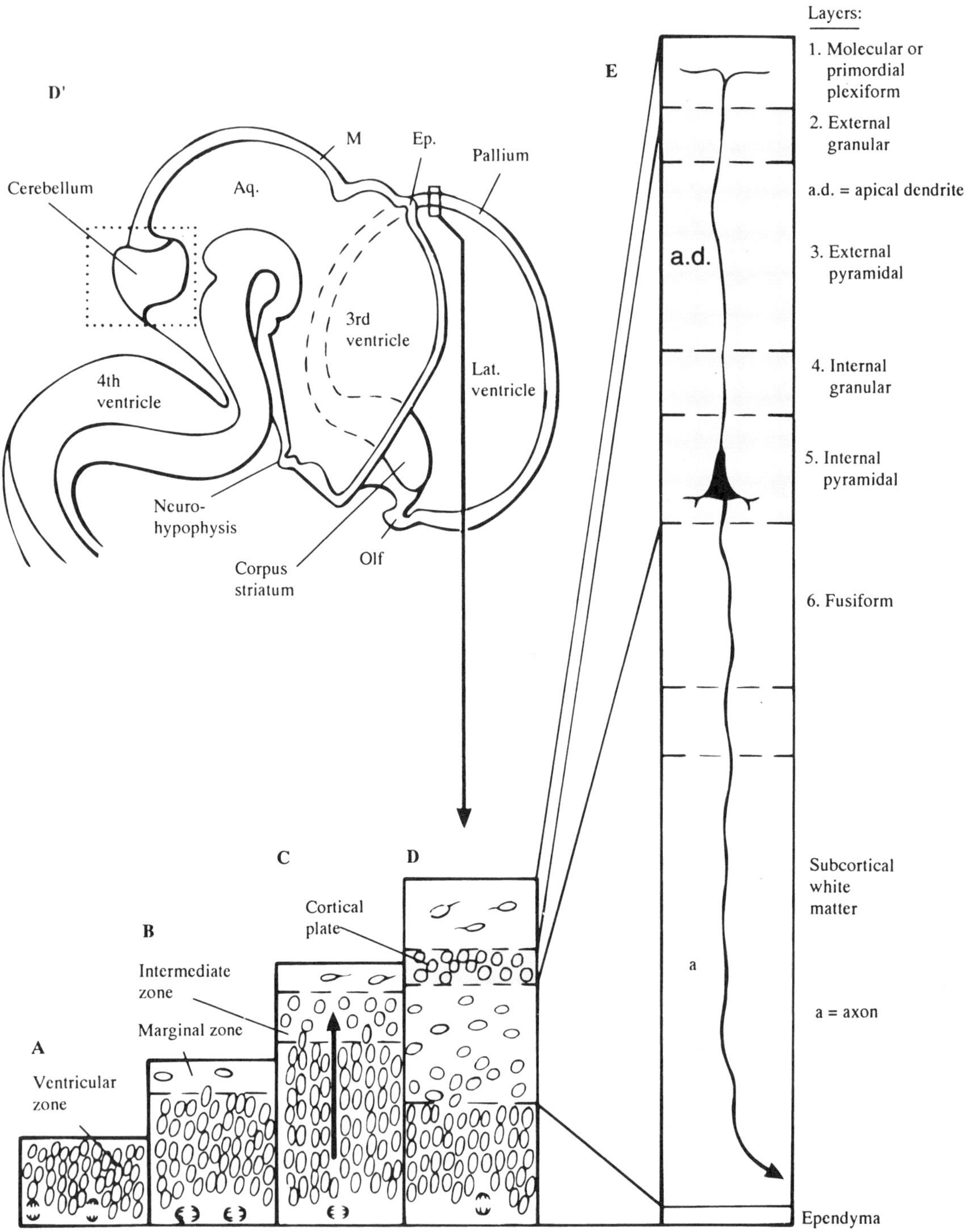

Fig. 2–14. Development of the Cerebral Cortex Schematic drawing showing the development of the cerebral cortex from the earliest stage of rapidly multiplying cells in the ventricular zone (**A**) to the definitive cortex (**E**). Layers 2–5 develop according to an "inside-out" sequence, as they derive from the cortical plate (*blue* in **D**), which begins to form inside the primordial plexiform layer towards the end of the embryonic period. (From O'Rahilly, R. and F. Müller, 1991. Human Embryology and Teratology. With permission from Wiley-Liss, a division of John Wiley and Sons, Inc., New York.)

layers 2–5[3]. This process is characterized by the fact that newly arrived cells migrate outward past their predecessors in the cortical plate. In other words, cortical layers 2–5 are gradually taking shape as newly arrived cells are occupying increasingly more superficial positions, i.e., they develop according to an "inside-out" sequence. The cells in layers 1 and 6, on the other hand, which are derived at an earlier stage, are developed according to an "outside-in" sequence. As in the rest of the neural tube, the ventricular zone becomes the ependyma. Although the cortical cytoarchitecture, as revealed in the laminar pattern, has developed its mature form at birth, the development of axonal and dendritic patterns and concomitant synapse formation continue well beyond the neonatal period.

The Ventricle System

The cavity in the rostral part of the neural tube develops into the ventricle system of the brain. The greatest morphologic changes take place in the cerebral hemisphere, in which a long curved cavity, the *lateral ventricle*, develops (Fig. 2–15). Its shape can best be understood if one recalls the development of the temporal lobe as a semicircular expansion of the ventral part of the hemisphere (Fig. 2–13). The different parts of the lateral ventricle are referred to as the *anterior horn*, *central part*, *posterior horn*, and *temporal horn*. The lateral ventricle is continuous with the *third ventricle* in the diencephalon through the *interventricular foramen* (of Monro). The third ventricle, which contains three prominent recesses, the optic, infundibular, and pineal recesses (Fig. 2–15B), is in turn continuous with the *fourth ventricle* in the rhombencephalon through a narrow canal, the *cerebral aqueduct*.

The brain ventricles are filled with a clear, colorless fluid, the *cerebrospinal fluid* (CSF), which is produced by the choroid plexus (*see below*). The fluid leaves the ventricular system through three openings in the fourth ventricle, the *median aperture of Magendie* in the midline and the two *lateral apertures of Luschka* in the lateral recesses.

Malformations of the ventricular system or disease processes that interfere with the normal flow of CSF can give rise to pathologic enlargement of the ventricular system, which is called *hydrocephalus* (see Chapter 3).

The Choroid Plexus

In certain regions of the developing brain, i.e., the medial surface of the hemispheres and the roof of the diencephalon and rhombencephalon, the wall remains very thin, consisting only of an ependymal layer. This layer invaginates into the ventricular cavity together with a highly vascularized mesenchyme to form the choroid plexus (Fig. 2–16), which produces the CSF. Where the telencephalic vesicle attaches to the diencephalon, i.e., the *hemisphere stalk* (Fig. 2–16B), the invagination in the medial wall of the telencephalon is continuous with the invagination in the roof of the diencephalon. The choroid plexus of the lateral ventricle, therefore, is directly continuous with the choroid plexus of the third ventricle at the level of the interventricular foramen. In the subsequent development, more and more fibers connect the rapidly developing hemispheres with the diencephalon and brain stem. These projection fibers gather primarily in the ventral and caudal walls of the hemisphere stalk, which increase rapidly in thickness thereby producing a broad attachment between the telencephalon and diencephalon. This situation may give rise to the erroneous impression that the two major parts of the forebrain, the hemisphere and the diencephalon, have fused (Figs. 2–16B and C). Some fusion, however, may take place at the region where the thin telencephalic (ependymal) wall of the lateral ventricle is attached to the dorsal surface of the thalamus, hence the name *lamina affixa thalami* (Fig. 2–16C).

After further expansion, the two hemispheres finally meet in the midline over the diencephalon. The mesenchyme, which thus becomes trapped between the hemispheres and the diencephalon, forms a highly vascularized connective tissue sheath, the *tela choroidea* or velum interpositum (Fig. 2–16D). The tela choroidea is located in the *transverse fissure* underneath the corpus callosum and the fornix in the fully developed brain (see below). The choroid plexus of the lateral ventricle appears as an invagination from the lateral edge of the tela choroidea, whereas the choroid plexus of the third ventricle represents invaginations from the inferior surface of the tela (Fig. 2–16D). As already indicated, however, the plexus in the lateral ventricle is continuous with the plexus in the third ventricle at the interventriclar foramen (Fig. 2–16E). Further, the choroid plexus is continuous with the pia mater through the *choroid fissure*, which

3. The cerebral cortex can be subdivided on cytoarchitectonic grounds into isocortex (neocortex) and allocortex. By far the largest area of the human cerebral cortex is represented by neocortex, in which six layers can be identified. It is the neocortex that during its development can be defined by the presence of a cortical plate.

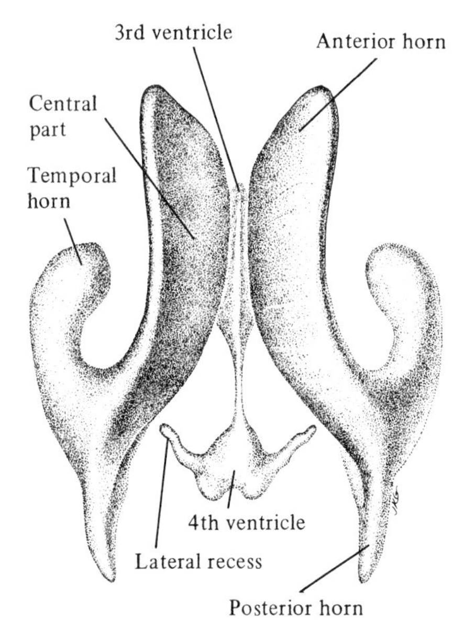

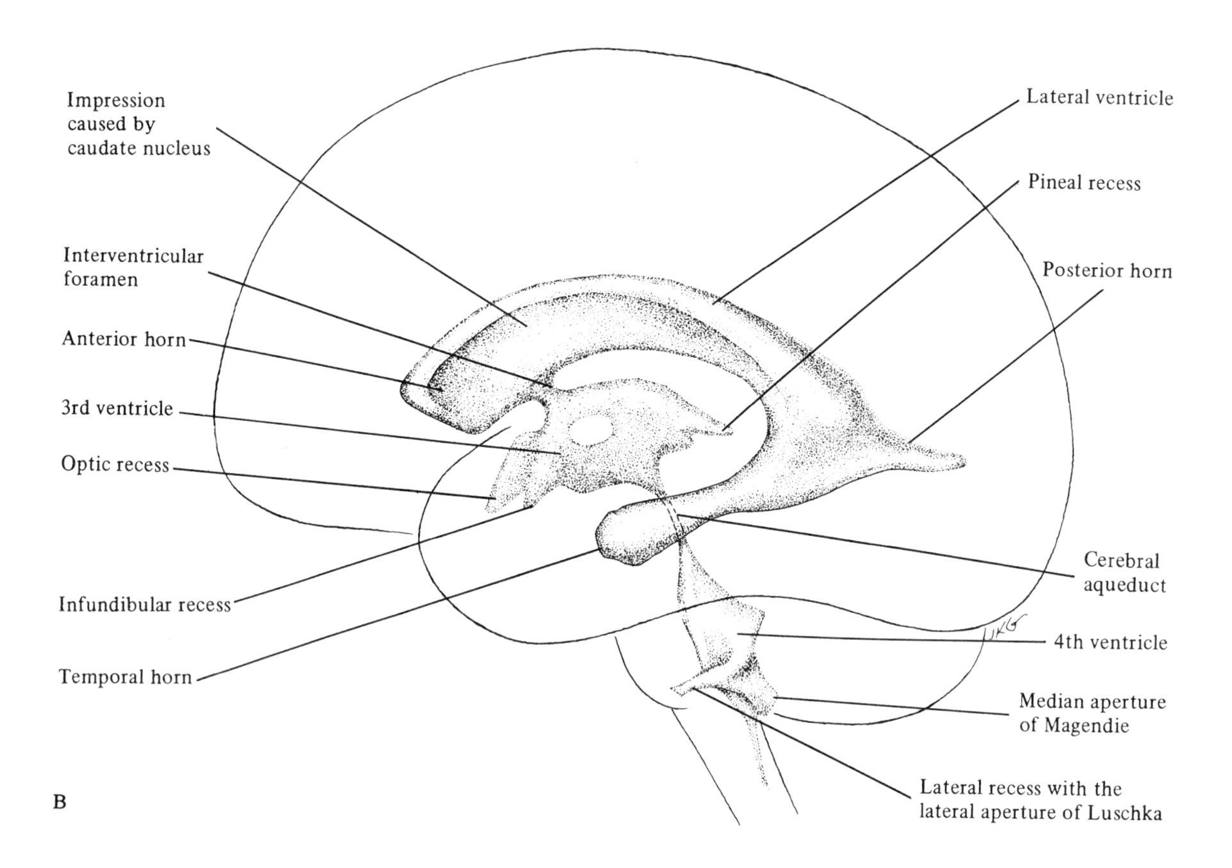

Fig. 2–15. Development of the Ventricle System With the appearance of the temporal lobe, the lateral ventricle develops into a long curved cavity. (**A**) Superior and (**B**) lateral views of the ventricle system. (Modified after Bailey, P., 1948. Intracranial Tumors. Charles C. Thomas, Publisher, Springfield, IL.)

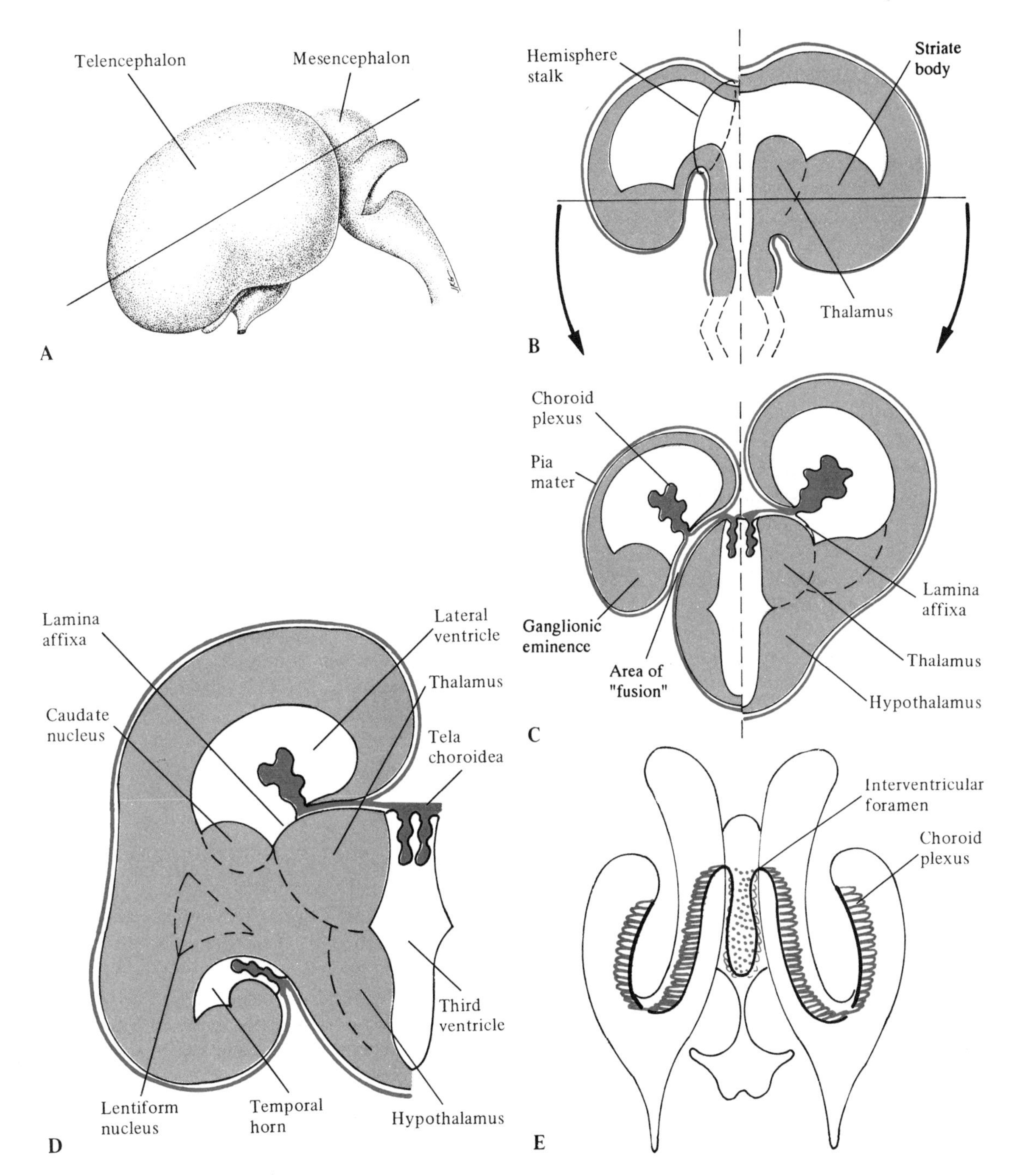

Fig. 2–16. Development of Choroid Plexus
A–C. In certain regions along the medial surface of the telencephalon and the roof of the diencephalon and rhombencephalon, the thin ependymal wall is pushed into the ventricular lumen by highly vascularized connective tissue, thus forming the choroid plexus. The line in **A** indicates the plane for the forebrain section shown in **B**, whereas the line in **B** illustrates the plane for the section in **C**. The sections to the right of the midline in **B** and **C** illustrate a somewhat later developmental stage than the sections to the left.

D and E. With the development of the temporal lobe, the choroid plexus of the lateral ventricle will eventually form a semicircular curve similar to that described by the lateral ventricle. (Modified after Kahle, W., 1978. Color Atlas and Textbook of Human Anatomy. Vol. 3: Nervous System and Sensory Organs. Georg Thieme Publishers. Stuttgart.)

indicates the site of invagination (black line in Fig. 2–16E). Since the choroid plexus in the central part of the lateral ventricle continues into the temporal horn, the choroid fissure forms a semicircular curve similar to that described by the developing temporal horn of the lateral ventricle.

Although the choroid plexus belongs to the circumventricular organs (Chapter 3), it has some unique features that distinguish it from the other circumventricular organs. In particular, it has an extensively folded surface consistent with its secretory role in the production of cerebrospinal fluid. The choroid epithelium is characterized by tight junctions, which provide for the blood–cerebrospinal fluid barrier (Chapter 3).

Corpus Callosum and Fornix

Fibers projecting from one developing hemisphere to the other, i.e., commissural fibers, cross the midline in a thickening in the embryonic lamina terminalis. This area increases rapidly in size to form the *commissural plate* (Fig. 2–17A). With the expansion of the hemispheres, the plate increases in size, overgrows the tela choroidea and the roof of the third ventricle, and emerges as the dorsally convex *corpus callosum* (Fig. 2–17B). Its curvature indicates the direction in which the hemispheres expanded. The different parts of the corpus callosum are referred to as the *rostrum*, *genu*, *body*, and *splenium*. The *anterior commissure*, which also develops from the original commissural plate, is another important bundle of commissural fibers that interconnect parts of the temporal lobes with each other. It crosses the midline just behind the upper part of the lamina terminalis (Fig. 2–17B).

The *septum pellucidum* is a thin wall that separates the anterior horns of the lateral ventricles behind the genu of corpus callosum. In the ventral free edge of the septum, a bundle of nerve fibers appears on each side of the midline. This is the *fornix*, which forms an arch over the thalamus. The two fornices diverge markedly as they sweep down as the *crura fornicis* behind the two thalamic nuclei into the temporal lobes. The last flattened part of the fornix, the *fimbria*, reaches the hippocampus in the medial wall of the temporal horn. The two crura fornicis meet to form the *body of the fornix* underneath the corpus callosum. However, they separate again and proceed forward and downward as the *columns of the fornix* in the ventral free edge of the septum to reach the *mammillary bodies* in the hypothalamus.

Commissural fibers connecting the hippocampal formations on the two sides cross over between the two crura fornices to form the *hippocampal commissure*. It is situated underneath the splenium of the corpus callosum to which it is intimately attached (Fig. 2–17C).

Striatum and Globus Pallidus

Soon after the hemispheres have appeared, their basal portion thickens rapidly through the proliferation of cells, thereby forming an especially thick part of the ventricular zone referred to as the *ganglionic eminence*. This structure first appears as a slight swelling in the telencephalic cavity (Fig. 2–18B) and gradually forms a longitudinal ridge in the floor of the lateral ventricle. The dorsal part of the hemisphere, which covers the ganglionic eminence like a mantle, develops into the *cerebral cortex*.

The ganglionic eminence contributes cells to a number of telencephalic subcortical structures including the *caudate nucleus* and *putamen*, together referred to as the *striatum*, as well as the *amygdaloid body* and the closely related *bed nucleus of the stria terminalis* (Chapter 20).

The *striatum*, which represents the most prominent part of the collection of nuclei referred to as the *basal ganglia* (Chapter 16), appears at the outset as a single cellular mass. With the rapid growth of the cerebral hemisphere, however, there is a concomitant increase in the number of fibers that connect the expanding hemisphere with the thalamus and the rest of the brain. These fibers project through the original mass of young neurons, which thereby acquires a striated appearance, hence the name *striate* body. Many of the fibers penetrating the striate body concentrate to form a sheet of white substance, the *internal capsule*, which divides the original striate body into caudate nucleus and putamen (Fig. 2–18B). The dorsomedially located

Fig. 2–17. Development of Corpus Callosum and Fornix

A. Median section in a 10-week embryo showing the commissural plate and its development into the corpus callosum.

B. Median section through the corpus callosum and diencephalon.

C. Three-dimensional drawing illustrating the relationships between corpus callosum, fornix, and hippocampus.

⟶

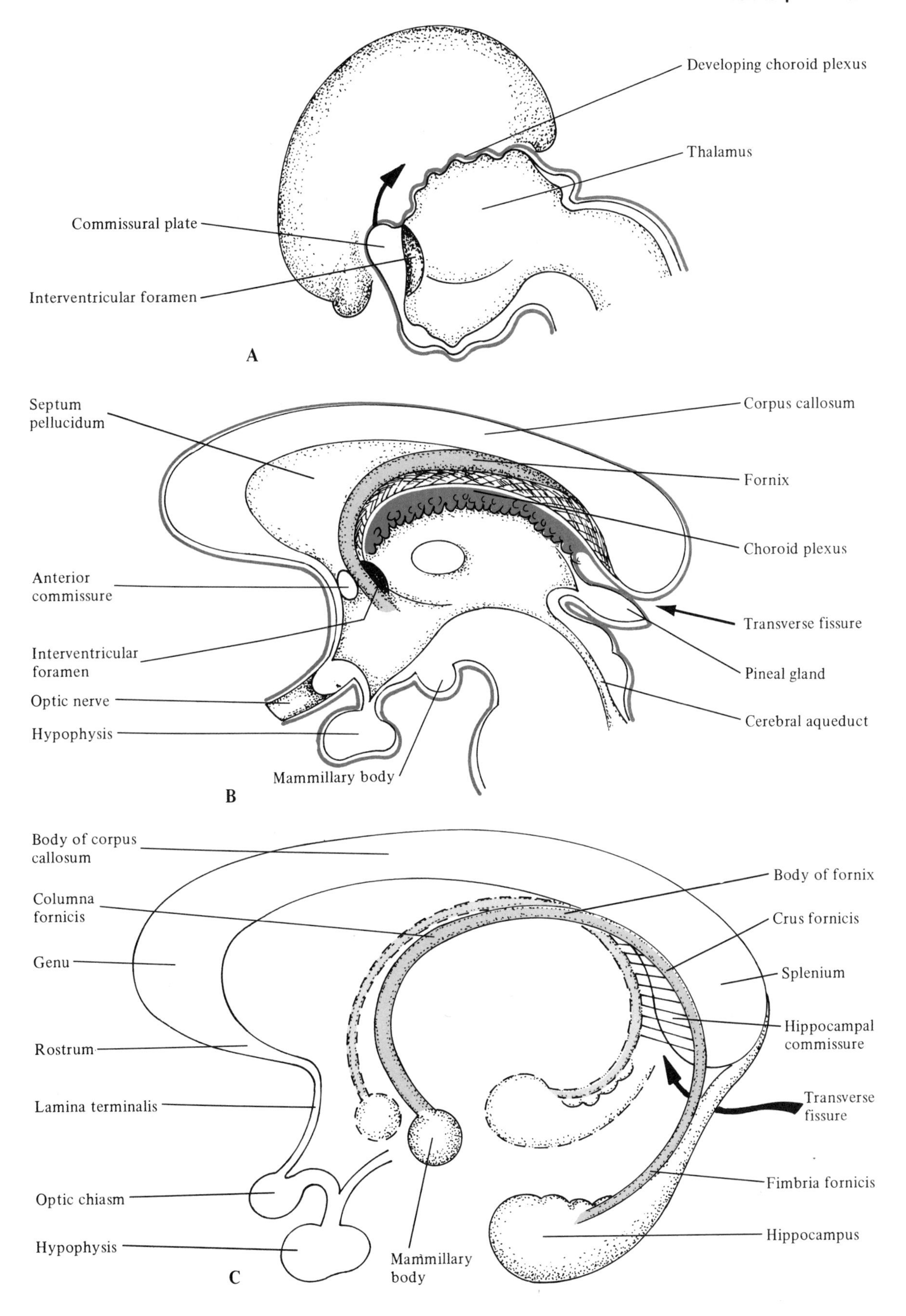
Developing choroid plexus
Thalamus
Commissural plate
Interventricular foramen
A
Septum pellucidum
Corpus callosum
Fornix
Choroid plexus
Anterior commissure
Transverse fissure
Interventricular foramen
Pineal gland
Optic nerve
Cerebral aqueduct
Hypophysis
Mammillary body
B
Body of corpus callosum
Body of fornix
Columna fornicis
Crus fornicis
Genu
Splenium
Hippocampal commissure
Rostrum
Transverse fissure
Lamina terminalis
Fimbria fornicis
Optic chiasm
Hippocampus
Hypophysis
Mammillary body
C

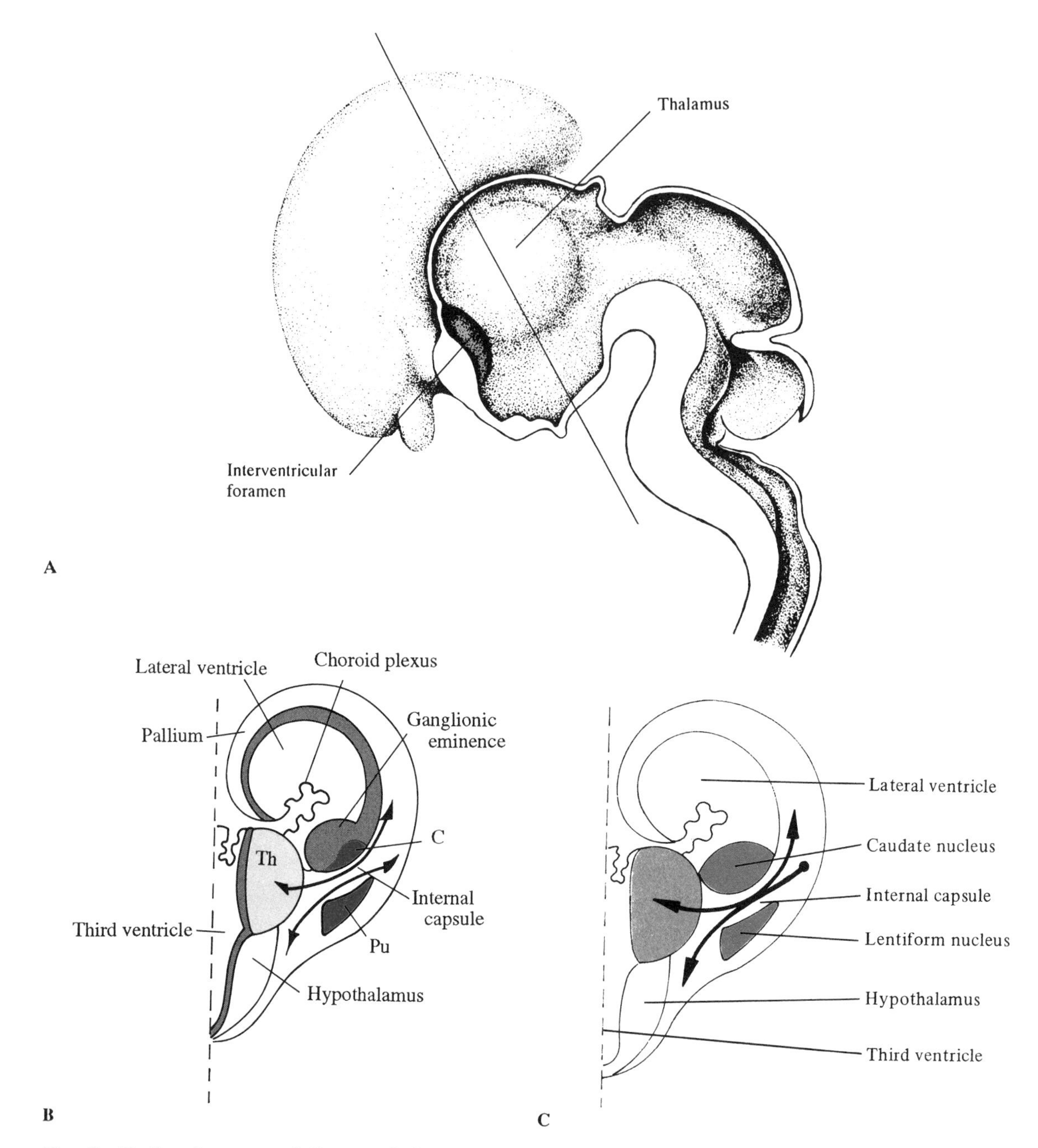

Fig. 2–18. Development of Corpus Striatum and Internal Capsule

A. The medial surface of the brain in a 43-mm embryo. The thalamus is developing in the diencephalon, and corpus striatum in the basal part of the telencephalon. Part of the ganglionic eminence can be seen through the interventricular foramen.

B. Transverse section through the developing brain. C = caudate nucleus, Pu = putamen, Th = thalamus. (Modified from Sidman, R. L. and P. Rakic, 1982. Development of the Human Central Nervous System. In W. Haymaker and R. D. Adams (eds.): Histology and Histopathology of the Nervous System. Charles C. Thomas, Springfield, Illinois.)

caudate nucleus bulges into the lateral ventricle, whereas the putamen is located close to the lateral surface of the hemisphere. In this process of gradual expansion of the cerebral hemisphere, the *hemisphere stalk*, which connects the hemisphere with the diencephalon (Fig. 2–16B), increases in thickness as more and more fibers accumulate in the basal and posterior walls of the *interventricular foramen*. As a result, the interventricular foramen becomes smaller and the hemisphere gradually acquires a broad attachment to the diencephalon.

On the medial side of the putamen another group of young neurons develops into the *globus pallidus*, another important part of the basal ganglia. The globus pallidus consists of two segments, the *external* and *internal segments*. There has been much speculation regarding the origin of the globus pallidus, especially in regard to the external segment, part of which may be of telencephalic origin. It is generally accepted, however, that a diencephalic germinal zone produces the internal pallidal segment, which during development is pulled into the basal telencephalon to join the external segment. Whereas the striatum represents the input side of the basal ganglia, the globus pallidus is the major output station (Chapter 16). This is clearly reflected by the fact that the two parts of the striatum, i.e., the caudate nucleus and the putamen, although fortuitously separated by the fibers of the internal capsule, nonetheless share similar histologic and histochemical characteristics. The globus pallidus has a very different cellular composition, which conforms to its function as an efferent structure in the basal ganglia. The putamen and the globus pallidus, although very different in character, form the *lentiform nucleus*.

Another event concerning the development of the caudate nucleus deserves special notice. The caudate nucleus is closely related to the lateral ventricle throughout its extent, and its caudal part, the *tail*, describes a semicircular curve similar to that displayed by the ventricle as it reaches into the temporal lobe. Since the caudate nucleus is situated in the floor of the central part of the lateral ventricle, it is easily appreciated that the tail of the caudate forms part of the roof of the temporal horn (Fig. 2–19A). The other part of the striatum, the putamen, which is situated close to the lateral surface underneath the future *insula*, is not affected to the same extent by the development of the temporal lobe.

Topography of Corpus Striatum, Thalamus, and Internal Capsule

The development of the corpus striatum in the telencephalon is matched by the growth of another nuclear mass, the *thalamus*, in the wall of the diencephalon (Figs. 2–10B and 2–18B). Whereas the thalamus is in direct contact with the caudate nucleus, it is separated from the lentiform nucleus by the *internal capsule*. These relationships can be appreciated in Figs. 2–18 to 2–20.

The internal capsule, which is V-shaped in horizontal sections, is located on the medial side of the lentiform nucleus (Fig. 2–19B). The anterior limb of the capsule is located between the head of the caudate and the lentiform nucleus, whereas the posterior limb separates the thalamus from the lentiform nucleus. The angle between the two limbs is referred to as the *genu*. The extent of the internal capsule becomes more evident in a three-dimensional drawing (Fig. 2–20). The posterior part of the internal capsule literally wraps around the posterior half of the lentiform nucleus, forming the *retrolenticular* and *sublenticular* parts of the internal capsule.

The fibers on the outside of the lentiform nucleus form the *external capsule* (Fig. 2–19B), whereas the *extreme capsule* separates another nucleus, the *claustrum*, from the cortex of the insula.

The shape of the lentiform nucleus resembles a somewhat compressed ice cream cone with the tip directed medially. It is not a homogeneous structure like the caudate nucleus but can be easily separated, even on macroscopic inspection, into *putamen* and *globus pallidus* (Fig. 2–19B). The putamen and caudate nucleus are characterized by a large number of densely packed small neurons, and their common origin is further revealed by the fact that the two structures are in continuity with each other through cell bridges located between the fiber bundles of the internal capsule. This is especially evident in the anterior limb of the internal capsule (Fig. 2–19B). The globus pallidus is characterized by large spindle-shaped cells that lie rather far apart. It also contains a large number of myelinated fibers, which explains its pale color. Considering these morphologic and developmental aspects, it seems reasonable to underscore the distinction between the *pallidum* (globus pallidus) and the *striatum* (caudate nucleus and putamen). *Corpus striatum* is used as a collective term for the caudate

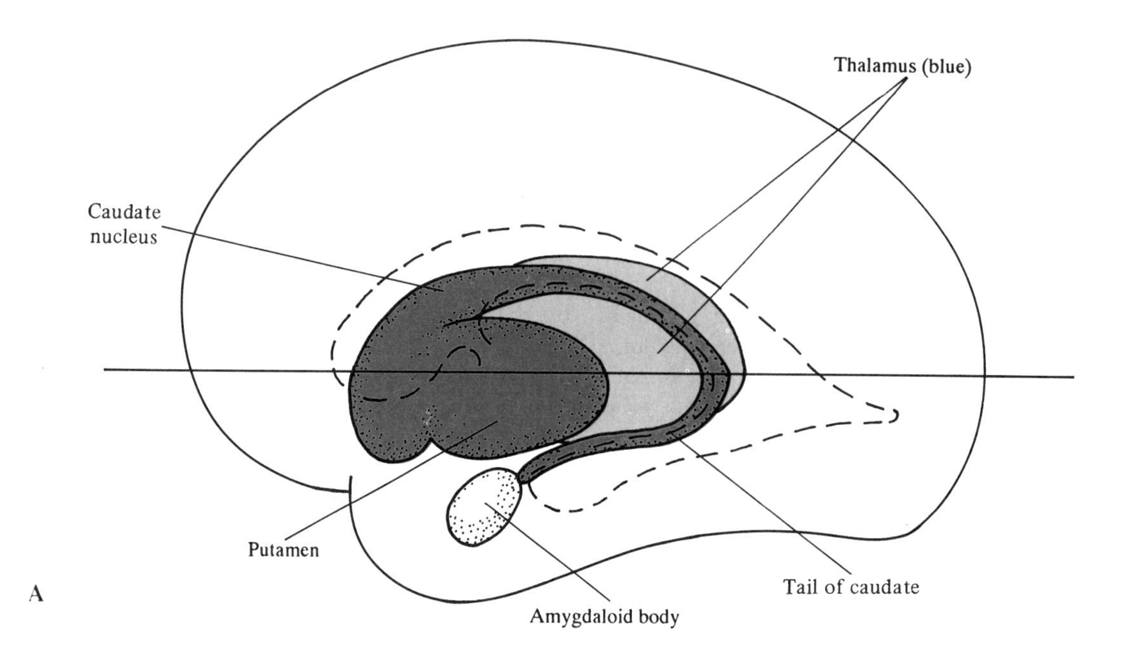
Thalamus (blue)
Caudate nucleus
Putamen
Tail of caudate
Amygdaloid body
A

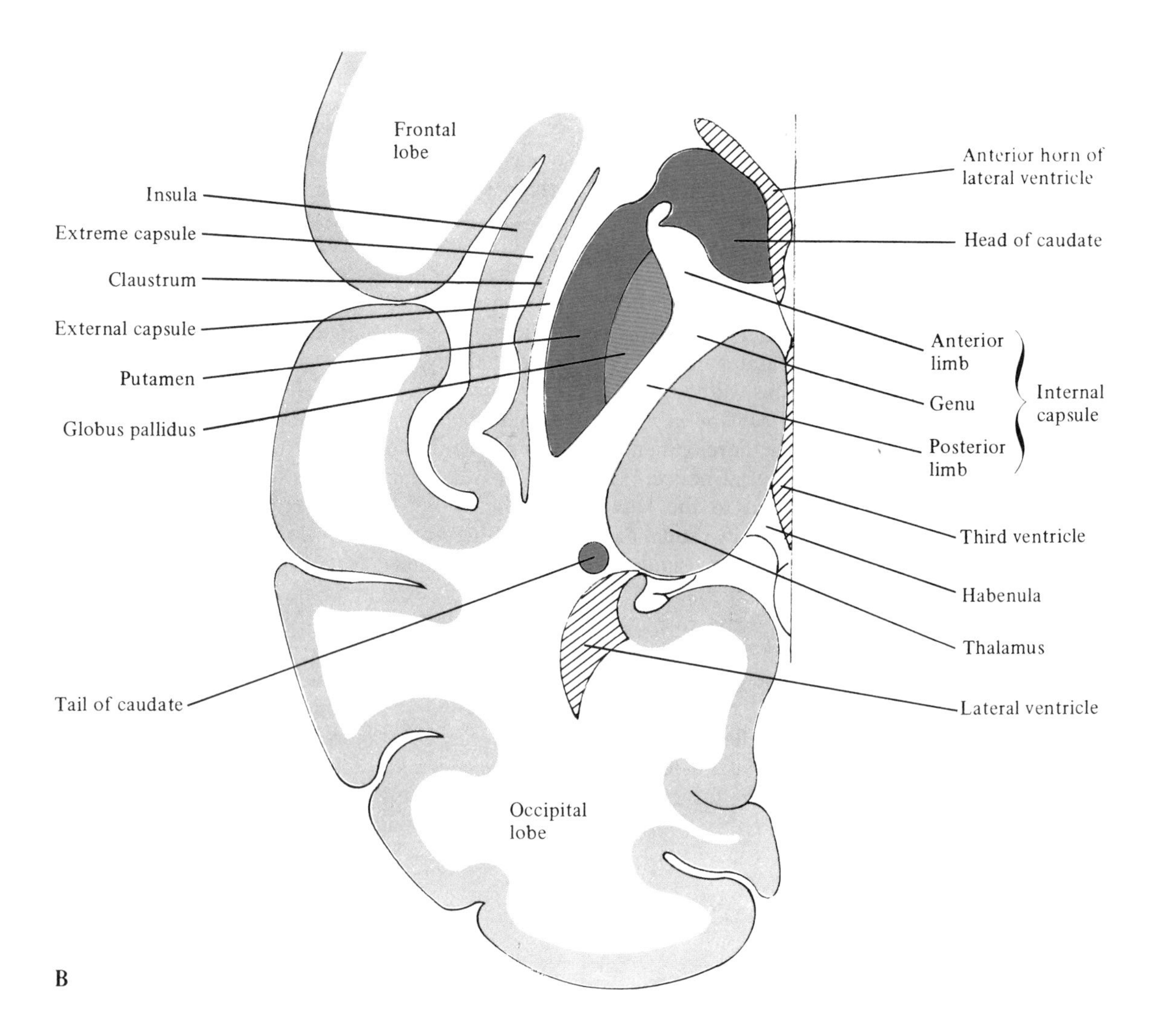
Frontal lobe
Insula
Extreme capsule
Claustrum
External capsule
Putamen
Globus pallidus
Tail of caudate
Anterior horn of lateral ventricle
Head of caudate
Anterior limb
Genu
Posterior limb
Internal capsule
Third ventricle
Habenula
Thalamus
Lateral ventricle
Occipital lobe
B

←

Fig. 2–19. Topography of Corpus Striatum, Thalamus, and Internal Capsule
A. Schematic drawing showing the location of the striatum and amygdaloid body and their relationship to the ventricle system in the cerebral hemisphere (side view).
B. Horizontal section (as indicated in **A**) through the corpus striatum, thalamus, and internal capsule.

nucleus, putamen, and globus pallidus, which together represent the main components of the basal ganglia (Chapter 16).

The topographic anatomy of the basal ganglia and the internal capsule is complicated and does not lend itself easily to verbal descriptions and illustrations. The relationships, however, are beautifully revealed in brain dissections.

Clinical Notes

The nervous system is characterized by a protracted maturation, and a variety of harmful environmental factors can derail its normal development. Although cell multiplication and migration usually are completed before birth, some of the granular cells in the hippocampus and cerebellum arise postnatally. Further, development of neuronal processes and establishment of synaptic connections continue into early childhood and probably even beyond. Most of the glial cells also seem to arise postnatally, and some of the fiber systems are not fully myelinated until the second or third year after birth.

Aging of the Brain

Although this chapter is devoted to developmental events and their relation to the anatomy of the CNS, and to some malformations that may be evident early in life, it is important to remember that the nervous system continues to change throughout life. Since elderly persons make up an increasingly large percentage of patients seen by most physicians, age-related changes are especially important, and the most consequential of these appear in the nervous system. Such changes start already in the third and fourth decades, but do not usually become noticeable until the fifth or sixth decade in the form of increased forgetfulness and perhaps some decline of intellectual abilities. Minor motor and sensory deficits may also become evident. These symptoms, which generally do

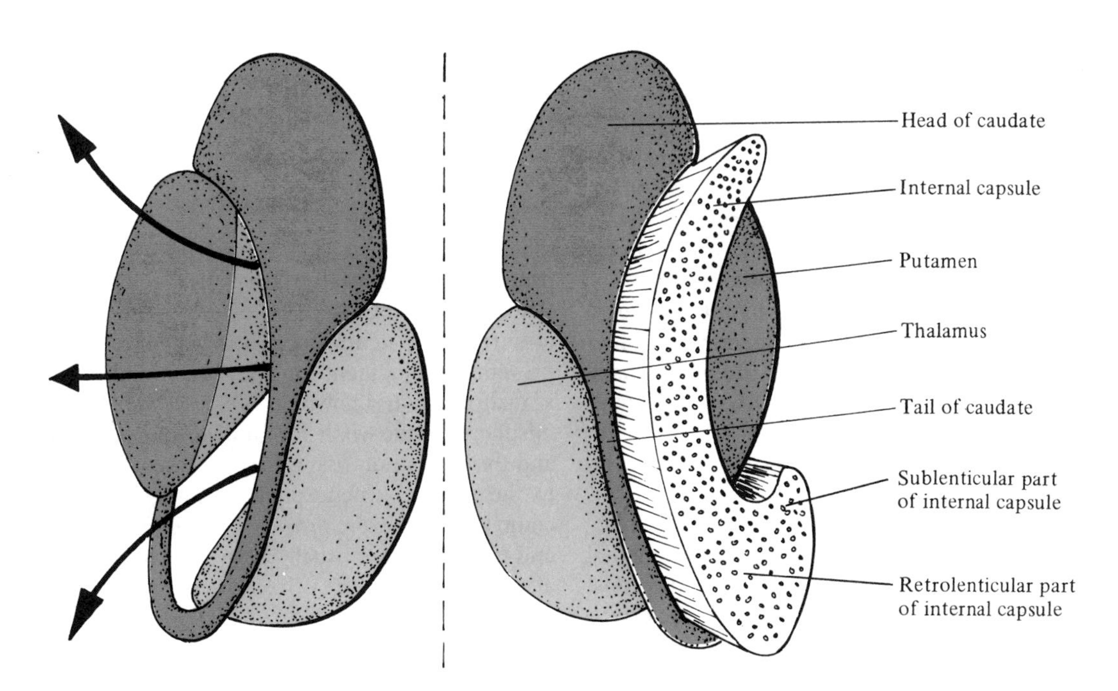

Fig. 2–20. Internal Capsule
Three-dimensional drawing showing the topography of the internal capsule, corpus striatum, and thalamus (Modified after Benninghoff, A., 1954. Lehrbuch der Anatomie des Menschen. Dritter Band: Nervensystem, Haut und Sinnes-organe. Urban & Schwarzenberg, München-Berlin.)

not reduce the quality of life unless they become pronounced, are due to degenerative neuronal changes, including a reduction in the number of neurons and synaptic connections. In fact, there is normally an estimated 10–15% reduction of brain weight from the third or fourth decade to the ninth decade. Another well-known phenomenon in old age is the increased accumulation of *lipofuscin* (aging pigment) in neuronal cell bodies. The yellowish-brown pigment granules represent lipid-containing residues of lysosomal and mitochondrial membranes. If the loss of intellectual abilities becomes severe enough to interfere with normal functioning, the person is said to suffer from dementia. The most common form of dementia is caused by Alzheimer's disease, in which there is a pathologic acceleration of the aging process (see Clinical Notes, Chapter 21).

Chromosomal Abnormalities and Environmental Factors

Chromosomal and Genetic Factors

Chromosomal anomalies and abnormal genes are responsible for many forms of *mental retardation*. A well-known example is *mongolism* (Down's syndrome or trisomy 21). The child has 47 chromosomes instead of 46, with an extra chromosome 21 from meiotic nondisjunction. It is recognizable at birth because of characteristic body features, including medial epicanthal folds with concomitant slanting of the eyes. The main disturbances in Down's syndrome and in some other metabolic and infectious disorders (e.g., phenylketonuria and congenital rubella, see below) are in large part a reflection of the faulty development and elaboration of neuronal processes and synaptic connections. Down's syndrome can be diagnosed in utero by cytogenetic examination of the amniotic fluid, a procedure now frequently recommended for older mothers, who have increased risk of chromosomal abnormalities. If the test is positive for Down's syndrome, a reasonable case can be made for abortion in order to prevent the birth of a child with mongolism.

A variety of *metabolic disorders*, including phenylketonuria, are also characterized by mental retardation. Phenylketonuria is one of the most common forms of aminoaciduria. It is characterized by deficiency of the hepatic enzyme phenylalanine hydroxylase, which is necessary for the conversion of phenylalanine to tyrosine.

Maternal Infections

A *congenital rubella* (measles) infection, especially in the first or second trimester, may cause mental retardation and malformations of the eyes, ears, and heart. Every female who has not had a rubella infection should therefore be vaccinated against the disease before reaching childbearing age.

Toxoplasmosis, which can cause mental retardation and severe malformations of the brain, is usually not recognized in the pregnant woman. It is therefore difficult to estimate the frequency with which the intracellular parasite *Toxoplasma gondii* affects the neonatal brain.

Radiation, Chemical Substances, and Drugs

Among the many toxins known to be harmful to the developing brain are *radiation*, *steroid hormones*, and various *chemical substances* including some drugs taken by the pregnant woman. *Thalidomide* is a well-known example, but there may be other potentially harmful drugs, including some *antiepileptic*, *antipsychotic*, and *antianxiety* drugs (tranquilizers).

Although we know too little about the long-term effects of "street drugs" such as marijuana, lysergic acid diethylamide (LSD), and phencyclidine ("angel dust"), it is a tautology to say that they are harmful. Excessive alcohol consumption is also a formidable health problem; it can cause a number of morbid conditions including irreversible brain damage, referred to as Wernicke's encephalopathy. Alcohol is also a fetal pathogen, and it should therefore be avoided by the pregnant woman.

Neural Tube Defects

The development of the neural tube is a critical process. If the neural groove fails to close, neural tissue remains exposed to the surface. A defective closure of the neural tube at the cranial end may prevent development of the brain and lead to a malformation known as *anencephaly*. The hemispheres or the whole forebrain are usually absent, and the little brain tissue that is present is exposed to the surface. Anencephaly is almost always accompanied by severe *myeloschisis* (cleft spinal cord), and the infants are usually born dead or die shortly after birth.

Spina bifida refers to a defect of the vertebral column. There are many variations ranging from *spina bifida occulta* (Fig. 2–21A), in which the abnormality is confined primarily to the vertebrae

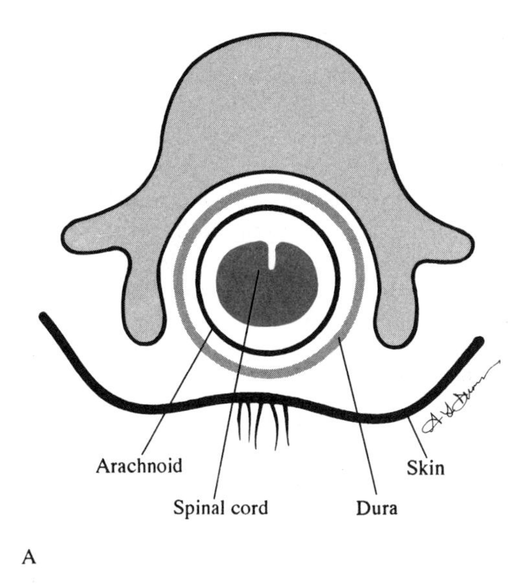

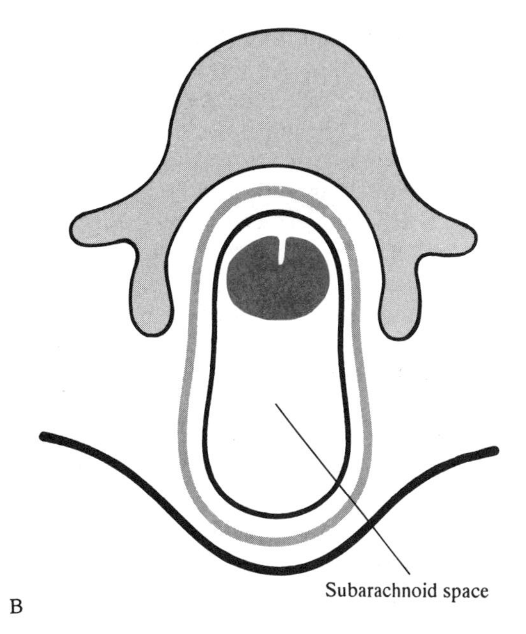

Placode

C

Fig. 2–21. Rachischisis and Spina Bifida
Diagrammatic features of various types of spina bifida. **A** = spina bifida occulta; **B** = meningocele; **C** = myelomeningocele. (From Escourolle, R. and J. Poirier, 1978. Manual of Basic Neuropathology. Translated into English by L. J. Rubinstein, 2nd. ed. W. B. Saunders Co., Philadelphia. With permission of Masson S. A. Paris.)

(unclosed posterior arch), to defects that involve the spinal cord and the meninges. Although spina bifida occulta can be completely asymptomatic, it may also be associated with spinal cord anomalies, e.g., splitting of part of the spinal cord, with concomitant anomalies in the lower limbs. This possibility should always be contemplated if a tuft of hair is present in the lumbar region.

Spina bifida is often accompanied by a posterior midline protrusion including some of the content of the vertebral canal. If only the meninges protrude through the defect, which is rare, the condition is referred to as a *meningocele* (Fig. 2–21B). A more severe and more frequent form is *myelomeningocele* (Fig. 2–21C), in which elements of the CNS or the neural placode protrude as well. This condition, which is often referred to as "spina bifida" in clinical jargon, is usually associated with neural deficits (e.g., bladder dysfunction or paralysis and lack of sensation in the legs), cerebellomedullary malformation, and hydrocephalus. Cerebello-medullary malformation, with or without myelomeningocele, is called Arnold–Chiari malformation (see Clinical Example). Whereas in the first half of this century babies born with spina bifida often succumbed to infections or progressive hydrocephalus, medical advances now make it possible for many of these children to survive.

Hydrocephalus

This is a condition characterized by an accumulation of CSF in the ventricle system. Congenital hydrocephalus often results from an obstruction occurring somewhere along the CSF pathway, especially where the passage is narrow, as in the cerebral aqueduct or in the region of the foramina of the fourth ventricle. Hydrocephalus causes enlargement of the cerebral ventricles and thinning of the cerebral substance. Congenital hydrocephalus is often apparent at birth or soon thereafter because of enlargement of the head, widening of the sutures, and bulging fontanelles (see Clinical Notes, Chapter 3).

Agenesis of Corpus Callosum

Disturbances in the development of the commissural plate may result in partial or complete lack of corpus callosum (see Clinical Example). This defect does not reveal itself easily during life, and the condition, if unaccompanied by other defects, is therefore often discovered by accident during a radiologic examination or at autopsy. However, agenesis of corpus callosum can be discovered by the application of special tests for interhemispheric transfer of information (see Chapter 22).

Microcephaly

An unusually small brain with a reduced number of brain cells can result from a variety of conditions, including environmental factors (e.g., hypoxia, radiation, and malnutrition) and chromosomal disorders. Microcephaly is responsible for many cases of mental retardation.

Craniopharyngioma

Epithelial rests of the adenohypophysial pouch in the region of the hypophysis sometimes give rise to a tumor that usually expands dorsally to invade the hypothalamus and the third ventricle. The clinical picture is characteristically diverse and dependent on the affected cell groups and pathways; involvement of the visual pathways is common. Hydrocephalus may occur if the flow of CSF through the third ventricle is impaired.

Disorders of the Eye

The development of the eye presents many aspects of great importance in clinical medicine. Of special significance is the fact that the retina is an outgrowth of the prosencephalon and that the optic nerve is a central brain tract. When clinicians look at the retina through the ophthalmoscope, they are actually looking at a part of the CNS.

Papilledema

The optic nerve is surrounded by the same membranes that cover the brain, and the subarachnoid space that surrounds the brain is in direct communication with a similar potential space around the optic nerve. An increase in the intracranial pressure, therefore, is directly transmitted to the subarachnoid space surrounding the optic nerve, with subsequent swelling of the optic disc, i.e., papilledema, which is one of the cardinal signs of increased intracranial pressure (see Fig. 13–15).

Coloboma

This malformation results from an incomplete fusion of the choroid fissure. It may be restricted to a cleft in the iris or may extend into the optic nerve.

Retinal Detachment

Retinal detachment is a pathologic separation of the retina from the pigment layer. It is in effect a reformation of the space originally present in the optic vesicle, and it can be caused by many conditions, including tumors and infections. It can also occur in response to trauma, especially in myopic eyes.

Clinical Example

Arnold–Chiari Malformation with Myelomeningocele (Chiari II)

A 36-year-old mother gave birth to a baby girl, who presented with a posterior midline protrusion in the lumbar region covered by delicate but intact skin. The girl's head was slightly enlarged, but her anterior fontanelle was soft and not bulging. Neurologic examination revealed weak dorsiflexion

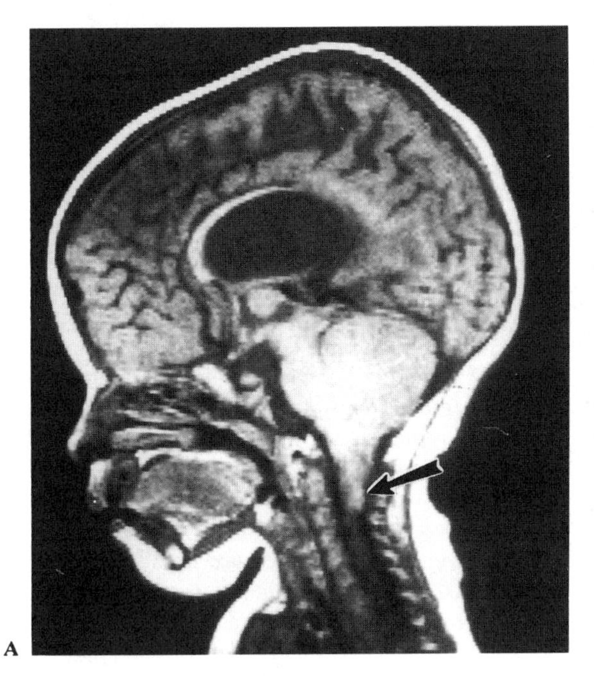

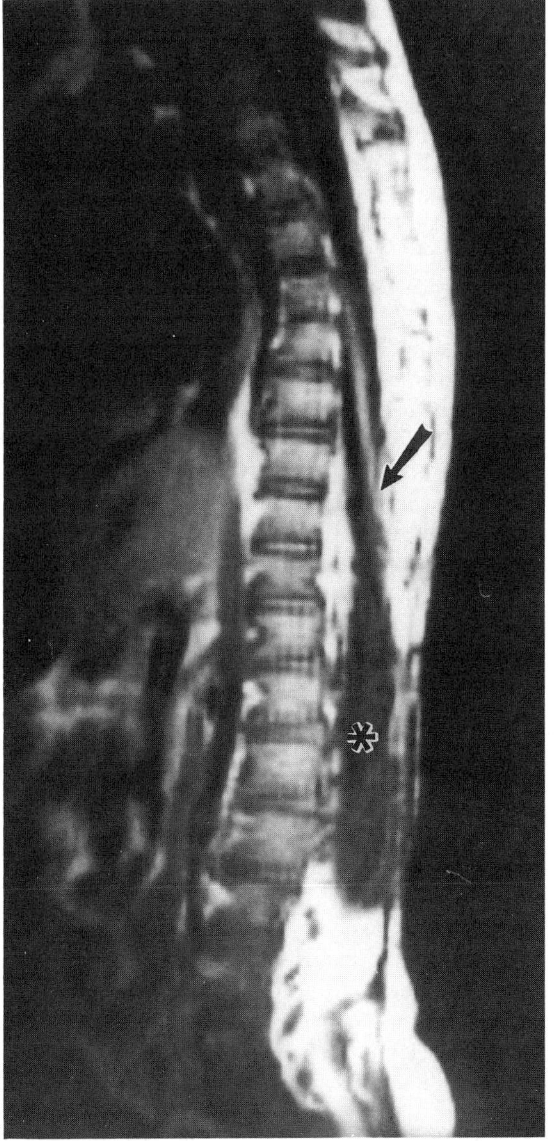

Fig. 2–22. Arnold–Chiari Malformation with Myelomeningocele

A. T1-weighted sagittal MRI of brain and cervical spinal cord shows elongation and downward displacement of the brain stem with tonsillar herniation (*arrow*). Note also agenesis of posterior portion of corpus callosum. (Photographs, courtesy of Dr. Paul Francel, University of Virginia.)

B. T1-weighted sagittal MRI of thoracolumbar spinal cord demonstrates feathering of cord (*arrow*), dilated thecal sac (*star*), and posterior bony defect.

and lack of plantar flexion, minimal anal wink (reflex contraction of the external anal sphincter when the surrounding perineal region is touched), and sensory loss below L5. A head ultrasound indicated abnormally large ventricles, but there was no evidence that the CSF was under increased pressure.

The patient was operated for her myelomeningocele (Fig. 2–22B) the day she was born; the neural placode was dissected free from the surrounding epithelium and was allowed to sink back into the widely opened vertebral defect, which was then covered by lyodura (lyophilized cadaver dura), a myofascial layer and skin. Although the child probably had a mild hydrocephalus (Chapter 3), no surgical intervention to reduce the amount of intraventricular CSF was undertaken, in the hope that the condition would correct itself. This, however, was not the case. Instead, signs of increased intracranial pressure, including irritability and bulging fontanelle, appeared several days after the operation, and a new ultrasound examination showed a further widening of the cerebral ventricles. The decision was therefore made to put in place a ventriculoperitoneal shunt. The shunt, in the form of a small plastic tube, was inserted into the right occipital horn with the tip of the shunt reaching to the region of the interventricular foramen. Located inside the tube is a small valve to control the pressure of the fluid in the skull and the direction of the flow of the CSF, which drains through the shunt into the peritoneal cavity or sometimes through a vein into the heart. Postoperatively, the ventricles decreased in size and the growth of the head progressed in normal fashion.

The patient was followed on a regular basis in a neurosurgical clinic, and she recovered some of the strength in her legs. A few months later, however, the infant appeared to have difficulties in swallowing manifested by poor feeding and prolonged feeding time. She also showed signs of respiratory distress with transient occurences of "noisy breathing" or laryngeal stridor (a harsh vibrating sound heard during respiration as a result of obstructed airways) and a few episodes of apnea (transient cessation of respiration). A neurologic examination revealed that the infant had weakness in the upper arms and a diminished gag reflex (constriction of the pharynx when its posterior wall is touched). Laryngoscopic examination indicated a laryngeal palsy with vocal cord paralysis on one side. Considering these symptoms, the most likely explanation seemed to be an Arnold–Chiari or cerebello-medullary deformity, which often accompanies myelomeningocele. The presence of an Arnold–Chiari deformity was evident from the midsagittal magnetic resonance image (MRI) of the head (Fig. 2–22A). Characteristically, the brainstem is displaced downward with the medulla oblongata and cerebellar tonsils herniating through the foramen magnum.

The patient underwent a posterior fossa-cervical decompression by removal of the arches of vertebrae C1, C2, and C3 and by a great enlargement of the foramen magnum. Following the operation, there has been a gradual but slow improvement; the patient still has some difficulty swallowing but the strength in the upper extremities has returned to normal, and there is no evidence of respiratory disturbances.

1. Why does a myelomeningocele often cause weakness and sensory disturbances in the lower parts of the leg?
2. Why did the hydrocephalus not become evident until several days after the child's birth?
3. Why did the infant have difficulties with swallowing and respiration?

Discussion

Over one-half of the myelomeningoceles in the thoracolumbar and lumbar regions, which are the most common locations, are associated with an Arnold–Chiari or cerebellomedullary deformity. It is not unusual for these patients to have an assortment of intracranial anomalies, involving not only the brain, (e.g., agenesis of corpus callosum, Fig. 2–22A) but also the dural sinuses and the skull. Hydrocephalus, which is often associated with Arnold–Chiari deformity, is probably the result of a number of factors including a narrowing of the cerebral aqueduct, occlusion of the cisterna magna by a displaced cerebellum, and opening of the lateral and median apertures into the cervical canal, where the brain stem is surrounded by fibrotic arachnoid tissue. Hydrocephalus with increased pressure usually becomes evident first after the operation for the myelomeningocele, when the seeping of CSF through the lumbar defect has been stopped.

The neurologic symptoms seen in the newborn girl, i.e., sensory disturbances and loss or weakness of movements in the lower parts of the legs, are typical for myelomeningoceles in the lumbar region since the primary sensory and motor regions that subserve the lower parts of the leg are located in the lower part of the spinal cord, which is compromised by this malformation. Even if the myelomeningocele is operated successfully, the neurologic symptoms usually do not dissappear.

Symptoms from the Arnold–Chiari deformity may be evident at the time of the first shunt operation, or can develop with some delay, as in this case, e.g., because of a malfunctioning shunt with subsequent rise in intracranial pressure. Symptoms from involvement of the medulla oblongata and the lower cranial nerves are typical for the Arnold–Chiari deformity, in which the medulla oblongata and part of the cerebellum is squeezed down into the foramen magnum. The affected cranial nerves, therefore, are usually the ones that exit from the medulla, since they are drawn into the vertebral canal and are therefore making an abnormal detour into the skull before they exit. The symptoms, as in this case, are often related to involvement of the glossopharyngeal and vagus nerves as reflected by difficulty in swallowing, diminished gag reflex, and vocal cord paralysis. The two respiratory problems, i.e., the vocal cord paralysis and the apneic episodes, can easily become life-threatening and may require intubation or tracheotomy (the formation of an artificial opening in the trachea). The tracheotomy by itself, however, may not abolish the apneic episodes, since these are most likely due to pressure on the respiratory center in the medulla oblongata (Chapter 19). Finally, pressure on the descending supraspinal motor pathways that influence the functions of the motor neurons in the spinal cord is the most likely explanation for weakness in the arms.

Suggested Reading

1. Brown, M. C., W. G. Hopkins, and R. J. Keynes, 1991. Essentials of Neural Development. Cambridge University Press: Cambridge.
2. Cowan, W. M., 1992. Development of the Nervous System. In: A. K. Asbury, G. M. McKhann, and W. J. McDonald (eds.): Diseases of the Nervous System; Clinical Neurobiology. W. B. Saunders Company: Philadelphia, pp. 55–24.
3. Hall, Z. W., 1991. An Introduction to Molecular Biology. Sinauer Associates: Sunderland, Massachusetts, pp. 388–459.
4. Jacobson, M., 1991. Developmental Neurobiology, 3rd ed. Plenum Press: New York.
5. O'Rahilly, R. and F. Muller, 1992. Human Embryology and Teratology. Wiley-Liss: New York, pp. 253–303.
6. Sidman, R. L. and P. Rakic, 1982. Development of the Human Central Nervous System. In: W. Haymaker and R. D. Adams (eds.): Histology and Histopathology of the Nervous System. Charles C. Thomas: Springfield, Illinois, pp. 3–145.
7. Volpe, J. J., 1987. Neurology of the Newborn, 2nd ed. W. B. Saunders: Philadelphia, pp. 2–68.

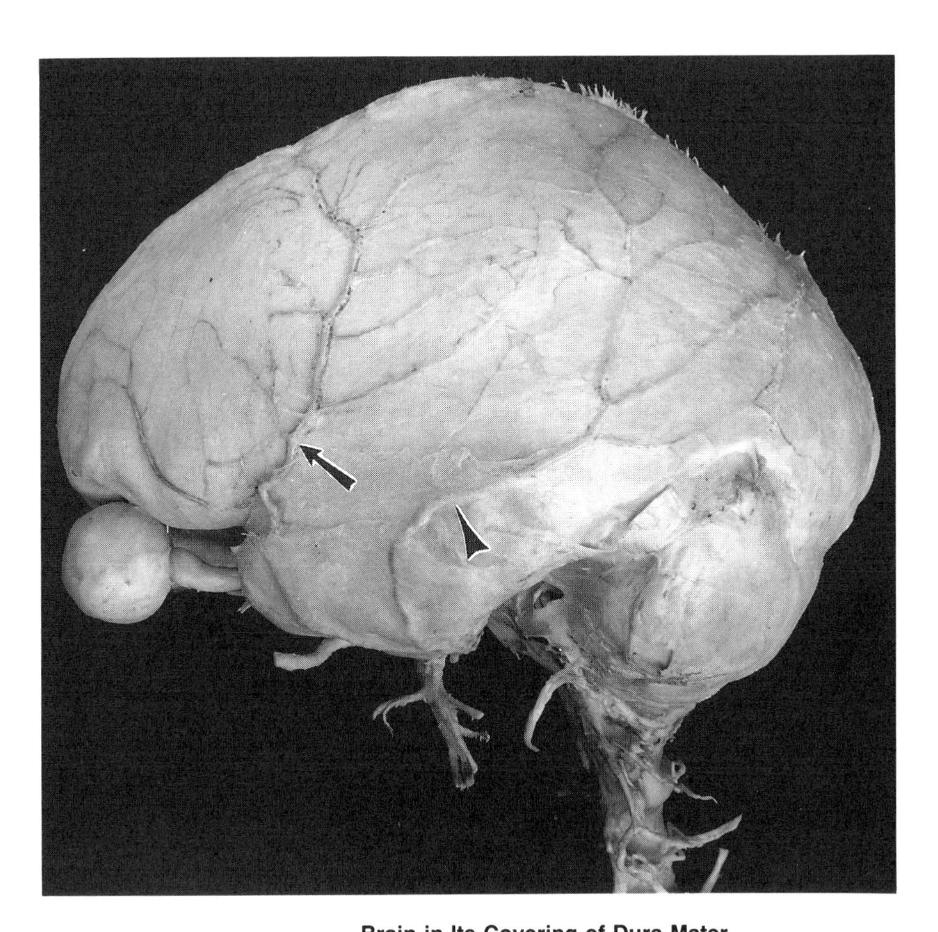

Brain in Its Covering of Dura Mater
Frontal branch of middle meningeal artery (*arrow*). Parietal branch of middle meningeal artery (*arrowhead*). (From Gluhbegoric, N. and T.H. Williams, 1980, The Human Brain: A Photographic Guide. JB Lippincott Co., Philadelphia. With permission of the publisher.)

3

Meninges and Cerebrospinal Fluid

The brain and the spinal cord are surrounded by three mesodermal coverings or meninges: the dura mater, the arachnoid, and the pia mater. The space between the pia mater and the arachnoid, i.e., the subarachnoid space, is filled with cerebrospinal fluid (CSF), which serves as a cushion between the delicate CNS and the rigid skull.

The meninges and their various specializations have important protective and supportive functions. In addition, special meningeal features are related to the production, circulation, and absorption of CSF, and endothelial-lined sinuses of the dura mater convey the venous blood to the internal jugular veins.

The meninges and their relationships to the brain and the CSF system are of great relevance in a number of clinical conditions including head injuries, intracranial hemorrhages, infections, and hydrocephalus.

Dura Mater

The tough and fibrous dura mater, or pachymeninx, is partly attached to the inner surface of the cranial cavity. Two large sheets of dura, the falx cerebri and the tentorium cerebelli, divide the cranial cavity into three communicating compartments, i.e., two supratentorial compartments and one infratentorial compartment (Figs. 3–1A and B).

Dural Folds

1. The *falx cerebri* is suspended vertically in the longitudinal cerebral fissure between the two cerebral hemispheres, thereby dividing the supratentorial compartment into a right and a left half (Fig. 3–1B). The falx cerebri is attached to the crista galli of the ethmoid bone in the front and to the horizontally suspended tentorium cerebelli in the back (Fig. 3–1C).

2. The *tentorium cerebelli* is a tent-like partial partition between the middle and posterior cranial fossae. It divides the cranial cavity into supratentorial and infratentorial compartments (Fig. 3–1B), and separates the cerebellum from the occipital lobes. Its free anterior edge forms the major part of a large oval opening, the *tentorial notch* or *incisure* (Fig. 3–1C), which surrounds the midbrain.

3. The *falx cerebelli* is a small midline dural fold in the posterior fossa. It is attached to the posterior inferior surface of the tentorium cerebelli and projects forward into the posterior cerebellar notch, which partly separates the two cerebellar hemispheres.

4. The *diaphragma sellae* is a small circular dural fold that bridges over the sella turcica and covers the pituitary gland. The stalk of the pituitary, which is attached to the base of the brain, passes through an opening in the center of the diaphragma sellae (Fig. 3–1C).

Dural Venous Sinuses

The cerebral veins are divided into external and internal veins, all of which eventually drain into venous sinuses. The sinuses are endothelium-lined spaces between two layers of dura. They occur along the attachments of the dural folds (Figs. 3–1B and 3–5).

Many of the external veins that drain the cerebral cortex on the convexity of the hemispheres empty into the superior sagittal sinus (Fig. 3–5). The external veins on the ventral surface of the brain connect with sinuses located in more basal parts of the skull cavity. The blood finally reaches the internal jugular vein through the sinuses.

1. The *superior sagittal sinus* runs along the attachment of the falx cerebri (Fig. 3–2). It widens as it approaches the internal occipital protuberance to form the *confluence of sinuses*, and next continues as one of the transverse sinuses, usually the right transverse sinus. The superior sagittal sinus receives tributaries from the superior cerebral veins on the convexity of the cerebral hemisphere (Fig. 3–5).

2. The *inferior sagittal sinus* is situated along the inferior free margin of the falx cerebri (Fig. 3–1B). It receives blood from the medial aspects of the hemisphere and from the falx cerebri.

3. The *straight sinus*, which represents the continuation of the inferior sagittal sinus, is situated along the attachment of the falx cerebri to the tentorium cerebelli (Fig. 3–2). The straight sinus receives the *great cerebral vein of Galen* (v. cerebri magna), which collects the venous blood from the left and right *internal cerebral veins* (Fig. 3–2) as well as from the two basal veins. The two internal cerebral veins drain the central parts of the brain. The left and right basal veins, which pass dorsally around the cerebral peduncles to enter the great cerebral vein of Galen, drain the basal parts of the brain. The straight sinus is usually continuous with the left transverse sinus at the confluence of the sinuses.

4. The *transverse sinus* (lateral sinus) runs horizontally along the bony attachment of the tentorium cerebelli, which extends from the internal occipital protuberance to the base of the petrous portion of the temporal bone (Fig. 3–3).

5. The *sigmoid sinus*, which is a continuation of the transverse sinus, drains into the internal jugular vein passing through the jugular foramen (Fig. 3–3).

6. The *cavernous sinus* is located on the lateral surface of the body of the sphenoid bone and is in close relation to the internal carotid artery (Fig. 3–3). It is connected with its counterpart on the opposite side through intercavernous sinuses. The cavernous sinus, which receives blood from the pituitary, from the orbita through the ophthalmic veins, and from the middle cerebral veins, is drained by the superior and inferior petrosal sinuses.

7. The *superior and inferior petrosal sinuses* lie upon the superior and posterior margins of the petrous temporal bone (Fig. 3–3). The superior petrosal sinus connects the cavernous sinus with the junction of the transverse and sigmoid sinuses. The inferior petrosal sinus is continuous with the internal jugular vein.

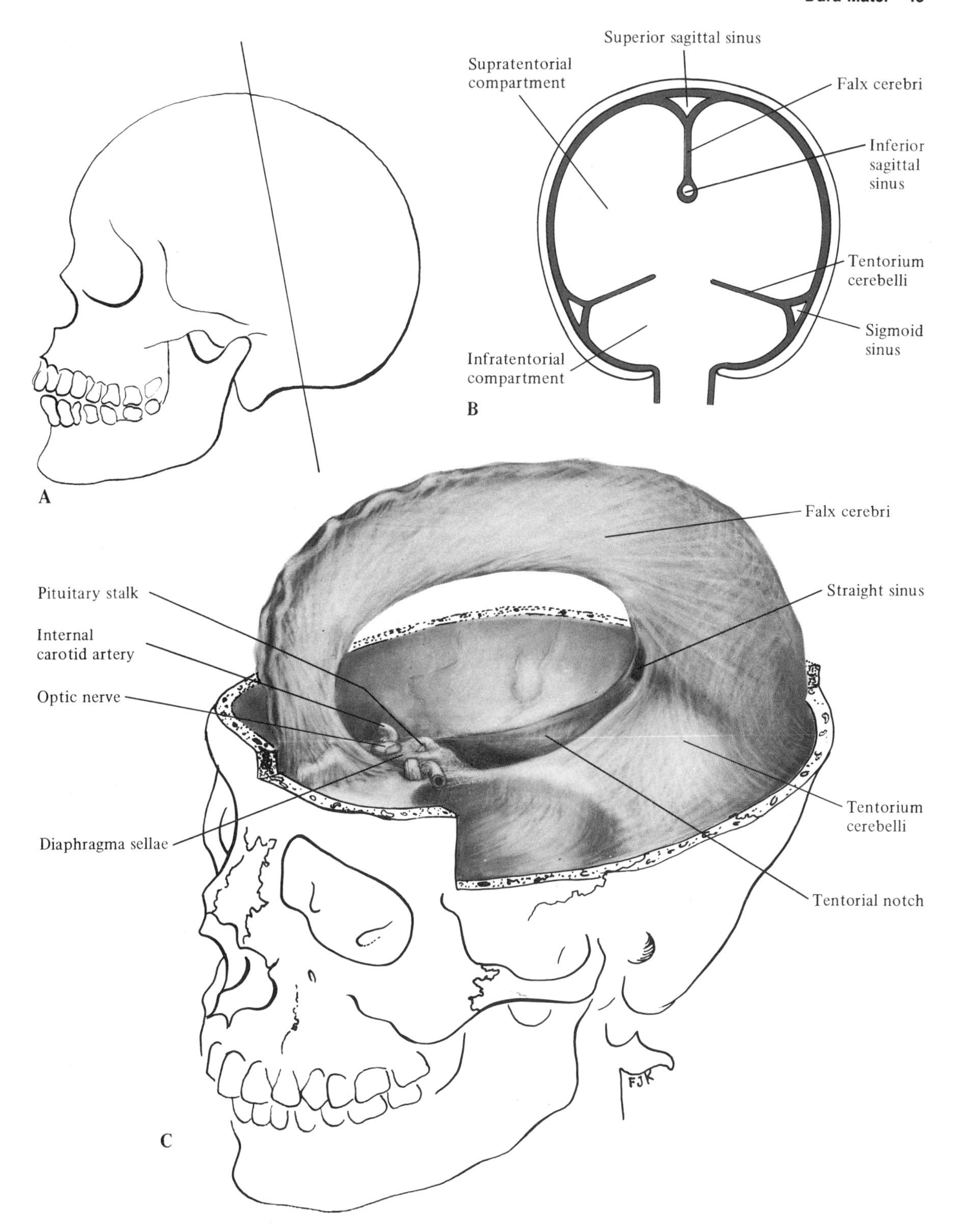

Fig. 3–1. Falx Cerebri and Tentorium Cerebelli
A illustrates the plane for the cross-section in **B**, which shows how the falx cerebri and tentorium cerebelli divide the skull cavity into supratentorial and infratentorial compartments. The thickness of the dura is exaggerated. **C** shows the attachment of falx cerebri to the crista galli and the tentorium.

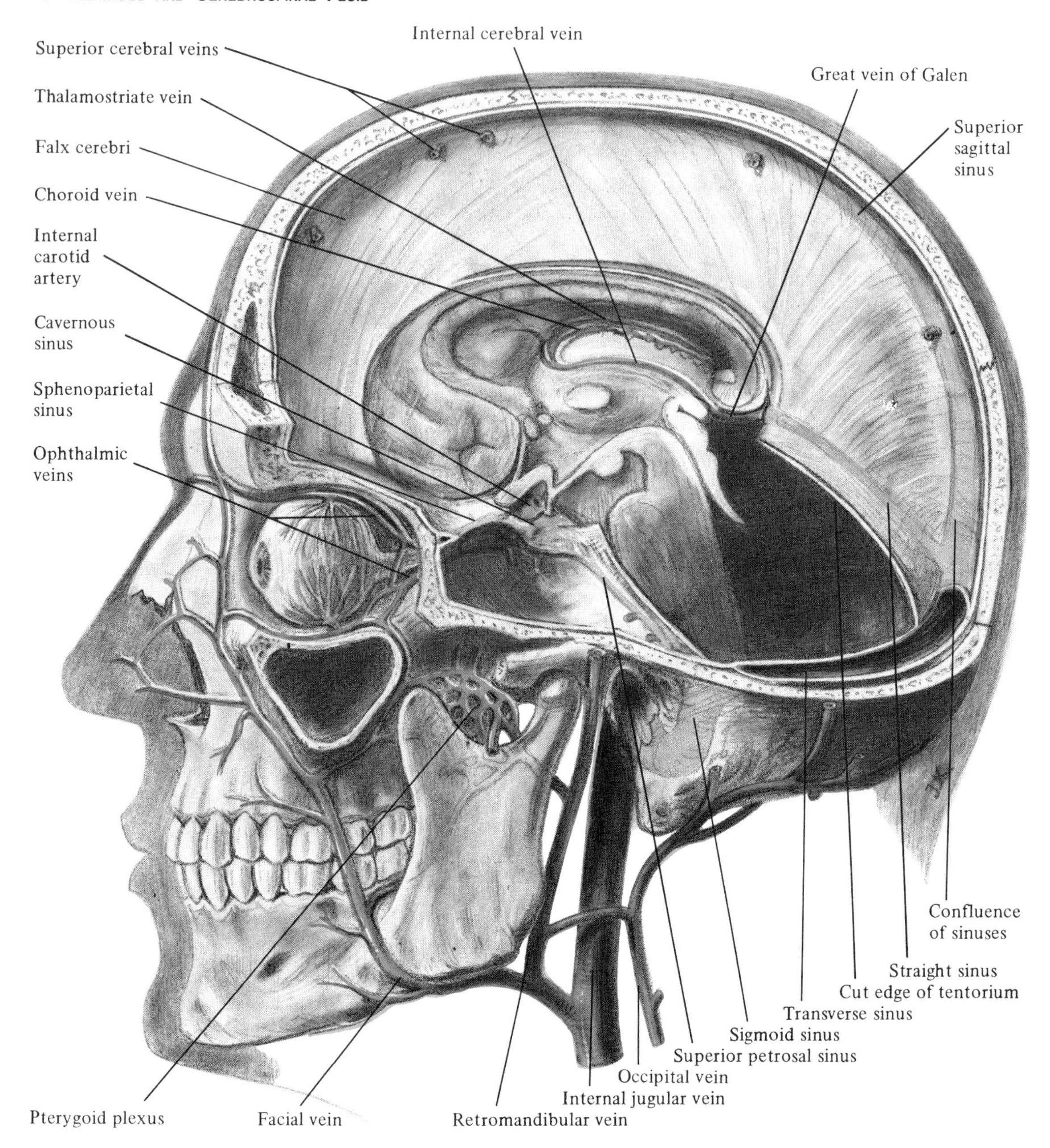

Fig. 3–2. The Venous Sinuses and the Veins of the Head

Dissection of the dural venous sinuses and the deep veins of the brain. The left half of the brain and the entire cerebellum have been removed. The deep veins in the floor of the right lateral ventricle have been exposed by removing the septum, the fornix, and part of the corpus callosum.

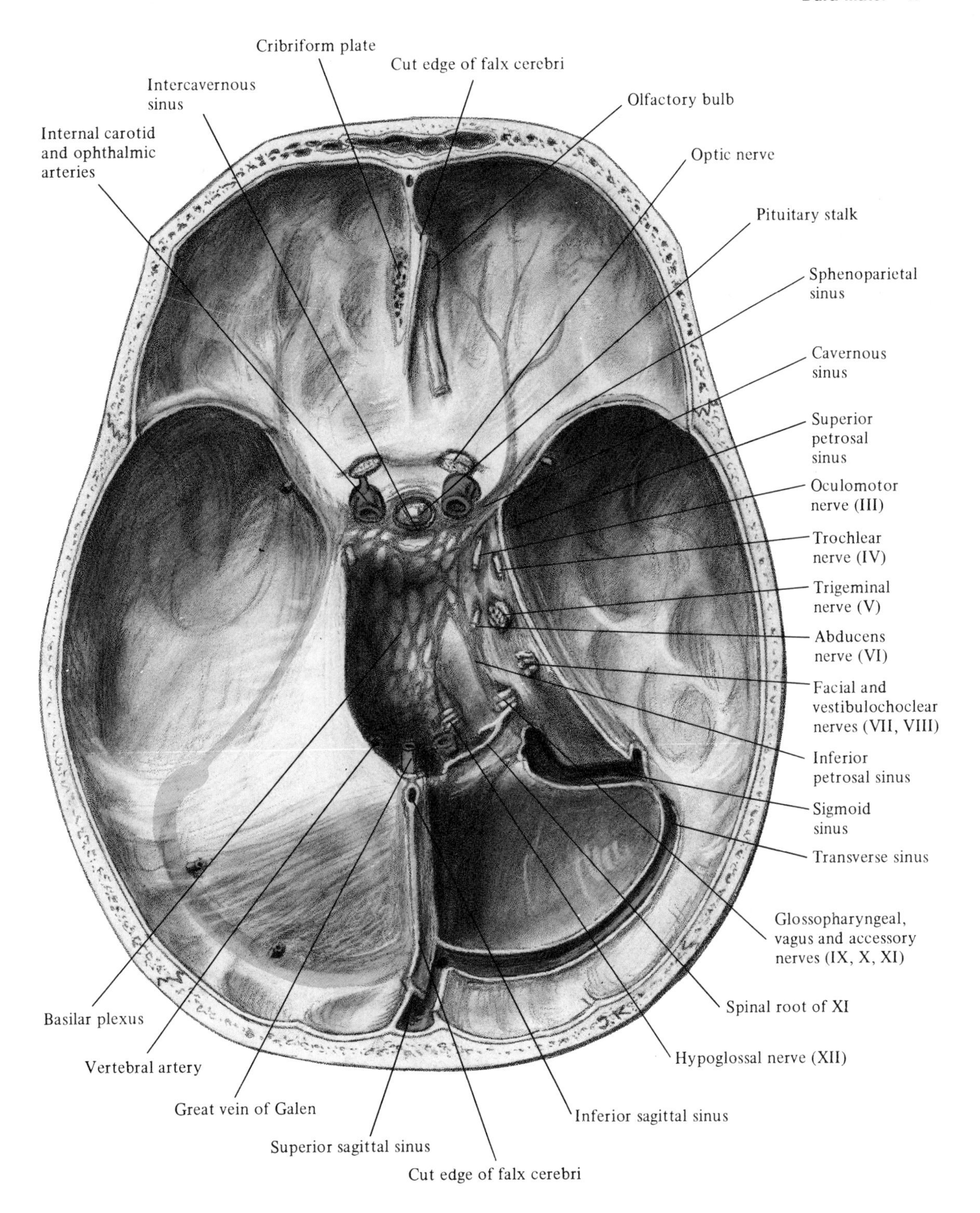

Fig. 3–3. Venous Sinuses on the Base of the Skull Cavity
The tentorium cerebelli has been removed on the right side to expose the sigmoid and inferior petrosal sinuses as well as the exits of the cranial nerves.

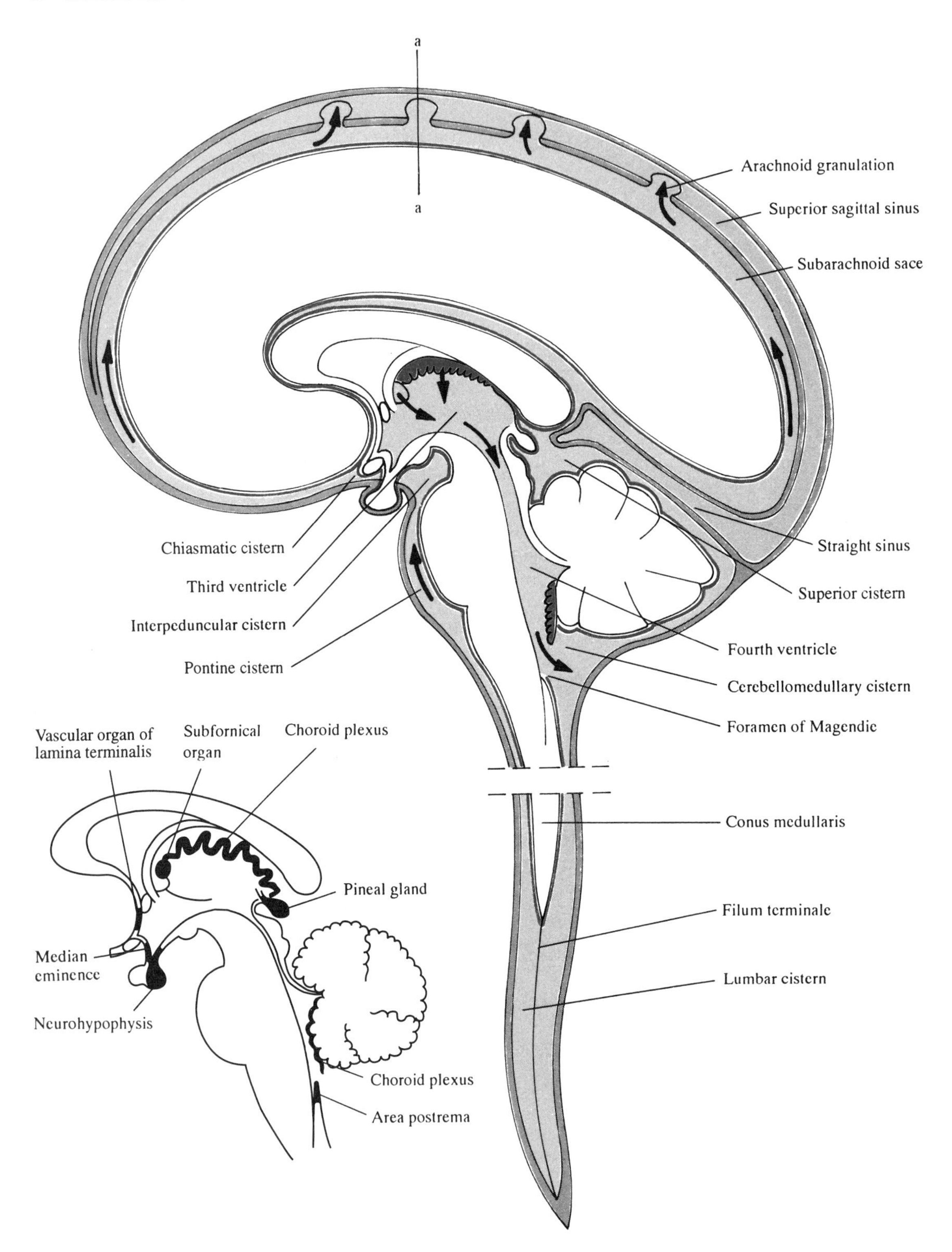

Fig. 3–4. Subarachnoid Cisterns
Schematic illustration showing the subarachnoid space and the flow of cerebrospinal fluid (CSF).
Inset: Drawing of median sagittal section through the human brain showing the locations of the circumventricular organs in *black*. (Modified from Weindl, A., Neuroendocrine aspects of circumventricular organs. In W. F. Ganong and L. Martini (eds.): Frontiers in Neuroendocrinology. Oxford University Press: New York, 1973, p. 4. With permission of the publisher.)

Pia–Arachnoid

The very thin pia mater is intimately associated with the surface of the brain, dipping into its fissures and sulci. The arachnoid, however, bridges over the various irregularities of the brain. The space between the pia and the arachnoid, the *subarachnoid space*, contains CSF, which serves as a protective "water cushion" for the brain and spinal cord. The pia and the arachnoid, together known as the leptomeninges, are connected by fine connective tissue strands or trabeculae. The relationships between the two membranes are in many areas so intimate that they can be regarded as one entity, the *pia–arachnoid*. In certain regions, however, the subarachnoid space widens to form cavities, the subarachnoid cisterns (Fig. 3–4):

1. *Cerebellomedullary cistern* (cisterna magna), between the cerebellum and the medulla oblongata
2. *Pontine cistern*, surrounding the pons
3. *Interpeduncular cistern*, between the two cerebral peduncles
4. *Superior cistern* between the splenium and the superior surfaces of the midbrain and the cerebellum. The superior cistern and the interpeduncular cistern are continuous with the
5. *Cisterna ambiens* on the sides of the midbrain
6. *Cistern of the lateral fossa* on the lateral side of the hemisphere where the arachnoid bridges over the lateral fissure
7. *Chiasmatic cistern* below and anterior to the optic chiasm.

Cerebrospinal Fluid

The CSF fills the ventricular system and the subarachnoid space. It is a clear and colorless water-like fluid that is formed primarily by the choroid plexus in the lateral ventricles and, to a lesser degree, by the choroid plexus of the third and fourth ventricles. The formation of CSF is complex and includes both passive filtration and active secretory mechanisms.

CSF produced in the lateral ventricles enters the third ventricle through the interventricular foramen and flows through the cerebral aqueduct into the fourth ventricle (Fig. 3–4). It then reaches the subarachnoid space through three openings, the *median aperture* (foramen of Magendie) in the posterior medullary velum and the two *lateral apertures* (foramina of Luschka) in the lateral recesses of the fourth ventricle. From the cerebellomedullary cistern and the pontine cistern, the CSF flows through the cisterns on the basal surface of the brain and upward over the lateral surfaces to the region of the superior sagittal sinus. Here most of the fluid is absorbed into the venous system through the *arachnoid villi*. Collections of microscopic arachnoid villi form visible *arachnoid granulations*, that protrude into the lateral expansions of the superior sagittal sinus through openings in the dura mater (Fig. 3–5).

The flow of CSF is fairly rapid. The total volume of CSF in the ventricle system and the subarachnoid space is only about 125 ml, but it is estimated that more than 4 times that amount, or about 500 ml, is formed during a 24-h period. A small amount of CSF seeps down around the spinal cord, but little is known about its circulation and absorption in the spinal subarachnoid space.

Functions of the Cerebrospinal Fluid

Besides serving as a cushion for the brain within the rigid skull, the CSF helps to maintain a suitable environment for the nervous tissue. CSF extends into the perivascular spaces (Virchow–Robin spaces) that follow the blood vessels into the deeper parts of the brain (Fig. 3–5). Here, metabolites and small solutes can diffuse quite freely between the extracellular fluid of the brain and the CSF in the perivascular space. Harmful metabolites, for instance, can be removed from the extracellular fluid by diffusion across the pia mater into the perivascular spaces and eventually be absorbed into the blood; in this sense, the perivascular spaces are reminiscent of the lymphatic system in the rest of the body.

Blood–Cerebrospinal Fluid and Blood–Brain Barriers

The capillaries in the brain, including those of the choroid plexus, have endothelial cells that are in general not fenestrated and furthermore are joined together by tight junctions. This prevents the free movement of compounds from the blood to the CSF or to the brain. These effective barriers, the blood–CSF barrier and the blood–brain barrier, allow the passage of small, lipid-soluble molecules but prevent some toxic substances as well as plasma proteins and other large molecules from entering the brain. Other compounds, such as glucose and some amino acids, are transported from the blood into the brain by active transport mechanisms.

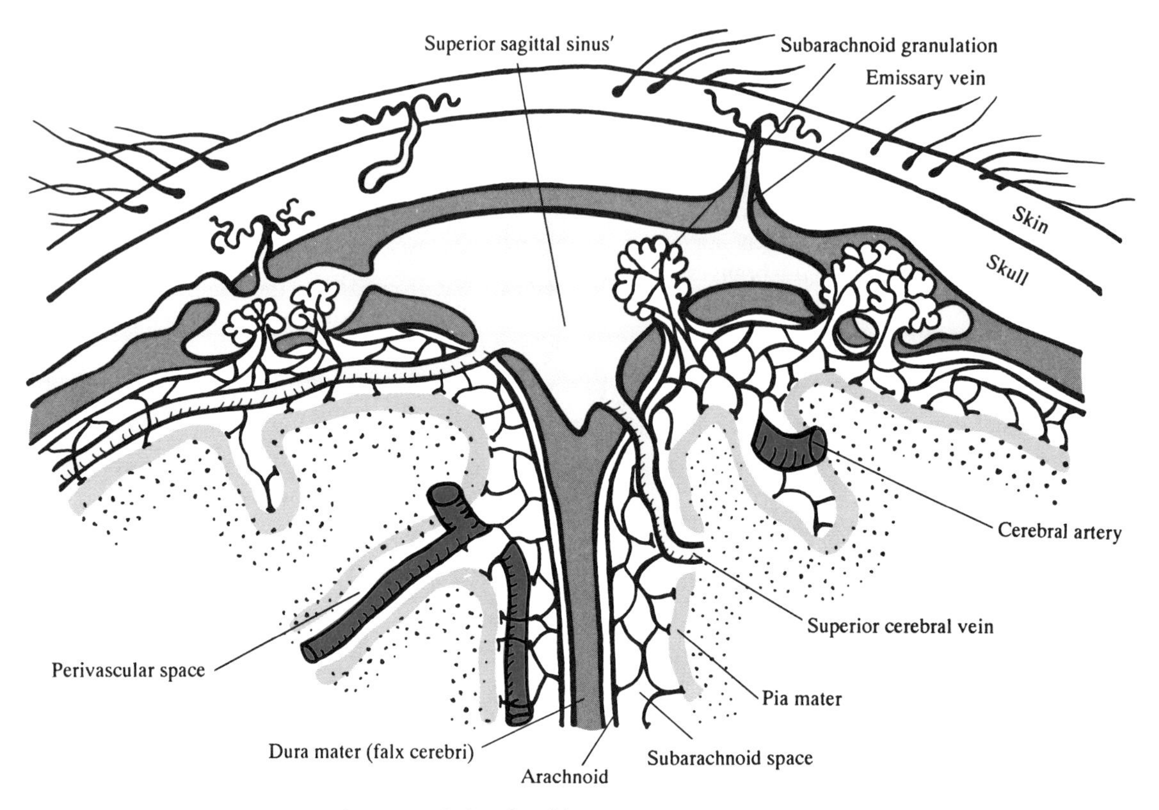

Fig. 3–5. Superior Sagittal Sinus and Arachnoid Granulations
Transverse section (indicated in Fig. 3–4 through the superior sagittal sinus showing the arachnoid granulations and the relationships between the different meninges.

Circumventricular Organs

The choroid plexus belongs to a group of structures that are called *circumventricular organs* (CVOs) because they are located along the surface of the ventricular system (Fig. 3–4, inset). Other structures within the group are the subfornical organ, the vascular organ of the lamina terminalis, the median eminence, the neurohypophysis, the pineal gland, and the area postrema. The CVOs are in general characterized by the lack of a blood–brain barrier due to fenestrated endothelium in their capillaries. As indicated above, however , the capillary endothelium in the choroid plexus does serve as a barrier for certain substances.

The subfornical organ and the vascular organ of the lamina terminalis are situated in the midline along the anterior wall of the third ventricle, the subfornical organ immediately ventral to where the two fornix columns come together at the level of the interventricular foramen, and the vascular organ alongside the lamina terminalis, midway between the anterior commissure and the optic chiasm. The lack of a blood–brain barrier makes it possible for blood-borne chemicals to activate these structures. The neurons in the subfornical and lamina terminalis organs, which have direct connections with each other and with many other brain regions, including neurosecretory cells and autonomic nuclei in the hypothalamus and brain stem, are of great importance for the regulation of fluid balance and blood pressure. The neuronal circuit involving the subfornical organ and the vascular organ of the lamina terminalis are described in Chapter 18, where the median eminence and neurohypophysis are also discussed; the area postrema is described in Chapter 10.

Meninges of the Spinal Cord

Dura Mater Spinalis

The spinal cord is enveloped by membranes that are directly continuous with those covering the brain. Unlike the situation in the skull cavity, in which the dura mater is attached to the periosteum of the skull, the *spinal dura mater* forms a tubular sac that is firmly attached to the bone only at the margin of the foramen magnum. The dural sac ends at the level of the second sacral vertebra. A filamentous extension of the spinal cord, the *filum terminale*, stretches from the caudal tip of the spinal cord to the bottom of the dural sac, where it is closely united with the dural sheath to form a cord, which stretches further caudally to become continuous with periosteum of the coccygeal bone (Fig. 3–8). Between the spinal dura mater and the periosteum of the vertebral canal is the *extradural space*, which is filled with a venous plexus and adipose tissue. The anterior and posterior spinal nerve roots are surrounded by a common dural sleeve that is continuous with the epineurium of the spinal nerve at the level of the intervertebral foramen.

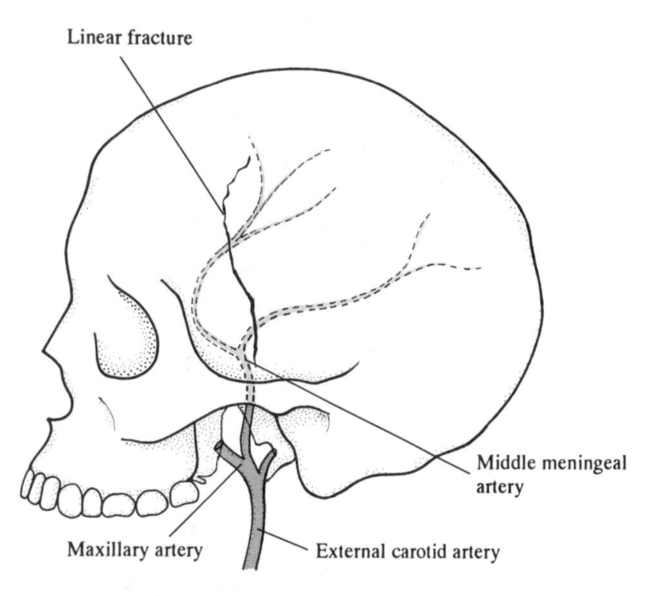

Fig. 3–6. Middle Meningeal Artery and Extradural Hematoma
Diagram showing the location of the middle meningeal artery on the inside of the skull. Extradural hematoma is often caused by a linear fracture in the parietotemporal region associated with rupture of the middle meningeal artery.

Pia Mater and Arachnoidea Spinalis

Whereas the *pia mater* is intimately attached to the surface of the spinal cord, the *arachnoid* is closely associated with the spinal dura mater. The subarachnoid space, which is filled with CSF, is particularly spacious below the conus medullaris, where it is referred to as the *lumbar cistern*. This cistern is occupied by the filum terminale and the nerve roots of the cauda equina (Fig. 2–4). Throughout its length the spinal cord is suspended in the dural sac by 21 pairs of lateral extensions of the pia, the *denticulate ligaments*.

Clinical Notes

Head Injuries

The brain is well protected by the skull and the dura mater. Nevertheless, brain damage is a frequent complication in head injuries, which are common in motorized and industrial societies. The brain injury may take the form of a "knock-out" or *concussion*, with transient loss of consciousness but no obvious injury to the brain, or there may be ecchymosis in the brain tissue, referred to as *contusion*. Finally, hemorrhage or laceration may occur and cause severe and permanent symptoms.

The most important consideration in the case of a skull fracture is the extent and nature of damage to the brain. Fractures of the base of the skull, which are difficult to detect in X-ray films, are more apt to cause severe damage to the brain or cranial nerves than are simple linear fractures of the vault. Leakage of CSF from the nose or bleeding from the auditory canal indicates the presence of a basal fracture. Although the brain can be severely damaged without any evidence of skull fracture, fractures of the cranial cavity always increase the risk of meningeal infections and bleeding.

Subdural and Extradural Hemorrhage

These are serious and not infrequent complications of head injuries. A *subdural hematoma* is a common complication that may result from even trivial head trauma. The subdural bleeding, which is most often located over the convexity of the frontal or parietal lobes, is usually attributable to the tearing of veins that bridge the space between the brain and the superior sagittal sinus (Fig. 3–5). The condition may develop slowly, and recovery following prompt evacuation is usually good. Rapidly developing subdural hematoma is usually associated with cerebral contusion and venous and arterial bleeding. Such an acute subdural hematoma is often fatal.

A skull fracture may be associated with rupture of one of the meningeal arteries or dural sinuses. The middle meningeal artery (Fig. 3–6), the largest of the meningeal arteries, is often damaged as a result of a linear fracture in the parietotemporal region. The subsequent bleeding between the dura and the skull results in a rapidly expanding mass, an *extradural hematoma*, that can cause irreversible damage to the brain if it is not removed.

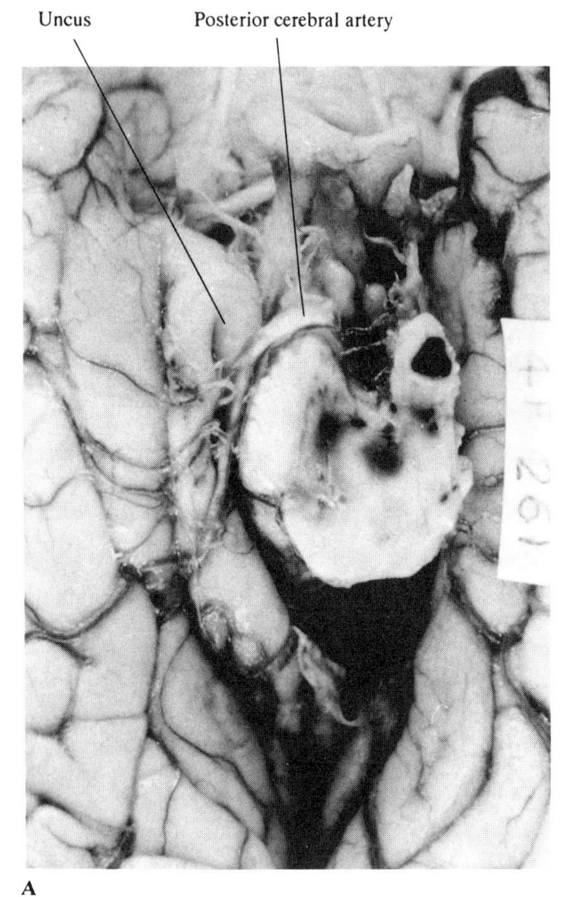

A

B

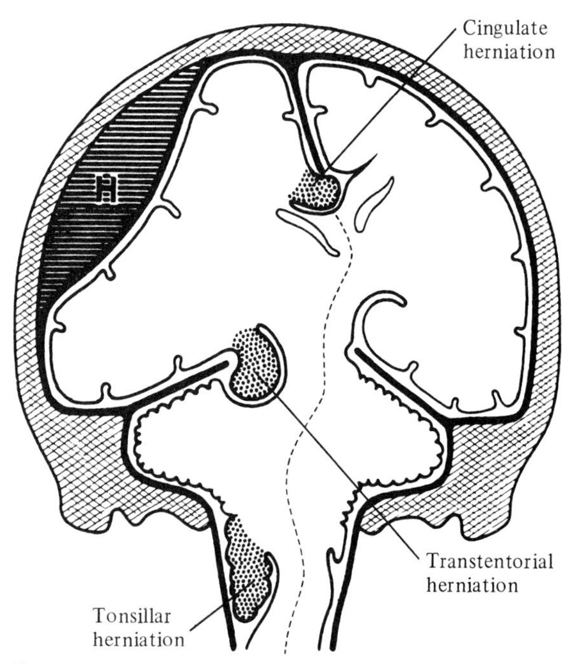

C

Fig. 3–7. Transtentorial Uncal Herniation

A. Expanding supratentorial lesions may cause downward transtentorial herniation of the uncus and parahippocampal gyrus (*left side of figure*) with compression of the brain stem and injury to the IIIrd cranial nerve. Note the distortion and bleeding in the mesencephalon. (Photograph courtesy Dr. Scott Vandenberg, University of Virginia.)

B. Third nerve damage may lead to dilation of the ipsilateral pupil and ptosis of the upper lid.

C. Summary diagram of the major types of cerebral herniations. (From Escourolle and Poirier, 1978. Manual of Basic Neuropathy, 2nd ed. Translated into English by L. J. Rubinstein. W. B. Saunders Co., Philadelphia. With permission of Masson et Cie, Paris.)

Transtentorial Herniation

Intracranial pathologic processes, such as hemorrhages, tumors, or infections, tend to cause a displacement and deformation of brain tissue with subsequent rise in intracranial pressure. Herniations usually develop along lines of decreased resistance to expansion. Herniation of part of the temporal lobe through the tentorial notch, for instance, is a very serious complication of an expanding supratentorial lesion, e.g., an extradural or subdural hematoma. The herniating parts of the temporal lobe, the uncus and adjacent parts of the parahippocampal gyrus, compress the midbrain as well as other structures within the tentorial incisure (Fig. 3–7A).

Although the uncal herniation syndrome can be fully appreciated only on the basis of a more comprehensive understanding of the functional anatomy of the brain (see Clinical Example: Transtentorial Herniation, Chapter 11), the main pathologic events will nevertheless be summarized at this point. The syndrome is characterized by the following classic triad: deterioration of consciousness, ipsilateral pupillary dilation with loss of the light reflex, and hemiparesis.

The alarming deterioration of consciousness to a state of deep coma (unconsciousness), which often occurs within minutes or hours, is due to the interference with a brain stem arousal system called the ascending reticular activating system (Chapter 10). It can also be due to the effect of the mass lesion on the cerebral cortex.

The pupillary dilation and loss of the light reflex are due to involvement of the parasympathetic fibers to the pupillary sphincter, which travel in the IIIrd cranial nerve, the oculomotor nerve. The parasympathetic fibers to the smooth muscle, the pupillary sphincter that constricts the pupil, are apparently very sensitive to compression. When the transtentorial herniation has progressed to the point where it directly affects the integrity of the mesencephalon, both parasympathetic and sympathetic functions are lost, and the pupil is "fixed" in midposition.

Pressure on the cerebral peduncle is likely to affect the descending corticofugal fiber bundles including the corticospinal (pyramidal) tract, which is of crucial importance for motor functions. Since the corticospinal tract crosses over to the opposite side in the pyramidal decussation in the lower part of the medulla oblongata, there will be a loss of motor function on the opposite side, i.e., contralateral hemiparesis. However, the contralateral hemiparesis can also be due to destruction of the cortical motor fields or their corticofugal fibers by the mass lesion. Sometimes the cerebral peduncle contralateral to the herniation can also be compressed against the free edge of the tentorium. This may cause loss of motor function on the same side as the herniation, i.e., ipsilateral hemiparesis ("false localizing sign"), which is much less common. Compression of the posterior cerebral artery may cause obstruction of blood circulation with subsequent necrosis of the tissue (infarction) in the distribution area of the posterior cerebral artery. With further herniation hemorrhages in the brain stem decrease the chance of survival.

Herniations of brain tissue can also occur underneath the falx cerebri, i.e., *cingulate herniation*, or down through the foramen magnum, *tonsillar herniation* (Fig. 3–7C).

Meningitis

Although the skull, the meninges, and the blood–brain barrier provide good protection from infectious agents, once an infection has established itself in the CNS, it is likely to be destructive. Meningitis, which is characterized by inflammation of the pia–arachnoid, is the most common infection of the CNS, and is usually caused by a bacterium or a virus. Since the subarachnoid space is continuous not only around the brain and the spinal cord but also with the ventricular system, the infection rapidly reaches most parts of the CNS. Fever, headache, and stiff neck are common symptoms in patients with meningitis, and the diagnosis is made by performing a lumbar puncture and examining the CSF for the presence of the pathogenic organism and for other abnormalities, including in most cases an increased number of white blood cells and elevated protein. It is especially important to recognize those infections that can be successfully treated with antibiotics before the patient dies. The risk of complications increases if therapy is delayed. One example of a serious complication is the formation of fibrous adhesions between the pia and the arachnoid, which may interfere with the circulation of CSF and cause hydrocephalus.

Hydrocephalus

This condition is characterized by an excessive amount of CSF in the ventricular system. This can happen because of increased production of the fluid, or because of disturbances in its circulation or absorption into the dural venous sinuses. Many cases of hydrocephalus with increased intracranial

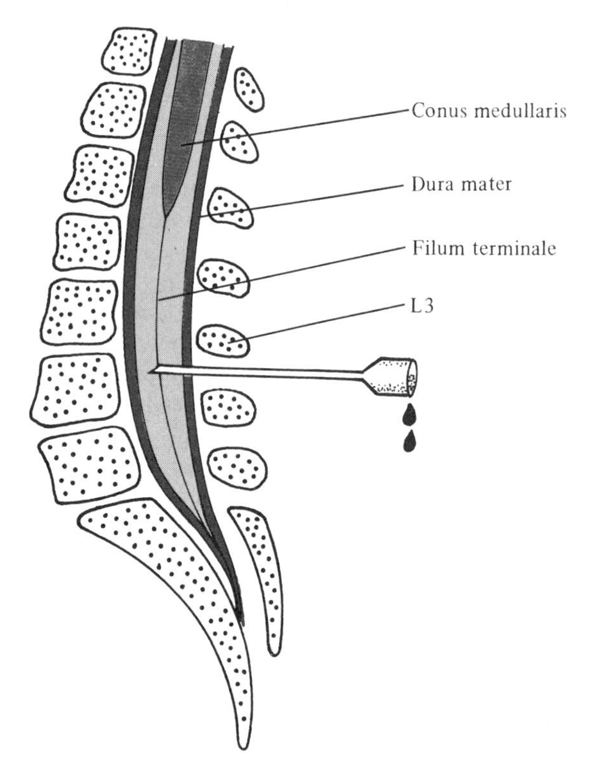

Fig. 3–8. Lumbar Puncture
CSF is obtained by inserting a needle into the lumbar cistern between the 3rd and 4th or the 4th and 5th lumbar spinal processes.

pressure are due to an obstruction in the circulation of CSF. The ventricular system contains several narrow passages, such as the interventricular foramen and the cerebral aqueduct, where a pathologic process can easily block the flow of CSF. If hydrocephalus occurs in childhood before the sutures have closed, the head may reach an enormous size unless the condition is treated. In older children and adults, in whom the sutures are closed, there is a gradual ventricular dilation with compression and thinning of brain tissue.

Lumbar Puncture

The composition of CSF, as well as its pressure, varies in different pathologic conditions. The analysis of the CSF can be of great help in establishing a diagnosis, especially in the case of meningitis and other forms of inflammatory conditions. The simplest and safest method of obtaining CSF is to insert a lumbar puncture needle into the lumbar cistern with the patient lying sideways. The lumbar cistern can be reached by placing the needle in the sagittal plane between the third and fourth, or the fourth and fifth lumbar spinal processes (Fig. 3–8). The spinal cord ends above L2; thus, there is no risk that the spinal cord will be damaged.

Normal CSF contains very little protein (15–45 mg/ml) and only a few lymphocytes per cubic millimeter. Its sugar content is somewhat lower than that of the blood, although the concentration of NaCl is somewhat higher. The amount of the different components, however, may change drastically when inflammation or bleeding involves part of the ventricular system or the meninges. An increase of red blood cells, for instance, may indicate subarachnoid bleeding. The pressure of the CSF, which is a good indicator of the intracranial pressure, can be estimated by attaching a manometer to the lumbar puncture needle. The normal pressure is between 6 and 14 cm CSF or water. Lumbar puncture is occasionally performed for the purpose of introducing therapeutic substances or local anesthetics into the subarachnoid space.

There are certain instances in which removal of CSF by lumbar puncture is dangerous. If, for instance, the intracranial pressure is greatly increased as a result of a brain tumor or bleeding, a sudden withdrawal of CSF from the spinal canal may cause downward herniation of the cerebellar tonsils through the foramen magnum (Fig. 3–7C). A *tonsillar herniation* ("cerebellar pressure cone") will cause compression of the medulla oblongata, which may compromise vital respiratory and circulatory functions (Chapter 19).

Clinical Example

Extradural Hematoma

A fifteen-year-old male riding his bicycle fell and struck the right side of his head on the curb. He had an immediate loss of consciousness, but within 5 minutes he was responding and apparently fully conscious. He was brought to the emergency room 45 minutes later, and while he was there he began to complain of headache. Routine skull X-ray films revealed a right temporal fracture (Fig. 3–9A). Soon after the films were obtained, the patient was examined by a neurosurgeon, who found that he was becoming agitated and confused, and that his pulse had dropped from the initial rate of 80 to 50. His respiratory rate had also decreased from 20 to 10, and he had become hypertensive.

While a computed tomographic (CT) scan was being ordered, it was noted that the patient's right pupil was larger than his left and was unresponsive

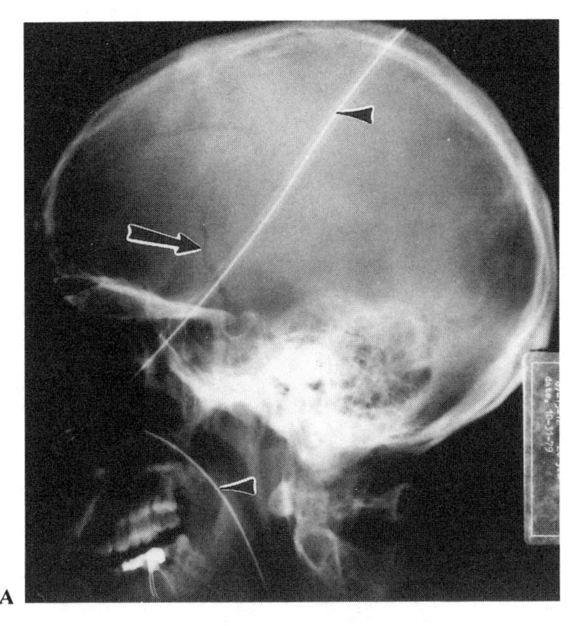

A

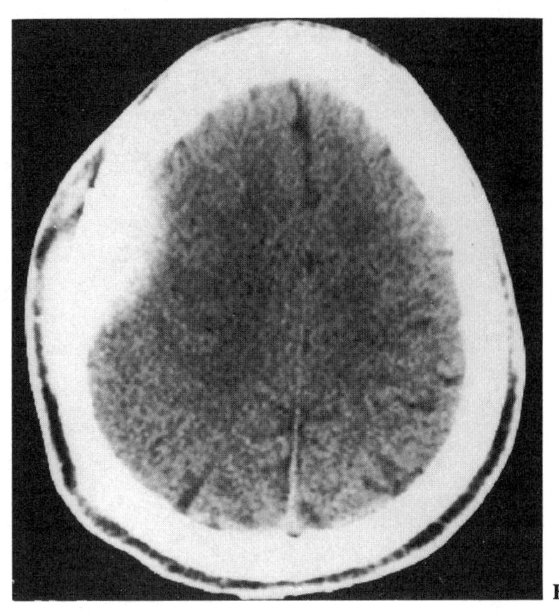

B

Fig. 3–9. Extradural Hematoma
A. Lateral skull radiograph showing temporal region fracture (*arrow*). The thin white lines (*arrowheads*) represent the nasogastric tube.
B. CT scan demonstrating an extradural hematoma. (Photographs courtesy Dr. John Jane, University of Virginia.)

to light. He also had a slight left hemiparesis. His agitation continued and the pulse rate diminished to 40. The neurosurgeon made a clinical diagnosis of an extradural hematoma and decided to operate without obtaining a CT scan.

1. On what basis did the neurosurgeon make the diagnosis?
2. Why did the patient have a left-sided hemiparesis, but a dilated pupil on the right side?
3. Why did the neurosurgeon decide to operate rather than wait for a CT scan, which would confirm the diagnosis?

Discussion

The neurosurgeon made a clinical diagnosis of an extradural hematoma based upon the history of a lucid interval, the right temporal skull fracture, the ipsilateral dilated pupil, and the contralateral hemiparesis. A lucid interval is characteristic both for epidural and chronic subdural hematoma, but it is usually longer in subdural hematomas, i.e., many days or even weeks. The agitation was felt to be due to increasing intracranial pressure. Decreasing pulse and respiratory rates and increasing blood pressure, besides agitation and confusion, are also signs of increased intracranial pressure.

Hemipareses due to extradural hematomas are caused either by involvement of the cortical motor areas or pressure on the descending corticospinal fibers in the base of the cerebral peduncle. In the first case, the motor symptoms will appear on the contralateral side of the hematoma, since the pyramidal tract crosses over to the other side in the pyramidal decussation. In the latter situation, the paralysis can be ipsilateral due to compression of the contralateral cerebral peduncle against the tentorial edge. This affects the corticospinal fibers which will cross further down in the pyramidal decussation.

Having made the diagnosis of extradural hematoma, and mindful of the rapidly deteriorating symptoms, the neurosurgeon decided to operate immediately in order to prevent a full-blown transtentorial herniation with loss of consciousness, respiratory failure, and death. The surgical procedure consisted of drilling a hole in the right temporal bone, which revealed a large epidural hematoma. The bleeding point was one of the branches of the middle meningeal artery (Figs. 3–6 and 3–9B), and the bleeding was stopped by cautery (a branding iron). Since the middle meningeal artery enters the skull through the foramen spinosum, another therapeutic measure is to plug the foramen with wax. The patient had an uneventful recovery.

Suggested Reading

1. Andrews, B. T. and L. H. Pitts, 1991. Traumatic Transtentorial Herniation and its Management. Futura Publishing Company: Mount Kisco, New York.
2. Browder, J. and H. A. Kaplan, 1976. Cerebral Dural Sinuses and Their Tributaries. Charles C Thomas: Springfield, Illinois.
3. Fishman, R. A., 1980. Cerebrospinal Fluid in Diseases of the Nervous System. W. B. Saunders: Philadelphia.
4. McKinley, M. J. and B. J. Oldfield, 1990. Circumventricular Organs. In G. Paxinos (ed.): The Human Nervous System. Academic Press: San Diego, pp. 415–438.
5. Rowland, L. P., M. E. Fink, and L. Rubin, 1991. Cerebrospinal Fluid: Blood–Brain Barrier, Brain Edema, and Hydrocephalus. In E. R. Kandel, J. H. Schwartz, and J. M. Jessell (eds.): Principles of Neural Science, 3rd ed. Elsevier: New York, pp. 1050–1060.

2

DISSECTION OF THE BRAIN

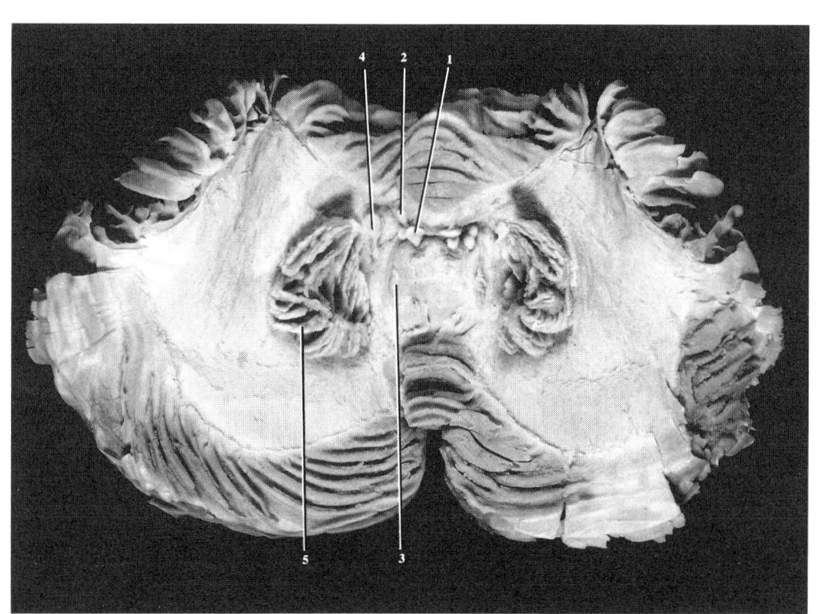

1. Fastigial nucleus
2. Anterior globose nucleus
3. Posterior globose nucleus
4. Emboliform nucleus
5. Dentate nucleus

This beautiful dissection of the intracerebellar nuclei in the human brain was performed by the Dutch neuroanatomist Nedzad Gluhbegovic from Utrecht.

4

Dissection of the Brain Introduction

The dissection of the human brain is a rewarding experience for both teachers and students if it proceeds in a systematic fashion. How to arrange the dissection course in the most practical manner depends to a certain extent on the time and the number of brains available. Although a general goal should be to have the students participate actively in the dissections, an insufficient number of brains may restrict the laboratory activities to a series of prosections. The following dissection guide is suitable for a 15- to 20-hour dissection course. Although it would be ideal to have two brains available, the dissections can be completed with one whole brain in addition to one half brain obtained by subdividing a brain in the midsagittal plane.

The dissection course will be considerably more rewarding if a group of 10–15 students can be assigned a teaching assistant. If arrangements are made ahead of time, neurology, psychiatry, or neurosurgery residents may be willing to assist in the dissections. This makes it especially enjoyable for the students because the residents tend to relate the anatomy of the central nervous system to clinically relevant problems. In return, the residents get a welcome review of brain anatomy.

To make it easier for the students to proceed with the dissections in a systematic fashion, it is advisable to have the assistants perform prosections of the different portions. This can, for instance, be arranged in the form of a preview of next day's session. As there is considerable variation in background and experience among the different teaching assistants, it may even be advisable to prepare the assistants by arranging a few practice sessions before the course starts. This ensures a certain uniformity in the teaching.

Following the autopsy, the brains are usually stored in strong formalin. To avoid the formalin fumes and reduce the risk of eczema and conjunctivitis, the brains should be soaked in water for a few days before use. It is recommended that a pair of surgical gloves or thin plastic gloves be used. To prevent the brain specimens from drying between sessions, it is recommended that they be stored in water or dilute alcohol in covered buckets or wrapped in a wet cloth in a plastic bag. The following instruments are needed for the brain dissection: brain knife, scalpel, scissors, forceps with fine terminals, anatomic forceps, and probe.

The dissection described in this guide requires the use of at least one and a half brains. The whole brain will be used for the study of the surface anatomy and for inspection of the meninges and the superficial arteries. It will also be used for the exploration of the ventricular system and for the study of the basal ganglia and the internal capsule. The half brain will be used for the blunt dissection of major fiber tracts, and the dissection can be facilitated by placing the brain in a freezer for several days. While the brain is in the freezer, it is advisable to store it in a plastic container filled with 10% formalin. Before the dissection, the brain is thawed in running cold water overnight. This deep-freezing procedure, which was developed by the Swiss neuroanatomist Josef Klingler, can with advantage be repeated several times.

First Dissection

Meninges and Subarachnoid Cisterns; Superficial Arteries; Vertebral–Basilar System; Internal Carotid System; Basal Surface and Cranial Nerves

Meninges and Subarachnoid Cisterns

The brain and the spinal cord are surrounded by three membranes, or meninges, which have protective and supportive functions.

Dura Mater (Pachymeninx)

The dura mater, which is partly attached to the inner surface of the cranial cavity, usually remains in the skull when the brain is removed at autopsy, and is therefore best studied in relation to the dissection of the head.

Arachnoidea and Pia Mater (Leptomeninx)

These two membranes are in many places so closely attached to each other by delicate fibrous strands, trabeculae, that they cannot be separated, and it is common practice to refer to them as an entity, the *pia–arachnoid.* However, there is a significant difference between the pia mater and the arachnoid in the sense that the pia closely follows the contour of the brain and dips into all irregularities of its surface, whereas the arachnoid follows the dura mater and therefore bridges over the irregularities. In places where the irregularities are pronounced, especially on the base of the brain, large cisterns are formed in the subarachnoid space between the pia and the arachnoid.

Subarachnoid Space and Subarachnoid Cisterns

The space between the pia and the arachnoid, the subarachnoid space, is filled with *cerebrospinal fluid* (CSF), which serves as a cushion for the delicate brain. This fluid comes from the ventricles of the brain and enters the subarachnoid space through three apertures in the fourth ventricle (see below). When the brain is removed from the skull, the fluid escapes and the subarachnoid space collapses. However, it should still be possible to identify some of the major cisterns on one of the brains before the meninges and the blood vessels are removed. Review the location of the following important subarachnoid cisterns (Fig. 3–4):

1. *Cistern of the lateral fossa* is an elongated expansion of the subarachnoid space at the place where the arachnoid bridges over the cleft formed by the lateral sulcus.
2. *Chiasmatic cistern* is situated below and anterior to the optic chiasm.
3. *Interpeduncular cistern* is located in the interpeduncular fossa in front of pons.
4. *Superior cistern* (cistern of the great cerebral vein of Galen) is situated between the splenium of corpus callosum and the superior surface of mesencephalon. It is continuous with cisterna ambiens on the sides of the midbrain.
5. *Pontocerebellar cistern* is situated in the angle between the pons and the cerebellum (see Clinical Example, Chapter 14).
6. *Cerebellomedullary cistern* (cisterna magna) is situated in the angle between the cerebellum and the medulla oblongata. It receives CSF through the median aperture (Magendie) and the two lateral apertures (Luschkae) of the fourth ventricle (see Cerebrospinal Fluid, Chapter 3).
7. *Lumbar cistern* of the spinal subarachnoid space is a large cavity, which extends from the second lumbar vertebra to the sacrum. It contains the cauda equina, and it can easily be reached by the aid of a lumbar puncture needle, if one needs to obtain CSF for diagnostic purposes (see Fig. 3–8).

Superficial Arteries

The brain holds a unique position in the human body in regard to blood supply. Although it is responsible for only 2% of the body weight in the adult, about one-fifth of the oxygenated blood that comes from the heart is carried to the brain. The blood reaches the brain through two pairs of arterial trunks, the *internal carotid arteries* and the *vertebral arteries* (see Fig. 24–1). Each internal carotid artery, which is a terminal branch of the common carotid artery, enters the skull through the carotid canal. The two major branches of the internal carotid artery, the *middle* and *anterior cerebral arteries*, supply the rostral parts of the brain including most of the basal ganglia and the internal capsule. The caudal parts of the brain, including the cerebellum and most of the brain stem, receive blood from the two vertebral arteries. The vertebral artery arises from the subclavian artery and ascends through the transverse processes of cervical vertebrae 1–6, whereupon it enters the skull through the foramen magnum. The vertebral arteries of the two sides unite at the caudal border of the pons to form the *basilar artery*, which in turn divides to form the two *posterior cerebral arteries* at the rostral border of the pons. Study the following arteries and arterial formations related to the vertebral–basilar system and the internal carotid system.

Vertebral–Basilar System

Vertebral Artery (Fig. 4–1)

1. *Posterior inferior cerebellar artery* (PICA) is the largest branch of the vertebral artery. It pursues a tortuous course between the medulla oblongata and cerebellum, and supplies the dorsolateral part of the medulla oblongata, the choroid plexus of the fourth ventricle, and the posterior and inferior parts of the cerebellum.
2. *Anterior spinal artery* is a single artery formed by a contribution from each vertebral artery. It supplies the median and paramedian parts of the medulla oblongata before it descends into the vertebral canal, where it is reinforced by radicular branches as it continues along the length of the spinal cord.

Basilar Artery

The unpaired basilar artery is formed at the inferior border of the pons by the combination of the two vertebral arteries (Fig. 4–1). It lies in the midline on the ventral surface of pons and divides into the two posterior cerebral arteries at the upper margin of the pons. The basilar artery gives off the following branches:

1. *Anterior inferior cerebellar artery* (AICA) arises from the caudal end of the basilar artery. It supplies the upper medulla and lower pons before it reaches the inferior surface of the cerebellum.
2. *Internal auditory artery (labyrinthine artery)* is a branch of the basilar artery or the anterior inferior cerebellar artery. It reaches the membranous labyrinth of the inner ear through the internal auditory canal.
3. *Pontine arteries*. Numerous median and paramedian branches enter directly into pons.
4. *Superior cerebellar artery* arises from the rostral end of the basilar artery and passes posteriorly along the upper border of the pons to reach the superior surface of the cerebellum. Note that the IIIrd cranial nerve, the oculomotor nerve, passes between the superior cerebellar artery and the posterior cerebral artery.

Posterior Cerebral Artery

Each of the two posterior cerebral arteries, which are terminal branches of the basilar artery, curves around the lateral aspect of the midbrain to reach the medial and inferior surfaces of the temporal and occipital lobes of the cerebral hemisphere (Fig. 4–2). Branches from the proximal segment of the posterior cerebral artery penetrate the posterior perforating space to reach structures in the interior of the brain, including the thalamus. Cortical branches from the posterior cerebral artery supply the visual cortex in the occipital lobe. The posterior cerebral artery is connected to the internal carotid artery on the same side through the *posterior communicating artery* (Fig. 4–1).

Internal Carotid System

Internal Carotid Artery

The internal carotid artery is a terminal branch of the common carotid artery, and it enters the skull through the carotid canal and gives off several collateral branches before it divides into the two terminal branches, the *middle cerebral artery* and the *anterior cerebral artery*. The *hypophysial arteries* and the *ophthalamic artery* come off the internal carotid artery as soon as the artery has entered the skull and they are usually not present on the part that is

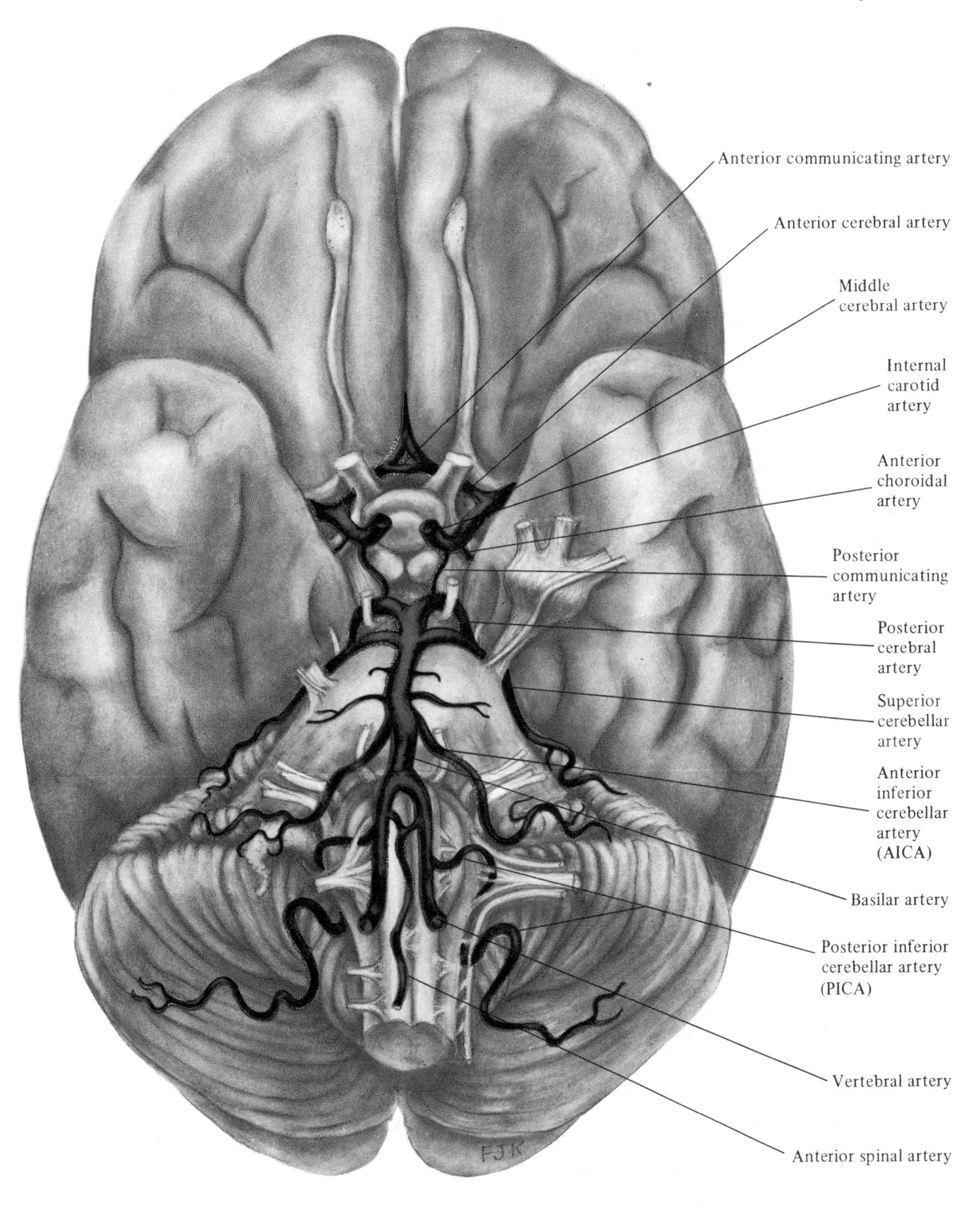

Fig. 4–1. Major Arteries on the Basal Surface of the Brain

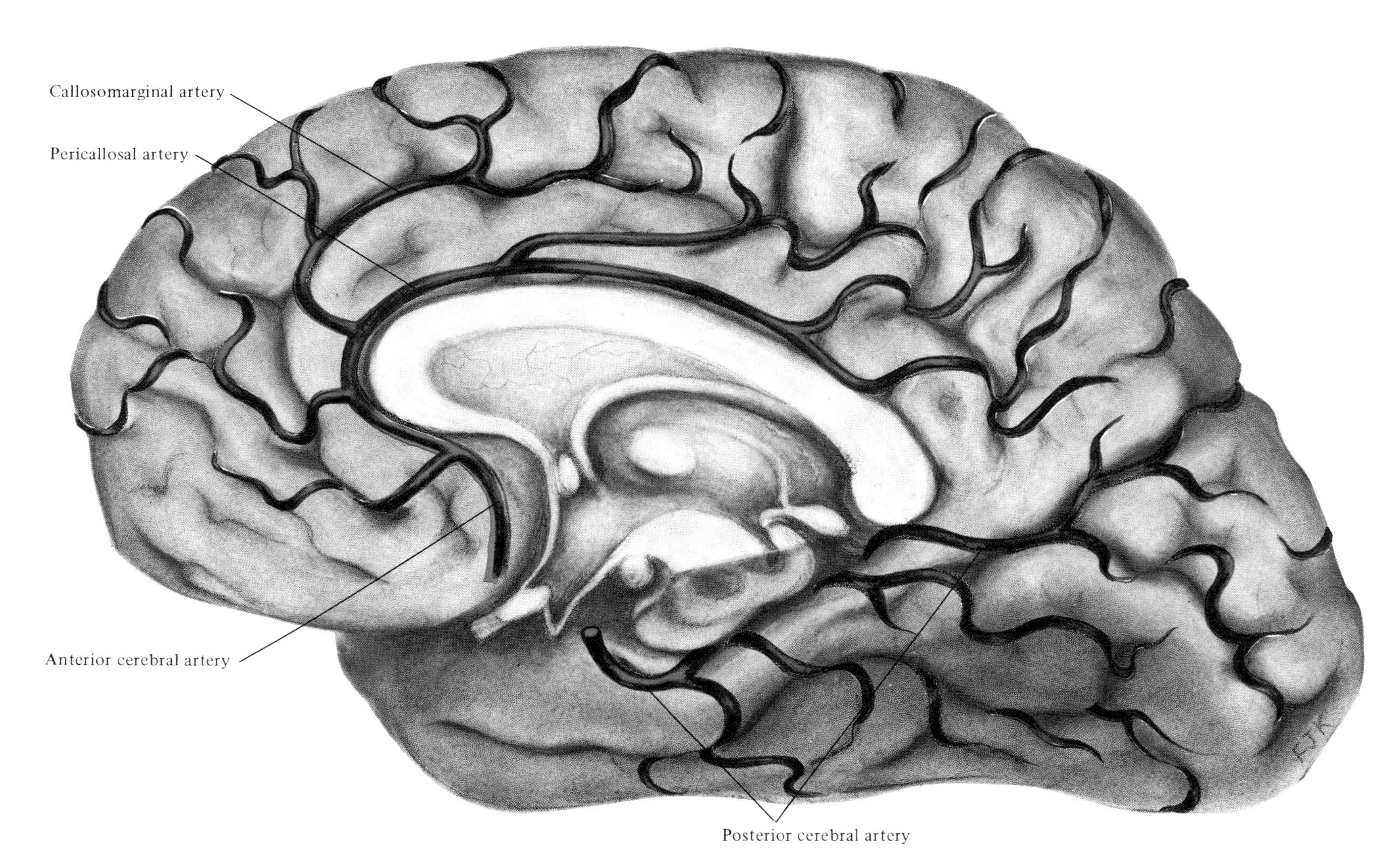

Fig. 4–2. Cortical Distribution of Anterior and Posterior Cerebral Arteries

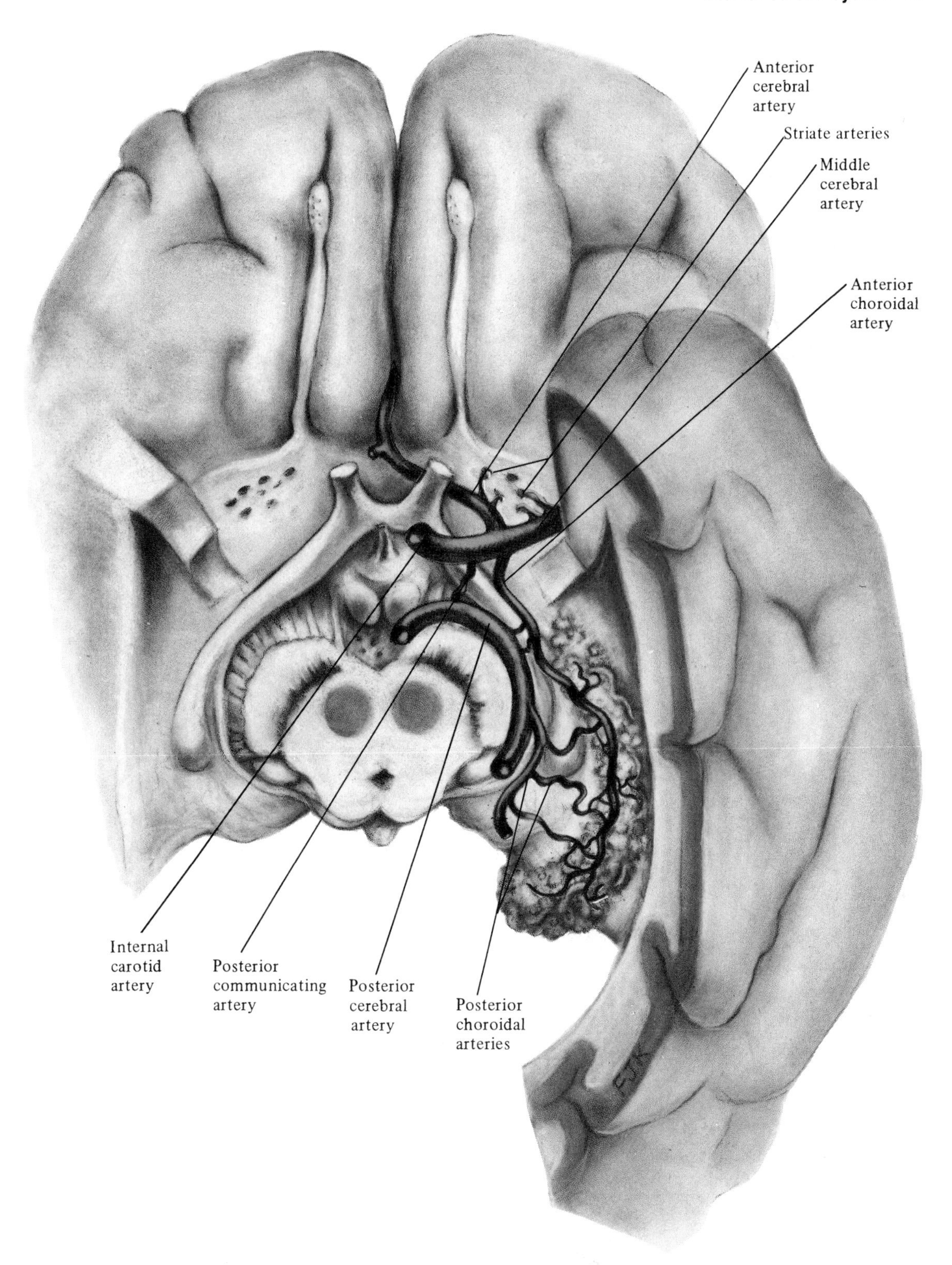

Fig. 4–3. Choroidal Arteries
The medial parts of the temporal lobes have been removed to illustrate the blood supply of the choroid plexus in the lateral ventricle.

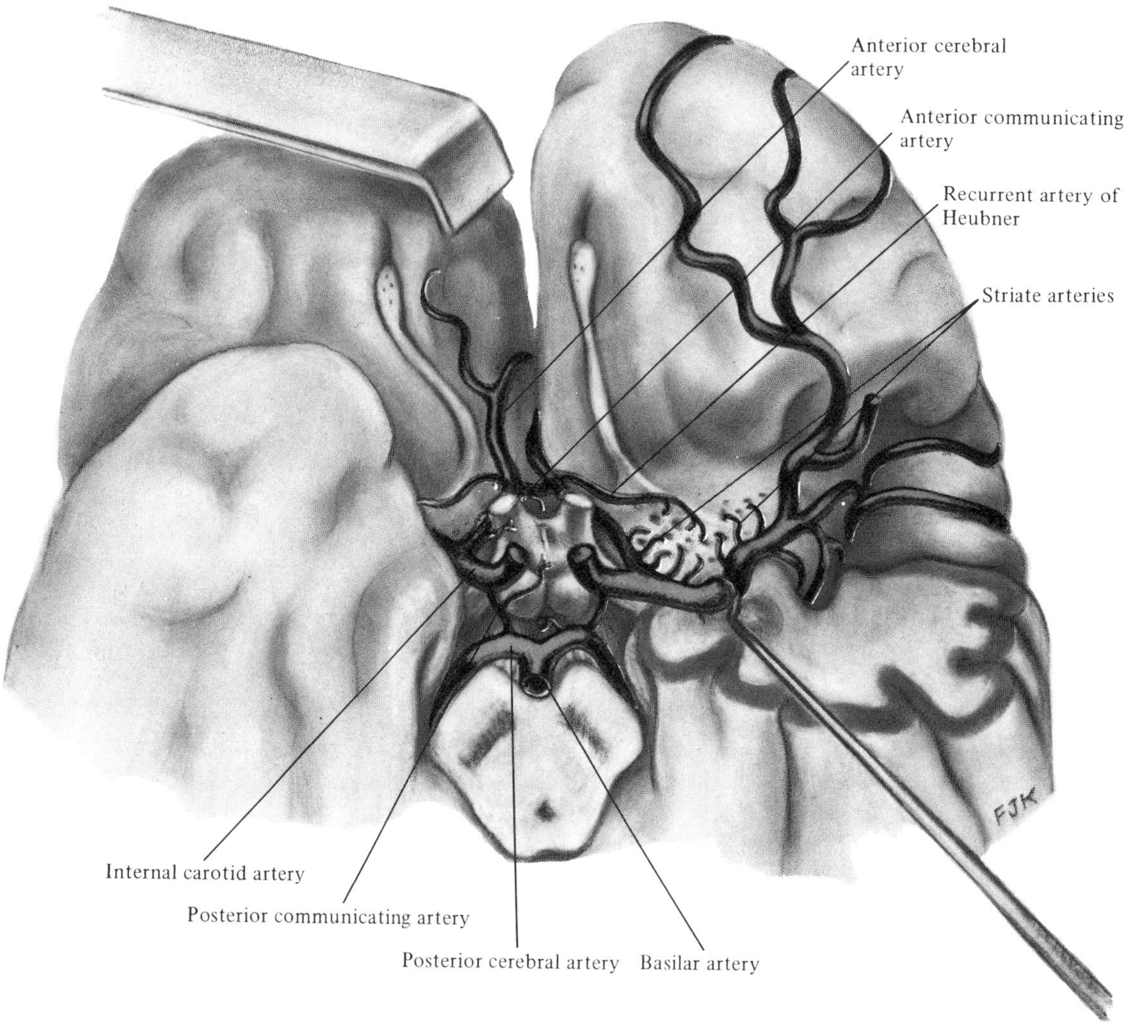

Anterior cerebral artery
Anterior communicating artery
Recurrent artery of Heubner
Striate arteries
Internal carotid artery
Posterior communicating artery
Posterior cerebral artery
Basilar artery
FJK

Fig. 4–4. Circle of Willis, Striate Arteries, and Anterior Perforated Space
The middle cerebral artery is retracted to show the striate arteries.

attached to the brain specimen. Identify the following two branches of the internal carotid artery (Fig. 4–3).

1. *Posterior communicating artery* varies considerably in length and caliber. It runs in a caudal direction to join the posterior cerebral artery.
2. *Anterior choroidal artery*, which sometimes comes off the middle cerebral artery, passes in a caudolateral direction beside the optic tract before it enters the choroid fissure to reach the choroid plexus in the inferior horn of the lateral ventricle. The choroid plexus also receives one or several posterior choroidal arteries from the posterior cerebral artery (Fig. 4–3).

Middle Cerebral Artery (Figs. 4–4 and 4–5)

This is the largest branch of the internal carotid artery, and it runs in a lateral direction between the temporal lobe and the frontal lobe toward the lateral sulcus. Many perforating branches, *striate arteries*, emerge from the initial segment of the middle cerebral artery, and penetrate the anterior perforated space to supply important structures in the interior of the brain (Fig. 4–4). On reaching the lateral sulcus, the middle cerebral artery divides into several large branches on the surface of the insula. These branches continue laterally and emerge on the lateral surface of the hemisphere between the lips of the lateral sulcus. Practically the whole lateral surface of the hemisphere, including

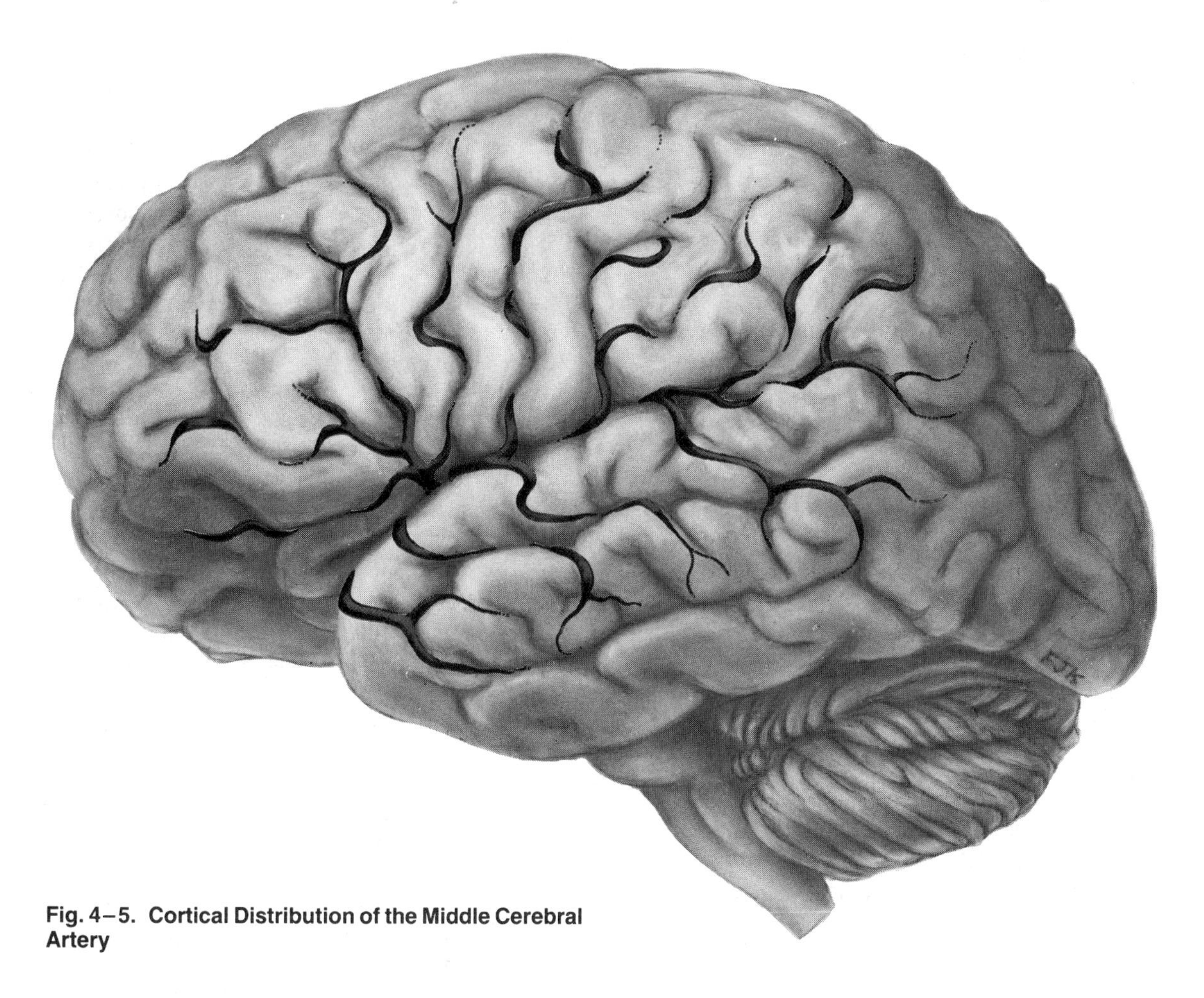

Fig. 4–5. Cortical Distribution of the Middle Cerebral Artery

important motor and sensory areas of the cortex, is supplied by branches of the middle cerebral artery (Fig. 4–5).

Anterior Cerebral Artery

This terminal branch of the internal carotid artery runs in a medial and rostral direction between the optic nerve and the anterior perforated space to the region of the longitudinal fissure, where it is connected with the corresponding artery on the opposite side by the short *anterior communicating artery* (Fig. 4–4). Perforating arteries to the hypothalamus and to other important structures in the basal part of the brain arise from the proximal part of the anterior cerebral artery.

Distal to the anterior communicating artery, the two anterior cerebral arteries ascend in the longitudinal fissure, where they curve upward and backward above the corpus callosum. Although the two arteries lie close together during their course in the longitudinal fissure, they are separated by the falx cerebri. The terminal ramifications of the anterior cerebral artery on the medial surface of the frontal and parietal lobes cannot be seen until the medial aspect of the hemisphere is exposed (Fig. 4–2).

Circle of Willis

The anterior and posterior communicating arteries help form an arterial circle, the circulus arteriosus cerebri or circle of Willis, which is of great clinical importance (see below). The very short anterior communicating artery connects the two anterior cerebral arteries in front of the optic chiasm, and the two posterior communicating arteries complete the posterior part of the circle by connecting the internal carotid artery with the posterior cerebral artery on each side. Small perforating arteries, which emanate from the different components of the circle of Willis, supply the basal parts of the diencephalon and mesencephalon. The initial segment of the anterior cerebral artery also gives rise to the *recurrent artery of Heubner*, which supplies the rostral limb of the internal capsule and adjacent parts of the basal ganglia (Fig. 4–4).

Clinical Notes

Stroke

Diseases of the cerebrovascular system are the most common of all neurologic problems, and stroke, i.e., a sudden onset of neurologic deficits, is one of the most frequent causes of death. Most strokes can be blamed on atherosclerosis, usually in the internal carotid artery or one of the large arteries on the base of the brain. The neurologic deficits that signify the stroke depend on the size and especially on the location of the destroyed brain tissue, i.e., the infarction, that results from the lack of sufficient blood supply. Based on knowledge of the blood vessels and their vascular territories in the brain, it is usually possible to determine both the location of the lesion and the vessel involved. Considering the great clinical significance of stroke, the functional anatomy of the cerebrovascular system will be treated more fully in Chapter 24 following the presentation of the different functional–anatomic systems.

Circle of Willis

The circle of Willis is of great clinical importance because it provides for the possibility of collateral circulation in the event of occlusion in one of the major arteries proximal to the circle. A thrombosis in the internal carotid artery on one side, for instance, may go unnoticed if sufficient collateral circulation can be established, either from the opposite internal carotid artery via the anterior communicating artery, or through the basilar artery via the posterior communicating artery. There are, however, great variations in the configuration of the circle, and an effective collateral circulation may not always be possible, because one of the anastomotic segments is absent or hypoplastic.

Other potential pathways for collateral circulation exist between the extracranial and the intracranial vessels, e.g., in the orbit between the arteries of the face and the ophthalmic artery, and between the various end branches of the superficial cerebral arteries.

Cerebral Angiography

The vessels of the brain can be visualized by injecting a contrast medium into one of the large arterial trunks carrying blood to the brain. This is usually done by puncturing the femoral artery and passing a catheter to the top of the aortic arch, from where the carotid or vertebral arteries can be reached by the tip of the catheter (Fig. 24–5). In the past, cerebral angiography was widely used, not only for demonstrating abnormalities in the blood vessels, but also for the indirect study of expanding lesions (e.g., tumors and hematomas) by observing displacement of vessels. However, since cerebral angiography is difficult and not without risk, it has to a large extent been replaced by computed tomography (CT) and magnetic resonance imaging (MRI), but is still used for the diagnosis of cerebrovascular disorders and some other conditions.

Subarachnoid Hemorrhage (SAH)

Bleeding in relation to the meninges is a serious and often fatal disorder. Whereas clinically significant hemorrhage related to the dura mater, i.e., subdural and extradural hematomas, are usually caused by head injuries (see Clinical Notes, Chapter 3), subarachnoid hemorrhage of clinical significance is usually of nontraumatic origin. A subarachnoid bleed often results from a ruptured arterial aneurysm in the circle of Willis or one of its major branches. Because most aneurysms lie in one of the large subarachnoid cisterns on the base of the brain, the extravasated blood tends to spread widely in the subarachnoid space. A subarachnoid bleed is a medical emergency, which overwhelms its victim with a sudden excruciating headache. An SAH is usually confirmed with CT scanning. If the patient survives, a more detailed examination can be performed with angiography, possibly followed by surgical treatment.

Basal Surface and Cranial Nerves

After having studied the arrangement of the major arteries on the base of the brain, cut through the superior cerebellar arteries as well as the posterior and middle cerebral arteries at their origin, and the anterior cerebral arteries where they disappear into the longitudinal fissure. Remove the circle of Willis together with the basilar and the vertebral arteries with their branches. By leaving the anterior, the middle, and the posterior cerebral arteries intact, it will be possible to study the distribution of these arteries on the medial and lateral side of the hemisphere in more detail later. Remove the remaining parts of the pia–arachnoid on the basal surface. The pia, which is attached to the cranial nerves, should be cut with scissors around the exit of the nerves. Identify the following structures (Fig. 4–6):

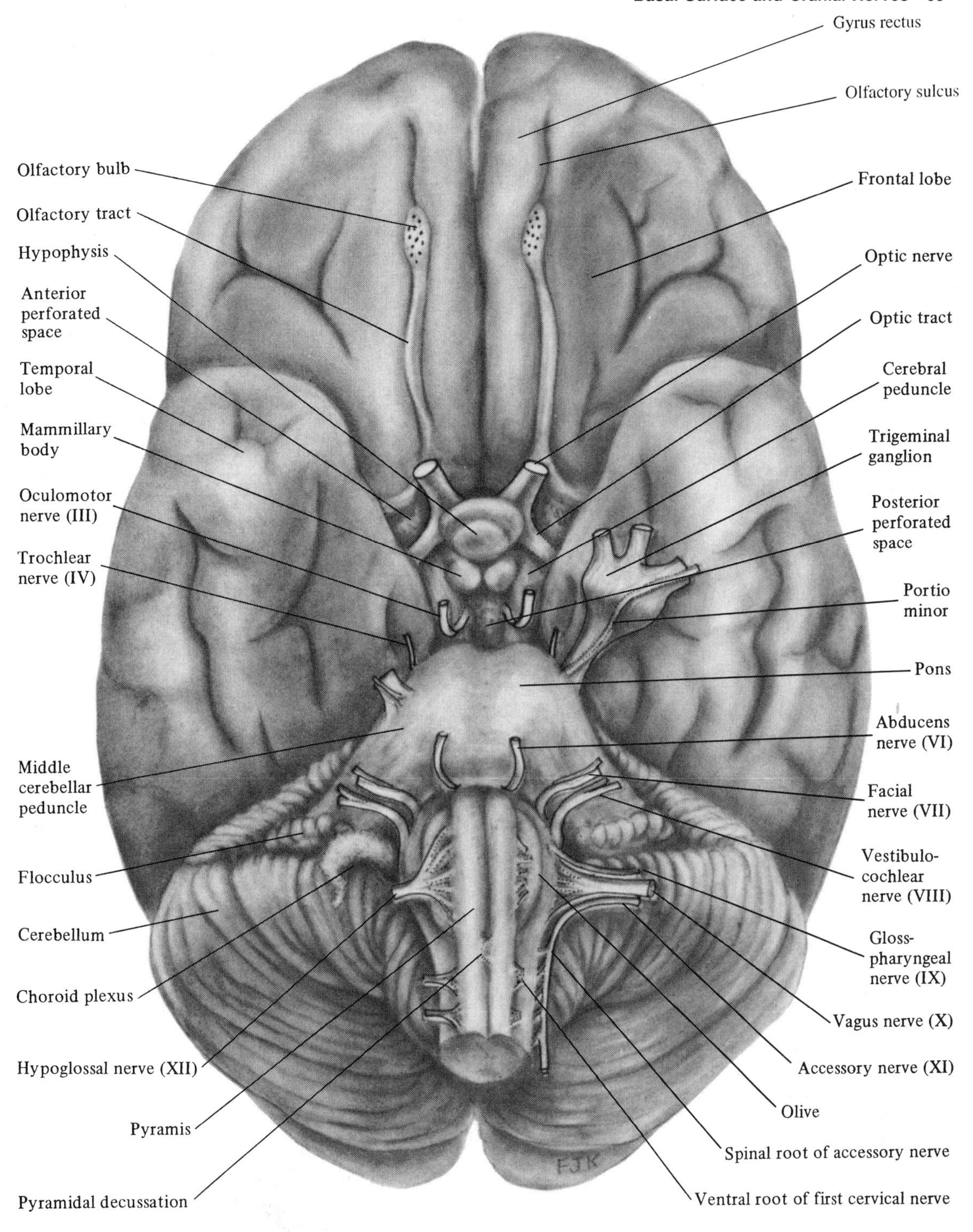

Fig. 4–6. Basal Surface of the Brain and the Cranial Nerves

Medulla Oblongata

1. *Pyramis* is a cone-shaped eminence formed by the corticospinal (pyramidal) tract. The corticospinal tract contains fibers passing from the cerebral cortex to the spinal cord.

2. *Pyramidal decussation* is situated at the lower end of the bulb. Here, most of the fibers in the corticospinal tract cross the median plane. The pyramidal decussation explains why the cortical

motor areas for voluntary control of motor functions primarily affect the contralateral side of the body.

3. *Olive* derives its name from the *inferior olivary nucleus*, which appears as a large wrinkled bag-shaped mass just beneath its surface. The inferior olivary nucleus is closely related to the cerebellum.

Cerebellum

Cerebellum, which is an important structure for control of motor functions, is divided into a medial part,

1. *Vermis*, and two lateral lobes or
2. *Cerebellar hemispheres*
3. *Choroid plexus* of the fourth ventricle protrudes through the lateral *apertures of Luschka* into the subarachnoid space in the angle between the medulla oblongata and the cerebellum.

Pons

The *middle cerebellar peduncles* connect the pons with the cerebellum. The transverse striations on the ventral surface of the pons represent pontocerebellar fibers connecting the base of the pons with the cerebellum. The other two pairs of cerebellar peduncles, the *inferior* and the *superior cerebellar peduncles*, cannot be seen from the basal side. They will therefore be studied later in relation to the dissection of the cerebellum.

Mesencephalon

1. *Cerebral peduncles*, one on each side, form the ventral surface of the mesencephalon. The basal part of the cerebral peduncle, the basis pedunculi, contains the majority of the fibers that descend from the cerebral cortex to the brain stem and the spinal cord.

2. *Interpeduncular fossa* is a deep depression between the two peduncles. Many small blood vessels perforate the floor of the fossa to reach the interior of the brain. When the vessels are stripped away during the dissection, the area exhibits a perforated appearance, the *posterior perforated space*.

Hypothalamus

1. *Mammillary bodies* (so named because of their resemblance to a woman's breasts) represent characteristic landmarks in the posterior part of the hypothalamus.

2. *Tuber cinereum*, which is hidden behind the hypophysis in Fig. 4–6, is a small slightly elevated area of gray matter in front of the mammillary bodies.

3. *Optic chiasm* is formed by the junction of the two *optic nerves*. Fibers from the nasal half of each retina cross the median plane in the optic chiasm and enter the contralateral *optic tract*, which swings around the lateral side of the cerebral peduncle to reach the *lateral geniculate body* in the thalamus. The partial decussation of the optic nerve fibers in the optic chiasm, and the fact that many units in the lateral geniculate body and the visual cortex receive input from the same part of the visual field of both eyes, provides for a high degree of stereoscopic vision in primates.

4. *Hypophysis* (pituitary gland) is lodged in the sella turcica and covered from above by the diaphragma sellae. The hypophysis is usually ripped off from the base of the hypothalamus when the brain is removed from the cranium during autopsy. The *hypophysial stalk*, which still attaches to the brain, contains an opening representing the infundibular recess of the third ventricle.

Telencephalon

The major part of the basal surface of the brain is represented by the *frontal* and *temporal lobes*. A conspicuous structure on the ventral surface of the frontal lobe is the:

1. *Olfactory bulb*, which is situated in a groove, the *olfactory sulcus*, lateral to the *gyrus rectus*.

2. *Olfactory tract* is the caudal continuation of the olfactory bulb. It attaches to the base of the brain in front of the

3. *Anterior perforated space*, whose perforations represent points of entrance for numerous arterial branches, the striate arteries, that have been ripped off from the anterior part of the circle of Willis and the middle cerebral artery during the removal of the blood vessels and the pia–arachnoid (Fig. 4–4).

Cranial Nerves

Identify the cranial nerves and study the place of their exits from the brain and the skull (Fig. 4–6, Table 4–1). The morphology and function of the cranial nerves are discussed in Chapters 11–14.

Clinical Notes

The base of the brain presents a number of topographic relations that are of great clinical relevance. Reference has already been made to the main arteries and the functional significance of the circle of Willis, and to the clinically important aneurysms related to this circle and neighboring arteries.

Structures on the base of the brain, especially the basal surfaces of the frontal and temporal lobes, are often damaged in head injuries. If the injury results in fractures through the base of the skull, some of the cranial nerves can easily be damaged at their point of exit through the base of the cranium.

Knowledge of the anatomic relations at the base of the brain is also essential for understanding the symptoms of tumors and other expanding lesions, which are fairly common in this region and often involve one or several of the cranial nerves either directly or indirectly. Most cranial nerves can be reliably identified with MRI as they pass through the basal cisterns to their exits from the skull.

The Suprasellar Region and Pituitary Adenomas

Tumors in the region of the sella turcica are fairly common, and an adenoma arising from the adenohypophysis is the most typical example. If the tumor is not discovered early on the basis of hormonal disturbances, it usually expands in the direction of less resistance, which is dorsally towards the base of the brain in the region of the optic chiasm (Fig. 4–8). Involvement of the fibers that cross in the chiasm, i.e., the fibers from the nasal halves of the retina, is likely to result in bitemporal visual field defects (see Clinical Example, Chapter 18).

The Cerebellopontine Angle and Vestibular Schwannomas

The cerebellopontine (CP) angle is the space between the lateral side of the pons, the cerebellum, and the posterior surface of the petrous bone. It contains the pontocerebellar cistern and several other important structures including the vestibulocochlear and facial nerves and the anterior inferior cerebellar artery. MRI sections in the axial plane can demonstrate the full extent of both the facial and the vestibulocochlear nerves as they pass through the pontocerebellar cistern and the internal auditory canal (see Clinical Example, Chapter 14).

Tumors in this region are referred to as cerebellopontine angle tumors, and the most common are schwannomas arising from the vestibulocochlear nerve in the internal auditory canal. Although schwannomas usually arise from the vestibular nerve, the initial symptom is often unilateral hearing loss, and the tumors are commonly referred to as acoustic neurinomas. If the tumor expands in the CP angle, the facial and intermediate nerves, as well as the Vth, IXth, and Xth cranial nerves, can also be involved. Large CP angle tumors may even cause cerebellar symptoms (see Clinical Example, Chapter 14).

Table 4–1. Exits of Cranial Nerves from the Brain and the Cranial Cavity.

Name	Its Ganglion	Exit from Brain	Exit from Skull
I. Olfactory nerve	Corresponds to the olfactory receptor cells in the olfactory mucosa	Olfactory bulb	Cribriform plate
II. Optic nerve	Corresponds to the ganglion cells in the retina	Optic chiasm	Optic canal
III. Oculomotor nerve	Ciliary ganglion (parasympathetic)	In front of the pons in the interpeduncular fossa	Superior orbital fissure
IV. Trochlear nerve	—	Dorsally, behind the inferior colliculus	Superior orbital fissure

Table 4–1. *Continued*

Name	Its Ganglion	Exit from Brain	Exit from Skull
V. Trigeminal nerve 1. Ophthalmic nerve 2. Maxillary nerve 3. Mandibular nerve	Trigeminal ganglion (som.) pterygopalatine ganglion, otic ganglion, and submandibular ganglion (all parasympathetic) are topographically related to the Vth nerve	Anterolateral part of the pons	1. Superior orbital fissure 2. Foramen rotundum 3. Foramen ovale
VI. Abducens nerve	—	Between the pons and the pyramid of the medulla oblongata	Superior orbital fissure
VII. Facial nerve	Geniculate ganglion (somatic)	Between the pons and the olive in the cerebellopontine angle	Internal acoustic meatus, facial canal, and stylomastoid foramen
VIII. Vestibulocochlear nerve 1. Cochlear nerve 2. Vestibular nerve	 Spiral ganglion Vestibular ganglion	Lateral of VII in the cerebellopontine angle	Internal acoustic meatus
IX. Glossopharyngeal nerve	Superior ganglion Inferior ganglion	On the lateral side of the olive	Jugular foramen
X. Vagus nerve	Superior ganglion	On the lateral side of the olive, behind IX	Jugular foramen
XI. Accessory nerve	—	1. On the lateral side of the olive, behind X 2. From the 5 or 6 first segments of the cervical part of the spinal cord	Jugular foramen
XII. Hypoglossal nerve	—	Between the pyramid and the olive	Hypoglossal canal

Second Dissection

Midsagittal Section; Lobes, Sulci, and Gyri

Midsagittal Section

If two whole brains are available, place the brain knife on the corpus callosum of one of them and bisect it by gently pulling the moistened knife through the brain. Do not press the knife into the tissue. Save one half for the preparation of frontal brain sections at the end of the course.

Study the distribution of the anterior cerebral artery on the medial surface of the hemisphere (Fig. 4–2). The artery ascends toward the genu of the corpus callosum and describes a wide curve backward on the dorsal side of this massive commissural plate, where it is sometimes referred to as the *pericallosal artery*. Another branch is called the *callosomarginal artery*. Study the distribution of the posterior cerebral artery on the medial side of the temporal and occipital lobes. Remove the pia–arachnoid on the medial surface of the hemisphere, and identify the following structures (Figs. 4–7 and 4–8).

Myelencephalon (Medulla Oblongata)

The medulla oblongata continues caudally without sharp boundary into the spinal cord. The *central canal* of the spinal cord continues up into the lower part of the medulla oblongata but gradually moves dorsally and opens into the fourth ventricle when it reaches the upper half of the medulla.

Metencephalon

1. *Pons* consists of a large ventral or basal part and a smaller dorsal part. The ventral part is characterized by many fiber bundles and scattered gray matter, the *pontine nuclei*.

2. *Cerebellum* is characterized by deep *fissures* separating narrow folds, *folia*. The tree-like outline of the white matter is referred to as the arbor vitae. The cerebellum is an important center for coordination of movements. It is separated from the pons and the medulla oblongata by the fourth ventricle.

3. The *fourth ventricle* is continuous caudally with the central canal and rostrally with the cerebral aqueduct. The rostral part of the roof of the fourth ventricle is formed by the *superior medullary velum*, which fills the angular space between the superior cerebellar peduncles. The caudal portion of the roof, the *inferior medullary velum*, is very thin and it has a large hole, the *median aperture of Magendie*, in its caudomedial part. Through the median aperture and the two *lateral apertures of Luschka* (Fig. 4–6) in the lateral recess, the CSF escapes from the ventricular system into the subarachnoid space. The *choroid plexus* of the fourth ventricle is attached to the ventricular surface of the inferior medullary velum.

Mesencephalon (Midbrain)

1. *Cerebral aqueduct* connects the third and the fourth ventricles. Obstruction of this narrow passage can easily block the flow of CSF (see Hydrocephalus in Clinical Notes, Chapter 3).

2. *Tectum* contains the *superior* and *inferior colliculi*. The superior colliculus contains visual reflex centers, whereas the inferior colliculus is an important relay center in the auditory pathways.

3. Ventral to the aqueduct are the *cerebral peduncles*, one on each side. The mesencephalon, pons, and medulla oblongata are usually referred to as the *brain stem*. Occasionally, the thalamus or even the whole of the diencephalon is included. The cerebellum, however, is not included in the term.

Diencephalon

1. *Third ventricle*, whose lateral wall is subdivided by the *hypothalamic sulcus* into an upper part,

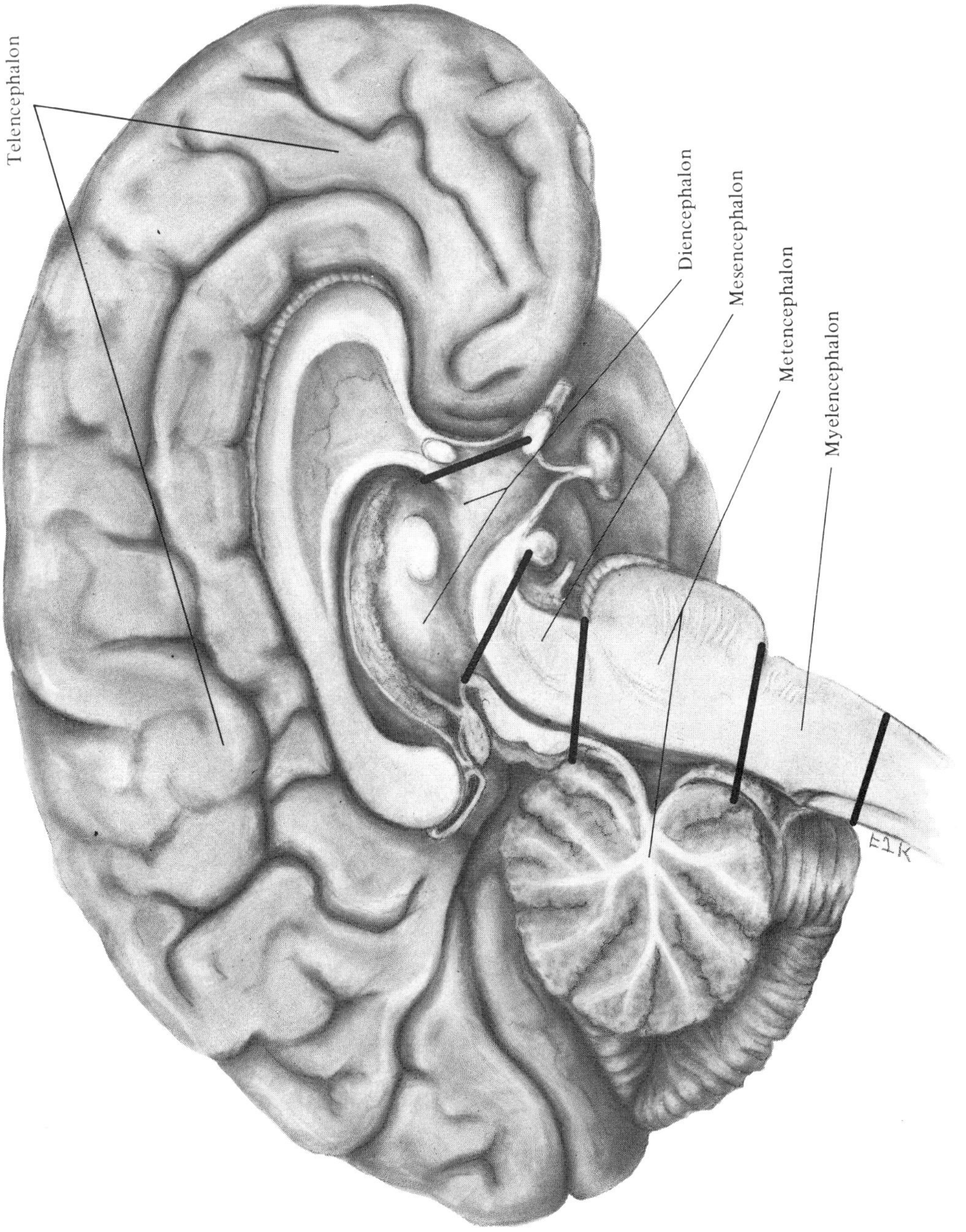

Telencephalon
Diencephalon
Mesencephalon
Metencephalon
Myelencephalon

Fig. 4–7. Major Subdivisions of the Brain

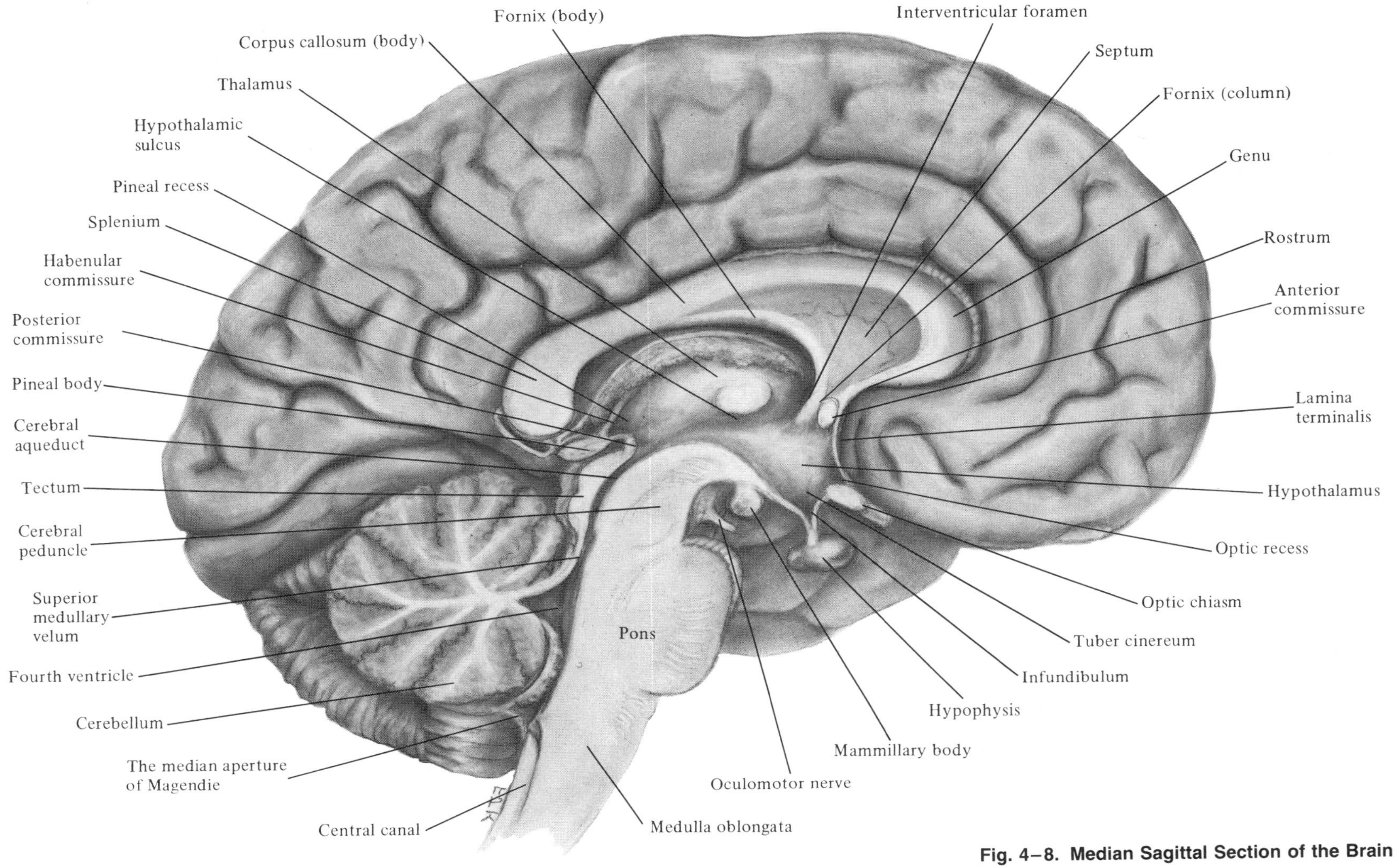

Fig. 4–8. Median Sagittal Section of the Brain

2. *Thalamus*,[1] and a lower part,

3. *Hypothalamus*, in which the *tuber cinereum* and the *mammillary body* can be recognized.

4. *Hypophysis* or *pituitary gland* is attached to the hypothalamus through the hypophysial stalk. Note the close relation to the ventral hypothalamus and the optic chiasm (see Clinical Example, Chapter 18).

5. *Optic chiasm* is located in the rostral part of the hypothalamus. Above the optic chiasm is a recess of the third ventricle, the *optic recess*. Another recess, the *infundibular recess*, extends into the *infundibulum*. Behind the thalamus is the *pineal recess*.

6. *Pineal body* or *epiphysis* protrudes from the pineal recess over the upper part of the mesencephalon between the two superior colliculi.

7. *Habenular commissure* is located above the pineal recess.

8. *Posterior commissure* is below the pineal recess. The fiber components in the habenular and the posterior commissures connect some diencephalic and mesencephalic regions on the two sides. The habenular and posterior commissures, like most commissures in the brain, contain both commissural and decussating fibers.

Telencephalon

1. *Corpus callosum* is the main commissural bundle between the two hemispheres. It is the largest fiber tract in the human brain, and its functions have been enigmatic until recently, when so-called "split brain" experiments (transection of corpus callosum) showed that this massive commissure is important for the transfer of experience from one hemisphere to the other. The corpus callosum consists of
(a) *Splenium* (the wide caudal part)
(b) *Body* (the large midportion)
(c) *Genu* (the anterior part) and
(d) *Rostrum*, which is continuous with the

2. *Lamina terminalis*, which forms the rostral wall of the third ventricle.

3. *Anterior commissure* contains fibers connecting the two temporal lobes. It is located behind the upper part of the lamina terminalis.

4. *Septum pellucidum*, which stretches between the anterior part of the corpus callosum and the fornix, separates the anterior parts of the two lateral ventricles from each other. Its dorsal part is a thin paired lamina that contains fibers and glia cells. Its thicker ventral part, the *precommissural septum*, contains the septal nuclei. The precommissural septum is located in the *subcallosal area* in front of the anterior commissure (Fig. 4–12).

5. *Fornix* constitutes a major projection system between the hippocampus in the temporal lobe and some areas in the basal telencephalon and diencephalon, particularly the mammillary body. The fornices of the two sides are joined for a short distance underneath the corpus callosum where they form the *body* of the fornix. As the two bundles curve forward and downward, they separate and form on each side a *column* that disappears into the wall of the third ventricle just in front of the *interventricular foramen*. Remove the gray matter of the hypothalamus by blunt dissection and follow the column of fornix to the mammillary body, were most of the fornix fibers terminate (Fig. 18–4). The posterior part of the fornix, the *crus fornicis*, deviates laterally underneath the corpus callosum and sweeps down behind the thalamus to reach the temporal lobe. This part of the fornix will be identified during the dissection of the hippocampus.

6. *Mammillothalamic tract* can usually be recognized as it leaves the mammillary body if the hypothalamic gray matter just above and slightly behind the mammillary body is carefully removed (Fig. 18–4).

Lobes, Sulci, and Gyri

The folding of the cortex results in a dramatic increase in surface area; compared to exposed cortex, it is estimated that almost three times more cortex is buried in the sulci. The complicated pattern of convolutions (*gyri*) and grooves (*sulci*) can be studied in detail following removal of the pia–arachnoid from the surface of the hemisphere. Although the appearance of the different folds and furrows varies between individuals, all brains have basic similarities. The subdivision of the hemisphere into lobes and gyri provides clinicians and students of the brain with a useful anatomic reference system.

The cerebral cortex has also been subdivided on the basis of its histologic appearance, which varies from region to region. The best known of these cytoarchitectural maps is the one published by the German histologist Brodmann almost 100 years ago. Brodmann identified more than 50 different regions, only a few of which can be sharply delineated from surrounding areas. Nonetheless, Brodmann's map is widely used by clinicians and

1. A median nuclear mass, massa intermedia, connects the two thalami across the third ventricle in about ⅔ of the population.

scientists, and many of his areas will be identified in a subsequent chapter on the Cerebral Cortex (Chapter 22). Although a few of the sulci, e.g., the central sulcus, demarcate areas with different histology, the various gyri do not usually coincide with specific cytoarchitectonic or functional areas of the cerebral cortex. For example, whereas the postcentral gyrus (see below) corresponds roughly to primary somatosensory cortex (Brodmann's areas 3, 1, and 2), the primary motor cortex (Brodmann's area 4) is mostly hidden in the depth of the central sulcus, and only in the dorsal part of the hemisphere does it cover more fully the exposed surface of the precentral gyrus (Fig. 15–8).

On the basis of some of the major grooves, the hemisphere is subdivided into five *lobes*: the frontal, parietal, occipital, and temporal lobes and the insula (Figs. 4–9 and 4–10).

The *frontal lobe* occupies the area in front of the *central sulcus* and above the *lateral sulcus*. The medial side of the frontal lobe is bounded posteriorly by an imaginary line from the central sulcus to the dorsal surface of the corpus callosum.

The *parietal lobe* is bounded by the central sulcus in the front, and by the lateral sulcus and its extension below. An imaginary line between the *parieto-occipital sulcus* and *preoccipital notch* indicates its caudal border.

The *occipital lobe* is located behind the above-mentioned imaginary line. On the medial side of the hemisphere, the occipital lobe is bounded anteriorly by the parieto-occipital sulcus.

The *temporal lobe* is situated underneath the lateral sulcus and in front of the above-mentioned line between the parieto-occipital sulcus and the preoccipital notch.

The *insula*, which is located in the bottom of the lateral sulcus, can be seen if the two lips of the sulcus are pulled apart. To expose it completely, the covering parts, i.e., the opercula of the frontal, parietal, and temporal lobes, must be removed (see below), but this should be done only after the surface topography of the other lobes has been studied.

Dorsolateral Surface of the Hemisphere

Frontal Lobe

The frontal lobe is the largest of the lobes, as it accounts for almost one-third of the cortical volume. It contains important motor and language areas in its posterior part, whereas its rostral part, the prefrontal cortex, is of great significance for many functions related to social behavior and higher mental activities.

Through the presence of three main grooves, the precentral sulcus, the superior frontal sulcus, and the inferior frontal sulcus, the frontal lobe is divided into four main gyri:

1. *Precentral gyrus* (corresponds in part to primary motor cortex or Brodmann's area 4)
2. *Superior frontal gyrus*
3. *Middle frontal gyrus*
4. *Inferior frontal gyrus*

The inferior frontal gyrus is subdivided into an orbital part, a triangular part, and an opercular part by the anterior and ascending ramus of the lateral sulcus. The motor speech area (Broca's area) is located in the triangular and opercular parts of the inferior frontal gyrus, usually on the left side (see Language Disorders, Chapter 22).

Parietal Lobe

Parallel to the central sulcus is the postcentral sulcus, and the convolution between these two grooves is the:

1. *Postcentral gyrus*, which coincides roughly with primary somatosensory cortex, areas, 3, 1, and 2. The intraparietal sulcus separates the
2. *Superior parietal lobule* from the
3. *Inferior parietal lobule*. Two important gyri are usually recognized in the last mentioned lobule, namely
4. *Supramarginal gyrus*, which caps the posterior tip of the lateral sulcus, and
5. *Angular gyrus*, which caps the posterior tip of the superior temporal sulcus. These two convolutions together with the adjoining parts of the occipital and temporal lobes are of special importance for language functions. They are part of the sensory language area, which is usually located in the left hemisphere.

Occipital Lobe

The gyri on the lateral and inferior sides of the occipital lobe are collectively referred to as *gyri occipitales*. The posterior part of the *calcarine sulcus*, which is the major groove on the medial side (see below), sometimes curves around the occipital pole to the lateral surface. The occipital lobe is primarily related to visual functions. Indeed, regious subserving visual functions extend beyond the boundaries of the occipital lobe into both parietal and especially the temporal lobes (see Chapter 13).

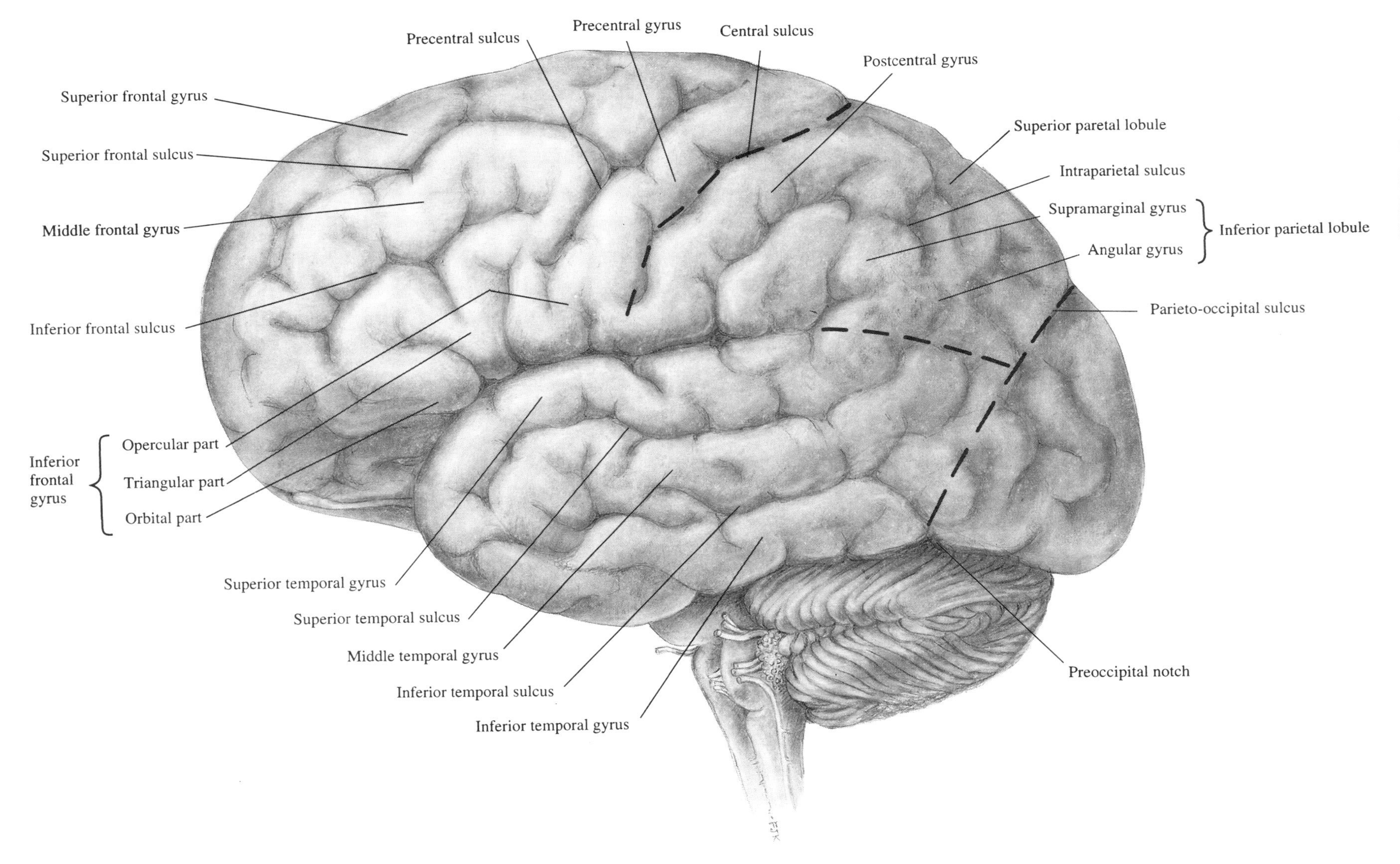

Fig. 4–9. Lateral Surface of the Brain

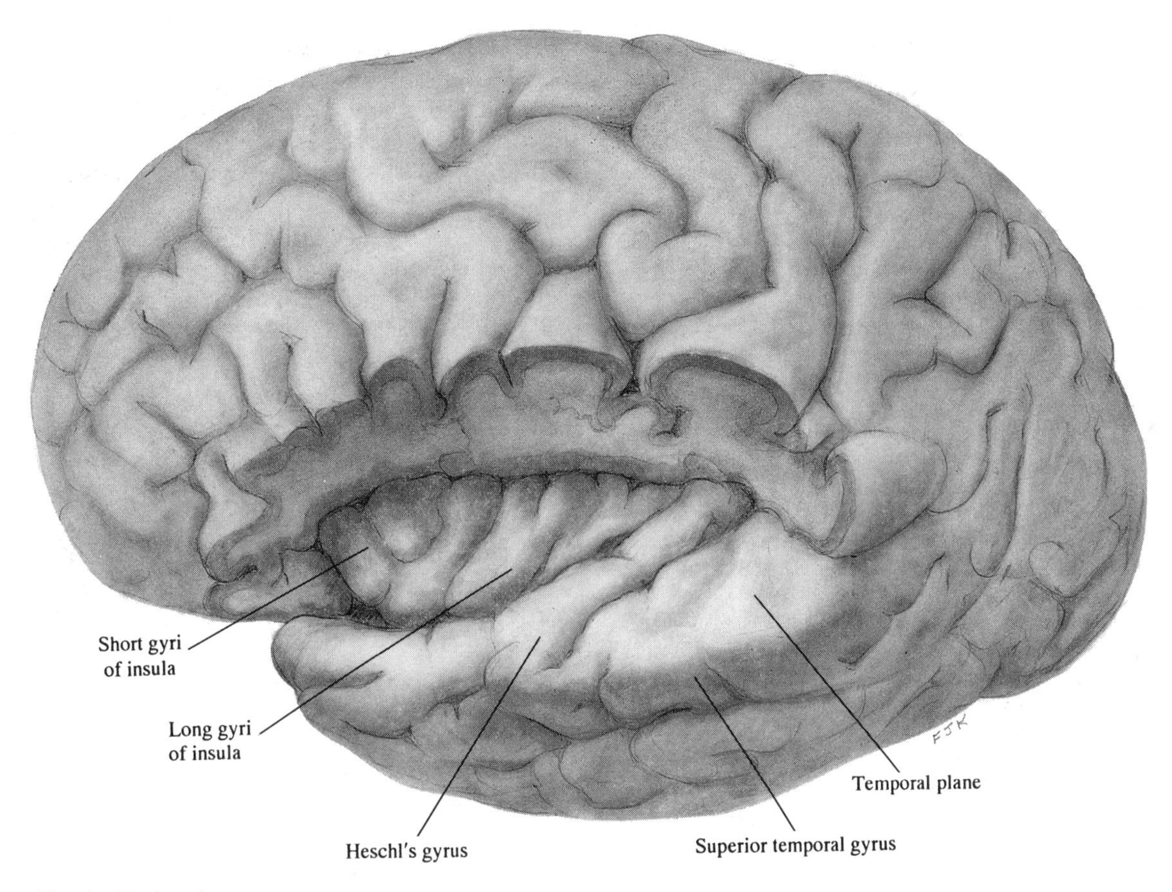

Fig. 4–10. Insula
The major part of the insula has been exposed by removing the opercular parts of the frontal and parietal lobes. The hemisphere is slightly tilted to provide a good view of Heschl's gyrus and the temporal plane on the dorsal surface of the superior temporal gyrus.

Temporal Lobe

The temporal lobe, which is almost as large as the frontal lobe, contains many different regions, including sensory areas for auditory and olfactory functions. Two large structures, the amygdaloid body and the hippocampus, are hidden from sight in the interior part of the lobe and will be exposed at the end of the Fourth Dissection.

The inferior, superior and temporal sulci subdivide the lateral surface of the temporal lobe into:

1. *Inferior temporal gyrus*
2. *Middle temporal gyrus* and
3. *Superior temporal gyrus*

Situated on the extensive dorsal surface of the superior temporal gyrus are the

4. *Transverse temporal gyrus* (Heschl's gyrus), represented by one or two small convolutions, which indicates the region of the primary auditory cortex (Brodmann's areas 41 and 42), and
5. *Planum temporale*, part of the auditory association cortex, behind the transverse temporal gyrus.

These two structures can be exposed if the lateral sulcus is spread apart, or if the opercular parts of the frontal and parietal lobes are removed as indicated in Figs. 4–10 and 4–11. This procedure also exposes the insula.

Insula

The insula can be exposed if the opercular parts of the frontal, parietal, and temporal lobes are removed as illustrated in Fig. 4–10. Use the side of the brain from which the brain stem and cerebellum were removed. Before the incision is made, however, it is advisable to spread apart the lateral sulcus to observe the location and the extent of the insula. The insula is subdivided by a central insular sulcus into several

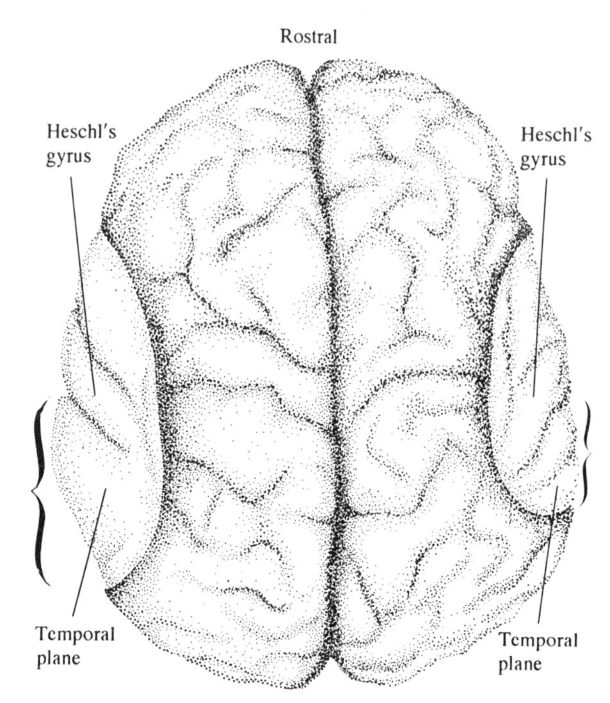

Fig. 4–11. Dorsal View of the Brain
Dorsal view of the brain with frontoparietal opercula removed to show the temporal planes on the dorsal sides of the superior temporal gyri. Note the large temporal plane on the left side. However, see Chapter 14.

1. *Short gyri* in front of the sulcus, and one or two
2. *Long gyri* behind the central insular sulcus.

The insula is surrounded by a circular sulcus. The anterior basal part of the insula, which separates the lateral surface of the insula from the anterior perforated space, is known as the *limen insulae*.

Medial and Basal Surfaces of the Hemisphere (Figs. 4–12 and 4–13)

1. *Cingulate gyrus* begins in the *subcallosal area* underneath the rostrum of the corpus callosum and continues above the corpus callosum into the temporal lobe, where it is continuous with the

2. *Parahippocampal gyrus* through the narrow *isthmus*. The parahippocampal gyrus and related parts of the temporal lobe will be examined in more detail in the fourth dissection. The rostromedial part of the parahippocampal gyrus hooks sharply backwards as the uncus. Part of the uncus represents the primary olfactory cortex (Chapter 12). The peripheral boundary of the cingulate gyrus is represented by the cingulate sulcus, which usually gives off a paracentral sulcus on the medial surface of the frontal lobe, and then divides into a marginal and a subparietal sulcus on the medial surface of the parietal lobe. The region between the paracentral sulcus and the marginal sulcus is referred to as

3. *Paracentral lobule.* The anterior part of this lobule is continuous with the precentral gyrus on the lateral side of the hemisphere, and the posterior part is an extension of the postcentral gyrus. Behind the marginal sulcus is

4. *Precuneus*, which is continuous with the superior parietal lobule on the lateral surface. The parieto-occipital sulcus separates the precuneus from

5. *Cuneus*, which belongs to the occipital lobe. A deep horizontal fissure, the *calcarine sulcus*, is an important landmark in the occipital lobe. The area surrounding the calcarine sulcus represents primary visual cortex (Brodmann's area 17).

The labeling of folds on the basal surface has always created problems. However, it is usually possible to identify two more or less uniform longitudinal convolutions, which extend from the occipital lobe into the temporal lobe, without indicating any boundary between the two lobes, hence the names:

6. *Medial occipitotemporal (lingual) gyrus*, and

7. *Lateral occipitotemporal (fusiform) gyrus.* The medial occipitotemporal gyrus, which is continuous rostrally with the parahippocampal gyrus, is separated from the lateral occipitotemporal gyrus by the *collateral sulcus*. The *occipitotemporal sulcus* separates the lateral occipitotemporal gyrus from the inferior temporal gyrus.

8. *Gyrus rectus* is located medial to the *olfactory sulcus* on the basal surface of the frontal lobe. The *olfactory bulb* and *olfactory tract* are lodged in this sulcus. Lateral to the olfactory sulcus are many irregular

9. *Orbital gyri.*

Clinical Notes

The cerebral cortex represents the crowning achievement of the evolutionary process. It is the single largest structure in the CNS, and as discussed in more detail in Chapter 22, it is critically involved in the perception of sensory information and in the control of voluntary movements, as well as in mental activities in which humans normally excel, e.g., functions related to memory and learning, reasoning, and language.

The fact that various functions are related to different parts of the cerebral cortex has promoted

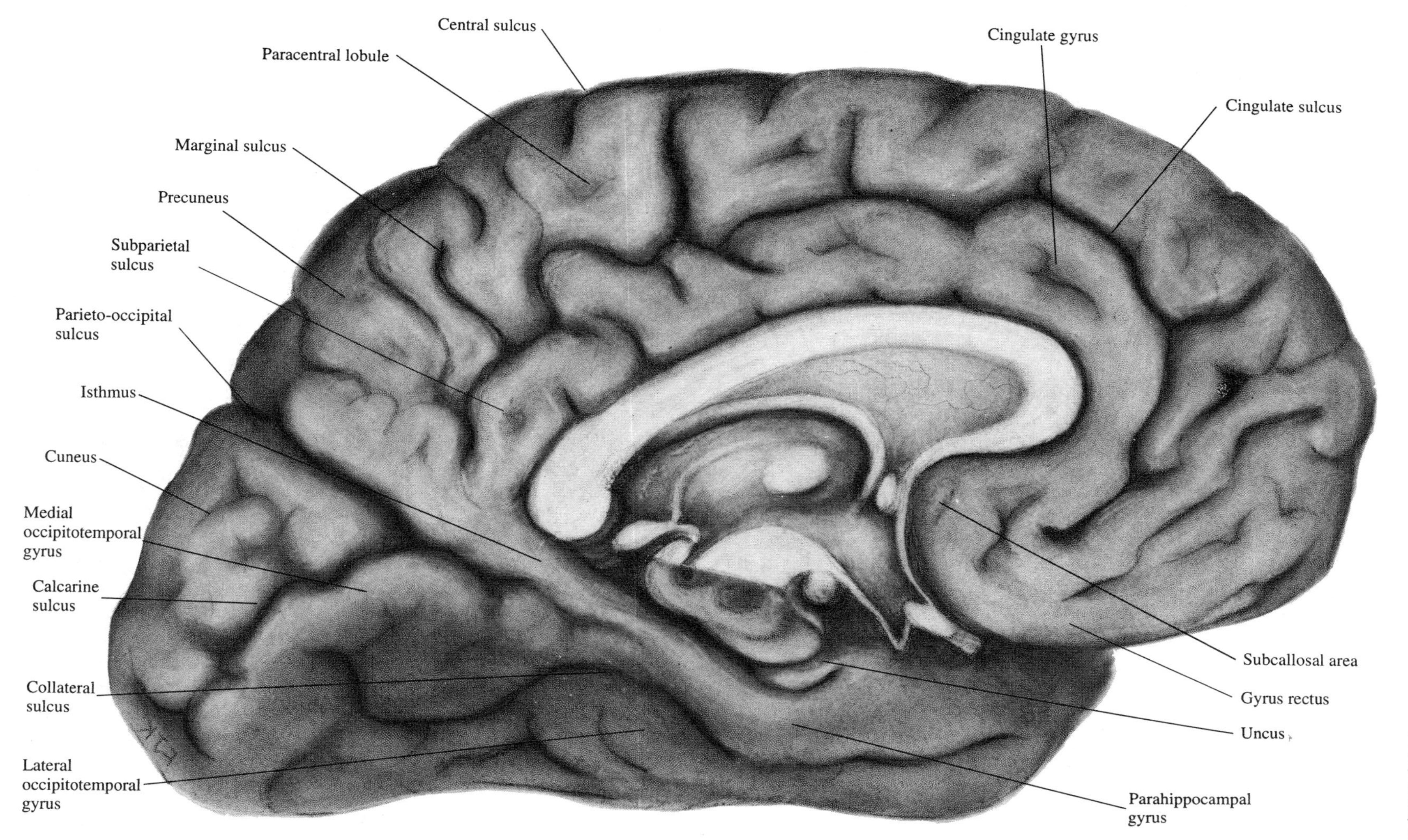

Fig. 4–12. Medial Surface of the Cerebral Hemisphere
The brain stem has been removed to expose the medial surface of the temporal lobe.

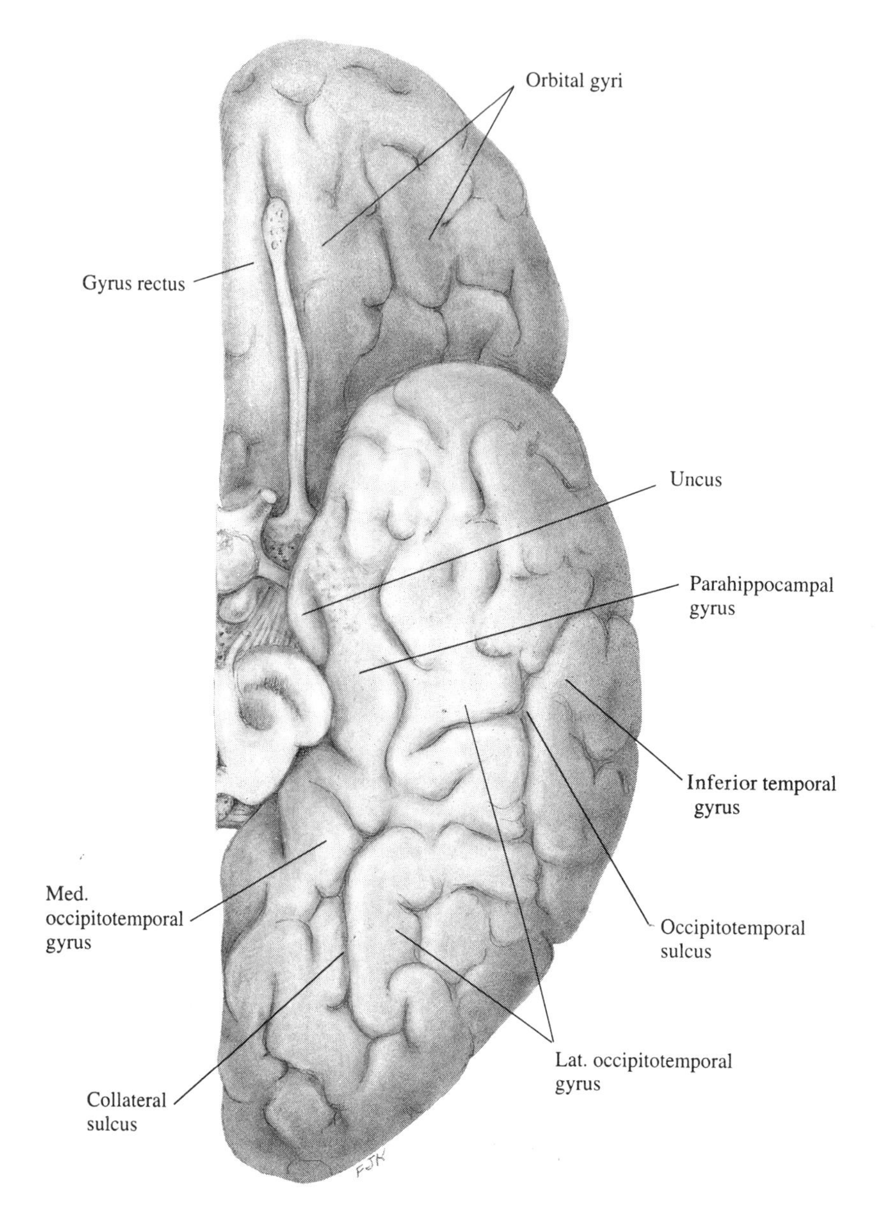

Fig. 4–13. Basal Surface of the Cerebral Hemisphere

the traditional notion of "cortical localization" of functions. However, it is important to realize that a functional center does not exist in isolation from the rest of the nervous system. For the sake of simplicity, a functional "center" can be conceived of as a "distribution center," in the sense that it is connected to many other parts of the nervous system. It derives its importance from the fact that it represents an essential part in a neuronal circuit, or in a network of circuits, that is important for the function in question. Therefore, if a "center" is damaged, the function it subserves is impaired to a lesser or greater extent. Herein lies the clinical significance of cortical "centers." Focal lesions in the cerebral cortex, e.g., vascular accidents or tumors, tend to produce specific functional deficits, and by observing the signs and symptoms produced by the lesion, the physician can usually determine the most likely location of the lesion. Since modern neuroimaging has dramatically improved our ability to determine the extent and location of brain lesions (Fig. 5–5), lesion analysis in the living human brain has become an important tool for the study of higher cortical functions (see Cortical Functions, Chapter 22).

Third Dissection

White Matter; Blunt Dissection of Major Fiber Systems

White Matter

The massive medullary center of the cerebral hemisphere consists of myelinated fibers, which can be divided into three different groups: *association, commissural*, and *projection fibers*. As an introduction to the blunt dissection of the major fiber bundles of the hemisphere, it is helpful to review and examine the location of the major bundles on a schematic drawing (Fig. 4–14).

Association Fibers

Association fibers interconnect cortical regions within the same hemisphere. Many of them form bundles, fasciculi, which contain long pathways as well as short fibers that enter and leave the bundles along their course. The main association bundles are summarized below.

1. *Arcuate fibers* (Fig. 4–15) are short fibers interconnecting adjacent gyri.
2. *Cingulum*, which is located in the cingulate gyrus and the parahippocampal gyrus, connects frontal, parietal, and temporal cortical areas on the medial side of the hemisphere.
3. *Superior longitudinal fasciculus* (*arcuate fasciculus*) is located in the more lateral parts of the hemisphere above the insula. It interconnects areas within the frontal, parietal, occipital, and temporal lobes. Since many of its fibers form an arch-shaped structure as they sweep down into the temporal lobe behind the insula, the bundle is also referred to as the arcuate fasciculus.
4. *Superior occipitofrontal fasciculus* is located in the medial part of the hemisphere underneath the lateral extension of the corpus callosum. It runs alongside the dorsal border of the caudate nucleus and interconnects the frontal lobe with more posterior parts of the hemisphere.
5. *Inferior occipitofrontal fasciculus* runs between the occipital and the frontal lobe along the ventral aspect of the insula in the lateral part of the hemisphere. Many of the fibers in the frontal part sweep down into the temporal lobe as the
6. *Uncinate fasciculus* (Fig. 4–16), which makes a sharp bend around the lateral sulcus to interconnect the orbital surface of the frontal lobe with the rostral part of the temporal lobe.

Commissural Fibers (Fig. 4–8)

The commissural fibers cross the midline and interconnect areas in the two hemispheres with each other. Whereas most of the fibers interconnect similar (homotopic) regions, others link dissimilar (heterotopic) areas on the two sides.

1. *Corpus callosum*, which contains about 300 million fibers, interconnects neocortical areas in all lobes. Although most regions are interrelated through commissural fibers, some regions, notably the primary visual cortex and the hand and foot regions of the somatosensory cortex, do not seem to receive commissural fibers.
2. *Anterior commissure*, which crosses the midline as a compact bundle in front of the fornix columns, interconnects areas in the two temporal lobes. A small anterior or olfactory part, which cannot be readily distinguished from the main posterior part in gross dissections, contains fibers that interconnect the anterior olfactory nucleus at the base of the olfactory tract with the anterior olfactory nucleus and the olfactory bulb on the other side.
3. *Hippocampal commissure* is situated underneath the splenium of the corpus callosum, from which it is difficult to separate. Its fibers interconnect the two hippocampus formations. The hippocampal commissure will be exposed during the next dissection.

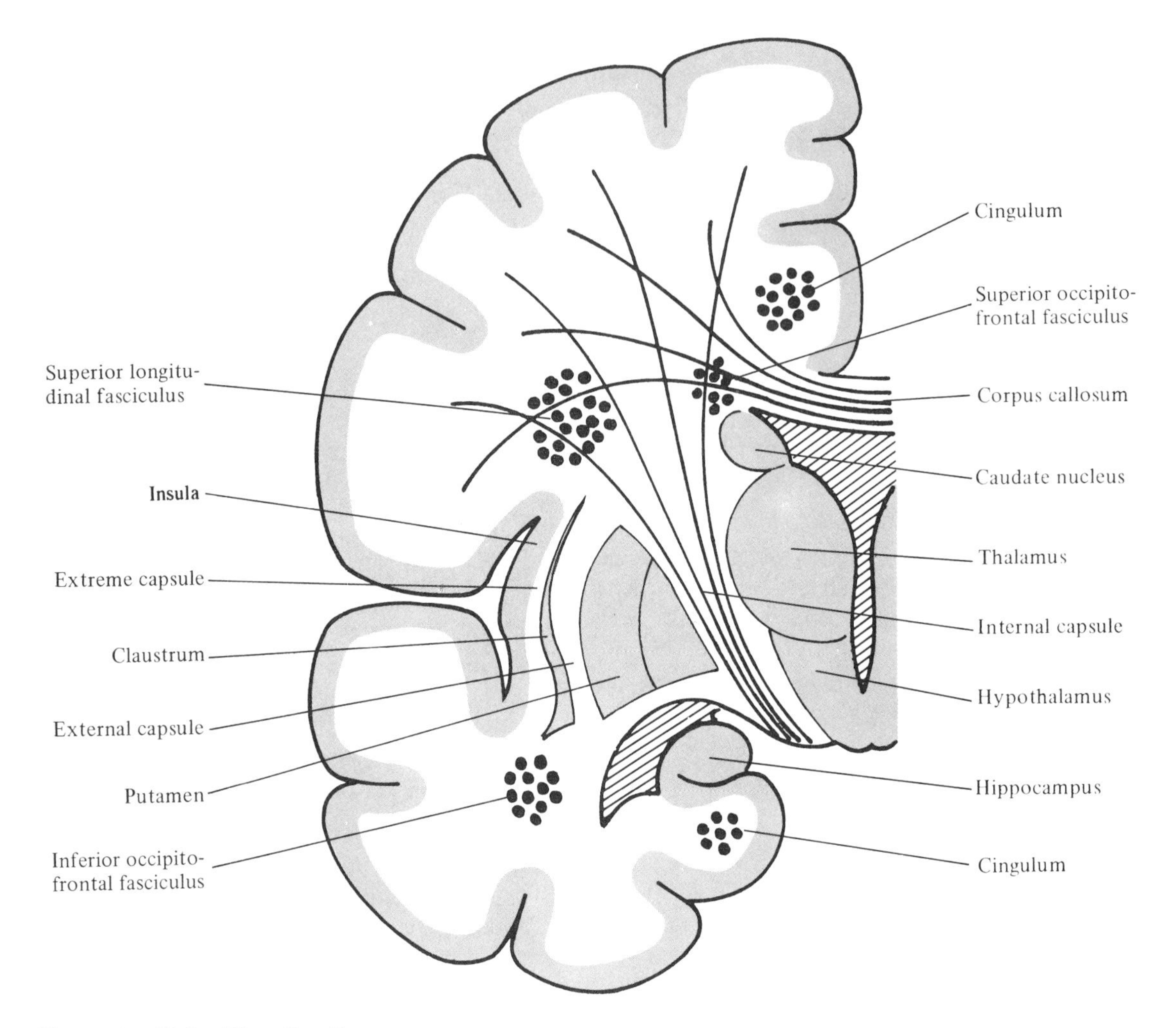

Fig. 4–14. Major Fiber Bundles
Diagram illustrating the position of some of the major fiber bundles in a frontal section of the brain.

4. *Posterior commissure* is located at the boundary between diencephalon and mesencephalon immediately rostral to the superior colliculus. It is a complex fiber bundle containing different components related to several nearby regions including the pretectal area and part of the oculomotor complex.

5. *Habenular commissure* is located above the pineal recess and contains stria medullaris fibers through which the habenular complex and some other subcortical regions are connected to structures on the opposite side.

Projection Fibers (Fig. 4–14)

Many of the fibers in the white matter of the hemisphere connect the cerebral cortex with subcortical regions. These are the projection fibers, which to a considerable extent are related to the *internal capsule.* The projection fibers can be divided into *corticopetal* and *corticofugal fibers*, depending on whether they transmit impulses to or from the cortex. The corticopetal fibers come primarily from the thalamus, form part of the internal capsule, and diverge toward the cerebral cortex. The corticofugal fibers, which originate in different parts of the cerebral cortex, converge toward the basal ganglia, the thalamus, and the internal capsule. The large majority of the corticopetal and corticofugal fibers form a fan-shaped fiber mass, *corona radiata*, as they emerge from the internal capsule. Some of the important fiber systems in the internal capsule are:

1. *Thalamic radiation*, which consists of both corticothalamic and thalamocortical projection fibers.
2. *Corticopontine fibers*, which originate in wide areas of the cerebral cortex and project to the pontine

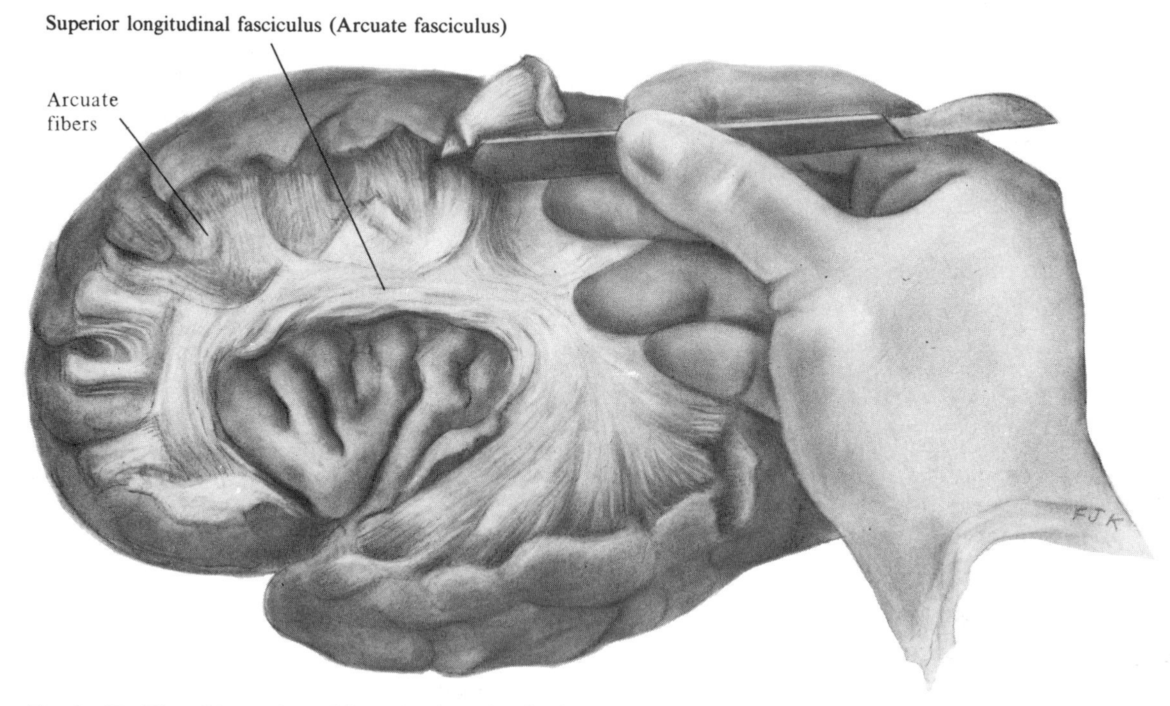

Fig. 4–15. Blunt Dissection of Superior Longitudinal Fasciculus and Arcuate Fibers

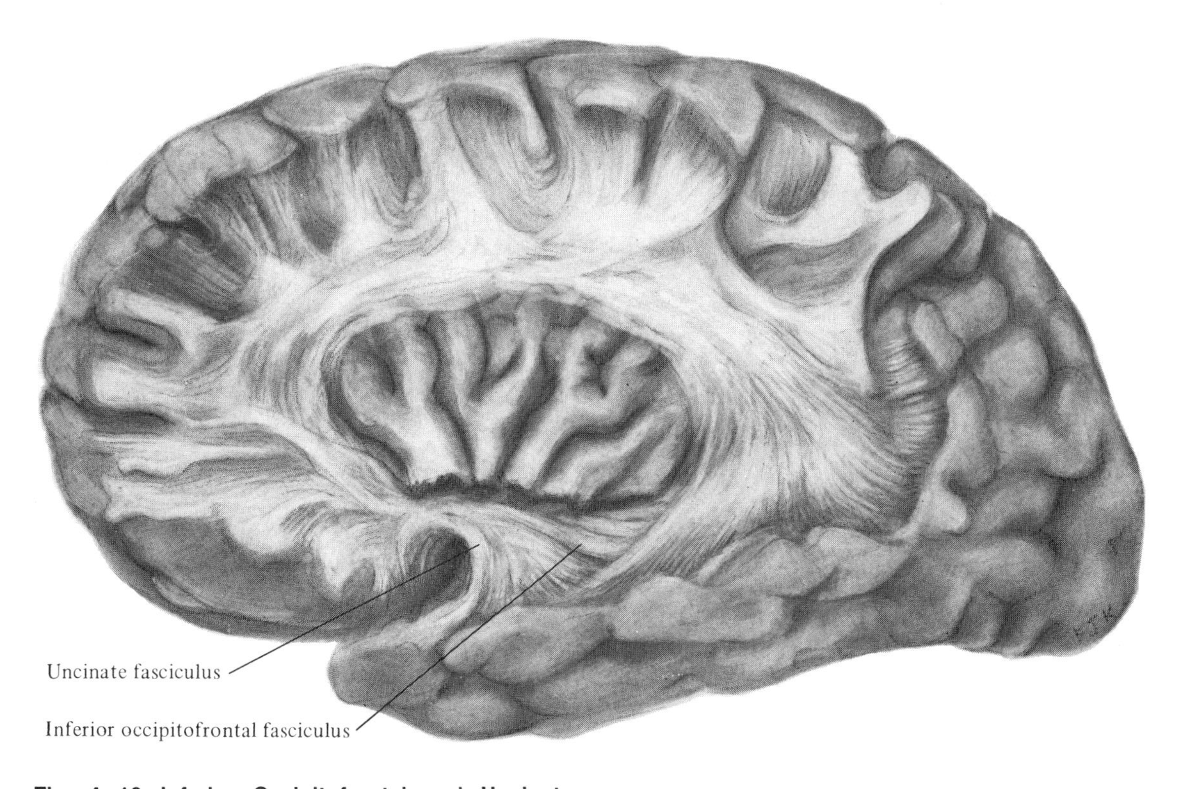

Fig. 4–16. Inferior Occipitofrontal and Uncinate Fasciculi

nuclei in the basal part of the pons. The pontine nuclei in turn project to the cerebellum.

3. *Corticobulbar and corticospinal fibers*, together often referred to as the pyramidal tract, originate in motor and to a lesser extent in somatosensory cortex, and project to cranial nerve nuclei in the brain stem and to cell groups throughout the length of the spinal cord.

4. *Corticoreticular fibers* originate primarily in motor and somatosensory cortex and project to the brain stem reticular formation.

Blunt Dissection of Major Fiber Systems

If the brain substance is peeled away in a methodical fashion from the lateral and the medial sides of the hemisphere, many of the major fiber systems of the white matter can be visualized. By using the blunt end of a scalpel handle on a well-fixed brain, the myelinated fiber bundles can be torn and split apart much like the lamellae of a mushroom. Although the blunt dissection is very useful to gain an appreciation of the structure of the white matter, it is important to realize that information regarding origin and termination of specific fiber tracts cannot be obtained by this type of dissection. Such information is obtained in microscopic analysis of histologic preparations. For the blunt dissection, use the hemisphere on which the insula was explored, and compare the appearance of the different fiber bundles on your specimen with the location of these bundles in a frontal section through the hemisphere (Fig. 4–14).

Dissection from the Lateral Side

1. Reveal the *superior longitudinal fasciculus* (arcuate fasciculus) by peeling away the cortex in the circular sulcus surrounding the insula. Expose the whole bundle by extending the dissection into the frontal and occipital lobes, and follow the part that sweeps down behind the insula into the temporal lobe (Fig. 4–15).

2. Continue the dissection underneath and deep to the basal part of the insula to expose the association system that connects the basal parts of the frontal lobe with the temporal and occipital lobes (Fig. 4–16). To do this, the ventral part of the insular cortex must be removed. The dorsal component of this system, the *inferior occipitofrontal fasciculus*, extends in a longitudinal direction between the frontal and the occipital lobes, whereas the more ventral component, the *uncinate fasciculus*, hooks around the lateral fissure to connect the orbitofrontal cortex with the anterior part of the temporal lobe.

3. At this point it is advisable to turn attention to the *geniculocalcarine tract*, or *optic radiation*,[1] which is the last link in the optic pathway from the retina to the visual cortex in the occipital lobe. This important projection system is usually exposed in efforts to follow the inferior occipitofrontal fasciculus in a posterior direction underneath the caudal part of the insula.

The geniculocalcarine tract originates in the lateral geniculate body of the thalamus and proceeds in a more or less lateral direction through the posterior (retrolenticular and sublenticular) part of the internal capsule before its fibers turn in a caudal direction toward the calcarine fissure in the occipital lobe (Fig. 4–17). It comes into view behind and underneath the caudal part of the insula and sweeps caudally in the form of a broad band, which disappears underneath the arcuate fasciculus (Fig. 4–16). If the arcuate fasciculus and the lateral part of the occipital lobe are dissected away, the geniculocalcarine tract can be followed all the way to the primary visual area in the region of the calcarine fissure (Fig. 4–12).

Although it is usually easy to follow the last part of the geniculocalcarine tract into the occipital lobe, it is more difficult to appreciate the first part of the tract, and it may be especially difficult to separate its ventral part from association fibers in the inferior occipitofrontal fasciculus underneath the insula. The more ventrally located fibers in the geniculocalcarine tract are first directed forward and laterally above the inferior horn of the lateral ventricle before they sweep in a caudal direction toward the occipital lobe. The part of the geniculocalcarine tract that makes a rostrolateral detour into the temporal lobe is referred to as *Meyer's loop* (Fig. 4–17).

4. Remove the rest of the superior longitudinal fasciculus and carefully dissect away the cortical layer of the insula (compare Fig. 4–14). Underneath the insular cortex is a thin lamella of white matter, the *extreme capsule*, which is often difficult to identify. Deep to the extreme capsule is a thin sheet of gray matter, the *claustrum*, which should be carefully stripped away to expose the *external capsule*, which is yet another thin layer of white matter

1. Although the optic radiation is often used as a synonym for the geniculocalcarine tract, one should be aware of the fact that it is also used for the system of sagittally running fibers between the thalamus and the occipital lobe. This massive system of fibers contains more than just the geniculocalcarine tract, e.g., fibers from the visual cortex to the lateral geniculate nucleus and pathways between the pulvinar and the occipital lobe.

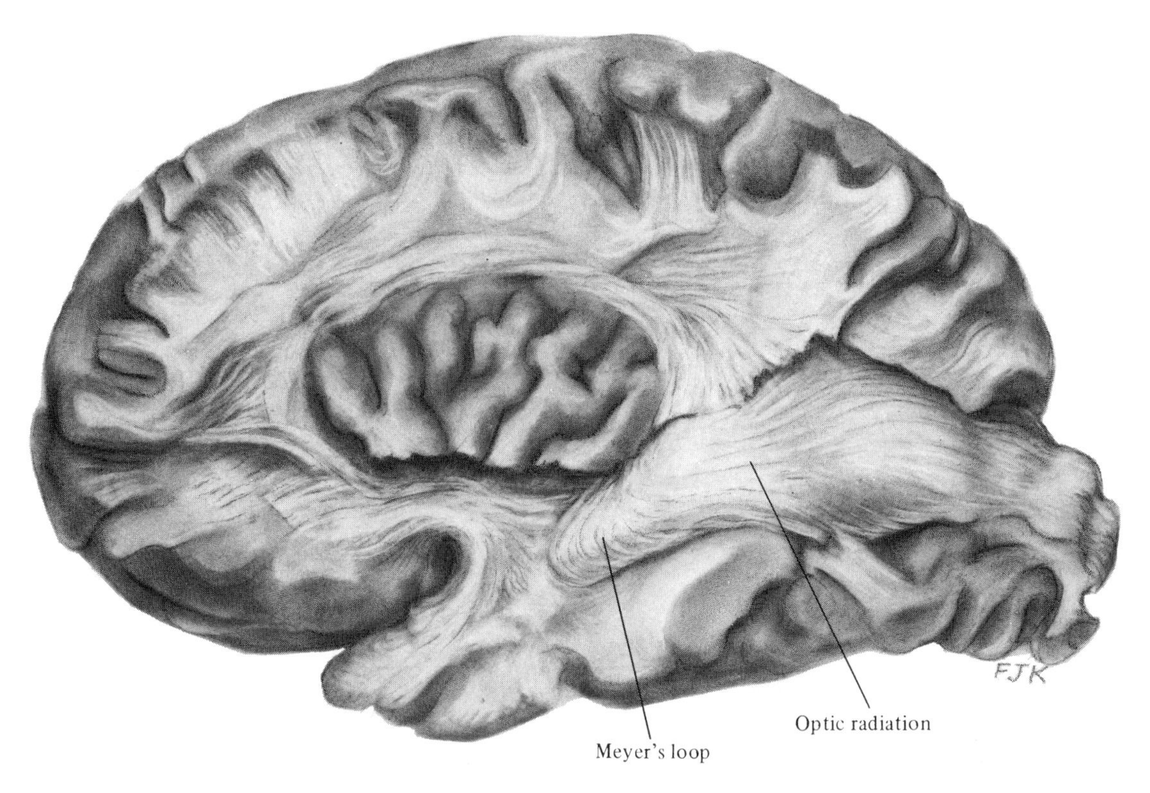

Fig. 4–17. Geniculocalcarine Tract

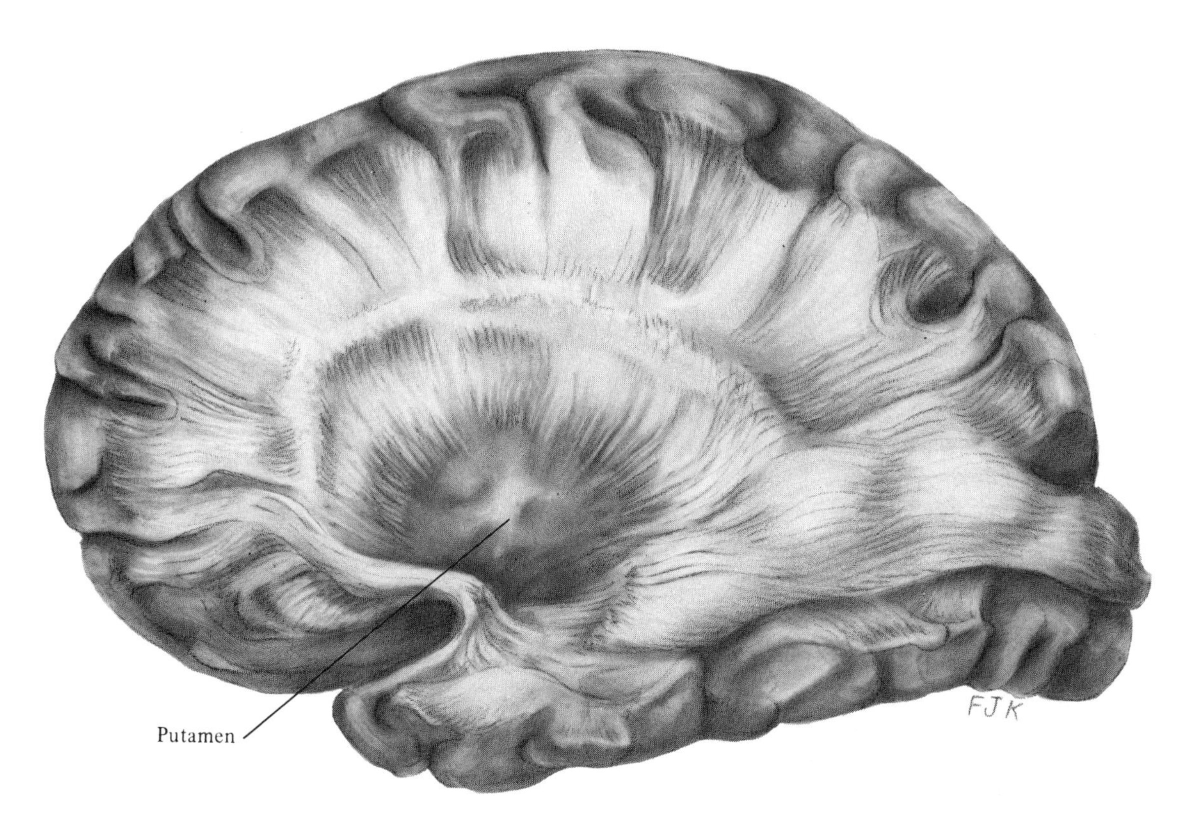

Fig. 4–18. Exposure of Putamen

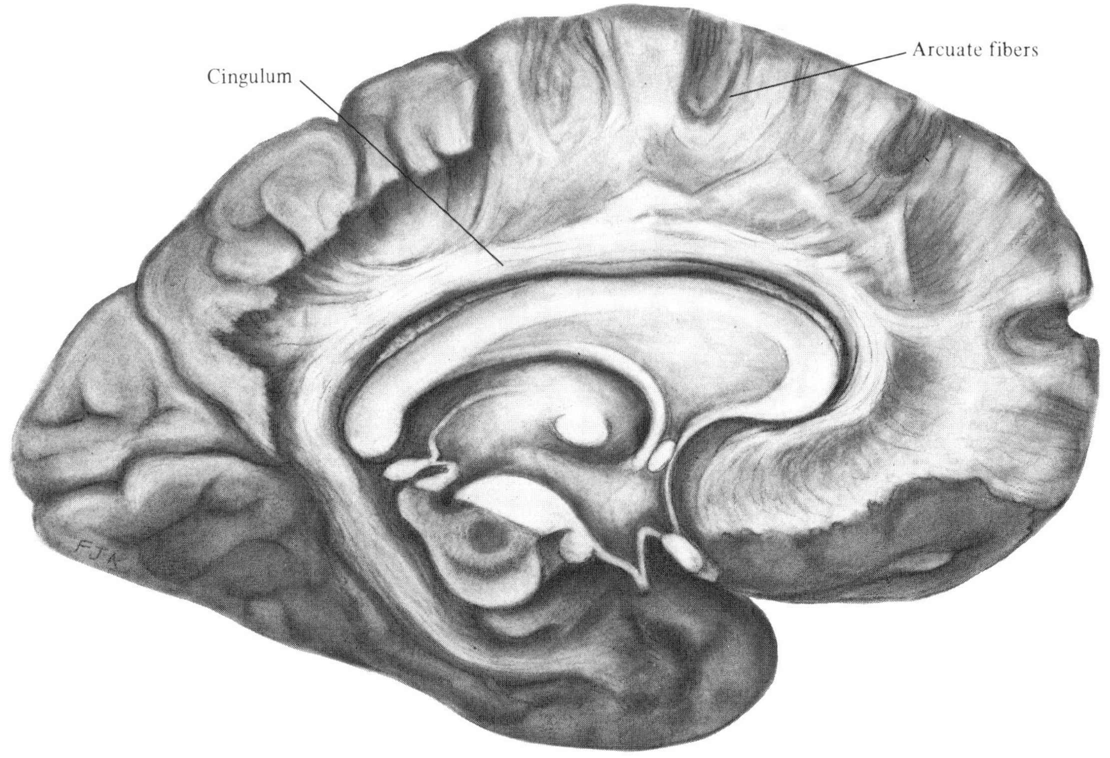

Fig. 4–19. Cingulum Exposed by Blunt Dissection from the Medial Side

composed of different kinds of fibers, i.e., association fibers, projection fibers and commissural fibers. Although it is sometimes difficult to identify the dorsal part of the claustrum, the more voluminous ventral part can usually be recognized.

5. Carefully strip away the external capsule to expose the lateral surface of the *putamen* (Fig. 4–18). The putamen can be easily recognized by its gray color. It represents the lateral part of the lentiform nucleus, an important component of the basal ganglia.

Dissection from the Medial Side

1. Expose the *cingulum* by peeling away the cortex of the cingulate gyrus. Follow this association bundle as it arches around the genu of the corpus callosum and continues in the direction of the subcallosal area. Complete the dissection in the caudal direction and follow the cingulum as it sweeps down into the parahippocampal gyrus behind the splenium of the corpus callosum (Fig. 4–19).

2. Strip away the remaining cortex on the medial side and identify numerous U-shaped fibers, i.e., *arcuate fibers*, which interconnect adjacent gyri.

3. Insert the handle of the scalpel between the cingulum and the corpus callosum and remove the cingulum by lifting the scalpel. This exposes the commissural fibers which radiate from the *corpus callosum* into the cerebral cortex (Fig. 4–20).

4. Cut off the free medial part of the corpus callosum as shown in Fig. 4–21, and remove it together with the septum pellucidum and the main part of the fornix. This can be done by transecting the columns of the fornix in front of the interventricular foramen and the crus fornicis where it sweeps down behind the caudal part of the thalamus. The removal of the corpus callosum exposes the lateral ventricle. Identify the thalamus in the floor of the lateral ventricle.

If the free edge of the corpus callosum is completely trimmed away, even the more laterally placed caudate nucleus can be identified. By removing also the temporal part of the cingulum

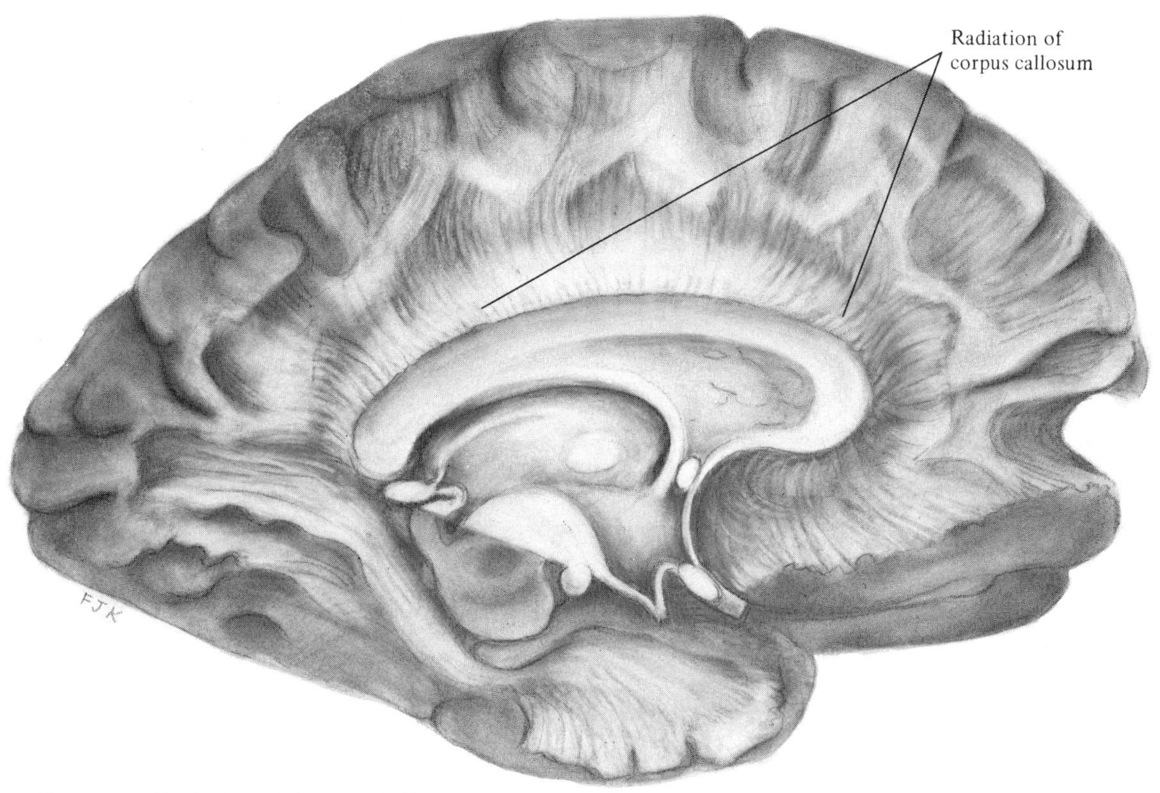

Fig. 4–20. Radiation of Corpus Callosum

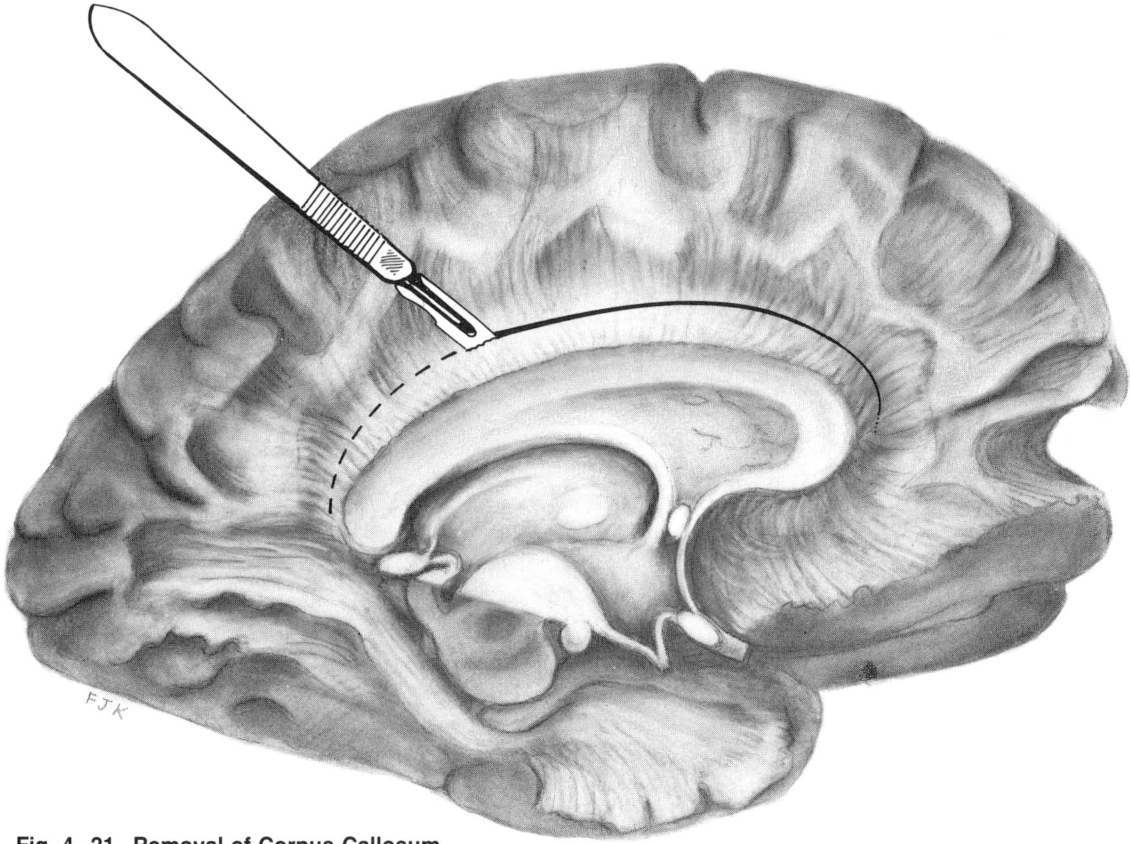

Fig. 4–21. Removal of Corpus Callosum

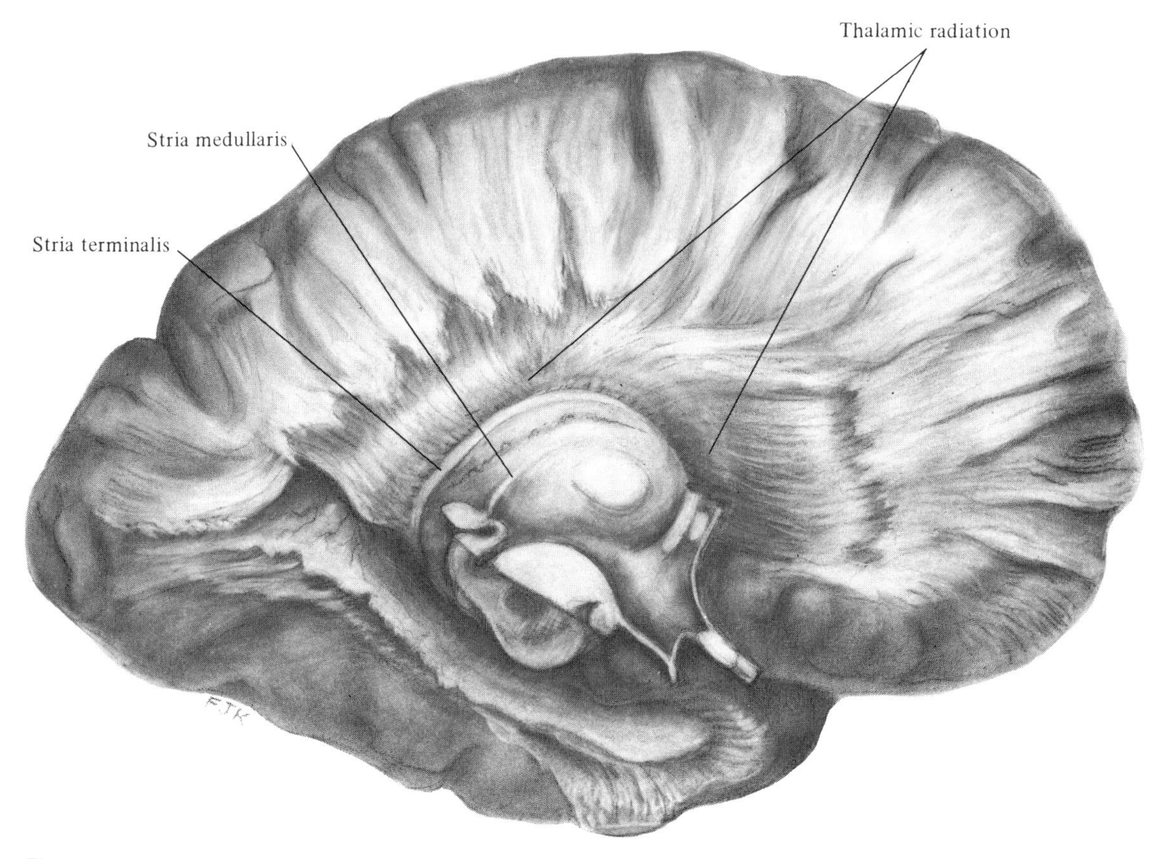

Fig. 4–22. Thalamic Radiation

together with the parahippocampal gyrus and the hippocampus formation, the whole extent of the lateral ventricle is revealed. Note the highly vascularized *choroid plexus*, which produces the CSF.

5. Dissect away the gray matter of the caudate nucleus and expose the fibers of the *internal capsule* and the *corona radiata*. The fibers that first come into view belong to the *thalamic radiation*, which connect the cortex with the thalamus (Fig. 4–22).

6. Continue the blunt dissection by removing the thalamus together with the more superficial fibers of the corona radiata. By doing so, other fibers of the corona radiata can be followed as they converge toward the internal capsule, and if the dorsal part of the mesencephalon is removed, the projection fibers can be followed all the way to the base of the cerebral peduncle in the mesencephalon. This fanshaped fiber system (Fig. 4–23), which bypasses the thalamus, contains primarily *corticofugal fibers*, i.e., corticopontine, corticoreticular, corticobulbar, and corticospinal fibers.

7. Locate the cut surface of the *anterior commissure* and expose the string-like commissural bundle by stripping away surrounding tissue. The bundle projects in a lateral direction and gradually curves backwards and downwards to reach the anterior part of the temporal lobe.

Clinical Notes

The Geniculocalcarine Tract

The geniculocalcarine tract forms a massive sheath of white substance, and it has a very characteristic course that is of great importance to the clinician. The fibers in the dorsal part of the radiation make a lateral-caudal sweep on the outside of the lateral ventricle and then proceed in the caudal direction to the region of the calcarine sulcus in the occipital lobe. The most ventral fibers, make a rostrolaterally directed detour above and around the temporal horn of the lateral ventricle (Fig. 4–17), before they turn caudally on the outside of the temporal horn in order to reach the region below the calcarine

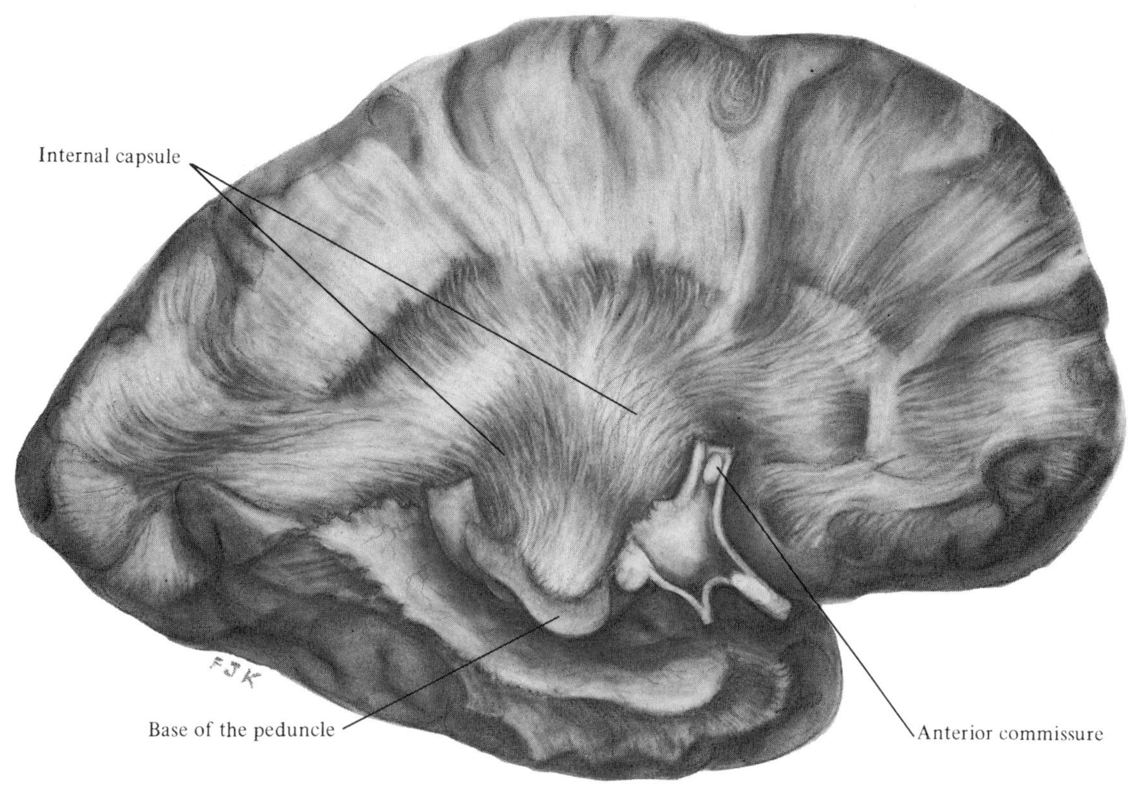

Fig. 4–23. Internal Capsule and Corona Radiata

sulcus. The rostrally directed loop is referred to as Meyer's loop.

Because of its extensive spread, part of the geniculocalcarine tract can easily be damaged by lesions in the posterior part of the hemisphere. Such lesions often result in incomplete homonymous defects, i.e., defects related to the same half of the visual fields of both eyes. Even lesions in the rostral part of the temporal lobe, especially if they reach more than 5 cm behind the rostral pole of the temporal lobe, tend to cause visual deficits, usually in the form of contralateral upper homonymous defects (see Chapter 13).

Corpus Callosum and "Split-Brain" Patients

As a result of a developmental disturbance, the corpus callosum can be partially or completely absent, a condition known as dysgenesis or agenesis of the corpus callosum. Often, such conditions occur together with other malformations of the brain, in which case the consequences may be serious. However, if dysgenesis of the corpus callosum is the only significant malformation, it may not result in remarkable symptoms, and the condition is sometimes discovered incidentally at autopsy or during an MRI examination of the brain. As a result, the functional significance of the corpus callosum remained obscure until the middle of this century[2] when scientists discovered that animals whose corpus callosum had been transected in the midline, so-called split-brain animals, were unable to transfer information from one hemisphere to the other. Soon thereafter, neurosurgeons introduced commissurotomy as a form of treatment in patients with severe and disabling epilepsy. In this procedure, the corpus callosum, together with the hippocampal commissure and sometimes the anterior commissure, are transected in the midline in order to prevent the spread of epileptic discharge

2. Some clinicians, however, were aware of the significance of the corpus callosum already at the turn of the century (see Chapter 22).

from one hemisphere to the other. This produced a number of split-brain patients, many of whom have been extensively studied. Although split-brain patients can function surprisingly well—there are no obvious changes in intellect and behavior—they are unable to perform certain tasks.

As an example we will consider a split-brain patient whose language centers are located in the left hemisphere, which is usually dominant in regard to language. If an object is placed in the patient's right hand (with eyes closed), the patient will be able to name the object, because the sensory information reaches the language centers in the dominant left hemisphere through the ascending somatosensory pathways that cross the midline in the lower part of the brain stem. However, if the objects is placed in the patient's left hand, the patient will be unable to name the object, because the right hemisphere does not have access to the memory for language in the left hemisphere.

Fourth Dissection

Ventricular System, Hippocampus Formation, Amygdaloid Body, Thalamus

Ventricular System

Before starting this dissection, it is useful to study the location of the brain ventricles on Figs. 2–15A and B. Afterward, remove the meninges and blood vessels from the dorsal side of the second brain. Then, separate the brain stem and the cerebellum from the rest of the brain by a transverse cut through the mesencephalon.

Exploration of the Lateral Ventricles

1. Pull the two cerebral hemispheres apart gently to study the location of the corpus callosum before cutting a series of horizontal sections down to the level of the corpus callosum (Fig. 4–24). Remove the medial parts of the cingulate gyri, and expose the dorsal surface of the corpus callosum as shown in Fig. 4–25. If the anterior cerebral arteries have not been removed, they will appear on the dorsal surface of the corpus callosum.

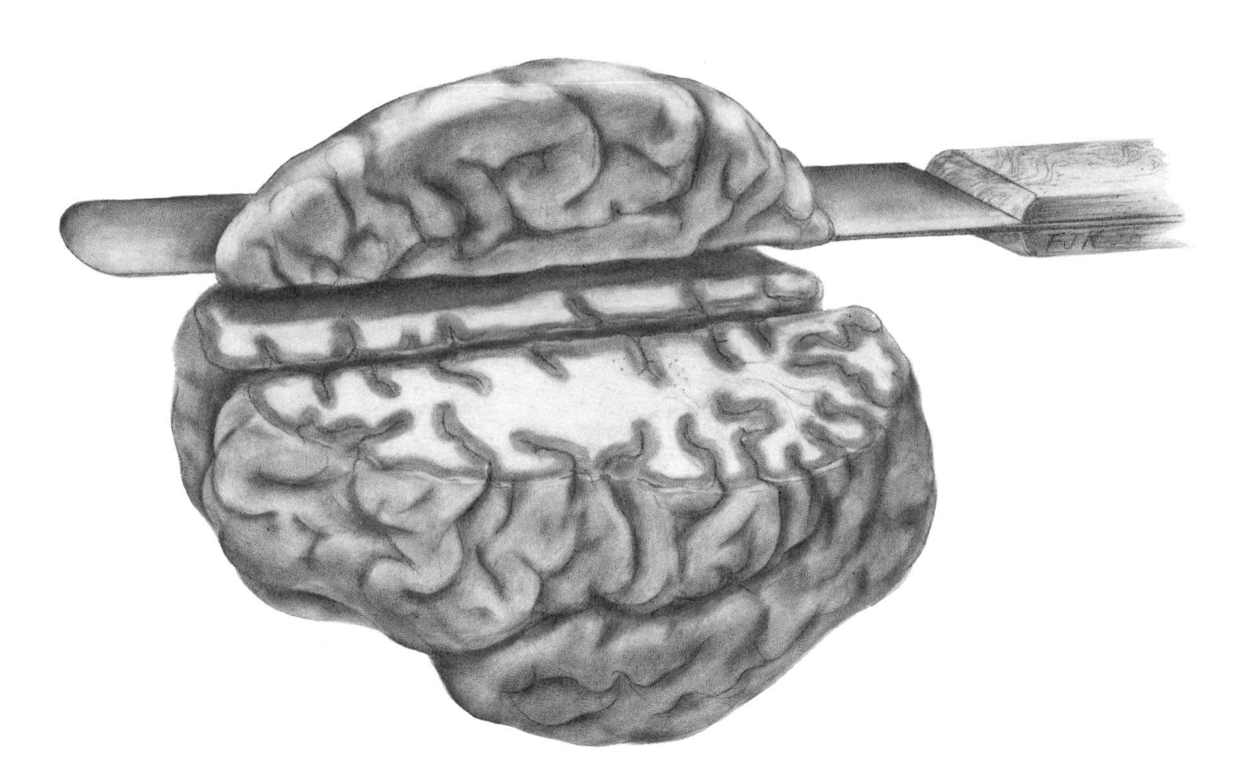

Fig. 4–24. Exploration of the Lateral Ventricles
To prepare for the exploration of the lateral ventricles, the dorsal parts of the brain are removed by cutting horizontal slices down to the level of the dorsal surface of the corpus callosum with the aid of a brain knife.

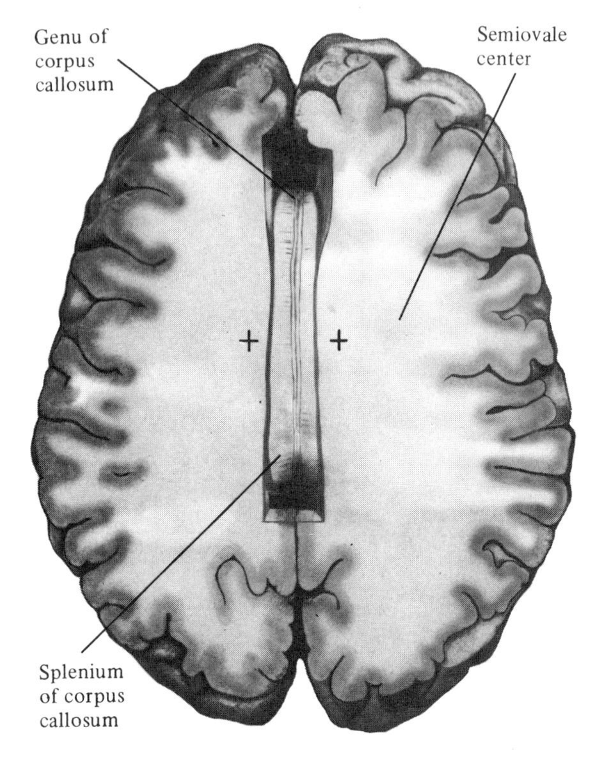

Fig. 4–25. Semioval Center and Corpus Callosum The *crosses* mark the points where access can be gained to the lateral ventricles. (From Rauber-Kopsch, 1955. Lehrbuch und Atlas der Anatomie des Menschen, Band 11, 19 Auflage. Georg Thieme Verlag, Stuttgart.)

The white matter in the center of the horizontal section, the *semioval center*, contains a mixture of association, commissural, and projection fibers. The highly convoluted surface of the hemisphere is covered by a 2–4 mm thick layer of gray matter, the *cortex cerebri*.

2. Palpate the semioval center just lateral to the exposed corpus callosum, and localize an area (+ in Fig. 4–25) where the tissue easily yields to the pressure of the finger. By cutting a small hole in the tissue at this point, access is gained to the lateral ventricle.

3. By gradually expanding the hole in a rostral and caudal direction according to Fig. 4–26, the whole roof of the lateral ventricle can be removed. Repeat the procedure on the contralateral side.

The most prominent structure related to the *anterior horn* of the lateral ventricle is the *caudate nucleus*, whose massive anterior part, the head of the caudate nucleus, bulges into the anterior horn from the lateral side.

The *central part* of the lateral ventricle is quite narrow. The *choroid plexus* and the *thalamus* are the most prominent structures in the floor of the central part. It may be surprising to realize that the thalamus, which is a diencephalic structure, forms part of the floor in the lateral ventricle, a telencephalic structure. During development, however, part of the telencephalon fuses with the diencephalon. In fact, the dorsal surface of the thalamus facing the lateral ventricle is covered by a thin ependymal layer, the *lamina affixa*, which belongs to the telencephalon (Fig. 2–16C).

The *posterior horn* extends for a variable distance into the occipital lobe. On the medial wall of the posterior horn is an eminence, the *calcar avis*, which is caused by the calcarine sulcus.

4. Continue the dissection on the right hemisphere by exposing the *temporal (inferior) horn*. To do this, cut away the frontal and parietal opercula covering the insula, and open up the temporal horn from the lateral side by cutting into the temporal lobe as indicated in Figs. 4–26 and 4–27. Make the cut in the direction of the superior temporal sulcus and remove, piece by piece, the superior temporal gyrus, until the whole extent of the temporal horn is exposed. The beautifully curved structure in the medial wall of the temporal horn is the *hippocampus* (Fig. 4–28), which will be examined in more detail below.

5. Inspect the location of the choroid plexus and its extension into the temporal horn. An expanded portion of the choroid plexus, the *glomus choroideum*, is located in the region of the atrium, i.e., where the temporal horn, the posterior horn, and the central part of the lateral ventricle come together.

The choroid plexus in the lateral ventricle receives its blood supply from the anterior and posterior choroidal arteries (Fig. 4–3).

Hippocampus Formation and Amygdaloid Body

1. Make a transverse cut behind the genu of the corpus callosum and continue the cut through the septum toward the interventricular foramina. As a consequence, the fornix columns are also transected and the trunk of the corpus callosum together with the body of the fornix can be bent backward as shown in Fig. 4–29. Note that in order to keep the tela choroidea intact during this maneuver, the fornices must be carefully separated from the underlying tela. With the fornices and the corpus callosum flipped backwards, it is

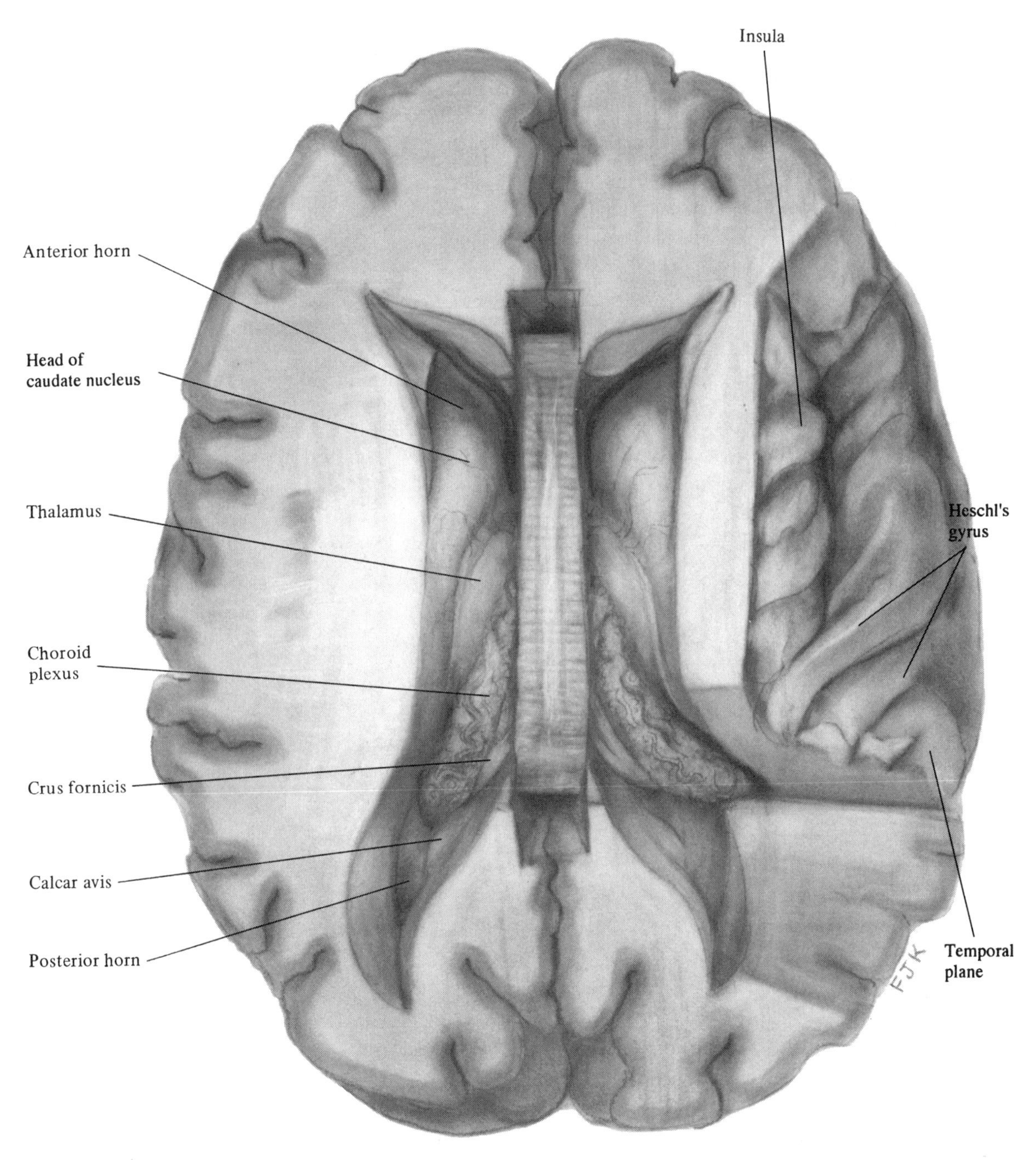

Fig. 4–26. Lateral Ventricles
The lateral ventricles have been exposed by removing the white substance overlying the ventricles.

possible to appreciate the transversely running fibers between the two crura fornicis, i.e., the hippocampal commissure, which connects the hippocampus formations on the two sides.

2. Follow the right fornix as it sweeps behind the thalamus in a lateral and downward direction into the temporal lobe. The last part of the fornix attaches to the hippocampus as a flattened band, the fimbria hippocampi. The relationships between the fornix and the hippocampus will be studied in more detail after the right temporal lobe has been separated from the rest of the brain.

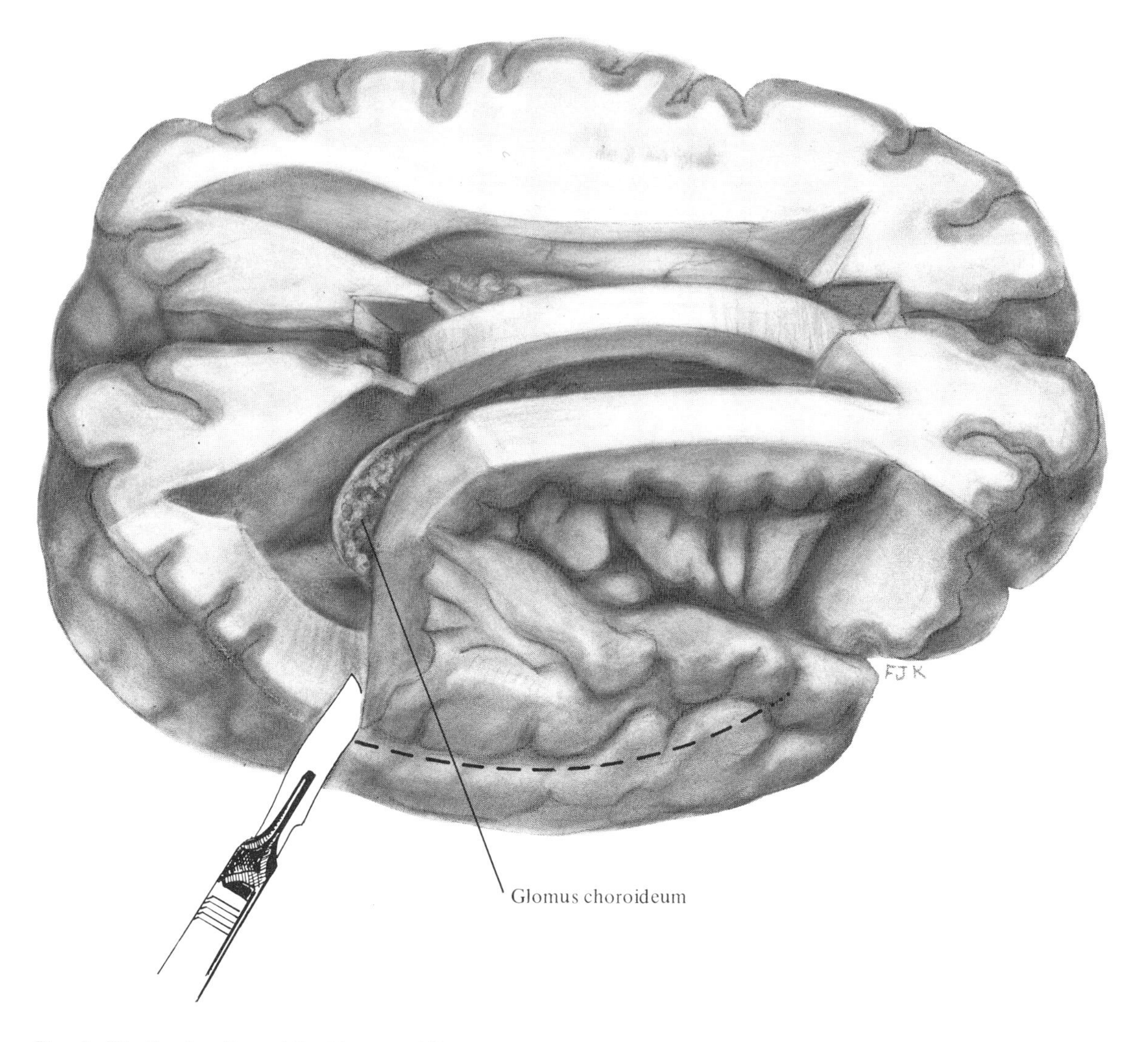

Fig. 4–27. Exploration of the Temporal Horn
Access to the temporal horn can be gained by making a cut in the direction of the superior temporal sulcus and removing the superior temporal gyrus piece-by-piece.

3. Be careful not to destroy the tela choroidea overlying the third ventricle in the process of separating the right temporal lobe from the rest of the brain. To remove the temporal lobe, locate the region of the limen insulae, where the temporal lobe is attached to the frontal lobe. Make a circular cut into the limen insulae by starting on the medial side above the uncus and proceed with the circular movement forward and then backwards on the lateral side towards the dorsal side of the hippocampus. It is important to make the cut as far dorsal as possible in order to preserve most of the amygdaloid body in the temporal lobe specimen, which will now be used for a closer inspection of the hippocampus and the amygdaloid body.

4. The hippocampus, which is of crucial importance for memory functions (Chapter 21), is located in the floor and medial wall of the temporal horn of the lateral ventricle. The anterior part of the hippocampus is marked by shallow grooves, which give it the appearance of a paw, hence the term pes hippocampi.

5. The fimbria hippocampi, the last flattened part of the fornix, is located on the dorsal side of the hippocampus; it becomes gradually thinner as it extends towards the rostral part of the hippocampus (Vignette, Chapter 21).

6. The dentate gyrus is located under the fimbria hippocampi. Its serrated surface can be nicely

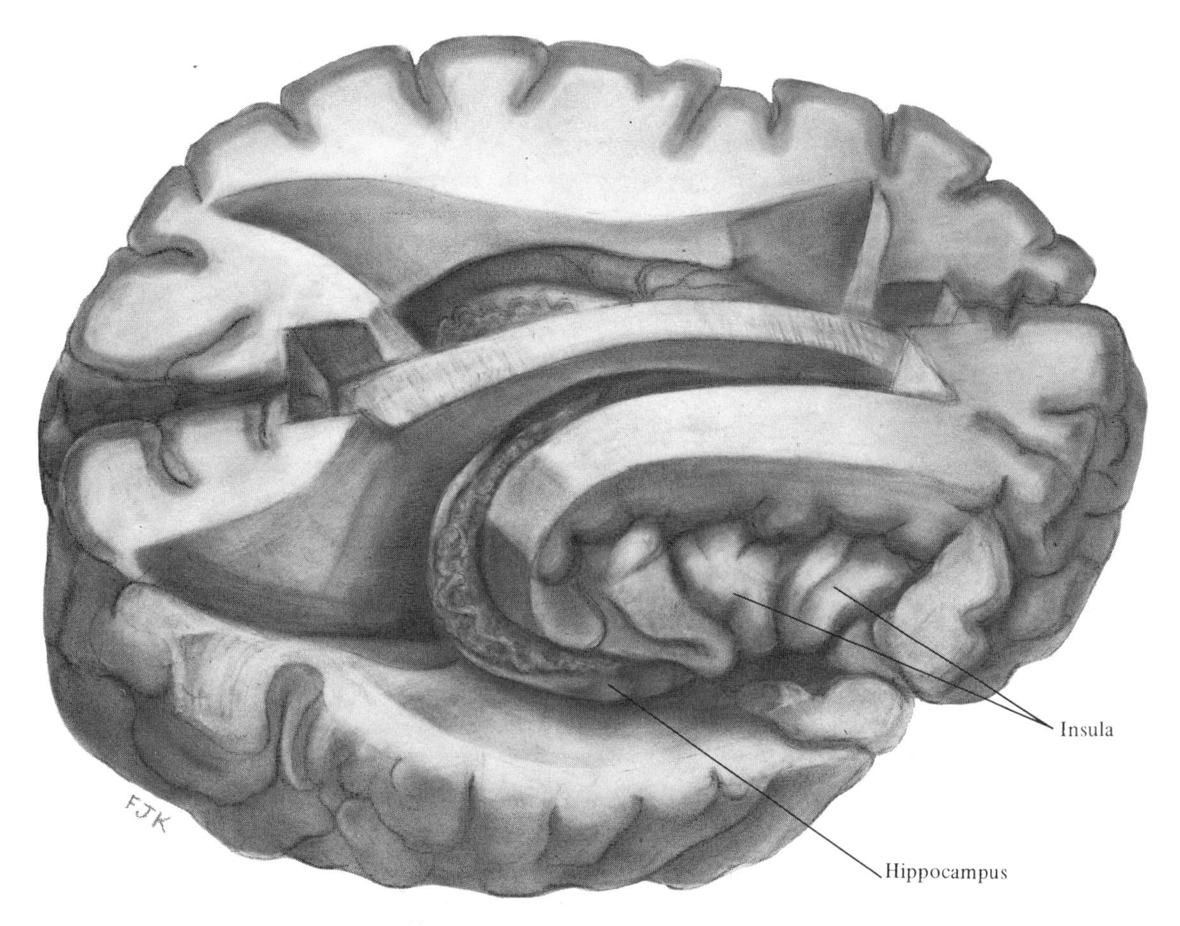

Fig. 4–28. Temporal Horn and Hippocampus
In this lateral view of the brain the temporal horn has been exposed by removing a large part of the temporal lobe.

exposed by lifting the medial edge of the fimbria and removing the pia and the small blood vessels in the groove between the fimbria and the superior surface of the parahippocampal gyrus.

7. The parahippocampal gyrus on the basal surface of the temporal lobe also merits special attention at this time. Most of its rostral part is occupied by the entorhinal area (Brodmann's area 28), which is closely related to the hippocampus. The entorhinal area (Fig. 4–30) can usually be identified because of its irregular surface caused by tiny wart-like bumps (similar to orange peel in texture), which indicate the location of islands of cells in the underlying cortex.

8. The amygdaloid body is a large subcortical structure located just in front of the hippocampus and the temporal horn of the lateral ventricle. The extent of the amygdaloid body can be appreciated by cutting a couple of horizontal slices through the remaining part of the temporal lobe in front of the hippocampus (Vignette, Chapter 20). The amygdaloid body is continuous with the cortex on the medial side of the temporal lobe in the region of the uncus. The amygdaloid body is well known for its role in emotional experience and as a modulator of emotional expression (Chapter 20).

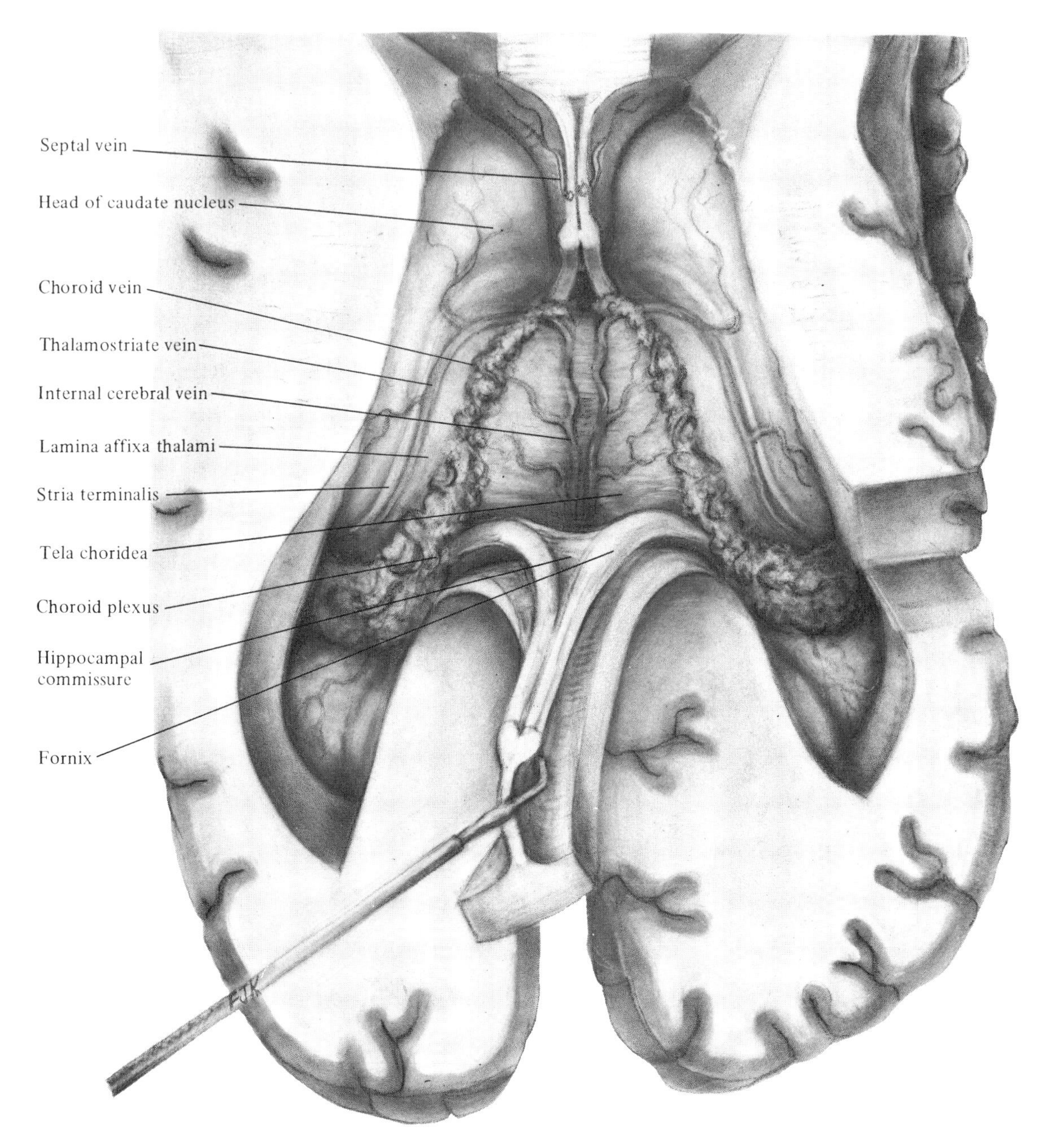

Fig. 4–29. Fornix, Hippocampal Commissure, and Tela Choroidea
The corpus callosum, septum, and fornix have been transected and flipped backwards to expose the tela choroidea.

Tela Choroidea of the Third Ventricle

The roof of the third ventricle is formed by the *tela choroidea* (velum interpositum), which extends between the two thalamic nuclei. The tela choroidea, which is located in the transverse fissure, constitutes a folding of the pia mater, and is continuous underneath the splenium of the corpus callosum with the pia–arachnoid on the dorsal surface of the mesencephalon. This highly vascularized part of the pia mater, which reaches foward above the two thalamic nuclei to the level of

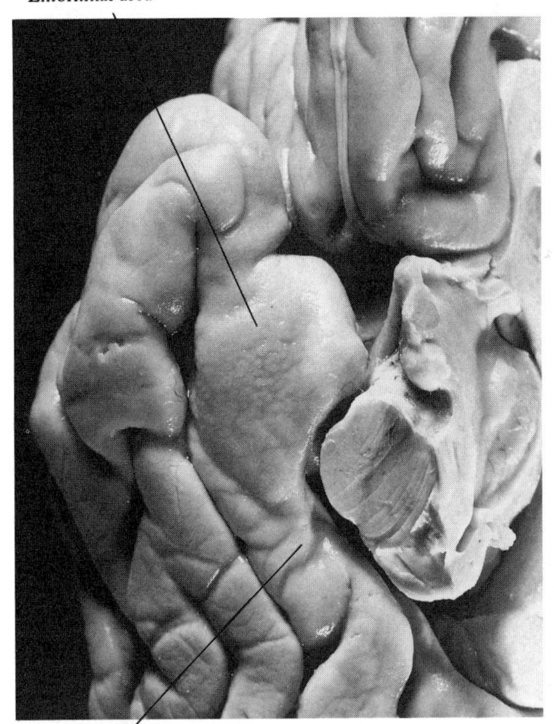

Fig. 4–30. The Entorhinal Area
The entorhinal area can usually be identified by the naked eye on account of its irregular surface caused by tiny wart-like bumps (see text). (Photography courtesy of Dr. Gary van Hoesen, Iowa University.)

the interventricular foramina, is separated from the third ventricle by a layer of ependymal cells. The choroid plexus, which has penetrated into the lateral ventricle from the tela choroidea, is also covered by a layer of ependymal cells. Two smaller vascular fringes, which project into the third ventricle from the underside of the tela choroidea, are continuous with the choroid plexuses of the lateral ventricles through the interventricular foramina (Fig. 2–16).

The biggest vessels in the tela choroidea are the two *internal cerebral veins*, which unite in the caudal part of the tela choroidea to form the *great cerebral vein*, which is continuous with the *straight sinus*. The internal cerebral vein receives blood from the *choroid vein*, the *septal vein*, and the *thalamostriate vein* (Fig. 4–29).

Exploration of the Thalamus, Epithalamus, and Third Ventricle

1. Remove the roof of the third ventricle by carefully stripping off the tela choroidea with attached choroid plexuses from the level of the interventricular foramina and backwards. As a consequence, the ependymal layer that covers the ventricular surface will also be removed. The *tenia choroidea* marks the place where the choroid plexus projected into the lateral ventricle, whereas the *tenia thalami* indicates the line where the roof of the third ventricle was attached to the thalamus (Fig. 4–31).
2. Continue the dissection by freeing the posterior part of the tela choroidea to expose the epithalamus, i.e., the *habenula* and the *pineal body*, as well as the *quadrigeminal plate* with the *superior* and *inferior colliculi*.
3. Review the remaining walls of the third ventricle on available preparations.

The *anterior wall* is formed by the lamina terminalis, the anterior commissure, and the fornix columns.

The *lateral wall* features the medial surfaces of the thalamus and the hypothalamus. Identify the stria medullaris along the mediodorsal edge of the thalamus.

The *posterior wall* of the third ventricle tapers in a funnel-shaped fashion into the cerebral aqueduct. Other formations in the posterior wall are the habenular commissure, the pineal body, and the posterior commissure. The *pineal recess* is that part of the third ventricle that bulges into the pineal stalk.

The *floor* is formed by the optic chiasm, the infundibulum, the tuber cinereum, and the mammillary bodies. It contains two recesses, the *optic recess* and the *infundibular recess*.

Thalamus, Basal Ganglia, and Internal Capsule

At this point it is practical to review the topographic relationships between the major forebrain nuclei and the white matter on the remaining part of the second brain. For this purpose cut a couple of horizontal sections at different levels in the two hemispheres and compare the sections (Figs. 4–32 and 4–33) with the picture of a frontal section at rostral or midthalamic level (Fig. 4–34), and with the drawings in Figs. 2–19 and 2–20. Identify the following structures:

1. *Thalamus* is a collection of nuclei, many of which constitute important "relay" stations for afferent impulses to the cerebral cortex. A vertical lamina of white matter, the *internal medullary lamina*, divides the thalamus into a medial and a lateral part.

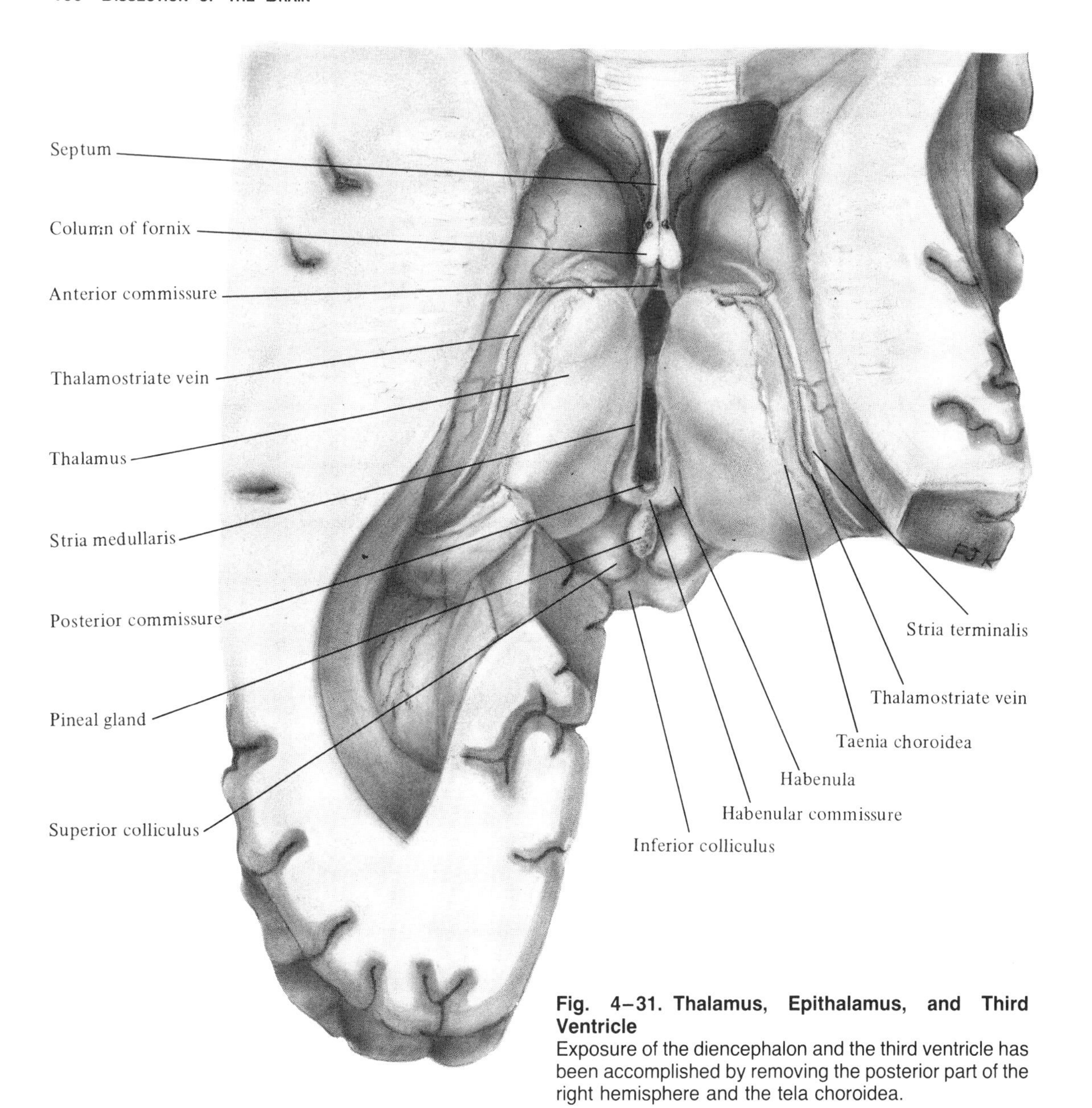

Fig. 4–31. Thalamus, Epithalamus, and Third Ventricle
Exposure of the diencephalon and the third ventricle has been accomplished by removing the posterior part of the right hemisphere and the tela choroidea.

Rostrally, and dorsally, the medullary lamina is divided into two layers, which surround a rostral group of nuclei. The extensive caudal part of the thalamus is called the *pulvinar*, under which the two geniculate bodies are located. The *medial* and *lateral geniculate bodies* serve as relay nuclei in the auditory and visual pathways.

2. *Caudate nucleus* is an elongated mass of gray matter, whose thick anterior portion, the *head* of the caudate, bulges into the anterior horn of the lateral ventricle from the lateral side. At the level of the interventricular foramen the head narrows into the *body* of the caudate, which lies adjacent to the thalamus. The body forms part of the floor of the central part of the lateral ventricle. The body of the caudate graudally tapers off and continues behind the thalamus as the *tail* of the caudate, which makes a sweep in a downward and forward direction into the temporal lobe.

3. *Lentiform nucleus*, which is located lateral to the

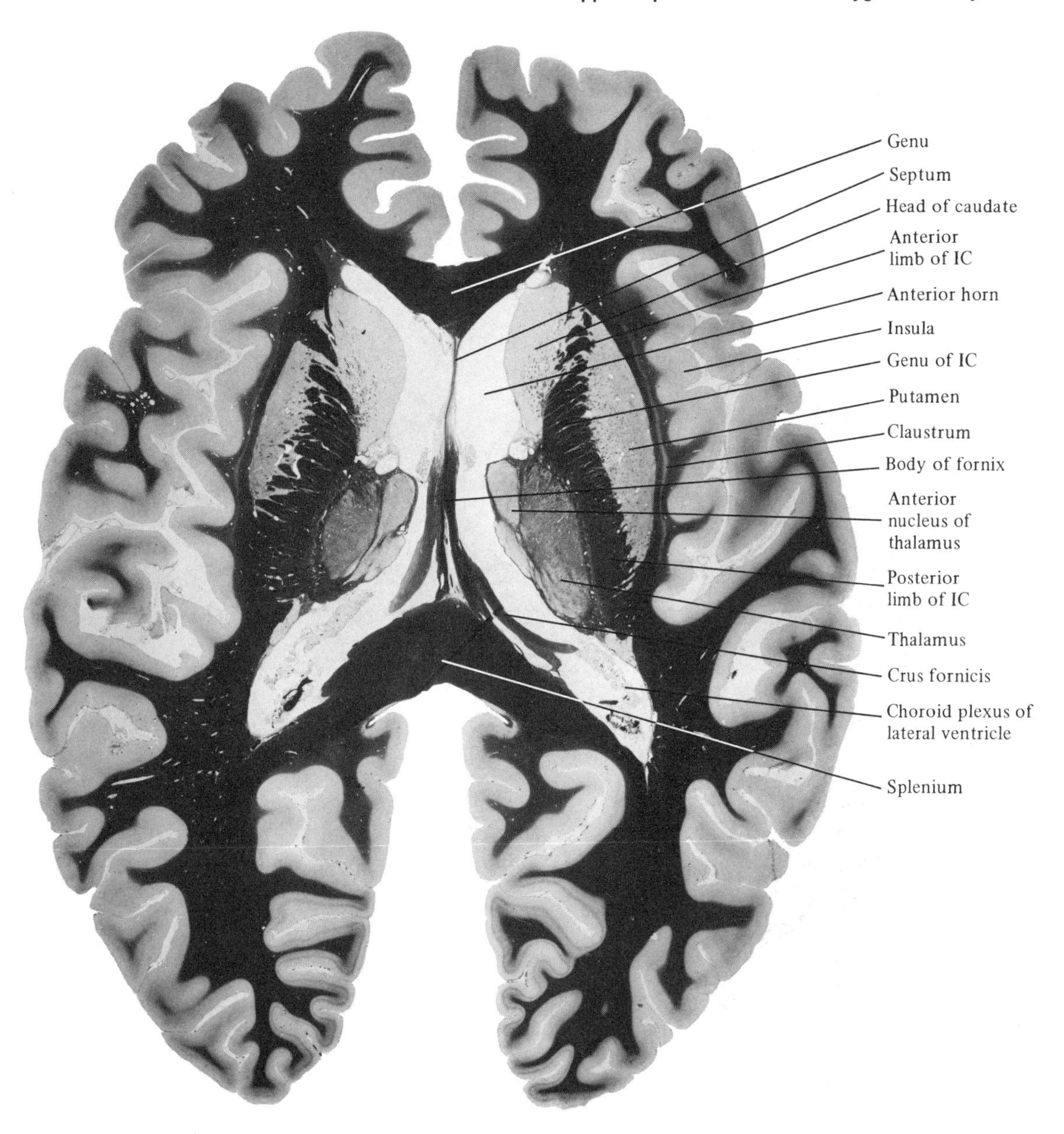

Fig. 4–32. Horizontal Section
(From the Yakoklev Collection.)

internal capsule, has the form of a short ice cream cone with the broad base facing the insula. However, it is separated from the insula by the *external capsule*, the *claustrum*, and the *extreme capsule*. The lentiform nucleus is divided into a lateral part, the *putamen*, and a medial part, the *globus pallidus* (pallidum). The putamen is continuous with the head of the caudate through cell bridges that split apart the white matter, especially in the anterior limb of the internal capsule, giving it a striated appearance. The putamen and the caudate nucleus, together referred to as the striatum, have a similar histologic structure, which is characterized by a large number of densely packed, medium-sized neurons. The globus pallidus, on the other hand, contains large cells which lie rather far apart. In addition, it contains many myelinated fibers, which give the structure a pale appearance on a freshly cut brain.

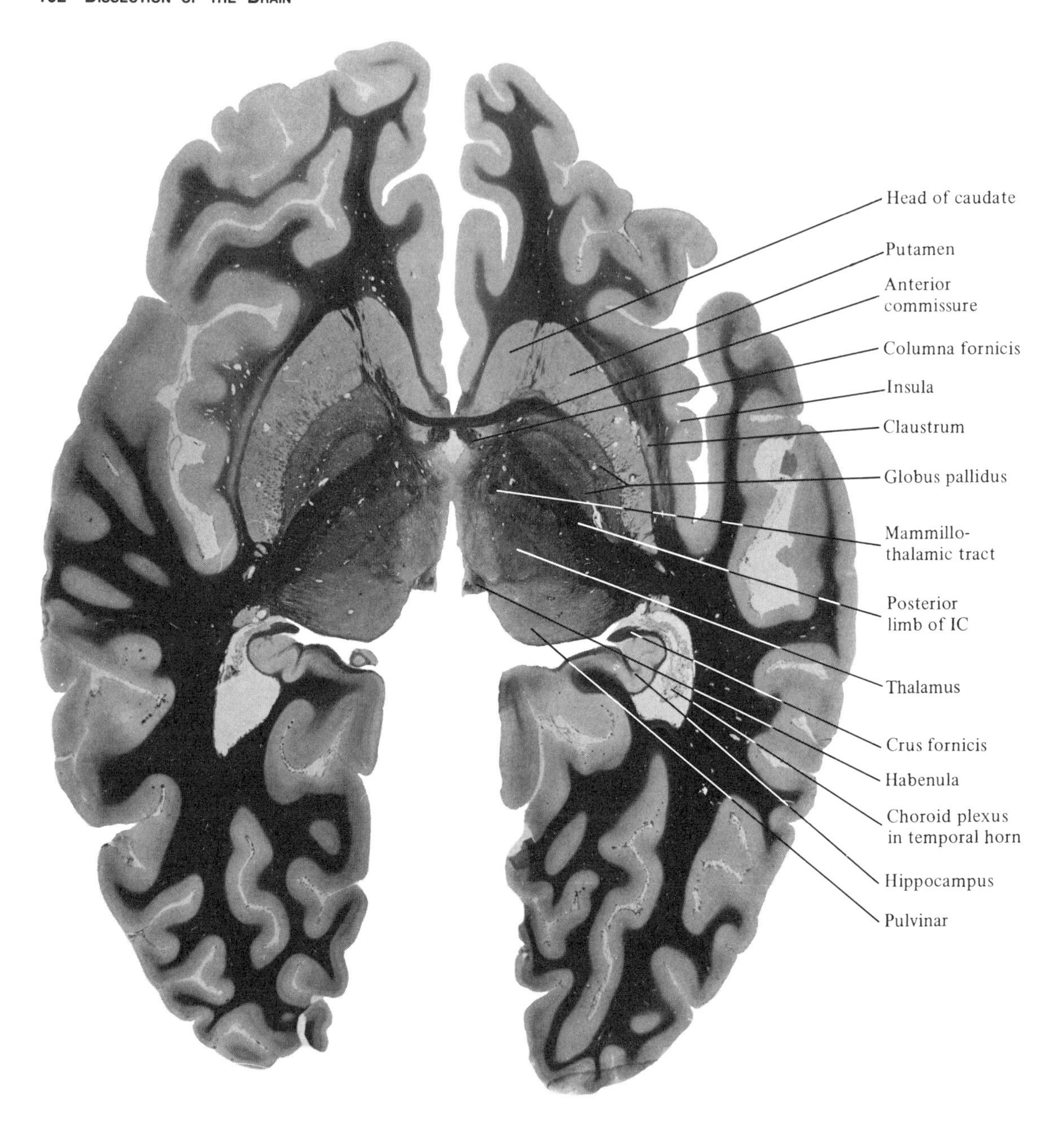

Fig. 4–33. Horizontal Section
This section is about 2 cm closer to the base of the brain than the section in Fig. 4–32. (From the Yakoklev Collection.)

The caudate nucleus and the lentiform nucleus are two of the major components in the basal ganglia, which are best known for their relationships to motor functions (Chapter 16).

4. *Internal capsule.* This massive layer of white matter contains projection fibers connecting the cerebral cortex with various subcortical structures. It is located medial to the lentiform nucleus, where it forms the internal part of the capsule surrounding the lentiform nucleus, hence the name internal capsule. The lamina of white matter lateral to the lentiform nucleus is referred to as the external capsule.

The internal capsule has a typical V-form on a horizontal section with the apex of the V pointing medially (Figs. 4–32, 4–33). The part of the capsule that is located between the head of the caudate medially, and the lentiform nucleus

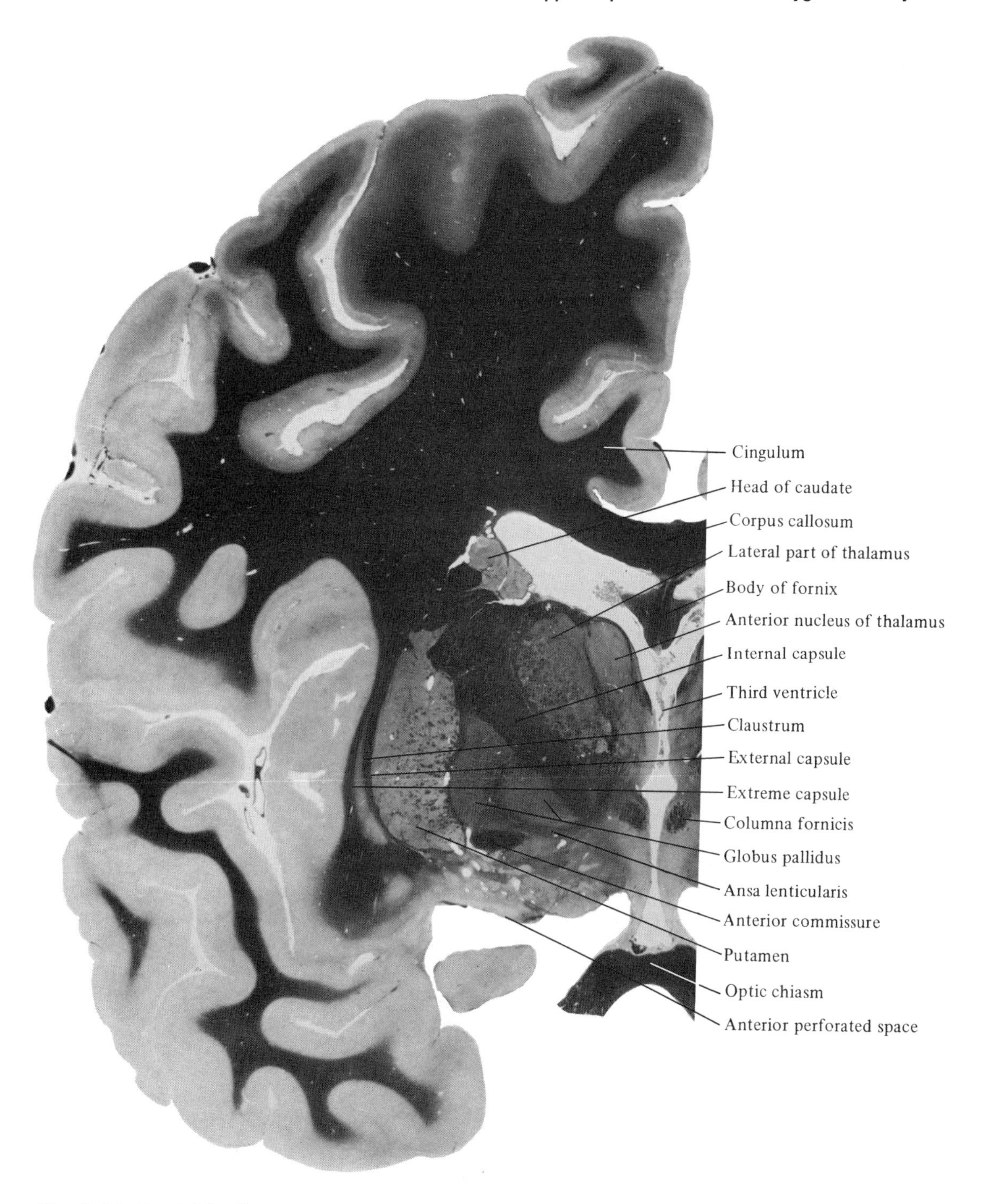

Fig. 4–34. Frontal Section
The upper part of the section is tilted somewhat in a posterior direction to cut through both the optic chiasm and the anterior thalamus. (From the Yakoklev Collection.)

laterally, is called the *anterior limb*. It is broken up by bands of gray matter which form bridges between the putamen and the head of the caudate nucleus. The anterior limb contains the frontal corticopontine tract as well as fibers connecting the frontal cortex with the thalamus in both directions, i.e., the anterior thalamic radiation. The angle of the internal capsule is known as the *genu*.

The *posterior limb* of the internal capsule

separates the lentiform nucleus on the lateral side from the thalamus on the medial side. The posterior limb contains the corticospinal tract (pyramidal tract) and the sensory radiation, which carries sensory impulses to the somatosensory cortex in the postcentral gyrus. Other fibers in the posterior limb belong to the corticopontine and corticoreticular systems as well as the thalamic radiation.

Those parts of the posterior limb which are located behind and underneath the lentiform nucleus are known as the *retrolenticular* and the *sublenticular* parts of the internal capsule. The optic radiation, which originates in the lateral geniculate body and terminates in the visual cortex, is located in both the retrolenticular and the sublenticular part. The auditory radiation, which comes from the medial geniculate body, projects through the sublenticular part of the internal capsule to the auditory cortex in the superior temporal gyrus.

In a frontal section through the hemisphere at midthalamic level one can usually trace the internal capsule in continuity with the base of the cerebral peduncle in the mesencephalon. The internal capsule fibers that proceed uninterrupted to the base of the cerebral peduncle belong to the corticospinal tract and the corticopontine system.

Clinical Notes

Ventricular System

Using modern imaging methods, the ventricular system can be visualized in great detail, and if need be, reconstructed in three dimensions (Vignette, Chapter 1). Since the ventricular system has a very characteristic appearance that varies little from individual to individual, alterations in its position, shape, and size can give valuable information regarding various pathologic processes.

Familiarity with the morphology of the ventricular system and with the formation and flow of CSF is also important for understanding one of the most common causes for *increased intracranial pressure*. This serious condition is often seen as a result of obstruction of the CSF circulation through the ventricular system. Narrow passages such as the interventricular foramen, the cerebral aqueduct, or the outlet foramina of the fourth ventricle can readily be obstructed by a pathologic process, e.g., a congenital malformation, a tumor, or an inflammatory lesion. Since CSF is continuously produced in the choroid plexus, the obstruction increases the amount of CSF in the cranial cavity, which causes an increase in the intracranial pressure. The cardinal symptoms of increased intracranial pressure are headache, vomiting, and papilledema (choked discs). Excessive amount of CSF in the cranial cavity is referred to as *hydrocephalus* (see Clinical Notes, Chapter 3).

Hippocampus and Alzheimer's Disease

Structures in the medial part of the temporal lobe seem to be especially vulnerable to various types of injuries, and pathologic changes in this part of the brain are prominent in several well-known and common brain disorders including Alzheimer's disease and certain forms of epilepsy. In Alzheimer's disease the most pronounced changes, including cell death and neurofibrillary degeneration (see Clinical Example, Chapter 21), usually occur in the hippocampus and entorhinal area. Since these structures are of crucial importance for memory, it can be easily appreciated that loss of memory is one of the cardinal symptoms in Alzheimer's disease.

Thalamus, Basal Ganglia, and Internal Capsule

These structures are in close proximity to each other, and they are located in the central and basal parts of the brain. Since they are surrounded by ventricular cavities and broad white fiber bundles, which provide excellent contrast, they can be easily recognized on tomographic images.

Although most cerebrovascular accidents are caused by atherosclerosis in the internal carotid artery or one of the large brain arteries (Chapter 24), the lenticulostriate arteries supplying the basal ganglia and internal capsule, or the thalamoperforating arteries of the posterior cerebral artery, are also favorite sites for pathological processes, often in patients with hypertension. Since the territories of these small arteries are restricted, vascular occlusions in this part of the brain often result in correspondingly small infarcts, usually covering a volume of one or two cubic centimeters, referred to as lacunar infarcts. However, even a small lesion can produce severe symptoms, for instance, if it is strategically placed to interfere with descending pathways for voluntary movements in the internal capsule (see Clinical Example, Chapter 15), or with sensory relay stations in the thalamus. The basal ganglia can also be specifically affected by cerebrovascular accidents, but their clinical significance is more often related to degenerative brain disorders, especially Parkinson's disease (see Clinical Notes, Chapter 16).

Fifth Dissection

Fourth Ventricle; Cerebellum; Brain Stem

Required for this dissection are one complete brain stem–cerebellum preparation, and one divided by a midsagittal section.

The cerebellum is attached to the brain stem by three pairs of compact fiber bundles: the superior, middle, and inferior cerebellar peduncles. The cerebellum and the brain stem are thus two closely related structures, which lie in the posterior fossa covered by the tentorium cerebelli. Note that in the normal position in the skull, the cerebellum is behind rather than above the brain stem (Fig. 4–35A), and a horizontal section through the brain stem–cerebellum preparation is nearly perpendicular to the long axis of the brain stem (Fig. 4–40).

Fourth Ventricle

The fourth ventricle is best studied on a median section through the brain stem and cerebellum (Fig. 4–35A). The upper part of its tent-shaped roof is formed by the *anterior medullary velum*, a thin plate, that extends between the two superior cerebellar peduncles. Rostrally, where it is attached to the quadrigeminal plate, it contains the decussating fibers of the trochlear nerve, which exits laterally. The *lingula*, a midline part of the cerebellar cortex, is fused to the outer surface of the anterior velum. The lower part of the roof is formed by a thin pia–ependymal membrane, from which the choroid plexus of the fourth ventricle is suspended. Although this membrane is often referred to as the *posterior medullary velum*, its medial part, which is related to the choroid plexus, does not seem to contain any myelinated fibers. The anterior medullary velum is continuous with the posterior velum at the apex of the fourth ventricle, the *fastigium*. The CSF leaves the fourth ventricle to enter the subarachnoid space through the median aperture, the foramen of Magendie, and the two lateral apertures, the foramina of Luschka. The relatively wide *median aperture*, which opens into the cerebellomedullary cistern, can be seen on the dorsal side of the brain stem in the angle between the medulla oblongata and the cerebellum if the two structures are gently parted (Fig. 4–35B). The *lateral aperture* can be identified on one of the half brain stem–cerebellum preparations by gently probing in the direction of the tubular *lateral recess* of the fourth ventricle as it curves over the dorsal aspect of the brain stem. The lateral aperture, which is located in the *cerebellopontine angle*, is usually easy to identify from the outside because part of the choroid plexus projects out into the subarachnoid space through the opening.

Cerebellum

Although the cerebellum is largely hidden under the caudal parts of the cerebral hemisphere, it is nevertheless an impressive structure of great significance in clinical medicine. The cerebellum receives information from different sensory organs and from many parts of the brain and spinal cord, and based on this multifarious input, the cerebellum responds by sending messages to other motor structures through which it controls posture and movement.

Major Subdivisions

The cerebellum, which weights about 150 g, is divided into a median part, the *vermis*, and two large lateral parts, the *cerebellar hemispheres*. The vermis is clearly separated from the hemispheres only on the inferior surface (Fig. 4–36). The cerebellum is characterized by an extensively folded cortex forming narrow elongated *folia* separated by furrows or *sulci*, which generally run in a transverse direc-

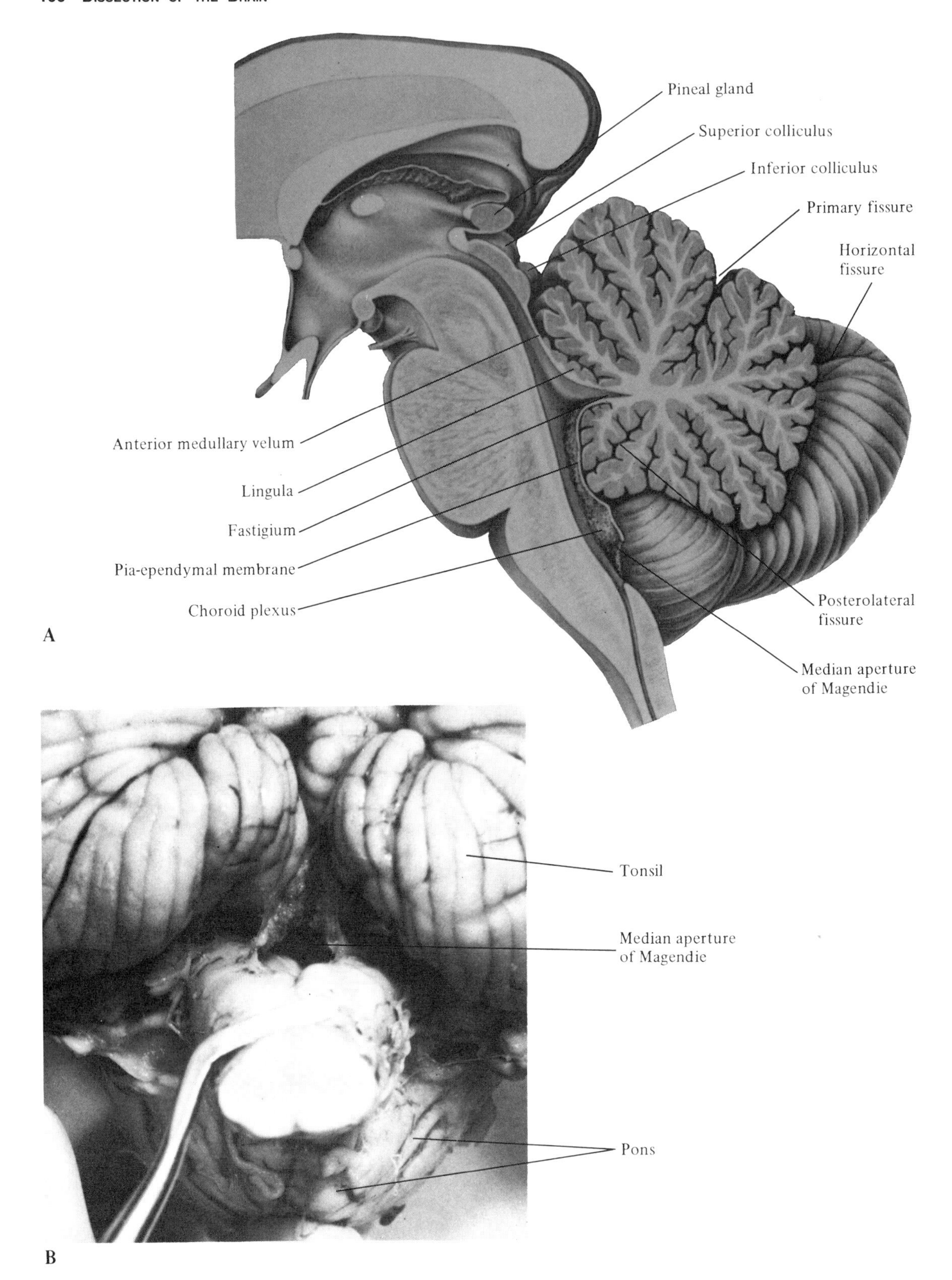

Fig. 4–35. Fourth Ventricle and Foramen of Magendie

A. Midsagittal section through the brain stem and cerebellum. (From Nieuwenhuys, R., J. Voogd, and C. van Huijzen, 1978. The Human Central Nervous System: A Synopsis and Atlas. With permission of the authors and Springer-Verlag, Berlin-Heidelberg.)

B. Photograph showing the median aperture of Magendie.

tion. Several deep clefts or fissures divide the cortex into various lobes and lobules, which have been given various names by different authors. The many nomenclatures are confusing and of interest only to the specialist. To appreciate the functional anatomy of the cerebellum and its relevance in clinical neurology, the medical student needs to pay attention only to a few major fissures and subdivisions (Figs. 4–35, 4–36). The main subdivisions, which are arranged in a transverse fashion, can best be understood by imagining the exposed surface of the cerebellum projected on a flat surface (Fig. 4–37). If one eliminated all the cerebellar fissures in this process, the cerebellar surface would be very long in the main axis of the brain but rather narrow from side to side, since the fissures run in a transverse direction. Identify the following fissures and lobes:

1. *Primary fissure* on the superior surface separates the *anterior* and *posterior lobes*.
2. *Horizontal fissure*, which extends from one middle cerebellar peduncle to the other within the posterior lobe, lies on the inferior surface close to its junction with the superior surface.
3. *Posterolateral fissure* separates the *posterior lobe* from the *flocculonodular lobe*, which is known for its close relations with the vestibular system. The flocculonodular lobe is located close to the brain stem on the inferior surface of the cerebellum, and it can be properly inspected only if the cerebellum is separated from the brain stem (see below). The main mass of the cerebellum, with the exclusion of the flocculonodular lobe, is referred to as the *corpus cerebelli*.

Although the vermis is clearly demarcated from the hemispheres on the inferior cerebellar surface, each part of the vermis is continuous with a corresponding part of the cerebellar hemisphere, just as on the superior surface. In other words, the *nodule* is continuous with the *flocculus*, the *uvula* with the *tonsil*, and the *pyramis* with the *biventer* and *gracile* lobules.

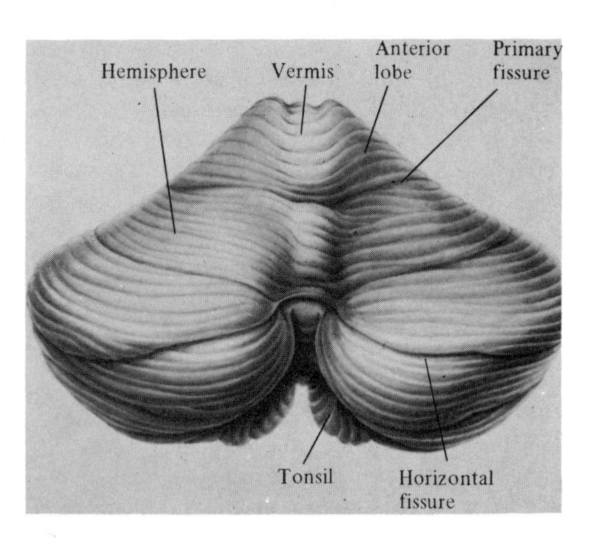

Fig. 4–36. Dorsal View of Cerebellum
(From Nieuwenhuys, R., J. Voogd, and C. van Huijzen, 1978. The Human Central Nervous System: A Synopsis and Atlas. With permission of the authors and Springer-Verlag, Berlin-Heidelberg.)

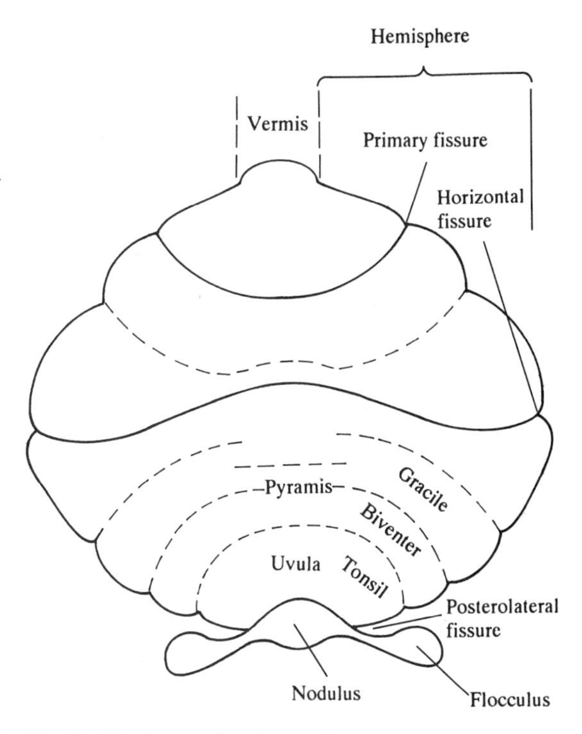

Fig. 4–37. Cerebellar Subdivisions
A highly schematic diagram of the exposed surface of the cerebellum projected on a flat surface showing the major fissures and subdivisions of the cerebellar cortex.

Cerebellar Peduncles

The cerebellum is attached to the brain stem by three pairs of compact fiber bundles: the superior, middle, and inferior cerebellar peduncles.

1. *Inferior cerebellar peduncle* (restiform body) contains both cerebellopetal and cerebellofugal fibers. It approaches the cerebellum from the posterolateral surface of the medulla oblongata.

2. *Middle cerebellar peduncle* (brachium pontis) is the largest of the peduncles. It occupies the most lateral position of the three peduncles, and can easily be identified as it extends laterally into the cerebellum from the base of the pons. It contains fibers that originate in the pontine nuclei in the contralateral basilar part of the pons.

3. *Superior cerebellar peduncle* (brachium conjunctivum) contains most of the efferent pathways that come from the intracerebellar nuclei and project to structures in the brain stem and diencephalon.

Blunt Dissection of the Cerebellar Peduncles

The blunt dissection of the cerebellar peduncles should be done only if there are extra brain stem–cerebellum preparations available, in which case it is well suited for a prosection. It can be properly performed only after some practice. If carefully done, however, the dissection reveals beautifully the topographic relations between the different peduncles. It is most practical to use a brain stem–cerebellum preparation cut in half by a median section.

Start from the lateral side of the middle cerebellar peduncle by removing the different cerebellar lobules, including the vermis, until the core of white matter is exposed (Fig. 4–38A). Identify the vestibulocochlear nerve and its attachment to the dorsolateral side of the brain stem and pull it gently in a dorsal direction. By doing so the ridge of gray matter, i.e., the cochlear nuclei, which extends from the attachment of the nerve transversely across the lateral aspect of the brain stem, is stripped away. This exposes the *inferior cerebellar peduncle*. Push a probe upward alongside the lateral aspect of the inferior cerebellar peduncle until the probe becomes visible on the opposite side between the superior and the middle cerebellar peduncles (Fig. 4–38A). Cut through the fibers of the *middle cerebellar peduncle* lateral to the probe and remove the group of fibers that have just been cut. This exposes the intracerebellar part of the inferior cerebellar peduncle, whose fibers sweep on the outside of the *superior cerebellar peduncle* and fan out toward the cerebellar cortex. The relationship between the inferior and the superior peduncles is best studied by pushing a probe in a superior direction alongside the medial aspect of the inferior cerebellar peduncle, and then separating the fibers of the two peduncles by forcing the probe in a posterior direction (Fig. 4–38B). The capsule of fibers that forms the "paddle," of the superior cerebellar peduncle (Fig. 4–38C) surrounds the largest of the intracerebellar nuclei, the dentate nucleus. By holding on to the "paddle," the superior cerebellar peduncle can be lifted from its bed to the point where it crosses the midline in the lower part of the mesencephalon. The characteristic corrugated form of the dentate nucleus can be observed by cutting a section through the "paddle."

Flocculonodular Lobe

The flocculonodular lobe can best be appreciated on the whole cerebellum following its separation from the brain stem (Fig. 4–39). This will also provide a brain stem preparation on which the floor of the fourth ventricle, the rhomboid fossa, can be inspected (see below). To separate the cerebellum from the brain stem, cut at right angle through the middle cerebellar peduncle posterior to the trigeminal nerve, and extend the cut toward the lateral aperture just in front of the flocculus. Now, lift up the inferior surface of the cerebellum from the brain stem, identify the inferior cerebellar peduncle by removing the tela choroidea of the fourth ventricle, and transect the peduncle. Likewise, lift carefully the anterior lobe of the cerebellum from the brain stem and cut through the superior cerebellar peduncle. Repeat the same procedure on the other side and remove the cerebellum.

The flocculonodular lobe, which constitutes only a small part of the whole cerebellum, consists of two divisions, the *nodule* in the vermis, and the laterally located *flocculus* which resembles a small fragment of cauliflower. The nodulus is tucked away under the main part of the cerebellum, whereas the flocculus lies adjacent to the vestibulocochlear nerve. The flocculonodular lobe, which receives its main input from the vestibular system, is a center for various vestibular reflexes.

Cerebellar Nuclei (Fig. 4–40)

The cerebellar nuclei, which receive input from overlying parts of the cerebellar cortex, give rise to the efferent cerebellar projections. The largest of the cerebellar nuclei, the dentate nucleus, can easily be identified by gross anatomic inspection if suitable sections are cut through the cerebellum. Use the half brain stem–cerebellum preparation and cut a nearly horizontal section along a line stretching from the horizontal fissure to slightly above the fastigium (Fig. 4–40A). The nuclei from lateral to medial are (Fig. 4–40B):

1. Dentate nucleus (red)
2. Emboliform nucleus (gray)

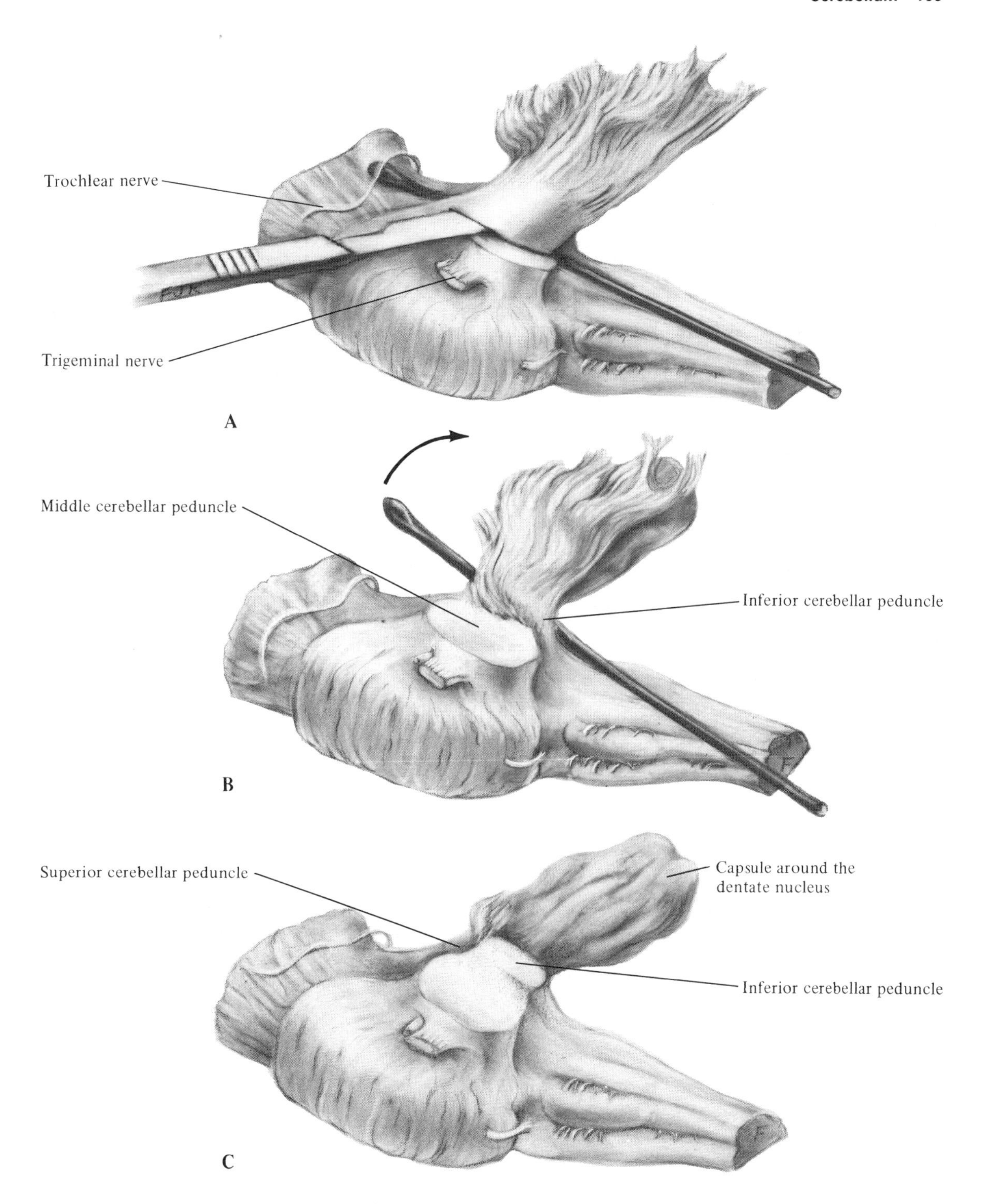

Fig. 4–38. Cerebellar Peduncles
A. After removal of the cerebellar lobules, a probe is pushed upwards alongside the inferior cerebellar peduncle until it becomes visible between the superior and middle cerebellar peduncles. The fibers of the middle cerebellar peduncle is then transected.
B. The relationships between the inferior and superior cerebellar peduncles can be studied by pushing the probe upwards alongside the medial aspect of the inferior cerebellar peduncle whereupon the fibers of the two peduncles are separated by forcing the probe in a posterior direction.
C. The inferior cerebellar peduncle has been transected to reveal the capsule of fibers around the dentate nucleus.

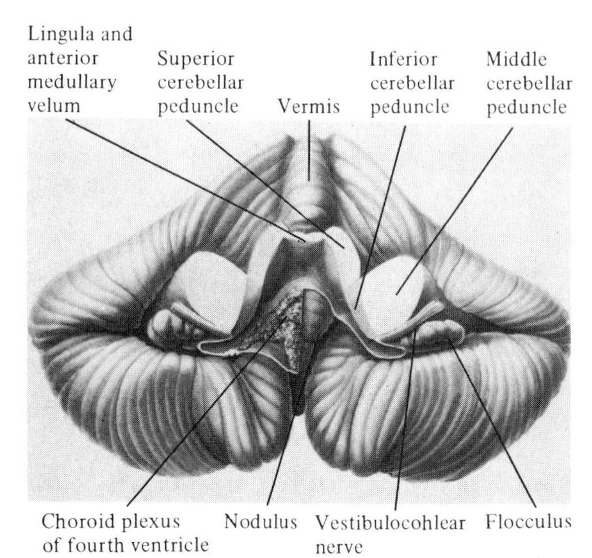

Fig. 4–39. Ventral View of Cerebellum
(From Nieuwenhuys, R., J. Voogd, and C. van Huijzen, 1978. The Human Central Nervous System: A Synopsis and Atlas. With permission of the authors and Springer-Verlag, Berlin-Heidelberg.)

3. Globose nucleus (gray)
4. Fastigial nucleus (blue)

The large, irregularly folded, bag-shaped dentate nucleus has its hilus directed ventromedially and rostrally. The more or less rounded fastigial nucleus close to the midline can usually also be recognized. The emboliform nucleus, which is continuous with the dentate nucleus, and the globose nucleus, which joins the lateral part of the fastigial nucleus, are difficult to identify. The extent of the dentate nucleus can be appreciated by cutting a series of sagittal sections through the still unused cerebellum from the second brain.

Brain Stem

Rhomboid Fossa

The floor of the fourth ventricle is shaped more or less like a rhombus and is accordingly referred to as the rhomboid fossa (Fig. 4–41). On each side of the *median sulcus*, which divides the fossa into two symmetrical halves, is a longitudinally running ridge, the *medial eminence*. The medial eminence is bounded laterally by the somewhat illdefined *sulcus limitans*. The small medullary fold that overhangs the opening of the central canal at the inferior end of the rhomboid fossa is called the *obex*. Identify the following structures in the rhomboid fossa:

1. *Striae medullares*, a group of medullated fiber bundles that connect the arcuate nuclei adjacent to the pyramis with the cerebellum. The pronounced elevation on the medial eminence rostral to the striae medullares is
2. *Facial colliculus*, which overlies the abducens nucleus and the genu of the facial nerve.
3. *Locus ceruleus*, a bluish gray area rostral to the facial colliculus in the area of the sulcus limitans. The area overlies the nucleus of the locus ceruleus, the cells of which contain melanin pigments and give rise to noradrenergic pathways, which are distributed widely in the brain and the spinal cord.
4. *Vestibular area* overlies the vestibular nuclei in the lateral part of the rhomboid fossa.
5. *Hypoglossal trigone* behind the striae medullares is formed by the underlying nucleus of the hypoglossal nerve.
6. *Vagal trigone*, lateral and posterior to the hypoglossal trigone, overlies the dorsal nucleus of the vagus nerve.
7. *Area postrema*, which like the other circumventricular organs lacks a blood–brain barrier (Chapter 3), bulges into the fourth ventricle alongside its lateral margins rostral to the obex. The area postrema has been implicated in many functions including food and water balance and cardiovascular regulation. It also appears to function as a trigger zone for the vomiting response.

Cross-Sections Through the Brain Stem

The structure of the brain stem is complicated, and many of its nuclear groups and pathways can be studied only in properly stained microscopic sections. However, the general topographic features

⟶

Fig. 4–40. Cerebellar Nuclei
By cutting a nearly horizontal section through a half brain stem–cerebellar preparation as indicated in **A**, it is usually possible to include some of the intracerebellar nuclei as illustrated in **B** and **C**. However, except for the dentate nucleus and possibly the fastigial nucleus, it is difficult to distinguish the other intracerebellar nuclei in a gross-anatomic preparation (**C**) unless special precautions are taken (see Vignette for Chapter 4). The emboliform nucleus cannot be seen in the dissected specimen in **C** because it is located behind the superior cerebellar peduncle. (**A** from Nieuwenhuys, R. J. Voogd, and C. van Huijzen, 1988. The Human Nervous System: A Synopsis and Atlas, 3rd ed. With permission of the authors and Springer Verlag, New York. **C** from Gluhbegovic, N. and H. Williams, 1980. The Human Brain: A Photographic Guide, JB Lippincott Co., Philadelphia. With permission of the publisher.)

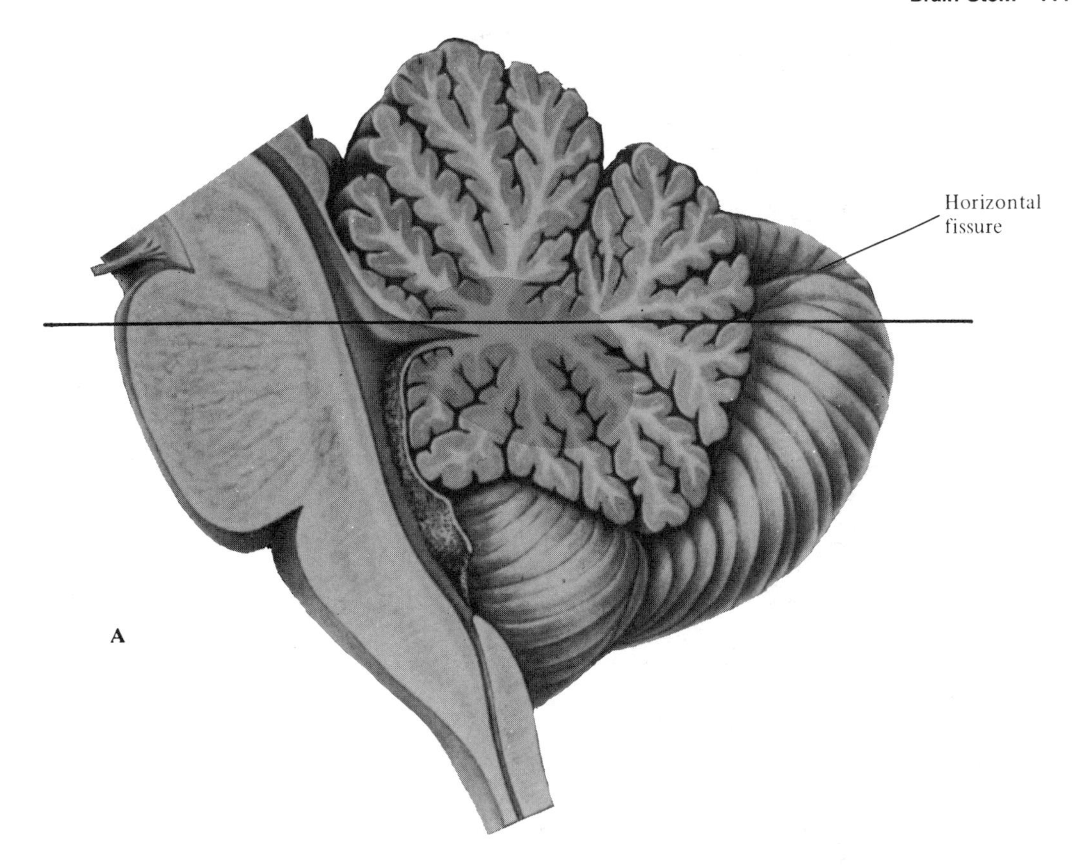
Horizontal fissure
A

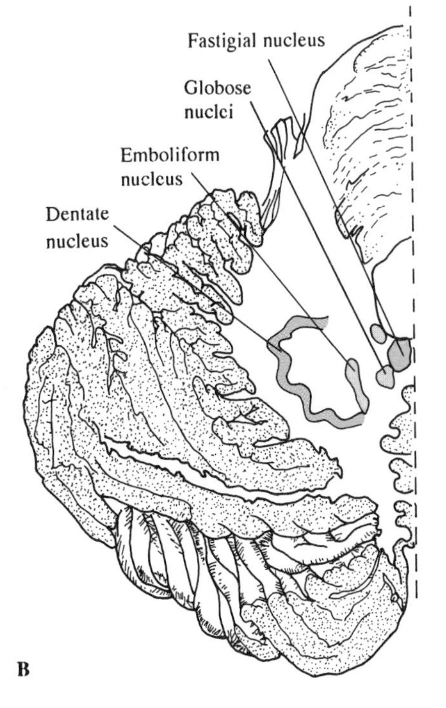
Fastigial nucleus
Globose nuclei
Emboliform nucleus
Dentate nucleus
B

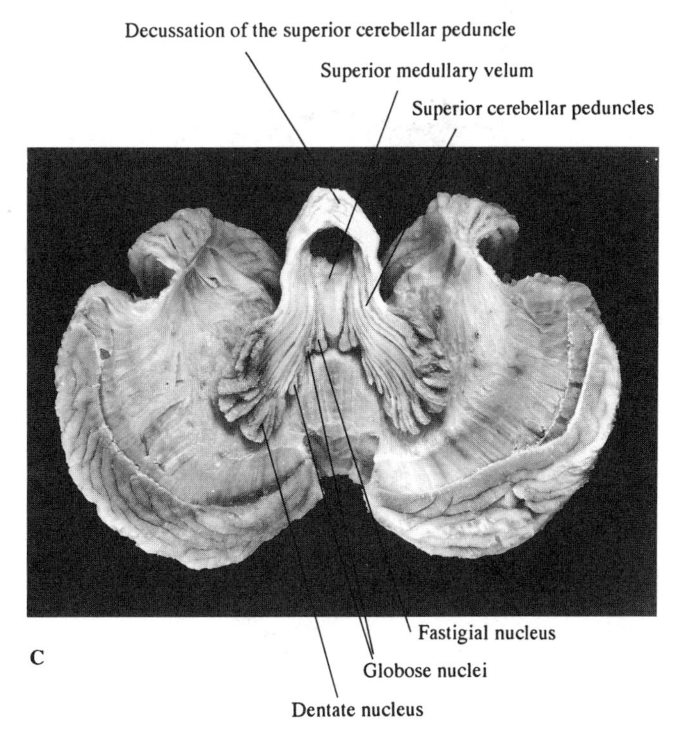
Decussation of the superior cerebellar peduncle
Superior medullary velum
Superior cerebellar peduncles
Fastigial nucleus
Globose nuclei
Dentate nucleus
C

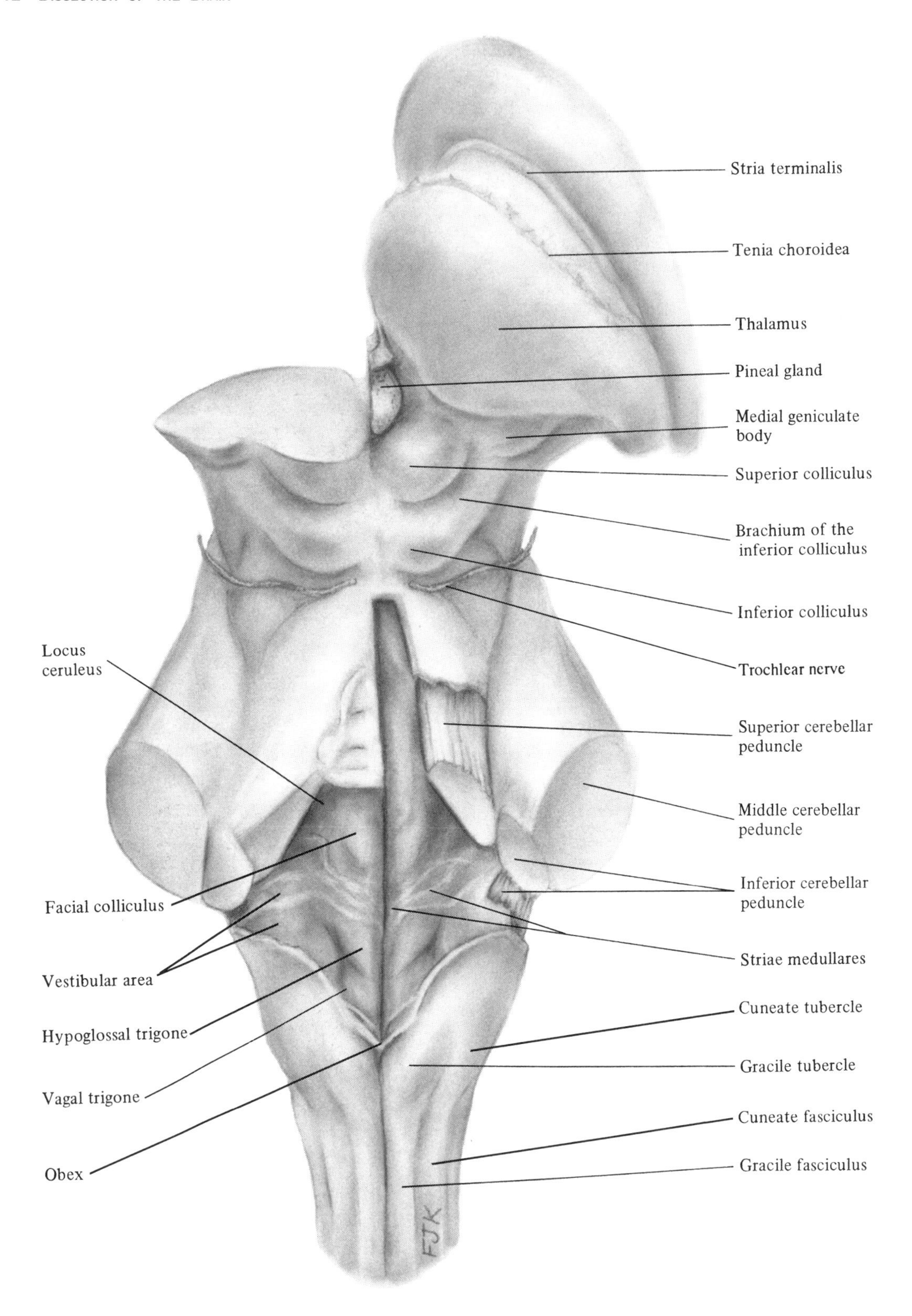

Fig. 4–41. Rhomboid Fossa
The floor of the fourth ventricle, the rhomboid fossa, has been exposed by removing the cerebellum.

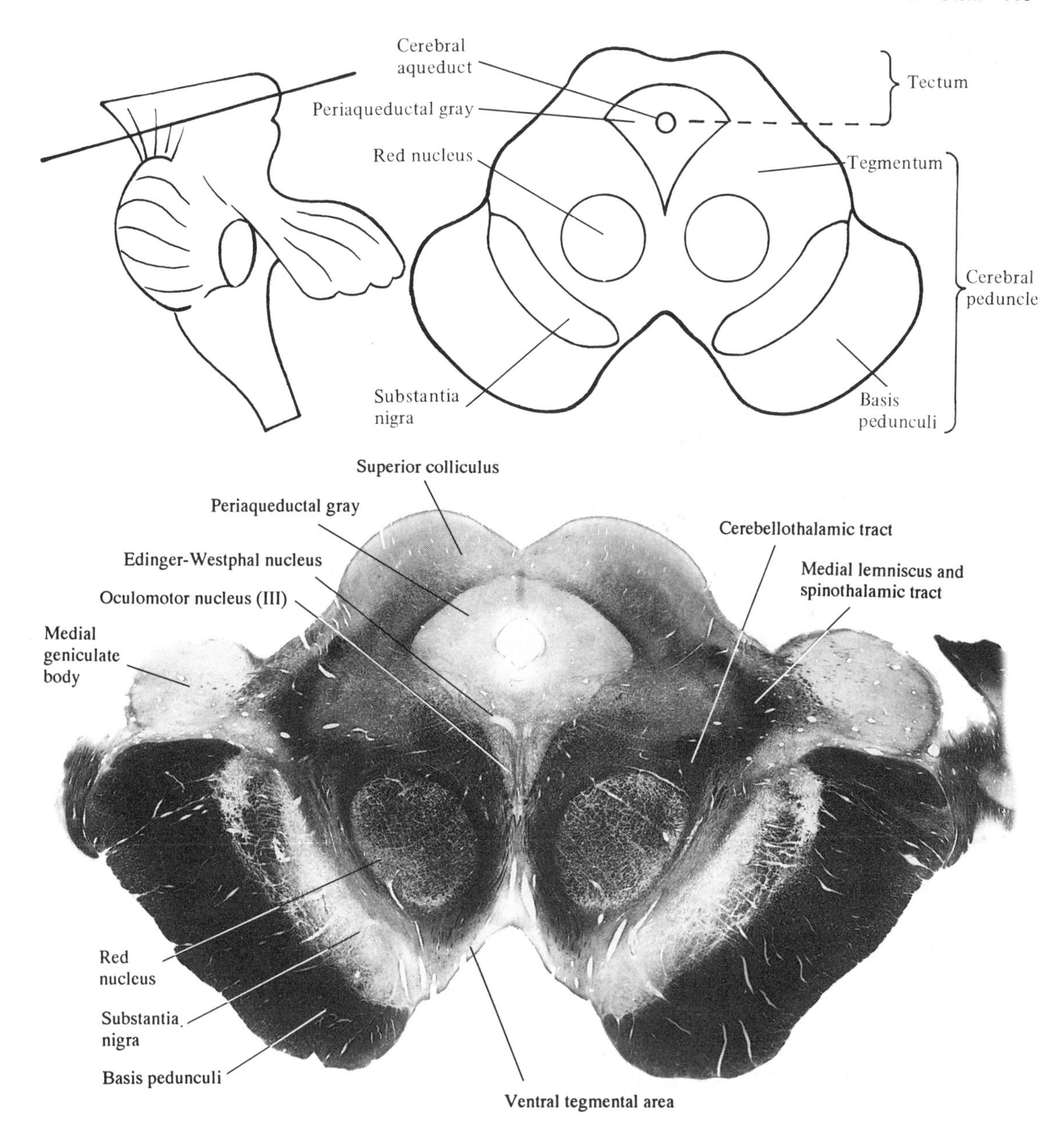

Fig. 4–42. Cross-Section of Mesencephalon
(From: Fundamental Neuroanatomy by W.H. Nauta and M. Feirtag. Copyright 1986 by WH Freeman and Company. Reprinted by permission.)

and some of the main nuclei and fiber systems can be identified with the naked eye in transverse sections cut during the brain dissection. The brain stem contains the cranial nerve nuclei and several other specific nuclei, many of which are embedded in the reticular formation (see Chapter 10). It also contains many ascending and descending pathways, through which impulses to and from the forebrain, the cerebellum, and the spinal cord are transmitted.

Mesencephalon (Fig. 4–42)

The mesencephalon is divided into the *tectum* (quadrigeminal plate) and the two *cerebral peduncles* by an imaginary plane through the cerebral aqueduct. Each cerebral peduncle in turn consists of the *tegmentum* dorsally and the *basis pedunculi* ventrally.

Tectum. The roof of the midbrain is composed of the *superior* and *inferior colliculi*, which contain

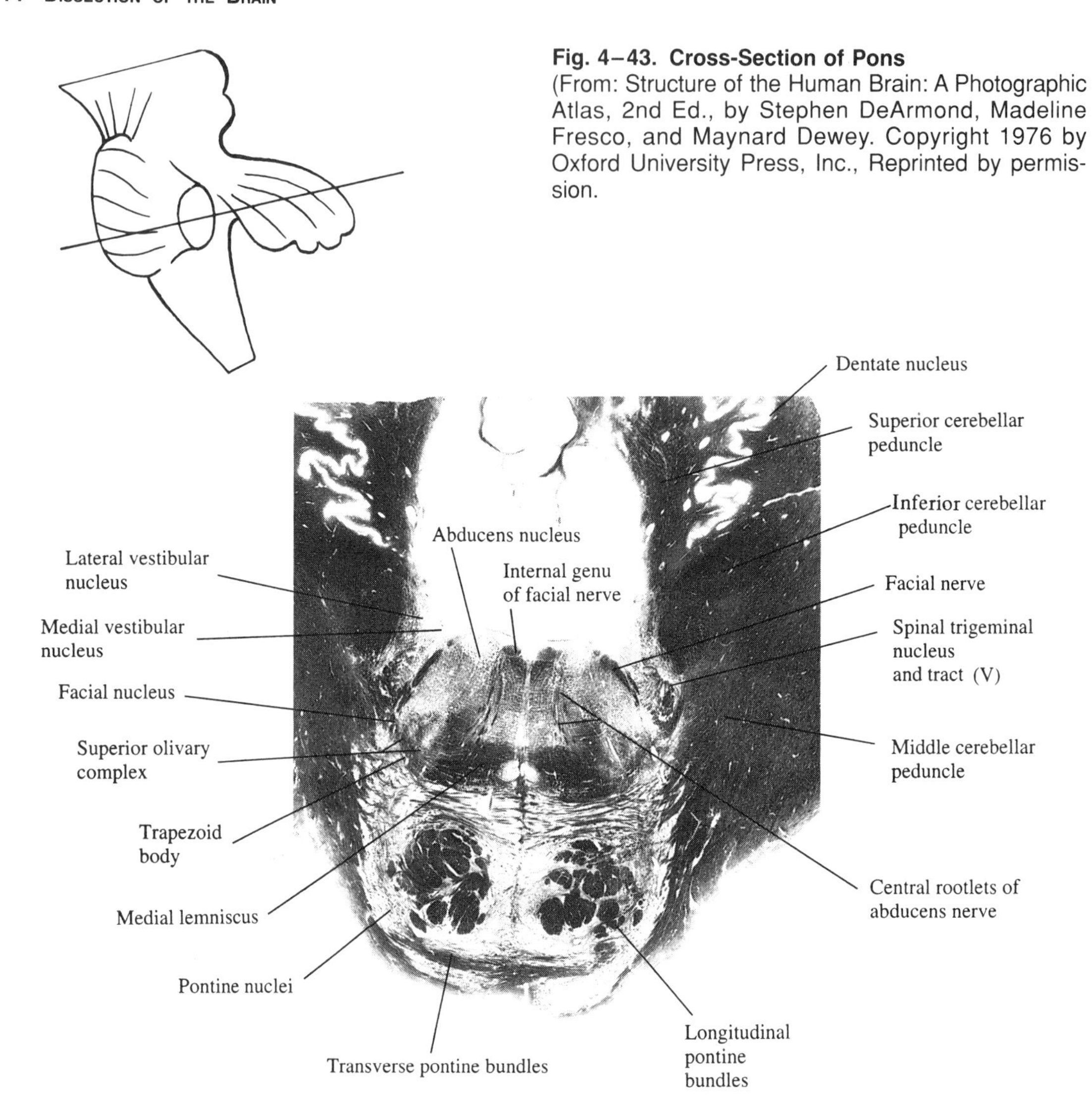

Fig. 4–43. Cross-Section of Pons
(From: Structure of the Human Brain: A Photographic Atlas, 2nd Ed., by Stephen DeArmond, Madeline Fresco, and Maynard Dewey. Copyright 1976 by Oxford University Press, Inc., Reprinted by permission.

important reflex centers for visual and auditory impulses. The inferior colliculus, furthermore, serves as a relay center in the auditory pathway from the cochlear nuclei to the medial geniculate body and the auditory cortex. *Periaqueductal gray* surrounds the cerebral aqueduct.

Tegmentum. The *nuclei* for the *oculomotor* (III) and *trochlear* (IV) *nerves* are located in the tegmentum at the border of the central gray substance. The oculomotor complex is located at the level of the superior colliculus, and the trochlear nucleus at the level of the inferior colliculus. The oculomotor centers cannot be recognized with certainty in the gross anatomic preparations. However, the most conspicuous structure in the tegmentum, the *red nucleus* (nucleus ruber), can usually be recognized as a round grayish-brown area (7–8 mm in diameter) surrounded by a capsule of white matter. The red nucleus is part of the motor system (Chapter 15).

The central part of the tegmentum throughout the brain stem is occupied by the *reticular formation*, which is characterized by a mixture of different cell aggregations separated by a wealth of nerve fibers running in all directions. The brain stem reticular formation influences a variety of functions; it controls motor activities and it is concerned with such basic activities as sleep and consciousness. The reticular formation in the lower pons and medulla contains important respiratory and cardiovascular centers. The somatosensory fiber systems, the *medial lemniscus* and the *spinothalamic tract*, are located in the lateral half of the tegmentum.

The Base of the Peduncle. The dorsal part of the basis pedunculi is characterized by a black or dark brown area, the *substantia nigra*, *pars compacta*, which gets its color from the presence of melanin-

Fig. 4–44. Cross-Section of Rostral Medulla Oblongata
(From: Structure of the Human Brain: A Photographic Atlas, 2nd Ed., by Stephen DeArmond, Madeline Fresco, and Maynard Dewey. Copyright 1976 by Oxford University Press, Inc., Reprinted by permission.

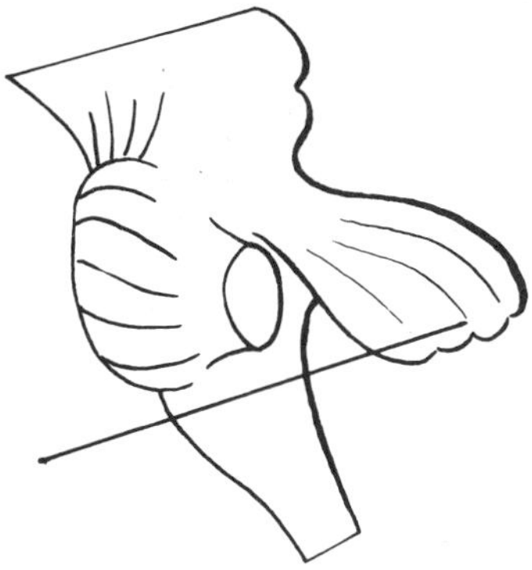

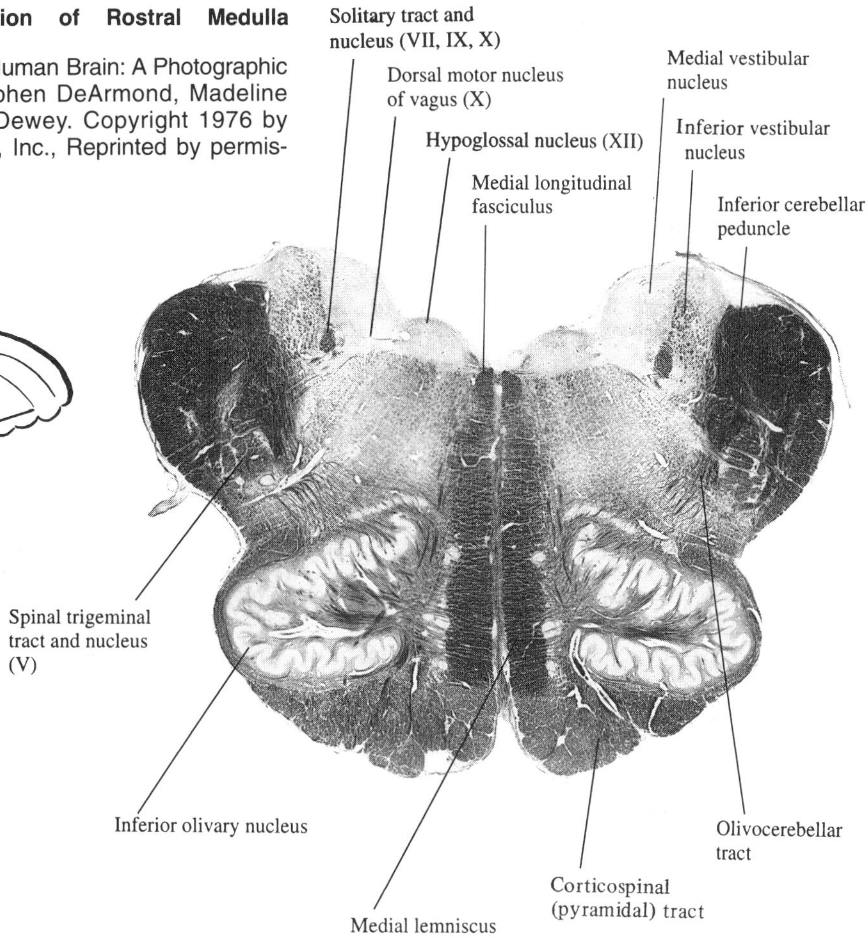

containing cells. The cells in the *substantia nigra, pars compacta*, use dopamine as their transmitter and they project to many areas in the forebrain (Fig. 10–7). The cells projecting to the striatum (putamen and caudate nucleus) constitute the nigrostriatal pathway, which is severely affected in Parkinson's Disease (see Clinical Notes, Chapter 16).

The ventral portion of the basis pedunculi, which is referred to as the *pes pedunculi*, contains descending pathways, that carry impulses from the cerebral cortex to the brain stem and the spinal cord. Sometimes the pes pedunculi is referred to as the base of the peduncle.

Pons (Fig. 4–43)

The pons consists of a large ventral or basal part and a smaller dorsal part, the pontine tegmentum.

Ventral Part. This part contains a large collection of cell groups, the *pontine nuclei*, as well as many transversely and longitudinally running myelinated fiber bundles, which gives the basal part a white color. The transverse fiber bundles, the *pontocerebellar tracts*, originate in the pontine nuclei on one side, cross the midline, and enter the cerebellum as the middle cerebellar peduncle. The pontocerebellar fibers form the second link in an important corticopontocerebellar pathway, through which cortical regions in one cerebral hemisphere can communicate with the opposite half of the cerebellum. The first link in this system is represented by longitudinally running fibers, the *corticopontine tracts*, which descend through the base of the cerebral peduncle and terminate in the pontine nuclei. The longitudinally running fiber bundles of the *corticospinal tract* come together at the lower end of the pons to form the two pyramids of the medulla oblongata (see below).

Tegmentum. The *pontine tegmentum* is continuous with the tegmentum of the midbrain and the medulla. Embedded in the pontine reticular for-

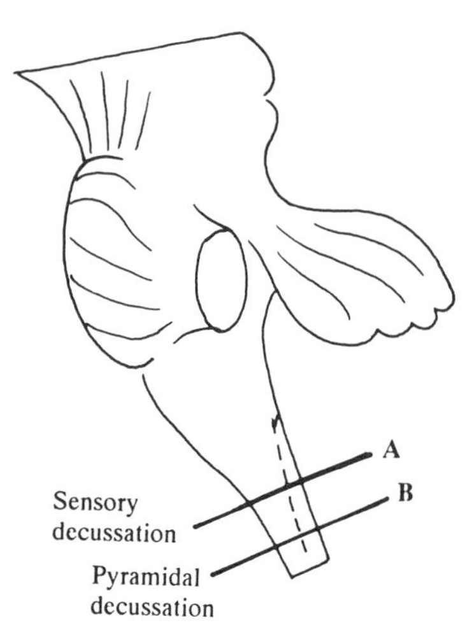

Fig. 4–45A. Sensory Decussation (Decussation of medial lemniscus)
4–45B. Pyramidal Decussation
These cross-sections of caudal medulla oblongata were kindly provided by Dr. T. H. Williams. (From Gluhbegovic, N. and T.H. Williams, 1980. The Human Brain: A Photographic Guide, JB Lippincott Co., Philadelphia. With permission of the publisher.)

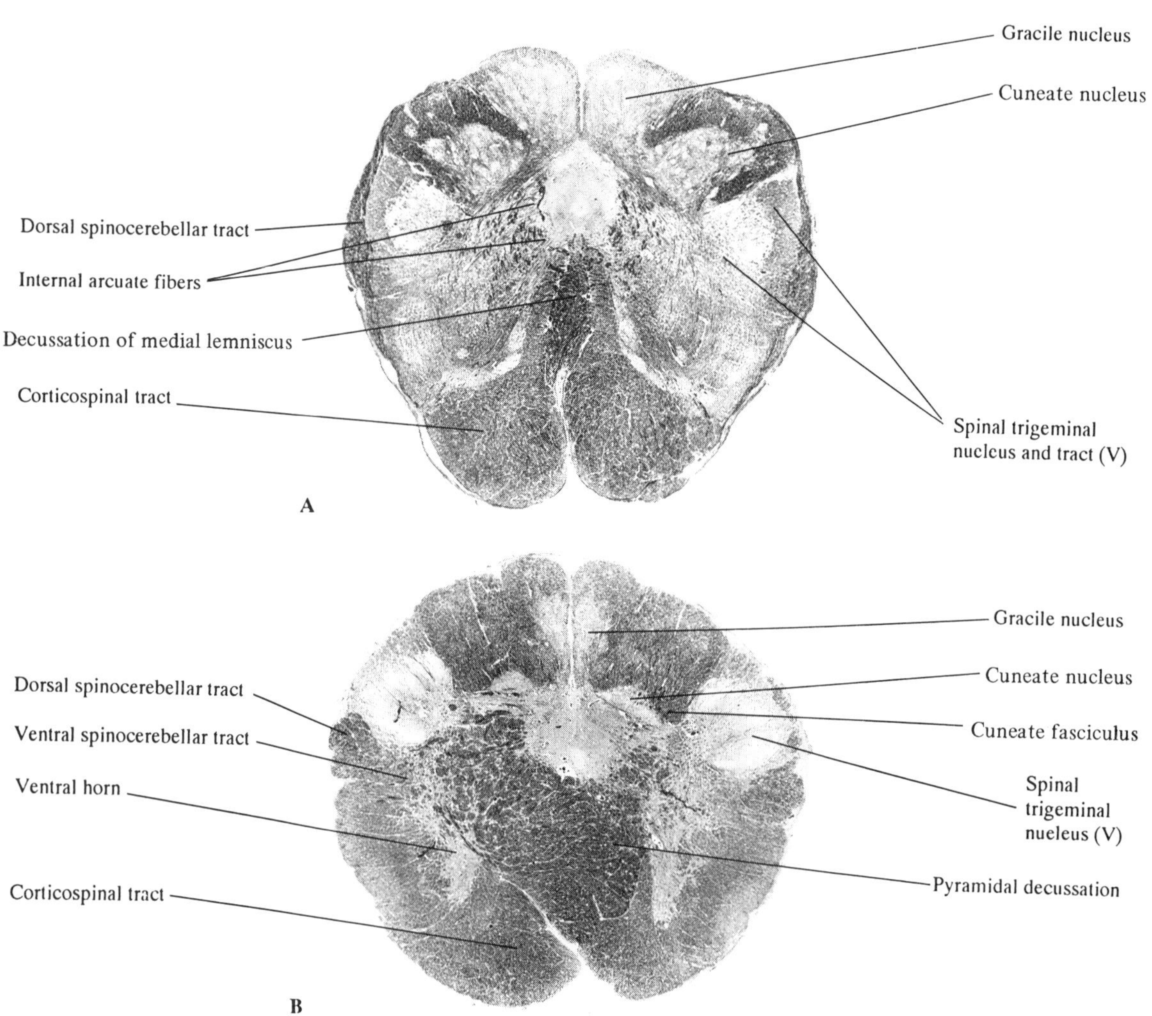

mation are many fiber tracts and several cranial nerve nuclei, namely those of the Vth, VIth, VIIth, and VIIIth cranial nerves. The *medial lemniscus* is situated close to the midline in the ventral part of the tegmentum, whereas the *spinothalamic tract* and the *lateral lemniscus* (secondary auditory fibers) are located more laterally in the tegmentum.

Medulla Oblongata (Figs. 4–44 and 4–45)

The medulla oblongata, which is the most caudal part of the brain stem, decreases in thickness as it approaches the spinal cord, and its boundary toward the spinal cord cannot be easily appreciated since the distinctive external features of the medulla gradually disappear in its lower part. The area of transition, however, is usually indicated on the ventral side by fibers crossing the median fissure. These fibers, which come from the pyramids, form part of the *pyramidal* or *corticospinal decussation* (Fig. 4–45B). The pyramidal decussation can best be appreciated by placing a probe in the median fissure close to the pons and pushing it in the direction of the spinal cord. Suddenly, the downward movement of the probe is obstructed by the decussating fibers. The pyramidal decussation, which is located about 2 cm from the lower margin of the pons, is the most conspicuous feature of the spinomedullary transition area.

One of the most characteristic structures in a cross-section through the upper half of the medulla oblongata is the *inferior olivary nucleus*, which looks like the dentate nucleus, i.e., like a crumpled bag with its hilus directed medially. It projects through the *inferior cerebellar peduncle* to the contralateral half of the cerebellum. The corticospinal tract forming the *pyramid* is located ventromedially to the inferior olive. The *medial lemniscus* is located close to the midline. Embedded in the gray matter underneath the floor of the fourth ventricle are several of the cranial nerve nuclei, namely those of the VIIIth, IXth, Xth, and XIIth cranial nerves.

The cross-section through the lower half of the medulla below the obex (Fig. 4–41) is similar to that of the spinal cord primarily because the dorsally located fourth ventricle has been replaced by the narrow *central canal*. The *dorsal funiculi* (posterior columns) represent additional spinal cord features. If the section was stained and examined in the microscope, however, one would immediately recognize some major departures from the typical spinal cord appearance. In contrast to the situation in the spinal cord, where the dorsal funiculi contain ascending somatosensory pathways in the form of heavy myelinated fiber bundles, i.e., the *gracile* and *cuneate fasciculi*, the dorsal funiculi in the lower half of the medulla oblongata are composed primarily of large nuclear masses, the *gracile* and *cuneate nuclei* (dorsal column nuclei), which form the *gracile* and *cuneate tubercles* on the dorsal surface of the bulb. These nuclei represent relay stations for the ascending somato-sensory fibers in the dorsal funiculi. Fibers arising in the gracile and cuneate nuclei on one side cross over to the opposite side anterior to the central canal to form the *medial lemniscus*. The crossing fibers are referred to as *internal arcuate fibers*, and the crossing itself as the *decussation of the medial lemniscus* (sensory decussation, Fig. 4–45A). The *pyramidal decussation* (Fig. 4–45B), which is situated immediately below the sensory decussation, marks the boundary between the medulla oblongata and the spinal cord.

Clinical Notes

The fact that the cerebellum and the brain stem are juxtaposed and closely related to each other anatomically has great clinical relevance. For instance, the three pairs of cerebellar peduncles are integral parts of both the cerebellum and the brain stem, and lesions in the brain stem can therefore interfere directly, not only with brain stem functions, but also with cerebellar functions. Furthermore, since both the brain stem and the cerebellum are contained within a restricted space in the posterior cranial fossa, a space-occupying lesion in any part of the fossa can easily cause symptoms from pressure on both the brain stem and the cerebellum. Tumors in the posterior fossa, furthermore, are notorious for their tendency to impede the flow of CSF through the fourth ventricle (see Medulloblastoma, below). Tonsillar herniation is another serious complication of an expanding lesion in the infratentorial compartment.

The brain stem is a small and compact part of the brain, which besides the cranial nerve nuclei, contains a large number of important structures and pathways. Even a relatively small lesion in the brain stem often damages several different cell groups and pathways, and specific brain stem syndromes can be recognized depending on the location of the lesion. These syndromes can be better appreciated following a review of the various functional–anatomic systems. Suffice it to say that the modern imaging methods, especially MRI, have dramatically improved the clinician's ability to identify anatomic structures and pathological processes in the infratentorial compartment, where the thick bone structures of the posterior cranial fossa often caused problems for earlier radiologic procedures.

Medulloblastoma

Almost two-thirds of all brain tumors in children are located in the infratentorial cavity. One of the most important tumors is the rapidly growing and very malignant medulloblastoma, which usually is located in the inferior vermis. Due to its strategic location in the region where CSF reaches the subarachnoid space from the fourth ventricle, the symptoms of increased intracranial pressure (papilledema, vomiting, and headache) usually occur early as a result of obstructive hydrocephalus. The tumor may occasionally spread throughout the subarachnoid space to the surface of the rest of the CNS, especially the spinal cord.

Tonsillar Herniation (Cerebellar Pressure Cone)

Herniation of parts of the cerebellar hemisphere, primarily the tonsils, through the foramen magnum can occur as a result of increased pressure in the posterior fossa. This is usually caused by an expanding lesion in the infratentorial compartment, or by general brain swelling that can occur in many disorders of the brain. It is also important to remember that a tonsillar herniation can be precipitated by a lumbar puncture in a patient suffering from increased intracranial pressure. A tonsillar herniation causes pressure on vital respiratory and cardiovascular centers in medulla oblongata and can easily result in respiratory failure and death.

Suggested Reading

1. England, M. A. and J. Wakely, 1991. Color Atlas of the Brain and Spinal Cord. Mosby Year Book: St. Louis.
2. Gluhbegovic, N. and T.H. Williams, 1980. The Human Brain: A Photographic Guide. JB Lippincott Co., Philadelphia.
3. Montemurro, D. G. and J. E. Bruni, 1988. The Human Brain in Dissection. 2nd ed. Oxford University Press: New York.
4. Smith, C. G., 1981. Serial Dissections of the Human Brain. Gage Publishing Limited: Toronto.

3

ATLAS OF THE BRAIN

Atlas: Myelin-stained Sections and matching MRI's in frontal, horizontal and sagittal planes

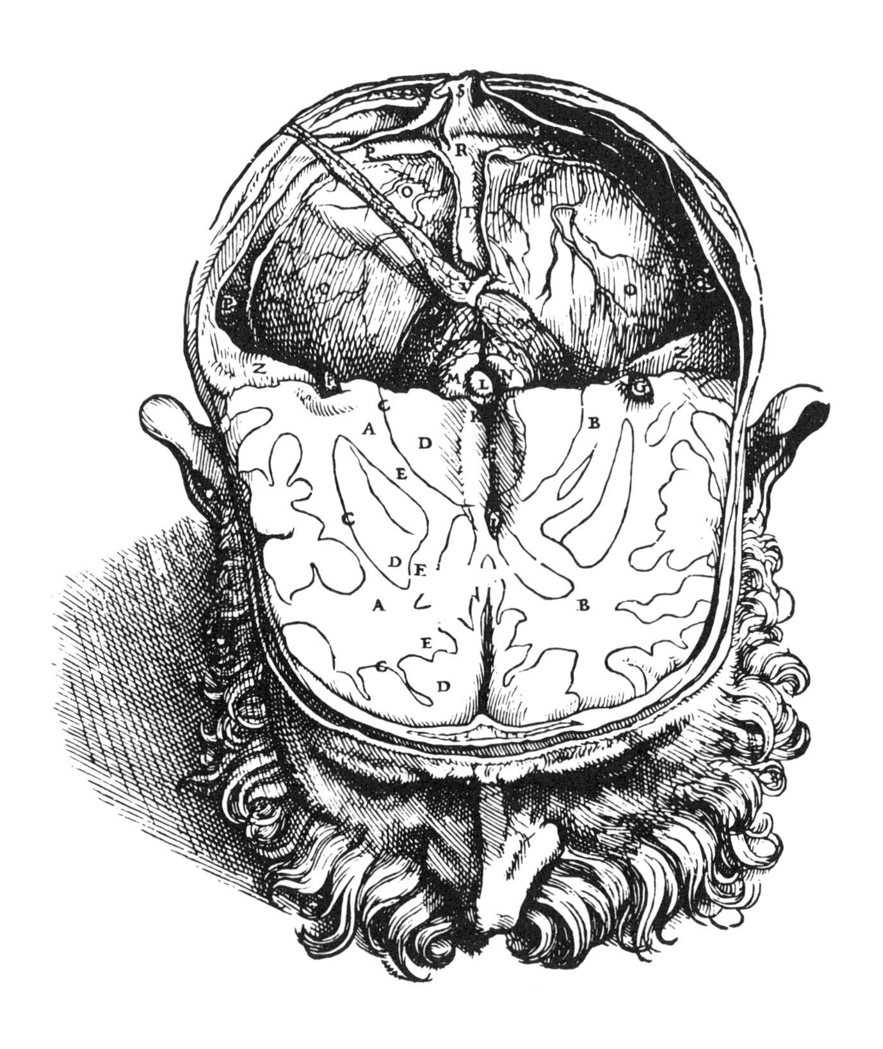

From Andreas Vesalius, 1543, De Humani Corporis Fabrica. Note the excellent rendition of the tentorium cerebelli and the sinuses in the posterior cranial fossa. Although the basal ganglia and the internal capsule are not specifically named, they are beautifully shown in the horizontal section through the cerebral hemispheres.

5

Atlas of the Brain

In order to take advantage of the extraordinary opportunities for in vivo topographic–anatomic analysis provided by modern imaging methods, it has become absolutely essential for the clinician to have a working knowledge of the functional–anatomic systems in the brain as presented in this and other textbooks. Since topographic–anatomic studies are an integral part of the process leading to a meaningful knowledge of anatomic systems, this section of the book provides the opportunity to compare high-quality myelin-stained sections of the human brain with corresponding MRI from a young adult man.

With the introduction of computed tomography (CT) in the early 1970s and the gradual development of magnetic resonance imaging (MRI) during the last several decades, it has now become possible to depict the anatomy of the brain and spinal cord in vivo. CT scanning, which is based on the same principle as radiography, i.e., the use of X-rays, is available in most hospitals and is still widely used, especially for evaluating vascular accidents and head injuries (see Clinical Notes, Chapters 3 and 15). It is also a relatively fast and comfortable procedure, which can be a determining factor when dealing with seriously ill patients. MRI, however, is in many instances more sensitive than CT scanning, and since it appears completely safe, it is rapidly becoming more available as the preferred modality, even if it is more expensive than CT. MRI exploits the effects of magnetic fields and radiofrequency waves on the body and measures their interaction with the nuclei of atoms present in the tissue. Although MRI may be performed using any element in the body with magnetic moment, practical considerations have restricted clinical application to the imaging of the single proton in the hydrogen nucleus, which makes it an excellent method for soft tissue imaging and for the differentiation of white and gray substance. The cortical bone of the skull, however, is all but invisible, since cortical bone possesses a low mobile proton density.

The fact that MRI provides a more detailed picture of brain morphology than CT scans has made it the preferred imaging method in the diagnosis of most brain disorders. Indeed, the most recent generations of MRI machines, which provide for a resolution of about ½ mm, have made it possible to study some of the more conspicuous intracortical structures in vivo (Damasio et al., 1993). Since pathologic alterations in cerebral cortex seem to be a hallmark of some of the most common neuropsychiatric disorders such as schizophrenia (Chapter 20) and Alzheimer's disease (Chapter 21), MRI may soon be able to play a more definitive role in the diagnosis of these illnesses.

The MRIs prepared to match the myelin-stained brain sections in the atlas below (Figs. 5–6 to 5–25) were acquired from a large number of coplanar slices using a special volume technique (Mugler and Brookeman, 1990). In this sequence arteries appear bright and veins dark (see midsagittal image, Fig. 5–1).

The Axial Plane

Since MRI can demonstrate brain structures with a high degree of contrast, neuroradiologists use a brain-based system of orientation rather than the traditional system based on bone structures of the skull, which is used for CT. The standard sections used by radiologists are transaxial or axial sections, i.e., parallel to the intercommissural plane, which is a plane through the anterior and posterior commissures, both of which can be easily identified on a midsagittal MRI section (Fig. 5–1). Transaxial sections, which produce horizontal sections through the cerebral hemispheres, are routinely presented with the posterior parts of the hemispheres pointing downwards. For certain conditions, patients are imaged in the coronal plane for CT, whereas with MRI coronal and sagittal planes are routinely used in addition to the more conventional axial plane.

Brain Stem Sections in MRI

MRI has for the first time made it possible to identify several brain stem structures in the living human brain, and the topographic anatomy of the brain stem is therefore of great interest to the clinician. Although it is still not possible to identify all of the important brain stem structures in MRI, their approximate location in relation to external and internal landmarks can be estimated provided one has a good mental image of the topographic anatomy of different brain stem levels. A good start to acquire such knowledge is to compare a few myelin-stained cross-sections through the brain stem with corresponding MRI sections. The plane of sectioning commonly used in anatomic atlases of the brain stem is perpendicular to its long axis. As indicated above, the routine plane used by neuroradiologists is the axial or intercommissural plane, but the orientations of these two planes are close enough to make such a comparison useful. However, since the axial MRI sections are presented with the posterior parts of the hemisphere pointing downwards, the posterior parts of the brain stem also face downwards in MRI (Fig. 5–2A). For those familiar with conventional anatomic atlases, in which the posterior part of the brain stem (dorsal surface) is facing upwards (Fig. 5–2B), the study of "inverted" brain stem sections may require a little bit of mental effort, however, the reversal soon becomes routine (Figs. 5–14 to 5–20; see also Fig. 4–3 and 10–1 to 10–5).

Variations on the MRI Theme

Although a technical description of MRI and its various techniques is beyond the scope of this book, it should be mentioned that a number of instrument parameters can be modified to produce different images. Routine MRI examinations are usually based on the spin-echo technique including T1-weighted and T2-weighted modes. The scans included in the atlas on the following pages are T1-weighted images, in which the white substance appears brighter than the gray substance, the CSF appears dark. In T2-weighted MRIs, on the other hand, the white substance is usually darker than the gray substance (Figs. 5–2A and 5–3), while CSF appears bright. Iron-rich substances, however, are also dark in the T2-weighted mode, which explains

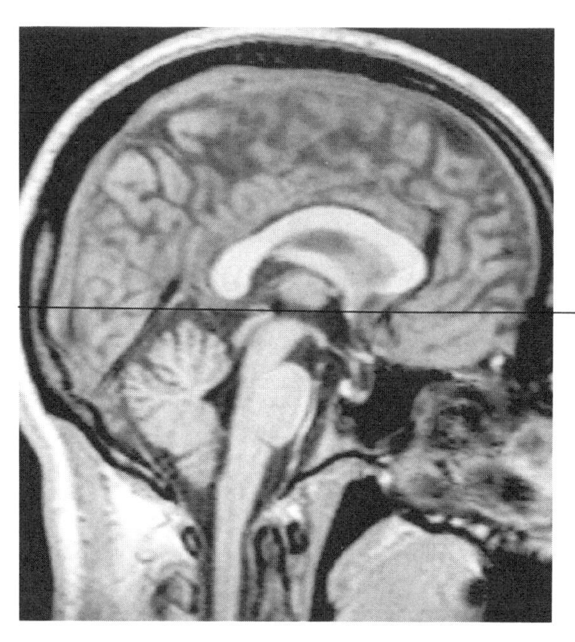

Fig. 5–1. Midsagittal MRI in T1-weighted Mode
Midsagittal MRI showing the basic reference plane for axial sections, i.e., the intercommissural plane through the anterior and posterior commissures. T1-weighted mode.

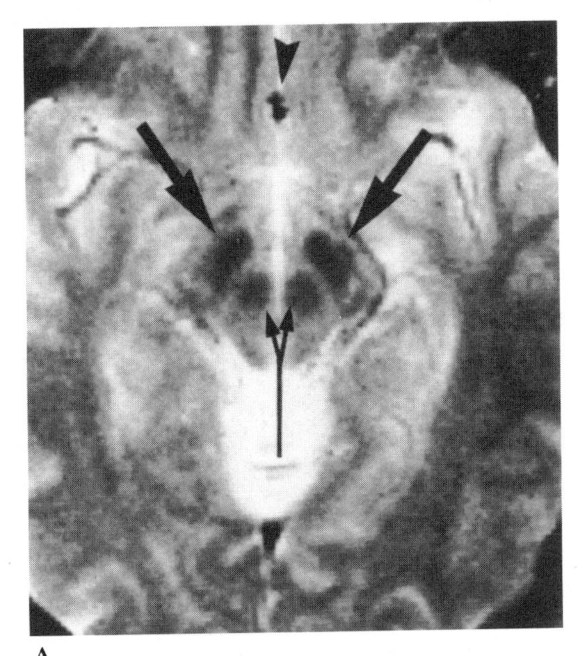

A

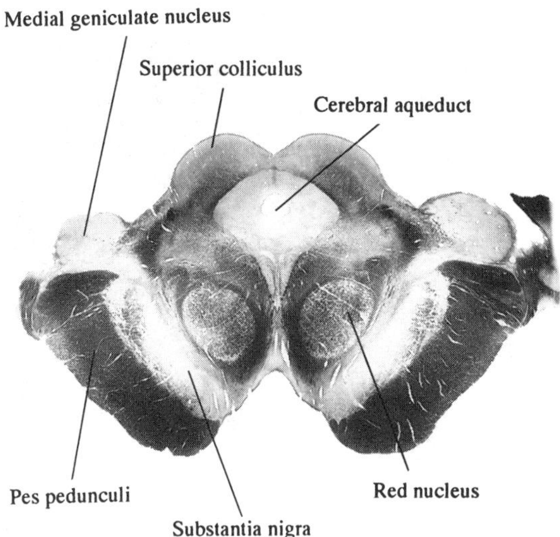

B

Fig. 5–2. The Midbrain in MRI and Myelin-stained Preparation
A. Normal axial MRI at the level of the midbrain. Iron-rich structures like the red nucleus and the substantia nigra are especially dark in T2-weighted MRIs. *Arrowhead* = anterior cerebral arteries. *Thick black arrows* = substantia nigra. *Forked arrow* = red nuclei (From Sartor, K., 1993. MR Imaging of the Skull and Brain. Springer-Verlag, Berlin. With permission of the author and the publisher.)
B. Transverse section through the midbrain as it usually appears in anatomic atlases, i.e., with the posterior parts of the section facing upwards. (The picture was kindly provided by Dr. Walle Nauta. From Nauta, W. J. H. and M. Feirtag, 1986. Fundamental Neuroanatomy. W. H. Freeman and Co., New York. With permission.)

the distinct rendition of the red nuclei and substantia nigra in Fig. 5–2A.

As indicated above, the MRI can be manipulated in a number of ways. One especially valuable method to enhance the contrast of the image is to make use of a paramagnetic contrast agent like gadolinium, which is injected intravenously in a chelated form to prevent the harmful effects of free metal ions. Some of the MRIs related to the Clinical Notes have been prepared with the aid of gadolinium enhancement (e.g., Fig. 17–9).

Fat tissue gives a bright signal in routine MRIs, and this may obscure the anatomic details, especially in areas such as the orbit. By the aid of a fat-saturation technique, however, the signal from the fat can be greatly reduced, thereby improving the visualization of the optic nerve and the other structures in he orbit (Fig. 5–4).

Three-Dimensional Reconstructions

Some of the MRIs in this book (e.g., vignettes for Chapters 3 and 13) represent three-dimensional perspectives of the living brain computed from a

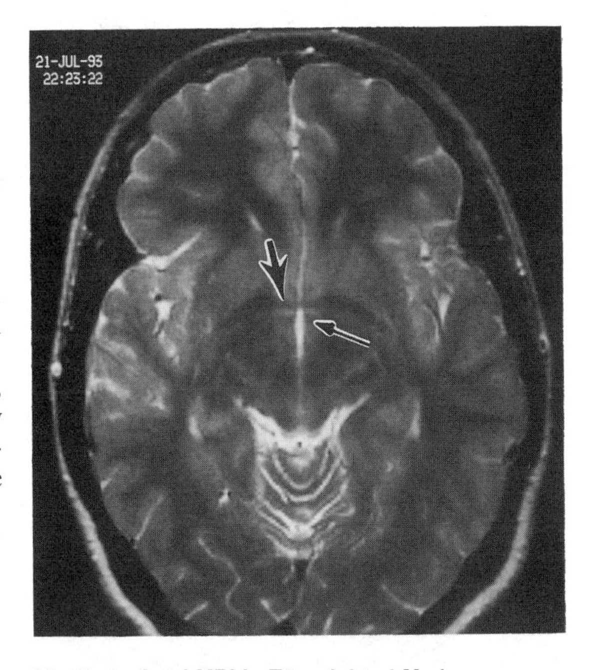

Fig. 5–3. Axial MRI in T2-weighted Mode
Axial MRI at the level of the crossing of the anterior commissure (*thick arrow*). *Thin arrow* = fornix. (Courtesy of Dr. Jim Berlino, University of Virginia.)

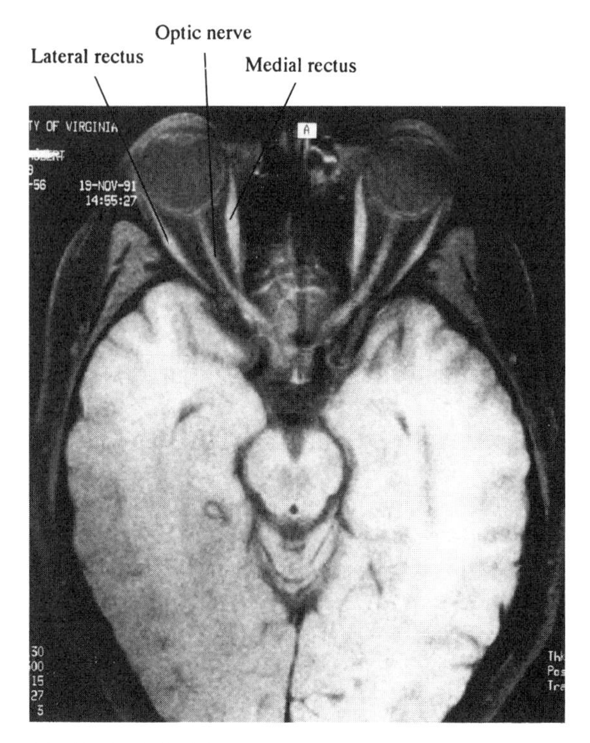

Fig. 5–4. Axial MRI using Fat Saturation Technique Axial T1-weighted MRI produced by Dr. Maurice Lipper, University of Virginia, with the aid of a fat-saturation technique to reduce the signal from the fat. This provides an especially clear image of the optic nerve and the eye muscles in the fat-rich orbit.

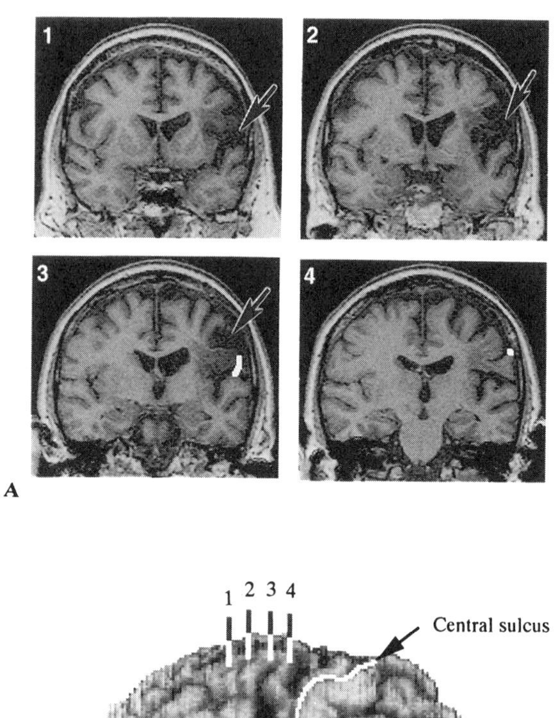

Fig. 5–5. Three-dimensional MRI
A. Four coronal T1-weighted MRIs from a patient with a hemisphere lesion.
B. Three-dimensional perspective of the brain in (A) computed from a series of coplanar slices. Note how easy it is to appreciate the location of the lesion in front of the central sulcus, indicated by a *white line.* (Courtesy of Dr. Hanna Damasio, University of Iowa.)

large number of coplanar slices. The ability to produce three-dimensional reconstructions from two-dimensional images has greatly facilitated in vivo anatomic analysis including the mapping of lesions in the brain. The series of MRIs presented in Figure 5–5 makes the point. From an analysis of the coronal slices (Fig. 5–5A), it would be impossible to ascertain that the lesion is located in front of the central sulcus, a fact that can be easily appreciated with the help of a three-dimensional view of the brain (Fig. 5–5B).

The three-dimensional reconstruction technique, which provides the opportunity to study the anatomy of the living human brain in a detail matching that of a postmortem examination, has a number of important clinical applications. For instance, it can be used by the neurosurgeon to map the exact location of a tumor or other pathologic structure before the operation. It also offers excellent possibilities for the study of cortical functions. The vignette for Chapter 13, for instance, demonstrates how the combination of three-dimensional MRI and positron emission tomography (PET) was used to locate a visual motion area on the lateral surface of the hemisphere. Another application relates to the lesion method in neuropsychology. Ever since Paul Broca first discovered that a lesion in the left inferior frontal gyrus causes dysfunction of language, or aphasia (see Vignette, Chapter 22), the study of cognitive deficits following cortical lesions in humans has been a favorite method for the analysis of cortical functions. The possibility of mapping the exact location of a lesion in a three-dimensional perspective has greatly improved the accuracy of such functional–anatomic correlations.

Suggested Reading

1. Damasio, H. and R. Frank, 1992. Three-dimensional in vivo mapping of brain lesions in humans. Arch. Neurol. 49:137–143.
2. Damasio, H., R. O. Kuljis, W. Yuh, G. W. Van Hoesen, and J. Ehrhardt, 1991. Magnetic Resonance Imaging of Human Intracortical Structure in Vivo. Cerebral Cortex 1:374–379
3. DeArmond, S. J., M. M. Fusco, and M. M. Dewey, 1989. Structure of the Human Brain. A Photographic Atlas, 3rd ed. Oxford University Press: New York.
4. Duvernoy, H. M., 1991. The Human Brain; Surface, Three-Dimensional Sectional Anatomy and MRI. Springer Verlag: Wien.
5. Fix, D., 1987. Atlas of the Human Brain and Spinal Cord. Aspen Publishers: Rockville.
6. Haines, D. E., 1995. Neuroanatomy. An Atlas of Structures, Sections, and Systems, 3rd ed. Williams & Wilkins: Baltimore.
7. Hayman, L. A. and V. C. Hinck, 1992. Clinical Brain Imaging; Normal Structure and Functional Anatomy. Mosby Year Book: St. Louis.
8. Kirkwood, J. R., 1990. Essentials of Neuroimaging. Churchill Livingstone: New York.
9. Mugler, J. P. and J. R. Brookeman, 1990. Three-dimensional magnetization-prepared rapid gradient-echo imaging (3D MP RAGE). Magnet. Res. Med. 15:152–157.
10. Naidich, T. P., D. L. Daniels, P. Pech, V. M. Haughton, A. Williams, and K. Pojunas, 1986. Anterior commissure: Anatomic-MR correlation and use as a landmark in three orthogonal planes. Radiology 158:421–429.
11. Sartor, K., 1993. MR Imaging of the Skull and Brain. Springer Verlag: Berlin.

Figs. 5–6 to 5–23
Myelin-stained sections (**A**) of the human brain cut in the frontal (Figs. 5–6 to 5–13), horizontal (5–14 to 5–20) and sagittal (5–19 to 5–23) planes have been matched as close as possible with corresponding MRIs (**B**). The horizontal myelin sections are close enough to the axial plane to serve as examples for the type of MRIs, routinely obtained by the radiologists. In order not to obscure the MR images, the labels are kept to a minimum, which also makes the comparison between the MRIs and the myelin-stained sections more challenging.

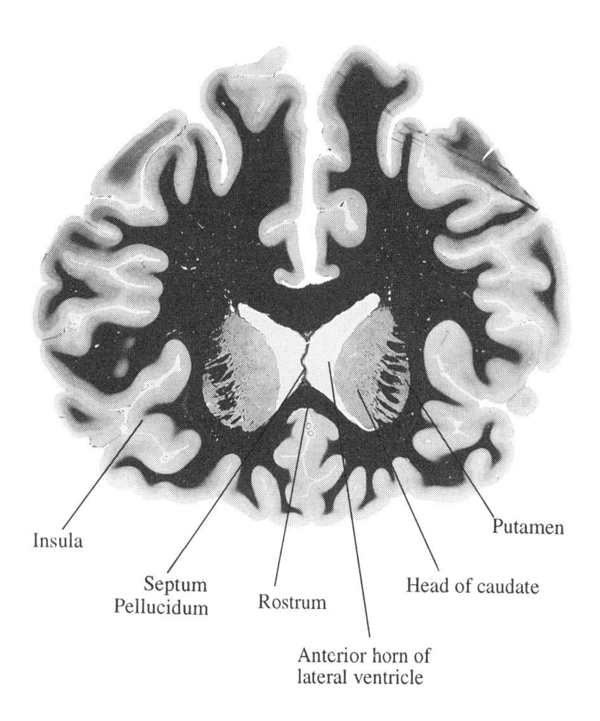

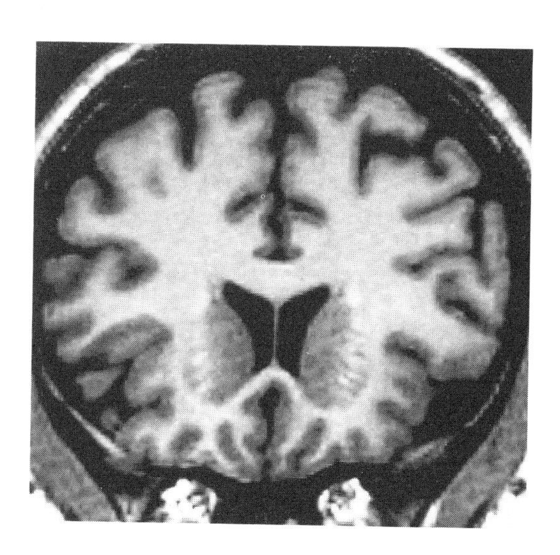

Fig. 5–6

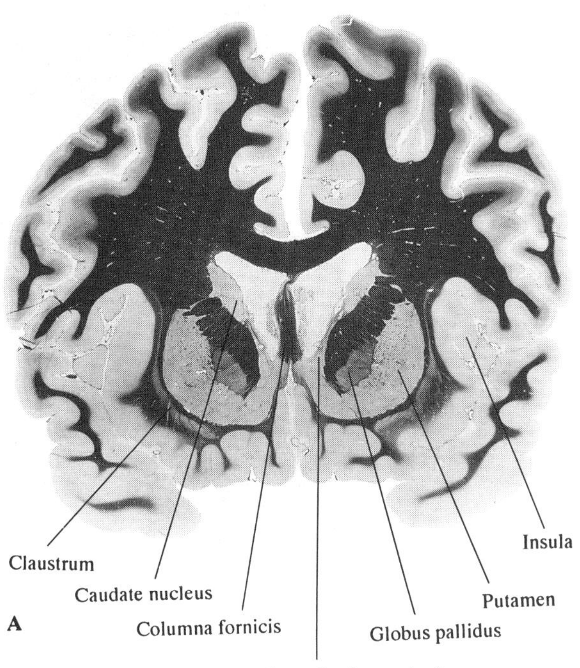

Accumbens
Fornix
Globus pallidus
Claustrum
B

Fig. 5–7

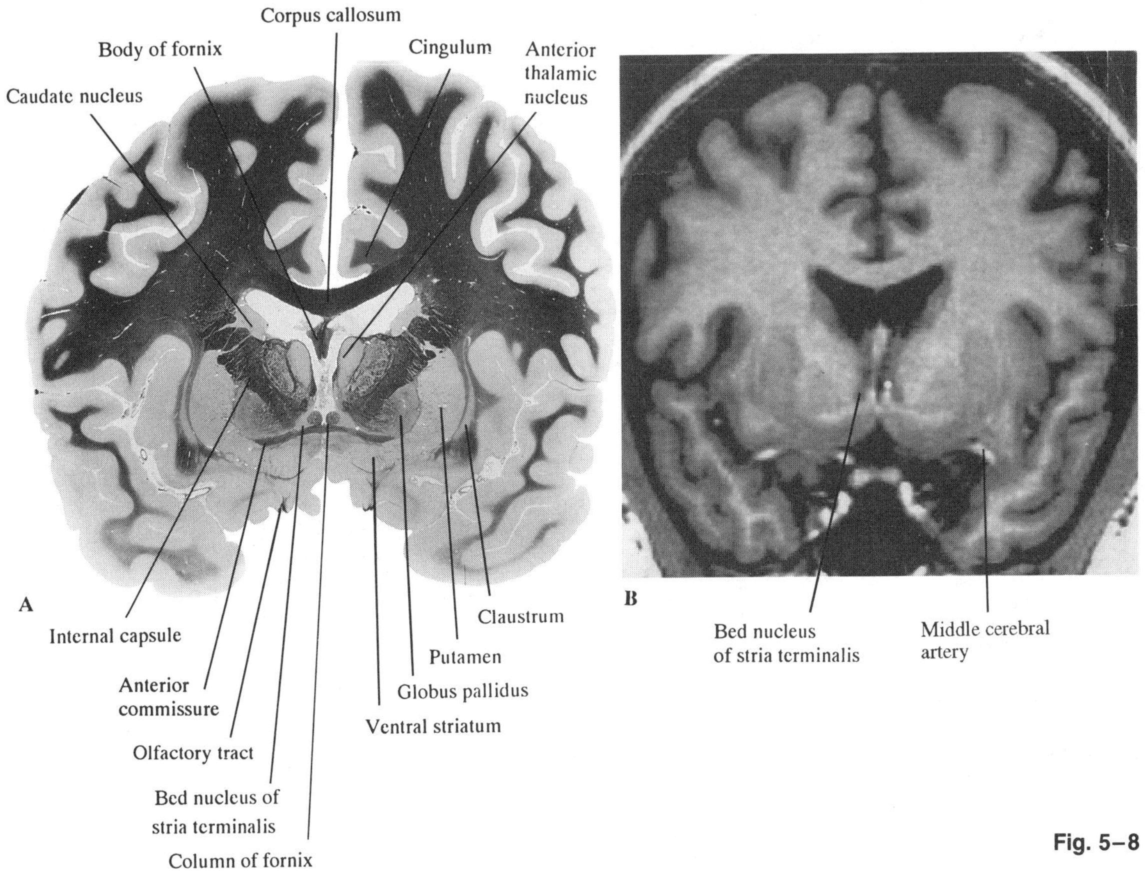

Fig. 5–8

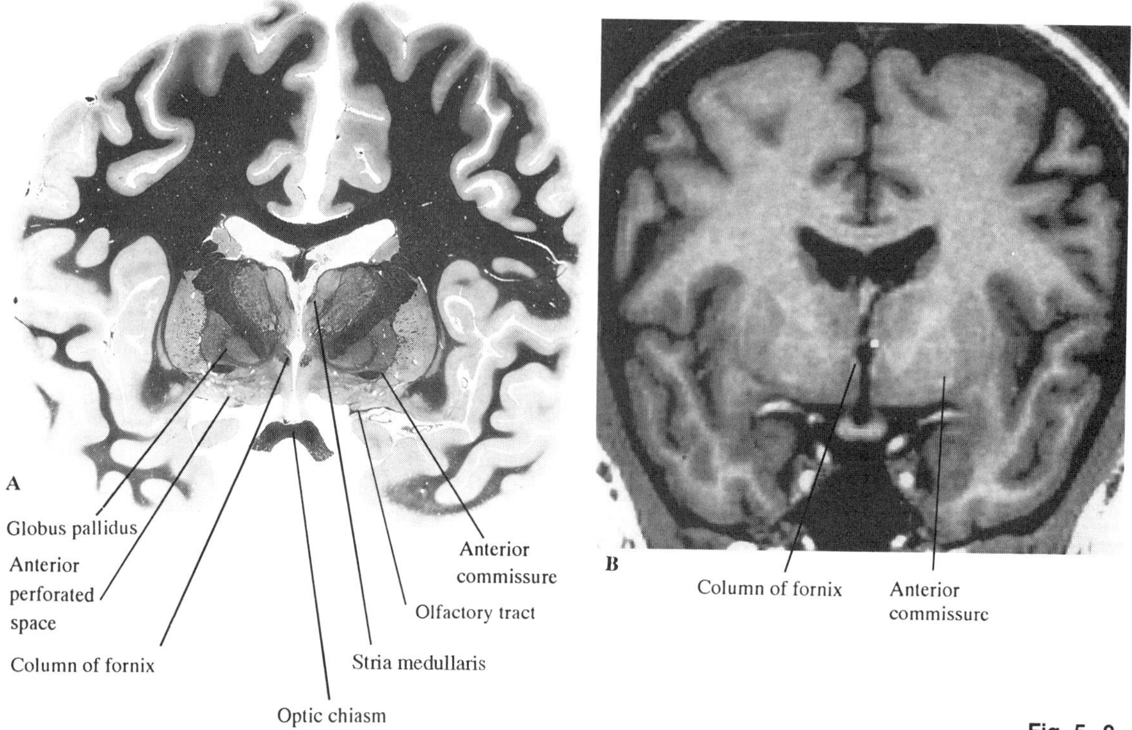

Fig. 5–9

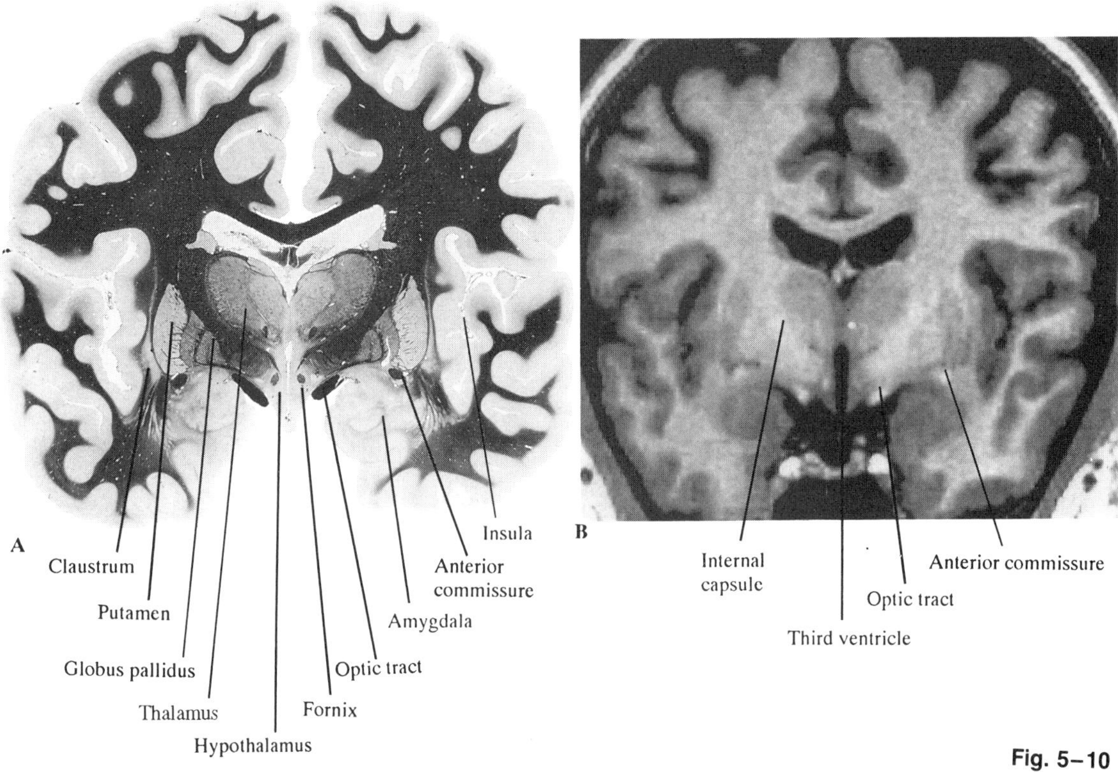

Fig. 5–10

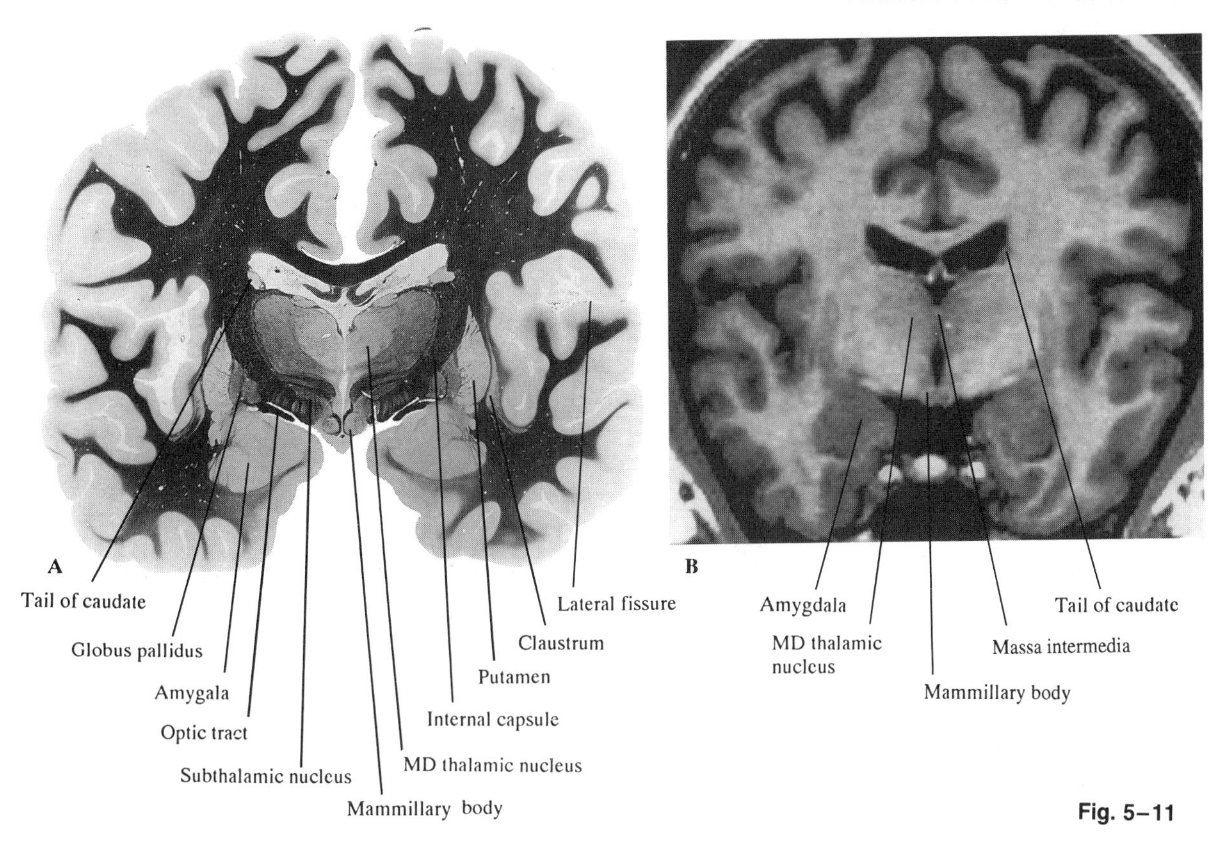

Fig. 5–11

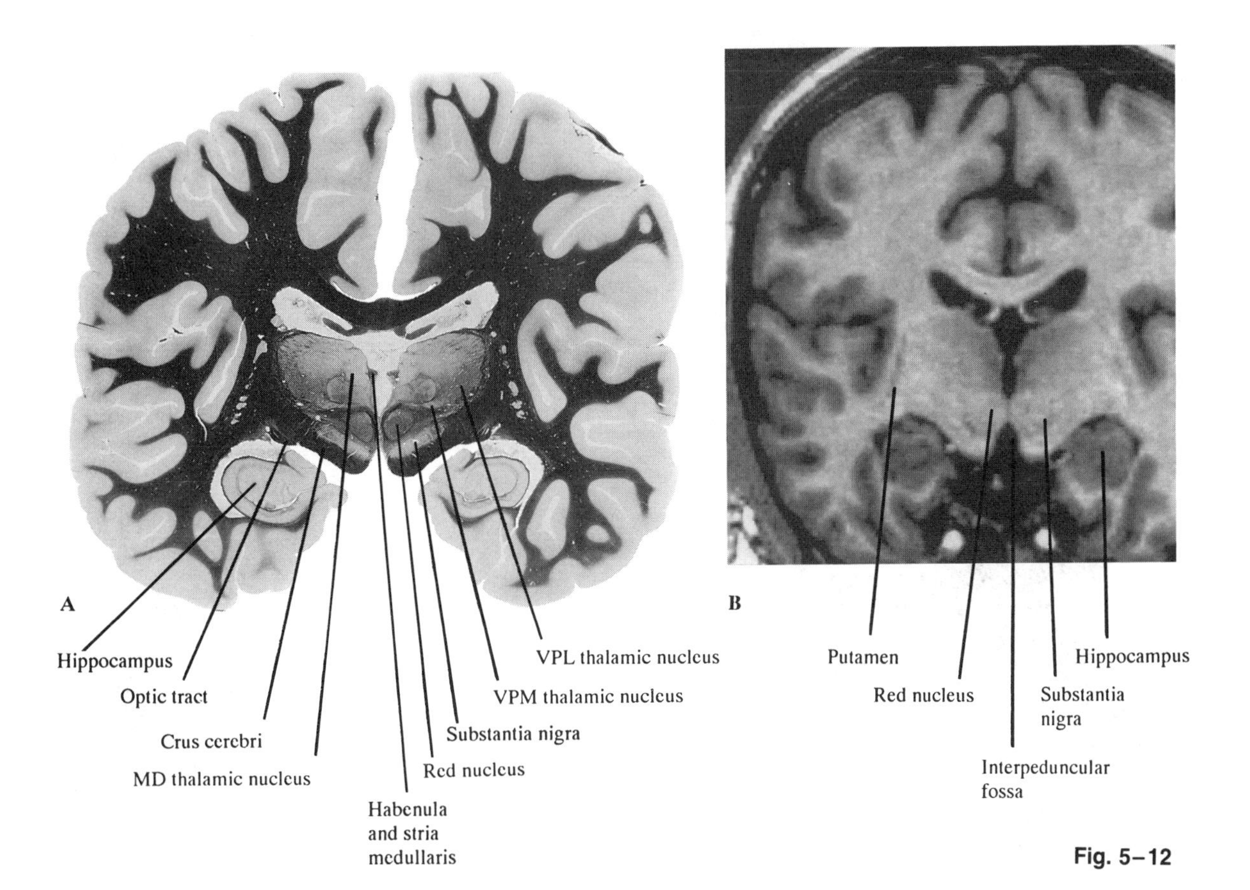

Fig. 5–12

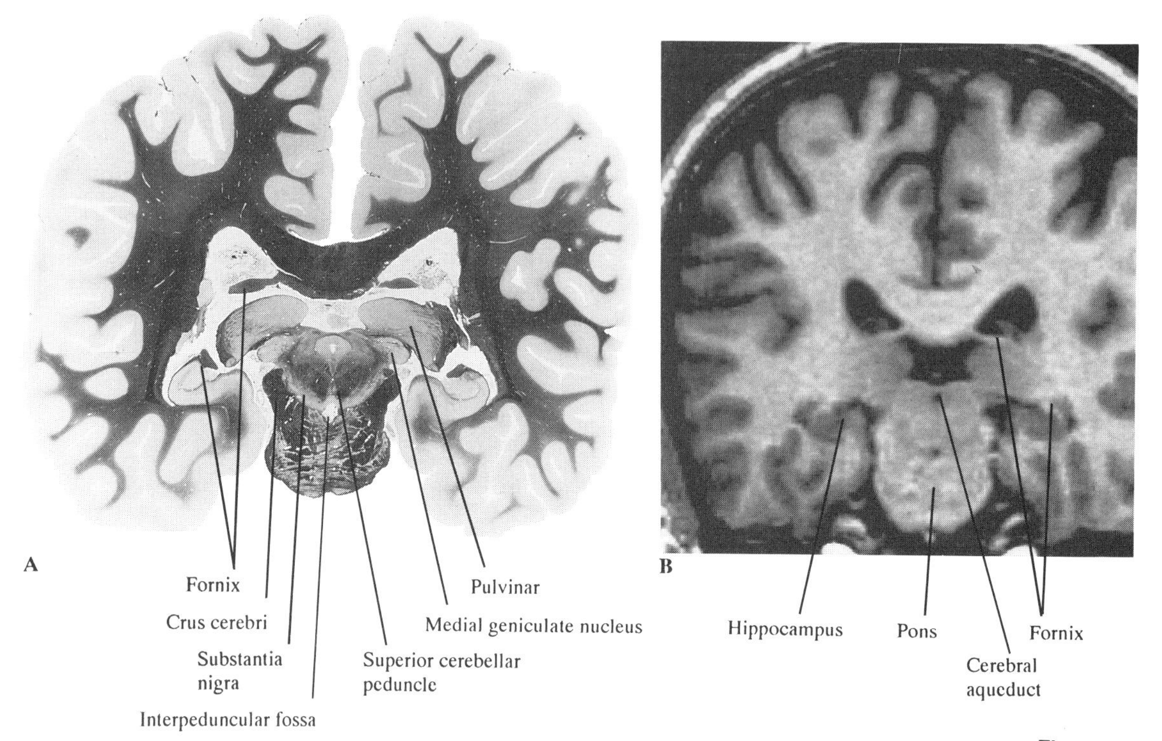

Fig. 5–13

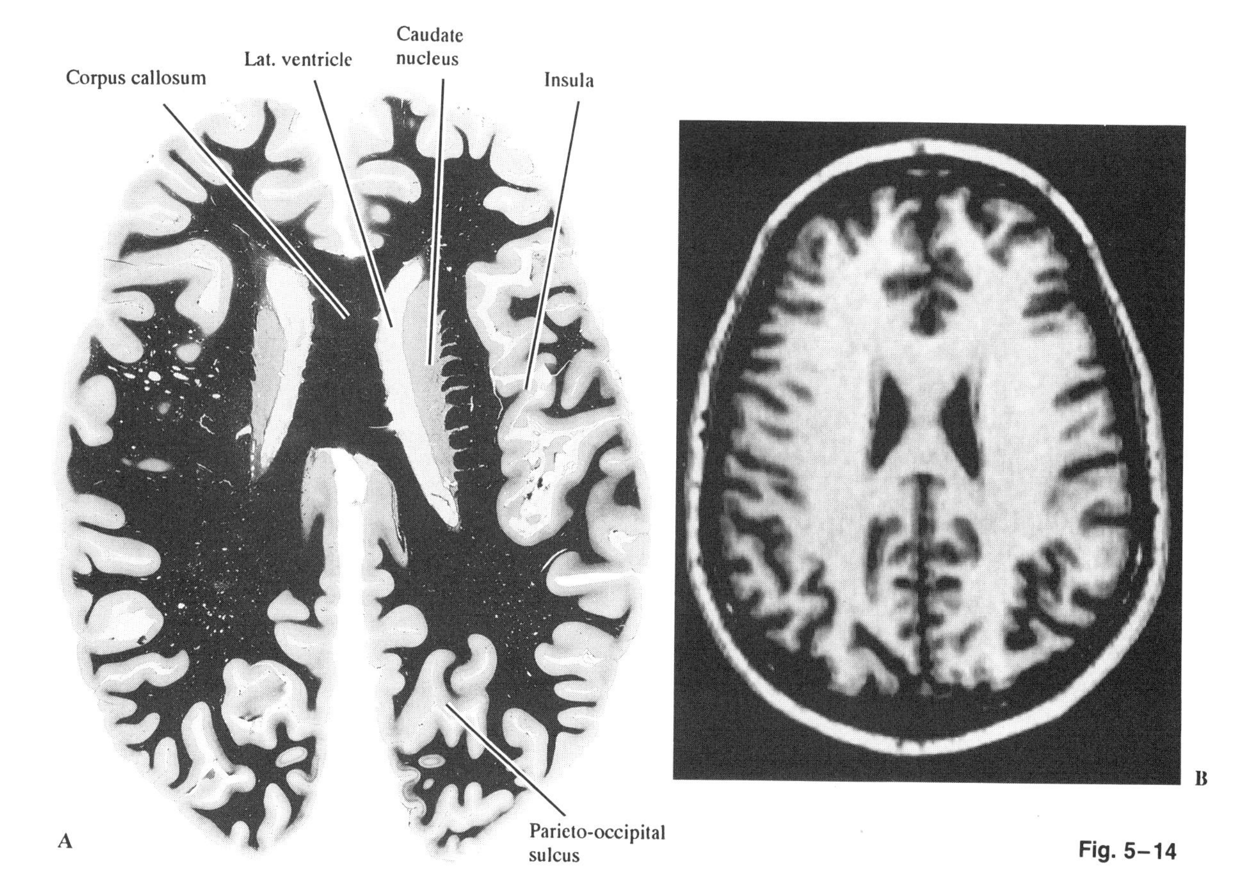

Fig. 5–14

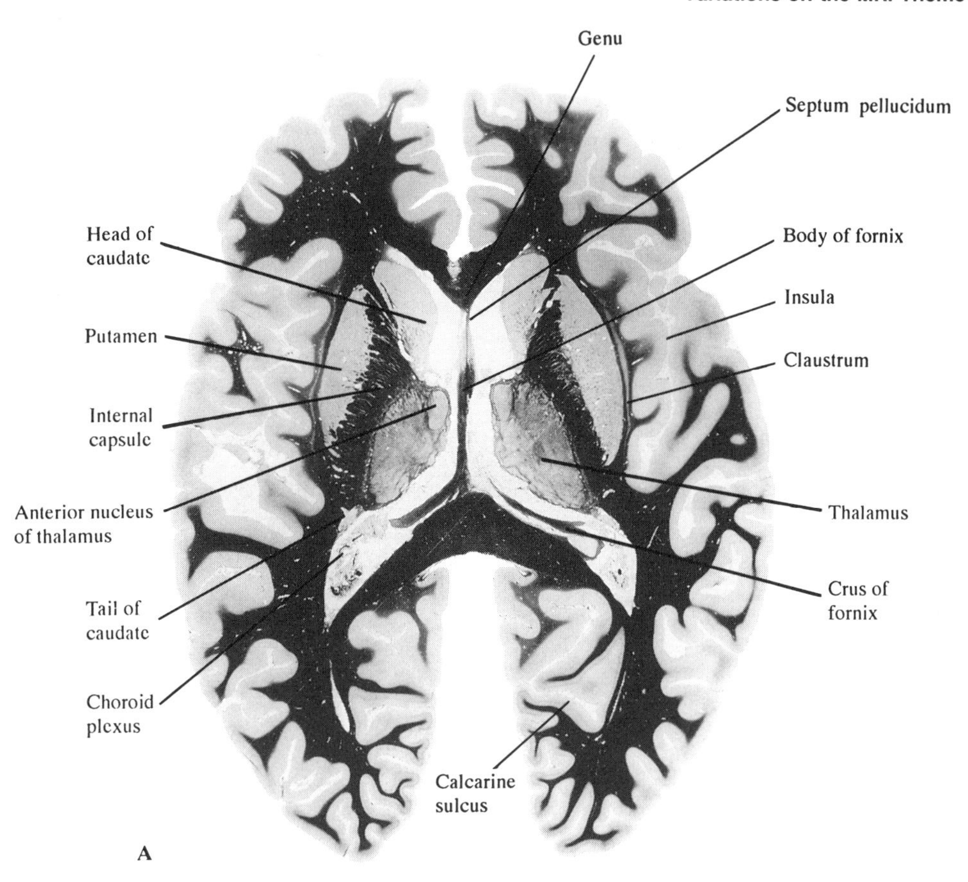
Genu
Septum pellucidum
Head of caudate
Body of fornix
Insula
Putamen
Claustrum
Internal capsule
Anterior nucleus of thalamus
Thalamus
Crus of fornix
Tail of caudate
Choroid plexus
Calcarine sulcus
A

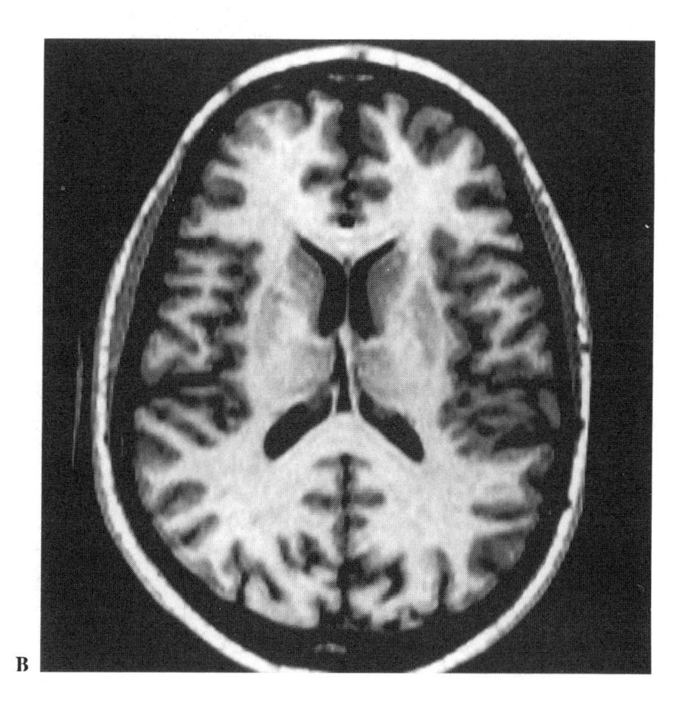
B

Fig. 5–15

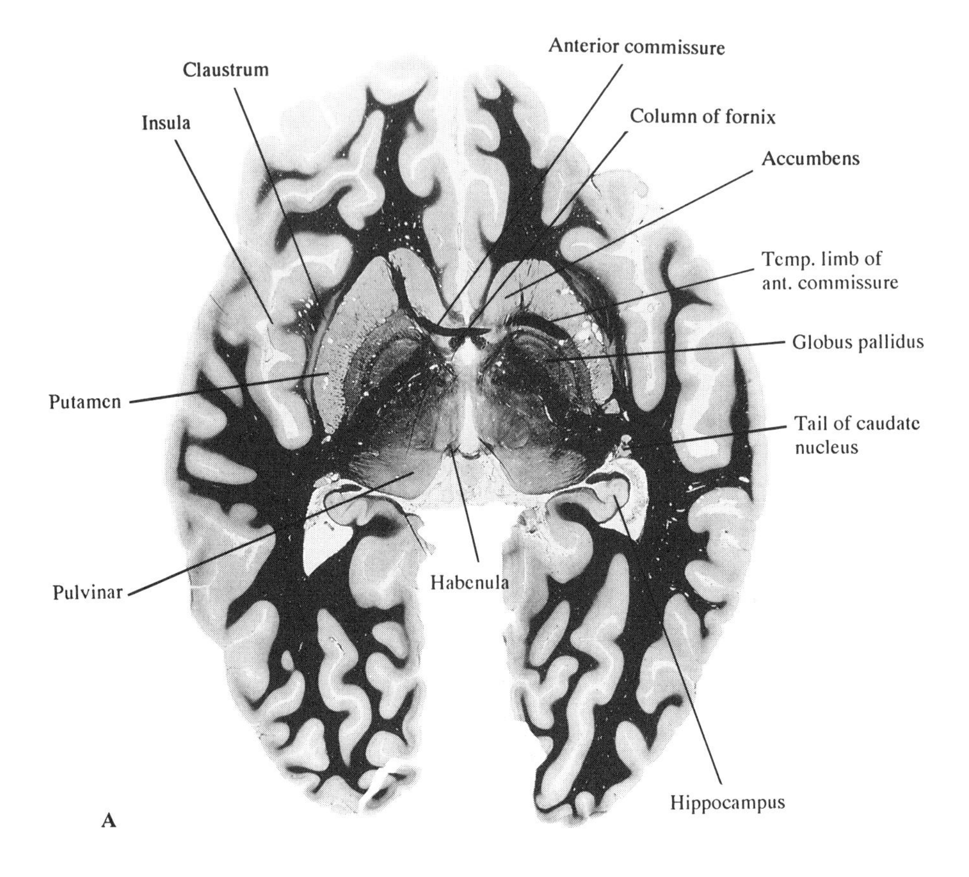

Anterior commissure
Claustrum
Column of fornix
Insula
Accumbens
Temp. limb of ant. commissure
Globus pallidus
Putamen
Tail of caudate nucleus
Pulvinar
Habenula
Hippocampus
A

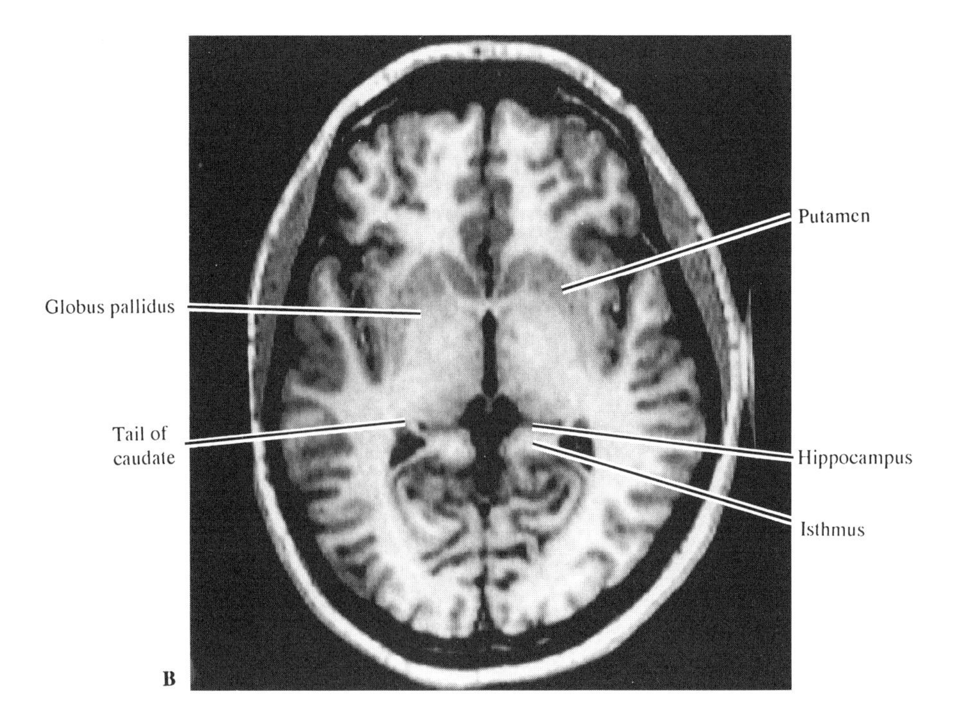

Putamen
Globus pallidus
Tail of caudate
Hippocampus
Isthmus
B

Fig. 5–16

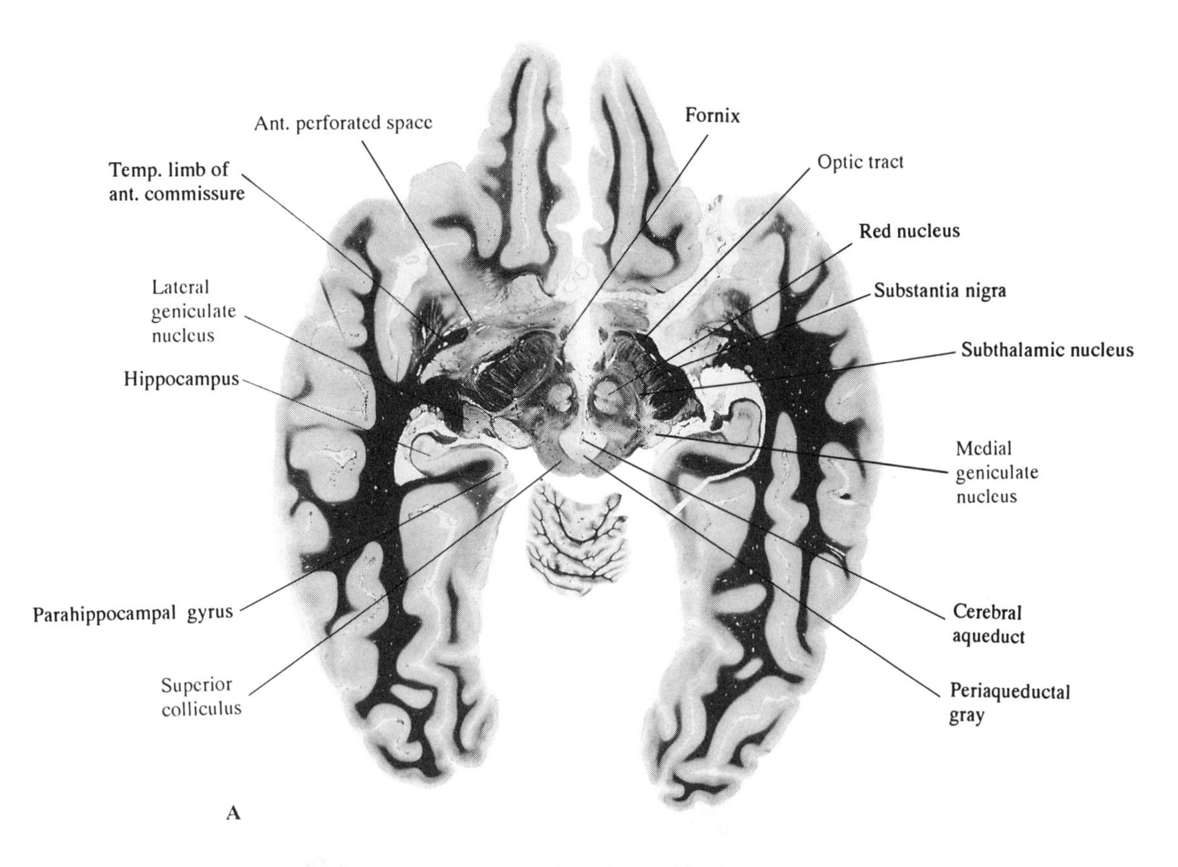
Ant. perforated space
Fornix
Temp. limb of ant. commissure
Optic tract
Red nucleus
Lateral geniculate nucleus
Substantia nigra
Subthalamic nucleus
Hippocampus
Medial geniculate nucleus
Parahippocampal gyrus
Cerebral aqueduct
Superior colliculus
Periaqueductal gray
A

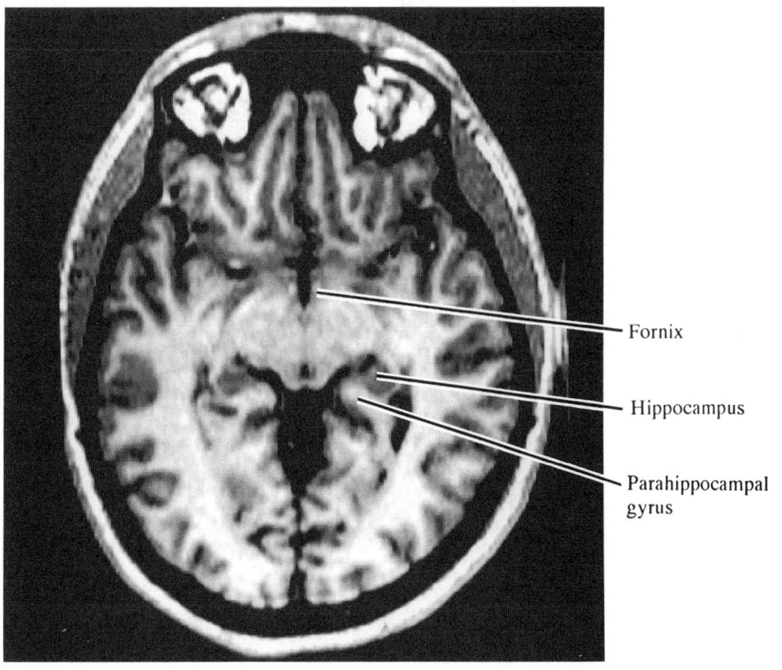
Fornix
Hippocampus
Parahippocampal gyrus
B

Fig. 5–17

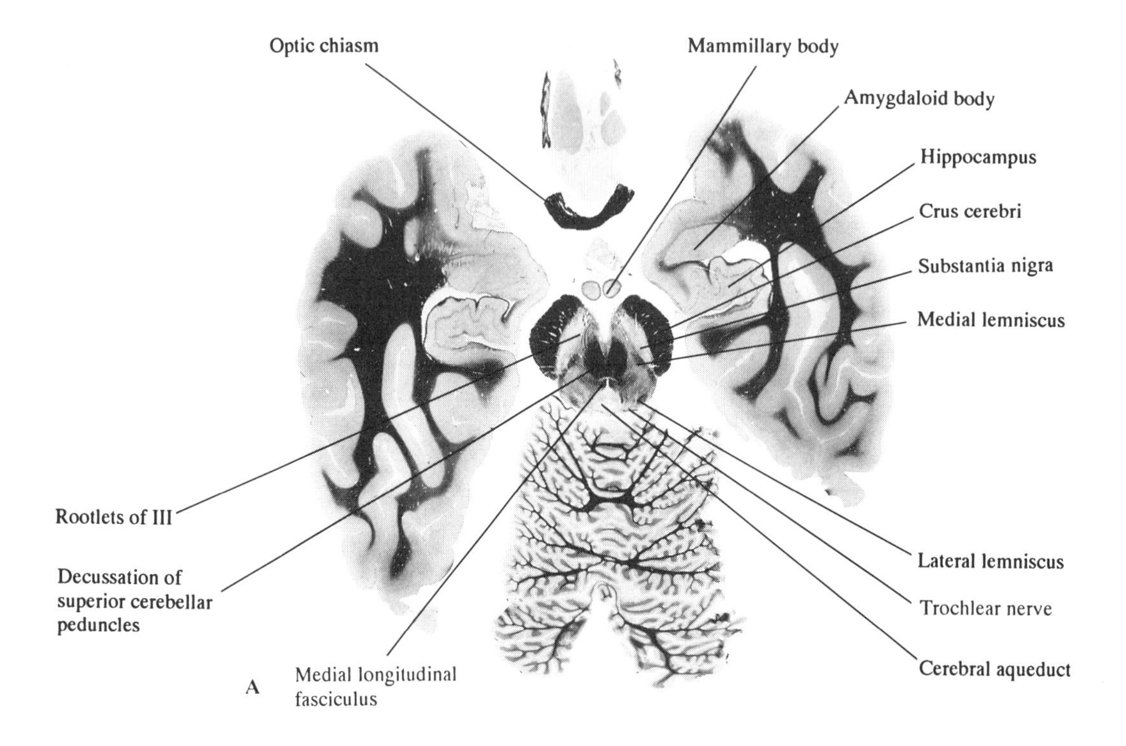
Optic chiasm
Mammillary body
Amygdaloid body
Hippocampus
Crus cerebri
Substantia nigra
Medial lemniscus
Rootlets of III
Decussation of
superior cerebellar
peduncles
Lateral lemniscus
Trochlear nerve
Cerebral aqueduct
Medial longitudinal
fasciculus
A

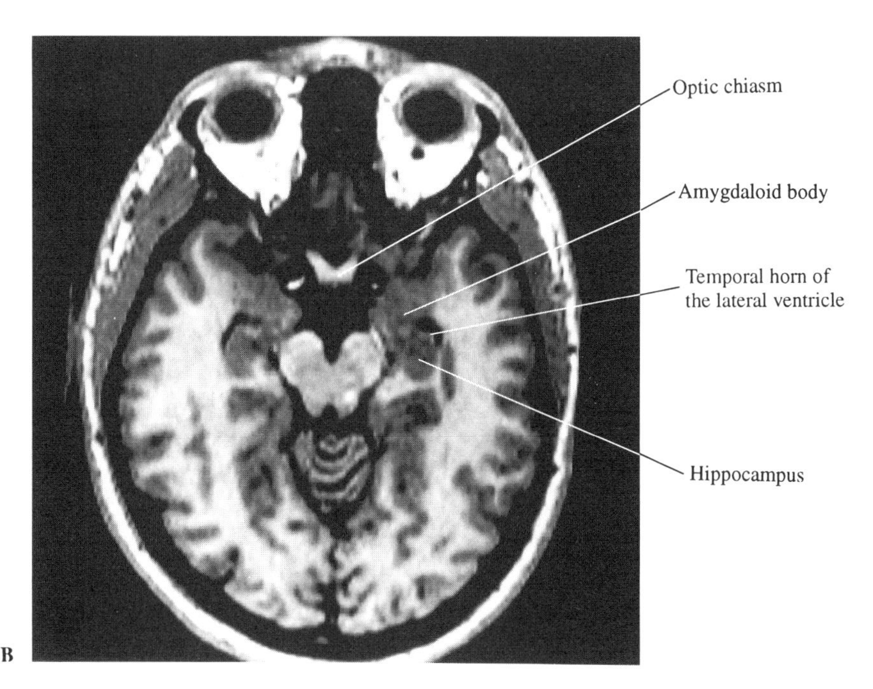
Optic chiasm
Amygdaloid body
Temporal horn of
the lateral ventricle
Hippocampus
B

Fig. 5–18

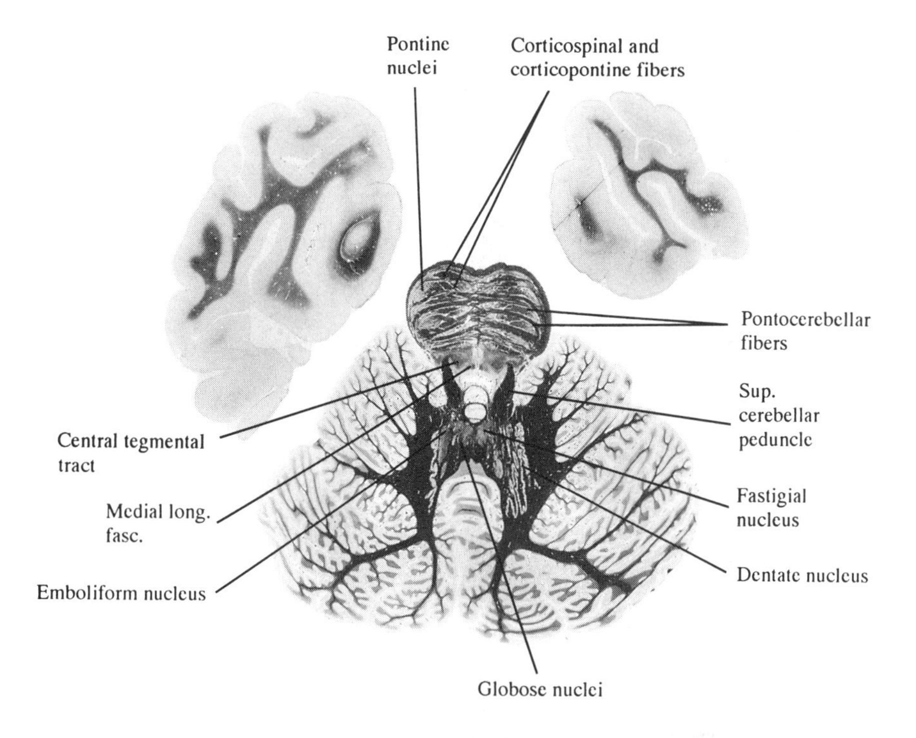
Pontine
nuclei
Corticospinal and
corticopontine fibers
Pontocerebellar
fibers
Sup.
cerebellar
peduncle
Central tegmental
tract
Medial long.
fasc.
Fastigial
nucleus
Emboliform nucleus
Dentate nucleus
Globose nuclei

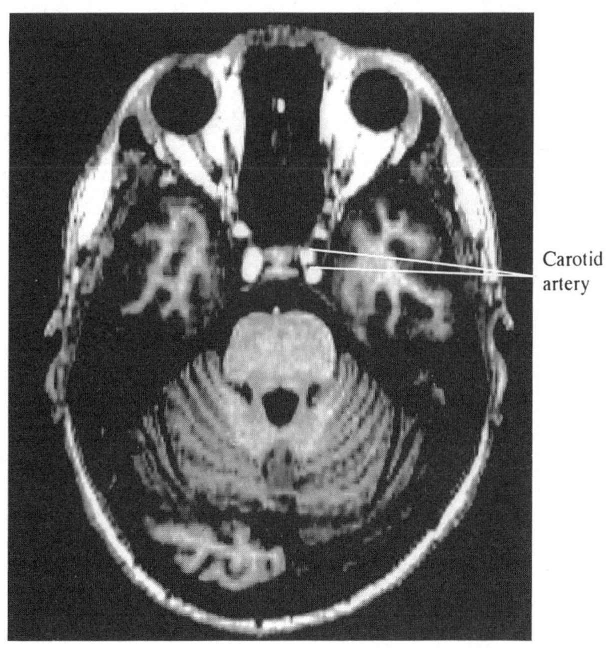
Carotid
artery

Fig. 5–19

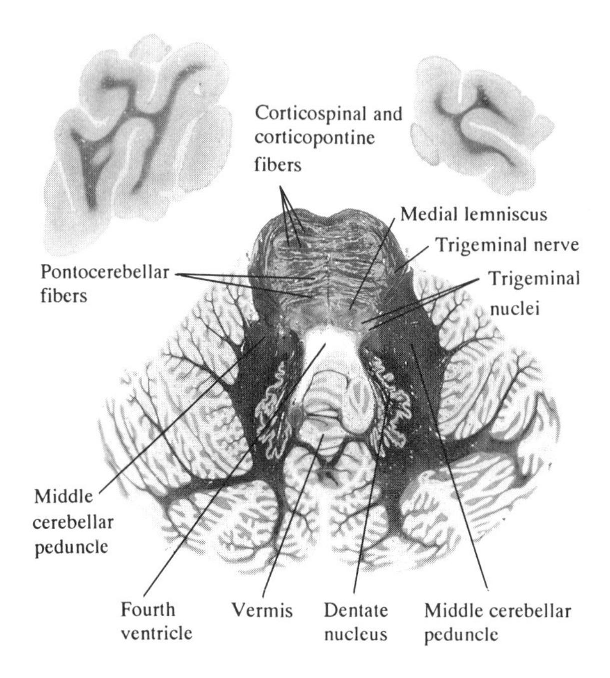
Corticospinal and
corticopontine
fibers
Medial lemniscus
Trigeminal nerve
Pontocerebellar
fibers
Trigeminal
nuclei
Middle
cerebellar
peduncle
Fourth
ventricle
Vermis
Dentate
nucleus
Middle cerebellar
peduncle

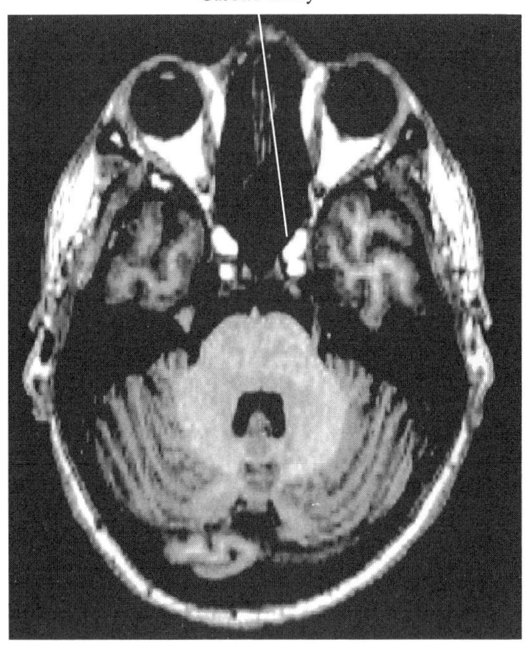
Carotid artery

Fig. 5–20

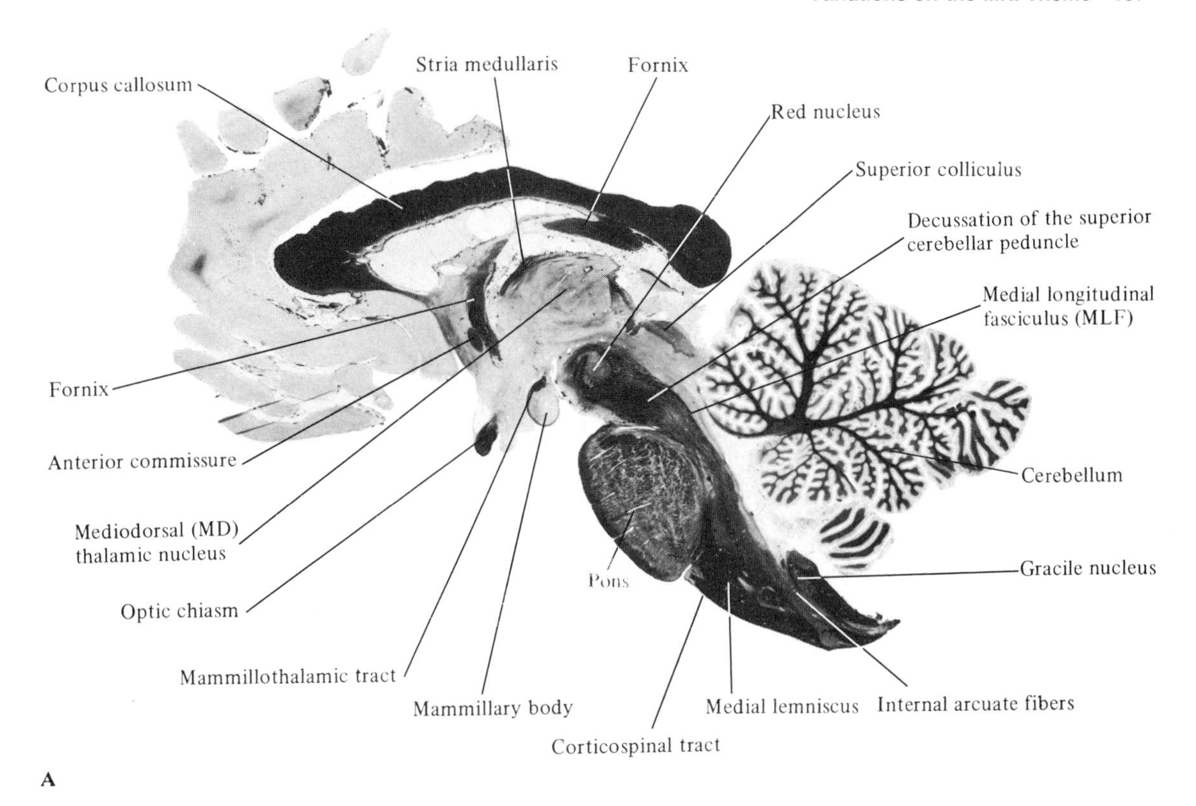

Ant. cerebral artery

Optic chiasm

Pituitary

Basilar artery

Straight sinus

Tonsil

B

Fig. 5–21A. Sagittal Sections of the Brain
(From the Yakovlev Collection.)

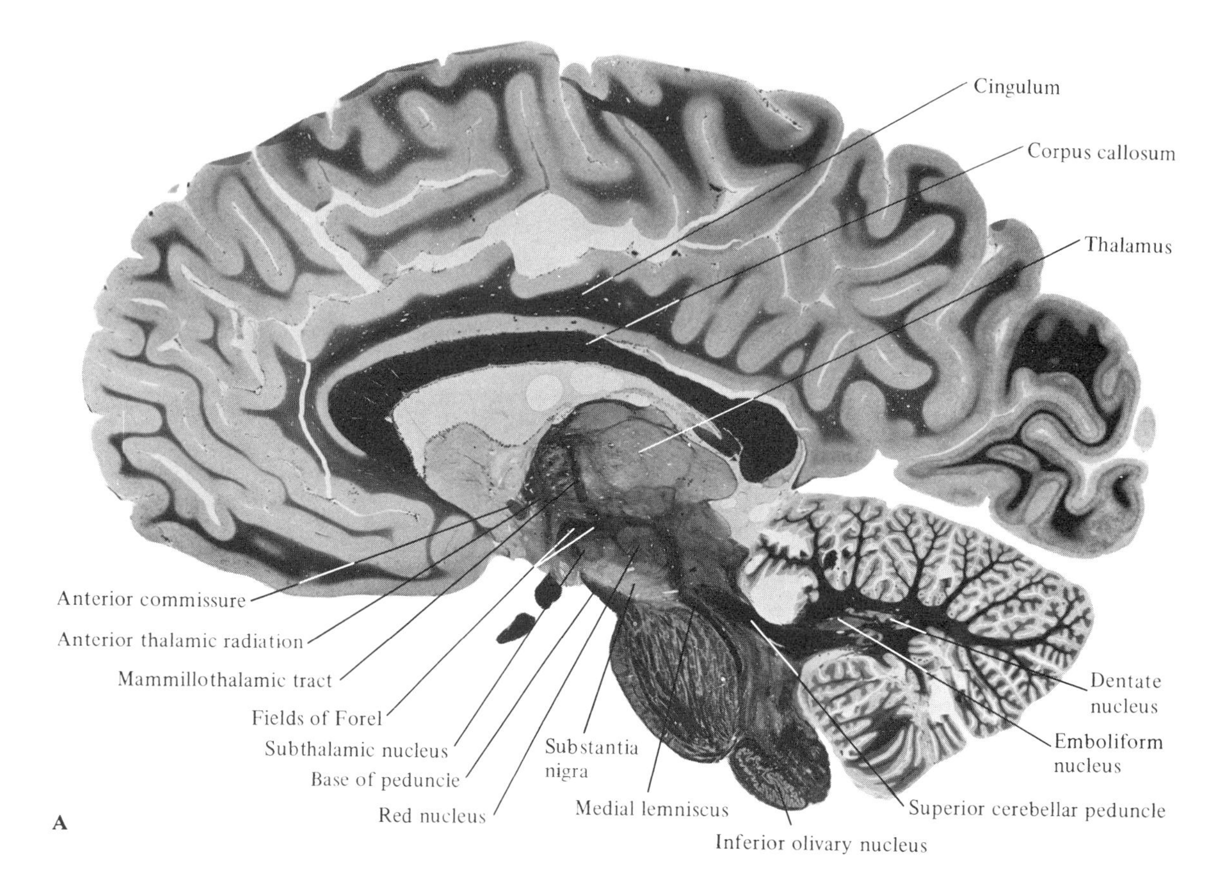

Cingulum
Corpus callosum
Thalamus
Anterior commissure
Anterior thalamic radiation
Mammillothalamic tract
Fields of Forel
Subthalamic nucleus
Base of peduncle
Red nucleus
Substantia nigra
Medial lemniscus
Inferior olivary nucleus
Superior cerebellar peduncle
Emboliform nucleus
Dentate nucleus
A

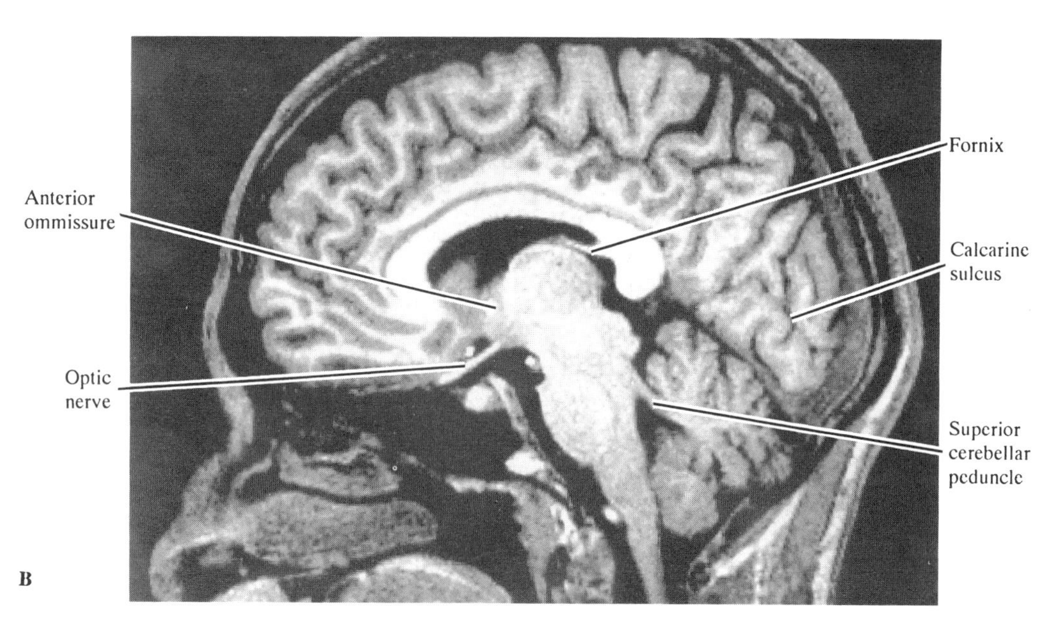

Anterior ommissure
Optic nerve
Fornix
Calcarine sulcus
Superior cerebellar peduncle
B

Fig. 5–22

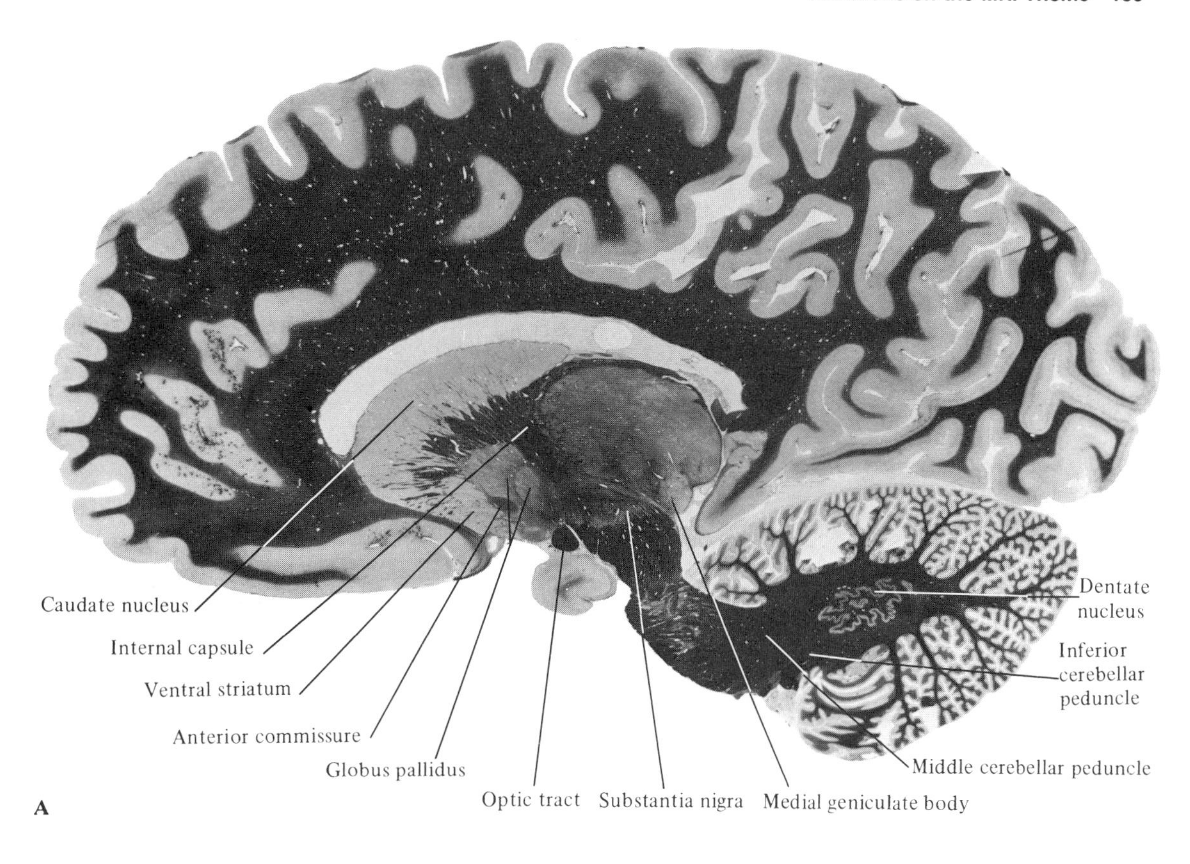
Caudate nucleus
Internal capsule
Ventral striatum
Anterior commissure
Globus pallidus
Optic tract
Substantia nigra
Medial geniculate body
Middle cerebellar peduncle
Inferior cerebellar peduncle
Dentate nucleus
A

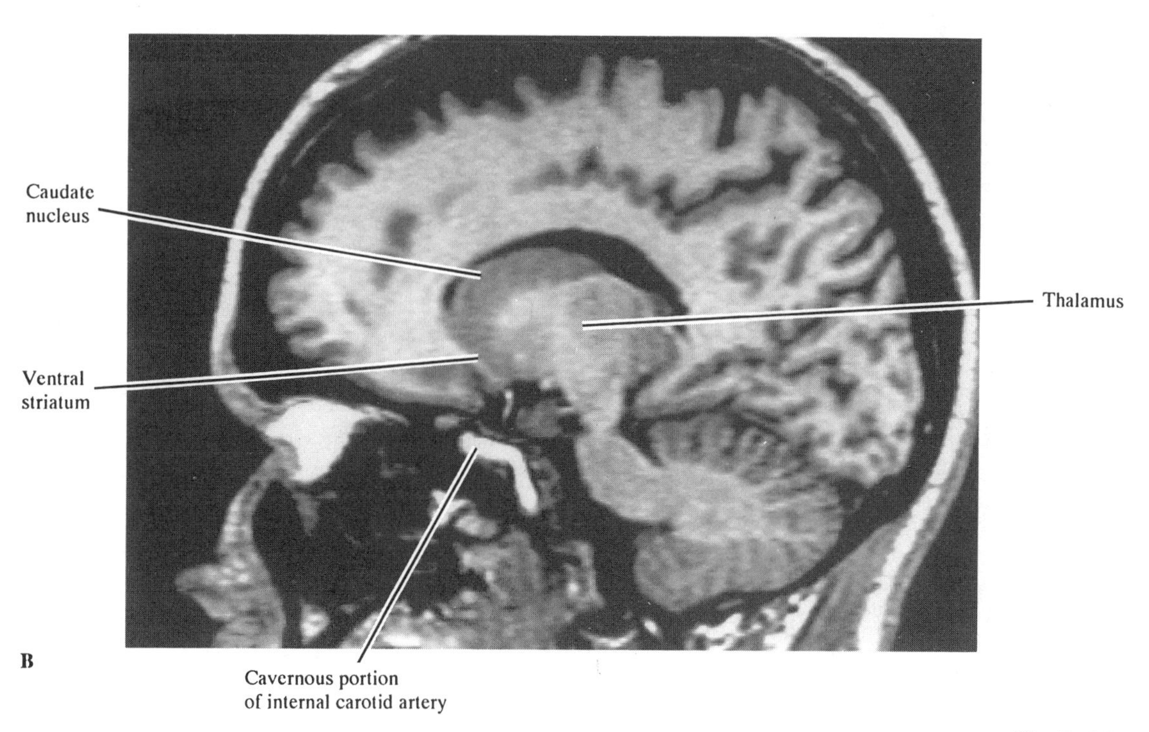
Caudate nucleus
Ventral striatum
Thalamus
Cavernous portion of internal carotid artery
B

Fig. 5–23

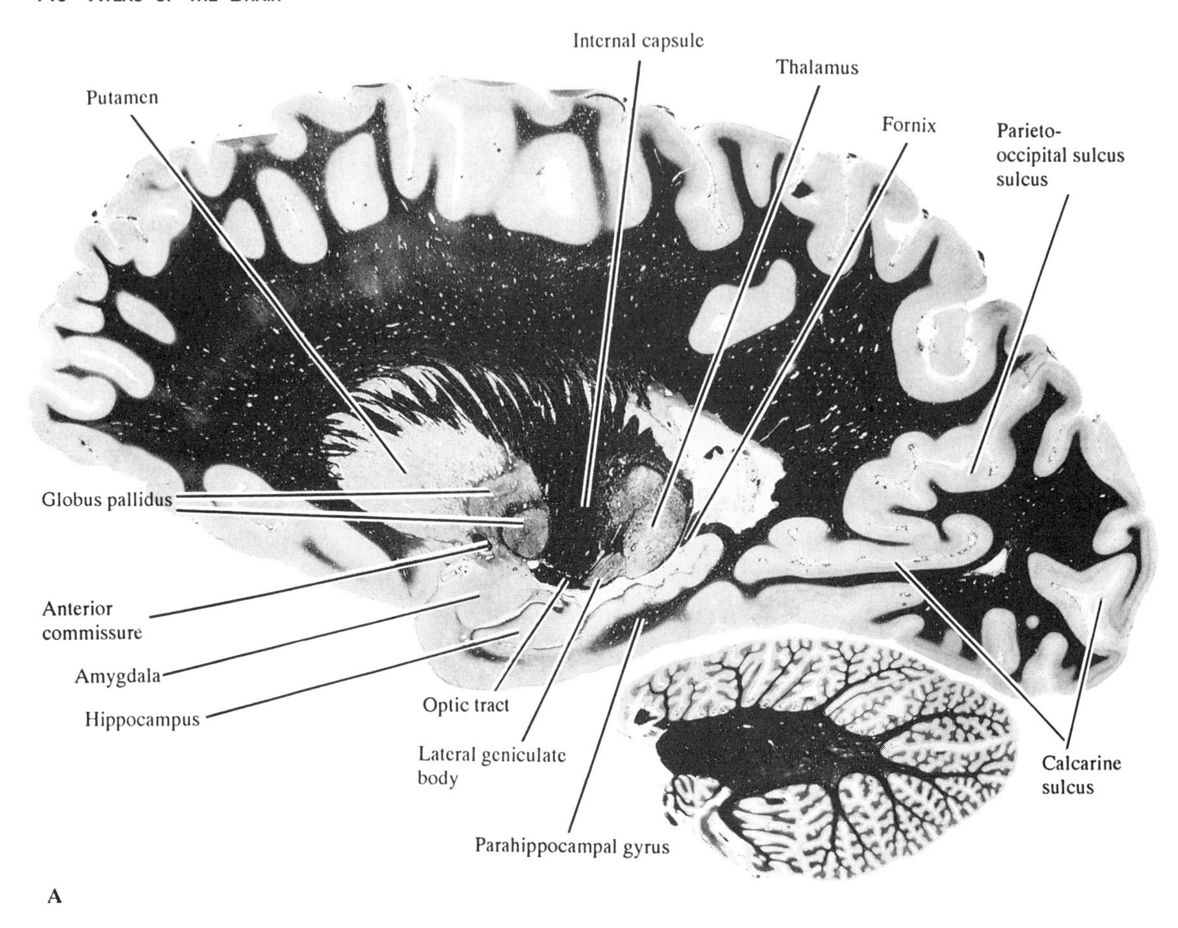

Internal capsule
Thalamus
Putamen
Fornix
Parieto-
occipital sulcus
sulcus
Globus pallidus
Anterior
commissure
Amygdala
Hippocampus
Optic tract
Lateral geniculate
body
Calcarine
sulcus
Parahippocampal gyrus

A

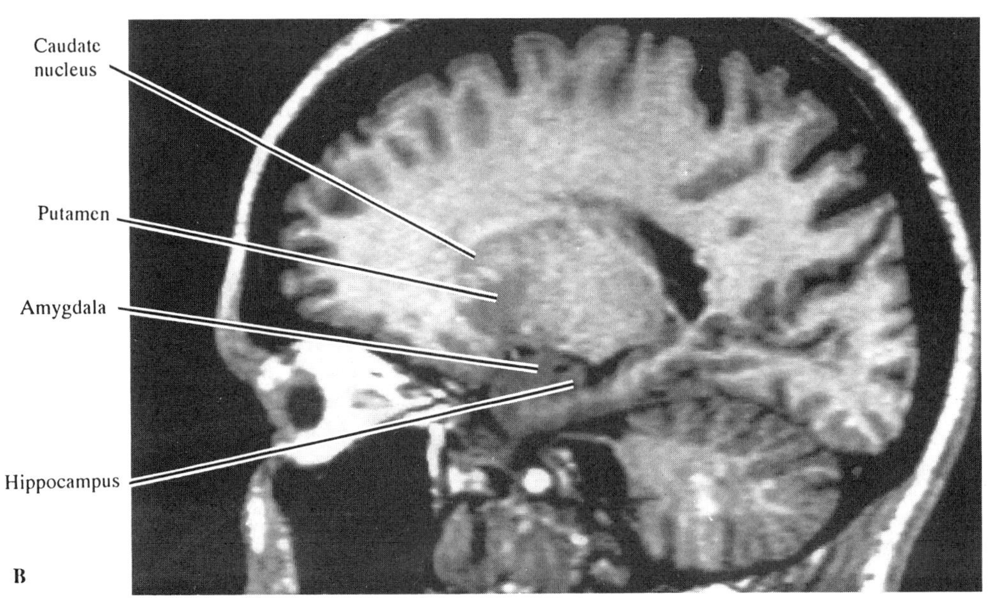

Caudate
nucleus
Putamen
Amygdala
Hippocampus

B

Fig. 5–24

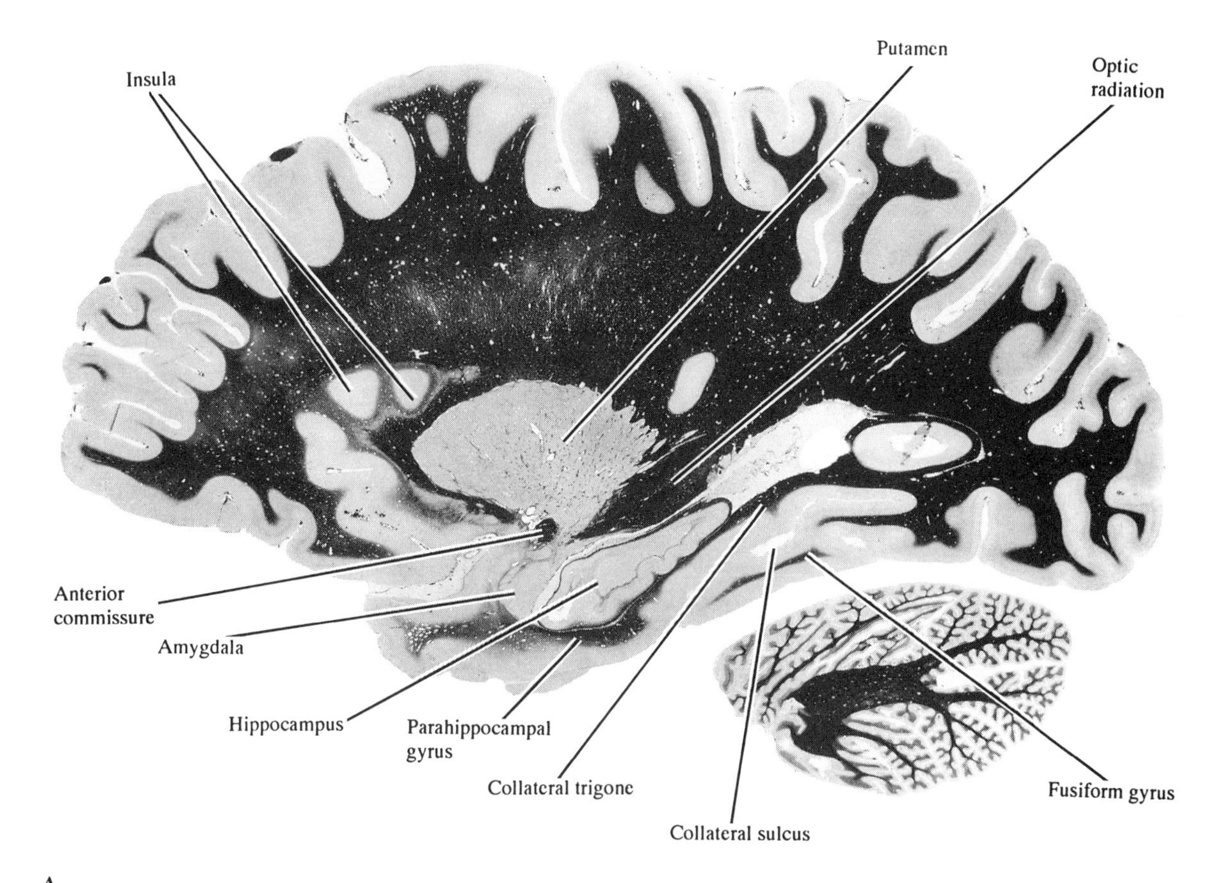

Insula
Putamen
Optic radiation
Anterior commissure
Amygdala
Hippocampus
Parahippocampal gyrus
Collateral trigone
Collateral sulcus
Fusiform gyrus

A

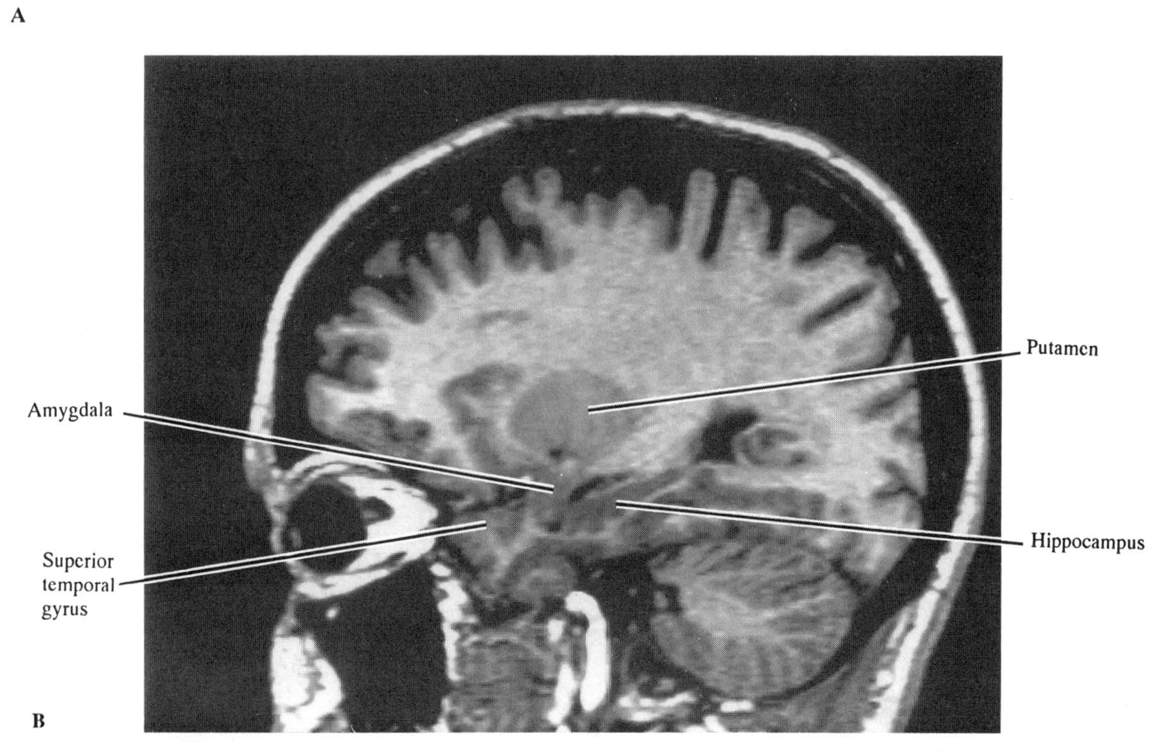

Amygdala
Superior temporal gyrus
Putamen
Hippocampus

B

Fig. 5–25

4

FUNCTIONAL NEUROANATOMY

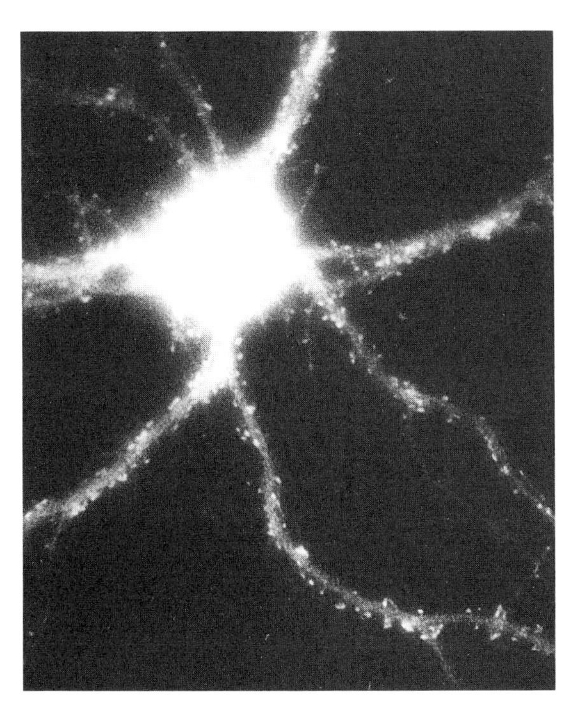

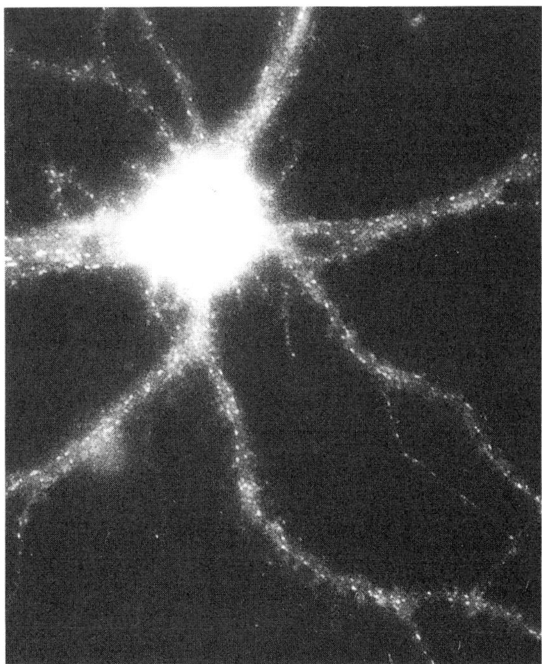

A cultured hippocampal neuron double-labeled with fluorescent antibodies against neurotransmitter receptors. The excitatory glutamate receptor GluR1 (A) is concentrated at postsynaptic sites on dendritic spines, whereas the inhibitory GABA receptor (B) is concentrated at postsynaptic sites on dendritic shafts. (Courtesy of Dr. Ann Marie Craig, University of Virginia.)

6

Neurohistology

The nervous system is composed of billions of nerve cells and neuroglial cells. The nerve cells, or neurons, which represent the basic information processing units, are connected to each other in highly specific patterns by means of neuronal processes, i.e., dendrites and axons of variable length and arrangement. The connections between neurons take place at synapses, which are specialized junctions for the transmission of nerve impulses. The functional significance of a neuronal microcircuit is a reflection of the morphology of its individual neurons, the way in which they are connected to each other, and the chemicals involved in the synaptic transmission.

Although basic information about the gross anatomic features of the brain and the spinal cord has accumulated through centuries of observation, it was not until the *light microscope* was refined in the nineteenth century that a successful study of the histologic structure of the nervous system was begun. Soon afterwards, techniques for the preparation and staining of histologic sections were developed that made it possible to visualize the different components of the nervous tissue and to study the distribution of neurons and the relationships between their processes.

Another great step forward in defining the structure of the nervous system occurred in the mid-twentieth century with the introduction of the *electron microscope*. In contrast to light microscopic techniques, which reveal only some components or aspects of the neuron or other nervous tissue elements, the electron microscope displays, with a high degree of magnification and resolution, all the different structures present in a section. The light and electron microscopic techniques are highly complementary, and the rapid advances in the elucidation of neuronal circuits and synaptic relations depend in large part on the successful combination of the two techniques.

Although the subjects of biochemistry, physiology, and molecular biology are essentially outside the scope of this text, it is nonetheless important to point out that the technical advances that have taken place in these fields have insured rapid advances in our understanding of how the cells in the nervous system develop and function; the molecular structure of a large number of specific proteins has been determined, and many of the physiologic and biochemical mechanisms responsible for synaptic transmission and signaling in the nervous system have been elucidated. As an increasingly more detailed picture of these events emerges, one is struck by their complexity and the means by which the neurons provide for great flexibility in their responses.

The Neuron

In spite of great variability in size and configuration, most neurons have in common certain morphologic features. To provide for efficient transmission of nerve signals and for the interaction of signals from various sources, the neurons have processes called *dendrites* and *axons*. The dendrites and the cell body usually receive input from other neurons or from receptor organs, and the axon, of which there is usually one per neuron, is specialized to conduct nerve impulses to other neurons or to effector organs (*muscles* or *glands*). The particular direction in the flow of information, i.e., from dendrites to axon, is known as the principle of dynamic polarization (see Neuron Doctrine, below). Although this classic view of the role played by axons and dendrites is true in most cases, there are important exceptions where dendrites may transmit the signal to another neuron, where axons may receive signals, or where hybrid processes perform both dendritic and axonal functions.

Transmission of the signal from one neuron to another occurs at specialized contact regions called *synapses*. This provides for a high degree of connectional specificity, but it is also in the specific arrangements of the synapses that much of the flexibility and potentials of the nervous system appear to reside.

Cell Body

The Cell Body Is the Trophic Center of the Neuron

The *cell body*, which consists of a large *nucleus* and the surrounding cytoplasm, the *perikaryon*, is the site for synthesis of most proteins. The function and survival of the neuronal processes are dependent on the integrity of the cell body.

The classic light microscopic method for the study of the cell body is the *Nissl method* (see below), which selectively stains the nucleus and one of the most characteristic cell components of the neuron, the *Nissl substance* (Fig. 6–1). The Nissl substance is basophilic and in the light microscope appears as dust-like material or as small or large granules. The Nissl substance is present in the cell body and the proximal dendrites, but is absent from the axon and in large part also from the axon hillock, especially in large neurons. The axon hillock is the conically shaped region from which the axon extends. Electron microscopic studies show that the Nissl substance (Fig. 6–1A) consists of concentrations of granular or *rough endoplasmic reticulum (RER)* and free *polyribosomes*. In the large Nissl bodies, the endoplasmic reticulum takes the form of parallel cisterns (Fig. 6–1B), "stacked one on top of another like pancakes" (Peters et al., 1991) and studded with polyribosomes. Ribosomal rosettes are also interspersed between the cisterns. The polyribosomes are responsible for the synthesis of proteins, which are in great demand for the maintenance of the elaborate neuronal processes. The granular endoplasmic reticulum is directly continuous with the agranular or *smooth endoplasmic reticulum*, which consists of

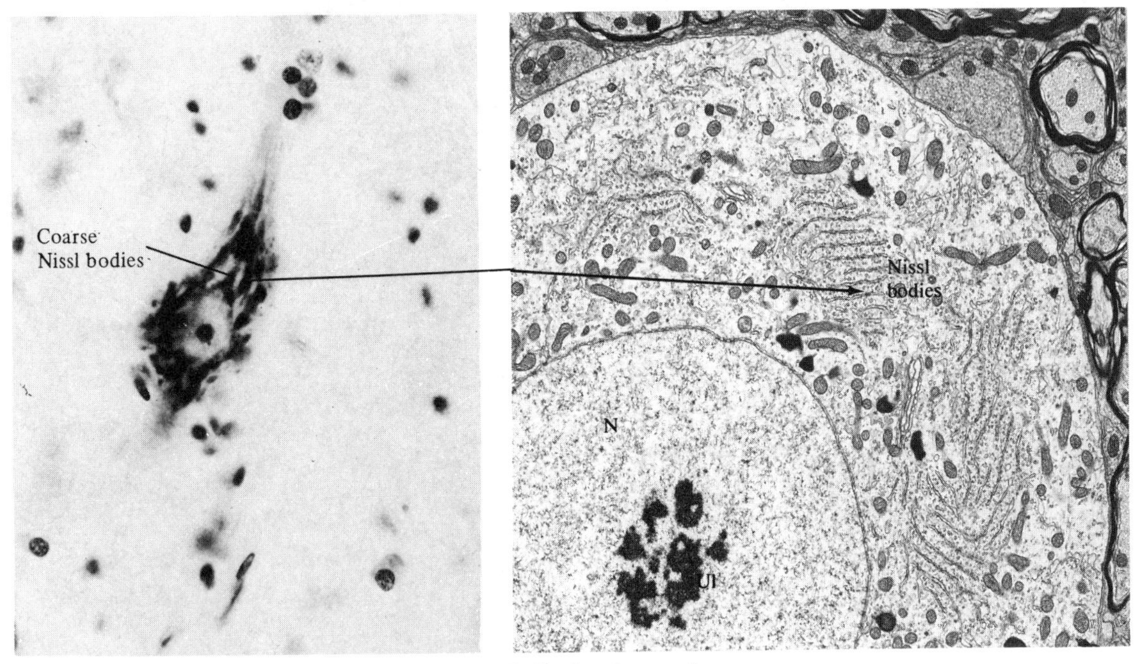

A Motor neuron

B Perinuclear region

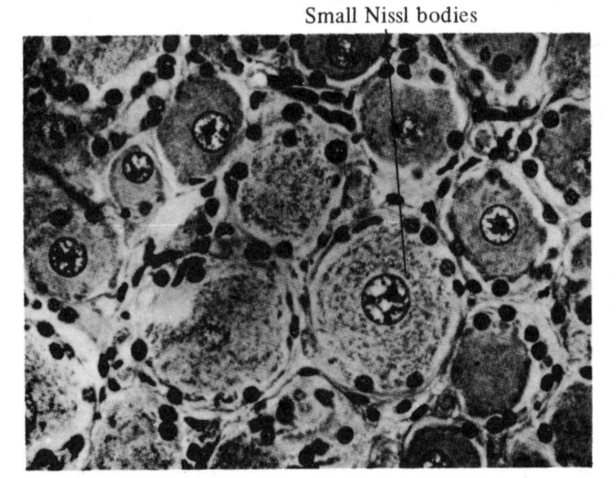

C Sensory neurons

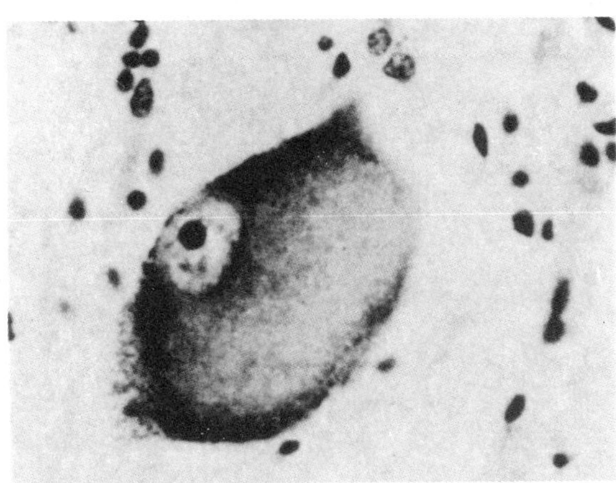

D Chromatolysis

Fig. 6–1. Neuronal Cell Bodies Stained with the Nissl Method

A. Motor neuron from the spinal cord.

B. Electron micrograph from the perinuclear region of a neuron shows that the Nissl substance consists of rough endoplasmic reticulum and free polyribosomes.

C. Sensory neurons from dorsal root ganglion.

D. Retrograde cell reaction with chromatolysis, cellular swelling, and peripheral displacement of the nucleus. (A. Photograph courtesy of H. Schroeder. C. from Bargman, W., 1977: Histologie und Mikroskopische Anatomie des Menschen. With permission of Thieme, Stuttgart. D. from Escourolle, E. and J. Boirier, 1978. Manual of Basic Neuropathology. W. B. Saunders Company. With permission of Masson et Cie, Paris.)

irregularly branching cisterns and tubules scattered throughout the neuron. Some cisterns have an unusually narrow lumen and are closely opposed to the plasma membrane, forming so-called *subsurface cisterns*.

The appearance of the Nissl substance varies between different types of neurons. Motor neurons have large Nissl bodies (Fig. 6–1A), whereas sensory neurons have smaller ones (Fig. 6–1C); small neurons often contain dust-like Nissl substance. The appearance of the Nissl substance also varies with the activity of the cell. The large Nissl bodies gradually diminish in size in neurons

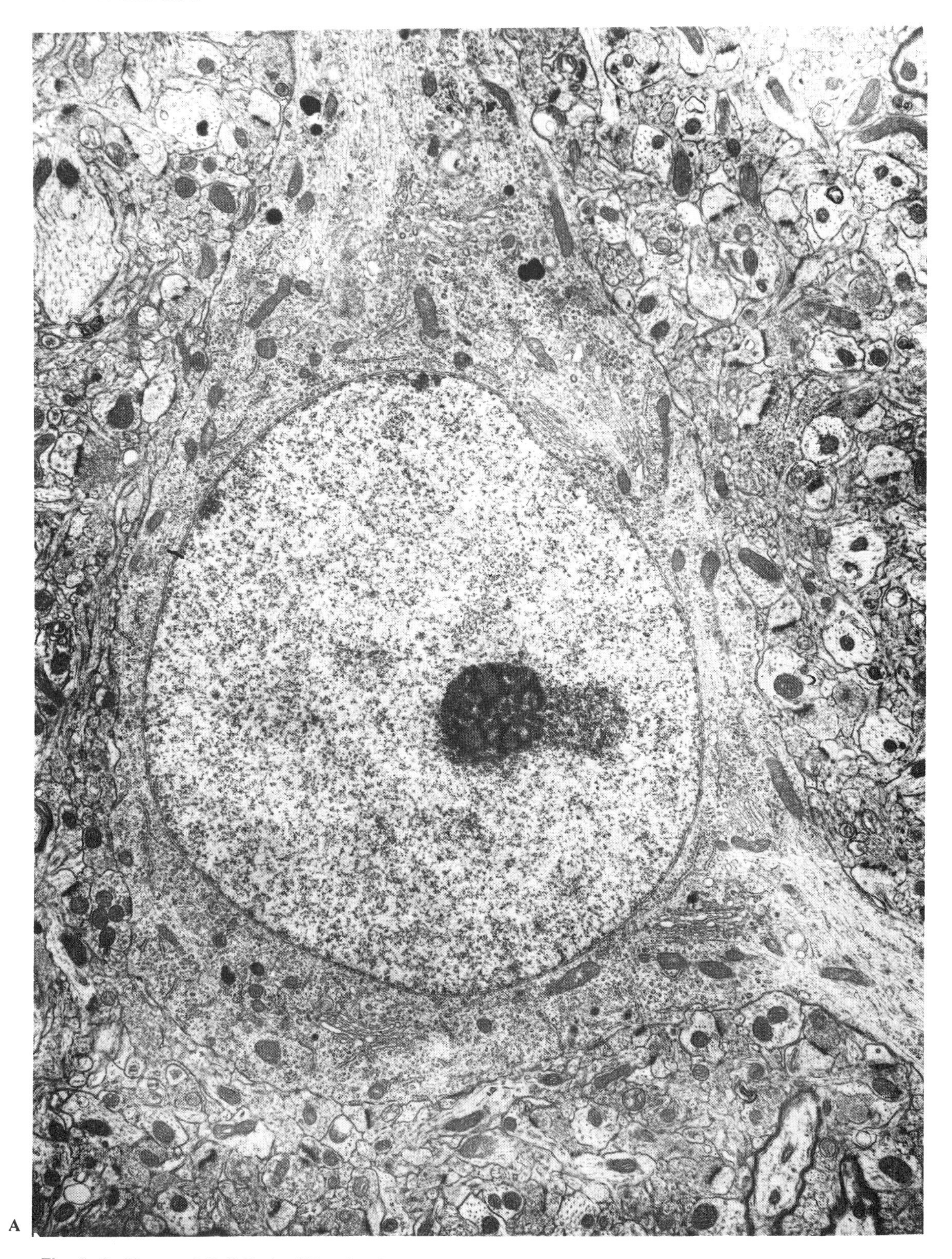

Fig. 6–2. Neuronal Cell Body: Ultrastructure

A. A pyramidal neuron of the cerebral cortex has been enlarged 10,000 times in this electron micrograph. The map in **B** identifies the various organelles. Mitochondria are colored *blue* and boutons from other neurons *red*. (Micrograph courtesy of A. Peters.)

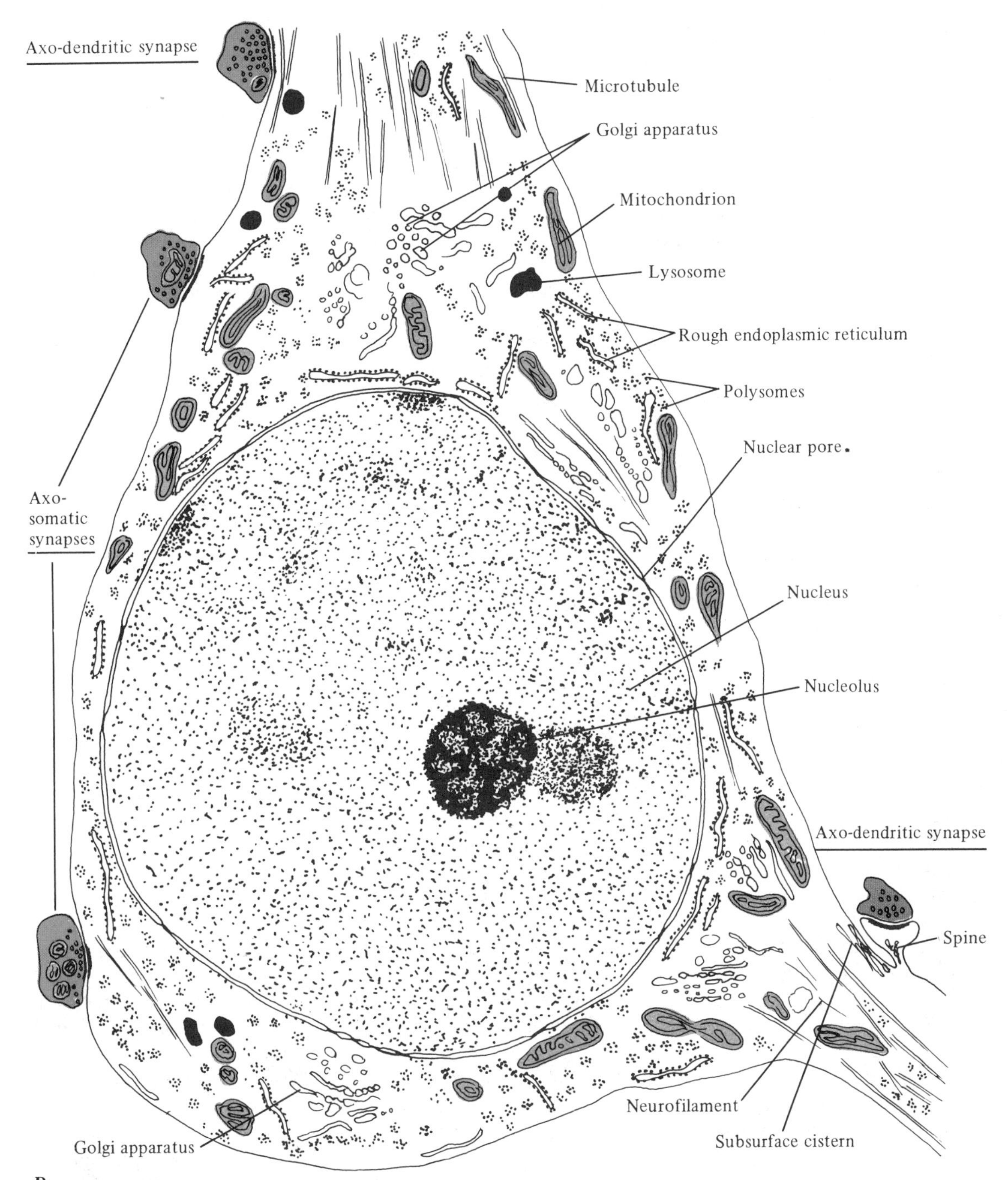
Axo-dendritic synapse
Microtubule
Golgi apparatus
Mitochondrion
Lysosome
Rough endoplasmic reticulum
Polysomes
Nuclear pore
Axo-somatic synapses
Nucleus
Nucleolus
Axo-dendritic synapse
Spine
Neurofilament
Subsurface cistern
Golgi apparatus

B

subjected to functional stress, and they disintegrate when the axon is injured or sectioned. The latter phenomenon, known as *chromatolysis*, may be part of a reconstructive process in response to injury; although the Nissl bodies disappear, the number of membrane-free polyribosomes as well as protein synthesis increases significantly. Therefore, the term *retrograde cell reaction* would be more appropriate for this process, which may also include swelling of the cell and displacement of its nucleus to the periphery of the perikaryon (Fig. 6–1D). The retrograde cell reaction was exploited in some of the very first experimental methods used in axon tracing (Chapter 7).

Neurons impregnated with one of the classic silver methods, the Bielschowsky method, display another characteristic structure, the *neurofibril*. The neurofibrils represent bundles of neurofilaments on which silver has been deposited. The neurofilaments (10-nm diameter), which can only be seen in the electron microscope (Figs. 6–2 and 6–5), give strength to the neuronal processes and support their shape. Collections of abnormal filaments, referred to as *neurofibrillary tangles*, are found especially in the large neurons in several brain regions in Alzheimer's disease (see Clinical Notes, Chapter 21) and some other neurodegenerative disorders. A small number of neurofibrillary tangles is also found in brains of normal elderly people.

The neurons display two other fibrillary structures, the *microtubules* and the *actin microfilaments*. As the name indicates, the microtubules are tubular structures (Figs. 6–3 and 6–5), with an outside diameter of about 25 nm. They play a key role in the intracellular transport of macromolecules (see Axonal Transport, below). The microfilaments (5–8 nm) or actin filaments, which are polymers of the globular protein actin, form a network underneath the plasma membrane. Actin filaments are especially prominent in growth cones (see Vignette, Chapter 2), where they form the basis for their motility. The neurofilaments, microtubules, and actin filaments form the main components of the *cytoskeleton*.

In addition to Nissl substance, neurofilaments, and microtubules, the nerve cells, like other cells, contain the Golgi apparatus, lysosomes, mitochondria, small peroxisomes, and various inclusions (Figs. 6–2A and B). Using his own silver method, Camillo Golgi (see Chapter 7) first discovered a perinuclear, net-like material in the cytoplasm, now generally referred to as the *Golgi complex* or *Golgi apparatus*. Electron microscopy reveals that this reticular complex consists of a system of flattened sacs surrounded by vesicles of varying size. The Golgi apparatus accumulates and segregates various secretory products, including those participating in the renewal of synaptic vesicles and the plasma membrane. Some of the neuropeptides that the neurons use as chemical messengers at the synaptic junctions are also packaged by the Golgi apparatus. The Golgi complex is also involved in the formation of the *primary lysosomes*, i.e., membrane-limited particles containing acid hydrolases. As part of the intracellular digestive system, the primary lysosomes fuse with material to be degraded, thereby turning into *secondary lysosomes* of varying size and shape. The large lipofuscin granules, the "wear-and-tear" or aging pigment, which appear with increasing age in nerve cells, represent insoluble residues of secondary lysosomes and mitochondria. *Mitochondria*, the principal energy source, are found in large numbers in the metabolically active nerve cells. These rodlet-shaped or spherical organelles vary in size and are distributed in a nearly random fashion throughout the neuron.

The chromatin-rich *nucleus*, which contains the genetic material (DNA) and a conspicuous *nucleolus*, is separated from the cytoplasm by a double-layered envelope (Fig. 6–2B). The two membranes of the *nuclear envelope* come together periodically to form *nuclear pores* for the passage of material, including ribosomal subunits and messenger RNA that carries the genetic information from the nucleus to the cytoplasm.

Surrounding the neuron is a thin, "excitable" *plasma membrane*, which provides the structural basis for the generation of electrochemical impulses and for the transmission of nerve signals. The plasma membrane appears in high-power electron micrographs as two dark bands separated by a lighter layer. It is in essence a lipid bilayer into which protein molecules are embedded (*membrane-spanning or integral membrane proteins*) or to which they are more or less loosely attached (*peripheral membrane proteins*). The proteins give the membrane its specific properties as reflected in a variety of ion channels, receptors, adhesion molecules, anchored proteins, and special enzymes.

Cell Processes

The configuration of the cell processes can be studied in detail by the *Golgi method*. A seemingly random and as yet unexplained impregnation of only a few out of hundreds of neurons has made the Golgi method an extraordinarily valuable technique. Because of the small number of nerve cells

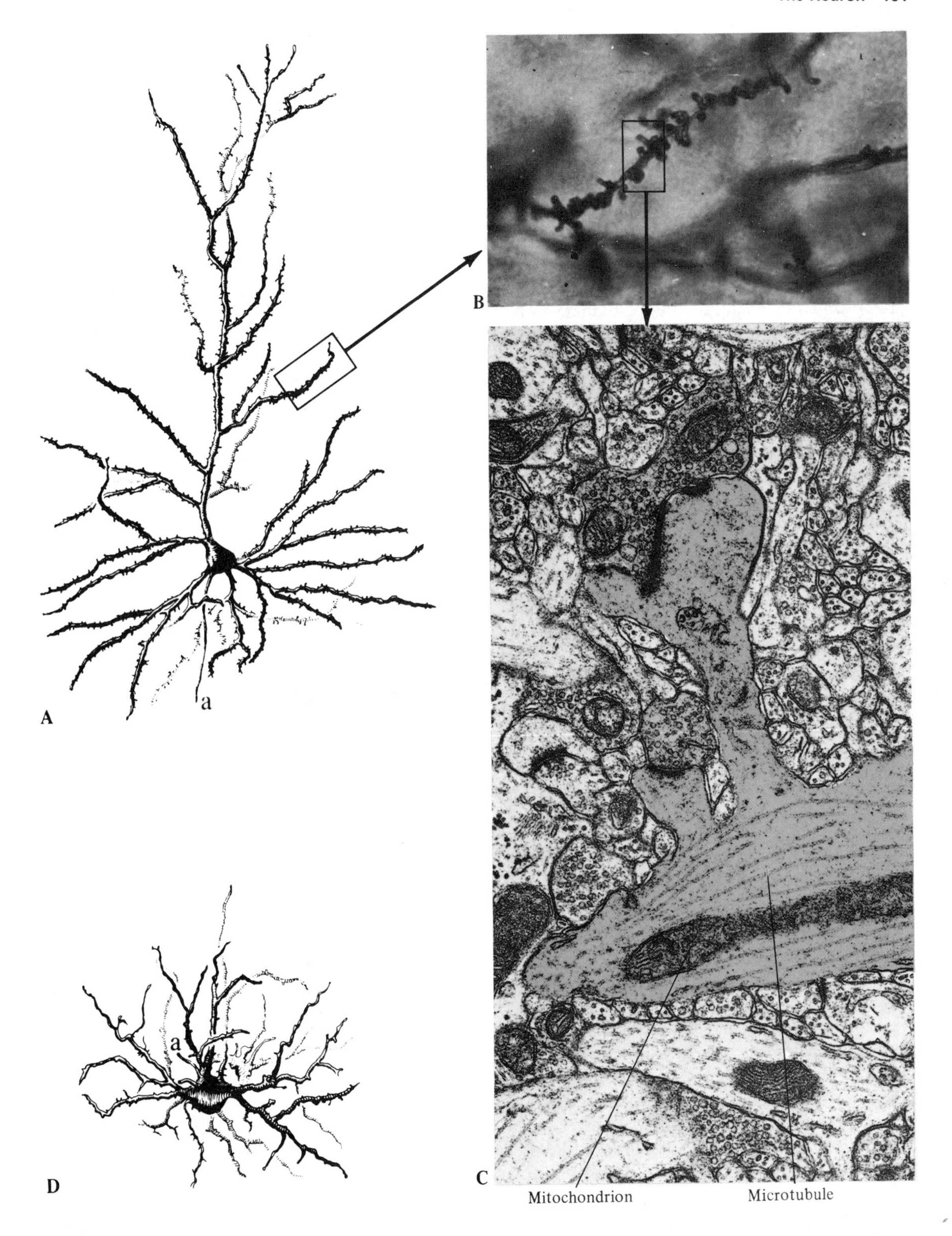

Fig. 6–3. Neuronal Processes

A. Pyramidal cell from the cerebral cortex stained with the Golgi method. *a*, axon.

B. Enlargement of a dendritic spine.

C. The electron microscope reveals that the dendritic spines are specialized for synaptic contacts. Dendrite and dendritic spines, *blue;* boutons, *red.*

D. Stellate cell from the dentate nucleus of the cerebellum. *a*, axon. (A and D from Peters, A., S. L. Palay and H. de F. Webster, 1976. The Fine Structure of the Nervous System. With permission of the authors and W. B. Saunders Company, Philadelphia.)

stained, the sections can be 100–300 μm thick, and this makes it possible to obtain a three-dimensional view of the cell and its processes (Figs. 6–3A and D).

According to the number of their processes, neurons can be classified as *multipolar*, *bipolar*, or *unipolar*. Most cells in the CNS are multipolar, and typical examples of such cells are shown in Figs. 6–3A and D. Bipolar cells are also found in many locations, including the cerebral cortex, thalamus, retina (Fig. 13–3), olfactory epithelium (Fig. 12–2), and the cochlear and vestibular ganglia. Unipolar cells are present in most sensory ganglia and have a single process, which divides in a T-like manner close to the cell body; one branch projects to the periphery and the other enters the CNS through the posterior root. These cells are sometimes referred to as *pseudounipolar* because they develop from bipolar neurons in which the initial parts of the two processes move toward each other and the parent cell body retracts, forming a single root process leading to the two branches.

Another classification of neurons is based on the length of their axons. Cells with long axons are referred to as *Golgi type I neurons* and those with short axons as *Golgi type II neurons*. Some neurons, such as the amacrine cells in the retina (Fig. 13–3) and the granule cells in the olfactory bulb (Fig. 12–2), do not fit this classification for the simple reason that they lack an axon. Their hybrid processes have both presynaptic and postsynaptic functions.

Dendrites

The dendrites constitute protoplasmic extensions of the cell body, and the initial parts of the dendrites contain essentially the same organelles as the rest of the cell body. The dendrites gradually taper off as they arborize in a tree-like manner in the vicinity of the cell body (Figs. 6–3A and D). The profuse branching explains why most of the neuronal surface area usually is associated with the dendrites. Large dendrites can usually be recognized in electron micrographs on the basis of their lack of a myelin sheath and the presence of a large number of microtubules (Fig. 6–3C) and scattered polyribosomes. Small dendritic branches, on the other hand, can be difficult to distinguish from small unmyelinated axons; both types of processes contain microtubules, but small axons usually contain more neurofilaments than small dendrites. Moreover, small axons usually do not receive synapses and they typically are collected in bundles (Fig. 6–9), whereas small dendrites often travel through the tissue as single structures, receive some synapses, and may have irregular contours.

The dendrites are eminently suited for the reception of a large number of axon terminals. Indeed, the dendrites of many neurons are studded with thorn-like processes, *dendritic spines* or *gemmules*, which are specialized for synaptic contacts (Figs. 6–3B and C). Occasionally, spines can also emanate from cell bodies. Although the spines may have different sizes and shapes to match the axon terminals, many spines are characterized by a rather slender stalk with a bulbous end (pedunculated spines); others are simple (sessile spines) or complex protrusions (excrescences). Isolated spines can usually be recognized in electron micrographs by their fluffy filamentous material and by the absence of other cytoplasmic organelles such as microtubules and mitochondria (Figs. 6–3C, 6–8, 6–9). Cisterns of smooth reticulum, however, often enter the stalk (Fig. 6–3C), and special accumulations of sacs and cisterns form the spine apparatus, which is the most conspicuous organelle of the spine. The spines appear to reach out for incoming axons, but their function is still somewhat of a mystery. They are very sensitive to functional and environmental disturbances; e.g., visual deprivation significantly reduces the number of spines on the cortical neurons that normally receive visual input. The development of spine synapses, like the elaboration of dendritic and axonal processes in general, is of obvious importance for the establishment of the appropriate neuronal circuitry. If these developmental events are disturbed, e.g., by genetic or metabolic disorders, the result may be a paucity of dendritic processes with a reduced number of strikingly abnormal spines, a likely concomitant of mental retardation.

The shape of the dendritic tree and its spines is often among the most characteristic morphologic features of a neuron. *Pyramidal cells*, which are found in all areas of the cerebral cortex, have cell bodies that are conical in shape (Fig. 6–3A), and are further characterized by two systems of spiny dendrites: a long apical dendrite, which is directed toward the surface, and several basal dendrites. The *Purkinje cells* in the cerebellum (Fig. 6–4A) have one of the most remarkable dendritic arborizations in the CNS. The dendritic tree, laden with over one hundred thousand spines, arises from one major dendritic trunk at the apical pole of the cell body. The very impressive dendritic arborization spreads out in a vertical plane at right angle to

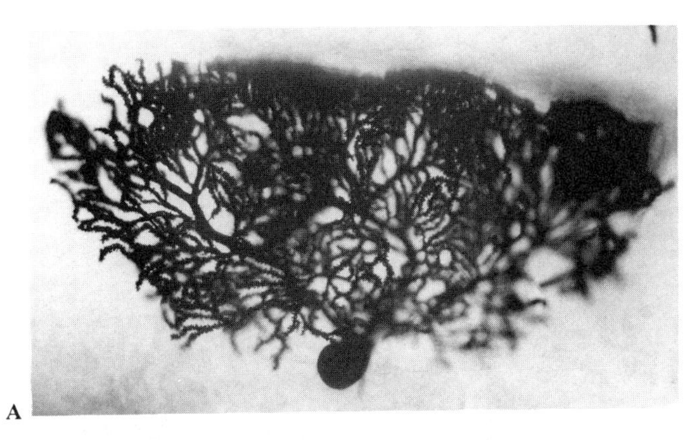

A

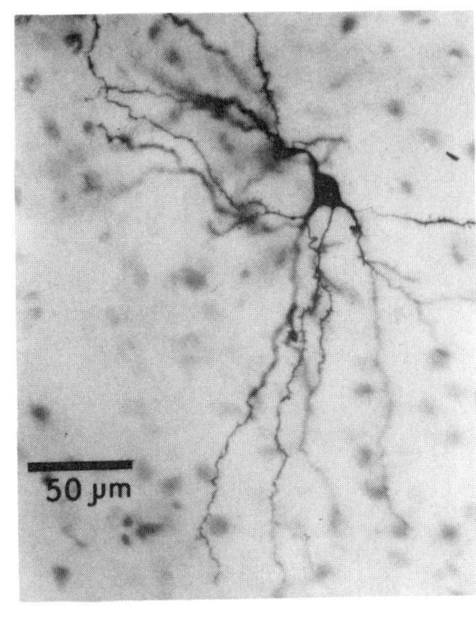

B

Fig. 6–4. Neuronal Processes
A. Purkinje cell from the cerebellum stained with the Golgi method. (Micrograph courtesy of S. Palay. From Palay, S. L. and V. Chan-Palay, 1974. Cerebellar Cortex. With permission of Springer-Verlag, New York–Heidelberg–Berlin.)

B. Neuron from the striatum visualized by a specific staining reaction following intracellular injection of horseradish peroxidase. (Micrograph courtesy of S. T. Kitai.)

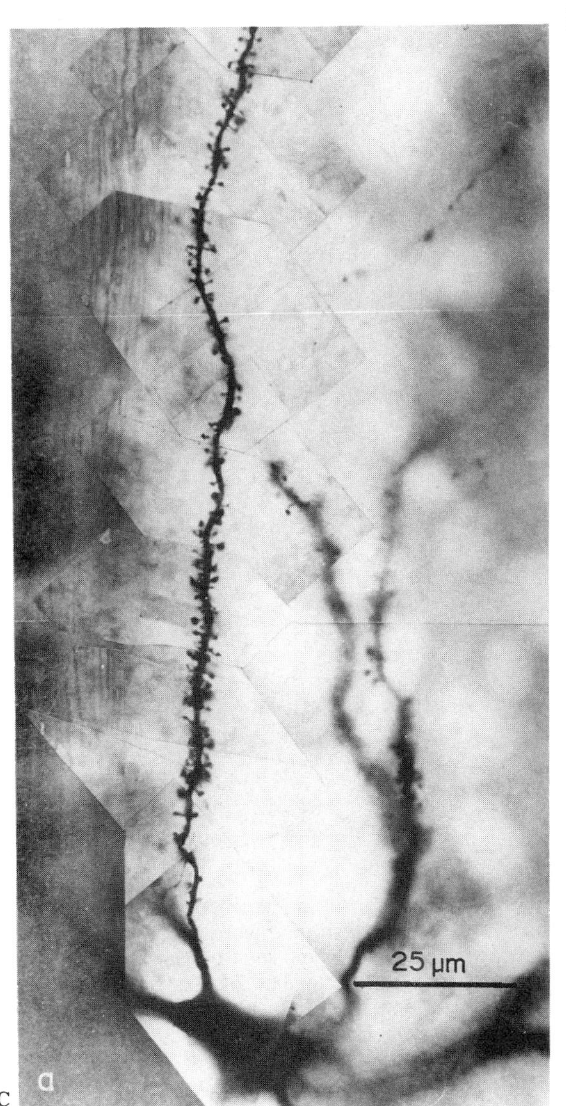

C

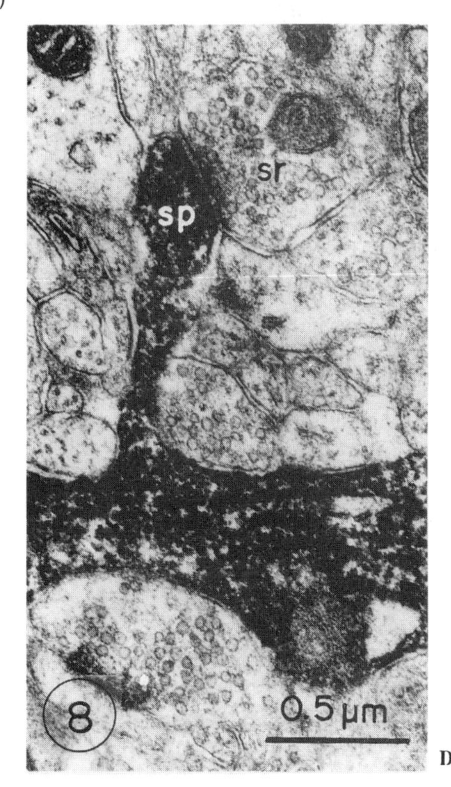

D

C. Photomontage of a single dendrite of a spiny striatel projection neuron intracellularly stained with HRP. (Ffrom Wilson, C.J. et al., 1983. J. Neuroscience, 3:383-398. With permission of Elsevier Sciene Publishers, New York.)
D. Electron micrograph of a dendritic spine (SP) from the dendrite shown in **C**. The spine is contacted by a containing small round synaptic vesicles. This kind of synapse is formed by most axons originating in cerebral cortex and thalamus. (Ffrom Wilson, C.Y. and Groves. P.M., 1980. J. Comp. Neurol. 194:599-615. With permission of Wiley-LIss, a division of John Wiley, New York.)

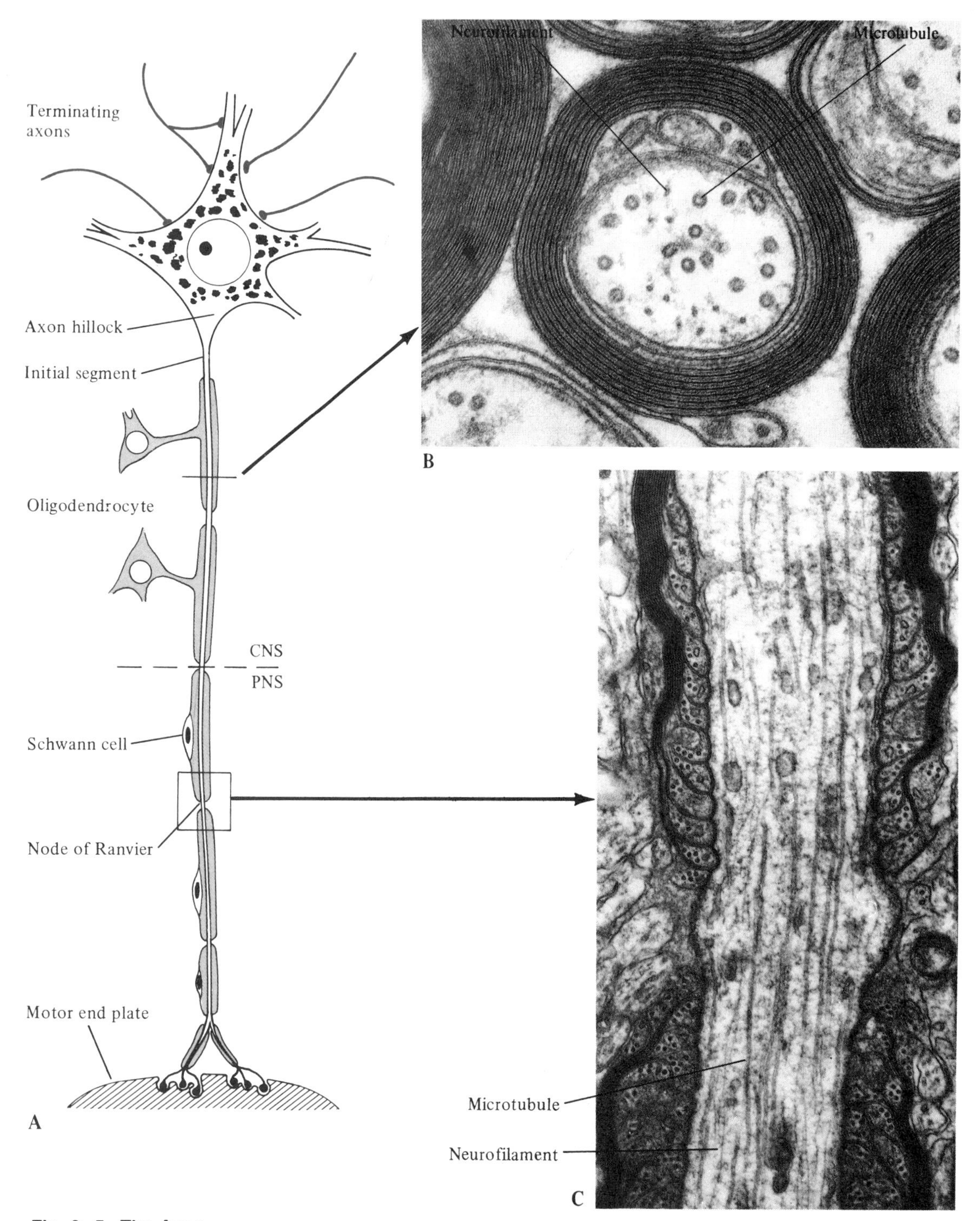

Fig. 6–5. The Axon

A. Schematic drawing of a motor neuron. The myelin sheath surrounding the axon is interrupted at regular intervals, and appears as a series of tubes in the light microscopic preparation.

B. Electron micrograph of a transversely sectioned axon indicates that the myelin sheath consists of thin lamellae, which form a regular pattern of concentric thin and thick lines. (Courtesy of Dr. Cedric Raine, Department of Neuropathology, Albert Einstein College of Medicine.)

C. Electron micrograph of a longitudinally cut axon at a node of Ranvier. When the myelin sheath approaches the node, it becomes increasingly thinner as successive lamellae terminate. Note that the axon contains both microtubules and neurofilaments. (C from Peters, A., S. L. Palay and H. de F. Webster, 1976. The Fine Structure of the Nervous System. With permission of the authors and W. B. Saunders Company, Philadelphia.)

the longitudinal axis of the cerebellar folium. The *stellate cells*, which appear in many parts of the CNS, have the dendrites extended about equally in all directions (Fig. 6–3D), and certain types of stellate cells are nearly devoid of dendritic spines. The most common neuron in the striatum is a medium-sized *multipolar neuron* with spine-laden dendrites (Fig. 6–4B). Other characteristic cell types are the *mitral cell* in the olfactory bulb (Fig. 12–2)—the name is derived from the fact that its cell body is reminiscent of a bishop's mitre—and the *cerebellar basket cell*, whose axon forms a series of axonal arborizations surrounding the Purkinje cells like baskets (Fig. 17–1).

The Axon

The axon arises from the cell body (Fig. 6–5) or from one of the main dendrites. Whereas the dendrites usually taper off gradually as they extend from the cell body, the axon acquires a relatively small diameter near its point of emergence from the axon hillock, and the rest of the axon has a rather uniform diameter. The first segment of the axon, the *initial axon segment*, is the site where the nerve impulse or action potential is automatically initiated, once the local potential exceeds threshold. Distinctive features of the initial axon segment are bundles of microtubules, bound to each other by cross-bridges, and the undercoating of the plasma membrane by electron-dense granulofibrillar material (Fig. 6–11B). A similar undercoating is present at the nodes of Ranvier (see below), and there is reason to believe that the undercoating marks the site of the high current density responsible for the initiation and propagation of the action potential in myelinated fibers. The axonal membrane at the initial axon segment and the node of Ranvier contains an extremely high density of sodium channels that may be held in place by the proteins of the undercoating. If the axon is myelinated, the myelin sheath begins just distal to the initial segment.

The length of the axon and the number of its collateral branches vary greatly. Neurons that project from one structure to another in the CNS, as well as the large motor neurons that innervate the striated muscles, have long axons (Golgi type I neurons); some of them reach from the lower end of the spinal cord to the muscles of the foot. Such neurons are referred to as *projection neurons*. The neurons that integrate information within a specific region of the CNS, the *local circuit neurons*, usually have short axons (Golgi type II neurons) that often arborize profusely in the vicinity of their cell bodies. Some axons proceed from one region of the CNS to another without emitting collateral branches, whereas the collateralization and branching patterns in other axons are quite elaborate. Recurrent collateral inhibition is a well-known physiological phenomenon, which is dependent on the presence of a special type of collateral, referred to as *recurrent collaterals*. Such collaterals, which emanate from the proximal part of the axon, activate inhibitory interneurons, which in turn innervate the neurons giving rise to the collaterals as well as their neighbors.

Near its point of termination, the axon ramifies to form preterminal branches, whose distal ends are enlarged to form *axon terminals* or *boutons terminaux*. The bouton contains the presynaptic part of the synapse (see below), and it is here that the nerve impulses cause the release of a neurotransmitter, which in turn affects another neuron or a target cell in the body. Unmyelinated axons and preterminal branches may also form a series of varicose terminals, *boutons en passant* (Fig. 6–6), along part or even most of their course.

Neurofilaments (Figs. 6–5B and C) are common in axons and, in general, are rare in dendrites. Notable exceptions are the large motor neurons in the spinal cord and the large pyramidal neurons (Betz cells) in the cerebral cortex, whose dendrites contain large numbers of neurofilaments. Ribosomes, on the other hand, are in general not present in the axoplasm, so that the proteins needed for the turnover of axonal components must be transported from the cell body.

Myelinated Axons

The *myelin sheath* is in effect a spiraling lipoprotein sheath. It serves as an electric insulator that considerably increases conduction velocity in larger axons. Axons larger than 1 μm are usually myelinated, and they are easily recognized in the electron microscope (Fig. 6–5B). The myelin sheaths, which are derivatives of the plasma membrane of oligodendrocytes in the CNS and of Schwann cells in the PNS, are interrupted at regular intervals by gaps, the *nodes of Ranvier* (Fig. 6–5). Each segment of myelin is therefore called an internode. The length and thickness of the internode vary with the thickness of the axon. Whereas in the peripheral nerves, one Schwann cell provides the myelin for one internode, many internodes in different axons are formed by a single oligodendroglial cell in the CNS (Fig. 6–7). This may,

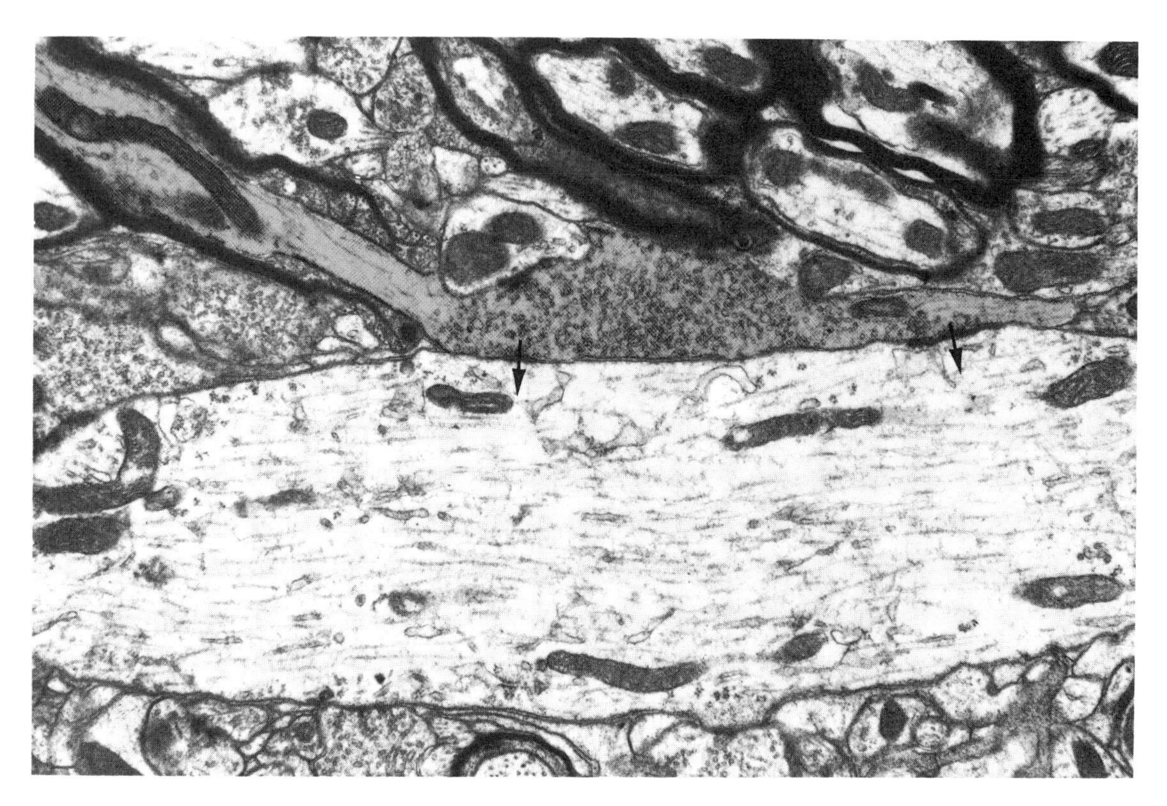

Fig. 6–6. Bouton en Passant
Axon (*red*) losing its myelin sheath and forming bouton en passant on dendrite with synaptic junctions (*arrows*). (Micrograph courtesy of E. Mugnaini.)

at least in part, explain the scarce remyelination following demyelinating disorders such as multiple sclerosis (see Clinical Notes).

Because the largest fibers have the longest internodes, and the nerve impulse jumps from node to node (saltatory conduction) in myelinated fibers, the length of the internode is of considerable functional importance. The largest fibers, which may have a conduction velocity of up to 100 m/sec, have internodes more than 1 mm in length. At its point of termination, a myelinated axon gradually loses its myelin sheath to form an axon terminal or a number of preterminal branches with boutons.

As indicated by the sensory and motor deficits that usually occur in patients with demyelinating disorders, the myelin sheath is of great importance for the functional integrity of the axon.

Axonal Transport[1]

Since the axoplasm is devoid of ribosomes, protein synthesis cannot take place in the axon, which is therefore dependent on a continuous delivery of various components from the perikaryon. The somatofugal movement of various substances occurs at different rates, and it has proved convenient to divide the transport system into two parts. The proteins needed for the continuous renewal of the axonal matrix, including microtubule and neurofilament proteins, are carried at a rate of less than 4–6 mm/day, referred to as *slow axonal transport*. Membrane-bounded structures, e.g., vesicles, mitochondria, and elements of the

1. Axonal transport was discovered in the 1940's by the Austrian-born scientist Paul Weiss and his colleagues, who noticed that if a peripheral nerve was ligated, there was a gradual thickening of the nerve proximal to the constriction. They correctly interpreted this phenomenon as a reflection of a continuous flow of axoplasm that had been halted by the constriction. Although they coined the term "axonal flow" for this phenomenon, the term "axonal transport" is now preferred.

The discovery of the axoplasmic transport phenomenon has greatly improved our understanding of many fundamental neurobiologic events, and it has served as a powerful stimulus for research on neuronal functions. Further, some of the most useful techniques for the tracing of neuronal pathways are based on the use of anterograde and retrograde axonal transport.

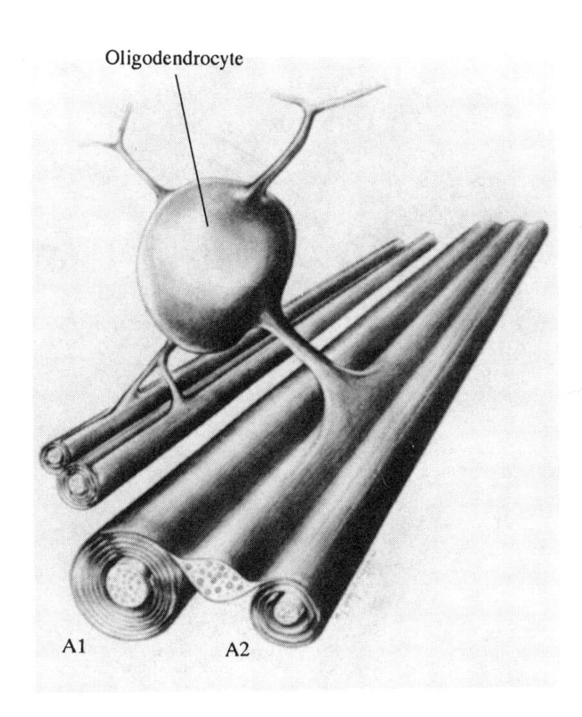

Fig. 6–7. Formation of Myelin Sheath
A possible three-dimensional relationship between the myelin sheaths A1 and A2 and their parent oligodendrocyte cell body. The shared outer tongue process lies between its two sheaths and is connected to the cell body by a single process. (From Friedrich, V. L. Jr., and E. Mugnaini, 1983. Myelin sheath thickness in the CNS is regulated near the axon. Brain Res. 274:329–331. With permission from Elsevier Biomedical Press: Amsterdam.)

agranular reticulum, are transported in a somatofugal direction at a rate of 50–400 mm/day, a process known as *fast anterograde axonal transport.* Rapid transport also occurs in the opposite direction: membrane material targeted for replacement is engulfed by lysosomes and undergoes *retrograde axonal transport* back to the cell body for degradation and recycling.

The Cytoskeleton and the Motor Proteins. The concept of the cytoskeleton is central to our current understanding of the axoplasm and axonal transport. The most prominent components of the cytoskeleton are the neurofilaments and microtubules, which are joined to each other by side arms or cross-bridges. This cytoskeletal matrix, in which the membrane-bounded cytoplasmic organelles are embedded, thus forms a rigid support structure for the neuron. The microtubules, which also serve as tracks for the rapid transport of organelles, are polar structures in which the opposite ends have distinct properties. For instance, one end of the microtubule, referred to as the plus end, grows faster than the other end, the minus end, during polymerization. It is estimated that the large majority of the microtubules in axons have their plus ends directed toward the peripheral end of the axon. The movement of organelles alongside the microtubules is thought to be accomplished by the aid of energy-transducing proteins, kinesin and dynein, which serve as unidirectional motors as they move in opposite directions along the microtubules. Dependent on the intended direction of movement, the organelles attach to one or the other of these motor molecules. Organelles bound to kinesin, which moves toward the plus end of the microtubule, are transported in the anterograde direction, whereas those bound to dynein, which moves toward the minus end, are transported in the retrograde direction.

The flow along the axon also has its downside in the sense that it can provide a mechanism for toxic and biologic agents, e.g., tetanus toxin or viruses, to reach the CNS from the periphery (see Clinical Notes).

The Synapse

The synapse consists of specialized *presynaptic* and *postsynaptic membranes*, separated by a 10- to 30-nm-wide *synaptic cleft.* The most characteristic features of the presynaptic element are the *presynaptic grid* and the accumulation of a large number of *synaptic vesicles*, some of which usually cluster in front of the presynaptic membrane. The most prominent feature of the postsynaptic membrane is the electron-dense material adhering to its cytoplasmic face. The presynaptic and postsynaptic membranes together with the synaptic cleft are referred to as the *synaptic junction.* The presence of dense material on the cytoplasmic sides of the opposing membranes makes the active zone or release site of the synapse easily identifiable. The

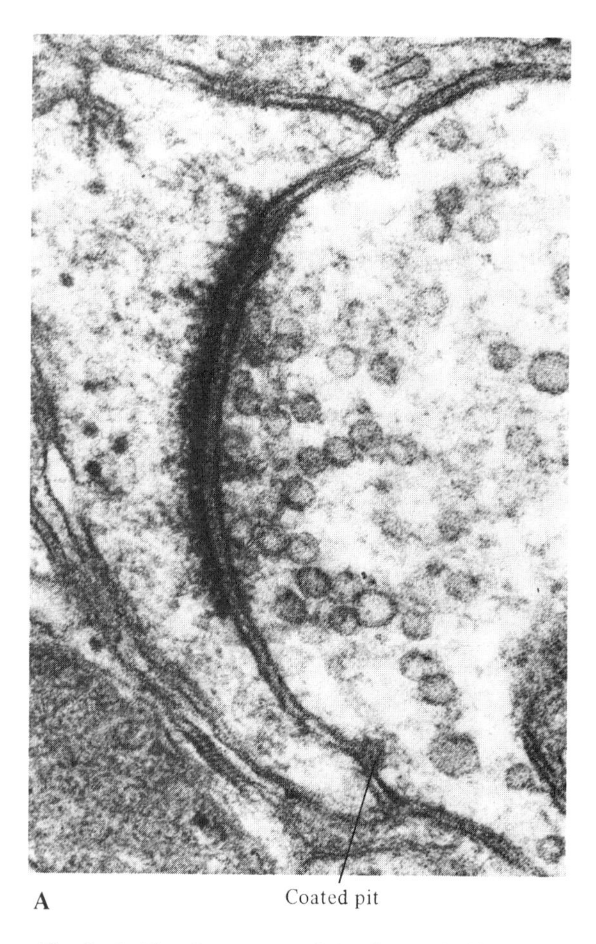

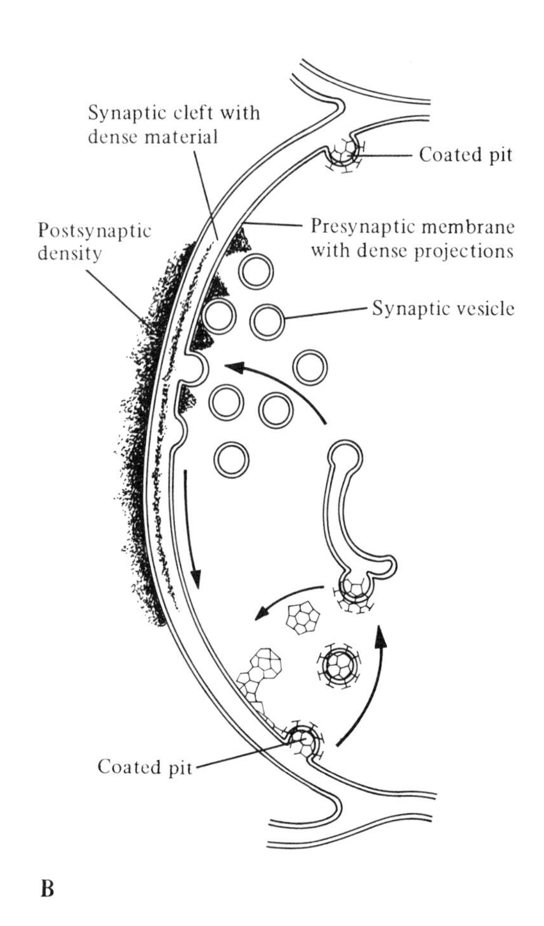

Fig. 6–8. The Synapse and the Synaptic Vesicles
A. The synapse consists of a presynaptic component, a synaptic left, and a postsynaptic component. The presynaptic component is characterized mainly by the accumulation of synaptic vesicles that contain the neurotransmitter.

B. Idealized drawing of the synapse in **A** illustrating the recycling of synaptic vesicle membrane. (Modified from The Journal of Cell Biology, 1973, 57:315–344 by copyright permission of The Rockefeller University Press.)

density is usually more prominent in relation to the postsynaptic membrane. Using special staining methods, the dense material on the presynaptic side manifests itself in the form of a presynaptic grid with pyramid-like projections (Fig. 6–8), which may play a role in docking the synaptic vesicles with the presynaptic membrane. As indicated in Fig. 6–8, even the synaptic cleft contains dense material, including filament bridges between the two membranes. The cleft material serves to bind the membranes together and may facilitate the rapid movement of the transmitter from the presynaptic to the postsynaptic membrane.

Intermediate junctions or *puncta adherentia* are another type of contact widely distributed in the nervous system. They are characterized by a symmetric disposition of dense material on the cytoplasmic surfaces of the opposing plasma membranes, but neither membrane is associated with accumulation of synaptic vesicles. Puncta adherentia are present between neurons, between glial cells, and between neurons and glial cells. As is the case in other epithelial tissue, puncta adherentia have purely adhesive functions, and are not involved in transmission of nerve signals.

Synaptic Vesicles

The synaptic vesicles come in different sizes and shapes. Most boutons are characterized by the presence of small vesicles that have a diameter of 30–50 nm and a clear center, *agranular vesicles*. At the neuromuscular junction and in many other

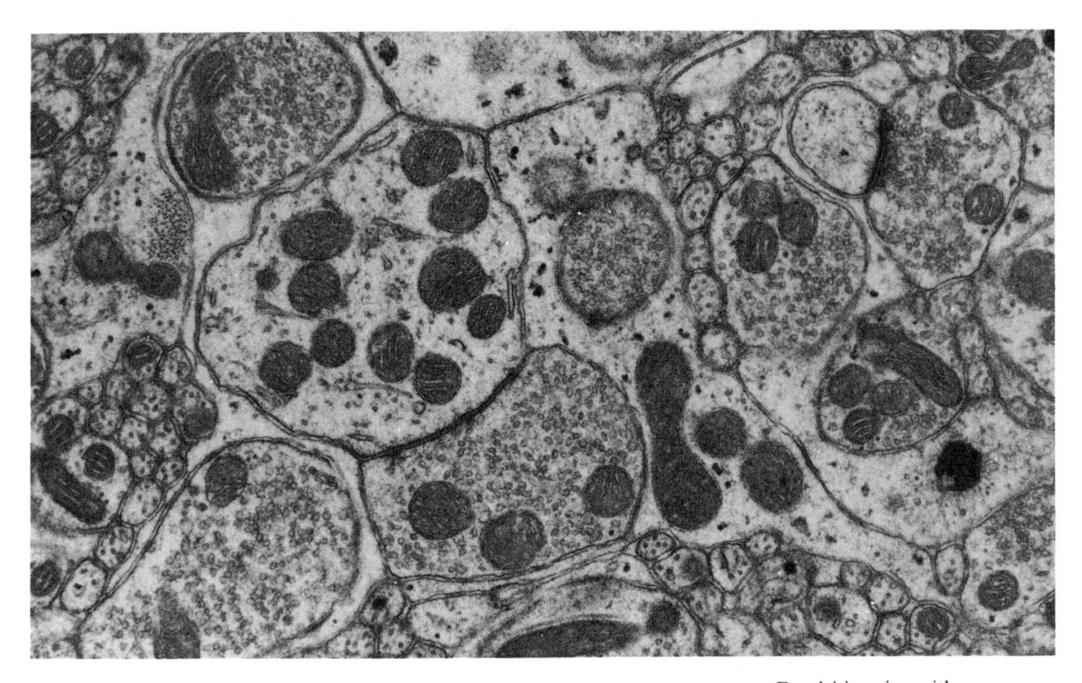

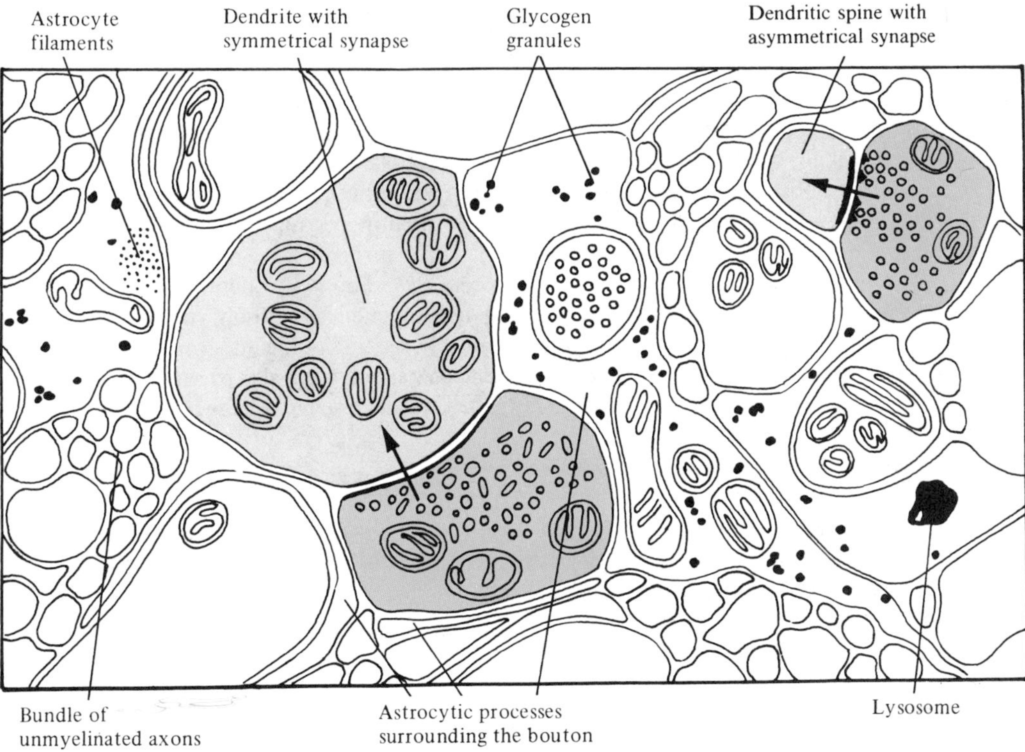

Fig. 6–9. Symmetric and Asymmetric Synapses
The symmetric synapse is characterized by a mixture of flattened and spherical vesicles, and the pre- and postsynaptic membranes are of about equal density. The asymmetric synapse is characterized by spherical vesicles and a prominent postsynaptic density. (The micrograph, which shows the neuropil in the dorsal cochlear nucleus of the rat, was kindly provided by E. Mugnaini.)

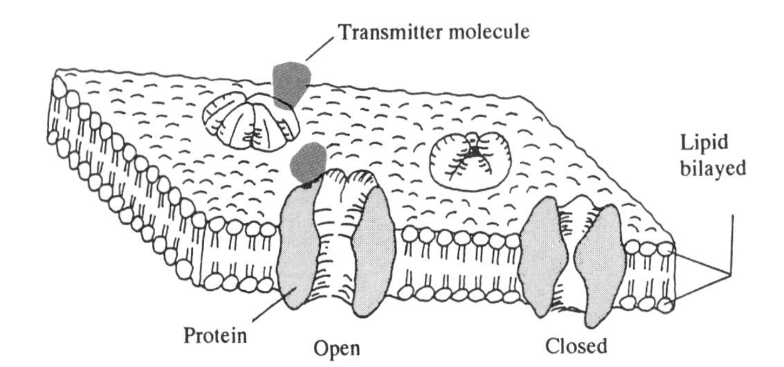

Fig. 6–10. Ion Channels
Schematic representation of a small part of the lipid bilayer of the cell membrane with interspersed ion channels. (From Brodal, P., 1992. The Central Nervous System. With permission from the author and Oxford University Press.)

parts of the nervous system such vesicles contain *acetylcholine (ACh)*; these synapses are called *cholinergic* synapses. Other vesicles may contain γ-aminobutyric acid (GABA), glycine, glutamate, or aspartate, all of them amino acid transmitters. The shape of the vesicle is to some extent dependent on the fixative used. Usually, the excitatory boutons contain round synaptic vesicles while inhibitory boutons, e.g., those containing GABA or glycine, have a tendency to become flattened or ellipsoid in aldehyde-fixed material (Fig. 6–9) and are often referred to as *pleomorphic* or *flat vesicles*.

Boutons containing catecholamines (noradrenaline, adrenaline, and dopamine) and probably also those containing serotonin have dense-cored vesicles called *granular vesicles*, which are somewhat larger (40–80 nm) than the agranular vesicles, but the core material may be seen only with certain preparative procedures. One or several large (80–100 nm) dense-cored vesicles, furthermore, can be found in almost every bouton, and there is evidence that these vesicles contain one or more neuropeptides. Whereas the small clear vesicles are released at the synaptic junctions, the large dense-cored vesicles can be released all along the axon. *Neurosecretory granules* or *vesicles*, finally, with diameters of up to 200 nm are associated with the neurosecretory neurons in the hypothalamus.

Ionic Channels

When the nerve impulse reaches the bouton, the small synaptic vesicles, according to the vesicle-exocytosis hypothesis, fuse with the presynaptic membrane in the intervals between the dense projections of the presynaptic grid, and, via exocytosis, release definite amounts (quanta) of the neurotransmitter. Influx of calcium into the terminal apparently plays a key role in the transmitter release. The positively charged calcium ions reach the interior of the axon terminal through passages formed by transmembrane proteins, i.e., *ionic channels*, in the lipid bilayer of the cell membrane (Fig. 6–10).

Having been released, the neurotransmitter diffuses across the synaptic cleft to combine briefly with specific receptors on the surface of the postsynaptic membrane[2]. The interaction between the chemical messenger and the receptor produces local (synaptic) potentials that can either depolarize or hyperpolarize the postsynaptic membrane, causing excitation or inhibition, respectively. These changes in the membrane potential are affected by the flow of Na^+ and K^+ ions through ion channels.

Many different types of ionic channels have now been identified, and as already indicated they are critically involved in a number of basic biologic processes. The ionic channels can be regulated or gated (opened or closed) not only by chemical substances including neurotransmitters (ligand-gated channels), but also by the membrane potential itself (voltage-gated channels). For instance, the

2. It is generally accepted that transmission of nerve impulses is achieved by the aid of transmitters at synaptic junctions, i.e., chemical synaptic transmission or wiring transmission, which provides for a high degree of connectional specificity and rapid changes. However, there is some reason to believe that an alternate way of parasynaptic or volume transmission exists, in which transmitters and neuroactive peptides would spread through the extracellular space and exert their action at a distance, and at places determined by the presence of appropriate receptors. This would provide a mechanism whereby populations of neurons rather than individual neurons could be regulated (see also the discussion of the neurotransmitters in Chapter 18). An often-cited reason for believing in parasynaptic transmission is the fact that the anatomic location for transmitter release does not always correspond to the location of the appropriate receptors. As a matter of fact, such "mismatches" may be the rule rather than the exception. Furthermore, receptors for a variety of transmitters have been localized to structures that do not receive classical synapses, including cerebral vessels, glial cells, and even cells of the immune system.

calcium channels in the presynaptic membrane referred to earlier are voltage-gated. The action potential, which results from the combined action of locally produced potentials in dendrites and the cell body, is also dependent on voltage-gated channels.

It should come as no surprise that disruption of neural transmission often has far-reaching consequences. Many well-known brain disorders, including Alzheimer's disease (Clinical Example, Chapter 21) and Parkinson's disease (Clinical Notes, Chapter 16), are characterized by transmitter disturbances, often in the form of reduction of certain transmitters. Other disorders, e.g., myasthenia gravis, derive from defects in some of the specific steps involved in synaptic transmission (see Clinical Notes).

Coexistence of Transmitters

In recent years it has become evident that ACh, GABA, or one of the monoamines can exist together with one, or sometimes several, neuroactive peptides. Since more and more examples of coexistence of classic neurotransmitters and neuropeptides are being discovered, coexistence may well be the rule rather than the exception in many parts of the nervous system. Although the functional significance of this coexistence has been clarified in only a few cases, it can hardly be denied that the presence of more than one chemical messenger tends to increase the complexity of synaptic transmission and provide for a more differentiated action. Elucidation of the physiologic and molecular-biologic aspects of the coexistence phenomenon may also open up new avenues for pharmacologic treatment of disorders of the nervous system.

Recycling of Synaptic Vesicle Membranes

During the process of exocytosis, the synaptic vesicle fuses with the presynaptic membrane. In order to compensate for this enlargement of the plasma membrane, one or several recycling mechanisms seem to be in place. One of the likely paths of membrane recycling is schematically illustrated in Fig. 6–8B. It has been shown that at least some of the vesicle components can be retrieved by the formation of *coated vesicles* by endocytosis, usually at some distance from the point of release. The coated vesicles subsequently lose their coat and coalesce to form cisterns, which divide to form new synaptic vesicles. Small-molecule, rapidly acting neurotransmitters including ACh, the monoamines, and amino acids such as GABA, glycine, glutamate, and aspartate, can be synthesized within the axon terminals, and their related vesicles are accordingly part of the recycling process. The neuroactive peptides, on the other hand, are synthesized and packaged in the cell bodies and moved to the terminals by fast axonal transport. Over 50 different peptides have now been identified as part of a growing list of chemical messengers or modulators in the nervous system.

Classification of Synapses

A useful anatomic classification of synapses is based on the position of the bouton on the postsynaptic neuron. Depending on the character of the postsynaptic component, the synapse can be *axodendritic*, *axosomatic*, or *axoaxonic* (Fig. 6–11). The postsynaptic element of an axodendritic synapse is either a dendritic shaft or a spine. Boutons on spines are referred to as axospinous, and they are often excitatory. Boutons on the soma, on the other hand, are often inhibitory. The postsynaptic component in an axoaxonic synapse is usually the initial segment of the axon or one of its boutons. Boutons on the initial axon segment are in a unique position to influence the generation of the axon potential, and they are often inhibitory. A bouton that terminates on another bouton cannot influence the firing of the nerve impulse of the postsynaptic neuron, but it can control the amount of transmitter released from the postsynaptic bouton. Synapses between boutons constitute the morphologic substrate for presynaptic inhibition and facilitation, and the change in synaptic function that can result from such mechanisms, i.e., synaptic plasticity, is generally believed to play a key role in memory functions.

Most synapses can also be divided into *symmetric* and *asymmetric* types (Fig. 6–9) according to the amount of dense material on the cytoplasmic face of the postsynaptic membrane. The asymmetric synapse is characterized by a prominent postsynaptic density and a widening of the intercellular space to about 30 nm, whereas the symmetric synapse has a less pronounced postsynaptic density and only a slight widening of the synaptic cleft, which measures 15–20 nm. In aldehyde-fixed material, asymmetric synapses are usually associated with spherical vesicles, whereas symmetric synapses are in general associated with pleomorphic vesicles, i.e., a mixture of flattened, ellipsoidal, and round vesicles. There seems to be a rather nice correlation between form and

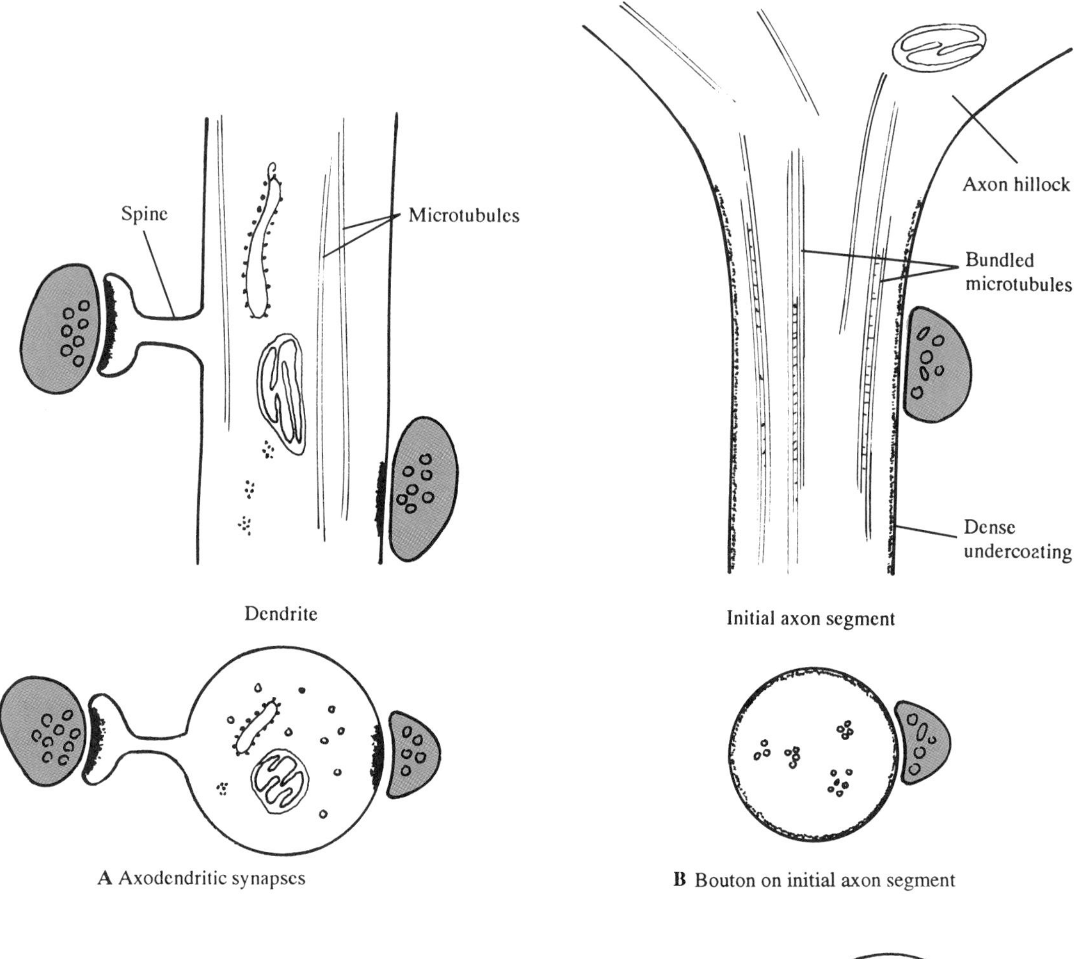

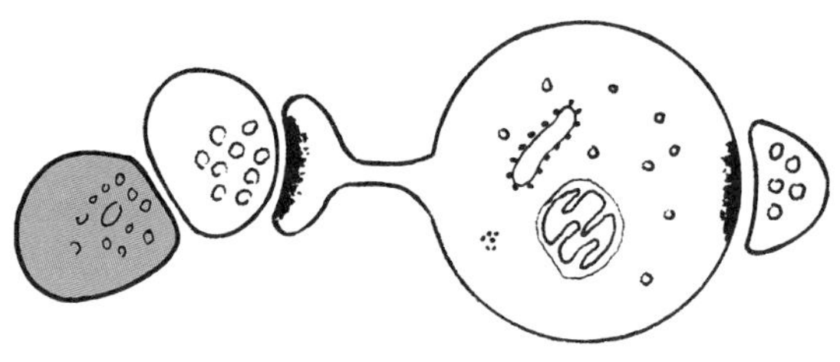

Fig. 6–11. Axodendritic and Axoaxonic Synapses
A. The postsynaptic element of an axodendritic synapse can be either a spine or a dendritic shaft.
B,C. The postsynaptic component of an axoaxonic synapse is usually the initial segment **(B)** or another bouton **(C)**. The initial segment of an axon can usually be distinguished from a dendrite by a characteristic undercoating, or dense layer, beneath the cell membrane and the presence of microtubules that are bundled together into fascicles.

function in that all symmetric synapses with pleomorphic vesicles are inhibitory, whereas asymmetric synapses with round vesicles are most likely to be excitatory. In other words, since most axospinous synapses are excitatory, they are usually of the asymmetric type. The inhibitory axosomatic synapses, on the other hand, are of the symmetric type.

In recent years, more unconventional synapses have been discovered in many parts of the brain and spinal cord, including primarily dendrodendritic synapses (see Chapter 12), but also occasionally synapses between dendrites and cell bodies.

The Neuron Doctrine

It is now generally accepted that the nervous system is composed of a large number of individual neurons that are related to each other by means of synapses operating by chemical signals. That the nervous system consists of cellular units (the neuron theory) rather than a continuous protoplasmic network (the reticular theory) was finally established by the aid of the electron microscope,

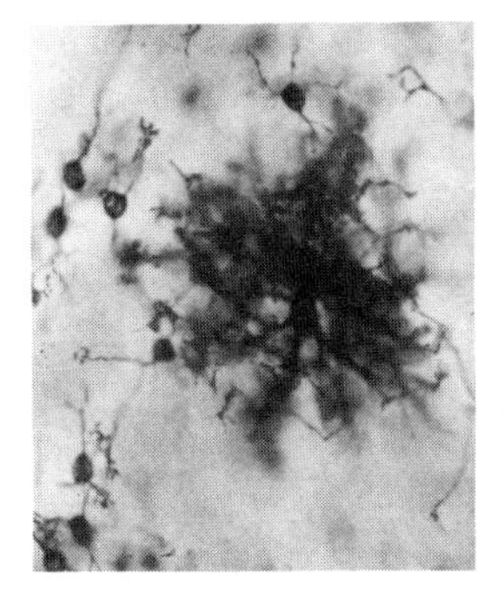

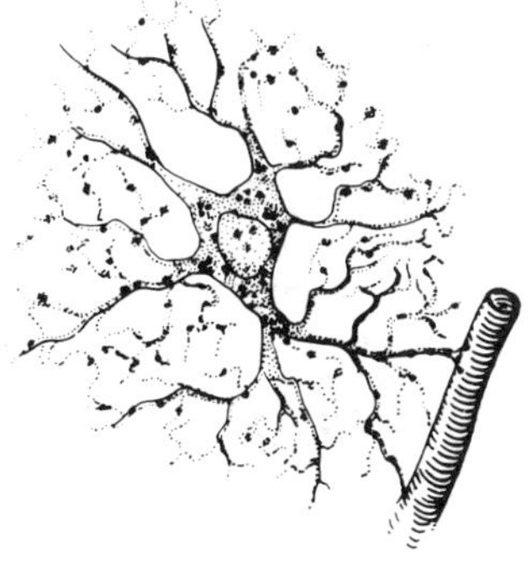

A. Protoplasmic astrocyte

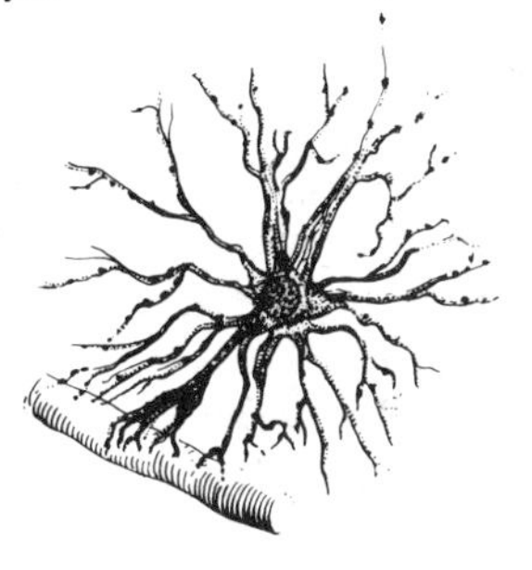

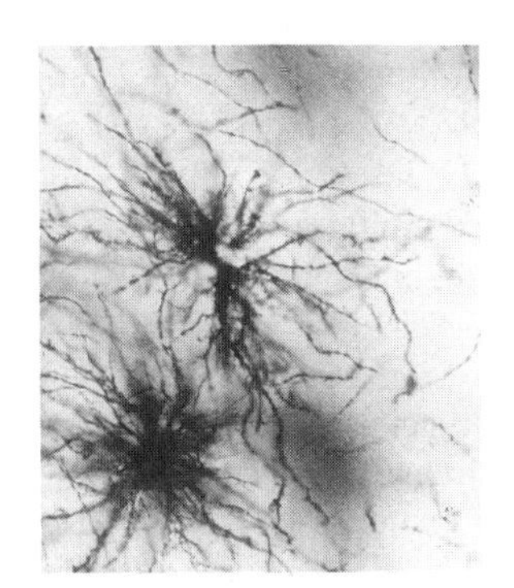

B. Fibrous astrocyte

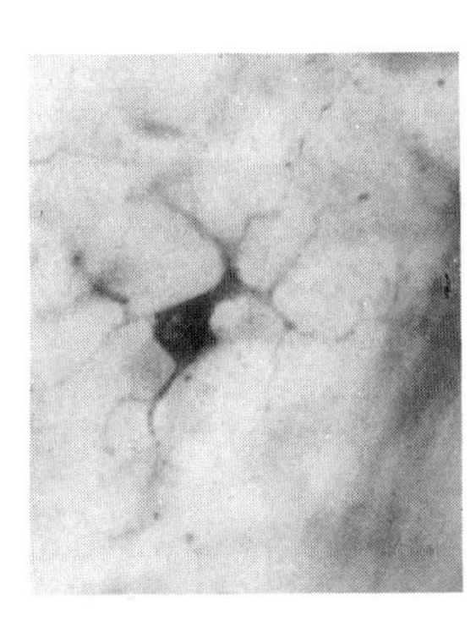

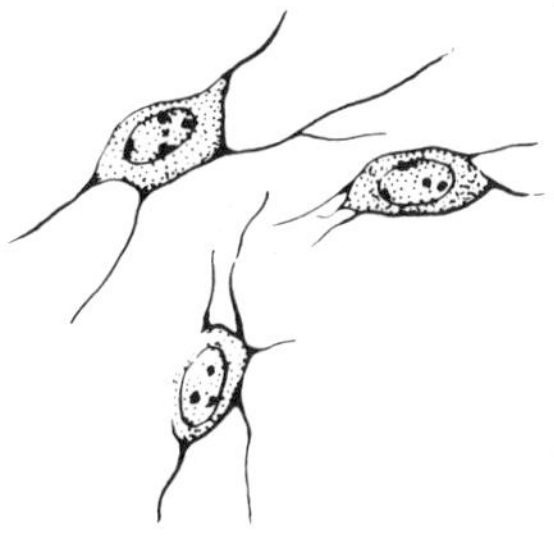

C. Oligodendrocytes

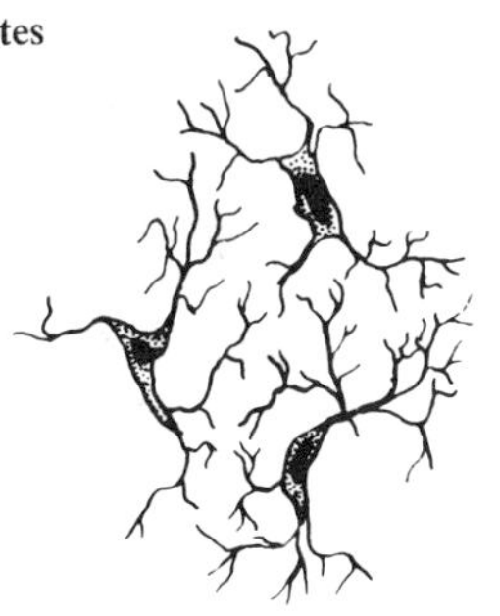

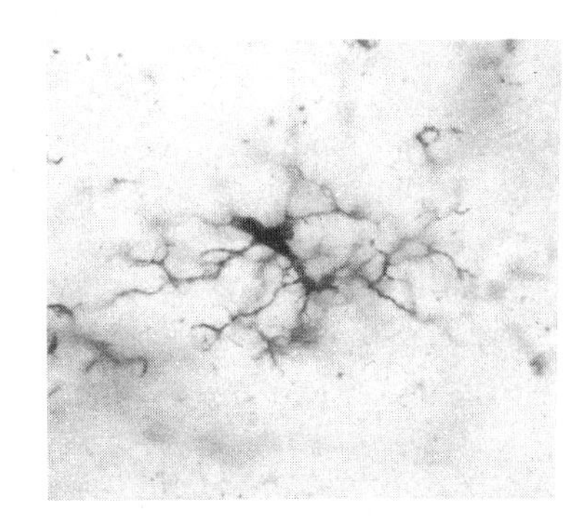

D. Microglia

E. Ependyma

Fig. 6–12. Neuroglia
Schematic drawings and photographs of histologic preparations showing the different types of glial cells. (Drawings from Jenkins, T. N., 1978. Functional Mammalian Neuroanatomy, 2nd ed. With permission of the author and Lea and Febiger, Philadelphia. Photographs; courtesy of Dr. Enrico Mugnaini.)

which permitted the detailed structure of the synapse to be first described[3].

According to the neuron theory, neurons are anatomic, functional, and trophic (nourishing) units. The theory also implies that the neurons are dynamically polarized in the sense that the dendrites and the cell body receive impulses, whereas the axon carries the impulses away from the cell body to other neurons or to the effector organs. Although this is generally true, it has been shown that dendrites or even cell bodies can transmit impulses to other neurons at chemical synapses. Such atypical synapses were first seen in the olfactory bulb (Fig. 12–1) and retina (Fig. 13–3), but they have now been discovered in many other parts of the brain and spinal cord. Further, some neurons communicate with one another via ultramicroscopic channels formed by special proteins called *connexins*, which are concentrated in specialized regions referred to as *gap junctions*. These intercellular channels provide electrotonic coupling and synchronized firing of neurons, and the gap junctions are often referred to as *electrical synapses*. The gap junctional channels also provide for metabolic coupling between neurons, since they allow molecules up to 1,200 daltons in molecular weight to move from one cell to the next.

Although the basic tenets of the neuron theory are still valid, certain aspects of the theory are being modified in the light of these discoveries. Furthermore, scientists are increasingly emphasizing different levels of organization, as expressed by local variations within individual neurons or by multineuronal circuits and systems, and they are turning their attention to the relations between these various levels of structural and functional organization, especially the oscillatory behavior of large neuronal assemblies.

Neuroglia

The neuroglial cells, which are more numerous than the nerve cells, are not directly involved in the transmission of nerve impulses, but they have many other important functions. Unlike neurons, they retain their capacity to divide in the adult nervous system. Neuroglial cells in the CNS are of four different types: *astrocytes*, *oligodendrocytes*, *microglia*, and *ependyma* (Fig. 6–12). Schwann cells and satellite cells in the peripheral nerves and ganglia are referred to as peripheral neuroglia.

Astrocytes

Astrocytes, the main interstitial cells, have many important functions. On the basis of morphology, astrocytes are divided into *protoplasmic astrocytes* and *fibrous astrocytes*. Protoplasmic astrocytes are located in the gray matter, and they have irregular and profusely branching processes that seem to conform to the surrounding neuronal elements. Fibrous astrocytes, on the other hand, are primarily found in myelinated fiber bundles, and they are characterized by fewer processes, which are more regular and which do not branch as often. Astrocytic processes typically form end-feet that cover the basal lamina around blood vessels and at the pia mater, where they form an almost continuous but patent sheath, the *glia limitans*. Astrocytes can be labeled with antibodies against glial fibrillary acidic protein (GFAP), which is the major component of astroglial filaments. They can also be easily recognized in electron micrographs (Fig. 6–9) because of their relatively light cytoplasm and the presence of glycogen granules and bundles of glial filaments, which are especially prominent in the fibrous astrocytes.

Astrocytes have a number of important functions, some of which are less well understood. During development, the fibers of the bipolar radial glial cells guide the migrating neurons to their destination. Radial glial cells, which disappear during the later stages of development, are believed to be the precursor cells to astrocytes; both radial glia and mature astrocytes stain positively with antibodies against GFAP. Besides guiding the neurons to their final destination, astrocytes provide a supporting framework for other elements in the CNS, and their processes are often seen to encircle boutons or groups of boutons as if to isolate them and prevent them from interacting functionally with surrounding structures (Fig. 17–2). In fact, glial–neuronal interactions may be much more dynamic than generally realized. Studies on the supraoptic nucleus in the hypothalamus, which is part of the hypothalamo-hypophysial system (Chapter 18), indicate that

3. The cell theory, enunciated in 1839 by the German zoologist Theodor Schwann, postulated that all animal tissues consist of elementary units or cells. The theory was generally accepted except for the nervous system, where it was difficult to elucidate the true nature of the cell processes. The controversy between the proponents of the neuron theory and those favoring the reticular theory attracted many of the leading neuroscientists during the second half of the last century and the debate continued well into the twentieth century. Theodor Schwann also discovered the myelin-forming cell in the PNS now generally known as the Schwann cell.

appropriate physiologic stimuli can result in rapid changes in glial structure, i.e., retraction of astrocytic processes, which in turn lead to a dramatic increase in the number of synaptic contacts in the supraoptic nucleus. Appropriate stimuli in this case include dehydration and lactation, which calls for increased activity in the neurosecretory cells of the supraoptic nucleus. Glial–neuronal interactions are not only structural in nature. For instance, astrocytes possess specific receptors for a number of transmitters, and it is known that astroglia can take up and catabolize neurotransmitters.

In various lesions of the CNS, the astrocytes take part in the repair by occupying the space left by degenerating neurons. The glioses include both hyperplasia (increase in number of cells) and hypertrophy (increase in the size of cells), and there is evidence that astroglia can also engulf and remove neuronal debris, especially degenerating synaptic terminals. The main scavengers in the CNS, however, are the microglial cells (see below).

The Astroglial Syncytium: a Mechanism for Metabolic Cooperation?

Electron microscopic studies have demonstrated numerous gap junctions between neighboring astroglial cells. Since gap junctions allow the passage of ions and small molecules from cell to cell, the interesting notion of a functional syncytium has been promoted as an effective mechanism for metabolic cooperation between individual units in the syncytium. Since astrocytes form gap junctions also with ependymal cells and oligodendrocytes, these two types of supporting cells may well be part of this metabolically coupled syncytium. It is well documented that the glial syncytium plays a role in ion buffering, especially of potassium, and it may well emerge as a critically important device for the dissemination of metabolites and other substances that would otherwise benefit mostly those astrocytes that are endowed with capillary and pial end-feet.

Oligodendrocytes

As the name indicates, oligodendrocytes emanate fewer first-order processes than the astrocytes; they are the myelin-forming cells in the CNS, and they are present in large numbers in white matter. They also occur as satellite cells around neuronal cell bodies and can be seen in other parts of the gray substance together with groups of myelinated fibers. It is usually easy to distinguish oligodendrocytes from astrocytes in electron micrographs, because the nucleus and cytoplasm are considerably denser in the oligodendrocytes. Glial filaments and glycogen granules, which are common in astrocytes (Fig. 6–9), are absent in oligodendrocytes. Oligodendrocytes can be labeled with antibodies to proteins specific for the myelin membrane and to special enzymes.

Microglia

Microglial cells, the main phagocytes, are small cells with delicate processes that give off spine-like projections (Fig. 6–12D). Their origin is still uncertain, but it is widely believed that at least some of them derive from the mesoderm and invade the brain during vascularization. Microglial cells, stained with Del Rio–Hortega's silver carbonate technique or with antibodies to elements of their plasma membrane, can easily be identified in the light microscope. Their identification in standard electron micrographs, however, is sometimes difficult because their cytoplasm has about the same density as that of the oligodendrocytes.

The microglial cells, which are present in both gray and white substance, undergo rapid proliferation in response to tissue destruction, and they migrate toward the site of injury where they act as scavengers by phagocytizing and removing tissue debris. Dense bodies, lysosomes, and lipofuscin granules are more common in microglial cells than in other types of glia. As already indicated, astroglial and oligodendroglial cells may also become phagocytic in response to injury, in which case there is likely to be an additional influx of phagocytic cells from the blood into the nervous system.

Ependyma

These epithelial cells (Fig. 6–12E) line the brain ventricles and the central canal of the spinal cord. The ventricular surface of most ependymal cells is covered with cilia, which seem to facilitate the movement of the CSF. Certain parts of the ventricular system, including the thin floor of the third ventricle and some of the other areas related to the circumventricular organs (see Chapter 3), are lined by a special type of ependymal cell called *tanycytes*. Such cells have an apical process that reaches into the nervous tissue and gives off a number of side branches with end-feet on nearby blood vessels, and it has been suggested that the tanycytes might be involved in the transport of substances from the CSF to the blood. There is also some evidence that substances may pass from the neuropil to the CSF via ependymal cells.

Clinical Notes

Multiple Sclerosis

The importance of the myelin sheath in the conduction of nerve impulses is well illustrated in cases of demyelinating diseases, i.e., conditions in which the destruction of the myelin sheath is the most striking feature. *Multiple sclerosis (MS)*, a widely known demyelinating disorder, commonly affects young adults in their prime. As the name indicates, multiple sclerotic (scar-like) plaques appear around the stripped axons (Fig. 6–13) in the white matter of the CNS. The symptoms very according to the location of the plaques. The optic nerve, periventricular areas, and spinal cord are favored sites.

MS is an autoimmune disease of unknown etiology, although both genetic and environmental factors seem to be involved. There is a definite geographic distribution in that MS is much more common in northern Europe and North America than in Asia and Africa. Unfortunately, there is little potential for repair of the demyelinated fibers, since the oligodendrocytes do not possess the remarkable capacity to reform myelin sheaths as do Schwann cells in the PNS (see below). Therefore, although there may be striking remissions, the disease usually causes a gradual deterioration of the patient's condition over many years.

Myasthenia Gravis

Myasthenia gravis (MG) is a disorder of neuromuscular transmission characterized by an excessive fatigability of striated musculature, especially the extraocular muscles, the levator palpebrae, and the muscles of facial expression. Recent advances in molecular biology have contributed greatly to an understanding of the pathophysiology of MG, which is now known to be an autoimmune disease affecting primarily the postsynaptic ACh receptors. Why certain people develop antibodies against their own receptors is still not known. Since the effect of the antibodies is to reduce the number of receptors, one of the treatment strategies is to prolong the action of the transmitter by giving the patient an acetylcholinesterase inhibitor, e.g., physostigmine or neostigmine. Thymectomy has also proven effective, because it alters some of the immunologic control mechanisms for production of antibodies.

Parkinson's Disease

This is a relatively common disorder characterized

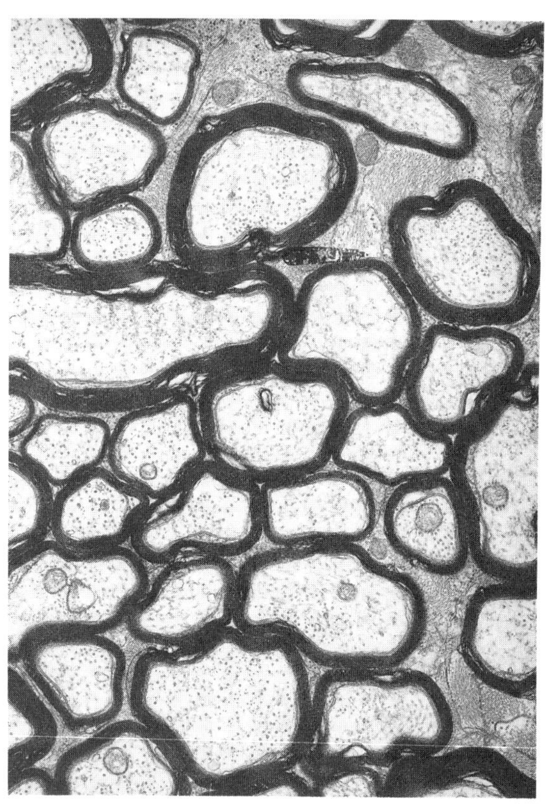

A

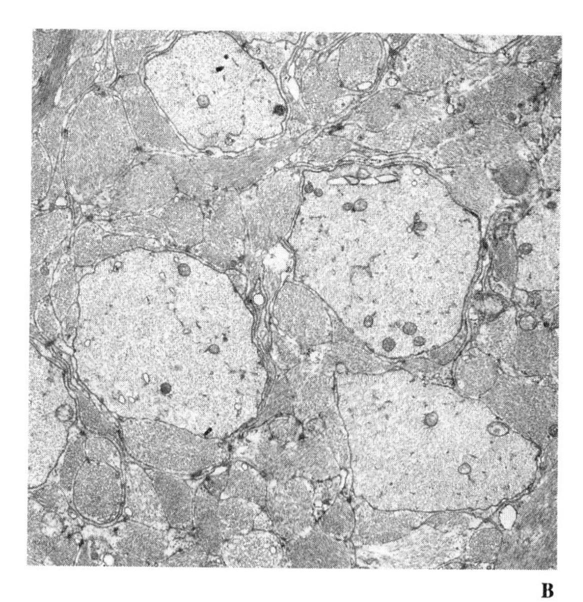

B

Fig. 6–13. Multiple Sclerosis
A. Normal axons in the white matter.
B. Tissue obtained post mortem from a patient with MS showing stripped axons in a sclerotic lesion in the white matter. (Courtesy of Dr. C. S. Raine, Albert Einstein College of Medicine, from 1991, The Journal of NIH Research, Washington, DC. Reprinted with permission.)

by a deficiency in a specific transmitter, i.e., dopamine, as a result of degeneration of the dopamine-synthesizing cells in the substantia nigra. The disease is discussed in more detail in Clinical Notes, Chapter 16.

Regeneration and Repair in the Nervous System

To what extent a function is recovered following an injury to the nervous system depends in large measure on the regeneration capacity of the damaged neurons. The capacity for regeneration in the PNS is impressive, whereas regeneration in the CNS is quite limited. Indeed, there is little evidence for regrowth of damaged axons following lesions in the human brain or spinal cord. On the other hand, damage to one part of the brain does not necessarily result in dramatic functional deficits, since nearby structures or parallel pathways might be able to take over the function performed by the damaged structure, or modify their own performance to offset some of the disturbed functions.

Regeneration of Peripheral Nerve Fibers

The degenerative changes that occur in a neuron whose axon has been transected are most pronounced in the part of the nerve fiber that is separated from the perikaryon. As mentioned previously, this part undergoes Wallerian degeneration, i.e., the axon including the myelin sheath breaks up into fragments and finally disappears. In the cell body of the damaged neuron, a process known as *chromatolysis* or retrograde cell reaction takes place (Fig. 6–1D). This is a reconstructive response to the injury, and promptly sprouts appear at the distal end of the proximal stump. The growing tips are likely to find their way into the spaces between the basal lamina and the proliferating Schwann cells, which play a crucial role by providing nerve growth factor (NGF) and other important molecules that promote axon growth. Eventually, the axon will be able to reinnervate the denervated target organ, especially following a crush injury, in which the basal lamina tubes are still intact.

If the nerve is completely transected and the gap between the proximal and distal stump is more than 1–2 cm, the growing axonal tips may not be able to cross the scar, in which case the sprouts and the scar tissue may form a painful neuroma. To prevent this from happening, it is important that the two ends of a nerve that has been interrupted by an injury be approximated as closely as possible, if necessary with the help of nerve transplants, which can provide chains of Schwann cells capable of promoting axonal growth.

Regeneration in the CNS

Damage to the CNS is usually irreversible, because the CNS lacks the guiding basal laminae sheaths, the Schwann cells, and the various types of growth-promoting proteins that promote axon growth in the PNS. Recent studies, however, indicate that axons in the CNS can regenerate if they have access to peripheral nerve segments that provide a suitable environment for growth. This opens the possibility of repairing lesions in the CNS by grafting pieces of peripheral nerves into the damaged area. However, the CNS is infinitely more complicated than the peripheral nerves, and this type of reconstructive brain surgery has not yet reached the stage where it can be used in patients.

Another promising method for repairing brain injury relies on the use of embryonic or neonatal tissue transplanted into the diseased brain. Such transplantations have proven remarkably successful in experimental animals, i.e., the transplants have survived and have been able to form functionally relevant connections in the host brain. This indicates that it may be possible to repair nervous pathways that have been destroyed by a lesion or a degenerative processs. In fact, some attempts to this effect have already been made in patients with Parkinson's disease (see Clinical Notes, Chapter 16), but with mixed success. Since the fetal tissue would be retrieved from aborted fetuses, the question of fetal transplantations and related research is closely tied to the abortion controversy, which besides its ethical and religious ramifications has become a highly charged political issue.

Spread of Viruses and Toxins

A number of viruses are neurotropic, i.e., they have an affinity for the nervous system, and like other infectious agents they produce common syndromes known as meningitis and encephalitis that may have serious consequences. Often, viruses spread from the port of entry to the nervous system through the blood vessels, as is the case with the human immunodeficiency virus (HIV) after it has entered the body through intravenous injections or invaded the vaginal or rectal mucosa during sexual intercourse. The virus can also reach the CNS by retrograde axonal transport in peripheral nerves. Herpes simplex virus (HSV) and rabies, as well as tetanus toxin, are likely to use this route to travel from the periphery to the CNS. An especially fulminant encephalitis, involving primarily the orbitofrontal cortex and the temporal lobe, is caused by HSV. It is believed that the olfactory pathway from the olfactory mucosa to the medial temporal lobe represents the primary route for the spread of this deadly disease.

The ease with which viruses reach the CNS through retrograde axonal transport and subsequent transneuronal transfer can be used to great advantage in the tracing of central nervous pathways (Chapter 7).

Suggested Reading

1. Calcium and Neuronal Excitability, 1988. Special Issue. Trends in Neuroscience, 11(10).
2. Fuxe, K. and L. F. Agnati (eds.), 1991. Volume Transmission in the Brain. Novel Mechanisms for Neural Transmission. Raven Press: New York.
3. Hillman, H. and D. Jarman, 1991. Atlas of the Cellular Structure of the Human Nervous System. Academic Press, London.
4. Jones, E. G., 1988. The nervous tissue. In L. Weiss (ed.): Cell and Tissue Biology. Urban & Schwarzenberg, Baltimore, pp. 279–350.
5. Kimelberg, H. K. and M. D. Norenberg, 1989. Astrocytes. Scientific American 260:44–55.
6. Nicholls, J. G. and A. R. Martin, 1992. From Neuron to Brain: A Cellular Approach to the Function of the Nervous System, 3rd ed. Sinauer Associates: Sunderland, Massachusetts.
7. Peters, A., S. L. Palay, and H. de F. Webster, 1991. The Fine Structure of the Nervous System. Oxford University Press: New York.
8. Shepherd, G. M., 1991. Foundations of the Neuron Doctrine. Oxford University Press: New York.
9. Shepherd, G. M., 1990. The Synaptic Organization of the Brain, 3rd ed. Oxford University Press: New York.
10. Steward, O., 1989. Principles of Cellular, Molecular and Developmental Neuroscience. Springer Verlag: New York.
11. Vallee, R. B. and G. S. Bloom, 1991. Mechanism of fast and slow transport. Annu. Rev. Neurosci. 14:59–92.
12. Walz, W., 1989. Role of glial cells in the regulation of the brain ion microenvironment. Prog. Neurobiol. 33:309–333.

Honoré Daumier, 1808–1879. Lithograph reproduced by the courtesy of the Museum of Fine Arts, Boston, Mass. Daumier's legend to this cartoon reads: "M. Babinet prévenu par sa portière de la visite de la comète."

7

Neuroanatomic Techniques

Although knowledge of the pathways in the brain and spinal cord cannot in itself explain how the nervous system works, the anatomic and chemical mapping of neuronal circuits and synaptic relations is a prerequisite for such understanding. Fortunately, in the pursuit of such information, the modern neuroscientist has a large number of sophisticated light and electron microscopic techniques to choose from.

"The gain in brain is mainly in the stain." This play on words by a contemporary neuroscientist takes note of the important fact that our knowledge of the brain is in essence a reflection of the methods of investigation. A booming dyestuffs industry in Germany in the second half of the last century facilitated the introduction of a number of new coloring agents in neurohistologic methods, many of which became widely used, e.g., Joseph von Gerlach's carmine method and Paul Ehrlich's methylene blue staining of neurons, Carl Weigert's potassium dichromate-hematoxylin stain for myelinated fibers, and the Nissl method for the study of cell bodies. Several silver methods for the staining of both normal and pathologic components of nervous tissue were also developed before and around the turn of the century, prominent examples being the Golgi method and the reduced silver methods by Bielschowsky and Ramón y Cajal.

A big step forward was taken in the 1950s, both with the introduction of the *electron microscope*, which made it possible to analyze in detail the synaptic contacts between individual neurons, and with the development of light microscopic methods to specifically stain degenerating axons and boutons with silver.

Another methodologic breakthrough came in the early 1960s with the development of a fluorescence method for the identification of monoamine-containing neurons and pathways. For the first time a method became available for the selective demonstration of neurons containing a known transmitter. This remarkable achievement, and the concurrent discovery that Parkinson's disease (see Clinical Notes, Chapter 16) is in essence a dopamine deficiency syndrome that can be successfully treated with L-dopa (a precursor to dopamine), ushered in a new era in basic and clinical neuroscience.

The rapid pace with which new discoveries have been made during the last few decades in all fields of neuroscience, from the mapping of neuronal connections to the elucidation of biochemical and molecular mechanisms, is a reflection of the large number of sophisticated methods that have become available during the last decades, including axonal transport methods and intracellular labeling techniques. Many of the new methods, including sensitive immunohistochemical methods for the identification of neuroactive substances, receptor binding techniques, and in situ hybridization for localizing the anatomic site for protein synthesis, can be applied both in experimental animals and in autopsied human brains. Since equally spectacular advances have taken place in the field of noninvasive imaging methods, it is now possible to compare data obtained in experimental studies and postmortem material with information gathered from correlated functional–anatomic studies of the living human brain.

The Nissl and Golgi Methods

The *Nissl method*, developed and perfected in the 1890s by the German psychiatrist Franz Nissl, is based on staining with basic aniline dyes, e.g., cresyl violet, thionine, or toluidine blue. In particular, these dyes stain the components containing nucleic acids including the nucleus and the ribosomes in the cytoplasm (Figs. 6–1A and B). The Nissl method reveals the size and shape of the cell bodies, and provides a map of cell assemblies and other cytoarchitectonic landmarks (Fig. 22–1).

The *Golgi method* is very different; it impregnates the whole neuron including its elaborate processes. In 1873 Camillo Golgi of Italy[1] discovered that some neurons became filled with a "black reaction," i.e., silver chromate precipitate, if brain tissue that had been hardened in potassium dichromate was subsequently treated in a silver nitrate solution. A fortunate yet unexplained quality of the Golgi method is the fact that it impregnates only a few percent of the neurons in a given area, without staining surrounding elements. With the black neurons in sharp relief against a light background (Figs. 6–3A and D), the method is especially valuable for studying the distribution of dendrites and unmyelinated axons. If all the nerve cells with their processes were impregnated in the section, the details would be lost in a dense tangle of fibers. The Golgi method, however, is capricious. It is not possible to predict which neurons will be stained, and many sections must be analyzed to obtain a

1. The Golgi Method and the 1906 Nobel Prize in Physiology or Medicine.

The unique advantages of the Golgi method have been exploited by many investigators, including one of Golgi's contemporaries, Santiago Ramón y Cajal of Spain. Cajal shared the 1906 Nobel Prize in physiology or medicine with Golgi, and his monumental work, aided in large part by Golgi's method, helped lay the foundation for modern neuroscience and continues to be a source of valuable information. Although Cajal and Golgi shared the Nobel Prize, they were on opposite sides in one of the great controversies at the turn of the century. Golgi was a staunch proponent of the nerve net or reticular theory, and it is somewhat ironic to realize that Cajal and many other pioneering neuroanatomists took advantage of the remarkable qualities of the Golgi method to turn the pendulum in favor of the neuron theory, which claimed that each nerve cell with its processes represents a functional unit.

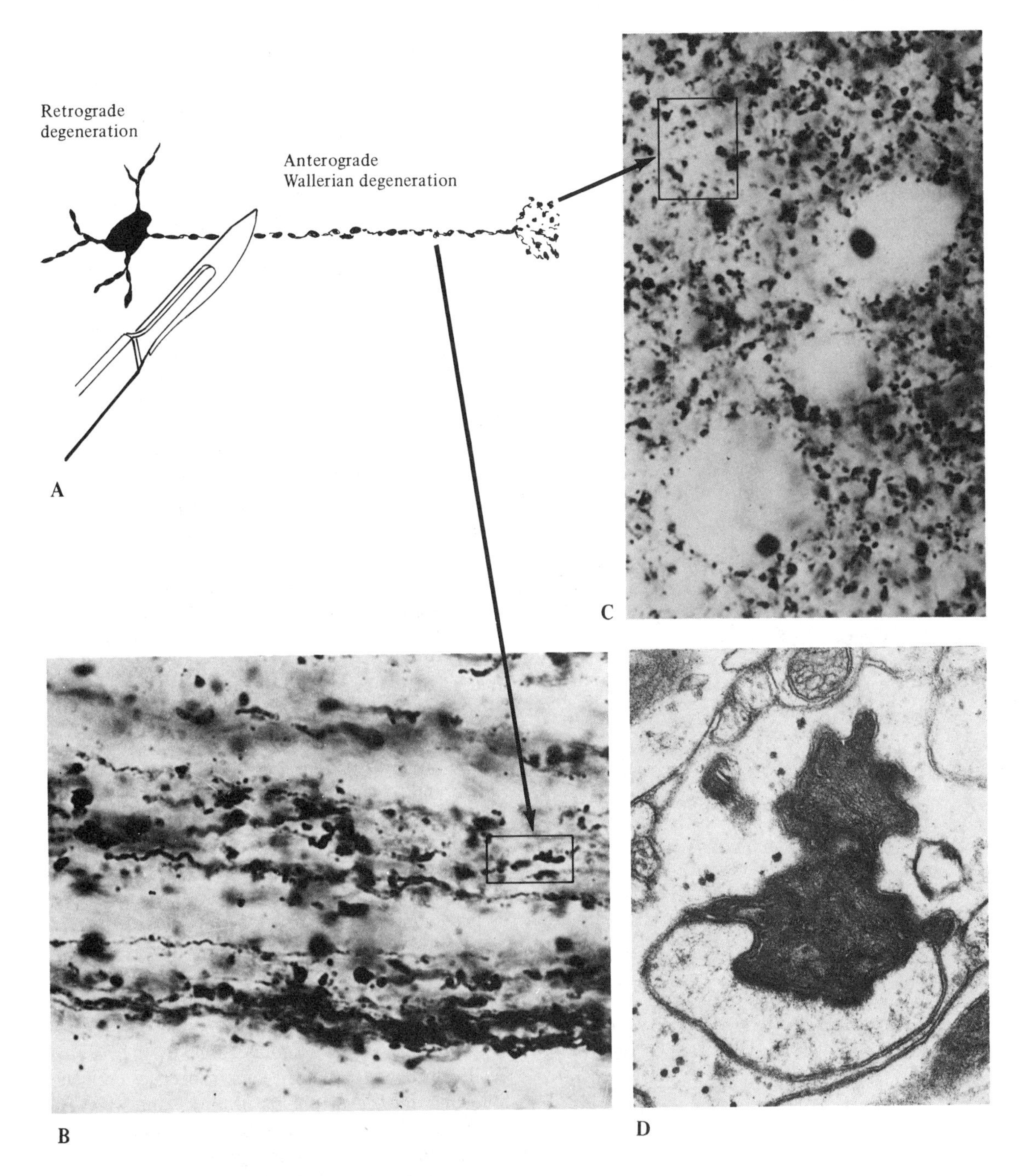

Fig. 7–1. The Study of Anterograde Degeneration with the Silver Method and Electron Microscopy
Following an experimental lesion (**A**) the reduced silver method can be used to reveal degenerating axons (**B**) en route to their destination (**C**). A silver impregnation study is often complemented by an electron microscopic investigation, in which the degenerating boutons and their postsynaptic structures can be identified (**D**). (A–C from de Olmos, J. S., S. O. E. Ebbesson, and L. Heimer, 1981. Silver methods for the impregnation of degenerating axoplasm. In: Heimer, L. and M. Y. Robards (eds.) Neuroanatomical Tract-Tracing Methods. With permission of Plenum Press, New York.)

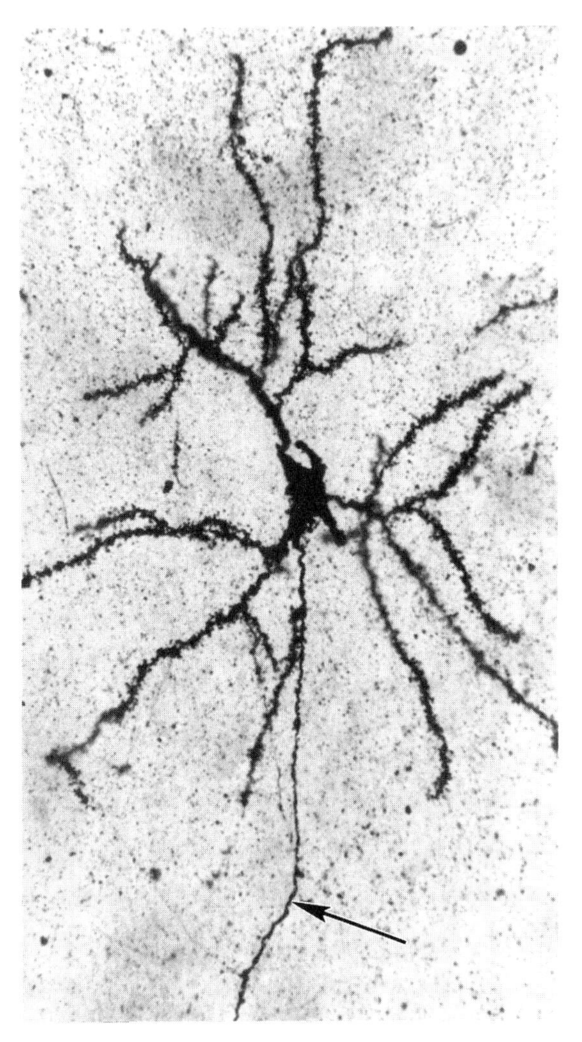

Fig. 7–2. Use of the Silver Method for Screening of Neurotoxic Effect
Degenerating neuron in parietal cortex following an injection of quinolinic acid into the striatum of a rat 24 hours before sacrifice of the animal. Signs of degeneration include an irregular cell body and a beaded axon (*arrow*). Note the dendritic spines and the large number of degenerating boutons in the background. Quinolinic acid, which is a derivative of tryptophan metabolism, may be a potential endogenous toxin in the human brain. (Micrograph courtesy of Drs. Carlos Beltramino and José de Olmos.)

satisfactory picture of the neuronal organization in any given area. This shortcoming of the Golgi method has in part been overcome by modern intracellular injection techniques, which make it possible to label specific neurons in a Golgi-like fashion (see below).

Axon Tracing Methods

Although individual axons can be traced with the Golgi method or by the aid of the intracellular injection technique, the tracing of long axons is usually performed with other techniques. Silver methods were developed around the turn of the century by several investigators including the German neuropathologist Max Bielschowsky and by Cajal, who discovered that most nerve fibers can be impregnated with silver if the tissue is treated in a silver solution followed by reduction of the silver ions to metallic silver in a reducing fluid. In the beginning, such methods, often referred to as reduced silver methods to distinguish them from the type of silver methods represented by the Golgi method, were used for the staining of normal fibers and for visualizing various pathologic phenomena in the human brain, e.g., neurofibrillary tangles (see Alzheimer's disease, Chapter 21). However, without some way of labeling a specific axon or a set of axons, it is almost impossible to trace axons in normal brains, because it is extremely difficult to follow a particular axon from section to section. One of the most successful tract-tracing methods in the past, therefore, was the Marchi method[2], which is based on the experimental induction of anterograde axonal degeneration.

The process of anterograde axonal degeneration is also known as Wallerian degeneration, as a tribute to Augustus Waller, the English physician–scientist, who in 1850 discovered that if an axon is severed from its cell body, it undergoes a gradual disintegration (Fig. 7–1). By transecting a pathway, or producing a surgical or chemical lesion in the area where a pathway originates, the Wallerian degeneration can be used to "mark" the pathway of interest, thus facilitating its recognition as it meanders through the brain or spinal cord. The experimental tracing of connections by the use of Wallerian degeneration became especially popular following the development of the Nauta–Gygax silver method in the middle part of this century.

Following transection of an axon, changes occur not only distal to the injury but also proximal to the

2. In the 1880s, Vittorio Marchi of Italy discovered that it was possible to stain selectively the breakdown products of the disintegrating myelin sheath with osmium tetroxide, if the tissue had been pretreated in potassium dichromate. Since the myelin sheath disintegrates together with the rest of the axon, the pathways containing myelinated fibers can be traced, but their terminal branches and the large number of thin unmyelinated axons cannot be detected by the Marchi method.

lesion, e.g., in the cell body, where the Nissl bodies gradually diminish in size and eventually disintegrate. This phenomenon, which is known as the retrograde cell reaction and which can be visualized by the aid of the Nissl method (Fig. 6–1D), has been used to identify cell bodies whose axons or axon terminals have been experimentally damaged. Since the appearance of the retrograde cell reaction depends on many variables and is less predictable than the anterograde Wallerian degeneration, methods based on the retrograde cell reaction have never gained widespread popularity. Nonetheless, both the Nissl method and the reduced silver method have been used for the study of retrograde cell degeneration.

Reduced Silver Methods

When the Dutch neuroanatomist Walle Nauta and the Swiss chemist Paul Gygax discovered that the impregnation of normal fibers could be suppressed if the sections were treated in phosphomolybdic acid and potassium permanganate before the silver staining, the stage was set for a dramatic expansion of our knowledge of fiber connections in the brain and spinal cord. By suppressing the staining of normal fibers, it is easy to follow the course of the degenerating fibers, and since the silver is deposited in the degenerating axoplasm rather than in the myelin as in the Marchi method, the degenerating fibers can be followed to their point of termination (Fig. 7–1).

An important advantage of the reduced silver methods is their ability to demonstrate the morphologic characteristics of degenerating neuronal components, which may allow morphologic differentiations of functional significance. This feature is also of special importance in the field of neurotoxicology (see below). However, since both cells of origin and fibers passing by the lesioned area are destroyed by the lesion, questions related to the origin of a specific pathway may not be easily resolved, especially when the experimental lesions involve subcortical structures. Although the "fibers of passage" problem can to some extent be solved by producing the lesion with a chemical substance, e.g., ibotenic acid, which destroys nerve cell bodies but not fibers of passage if properly done, this shortcoming was an important factor in the rapidly increasing popularity of the autoradiographic technique during the 1970s.

A seldom used but nevertheless important application of the reduced silver methods is the study of degenerating axons in human brains. Many of the powerful new tracing methods, which require intracerebral injections of the tracer, can for obvious reasons not be used in the human. Some of the reduced silver methods, on the other hand, have been specifically modified for use on autopsied human brains, and if the patient should happen to die within a relatively short (1–8 weeks) time period following a focal lesion in the brain, silver staining of degenerating fibers is likely to provide valuable information regarding the structures and pathways destroyed by the lesion. Since some products of axonal degeneration can remain for several months and sometimes for a year or more after damage to the human brain, postmortem silver staining can provide important information even if a person survives for a considerable period following a lesion in the CNS.

Although silver staining of degenerating fibers in the human brain has some obvious disadvantages, in the sense that the location and the size of the lesion as well as the survival time following the appearance of the lesion cannot be controlled, it nevertheless may clarify anatomic relations, which cannot be obtained in any other way. Unfortunately, the histotechnical aspects of silver methods, including the interpretation of the result obtained, are not without difficulties, and many neuropathologists are reluctant to make the reduced silver methods part of their technical arsenal.

The reduced silver methods are seldom used for the experimental tracing of connections nowadays, but they seem to be headed for a revival in the rapidly expanding field of *neurotoxicology*, which has become a matter of great public concern. Efforts have intensified to find reliable and convenient methods to identify chemicals that are potentially harmful to our health because of their toxic effect on the nervous system. Characterized by great sensitivity and distinct rendition of the morphology of degenerating neurons and their processes (Fig. 7–2), the reduced silver methods are increasingly being used for the screening of neuronal damage caused by neurotoxic substances including drugs of abuse and other chemical substances.

Axonal Transport Methods

Anterograde Tracing Methods

Autoradiographic Method. This method was introduced as a tracing technique in the late 1960s by several American research teams including Paul Weiss, Raymond Lasek, Stephen Goldberg, and their collaborators. It is based on the well-known biologic principle that all neurons synthesize proteins and transport some of them through the axon to the terminals (see Axonal Transport, Chapter 6). If radioactively labeled amino acids, usually tritiated leucine or proline, are injected into an area containing cell bodies, the amino acids are taken up by the cell bodies and incorporated into proteins synthesized by the neurons. The proteins that are transported in an anterograde direction to the axon terminals can be detected by autoradiography.

The unique advantage of the autoradiographic method is related to the fact that the label is taken up only by the cell bodies and not by fibers passing through the injection site. This gives a clear indication of the origin of the labeled pathway. However, the autoradiographic method is a lengthy procedure, and it does not provide the morphologic details of the labeled processes on the light microscopic level as does the recently introduced PHA-L method.

PHA-L Method. In 1984, two American neuroscientists, Charles Gerfen and Paul Sawchenko, discovered that the plant lectin *Phaseolus vulgaris*-leucoagglutinin (PHA-L) was a superior anterograde tracer. Soon after iontophoretic injection of PHA-L into a specific part of the nervous system, the tracer is incorporated into the cell bodies in the area of injection, and some of the lectin is transported at the slow rate (4–6 mm/day) of axonal transport to the axon terminals (Fig. 7–3). After a certain number of days, which depends on the length of the pathway being studied, the experimental animal is sacrificed and histologic sections are prepared in routine fashion from the brain or spinal cord. The neurons that have incorporated the tracer can then be detected by an immunohistochemical technique (see below). One of the great advantages of the PHA-L method is that the neurons whose cell bodies have taken up the lectin in the area of the injection become labeled in a Golgi-like fashion (Fig. 7–3B); their dendritic trees and axonal projections are displayed in great detail, and the axonal branches can be followed to their point of termination (Fig. 7–3C).

The PHA-L method and other tracer methods with similar qualities, e.g., the biocytin method, are now setting the standards for methods used in anterograde tract tracing on both the light and the electron microscopic level. Using a double-immunolabeling protocol, the PHA-L method can be combined with another immunohistochemical technique designed to reveal the specific transmitter in the neuron with which the PHA-L-labeled terminals establish their contacts (Fig. 7–4).

Retrograde Tracer Methods

In 1970–1971, the Swedish neuropathologist Krister Kristenson and his colleagues discovered that horseradish peroxidase (HRP) and the fluorescent dye Evans Blue (bound to albumin) were readily taken up by axon terminals and transported back to the corresponding cell bodies, where they could be detected (Fig. 7–5) either by a simple staining procedure (in the case of HRP) or by direct visualization in a fluorescence microscope (in the case of Evans Blue). In recent years many other excellent retrograde tracers, including the subunit B of cholera toxin and a number of fluorescent dyes, e.g., Fast Blue, Fluoro-Gold, and rhodamine beads, have been introduced. Fluorescence microscopy has the capacity to reveal substances in very low concentrations, and if two substances are chosen so that each fluoresces in a different color at a different excitation wavelength, it is possible to distinguish double-labeled cells in one and the same preparation following injections of the two substances in different projection regions of long branching axons (Fig. 7–6). Such double-labeling procedures have demonstrated that the axonal projections of many cell groups in the CNS are characterized by long branching axons. Another popular application for the retrograde fluorescent markers is their use in combination with fluorescent immunomarkers for the visualization of the neurotransmitter in the same neuron that is labeled by the retrograde fluorescent substance.

In spite of the great popularity of the retrograde fluorescent markers, which do not need any special histochemical staining procedure in order to be visualized, HRP is still widely used, especially in combined light and electron microscopic studies, since the HRP reaction product can be easily identified also at the electron microscopic level. The uptake and transport of HRP can be enhanced considerably by conjugating it with other substances such as plant lectins, e.g., wheat germ agglutinin (WGA). HRP, furthermore, is a favorite substance for intracellular labeling (see below).

In addition to being transported in a retrograde direction, HRP and several of the fluorescent tracers can also be taken up by cell bodies and transported in an anterograde direction to the axon terminals. However, the use of HRP and some of the fluorescent tracers as anterograde tracers is to some extent hampered by the difficulties of distinguishing anterogradely labeled terminal arborizations from axonal collaterals of retrogradely labeled cells.

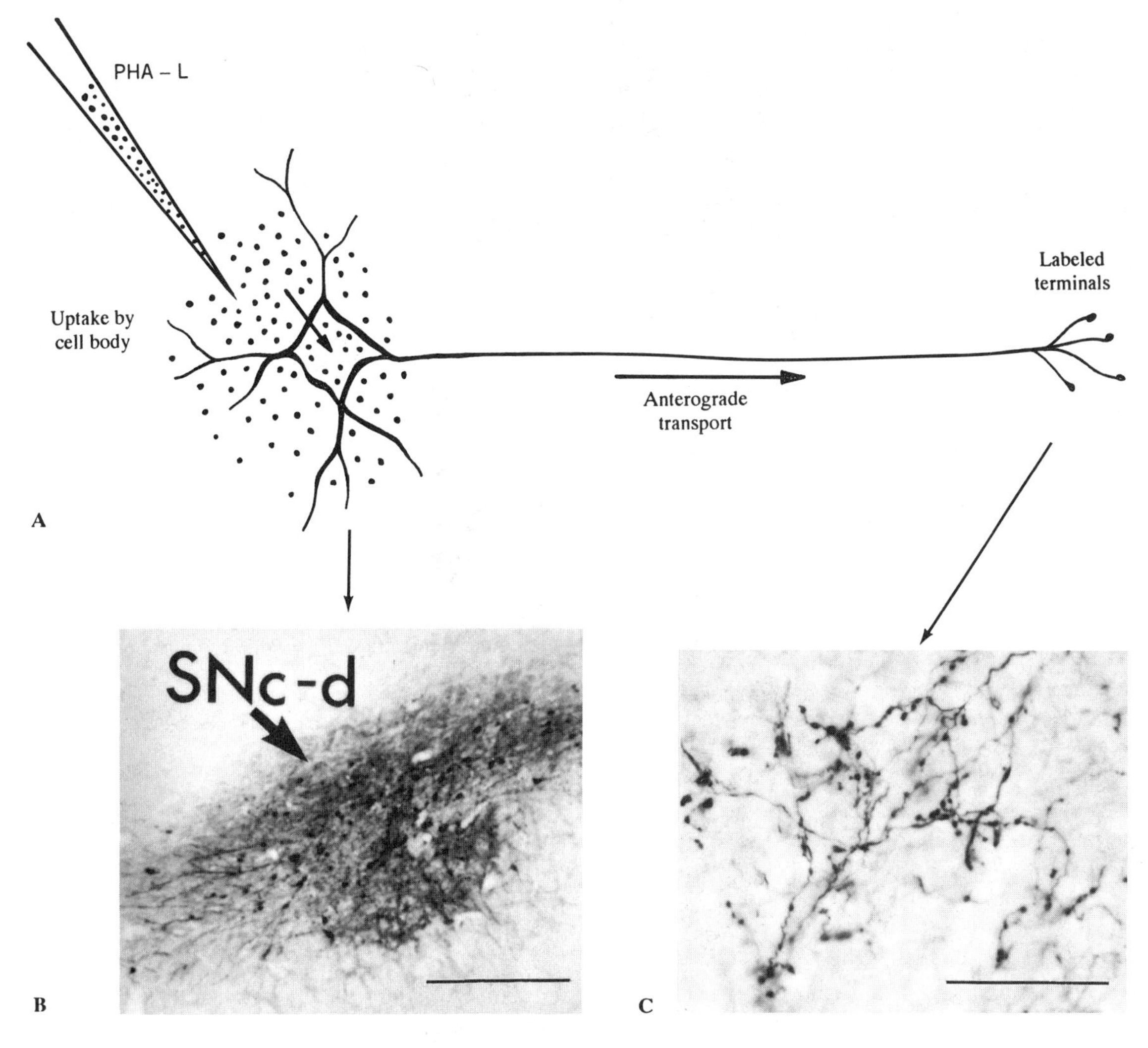

Fig. 7–3. The Anterograde Tracing Method Using PHA-L
If PHA-L is injected by iontophoresis through a fine pipette into a presumed area of origin of a pathway (in this case the substantia nigra) (**A**), the tracer is taken up by the cell bodies in the vicinity of the injection (**B**) and transported through the axons to their terminals in the striatum (**C**). (SNc-d: dorsal part of substantia nigra, pars compacta). Bar in **B** = 1 mm; in **C** = 0.1 mm. (Photographs courtesy of Dr. Charles Gerfen, National Institute of Health. From Gerfen, C. R., M. Herkenham, and J. Thibault, 1987. The neostriatal mosaic. II. Compartmental organization of mesostriatal dopaminergic and non-dopaminergic systems. J Neurosci 7:3915–3934. With permission of Oxford University Press.)

Transneuronal Tracer Methods

So far, we have discussed methods that are used for the tracing of direct monosynaptic projections from one region to another in the nervous system. Some of the tracers used in these methods can be transferred from neuron to neuron at synaptic junctions, in which case they can be used to analyze multineuronal pathways or systems of connected neurons. WGA-HRP, which was mentioned above as an excellent tracer, as well as tritiated proline (Fig. 13–12A), has been used with some success for this purpose. However, one of the problems in the tracing of multineuronal circuits is the dilution of the tracer at each synaptic transfer. This drawback has recently been eliminated through the use of live viruses.

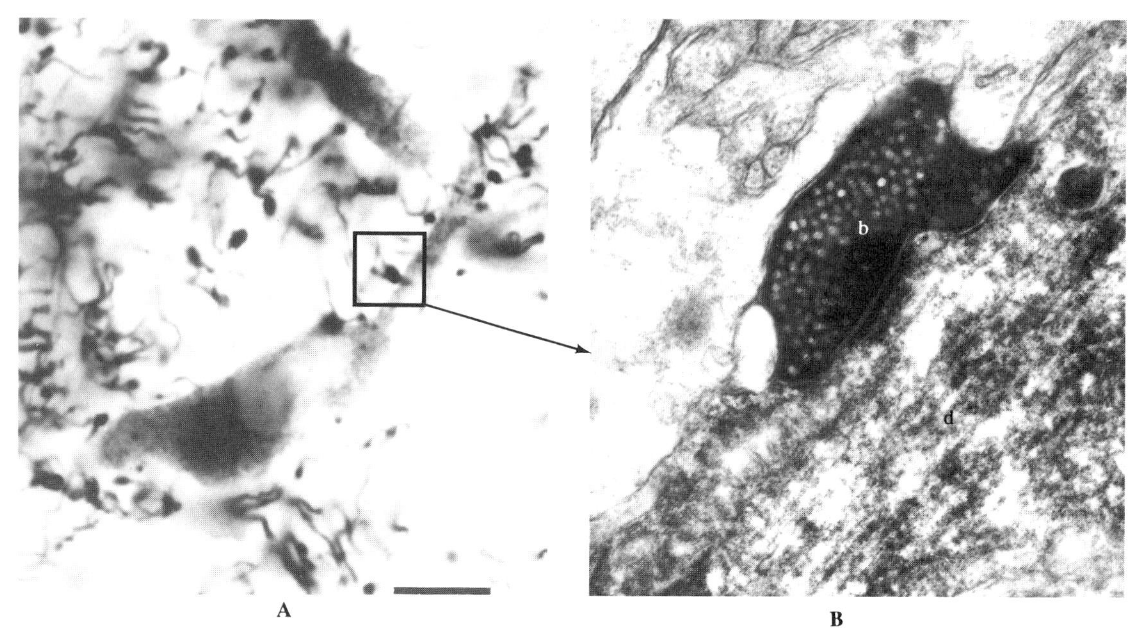

Fig. 7–4. PHA-L Method in Correlated Light and Electron Microscopic Studies
A. PHA-L-labeled axons and axon terminals surrounding a cholinergic neuron identified by immunohistochemistry using an antibody against ACh in the basal forebrain. Bar = 10 μm.

B. Electron micrograph showing one of the PHA-L-labeled boutons (b) in synaptic contact with the immunocytochemically labeled dendrite (d). (Micrograph courtesy of Dr. L. Zaborszky. From Brain Research, Vol. 570, 1992, pp. 92–101. With permission from Elsevier Science Publishers B. V.)

Viral Transport Method. The fact that a number of viruses can reach the CNS by retrograde axonal transport in peripheral nerves and then spread across synapses to other parts of the brain and spinal cord can be a serious health problem (see Clinical Notes, Chapter 6). However, axonal transport of viruses can also be exploited for experimental tract tracing in animals. The neurotropic viruses used in transneuronal tracing are especially herpes viruses, e.g., herpes simplex and pseudorabies virus, a herpes virus endemic to pigs and cattle. The advantage of using live viruses is that they replicate after the transneuronal transfer, which in turn amplifies the "tracer signal" in the recipient neuron. The virus antigen, which is synthesized when the virus reaches the cell nucleus, represents the marker that can be detected by immunohistochemical techniques. Viruses, like many other tracers, can be transported both in anterograde and retrograde direction, and it appears that the direction of transport is in large part dependent on the strain of the virus[3].

In spite of its great potential for revealing functionally relevant multineuronal circuits in the brain, the transneuronal viral transport method has not been widely used so far. Some of the reluctance to use the method is related to the fact that most of the neurotropic viruses are extremely virulent and likely to cause CNS infections in the experimental animal. They may also present a health hazard to humans, although this does not seem to be a great concern in the case of the pseudorabies virus, which in only a few instances has been suspected of causing infections in humans, although the disease caused by this virus in livestock is a source of significant loss to farmers in many parts of the world.

Although the transneuronal viral transport method has not gained widespread popularity, its

3. The neurotropic viruses may present a particular case where clinical and basic research are mutually reinforcing, since the neurotropic viruses are also under intense scrutiny as potential recombinant carriers of genetic repairs to specific sites in the CNS.

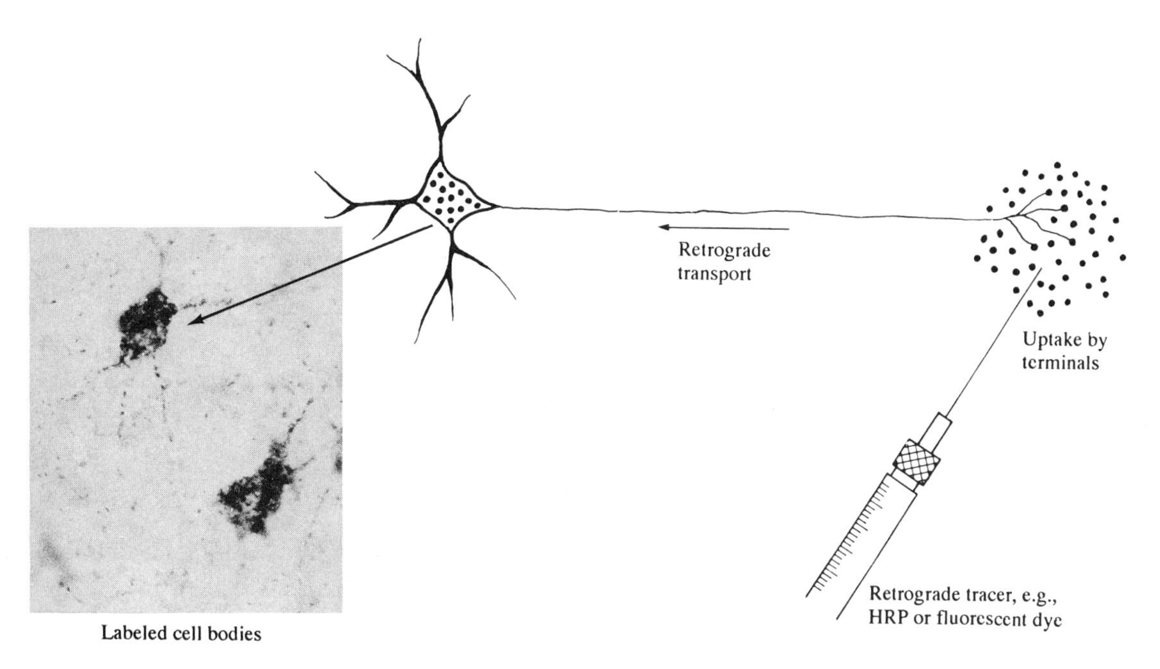

Fig. 7–5. Retrograde Tracing Methods
The retrograde tracer (e.g., horseradish peroxidase, HRP, or a fluorescent dye in solution) is injected into the terminal area of a projection where it is taken up by boutons. It is subsequently transported in retrograde direction to the cell bodies, where it can be detected by a simple staining procedure (in the case of HRP) or by direct visualization in a fluorescence microscope. Although HRP is readily transported both in retrograde and anterograde directions, it is most often used in retrograde tracing experiments.

remarkable capacity to identify functionally relevant multineuronal circuits has been well illustrated whenever the method has been used. The elucidation of the central autonomic network (Chapter 19) is a good example. To this end suspensions of the virus have been injected either into sympathetic ganglia or directly into visceral organs. Following a certain survival time, usually several days, the virus has had time enough to be retrogradely transported in the autonomic preganglionic fibers and then to be transneuronally transferred to the neurons that project to the preganglionic sympathetic or parasympathetic neurons in the CNS (Fig. 7–7). As indicated by the labeling in this figure, some of the most prominent parts of the central autonomic network in the forebrain are the paraventricular hypothalamic nucleus and the continuum formed by the central amygdaloid nucleus, the bed nucleus of the stria terminalis, and the cell columns between these two nuclei. This continuum is called the extended amygdala (see Chapters 16 and 20).

2-Deoxyglucose (2DG) Method. This method, which was developed in 1977 by the American scientist Louis Sokoloff and his collaborators as a means of studying cerebral glucose metabolism in vivo, can also be conceived of as a transneuronal tracing method with a distinct functional quality. In this method, a radioactively labeled substance, [^{14}C]2-deoxyglucose, is injected into the animal. Since 2DG is an analogue of glucose, it is taken up by the neurons as a source of energy. Unlike glucose, however, 2DG forms a metabolite that cannot easily cross the cell membrane, i.e., it is essentially trapped in the tissue. If the cells of a specific neuronal system are being stimulated, they increase their uptake of glucose and 2DG. This results in an increased intracellular accumulation of the radioactively labeled metabolite, which can be visualized by the autoradiographic technique following sacrifice of the animal (Fig. 13–12B). In other words, the method can be used to identify multineuronal functional systems in response to various stimuli or behavioral situations. However, the resolution of the technique is not good enough to identify the specific neurons involved.

The study of multineuronal pathways can also be accomplished by combining different tracer methods in one and the same experiment. Such experiments, although technically demanding, are becoming increasingly popular for the detailed

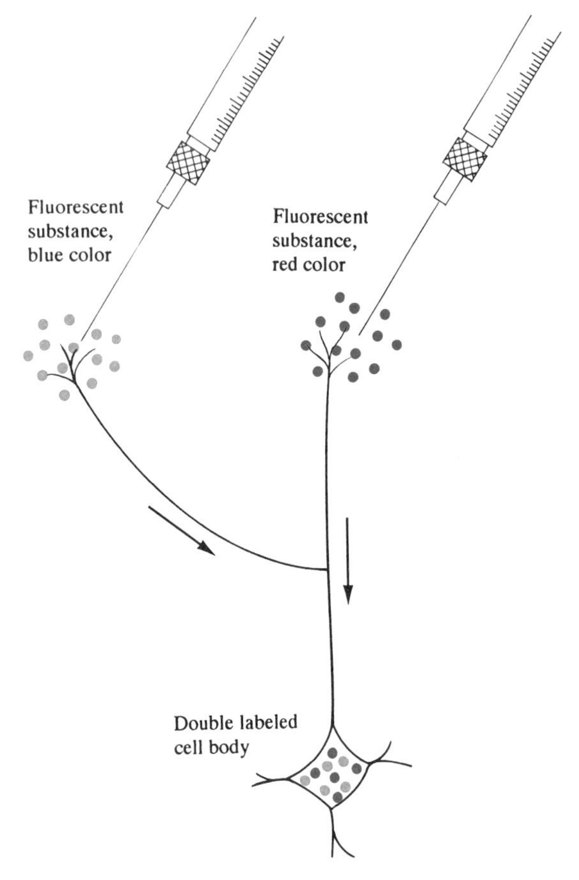

Fig. 7–6. Tracing of Collateral Projections
Two substances that fluoresce with different colors are injected in different projection areas of long branching axons. The fluorescent substances are transported in a retrograde direction to the same cell body, which thus becomes double-labeled. (From Steward, O. Horseradish peroxidase and fluorescent substances and their combination with other techniques, 1981. In: Heimer, L. and M. J. Robards. Neuroanatomical Tract-Tracing Methods. With permission of Plenum Press, New York.)

study of neuronal circuits (see Combined Techniques, below).

Intracellular Labeling Techniques

Intracellular injection of a marker substance is an excellent way to reveal the morphology of a neuron in a Golgi-like fashion (Fig. 6–4). It can be done in a number of ways: in the living animal, in unfixed brain tissue incubated in a slice chamber, or even in previously fixed brain slices. The usefulness of the intracellular labeling technique goes far beyond the morphological description of the injected neuron. One of the premium advantages of in vivo and in vitro (e.g., incubation of brain tissue in a slice chamber) cell labeling is that the neurophysiologic properties of the neuron can be studied before the marker is injected, which provides an excellent opportunity to directly correlate cellular physiology and morphology. Several substances, including HRP, biocytin, and Lucifer Yellow, are used as markers, and as described in some detail below, the technique can be combined with tract-tracing methods and immunocytochemical methods to determine the transmitter content of the intracellularly labeled neuron and of synaptic terminals contacting the labeled cell (see also Vignette, Chapter 12).

Transmitter Histochemistry

The Fluorescence Method for Monoamines

The fact that noradrenaline can react with formaldehyde to produce a fluorescent compound was already known, when in the early 1960s a Swedish research team, Nils-Ake Hillarp, Bengt Falck, Arvid Carlson, and collaborators, set out to develop a reliable fluorescence method for the demonstration of monoamine-containing neurons in the CNS. Their efforts were amply rewarded, and aided by increasingly sensitive modifications of the original method (the Falck-Hillarp method), and lately with the aid of immunohistochemical methods, the monoaminergic cell groups and pathways have been charted in great detail. Four different monoamines, i.e., dopamine, noradrenaline, adrenaline, and serotonin (5-hydroxytryptamine), serve as transmitters in the CNS, and the cell groups that give rise to the monoaminergic systems are located primarily in the brain stem (Chapter 10).

Immunohistochemical Methods

With the introduction of sensitive immunohistochemical methods in the 1970s, scientists were no longer restricted to the identification of monoaminergic transmitters; potentially, all transmitters and chemical messengers can now be identified. The immunohistochemical technique is based on the immune response, i.e., an organism's ability to develop antibodies against a foreign substance, the antigen. The antibodies are produced by injecting the antigen, e.g., a transmitter or enzyme, into a suitable species like the rabbit. The tissue section containing the specific antigen is then bathed with a solution of the antibodies, and the binding sites are revealed with fluorescent compounds such as rhodamine or fluorescein or with enzymatic reactions coupled to a chromogen such as diaminoben-

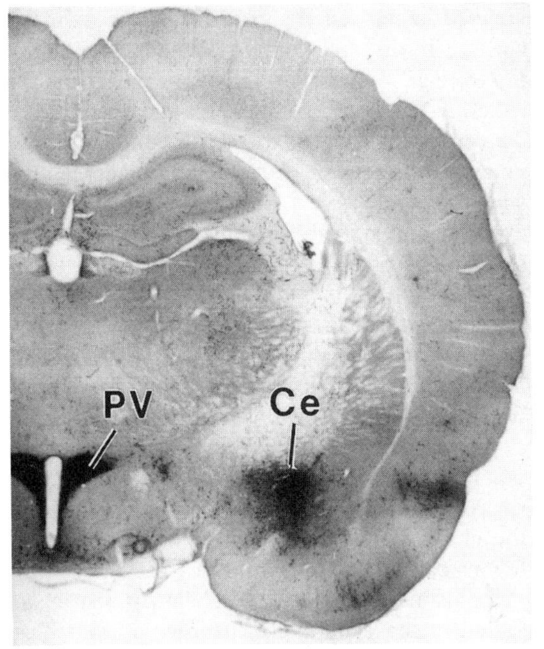

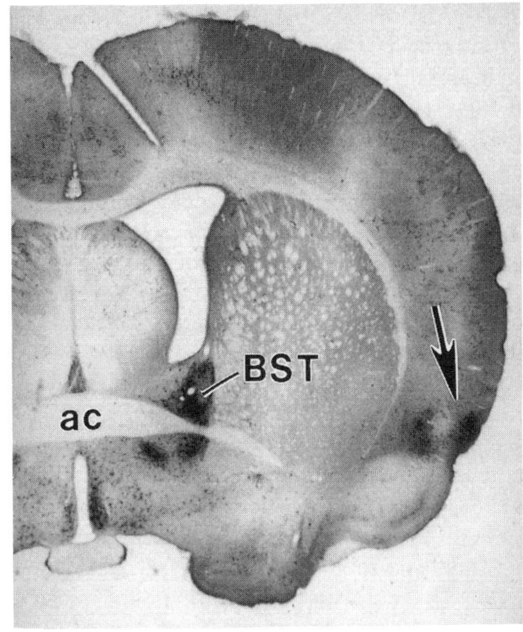

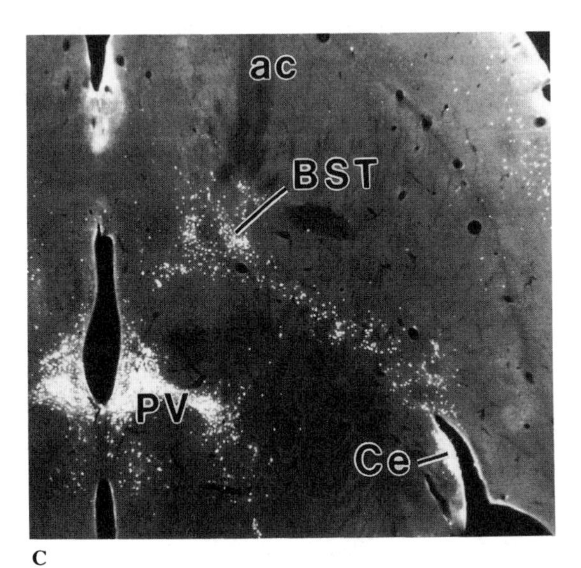

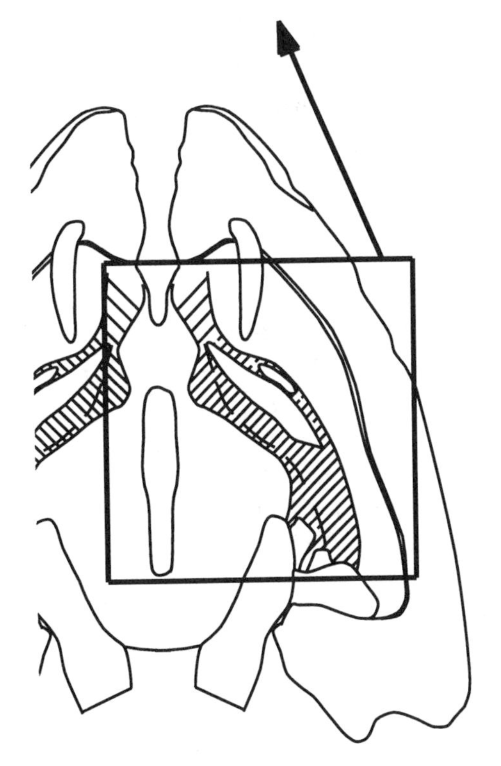

Fig. 7–7. Transneuronal Viral Transport Method
A and B. Retrograde viral labeling of the cells in the paraventricular hypothalamic nucleus (PV), central amygdaloid nucleus (Ce), and bed nucleus of the stria terminalis (BST) following injection of pseudorabies virus into the gastrointestinal tract of a rat, whose brain was cut into frontal sections (**A** is somewhat more caudal than **B**). The virus has been transsynaptically transported to the hypothalamus and forebrain via the solitary–vagal complex in the brain stem. The *arrow* in **B** points to the part of the rat cerebral cortex that corresponds to the insula in the human. ac = anterior commissure. (Courtesy of Dr. Richard Miselis.)

C and D. Horizontal brain sections from another rat treated in the same fashion as the previous animal. The section in **C**, which cuts through the most voluminous part of the PV, corresponds approximately to the *square* indicated in the drawing of the whole hemisphere section in **D**. Note the continuous labeling of neurons in part of the extended amygdala (*hatched area*). (Courtesy of Dr. George Alheid.)

zidine (DAB), so that the antigen–antibody complex can be detected in the microscope. In principle, any protein formed in the body can be detected by the immunohistochemical method. With the commercial production of a large number of specific antibodies, and the development of new and highly sensitive immunohistochemical methods for use at both the light and the electron microscopic levels (Fig. 17–2), the mapping of transmitter-specific cell groups and peptidergic systems has expanded enormously into a specialty known as "chemical neuroanatomy" (Chapter 23).

One of the most remarkable developments during the last decade has been the revival of the study of the human brain. For most of this century, the anatomic study of the human brain was restricted largely to the study of cytoarchitecture and major fiber bundles by the aid of the Nissl method, classic silver methods, and myelin stains; the detailed mapping of neuronal circuits using sophisticated methods requiring injections of chemically and biologically active tracers into the brain is for obvious reasons restricted to animal experiments. However, by the aid of immunohistochemical methods it is now possible to map transmitter-specific and peptidergic cell groups and pathways in human brains (Fig. 16–2C). To obtain additional perspectives, the results can be compared with what is already known about these structures in various animals including other primates.

Combined Techniques: The Tracing of Successive Links in a Neuronal Circuit

The most recent trend in experimental neuroanatomy is characterized by an approach in which a combination of techniques, including the intracellular injection technique and various axon tracing and immunohistochemical techniques, is used in the same experiment. In such experiments, synaptic relations can be studied between neurons that are also characterized in terms of their axonal projections and/or transmitter content. Although technically demanding, such experiments offer the great advantage of being able to correlate cell morphology, connectivity, and histochemistry in the same tissue specimen. This approach is exemplified in the schematic Fig. 7–8A, which illustrates the tracing of a pathway from the upper right-hand corner to the lower left-hand corner. Axonal degeneration is elicited by an experimental lesion as indicated by the scalpel in the figure. Alternatively, an anterograde tracer, e.g., PHA-L, can be injected in the cell group in the upper right-hand corner. If, in the same specimen, a retrograde tracer, e.g., fluorescent substance or HRP, is injected in the lower left-hand corner, the neuron in the center of the figure can be identified in terms of its axonal projection. After a certain survival time, which depends largely on the choice of tracers, the experimental animal is sacrificed. If suitable tissue slices are prepared, the retrogradely labeled neurons in the center of the figure can be identified (B) and injected with an intracellular tracer, e.g., Lucifer Yellow, which stains the neurons in a Golgi-like fashion (C). This makes it possible to identify their dendrites, where most of the synaptic contacts occur. If the light microscopic analysis reveals potential synaptic contacts between the participating neurons, the material can be prepared for study on the electron microscopic level, where synaptic relations can be studied in detail (D).

Transmitter immunohistochemistry can also be added to the experiment in order to characterize the participating neurons in terms of their chemical messengers. This situation is also illustrated in (D), which shows a Lucifer Yellow-labeled dendrite contacted by a degenerating bouton (DB) as well as by an immunohistochemically identified tyrosine hydroxylase-positive bouton (TH).

The possibilities to combine and exploit various techniques appear almost unlimited, and even if some of the combinations require a determined effort and considerable expertise, the technical repertoire is finally in place to disentangle the extraordinary complicated neuronal systems and microcircuits that form the basis for our behavior.

⟶

Fig. 7–8. Tracing of Successive Links in a Neuronal Circuit

A. This highly schematic and self-explanatory drawing illustrates some of the possibilities available to the neuroscientist who wants to identify successive links in a neuronal circuit. **B**, **C**, and **D**: Examples from experimental studies using the general approach outlined in **A**.

B. Cell body retrogradely labeled with a fluorescent tracer, which appears in the form of perinuclear granules following injection of the tracer in the area to which the neuron projects (see lower right-hand corner in A).

C. The cell in **B** after it has been intracellularly injected with Lucifer Yellow as indicated in **A**.

D. Electron micrograph showing an immunohistochemically identified tyrosine hydroxylase-positive bouton (TH) and a degenerating bouton (DB) in synaptic contact with a Lucifer Yellow-labeled dendrite, which is overlaid by a light blue color. (B and C from Buhl, E. H., 1993. Reprinted by permission of Wiley-Liss, a division of John Wiley and Sons, Inc., New York. Microscopy Research and Techn. 24:15–30; D. Micrograph courtesy of Luke Johnson and Paul Jays, Oxford University.)

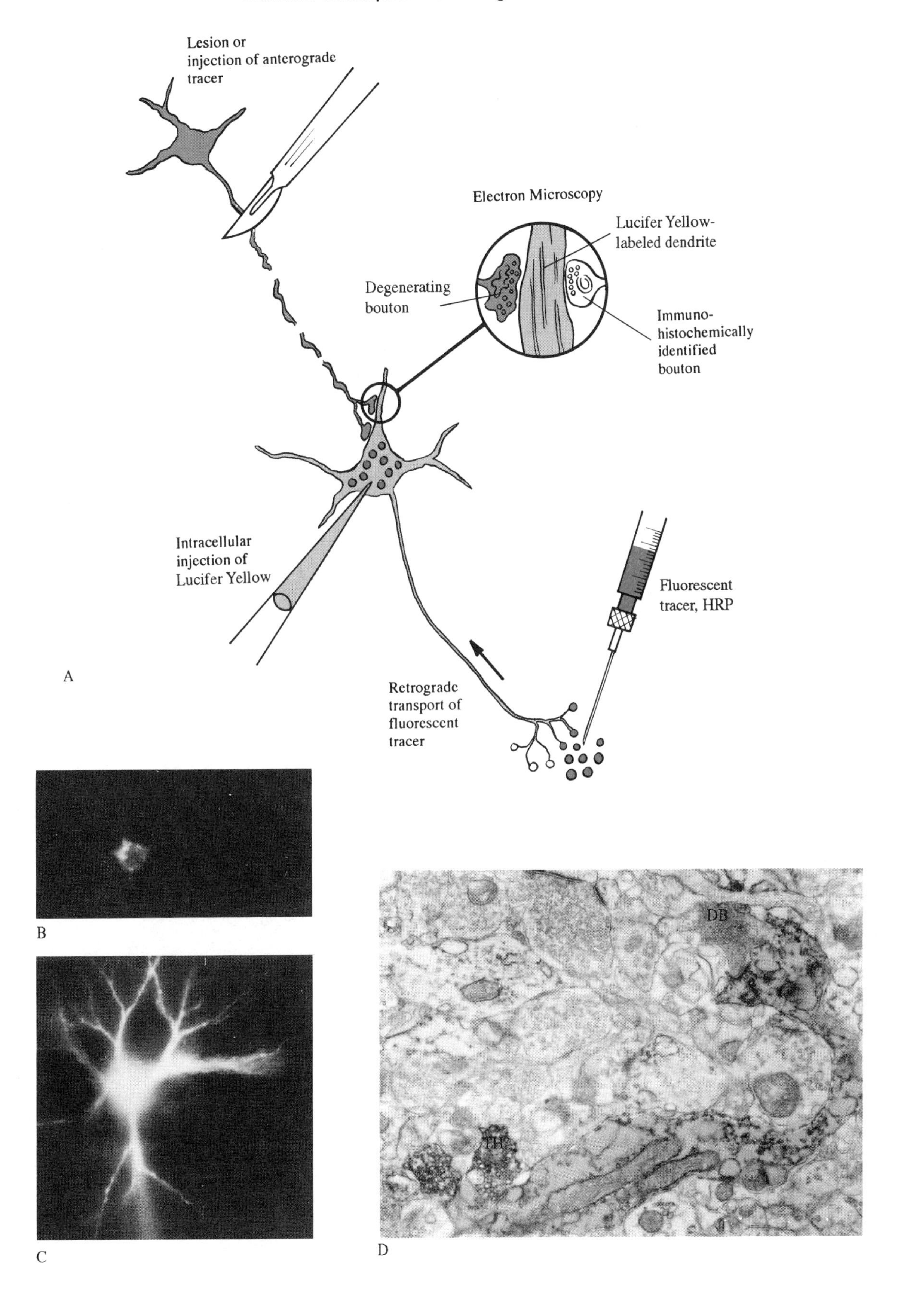
Lesion or
injection of anterograde
tracer
Electron Microscopy
Lucifer Yellow-
labeled dendrite
Degenerating
bouton
Immuno-
histochemically
identified
bouton
Intracellular
injection of
Lucifer Yellow
Fluorescent
tracer, HRP
A
Retrograde
transport of
fluorescent
tracer
B
C
DB
TH
D

In Situ Hybridization

The most recent addition to the technical arsenal is in situ hybridization, which offers the opportunity to localize the site where special mRNAs are synthesized. In situ hybridization is based on recombinant DNA technology, and it has been useful for locating the mRNAs associated with various neurotransmitter/neuropeptide systems (Fig. 16–2C) and receptors. Other exciting areas for exploration include the regulation of gene expression during development and changes in gene expression in response to stimulation.

Suggested Reading

1. Björklund, A., T. Hökfelt, F. G. Wouterlood and A. N. van den Pol (eds.), 1990. Handbook of Chemical Neuroanatomy. Vol. 8: Analysis of Neuronal Microcircuits and Synaptic Interactions. Elsevier: Amsterdam.
2. Bolam, J. P. (ed.), 1992. Experimental Neuroanatomy. A Practical Approach. Oxford University Press: Oxford.
3. Buhl, E. H., 1993. Intracellular injection in fixed slices in combination with neuroanatomical tracing techniques and electron microscopy to determine multisynaptic pathways in the brain. Microsc. Res. Techn. 24:15–30.
4. Heimer, L. and M. J. RoBards (eds.), 1981. Neuroanatomical Tract-Tracing Methods. Plenum Press: New York.
5. Heimer, L. and L. Záborszky (eds.), 1989. Neuroanatomical Tract Tracing Methods 2. Recent Progress. Plenum Press: New York.
6. Meredith, G. E. and G. W. Arbuthnott (eds.), 1993. Morphological Investigations of Single Neurons in vitro. John Wiley & Sons, Inc. Somerset, New Jersey.

Midline sagittal MRI of cervical spinal cord and brain structures in the infratentorial cranial compartment. (Photograph courtesy of Dr. Maurice Lipper.)

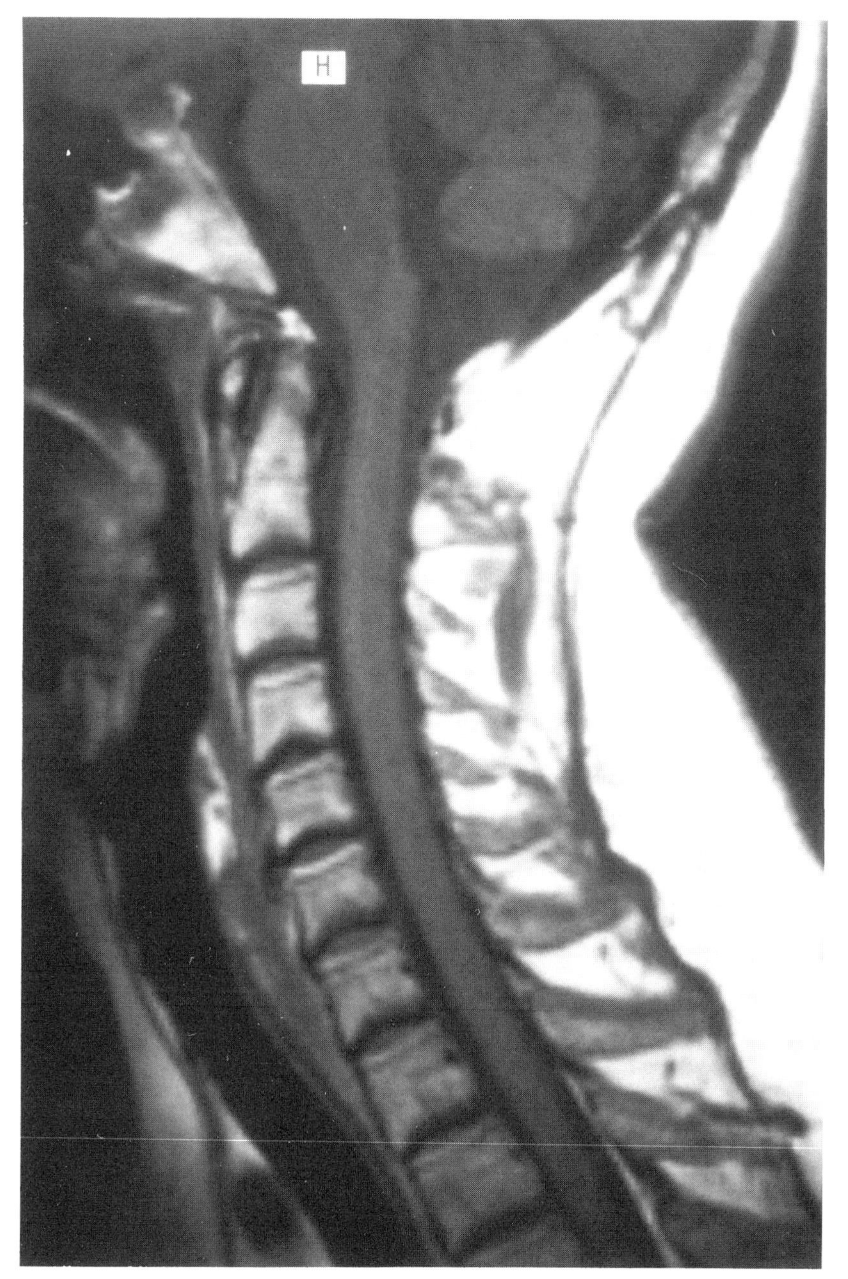

8

The Spinal Cord

The spinal cord is an almost 0.5-m-long cylindrical part of the CNS that occupies the upper two-thirds of the vertebral canal. It has a unique role as a conductor of impulses between the brain and the peripheral nerves; ascending pathways carry impulses to the brain from all parts of the body except the head, whereas various brain regions can influence the spinal cord motor neurons and modulate the activity in the spinal cord sensory relay nuclei through a number of descending pathways. But the spinal cord is much more than a conductor of ascending and descending impulses; it contains a wealth of intrinsic neuronal circuits subserving a multitude of spinal reflexes.

The frequency with which the vertebral column and its associated structures are damaged underscores the clinical importance of the spinal cord. Spinal cord disorders are often serious in nature, and their treatment presents a great challenge to the clinical staff. Although such disorders cannot always be cured, they can be understood and properly managed by people who are familiar with the functional-anatomic features of the spinal cord and its structural relationships to the vertebral column.

The vertebral canal and its contents are seldom dissected in medical school courses, but the anatomy of the spinal cord is of great importance in medical theory and practice. The gross anatomic features of the spinal cord and its particular relationship to the surrounding vertebral column is a subject of immediate interest for the practicing physician, and a basic knowledge of the internal organization of the spinal cord is a prerequisite for further inquiry into the functional-anatomic systems of the CNS.

The Spinal Cord and Its Relationship to the Vertebral Column

Gross Anatomic Features

The spinal cord forms a nearly cylindrical column that is almost 0.5 m long in the adult; its diameter varies between 1 and 1.5 cm. Two continuous rows of nerve roots emerge on each side of the cord (Fig. 8–1). The roots join distally to form 31 pairs of spinal nerves, which exit through the intervertebral foramina. The spinal nerves distribute sensory and motor fibers to all parts of the body.

The thickness of the spinal cord varies with the number of outgoing and incoming peripheral nerve fibers at different levels. The parts of the spinal cord that innervate the upper and lower limbs form spindle-shaped enlargements, the *cervical enlargement*, which extend between C3 and T1, and the *lumbar enlargement*, segments L1–S2 (see Fig. 8–3).

A cross-section of the spinal cord shows a characteristic division between the centrally placed gray substance and the surrounding white substance. The latter is formed by the long ascending and descending fiber tracts as well as intraspinal pathways. The cell groups seen in cross-sections actually form columns of cells that in some instances extend throughout the length of the spinal cord.

Spinal Nerves

The gray matter and white matter of the spinal cord are not segmented. However, the innervation of the body is orderly in the sense that one part of the spinal cord supplies a limited segment of the body with sensory and motor fibers. This occurs through the distribution of 31 pairs of segmentally arranged spinal nerves, which give the impression of a segmental organization of the spinal cord.

A ventral root and a dorsal root come together in the intervertebral foramen and form a *spinal nerve* (Fig. 8–1). The *ventral root*, which emerges as a series of root filaments along the ventrolateral sulcus, contains efferent somatic and, at specific levels, efferent visceral (sympathetic and parasympathetic) fibers.[1] These axons originate in motor nuclei of the spinal cord. A *dorsal root*, likewise, consists of several root filaments, which are attached to the spinal cord in the region of the dorsolateral sulcus. It contains primary afferent somatic and visceral fibers from cells in the *dorsal root ganglion* (spinal ganglion). The dorsal root ganglion appears as an oval swelling along the dorsal root close to its junction with the ventral root.

After the third fetal month, the vertebral column grows faster than the spinal cord. This results in a relative displacement of especially the lower parts of the spinal cord to higher levels of the vertebral column. The spinal nerves emerge from the vertebral canal before this unequal growth commences, and the dorsal and ventral roots, in particular the lumbosacral ones, are stretched out between their point of attachment to the cord and their entry into the intervertebral foramina. This growth or lengthening of the roots occurs within the vertebral canal, and the large collection of nerve roots below the first lumbar vertebra is called the *cauda equina* (Fig. 2–4).

Segmental Innervation of the Body

The peripheral distribution of the spinal nerves reflects an original segmental organization in which each spinal nerve is composed of fibers that are related to the region of the skin, muscles, or connective tissue that develops from one body segment (somite).

Innervation of the Skin: Dermatomes

The segmental organization is most pronounced in regard to the sensory innervation of the skin. The area of the skin that is supplied by sensory fibers of an individual spinal nerve is called a *dermatome*. Charts of the different dermatomes (Fig. 8–2) are important aids in the diagnosis and localization of pathologic processes that affect the spinal nerves or the spinal cord. However, one should be aware of the fact that there is considerable overlap between neighboring dermatomes. It is useful to remember a few key dermatomes.

C3: Neck
C5: Deltoid region
C6: Radial forearm and thumb
C8: Ulnar border of hand and little finger
T4–T5: Nipple
T10: Umbilicus
L1: Groin
L3: Knee region
L5: Dorsal side of foot and great toe
S1: Lateral side of foot and little toe
S3–S5: Genito-anal region

1. The ventral roots also contain a considerable number of mostly unmyelinated sensory fibers. Most of these fibers, however, seem to end blindly within the root and only some enter the spinal cord. Their function is unknown, but they may account for the occasional lack of success of dorsal rhizotomy in relieving pain.

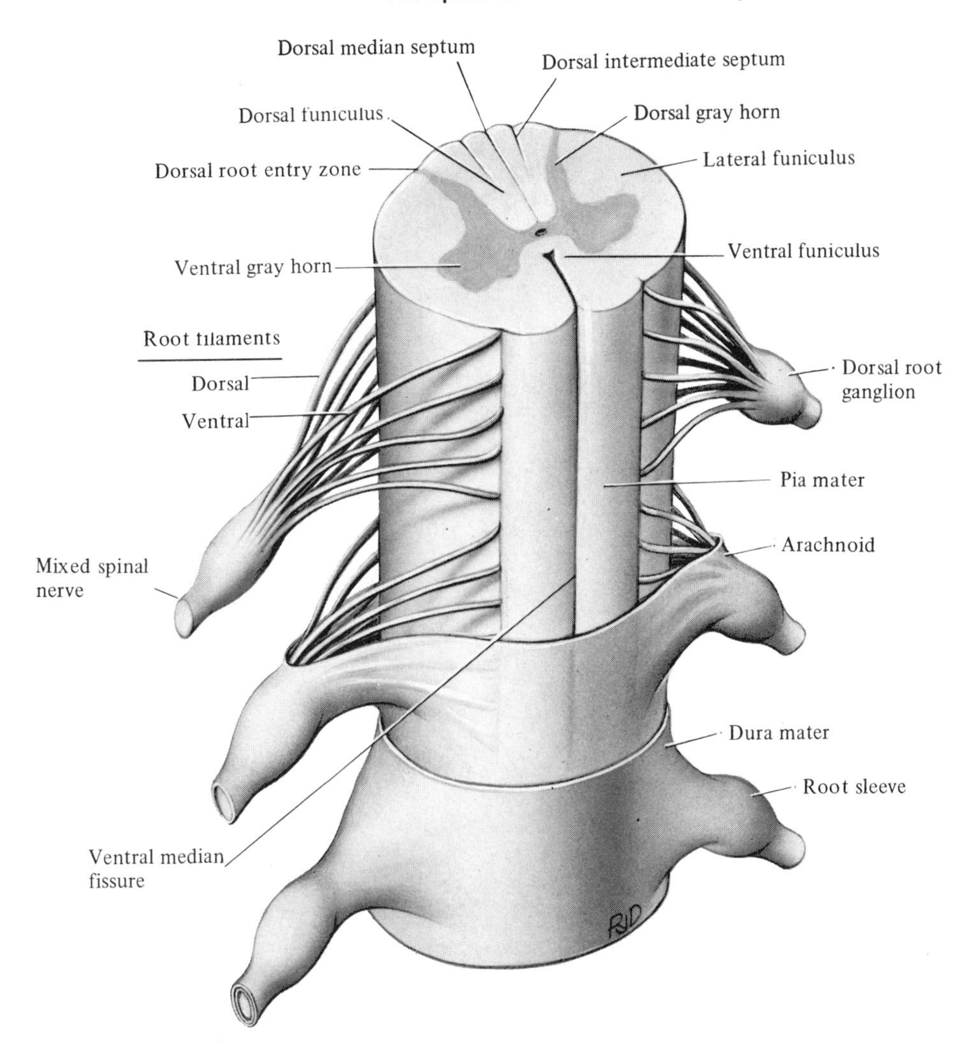

Fig. 8–1. Spinal Cord
The spinal cord forms a nearly cylindrical column surrounded by dura mater, arachnoid, and pia mater, all of which are continuous with the corresponding membranes of the brain. A series of dorsal and ventral rootlets emerges on each side of the cord. (Photograph courtesy of M. B. Carpenter. From Carpenter, M. B., 1978. Core Text of Neuroanatomy. 2nd. ed. With permission of the Williams & Wilkins Company, Baltimore.)

It is important to try to distinguish a spinal nerve or root lesion from a peripheral neuropathy. To do so one must know the cutaneous distribution of the peripheral nerves. The cutaneous areas supplied by the various nerves are illustrated in the Appendix.

Innervation of the Muscles

Although the segmental organization is less obvious in regard to the motor innervation, the muscles deep to a certain area of the skin are usually innervated by approximately the same segments as the overlying skin. The nerve supplies of the individual muscles are illustrated in the Appendix. A summary of the spinal cord levels for the motor neurons that innervate various muscle groups is presented below.

- C1–C4: Neck muscles
- C3–C5: Diaphragm
- C5–C6: Biceps
- C5–C8: Muscles of the shoulder joint
- C7–C8: Triceps and long muscles of the forearm
- C8–T1: Muscles for movements of the digits and the small intrinsic muscles of the hand

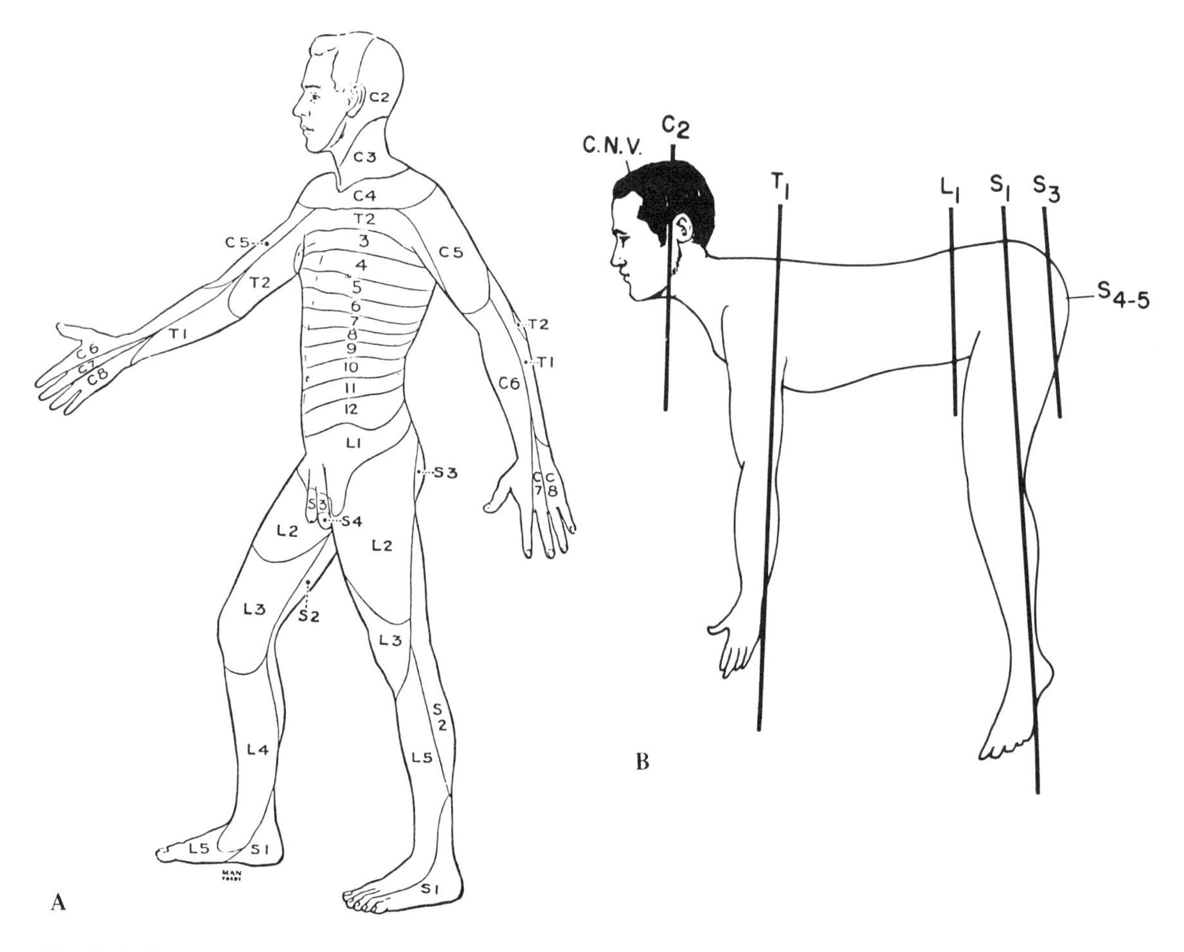

Fig. 8–2. Dermatomes
A. The segmental innervation of the skin, i.e., the dermatomes, from a lateral view.
B. Key dermatome boundaries (A. After Foerster, O., 1933. The dermatomes in man, Brain, 56:1–39, B. From Curtis, B. A., S. Jacobson, and E. M. Marcus. 1972. An Introduction to the Neurosciences. With permission of W. B. Saunders Company, Philadelphia.)

T2–T12: Axial musculature, intervertebral muscles, muscles of respiration, and abdominal muscles
L1–L2: Flexors of the thigh
L2–L3: Quadriceps femoris
L5–S1: Gluteal muscles
S1–S2: Plantar flexors of the ankle
S3–S5: Muscles of the pelvic floor, bladder, sphincters, and external genital organs.

Internal Organization of the Spinal Cord

The pattern of gray and white matter seen in transverse sections through the cord varies according to the level of the cord (Fig. 8–3). The largest cross-sections are present in the cervical and the lumbar enlargements. Here the dorsal and especially the ventral horns of the gray substance are particularly well developed. A characteristic feature of the gray matter in the thoracic and upper lumbar segments is the presence of a lateral horn in which the autonomic preganglionic motor neurons are located. The amount of white substance decreases in progressively caudal sections because the long ascending and descending pathways contain fewer axons at successively more caudal levels of the spinal cord.

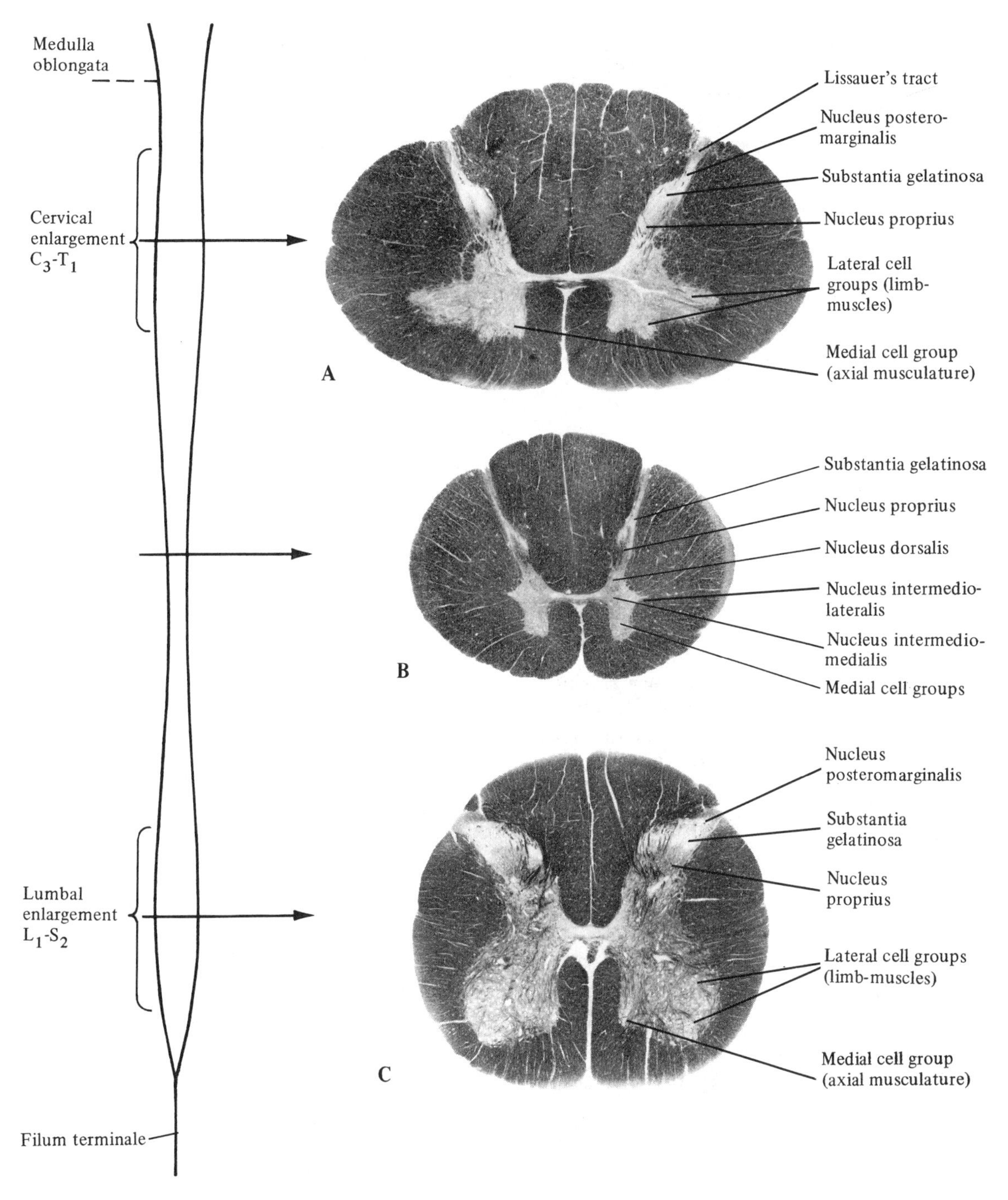

Fig. 8–3. Cross-Sections of the Spinal Cord
The cross-sections of the spinal cord are wider at the level of the cervical and lumbar enlargements than elsewhere. Note that the relative amount of gray and white matter is also different at different levels. The amount of white matter decreases gradually in the caudal direction, since the long ascending and descending fiber tracts contain fewer axons at successively more caudal levels of the spinal cord. The main nuclear groups in the gray matter have been indicated on the right halves of the spinal cord sections. (The myelin-stained sections of the spinal cord are reprinted from Gluhbegovic, N. and T. H. Williams, 1980. The Human Brain and Spinal Cord. JB Lippincott Co., Philadelphia. With permission of the publisher.)

Gray Matter

The gray matter has a characteristic butterfly appearance in cross-sections. It consists of a large number of neurons and their processes, in addition to an even larger number of neuroglial cells. The head of the dorsal horn, the substantia gelatinosa, is characterized by a large number of small neurons, and one generally gets the impression that the neurons increase in size as one moves from the apex of the dorsal horn in the direction of the ventral horn (Fig. 8–3). A more detailed analysis, however, reveals that most parts in front of substantia gelatinosa contain neurons of various sizes, although the largest neurons, i.e., the primary motor neurons, appear in the ventral horn.

Although the different parts of the gray matter are usually referred to as the *ventral* and *dorsal horns*, and the *intermediate gray* between the two horns, it is important to realize that the horns in reality represent columns of gray matter that extend throughout the entire length of the spinal cord.

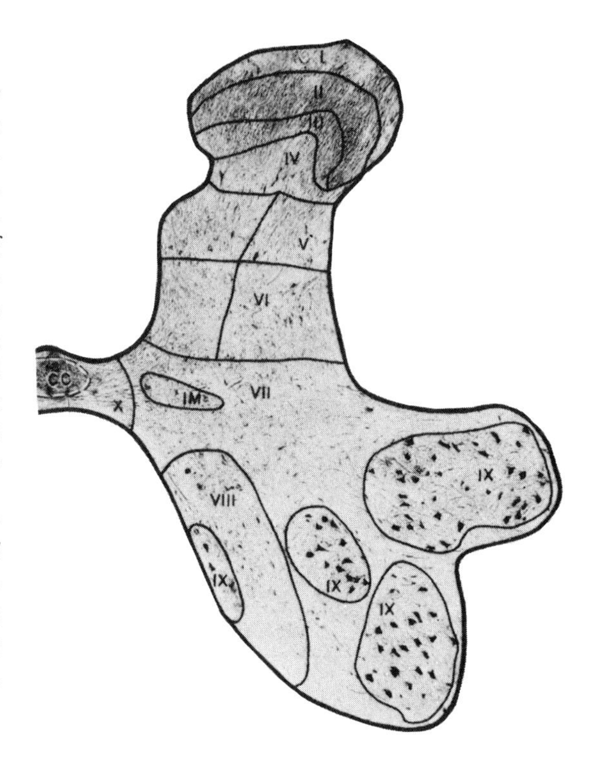

Fig. 8–4. Rexed's Laminae
Human spinal cord segment L5 stained with the Nissl method. On the basis of cytoarchitectonic features, the gray substance of the spinal cord can be subdivided into 10 zones or laminae (From Truex and Carpenter, 1969. Human Neuroanatomy. With permission of Dr. Carpenter and the Williams & Wilkins Company.)

Rexed's Laminae

In 1952, the Swedish neuroanatomist Bror Rexed divided the spinal cord gray matter into a series of 10 zones or laminae (Fig. 8–4) on the basis of cytoarchitectonic features. Soon after his pioneering histologic study, Rexed retired from the laboratory and spent the rest of his active career as the minister of social affairs in Sweden. In the meantime, however, Rexed's scheme became increasingly popular especially since it proved to be functionally relevant, i.e., functional properties of neurons tend to change in a predictable fashion from lamina to lamina.

The first six laminae subdivide the dorsal horn into more or less horizontal zones. Laminae I–IV comprise the head of the dorsal horn, whereas lamina V constitutes the neck of the dorsal horn. Lamina VI is present only in the enlargements. Lamina VII corresponds in general to the intermediate gray, whereas laminae IX and X occupy the ventral horn.

Lamina I, the *marginal zone*, consists of a thin layer of cells that caps the tip of the dorsal horn. The neurons are of varible size, and many of the axons originating in the marginal zone join the spinothalamic pathway (see Chapter 9); others project to other parts of the spinal cord.

Lamina II, the *substantia gelatinosa*, forms most of the head of the dorsal horn, and is especially well developed in the enlargements. It acquires its gelatinous appearance from a wealth of small neurons and unmyelinated fibers. The substantia gelatinosa can be easily recognized in myelin-stained sections because of its almost total lack of myelinated fibers. Inhibitory interneurons in the substantia gelatinosa have attracted much attention in recent years in connection with the gate theory of pain (see Chapter 9).

Laminae III and IV do not contain as many cells as lamina II, but some of the cells, especially in lamina IV, are larger than in the substantia gelatinosa. These laminae can be most easily identified in myelin-stained section, where especially lamina IV stands out because of its high concentration of myelinated fibers. Several of the cells in lamina IV, like those in the adjoining laminae V and VI, send their axons into the spinothalamic tract. Lamina IV corresponds in large part to *nucleus proprius*.

Laminae V and VI occupy the base of the dorsal horn and are usually considered together since they cannot be easily distinguished from each other. They contain mostly medium-sized neurons.

Lamina VII, which is populated by a large number of rather widely scattered medium-sized neurons, has poorly defined borders. However, it contains some well-known cell columns, e.g., the intermediolateral cell column, which is present from T1 to L2 and which contains the preganglionic sympathetic neurons. This column is responsible for the formation of the lateral horn in the thoracic part of the spinal cord. A corresponding cell column is located at levels S2–S4 for the sacral part of the parasympathetic division. Another distinctive cell column is Clarke's column, which is located in the dorsomedial part of the lamina at the base of the dorsal horn in segments about C8–L3. Axons from the nucleus of Clarke form the dorsal spinocerebellar tract (Chapter 17).

Lamina VIII differs from the previous laminae in two respects. It extends in a dorsoventral rather than a mediolateral direction in the enlargements, and it reaches its largest extent in the thoracic part of the spinal cord where it covers most of the ventral horn. It contains interneurons and propriospinal neurons, especially with long axons which project up and down for several segments.

Lamina IX contains the large multipolar motor neurons for the innervation of the skeletal muscles. The cells are arranged in several cell columns, which are especially prominent in the enlargements, where dorsolateral, ventrolateral, and central columns for the innervation of the extremities can be identified. Within these cell groups in the lateral part of the ventral horn, there is a further pattern of somatotopic localization in the sense that the neurons that innervate the distal limb musculature are located dorsal to those supplying the proximal limb muscles. A ventromedial cell column, which innervates the neck and trunk muscles, can be identified from segments C1 to L4.

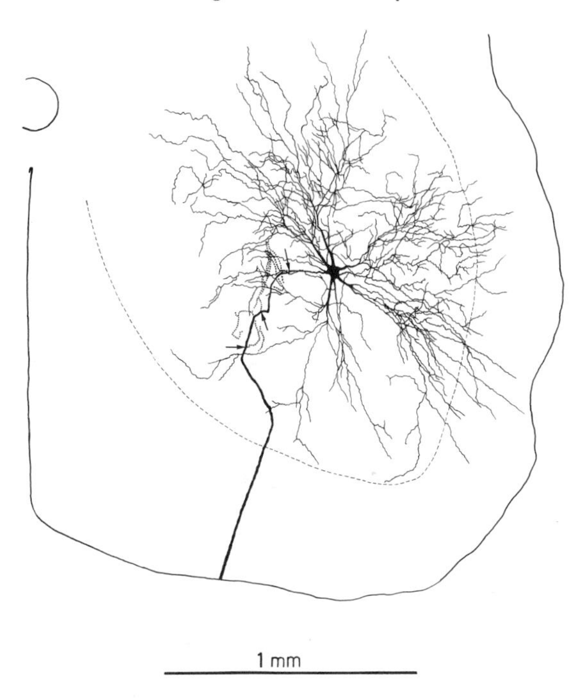

Fig. 8–5. Motor Neuron
Motor neuron in lamina IX labeled intracellularly with HRP. Note the extensive dendritic arborizations. (From Cullheim, S. and J. O. Kellerth, 1976. Combined light and electron microscopic tracing of neurons including axons and synaptic terminals after intracellular injection of horseradish peroxidase. Neurosci. Lett. 2:307–313. With permission from the authors and Elsevier, Amsterdam.)

Dendroarchitecture

It is important to realize that the Nissl method, on which Rexed's laminar scheme is based, only stains nerve cell bodies. The often elaborate dendritic trees, which receive the majority of the input, are not revealed. Since the dendrites often form a very elaborate three-dimensional pattern, the Nissl method leaves a false sense of simplicity. This becomes plainly evident when applying the Golgi method or an intracellular staining technique, which display the whole neuron including its dendritic and axonal processes. Figure 8–5, for instance, shows an example of an intracellularly labeled motor neuron in the dorsolateral column of lamina IX. The marker substance, which in this case was HRP, reveals an amazingly rich and multipolar, more or less spherical dendritic tree with an especially large spread in the transverse plane. The widely spread dendrites provide the opportunity for this type of cell to receive a multitude of afferent fibers from various directions.

The orientation of the dendrites and the dendritic spread varies from location to location within the spinal cord. For instance, lamina IV cells, many of which give origin to spinothalamic fibers, tend to have an antenna-like morphology, i.e., a cone-shaped dendritic tree with the apex directed ventrally and many of the dorsally directed dendrites reaching into the substantia gelatinosa, where they can be influenced by the synaptic input to this region. The dendritic orientation and spread, i.e., the dendroarchitecture, show a variety of different forms across the gray substance of the spinal cord, but it appears that cells within the same lamina tend to exhibit similar dendritic branching patterns, which again speaks to the validity of Rexed's laminar subdivisions.

Chemoarchitecture

The discovery, during the last few decades, of transmitter-specific neurons and pathways in the nervous system has had an enormous impact on all fields of neuroscience, not least on the clinical side, where it suggested the possibility of correcting transmitter disturbances in a systematic fashion by pharmacologic means. In the spinal cord, the opiates methionine enkephalin (M-ENK) and leucine enkephalin (L-ENK) have attracted special attention because of their potential importance in pain mechanisms (Chapter 9). But the opiates are only some of a large number of chemicals that have been identified in spinal cord neurons and terminals. These include a large number of neuropeptides besides the opiates, as well as the monoamines and several amino acids, e.g., glutamate, glycine, and GABA.

The dorsal horn is especially rich in neuropeptides, which in general have a laminar localization. Substance P-positive neurons, for instance, are present both in the dorsal root ganglia and in the dorsal horn, and substance P has attracted attention as a possible transmitter in the nociceptive pathway (Chapter 9). The fact that neuropeptides are present both in terminals and in different types of neurons suggests that they are involved in many functions besides nociception. The monoaminergic terminals that have been identified throughout most of the spinal cord are generally of supraspinal origin. Dopamine is found primarily in the dorsal horn, whereas serotonin and noradrenaline are abundant in the ventral horn, where they, like ACh, are implicated in motor functions (Chapter 10). Both GABA and glycine are important inhibitory transmitters in the spinal cord; GABA-ergic neurons and terminals are prominent in the dorsal horn, especially laminae I and II, whereas glycine is more abundant in the ventral horn, where for instance the inhibitory Renshaw cells (Chapter 15) contain glycine.

Interneurons

The majority of the neurons in the gray substance of the spinal cord as well as in the rest of the CNS are *interneurons* (association neurons, intercalated neurons), which serve important integrative functions as part of multisynaptic spinal reflexes and feedback circuits. Many of the axons of interneurons remain in the same segment (intrasegmental interneurons); others project to neighboring segments (intersegmental interneurons) or proceed to the opposite side of the spinal cord (commissural interneurons).

Propriospinal neurons form a special type of interneuron that send their axons to other spinal cord segments through fasciculi proprii adjacent to the gray substance. They coordinate the activity of various muscles during motor synergies.

It is hardly possible to overestimate the importance of the interneuronal system. The interneurons in the spinal cord by far outnumber the motorneurons and the long-axoned neurons that represent the origin for ascending pathways. The interneurons are intercalated in various reflex loops and intrinsic neuron circuits; they also serve as intermediaries in relationships between long descending pathways and spinal motor and sensory mechanisms, or between peripheral afferent fibers and spinal cord neurons giving rise to long ascending pathways. The multitude of interneurons provides an almost infinite number of circuits and synaptic contacts, where messages can be modified or fundamentally altered. In short, the interneuronal net is largely responsible for the amazing readiness and flexibility with which we respond to various stimuli. Furthermore, groups of spinal cord interneurons referred to as *pattern* or *locomotor generators*, handle many of the detailed interactions responsible for the finely tuned muscle contractions and relaxations required for rhythmic stepping movements (Chapter 15).

White Matter

The white matter on each side of the spinal cord can be divided into a *dorsal, lateral*, and *ventral funiculus* (Fig. 8–6A). The fibers are arranged functionally in more or less well-defined ascending and descending pathways, which are given names implying their origin and termination (Fig. 8–6B). There are also association pathways which connect

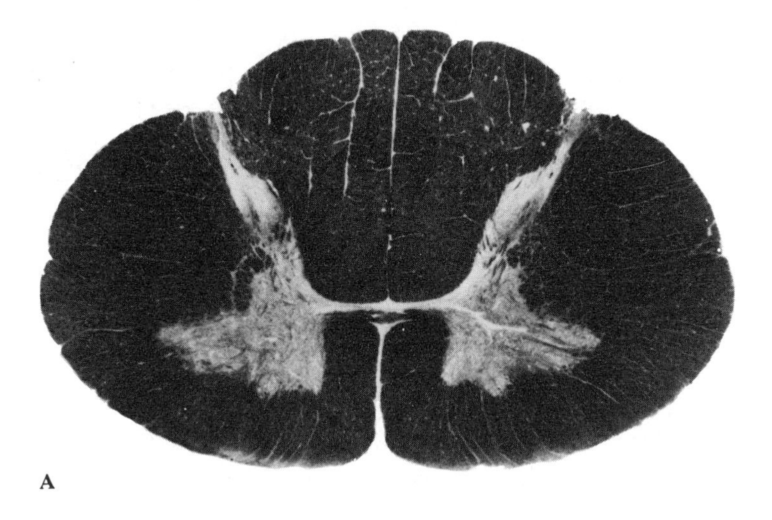

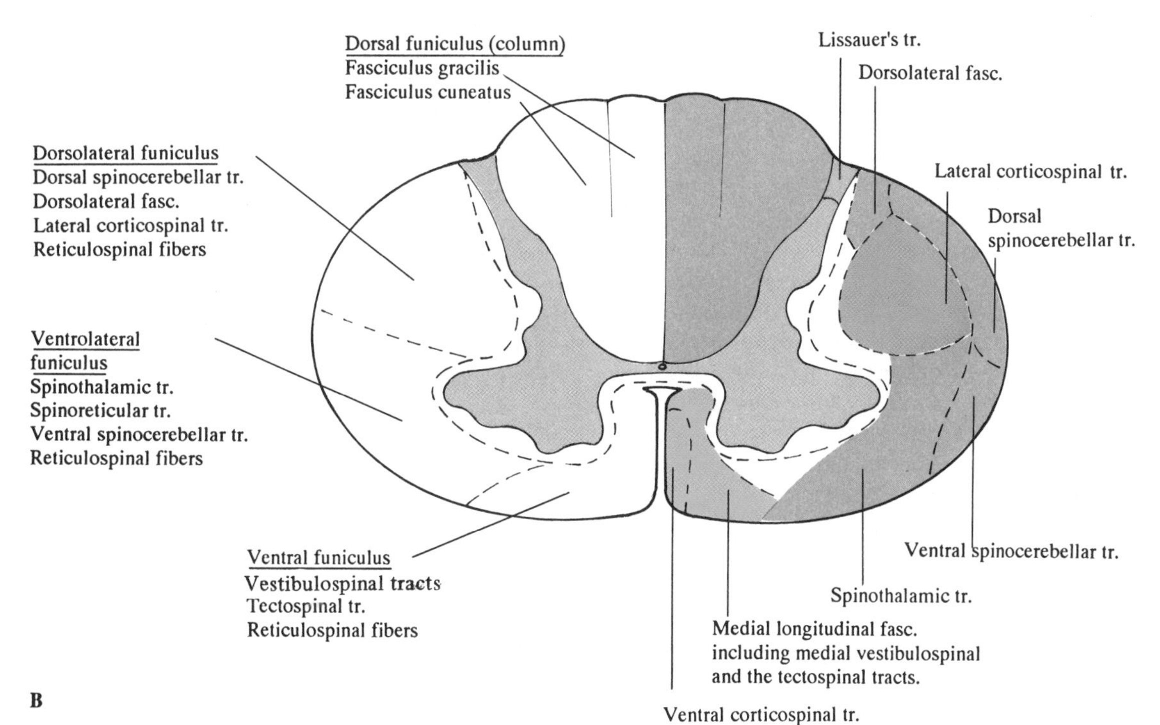

Fig. 8–6. White Matter of the Spinal Cord
A. Transverse myelin-stained section of the spinal cord through the cervical enlargement. The ascending and descending pathways form the white matter of the spinal cord, which appears black in this myelin-stained preparation. (From Gluhbegovic, N. and T. H. Williams, 1980. The Human Brain and Spinal Cord. JB Lippincott Co., Philadelphia. With permission of the publisher.)

B. Since there is considerable overlap between different fiber tracts, it is difficult to illustrate all spinal cord tracts in a schematic drawing. Nevertheless, the position of some of the most important ascending (*blue*) and descending (*red*) pathways is indicated on the *right side*. The reticulospinal fibers, which serve many important somatomotor and autonomic functions, descend in all funiculi except the dorsal funiculus.

different segments within the spinal cord. These pathways, the *fasciculi proprii*, are found in all three funiculi and they are usually located adjacent to the gray substance. *Lissauer's tract*, which is wedged between the dorsal horn and the surface of the spinal cord, is to a large extent a propriospinal pathway.

The ascending somatosensory pathways, i.e., the dorsal column–medial lemniscus system and the spinocervicothalamic and the spinothalamic tracts, are described in more detail in Chapter 9, and the descending pathways in Chapter 15. The spinocerebellar pathways are discussed in relationship to the functional anatomy of the cerebellum in Chapter 17.

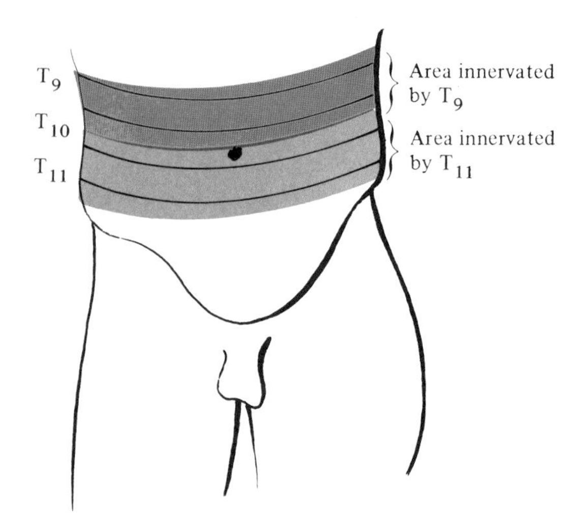

Fig. 8–7. Overlap of Dermatomes
There is considerable overlap between neighboring dermatomes, and the overlap is usually greater for touch than it is for pain. A small sensory deficit, therefore, is more easily detected by testing for pain rather than for touch.

Clinical Notes

Segmental Signs and Symptoms in Diseases of the Spinal Nerves or the Nerve Roots

In contrast with a peripheral nerve disease, in which the signs and symptoms appear within the distribution of the affected nerve, diseases of the spinal nerves or the nerve roots are characterized by *segmental loss* of function. Lesions of the spinal cord may also give segmental symptoms, but such lesions are usually accompanied by involvement of long ascending or descending pathways as well.

Spinal nerve disorders cause both motor and sensory disturbances, whereas root disorders produce either motor or sensory symptoms depending on whether a ventral or dorsal root is affected. The stretch reflexes are affected in both dorsal and ventral root lesions.

Sensory Symptoms

Compression or inflammation of a root manifests itself as pain localized to the affected dermatome, i.e., *radicular* or *root pain*. Other types of abnormal sensation, or *paresthesias*, are tingling, prickling, burning, and itching. Increased sensitivity to various stimuli is called *hyperesthesia*. When the fibers lose their capacity to conduct sensory impulses, the corresponding symptoms are referred to as *hypesthesia* (diminished sensitivity) or *anesthesia* (loss of all forms of sensation).

When testing for sensory loss one should remember that there is a considerable overlap between neighboring dermatomes. The upper half of dermatome T10, for instance, is innervated not only by fibers from T10, but also by fibers from T9, whereas the lower half of T10 receives additional innervation from T11 (Fig. 8–7). Therefore, interruption of one dorsal root may not be noticed. It is useful to remember that a small sensory deficit can be more easily detected by testing the sensibility for pain rather than for touch.

Motor Symptoms

A complete loss of motor function is referred to as *paralysis*; a lesser loss is called *paresis*. The motor deficits in spinal nerve or ventral root disorders are characterized by a flaccid (soft, flabby) paresis or paralysis of the affected muscles. There is also significant *atrophy* and reduced tonus, i.e., *hypotonus*, of the affected muscles, whose stretch reflexes are absent or weakened compared to the normal side. An affection of the peripheral motor neuron (see Chapter 15), therefore, is characterized by flaccid paralysis, atrophy, hypotonus, and hyporeflexia.

Disc Prolapse

This is the major cause of backache (lumbago) and radicular pain in the leg corresponding to the distribution of the sciatic nerve (sciatica). It is often seen in middle-aged people, in whom degenerative changes in the anulus fibrosus and the posterior longitudinal ligament are common. The condition is often precipitated by a flexion injury that forces the disc (or prolapsing nucleus pulposus) in a pos-

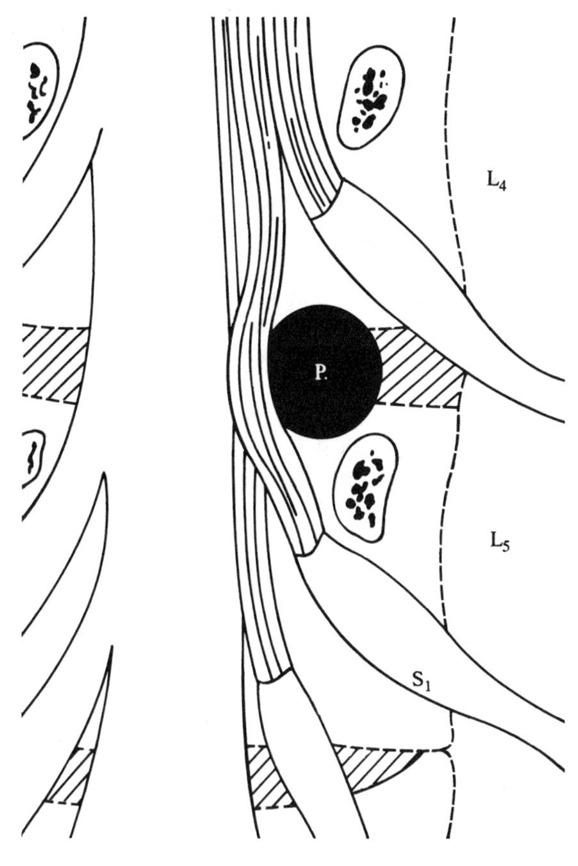

Fig. 8–8. Disc Prolapse
A protrusion (P) of the 4th intervertebral disc between L4 and L5 may compress either the 5th lumbar root or the 1st sacral root, or both. (Redrawn from Spurling and Grantham, 1940. Arch. Surg. 40:375–388. In: Brodal, A., 1981. Neurological Anatomy, 3rd ed. With permission of Oxford University Press.)

terolateral direction into the vertebral canal, where the disc compresses the nerve roots or the spinal nerve.

Flexion of the vertebral column mainly takes place in the lower lumbar region, and the discs between L4–L5 and L5–S1 are therefore most often affected. Intervertebral discs in the upper lumbar and lower thoracic regions are less commonly affected. Since the spinal cord ends at vertebral body L1, lumbar and sacral roots hang down below the cord at the cauda equina. Each set of roots proceeds obliquely toward its intervertebral foramen in such a manner that protrusion of an intervertebral disc may compress one or two roots, either right or left. For instance, protrusion of disc L4–L5 may compress either the 5th lumbar root or the 1st sacral root, or both (Fig. 8–8). The most common sites for disc protrusions in the cervical region are C5–C6 and C6–C7.

Compression of the 5th lumbar root usually produces radiating pain and sometimes corresponding sensory loss in the lateral calf, dorsal surface of the foot, and the first to third toes. Extension of the big toe and foot may be weakened.

Compression of the 1st sacral root produces radiating pain and sometimes corresponding sensory loss in the posterior surface of the leg and the outer plantar surface of the foot and the fourth and fifth toes. The plantar flexors of the foot may be weakened and the Achilles reflex is often diminished or absent (plantar flexion of the foot when the Achilles tendon is tapped, see Chapter 15).

Herpes Zoster

This disease is caused by a virus that is related to chickenpox virus. It causes inflammation of isolated posterior root ganglia, resulting in a skin lesion known as shingles. The symptoms are usually confined to one or two adjacent dermatomes, most often in the thoracic region. The initial symptom is radicular pain followed in a few days by reddening and the appearance of vesicles in a typical segmental distribution on one side of the body.

Spinal Cord Disorders

Spinal cord disorders are quite common and often produce distinctive syndromes, which can best be understood following a discussion of the long ascending and descending pathways of the spinal cord (Chapters 9 and 15). Therefore, only a brief summary of the main diseases of the spinal cord is presented here.

Like all types of nervous diseases, the signs and symptoms of a spinal cord disorder are dependent on the extent of the damage and the level of the injury. For instance, a complete *transverse lesion* of the spinal cord, which is often traumatic in origin, results in loss of all voluntary movements and all sensations below the level of injury. A *degenerative disease*, on the other hand, is usually characterized by selective involvement of certain functional-anatomic systems, e.g., the motor neuronal cell groups in *amyotrophic lateral sclerosis*, or some of the long ascending and descending tracts in *Friedreich's ataxia* and other s.c. *spinocerebellar degenerations*. Another typical spinal cord disease is *syringomyelia*, a disease of unknown etiology, which is characterized by a pathologic cavitation of the central gray substance often in the cervical part of the spinal cord. The lesion produces a characteristic segmental loss of pain and temperature sense (Fig. 9–8 and Clinical Example in Chapter 9) by involving the second-order neurons as they cross the midline.

Trauma

Automobile and diving accidents are two of the most common causes of severe *fracture-dislocations* of the spine and concomitant transverse lesions of the spinal cord. The most mobile parts of the vertebral columns, the lower cervical and the upper lumbar regions, are often damaged, and if the injury is severe enough, the patient will be suffering from a *paraplegia* (paralysis of both lower extremities) if the lower part of the spinal cord is damaged, or a *tetraplegia* (paralysis of all four extremities) if the lesion occurs in the cervical area (see Clinical Notes in Chapter 15).

Myelitis

An inflammatory disease of the spinal cord, i.e., a myelitis, is usually the result of a general infection of the nervous system. The cause of myelitis is often unknown; other times it is caused by a virus, e.g., *poliomyelitis, herpes zoster* (see above), or *rabies.*

Tabes dorsalis is a tertiary form of neurosyphilis which is characterized primarily by degeneration of the dorsal roots and the dorsal columns, resulting, typically, in lightning pains and ataxia (loss of muscular coordination). Neurosyphilis, which usually develops 15–20 years following the initial infection with *Treponema pallidum*, is today relatively uncommon because of successful therapy with penicillin.

Tumors

Since most of the tumors in the vertebral canal arise outside the spinal cord rather than within the substance of the cord, they can be removed surgically before they have produced irreversible damage, provided they are recognized at an early stage. Compression of a nerve root causes radicular pain and other sensory and motor disturbances, and the segmental signs and symptoms usually indicate the location of the tumor. Further, compression of the long pathways may result in motor and sensory symptoms, which are often of an asymmetic distribution.

Vascular Lesions

Vascular disorders of the spinal cord are uncommon compared with cerebrovascular diseases. Nevertheless, infarctions and bleeding do occur, most often in the territory of the anterior spinal artery (see Clinical Example in Chapter 24).

Clinical Example

Disc Prolapse

A 38-year-old male manual laborer suffered acute onset of back pain while operating a pneumatic hammer. The pain radiated down the posterior aspect of his right leg and into the ventral surface of his foot. He noticed that coughing made the pain much worse.

The neurologic examination was remarkable for percussion tenderness over his lower back with a sharp pain that radiated down the posterior aspect of his right leg. The right gastrocnemius muscle was weak and the Achilles reflex was absent on the right side. The motor deficits were matched by loss of pain and temperature sensitivity over the lateral surface of his right foot.

The man was treated with strict bed rest and his pain was alleviated with analgesics. However, his condition did not improve and a CT scan (Fig. 8–9B) was obtained before he was transferred to a neurosurgical unit.

1. What do you think is the nature of his disease process?
2. What nerve root is most likely to be involved?
3. Into which large peripheral nerve does this nerve root contribute fibers?
4. Compare the CT scan from this patient with a normal CT scan taken from a higher level (Fig. 8–9A). The arrow in the pathologic CT scan is pointing to disc material that should not be present in the vertebral canal. The herniated disc is pressing on an elliptically shaped area in the center of the picture. What does this area represent and what nervous structures are located there?

Discussion

The patient's history is typical for a disc prolapse (herniated disc), and the radiation of pain down the posterior aspect of the leg into the foot indicates compression of the S1 nerve root (see Fig. 8–8). This root contributes fibers to the sciatic nerve and one of its branches, the tibial nerve, which innervates the medial and lateral heads of the gastrocnemius muscle. Selective loss of gastrocnemius power coupled with a reduced or absent Achilles reflex (mediated by the gastrocnemius muscle) is typical of an S1 root lesion. The sensory deficit is also consistent with such a lesion.

This type of herniation is referred to as a central-lateral disc herniation. The elliptically shaped

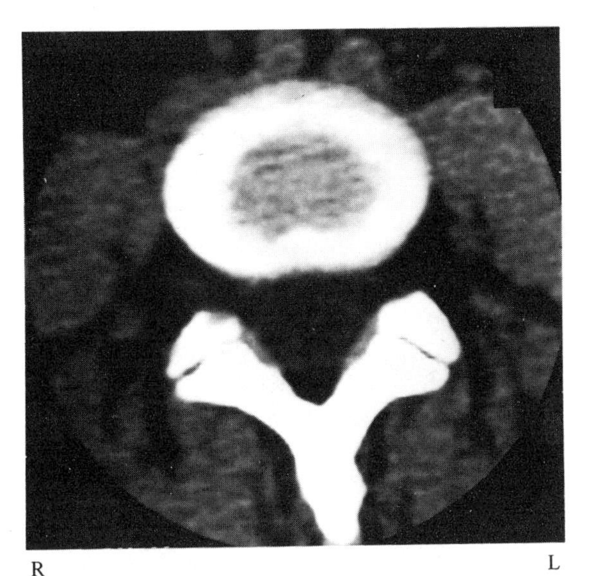

A. Normal

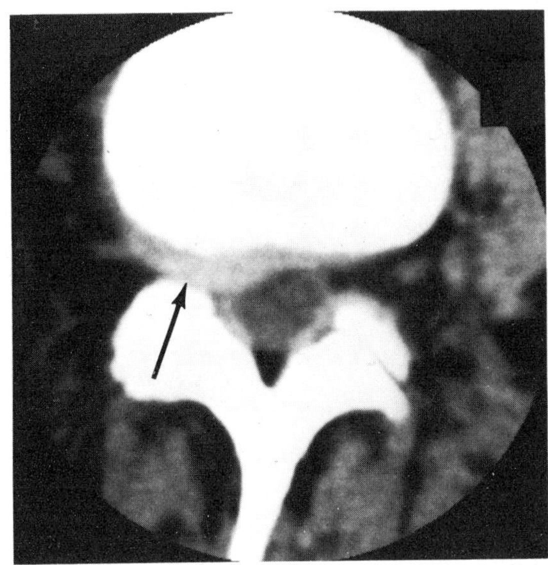

B. Disc Prolapse

Fig. 8–9. A and B.
A. Normal CT scan.
B. Disc prolapse.
(The CT scans were kindly provided by Dr. V. Haughton.)

structure against which the disc material is pressing is the dural sac, containing the cauda equina and leptomeninges. The ring of enhancement around the dural sac represents filling of the epidural venous plexus. Selective filling of this plexus with contrast dye (lumbar venogram) is also helpful in demonstrating herniated discs. Many patients with disc herniations improve with bed rest and analgesics, but some ultimately require surgery for removal of the extruded disc material.

Suggested Reading

1. Adams, R. D. and M. Victor, 1993. Principles of Neurology, 7th ed. McGraw-Hill: New York, pp. 171–196.
2. Schoenen, J. and R. L. M. Faull, 1990. Spinal cord: Cytoarchitectural, dendroarchitectural, and myeloarchitectural organization. In G. Paxinos (ed.): The Human Nervous System. Academic Press: San Diego, pp. 19–53.
3. Schoenen, J. and R. L. M. Faull, 1990. Spinal cord: Chemoarchitectural organization. In G. Paxinos (ed.): The Human Nervous System. Academic Press: San Diego, pp. 55–75.
4. Schoenen, J. and G. Grant, 1990. Spinal cord: Connections. In G. Paxinos (ed.): The Human Nervous System. Academic Press: San Diego, pp. 77–92.
5. Willis, W. D. Jr. and R. E. Coggeshall, 1991. Sensory Mechanisms of the Spinal Cord, 2nd ed. Plenum Press: New York.

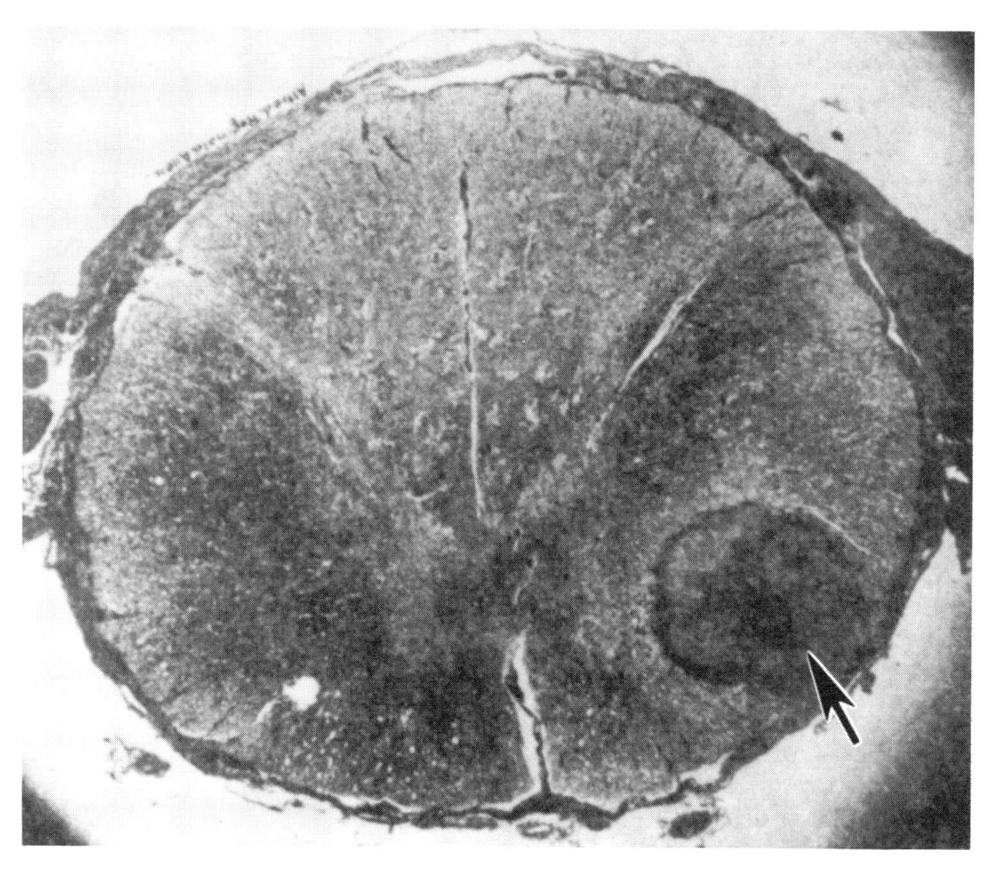

Cross section of the cervical part of the spinal cord showing a tuberculoma (arrow) which disrupted pain and temperature sensitivity on the contralateral side of the body below the level of the lesion. The picture, which was published by the illustrious American neurologist William Gibson Spiller in 1905 ("The location within the spinal cord of the fibers for temperature and pain sensation"; Journal of Nervous and Mental Disease, Vol. 32, pp. 318–320), did much to solidify the idea that the spinal pathway for pain and temperature is located in the anterolateral funiculus, and that the fibers carrying the impulses cross at the level of the spinal cord.

The incidence of tuberculosis has decreased in many parts of the world for several decades. However, the disease is again on the rise, because tuberculosis control programs are not dealing effectively with the problem. In addition, the emergence of multiple drug resistant strains of *Mycobacterium tuberculosis* in AIDS patients constitutes an additional concern.

Ascending Sensory Pathways

The classic view of two largely independent somatosensory channels, the dorsal column–medial lemniscus system for tactile sensitivity and position sense, and the spinothalamic system for pain and temperature sensitivity, has been modified in recent years through the discovery of additional spinal pathways for transmission of sensory impulses to the brain. This is highly significant, because the sensory symptoms that appear in patients with lesions in the CNS have sometimes been difficult to explain by the traditional dual-processing theory. They can now be better appreciated on the basis of a more dynamic view of the somatosensory pathways.

Although sensory and motor neurons are closely interconnected throughout the nervous system, the ascending sensory pathways are to a large extent separated from the structures and pathways that control motor functions. Therefore, nervous system disorders can in large measure affect sensory or motor functions independently.

The traditional view of the somatosensory pathways to the cerebral cortex is that the dorsal column–medial lemniscus pathway carries information necessary for discriminative touch, vibratory sensibility, and position sense, and that the spinothalamic pathway is mainly responsible for pain and temperature sensitivity. Clinical findings however could not always be explained on the basis of this dual-processing theory, and it is now clear that transmission of sensory information from the periphery to the brain can in many cases occur through several channels in the spinal cord. The changing view is related not so much to the pathway for pain and temperature but rather to the ascending systems that are responsible for tactile sensitivity and position sense.

The ascending sensory pathways can be divided into three groups:

1. Pathways for *pain* and *temperature*. The main pathway is the *spinothalamic tract*, which is accompanied in its course by fibers terminating in the reticular formation, the *spinoreticular tract*, and by fibers terminating in the mesencephalic tectum and periaqueductal gray matter, the *spinomesencephalic tract*. The spinothalamic, spinoreticular, and spinomesencephalic tracts are included in the so-called *anterolateral (AL) system*.

The traditional subdivision of the spinothalamic tract into a lateral spinothalamic tract for pain and temperature and a ventral spinothalamic tract for tactile stimuli is of little value.

2. Pathways for *tactile information*, *vibration*, and *position sense* (static position sense and kinesthesia). Impulses related to these modalities are carried via several pathways, particularly in the *dorsal funiculus* (column) and in the dorsal part of the lateral funiculus. Although some touch–pressure signals, and probably also proprioceptive impulses, can be mediated via the spinothalamic tract, the critical pathways for tactile information and position sense are located in the dorsal half of the spinal cord.

3. Pathways for *somatosensory impulses to the cerebellum* are described in Chapter 17.

Pathways for Pain and Temperature: The Anterolateral System

The sense of pain has several submodalities. Intense stimulation of the skin results in *superficial pain*, which is well localized. *Deep (aching) pain*, which arises in skeletal muscles, tendons, and joints, is poorly localized. Since the receptors for pain have a high threshold, they are excited only by stimuli that may cause tissue damage, i.e., harmful or noxious stimuli. Pain receptors, therefore, are referred to as *nociceptors*. *Visceral pain*, like deep pain, has an aching character and is poorly localized. *Itch* is closely related to pain and the two modalities are transmitted by the same pathways.

Thermal sensation can be divided into two discrete modalities: *warm* and *cold*. The temperature sensitivity on the skin is localized to cold spots and warm spots occupied by specific cold receptors, which usually are activated by temperatures below 33°C, and warm receptors, which discharge at an increasing rate with temperatures above 33°C.

Receptors and Peripheral Pathways

Most, if not all, *nociceptors*[1] (pain receptors) and *thermoreceptors* (cold and warm receptors) are represented by free nerve endings that are related to fine unmyelinated C fibers,[2] or thinly myelinated A-delta (Aδ) fibers. The cell bodies of the fibers lie in the dorsal root ganglion, and the central axonal processes proceed through the dorsal root to enter the lateral part of Lissaur's tract, where the small-diameter fibers divide into ascending and descending branches that send collaterals to the marginal zone (lamina I) and the substantia gelatinosa (lamina II). Other fibers proceed to termination in deeper parts of the dorsal horn and the intermediate gray substance.

Many of the fibers that transmit pain and temperature impulses take part in reflexes of various

---

1. It is common to divide the receptors into *nociceptors* (pain receptors), *thermoreceptors*, *mechanoreceptors*, *chemoreceptors*, and *photoreceptors*. Nociceptors and thermoreceptors represented by free nerve endings, respond to tissue damage and to cooling and warming respectively. There are several varieties of mechanoreceptors responding to touch pressure and vibration or to tension and stretch. Chemoreceptors, mostly associated with visceral sensitivity, respond to various metabolites, pH, carbon dioxide, oxygen, etc. If the receptors are located in the skin or the subcutaneous connective tissue, and respond to stimuli from the outside, they are referred to as *exteroceptors*. Those located in the visceral organs are called *interoceptors* (visceroceptors). *Proprioceptors* are located in muscles, tendons, and joints.

2. The classification of nerve fibers is somewhat confusing. In one classification the fibers are divided into A, B, and C fibers according to their diameter and physiologic properties. A fibers are somatic myelinated fibers, and they fall into four partly overlapping groups, α, β, γ, and δ, with decreasing diameter. B fibers are lightly myelinated preganglionic fibers in the autonomic nervous system, and C fibers include all unmyelinated fibers of both the somatic and autonomic (postganglionic fibers) system. Sensory fibers are either Aαβ fibers (conduction rate 120–30 m/sec), Aδ fibers (conduction rate 4–30 m/sec), or C fibers (conduction rate less than 2.5 m/sec).

Another classification that pertains to the afferent fibers in a muscle nerve recognizes four groups, I, II, III, and IV. Groups I and II correspond to Aαβ, group III to Aδ, and group IV to C fibers.

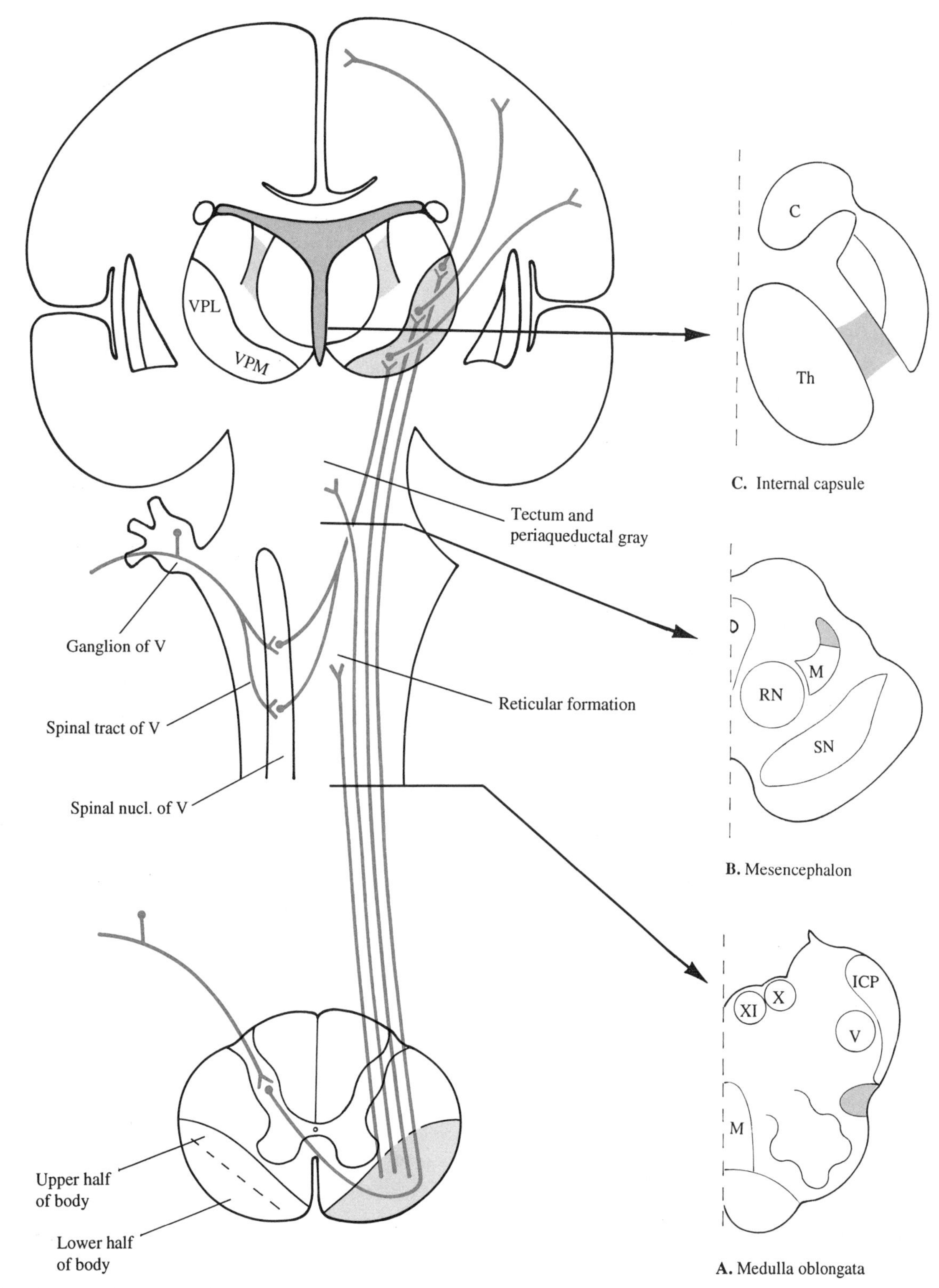

Fig. 9–1. Pathways for Pain and Temperature
Diagram illustrating the course and termination of the main pathways for pain and temperature, i.e., the spinothalamic and trigeminothalamic tracts. Many of the ascending fibers in the anterolateral system of the spinal cord as well as from the trigeminal system terminate in the reticular formation, periaqueductal gray, and tectum. The ascending pathways for pain and temperature are colored *blue* in the cross-sections through medulla oblongata (**A**), mesencephalon (**B**), and internal capsule (**C**). MD = mediodorsal thalamic nucleus; VPL = ventral posterolateral thalamic nucleus; VPM = ventral posteromedial thalamic nucleus; C = caudate nucleus; ICP = inferior cerebellar peduncle; M = medial lemniscus; RN = red nucleus; SN = substantia nigra; Th = thalamus

types, e.g., autonomic responses and the flexion withdrawal response (Chapter 15). Other afferent fibers establish synaptic contacts with cells whose axons form the ascending spinal pathway for pain and temperature.

Spinal Pathways (Fig. 9–1)

Spinothalamic Tract

The spinothalamic tract is the main spinal cord pathway for transmission of pain and temperature impulses. It also carries some touch–pressure sensations, including stimuli from the bowel and bladder, as well as sexual sensations. Its cells of origin are located primarily in the *marginal zone* (lamina I of Rexed) and the *nucleus proprius* (laminae IV and V). The cells, however, have long dendrites (Fig. 9–1), which usually spread across several neighboring laminae. The majority of the axons cross in the white commissure in the same or adjacent segment and ascend in the anterolateral funiculus on the opposite side of the cord.

Since crossing fibers are added to the inner aspect of the tract, fibers from successively more rostral levels will occupy increasingly deeper parts of the tract. This provides for a *rough somatotopic organization* in the sense that the lower parts of the body are represented laterally and the upper parts medially.

The spinothalamic tract ascends in the anterolateral white matter toward the brain stem, where it is located on the dorsolateral aspect of the inferior olivary nucleus in the medulla oblongata (Fig. 9–1A). Higher in the brain stem, it is situated dorsal to the medial lemniscus (Fig. 9–1B), which gradually moves laterally through its ascent in the brain stem. The spinothalamic fibers terminate in a somatotopic fashion primarily in the *ventral posterolateral nucleus* (*VPL*) and in the nearby *posterior nuclei* of the thalamus. The *intralaminar thalamic nuclei* (Chapter 22) also receive significant but diffuse and nonsomatotopically organized contributions from the anterolateral system.

Spinoreticular and Spinomesencephalic Tracts

The spinothalamic fibers are accompanied by fibers that terminate in the medial parts of the brain stem *reticular formation*, i.e., the spinoreticular tract, and in the *tectum* and *periaqueductal gray* of the midbrain, the spinomesencephalic tract. The spinoreticular fibers, which are both crossed and uncrossed, are involved in various reflex adjustments. Further, the spinoreticular tract represents the first link in a spinoreticulothalamic pathway to the intralaminar and midline thalamic nuclei, which in turn seem to be able to activate widespread areas of the cerebral cortex (Chapter 10). The spinomesencephalic tract terminates in the superior colliculus, and many of its fibers establish important synaptic relationships with a "pain inhibiting system" located in the periaqueductal gray and midbrain raphe nuclei, which gives origin to descending and mostly polysynaptic pathways to the dorsal horn (see below).

Pathways for Pain and Temperature from the Face

Pain and temperature impulses from the face travel along fibers in the different components of the trigeminal nerve. The cell bodies are located in the trigeminal ganglion, and the central processes reach the *nucleus of the spinal trigeminal tract* (Fig. 9–1; Note that only the fibers coming in through the mandibular nerve are indicated in the figure). Small fiber components of the facial, glossopharyngeal, and vagus nerves (from the external ear, auditory canal, and middle ear, as well as from the back of the tongue, pharynx, larynx, and esophagus) also terminate in the nucleus of the spinal tract.

Many of the fibers from the nucleus of the spinal tract cross to the opposite side of the medulla, join the medial lemniscus, and reach the *ventral posteromedial nucleus* (*VPM*) and intralaminar nuclei of the thalamus. Fibers also terminate in the reticular formation.

Thalamocortical Projections and the Perception of Pain

To what extent the perception of pain and temperature requires the cerebral cortex is not clear. Although clinical studies of patients with cortical lesions indicate that the cerebral cortex is not absolutely essential for conscious appreciation of pain, it is known that the thalamocortical projections from the VPL and VPM to the first somatosensory area, S-I, include axons from nociceptive as well as from other sensory-specific thalamic neurons. Recent PET studies, furthermore, indicate that not only the somatosensory cortex, but also part of the anterior cingulate area, participates in the perception of pain[3]. Therefore, it seems likely that the cerebral cortex is directly involved both in the

3. The apparent involvement of the anterior cingulate area in the perception of pain might provide some of the explanation for the beneficial effect obtained by cingulotomy, i.e., transection of the anterior part of the cingulum bundle, in patients with intractable pain.

perception of pain and in determining the localization of the painful stimulus. The mechanisms involved, however, are not well known. The cortical somatosensory areas are discussed in more detail in the next section on the dorsal column–medial lemniscus system.

Gate Control Hypothesis of Pain

According to this theory, which was proposed by Ronald Melzack and Patrick Wall in 1965, the substantia gelatinosa acts as a gating mechanism for the control of afferent input to the spinothalamic neurons (*blue* in Fig. 9–2). The activity in small pain-carrying C fibers keeps the gate open, and activation of large myelinated A fibers closes the gate. To accomplish a closing of the gate, small substantia gelatinosa cells (black in Fig. 9–2), which project to the spinothalamic neurons in the dorsal horn, would have to be excited by large afferents. This would produce presynaptic inhibition of the afferent input to the spinothalamic neurons, and the pain impulses, which travel along the small fibers, would not be able to reach the brain.

The gating principle seemed to provide an elegant explanation for some well-known phenomena, e.g., the tendency to rub a sore spot for relief from pain (the rubbing would activate the large fibers, thereby closing the gate). Although the gate theory of pain may have to be modified in the light of new research data, the fact remains that a mechanism is in place whereby activity in large myelinated afferents can inhibit the responses of dorsal horn neurons to painful stimuli. Furthermore, the gate theory of pain provides a striking example of how an imaginative hypothesis can stimulate scientific inquiry and even bring about a change in therapy (see Clinical Notes).

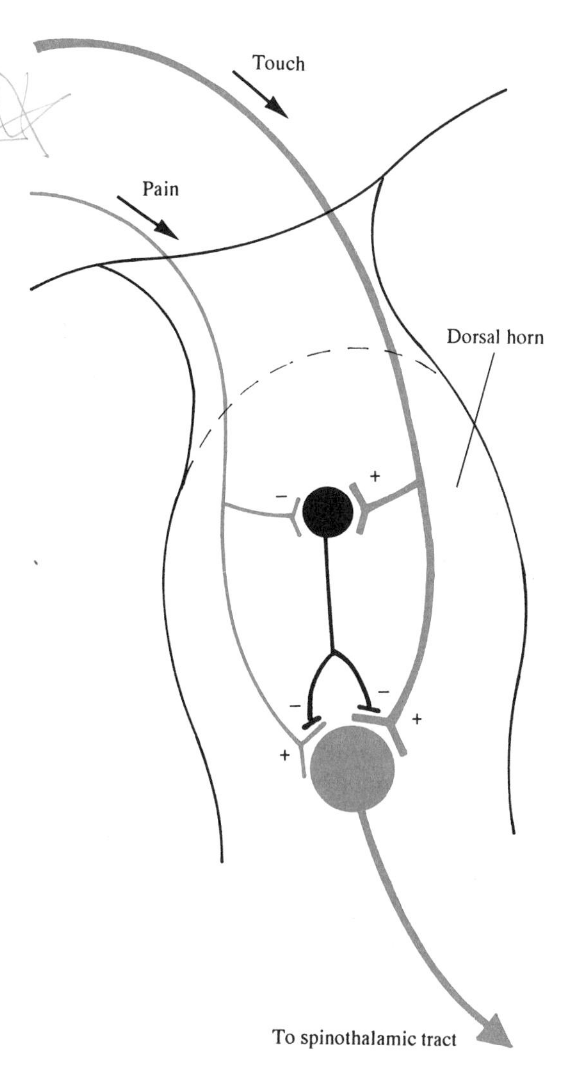

Fig. 9–2. The Gate Control Theory for Pain
Diagram explaining how activity in small C fibers (pain) would keep the gate open, whereas impulses reaching the dorsal horn through large A fibers in the medial portion of the dorsal root would close the gate. Although the gate control hypothesis has been severely criticized in recent years, it has stimulated much research and the development of therapeutic devices for pain control. *Black* neuron, inhibitory interneuron. *Blue*, primary afferent fibers and spinal projection neuron. (Modified after Melzack, R. and P. D. Wall, 1965. Pain mechanisms: A new theory. Science. 150:971–979. Copyright 1965 by the AAAS.)

Descending Control of Sensory Input to the Dorsal Horn

Centrifugal control of sensory input is a general principle that pertains to all sensory pathways. With respect to the sensory pathways discussed in this chapter, this regulatory effect is mediated by descending pathways that originate at various levels in the brain, including the cerebral cortex, basal forebrain, and several brain stem regions. Some of the fibers in the corticospinal tract, for instance, terminate on interneurons in the dorsal horn, and maybe even on terminals on incoming primary sensory fibers (Chapter 15).

The Endogenous Pain-Modulating System

Much attention in the recent past has focused on descending projections from the periaqueductal gray to the dorsal horn, especially after it was realized that electrical stimulation in the periaqueductal gray and adjoining areas in the reticular formation produce analgesia, i.e., stimulation-produced analgesia(SPA). The periaqueductal gray is also an important coordination center for various autonomic and somatomotor aspects of the defense reaction (see the Central Autonomic Network in Chapter 19), and it would seem to make sense that the pain-modulating system is closely related to areas involved in the fight–flight response.

Since neurons in the periaqueductal gray do not in general project directly to the spinal cord, the pain-modulating effects of stimulation in the periaqueductal gray are mediated by multisynaptic pathways, with relay nuclei located especially in the pontine reticular formation and in the medulla oblongata. One of the transmitters in the descending pathway from the dorsolateral pontine reticular formation is noradrenaline, whereas many of the descending fibers from the medulla oblongata originate in the serotonergic nucleus raphe magnus (Chapter 10). These pathways exert their analgesic effect by inhibiting the nociceptive impulse transmission in the dorsal horn. Part of this effect may be mediated by the release of enkephalin or endorphin ("the morphine within"), i.e., an endogenous substance with opiate-like activity. However, many other transmitters and various types of receptors are likely to be involved in the many circuits that are related to this system both in the brain stem and in the spinal cord. For instance, several circuits have been suggested to exist at the level of the dorsal horn, including both a direct and an indirect inhibition of the spinothalamic projection neurons (Fig. 9–3).

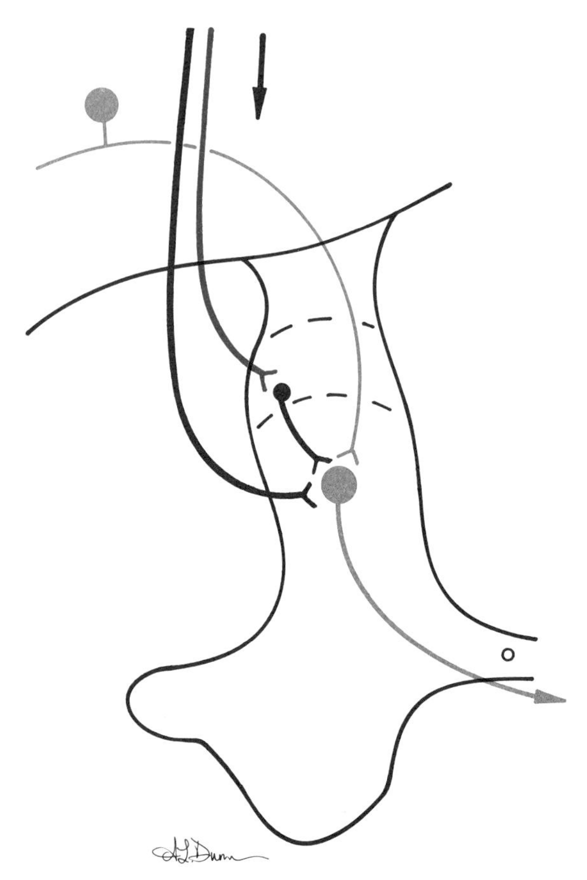

Fig. 9–3. Descending Control of Pain Transmission Modulation of pain transmission is believed to occur at the level of the dorsal horn by descending serotonergic pathways originating in the brain stem raphé system. Descending excitatory fibers (*red*) would activate enkephalinergic interneurons, which may exert postsynaptic and maybe even presynaptic inhibitory control over incoming fibers transmitting pain impulses (*blue*). Another likely scenario is that descending inhibitory fibers (*black*) establish direct contact with the spinothalamic projection neurons.

Pathways for Touch–Pressure, Vibration, and Position Sense: The Dorsal Column–Medial Lemniscus System

Touch is the experience of light stimulation of the skin, whereas activation of receptors deeper in the skin arouses the sensation of pressure. Touch and pressure, however, can be thought of as representing a continuum of stimulus intensity, *touch–pressure*.

Vibration is aroused by an oscillating stimulus, applied by the aid of a tuning fork, that activates receptors deep to the skin.

Position sense is composed of two submodalities: *static position sense* and *kinesthesia*. Static position sense (postural sense) signals the position of a limb in space, whereas kinesthesia reveals the movement of a limb.

Receptors and Peripheral Pathways

In contrast with the pain and temperature receptors, which are represented by free nerve endings, the receptors related to touch–pressure, vibration, and position sense are more or less specialized and often encapsulated (Fig. 9–4).

The tactile receptors signaling information about touch–pressure are represented by specialized epithelial cells of Merkel, by *encapsulated nerve*

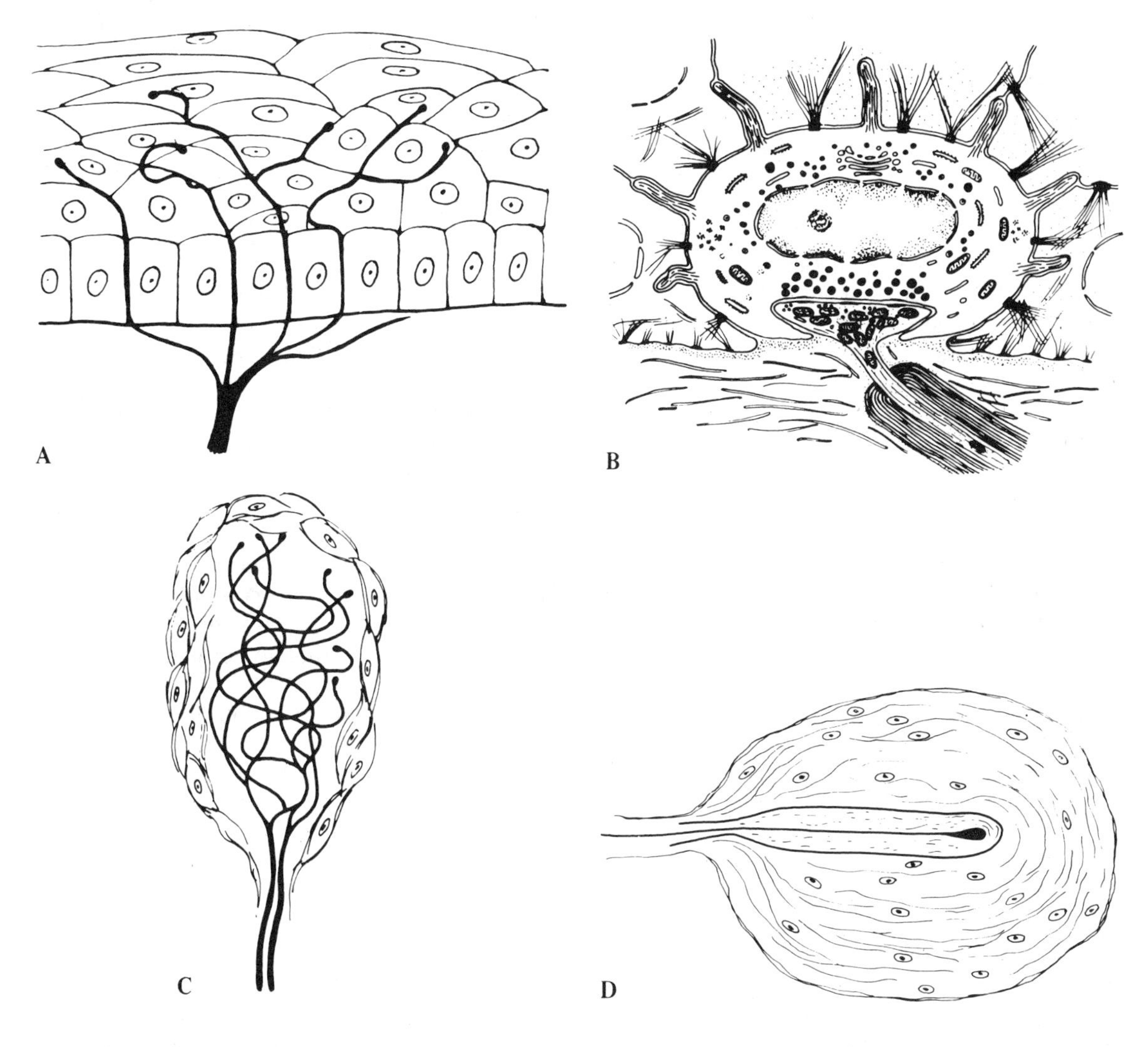

Fig. 9–4. Sensory Receptors
A. Free nerve endings. **B.** Merkel's specialized cell. **C.** Meissner's corpuscle. **D.** Pacinian corpuscle (**A**, **C**, and **D** from Geneser, F. 1981. Histologi. With permission of the author and Munksgaard, Copenhagen, Denmark. **B** from Bannister, L. H., 1976. Sensory Terminals of Peripheral Nerves. In: D. N. Landon (ed.): The Peripheral Nerve. With permission of Chapman and Hall Ltd.).

endings (e.g., Ruffini's and Meissner's corpuscles), and by hair follicle[4] receptors.

Vibration is mediated by highly specialized encapsulated nerve endings, the *Pacinian corpuscles*.

Position sense is apparently dependent on various *proprioceptive receptors* in joints and tendons and probably also in the muscles.

The cell bodies of the fibers related to these various receptors are located in the dorsal root ganglion, and their central processes, many of which are large and myelinated, proceed through the dorsal root to the spinal cord, where they divide in an ascending and a descending branch, both of which contain many collaterals that establish synaptic contacts with various cell groups in the spinal cord. Many of the ascending collaterals enter the dorsal column and some of them reach the medulla oblongata, where they terminate in an orderly somatotopic fashion in the dorsal column nuclei, also called the gracile and cuneate nuclei.

4. Although the specialists refer to the sensation generated by many of the hair follicle receptors (velocity receptors) as flutter, they are activated when the physician tests the sense of "touch" with a wisp of cotton, and for the sake of convenience they will be considered as touch receptors.

Spinal Pathways (Figs. 9–5 and 9–6)

Tactile Sensitivity

The Dorsal Column–Medial Lemniscus Pathway. This is the most important spinal pathway for touch–pressure, necessary for the finer analysis of tactile stimuli. Many of the fibers that enter the dorsal columns directly from the dorsal roots terminate at different levels in the spinal cord and only about 25% of the fibers reach the dorsal column nuclei, the *gracile* and *cuneate nuclei* in the medulla oblongata. Whereas the fasciculus gracilis carries information from the lower part of the trunk and the leg, the fasciculus cuneatus transmits impulses from the upper part of the trunk and the arm. When the axons reach the dorsal column nuclei, they establish synaptic contacts with cells whose axons cross over to the other side of the medulla and form the *medial lemniscus*. The crossing fibers, the internal arcuate fibers, form the *decussation of the medial lemniscus*, which is located in the lower part of the medulla (Fig. 9–5A) just above the pyramidal decussation. The medial lemniscus is located medially in the medulla, but it gradually moves laterally through its ascent in the brain stem to reach the VPL in the thalamus. Some of the fibers in the dorsal column–medial lemniscus system also terminate in the posterior thalamic nucleus.

Alternative Pathways. Although the dorsal column–medial lemniscus system is the major pathway for touch–pressure, other routes are available for tactile stimuli. For instance, fibers ascending in the dorsolateral fasciculus (Fig. 9–6) also convey tactile information. A special pathway in this part of the spinal cord is the *spinocervical tract*, which originates primarily in the nucleus proprius (laminae IV–V) and ascends on the same side to the *lateral cervical nucleus*, which is located on the lateral aspect of the posterior horn in the uppermost part of the cervical cord. This pathway apparently transmits signals primarily from hair receptors. The axons from cells in the lateral cervical nucleus cross the midline in the white commissure, and ascend in the brain stem in close relation to the medial lemniscus, to reach the VPL.

There is evidence both from animal experiments and clinical studies that touch–pressure information sufficient for contact recognition can reach the brain also through the spinothalamic tracts on both sides. In short, several possibilities are available for tactile information to reach the brain from the spinal cord, but the dorsal columns are essential for the finer analysis of tactile stimuli (Fig. 9–6A).

Vibration

The vibratory impulses are conveyed through the *dorsal column–medial lemniscus system* to the VPL (Fig. 9–6B).

Position Sense

The traditional view is that the dorsal column–medial lemniscus system is responsible for the proprioceptive information. This concept, however, has been difficult to reconcile with the clinical experience that dorsal column lesions at the thoracic or lumbar levels may produce a reduction in vibratory sensitivity but not in the position sense of the leg. Recent discoveries seem to have settled this apparent contradiction.

In brief, the situation seems to be as follows. Whereas the cuneate fasciculus serves position sense in the arm, the spinal pathway for position sense in the leg is primarily located in the dorsolateral funiculus (Fig. 9–6C). Since the fibers related to position sense in the dorsolateral funiculus project to a cell group (nucleus z) that is closely related to the gracile nucleus in the medulla oblongata; they are accordingly referred to as the *spinomedullary tract* (Fig. 9–6C). Axons from this cell group cross to the other side of the medulla and join the medial lemniscus in regular fashion.

Pathways from the Face

Fibers carrying tactile stimuli from the face reach the brain stem through the various components of the trigeminal nerve. The cell bodies of the fibers are located in the trigeminal (semilunar) ganglion, and the central processes terminate in the *chief sensory nucleus* of the trigeminal nerve (Fig. 9–5). Most of the axons from cells in this nucleus cross over to the other side and form part of the *trigeminothalamic tract*, which terminates in the VPM. Some of the fibers from the trigeminal nucleus reach the ipsilateral VPM.

The fibers responsible for position sense in the face are unique in that their perikarya are located within the CNS, specifically in the mesencephalic nucleus of the Vth cranial nerve. The central processes from these cells establish synaptic relationships with reticular formation cells, which in turn project to the VPM. They also establish contact with the trigeminal motor nucleus, thereby establishing a monosynaptic reflex arc for the jaw jerk or mandibular reflex (see Clinical Notes: Trigeminal Nerve in Chapter 11).

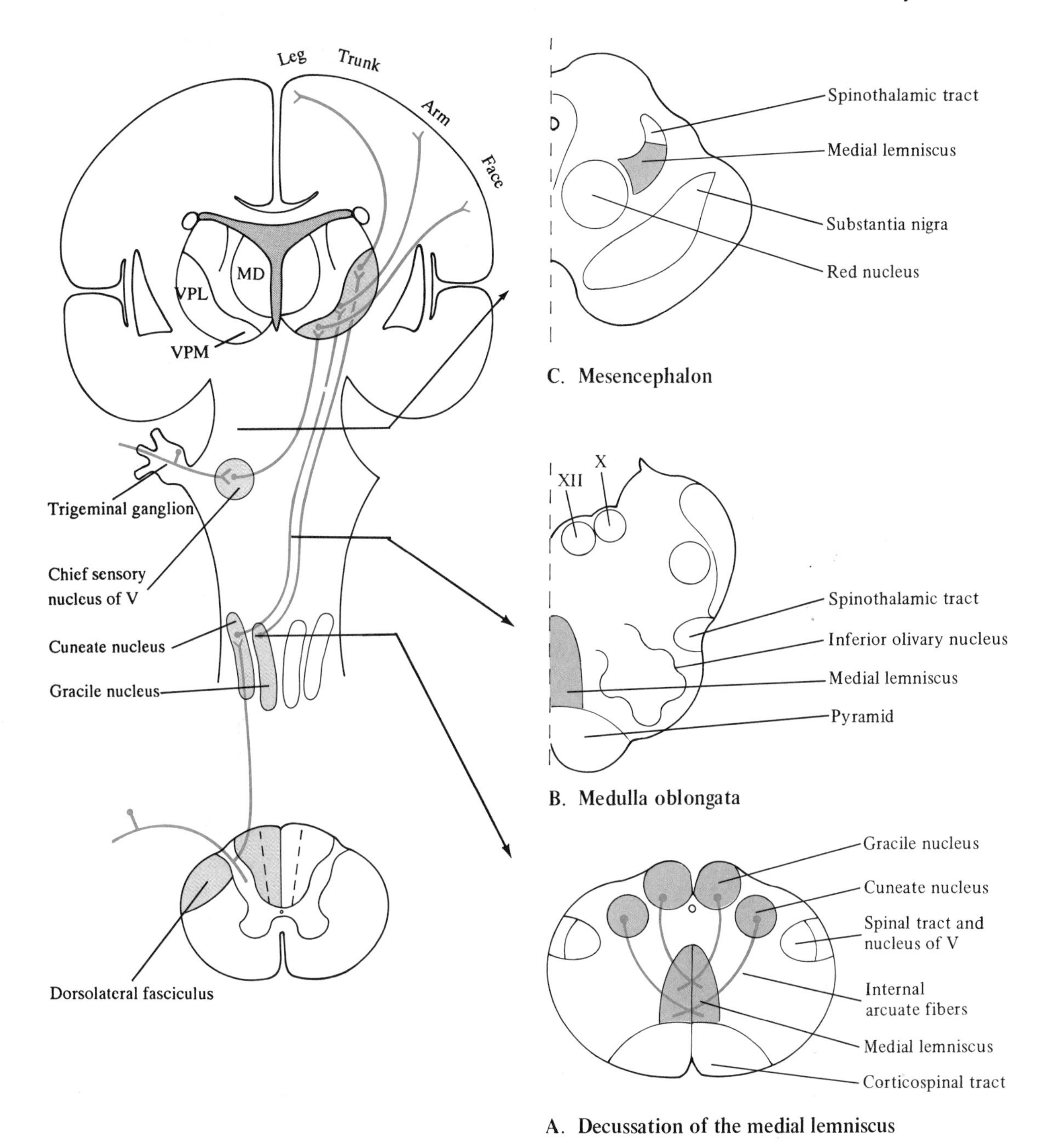

Fig. 9–5. Pathways for Tactile, Vibratory, and Proprioceptive Impulses
Diagram illustrating the dorsal column–medial lemniscal system mediating proprioception and stereognosis. Proprioceptive impulses and impulses for tactile discrimination are also transmitted in the dorsolateral fasciculus (see Fig. 9–6). **A–C.** Cross-sections showing the decussation of the medial lemniscus (**A**), and the position of the medial lemniscus in the medulla oblongata (**B**) and mesencephalon (**C**). MD = mediodorsal nucleus; VPL = ventral posterolateral nucleus; VPM = ventral posteromedial nucleus.

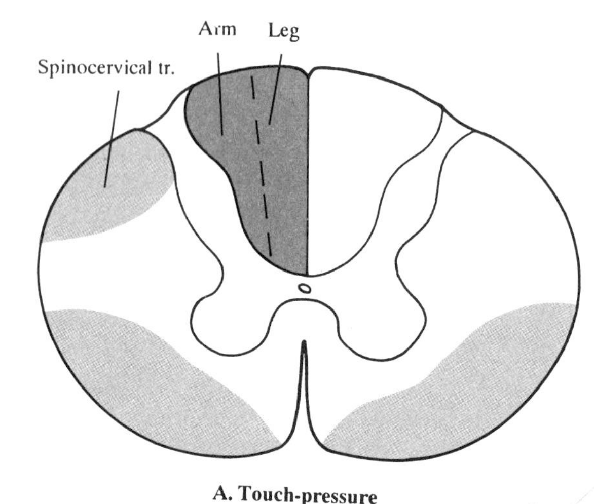

A. Touch-pressure

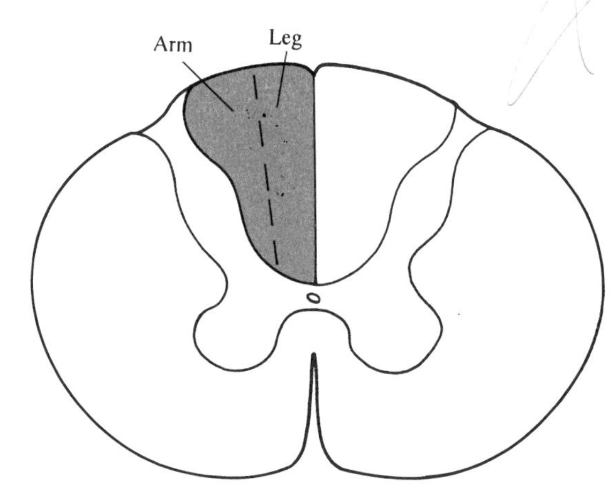

B. Vibration

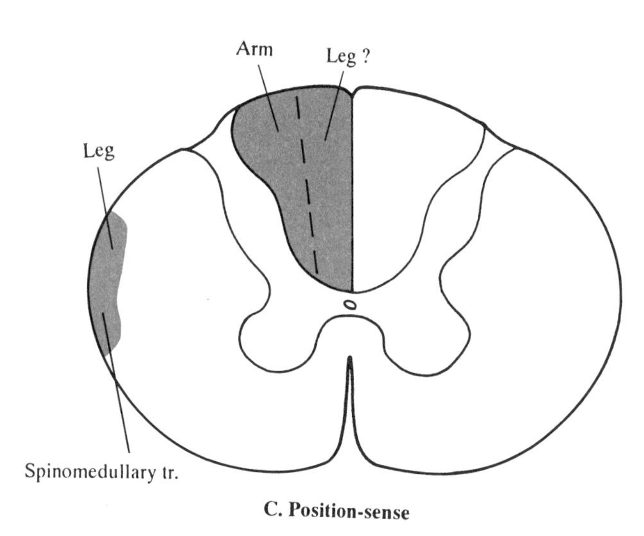

C. Position-sense

Fig. 9–6. Somatosensory Pathways in the Spinal Cord

Thalamocortical Projections and the Somatosensory Cortex

The medial lemniscus fibers and the trigeminothalamic fibers terminate in a somatotopic fashion in the VPL and the VPM, and this place specificity is maintained also in the thalamocortical projections, which proceed through the *posterior limb of the internal capsule* (Fig. 9–1C) to the primary somatosensory cortex, S-I, in the postcentral gyrus. A more limited projection reaches the secondary somatosensory cortex, S-II, in the superior lip of the lateral fissure (Fig. 9–7A).

S-I can be divided into four cytoarchitectural areas or strips along the posterior central gyrus: Brodmann's areas 3a, 3b, 1, and 2 (Fig. 9–7A). Area 3b, which receives the majority of the fibers from the ventral posterior (VP) nucleus, is mostly hidden from view in the depth of the central sulcus. This is the area for the basic processing of tactile information; it contains a complete map of the cutaneous receptors, and the different parts of the body are represented in a systematic fashion (somatotopic organization) in regions that are in proportion to their importance in somatic sensation (reflected by the density of innervation) rather than to their actual size (Fig. 9–7B). For instance, the areas serving the lips and the fingers are considerably larger than the areas related to the wrist and the neck.

Further processing of the tactile information takes place in areas 1 and 2, which receive afferents from area 3b and also to some extent from the ventral posterior thalamic nucleus. Like area 3b, areas 1 and 2 are somatotopically organized, and physiologic studies in primates indicate that area 1 is primarily concerned with the texture of objects, whereas area 2, which is related to both cutaneous and deep receptors, is more important for the discrimination of sizes and shapes. These physiologic findings are consistent with the fact that patients with extensive lesions in the postcentral gyrus suffer lasting sensory impairments, including inability to recognize the texture, size, and shape of an object (see Clinical Notes). Area 3a, finally, is activated by deep receptors and muscle spindles, and may be closely related to motor activities.

A second somatosensory area, S-II, is located in the upper bank of the lateral sulcus and adjoining insula (Fig. 9–7A). This area, which is also somatotopically organized, receives input from all S-I fields as well as from the ventral posterior. It is noted mostly for its participation in a multisynaptic corticocortical pathway through which somatosensory information reaches the hippocampus and amygdala.

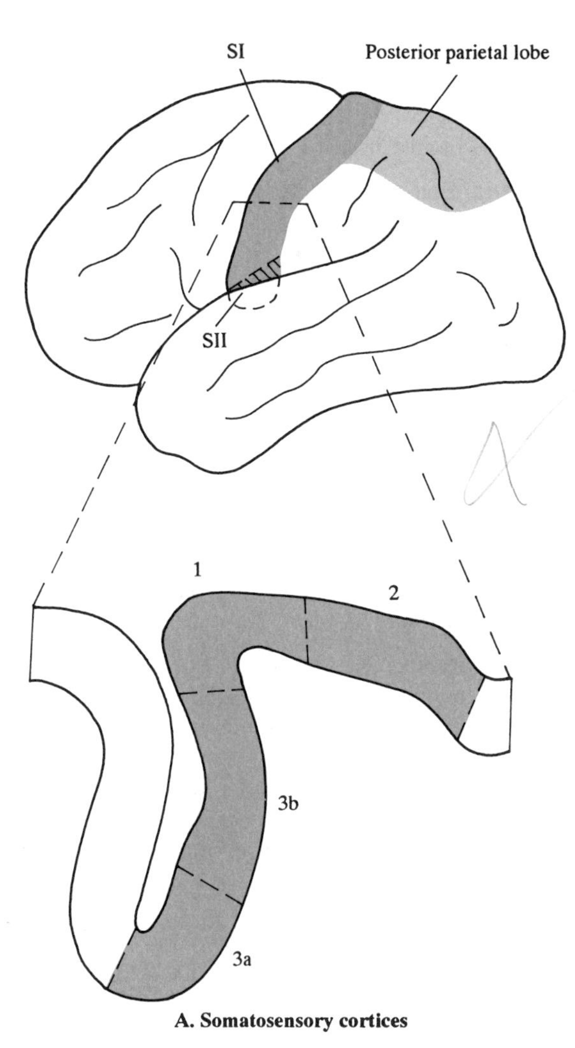

A. Somatosensory cortices

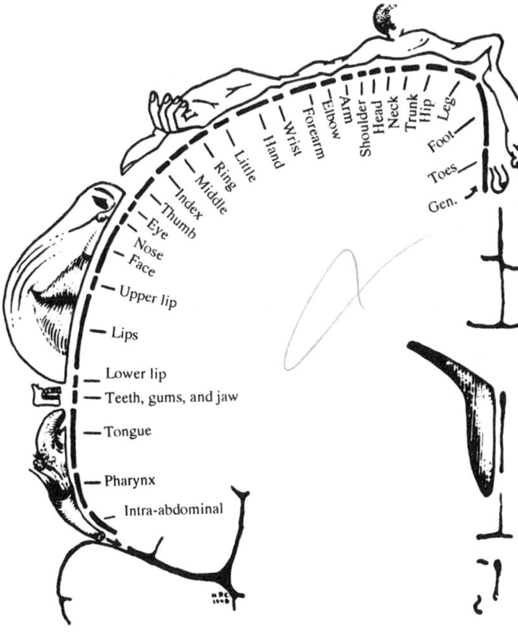

B. Sensory homunculus

Fig. 9–7. Somatosensory Cortices and the Sensory Homunculus
A. The primary somatosensory cortex, S-I, is located in the postcentral gyrus and in the depth of the central sulcus. It corresponds to Brodmann's areas 3, 1, and 2 (see Chapter 22). Another secondary sensory area is located in the upper lip of the lateral fissure. It was discovered more recently and is, therefore, referred to as the secondary somatosensory cortex, S-II. (Modified from Kaas, 1990 and Kandel et al., 1991.)
B. Within the primary somatosensory cortex various parts of the body are represented in regions that are related to that part's importance in somatic sensation rather than to its size. Note the large areas devoted to lips, thumb, and fingers. (From Penfield, N. and T. Rasmusson, 1955. The Cerebral Cortex of Man. With permission of Macmillan, New York.)

The posterior parietal cortex, especially its rostral part behind the primary somatosensory cortex, is also related to the somatosensory system, whereas its posterior part is more closely associated with the visual system. Lesions in the posterior parietal area tend to produce symptoms of neglect or inattention to somatosensory and visual impulses coming from the opposite side of the body or visual space (see Clinical Notes).

Descending Control of Dorsal Column Nuclei

Mention was made earlier of centrifugal control of sensory input on the level of the spinal cord. The dorsal column nuclei and other sensory relay nuclei in the brain stem, including the trigeminal complex, also receive descending fibers from other parts of the brain stem and from the cerebral cortex, primarily the sensorimotor cortex in the post- and precentral gyri. Many of these fibers appear to be collaterals of the descending corticospinal axons, and their effect on the sensory input is generally of an inhibitory nature. The descending control system is believed to play an important role in the changes of sensory transmission that occur during voluntary movements and in sleep.

Clinical Notes

The examination of patients with sensory disorders is difficult because the test procedures are rather crude and the responses are dependent on the patient's cooperation. Although the results obtained in a sensory examination may appear perplexing, they may accurately reflect complicated mechanisms of sensory transmission that are still not fully understood.

The symptoms following lesions of the sensory pathways are to a certain extent dependent on the location of the lesion. Lesions of the *spinal nerves* or the *nerve roots*, for instance, result in characteristic segmental syndromes.

Spinal cord lesions often give rise to sensory deficits of a dissociated character. For instance, a lesion in the anterior part of the spinal cord may abolish the sensitivity for pain and temperature, but it may not interfere substantially with transmission of tactile and propioceptive impulses, e.g., anterior spinal artery syndrome (see below). A lesion in the *brain stem* is often characterized by involvement of one or several of the cranial nerve nuclei. Lesions in the postcentral gyrus may result in loss of discriminative sensory functions, whereas damage to the posterior parietal lobe is characterized by more complicated sensory defects, including neglect syndromes (see Clinical Notes, Chapter 22).

Peripheral Nerve Lesions

The symptoms following interruption of a peripheral nerve will vary depending on whether a motor, a sensory, or a mixed nerve is affected. In regard to sensory symptoms, it is important to be familiar with the cutaneous distribution of the more important peripheral nerves (see Appendix) to differentiate between a peripheral nerve lesion and a lesion of the nerve roots or the CNS. Since there is a considerable overlap between neighboring nerves, the area of sensory loss following interruption of a cutaneous nerve is always smaller than its anatomic distribution.

Spinal Nerve and Nerve Root Lesions

Although the manifestations of spinal nerve and root lesions are similar to those accompanying peripheral nerve lesions, the distribution of various symptoms is different. Spinal nerve and root lesions have a typical segmental character.

Spinal Cord Syndromes

Segmental Symptoms with Lesions of the Spinal Cord

A segmental distribution of sensory symptoms is generally caused by diseases that affect the spinal nerves or the spinal roots, e.g., disc prolapse (see Clinical Example, Chapter 8), but can also occur in disorders that affect the center of the spinal cord. A typical example is the segmental sensory disturbances seen in patients with *syringomyelia* (syringomyelic syndrome), which is a degenerative disease causing a pathologic cavity involving the central region of the gray matter. Destruction of the decussating fibers in the anterior white commissure that carry impulses for pain and temperature results in a segmental anesthesia (loss of sensation) of dissociated type, i.e., loss of pain and temperature sense with preservation of tactile sensitivity (see Fig. 9–8 and Clinical Example).

Anterior Spinal Artery Syndrome

A thrombotic lesion in the anterior spinal artery may be secondary to atherosclerosis, an inflammatory disease, or compression of the vessel by a tumor. Since the artery supplies the anterior two-thirds of the spinal cord, there is a bilateral loss of pain and temperature sensation (spinothalamic tract) below the level of the lesion. Further, involvement of the corticospinal tract and the ventral horns results in disturbances of motor functions (see also Clinical Example, Chapter 24). However, the ascending pathways in the dorsal part of the spinal cord are usually spared and there is accordingly little or no interference with touch, vibration, and position sense.

Dorsal Column Syndrome

Dorsal column lesions are characterized by deficits in vibration and joint-position sense. The impairment of proprioceptive sensibility results in disturbances in coordination, i.e., sensory ataxia.[5] If the lower extremities are affected, the patient has difficulty maintaining balance when the eyes are closed (Romberg's sign). Interruption of the pathway for proprioceptive and tactile impulses also interferes with discriminative functions, such as the ability to recognize the shape and the form of an object by palpation.

There are several degenerative disorders that affect the dorsal columns. Well-known examples are tabes dorsalis, Friedreich's ataxia and subacute combined degeneration. Sensory ataxia and painful paresthesias (abnormal sensations) are the major symptoms in *tabes dorsalis*, the most common form of neurosyphilis, which usually appears 15–20 years after the initial *Treponema pallidum* infection. The disease is seldom seen today because the primary spirochete infection is usually treated successfully with penicillin. *Friedreich's ataxia*, which usually appears at or before puberty, is a hereditary spinocerebellar disorder in which the ascending sensory pathways in the dorsal columns and the

5. Although sensory ataxia is most often seen in dorsal column disorders, it can occur whenever the pathways for proprioceptive and tactile impulses are interrupted, e.g., as a result of a generalized lesion in the peripheral nerves (polyneuropathy) or following a lesion of the ascending lemniscal pathways in the brain.

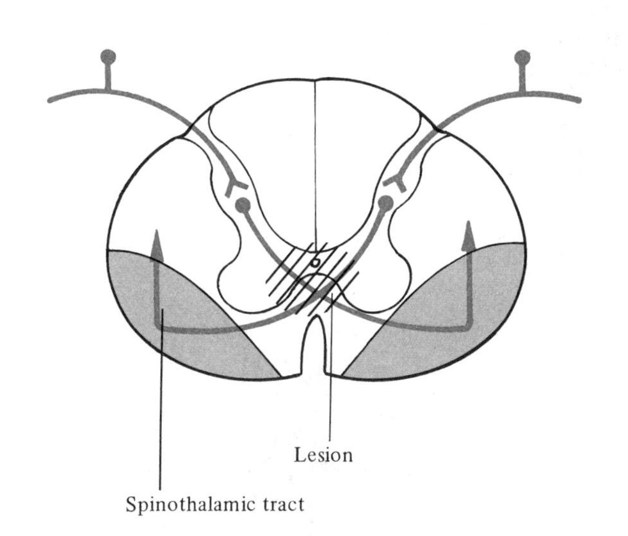

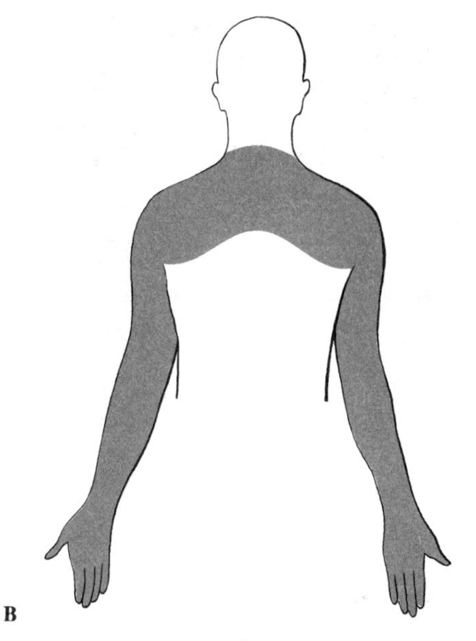

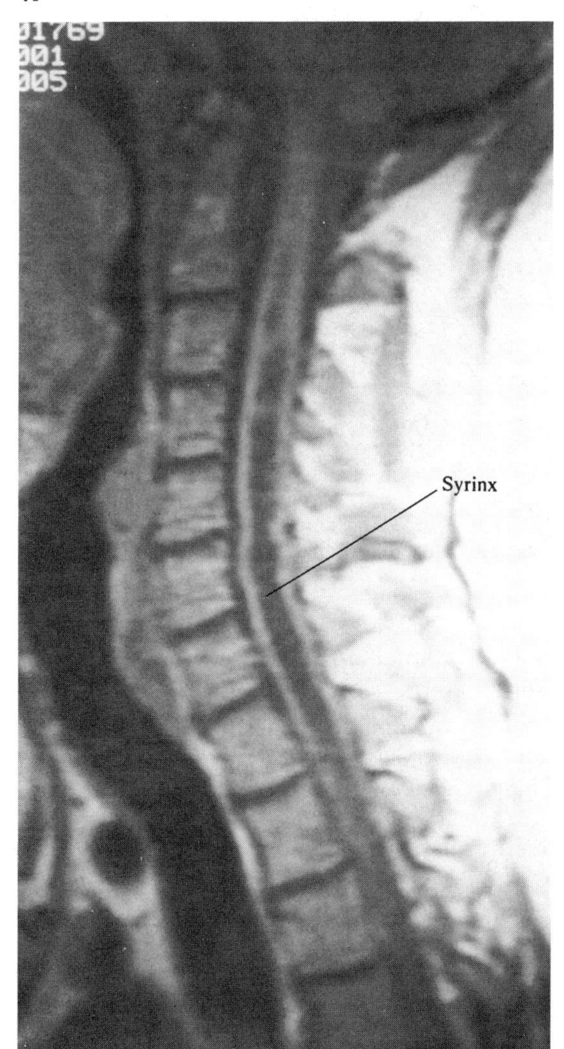

Fig. 9–8. Syringomyelia
A. Centrally located syringomyelic cavity in the spinal cord. The lesion interrupts the spinothalamic fibers as they cross the midline in the anterior white commissure. The sensory deficit following this type of lesion in cervical and upper thoracic region is shown in **B**.
B. Cape-like distribution of pain and temperature deficit caused by a syringomyelic cavity in the cervical and upper thoracic region. The lesion (**A**) and concomitant sensory deficits are usually not as symmetric as the ones indicated here (see Clinical Examples).
C. Midline sagittal MRI of the cervical spinal cord from a patient with syringomyelia. (Courtesy of Dr. Maurice Lipper).

spinocerebellar tracts undergo a slow but inevitable degeneration reflected in a gradually worsening ataxia. *Subacute combined degeneration* denotes a disorder in which there is a combination of dorsal column and corticospinal tract degeneration. It is usually caused by vitamin B_{12} deficiency.

Brown–Séquard Syndrome

This is the well-known, although rarely seen, syndrome caused by hemisection of the spinal cord. It is characterized by a contralateral loss of pain and temperature sensations which begin one or two segments below the level of the lesion (Fig. 9–9). There is also ipsilateral loss of proprioceptive sensation and ipsilateral motor paralysis below the level of transection. The sense of touch is usually preserved, since tactile impulses from one side ascend on both sides of the spinal cord (Fig. 9–6A).

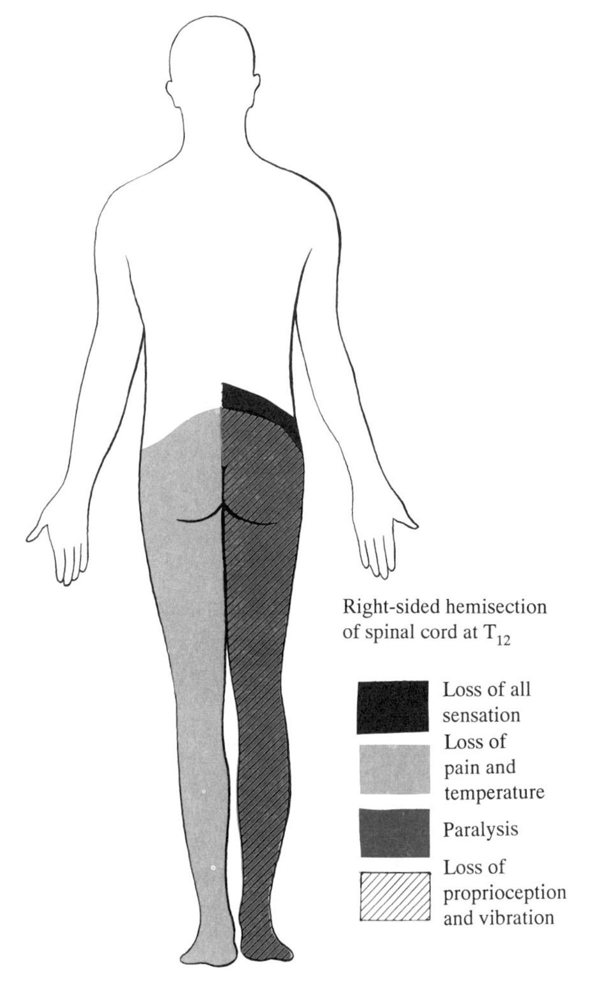

Fig. 9–9. Brown–Séquard Syndrome
Diagram illustrating the deficits following a hemisection of the right side of the spinal cord at the level of T12. A spinal cord lesion is hardly ever as precisely located to one half of the spinal cord as shown here, but the figure has been included for didactic purposes.

Lesions in the Somatosensory Cortex

Because the somatosensory pathways to the cerebral cortex are crossed, a lesion of the postcentral gyrus results in sensory loss in a part of the body on the opposite side. The primary modalities of pain, temperature, and touch may not be significantly impaired but the patient usually has difficulties localizing pain and touch stimuli. There is *loss of several discriminative sensory functions*, collectively called the cortical sensations.

Agraphesthesia, i.e., inability to recognize numbers or letters (larger than 4 cm) drawn on the skin, is dependent on the ability to localize tactile stimuli.

Two-point discrimination is the ability to separate two blunt points from one another. Normally, a double stimulus with a distance of about 5 mm can be recognized as two separate points on the fingertips, whereas the distance for two-point discrimination on most of the body surface varies between 4 and 8 cm.

Stereoanesthesia signifies the inability to appreciate the texture, size, and shape of an object.

Astereognosis means that the patient cannot identify an object by palpation in spite of the fact that primary sensory modalities, e.g., touch and vibration, are intact.

Lesions in the Posterior Parietal Cortex

Lesions in the parietal cortex behind the postcentral gyrus, especially in the right hemisphere, tend to produce inattention or neglect of sensory stimuli coming from the opposite side of the body or from the visual space on the opposite side (see Clinical Notes, Chapter 22).

Pain: A Clinical Problem with Many Dimensions

Patients suffering from pain are a daily experience for most physicians. Although momentary pain due to muscular cramp or some other factor is familiar to everybody, pain is also a pre-eminent sign of disease, in which case it serves as an effective alarm system. A variety of noxious (harmful) stimuli can activate pain receptors in various organs of the body, and the character of pain and its distribution is sometimes quite typical, with little doubt about the underlying disease. Skin pain, for instance, is often well localized, and involvement of peripheral nerves or nerve roots may produce a typical location of the pain (see Disc Prolapse in Clinical Example, Chapter 8). Deep pain related to the musculoskeletal system (e.g., arthritis) or the visceral organs is often poorly localized and has an aching character. Pain originating in visceral structures is often felt in the skin, not necessarily in the part overlying the diseased organ but in part of the skin that is innervated by the same spinal cord segment as the visceral organ. One of the most common examples is the pain felt along the ulnar side of the left arm in myocardial infarction. This phenomenon is known as "referred pain" (see Clinical Notes, Chapter 19).

A large number of diseases can produce chronic or intractable pain, which is usually disabling to the patient and a difficult challenge to the physician. Rheumatoid arthritis, cancerous growths and metastases in various parts of the body, or damage to sensory nerves and pathways, i.e., neuropathies, are typical examples. Involvement of peripheral nerves, i.e., peripheral neuropathies, is common and can be caused by a variety of conditions, e.g., trauma, infections (e.g., herpes zoster), diabetes, or nutritional disorders, to mention a few. One of the most common causes of chronic pain is disc prolapse. Spinal cord injuries and other lesions in the CNS, especially those involving the spinothalamic tract or its termination in the posterolateral thalamus, are also sources of neuropathic pain.

Phantom Limb. Following amputation of a limb, most amputees have a feeling that the limb or part of the limb is still present. Many have a tingling feeling, and some report intermittent or continuous pain in the missing limb. Phantom limb pain is a complicated phenomenon apparently resulting from several factors, including peripheral neuropathy as well as abnormalities in the CNS, e.g., hyperactivity in dorsal horn neurons because of deafferentation.

Pain Control

Although most pains can be managed with analgesics (pain-relieving drugs), the choice of drug is sometimes a delicate issue, especially in regard to opiates, which are the most effective pain killers but which also carry the risk of drug abuse. There are also a number of non-drug methods available for pain control, including behavioral methods such as relaxation, hypnosis, and biofeedback, with varying success rate.

Transcutaneous Nerve Stimulation (TNS). One method of pain amelioration is based on the gate control theory of pain (see above). Stimulating electrodes (stimulators) are pasted to the skin close to the area of the pain. The most successful results, apparently, have been obtained in traumatic peripheral nerve injuries, especially when the stimulator can be placed over an intact part of the nerve. The electrical stimulation, according to the gate control theory, activates fibers capable of closing the hypothetical gate, thereby preventing pain stimuli from reaching the brain.

Although stimulators can also be placed directly on the peripheral nerves and on the dorsal column (dorsal column stimulatory, DCS), or even within some parts of the central "pain-inhibiting system" discussed earlier, these procedures are not without risk and their long-term efficiency is still in doubt. Therefore, they are mostly in the experimental stage and not used on a routine basis.

Acupuncture (needling), i.e., the insertion of long needles into specific points of the body, has long been an accepted therapeutic method in the Orient for a variety of diseases. It is now being used increasingly also in some Western countries to treat chronic pain. Although the mechanism of pain relief is not completely understood, the most likely explanation is that the slightly painful stimuli caused by the needle activate peripheral nerves. These send impulses to the brain to release endorphins (morphine-like peptides), which can work at different sites in the CNS to inhibit pain transmission.

Neurosurgical Procedures, e.g., transection of posterior roots (rhizotomy) or the spinothalamic tract (anterolateral cordotomy) are used only in exceptional cases, usually in patients with terminal cancer and when other efforts to control pain have failed.

Clinical Examples

Syringomyelia

A 28-year-old woman complained of numbness in her right arm. Over the preceding 5 years she had gradually lost sensation in her right arm and hand, and over the last few months she had occasionally burned her right fingertips while cooking but had not been aware of the injuries at the time they occurred.

On examination, a Horner's syndrome was evident on her right side. She also had mild atrophy of her right shoulder muscles and marked atrophy of the intrinsic hand muscles and thenar and hypothenar muscles on the right side. Several poorly healing burn scars were noticed on her right fingers. Muscle tone was reduced, and the stretch reflexes were absent in the right arm. The left arm and both legs were normal.

Sensory examination revealed loss of pain and temperature sensation in the right arm and shoulder in a cape-like distribution, but position and vibratory sensations were normal.

An EMG study of the right arm showed pronounced fibrillations of all muscle groups innervated by spinal cord segments C5–T1.

1. Where is the lesion located?
2. Why do fibrillations occur in the muscles of her right arm and hand?
3. Why are the sensory deficits related only to one arm?
4. Why does the patient have a Horner's syndrome?

Discussion

Denervation of muscles and sensory loss in a limb can arise from a variety of pathologic processes involving the spinal cord, nerve roots, nerve plexuses, or peripheral nerves. In this case, preservation of position and vibratory sensation in the right arm militates against peripheral nerve, plexus, or root disease.

The symptoms and signs presented by this patient are typical of syringomyelia of the cervical spinal cord, in this case on the right side. This patient probably has a cervical syrinx similar to the one shown in Fig. 9–8A, but restricted largely to the right side of the spinal cord. The diagnosis can be easily confirmed with MRI (see Fig. 9–8C).

Extension of the syrinx cavity into the anterior gray matter results in progressive loss of anterior horn cells and denervation/atrophy of the corresponding muscles with concomitant fibrillations.

Ipsilateral spinothalamic fibers are disrupted as they approach the anterior white commissure (Fig. 9–8A), resulting in a "suspended" sensory level typically in a cape distribution involving an arm and part of the shoulder.

If the descending sympathetic fibers or the ciliospinal center in T1–T2 are involved in the lesion, an ipsilateral Horner's syndrome appears as in this case. If the cavity continues to enlarge, posterior column fibers or corticospinal tract fibers may also be damaged.

Suggested Reading

1. Brodal, A., 1981. Neurological Anatomy in Relation to Clinical Medicine, 3rd ed. Oxford University Press: New York, Oxford, pp. 46–148.
2. Fields, H. L., 1987. Pain, McGraw-Hill Book Co.: New York.
3. Fields, H. L., M. M. Heinricher and P. Mason, 1991. Neurotransmitters in Nociceptive Modulatory Circuits. Ann. Rev. Neurosci. 14:219–245.
4. Kaas, J. H., 1990. Somatosensory system. In G. Paxinos (ed.): The Human Nervous System. Academic Press. San Diego, pp. 813–844.
5. Macchi, G., A. Rustioni and R. Spreafico, 1983. Somatosensory Integration in the Thalamus. Elsevier: Amsterdam.
6. Rustioni, A. and R. J. Weinberg, 1989. The Somatosensory System. In A. Björklund et al. (eds.): Handbook of Chemical Neuroanatomy, Vol. 7. Elsevier: Amsterdam.
7. Willis, W. D. and R. E. Coggeshall, 1991. Sensory Mechanism of the Spinal Cord, 2nd ed. Plenum Press: New York.

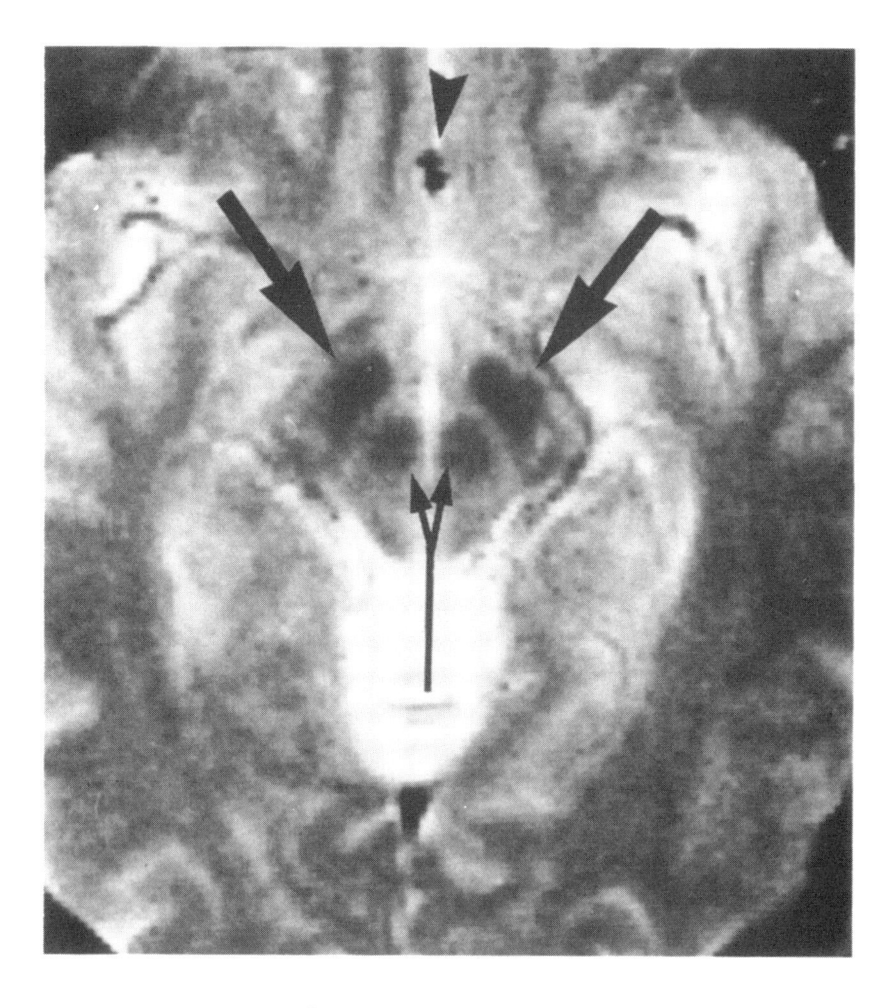

Axial MRI (T2-weighted image) showing normal midbrain in elderly adult. *Thick black arrows* = substantia nigra; *forked arrow* = red nuclei; *arrowhead* = anterior cerebral arteries. (From Sartor, K. 1992. MR Imaging of the Skull and Brain. Springer-Verlag Berlin. With permission of the publisher.)

10

Brain Stem, Monoaminergic Pathways, and Reticular Formation

The brain stem is densely packed with many vital structures such as long ascending and descending pathways, and specific nuclear groups including the nuclei of the cranial nerves. The core of the brain stem is occupied by the reticular formation, i.e., diffuse aggregations of cells surrounded by a wealth of interlacing fibers. The reticular formation is a complicated but highly organized part of the brain. It is crucially involved in a number of functions, including motor activities, respiration, cardiovascular functions, and mechanisms related to sleep and consciousness. Most of the monoaminergic cell groups in the brain are located within the reticular core of the brain stem, and their projections are widely distributed in the CNS.

On the basis of a skillful examination of the patient and knowledge of the anatomy of the nervous system, it is usually possible to make a correct anatomic diagnosis, i.e., to determine the location of the pathologic process in the nervous system. Nowhere is this more clearly demonstrated than in the brain stem, where lesions often involve cranial nerve nuclei and long ascending and descending pathways in different combinations.

After the presentation of a few representative brain stem sections, special attention will be paid to the reticular formation and the monoaminergic systems. The central nervous structures related to the cranial nerves will be discussed in the following chapters.

Cross-Sections Through the Brain Stem

The Spinomedullary Transition Area

The transition between the spinal cord and the medulla oblongata is characterized by a gradual change in surface anatomy and internal structure. The pyramidal and lemniscal decussations gradually cause a breakup of the typical H-shaped pattern formed by the centrally located gray substance in the spinal cord. The dorsal column nuclei, i.e., nucleus gracilis and nucleus cuneatus, which appear in the dorsal funiculi (Fig. 4–45A), will eventually replace the white substance represented by the ascending somatosensory fibers in the dorsal funiculi.

Pyramidal Decussation (Fig. 4–45B)

This is the place where the majority (about 90%) of the fibers in each pyramid cross over the midline and continue as the *lateral corticospinal tract* in the posterior part of the lateral funiculus on the opposite side of the spinal cord. The crossing fibers separate the ventral horns from the rest of the gray substance, and the ventral horns quickly disappear at a slightly higher level. The pyramidal decussation, which can be easily identified during the gross anatomic dissection, is a good marker for the area of transition between the medulla oblongata and the spinal cord.

Lemniscal Decussation (Fig. 4–45A)

The *internal arcuate fibers*, which cross the midline immediately above the pyramidal decussation, originate in the dorsal column nuclei. After the crossing, the fibers continue as the *medial lemniscus* through the brain stem to terminations in the thalamus. The *gracile* and *cuneate nuclei* are prominent structures, and their location on MRI can be inferred from studying the surface anatomy or by identifying the *spinal trigeminal complex*, which is located on the lateral side of the cuneate nucleus.

Upper Medulla Oblongata (Fig. 10–1)

A cross-section through the rostral half of the medulla oblongata looks very different from the previous two sections. Several new structures have appeared, and the central canal has opened up into the fourth ventricle. Three of the most prominent structures—the *pyramidal tract*, the *inferior olivary nucleus*, and the *inferior cerebellar peduncle*, which forms the characteristic lateral wing in this cross-section—can be easily identified on MRI. The inferior olivary nucleus is a large, irregularly folded, bag-shaped nucleus; many of its projection fibers cross the midline and reach the opposite half of the cerebellum through the inferior cerebellar peduncle.

The gray matter adjacent to the fourth ventricle is represented by a number of important nuclei and diffuse aggregations of cells, whose location in MRI can sometimes be inferred from the appearance of slight elevations in the floor of the fourth ventricle. The *hypoglossal nucleus* (XII), which represents the somatomotor column, is located close to the midline. Its root fibers traverse the medulla in a nearly sagittal plane and leave the brain stem between the inferior olive and the pyramid.

Further laterally, close to the ventricular surface, is the largest representative of the visceral motor column, the *dorsal motor nucleus of the vagus* (X). This nucleus flanks the medial aspect of a larger collection of cells, the *nucleus of the solitary tract* (VII, IX, and X), which is the main viscerosensory nucleus. These two nuclei, together with the *nucleus ambiguus*, which innervates the striated muscles of the soft palate, pharynx, and larynx, as well as the heart (cardioinhibitory fibers), are closely interrelated. They are connected with the periphery through the glossopharyngeal and vagus nerves, which can be identified on MRI as they bridge the subarachnoid space to reach the jugular foramen (Fig. 11–14).

The *area postrema*, one of the circumventricular organs (Chapters 3 and 18), which is closely related to the solitary nucleus, is located at a more caudal level than that shown in Fig. 10–1. To be more precise, it appears as two small prominences that bulge into the fourth ventricle on each side of the midline just rostral to the obex. The area postrema, which is strategically located in juxtaposition to the middle third of the solitary nucleus, is a chemoreceptor zone for substances in the blood and CSF. Although its functions are not well known, it is undoubtedly important for homeostasis, and is also well known as a trigger zone for vomiting.

Stretching from the vagal–solitary complex obliquely across the medulla oblongata is the *intermediate reticular zone* (area between broken lines in Fig. 10–1), which includes not only the nucleus ambiguus but also the region known as the *ventro-*

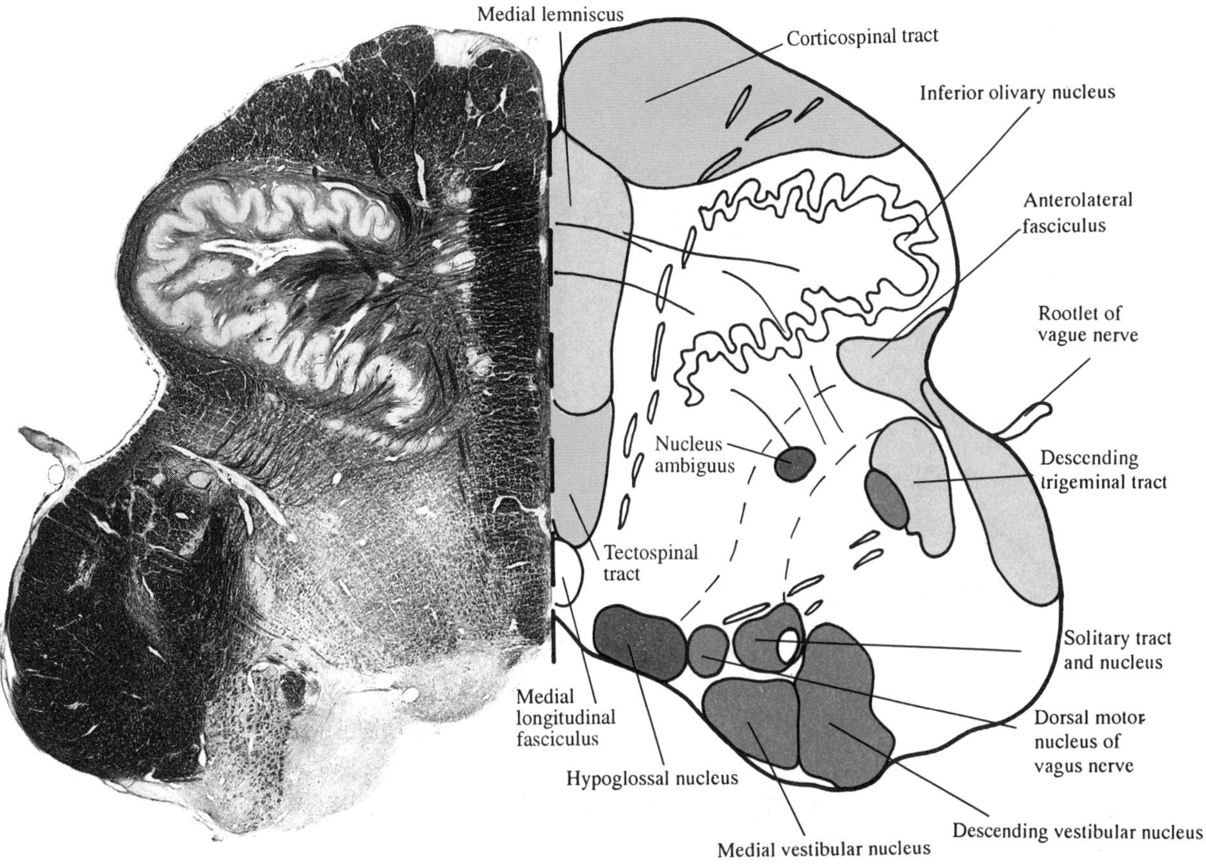

Fig. 10–1. Cross-Section through Upper Medulla Oblongata

Motor nuclei and pathways are colored *red*, sensory nuclei and pathways *blue*, and parasympathetic nuclei *gray*. The zone between the broken lines that extends diagonally across the medulla represents the "intermediate reticular zone" (see text). In Figs. 10–1 to 10–5, the posterior surface of the brain stem faces downwards as it does in axial MRIs (see Figs. 5–2 and 5–14 to 5–20). For brain stem images as they appear in conventional anatomical atlases see Figs. 4–42 to 4–45. (From: Fundamental Neuroanatomy by W.H. Nauta and M. Feirtag. Copyright 1986 by WH Freeman and Company. Reprinted by permission.)

lateral medulla oblongata close to the lateral surface. Although poorly defined in Nissl-stained preparations, the various parts of the intermediate reticular zone show great similarities in connectivity and histochemistry. For instance, the region stands out from the rest of the medulla oblongata by its content of catecholamine-containing neurons, in this case noradrenergic and adrenergic neurons, which are especially numerous in the region of the vagal–solitary complex and in the ventrolateral medulla. The intermediate reticular zone (the autonomic zone of the reticular formation), the vagal–solitary complex including the area postrema, and the nucleus ambiguus are part of the central autonomic network for the regulation of visceral, cardiovascular, and respiratory functions (Chapter 19).

Whereas adrenergic neurons in the medulla oblongata are restricted primarily to the intermediate reticular zone, the most prominent noradrenergic cell group, the *locus ceruleus*, is located close to the floor of the fourth ventricle in the rostral pons (Fig. 10–8). The medullary reticular formation also contains serotonergic neurons, especially in the region just dorsal to the main part of the inferior olivary nucleus, but the majority of the serotonergic neurons are located close to the midline in the *raphe nuclei*, which form a more or less continuous system throughout the brain stem (Fig. 10–6).

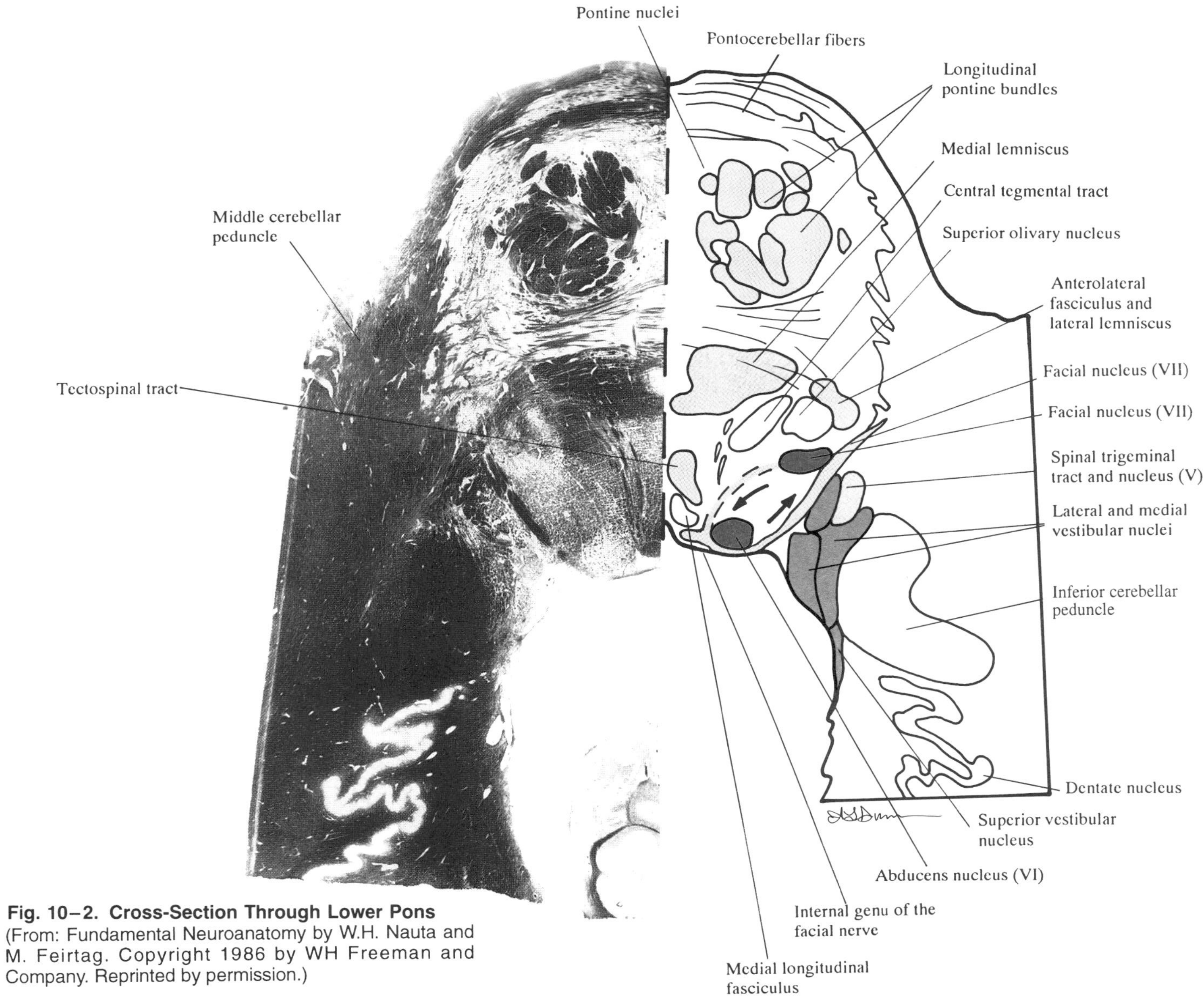

Fig. 10–2. Cross-Section Through Lower Pons (From: Fundamental Neuroanatomy by W.H. Nauta and M. Feirtag. Copyright 1986 by WH Freeman and Company. Reprinted by permission.)

Somatosensory nuclei are located in the more lateral parts of the medulla oblongata. These include the *spinal trigeminal nucleus* (V) and the *medial and inferior vestibular nuclei* (VIII). The *pyramidal tract* as well as the thick lamina of white substance close to the midline containing the *medial lemniscus*, *tectospinal tract*, and *medial longitudinal fasciculus* (MLF), can be recognized on MRI. The *spinothalamic tract* and *ventral spinocerebellar tract*, which are situated close to the lateral surface, cannot be separated at this level and are therefore often referred to as the *anterolateral fasciculus*.

Lower Pons (Fig. 10–2)

The large *middle cerebellar peduncles*, through which fibers from the pontine nuclei reach the cerebellum, constitute prominent landmarks in cross-sections through the pons. In a section through the lower half of the pons, several of the vestibular nuclei (VIII), i.e., the *lateral*, *medial*, and *superior vestibular nuclei*, are still present in a lateral position, as are the *spinal trigeminal nucleus and tract* (V). The *trigeminal motor nucleus* and the main *sensory nucleus of the trigeminus* (V) are located at a slightly more rostral level (Fig. 11–1).

The *abducens nucleus* (VI) is located close to the midline underneath the floor of the fourth ventricle, whereas the position of the *facial nucleus* (VII) corresponds to the location of the ambiguus nucleus in the medulla oblongata section. The abducens root fibers proceed in a nearly straight sagittal course to their exit at the border between pons and medulla oblongata, whereas the facial root fibers behave in a highly unusual way; they make a detour through the reticular formation toward the floor of the fourth ventricle, where they bend around the abducens nucleus and then pass obliquely across the lateral part of the tegmentum to their point of exit between the pons and medulla oblongata in the cerebellopontine angle. The loop of the facial root fibers around the abducens nucleus is referred to as the *internal genu* of the facial nerve, and this area is indicated by a slight elevation, the *facial colliculus*, in the rhomboid fossa (Fig. 4–41). The *salivatory nucleus*, which sends secretomotor fibers to the salivary glands, is located dorsal to the caudal pole of the facial nucleus.

An axial MRI scan through the caudal pons provides a good overview of the *cerebellopontine angle* (Figs. 11–2D and 14–6B) with excellent delineation of the *facial* (VII) and *vestibulocochlear* (VIII) *nerves* as they proceed towards the internal auditory meatus. The *abducens nerve* can often be identified in the same axial MRI.

The *medial lemniscus*, which is a good landmark on the axial MRI, is situated close to the midline but has changed its orientation from a vertical to a nearly horizontal lamina of fibers. Dorsolateral to the medial lemniscus is the *central tegmental tract*, a heterogeneous fiber system that contains both ascending and descending fibers and which is related to the reticular formation throughout the pons and mesencephalon. Many of the descending fibers terminate in the *inferior olivary nucleus*. The medial lemniscus is partly penetrated by the *trapezoid body*, which contains crossing secondary auditory fibers originating in the cochlear nuclei. The cochlear nuclei are located slightly more caudal on the dorsolateral side of the inferior cerebellar peduncle at the level where this peduncle disappears from view as it enters the cerebellum (Fig. 4–38). The trapezoid body continues in a rostral direction as the *lateral lemniscus* together with the *anterolateral fasciculus*. The *superior olivary nucleus* is part of the auditory system (Chapter 14).

The tectospinal tract and MLF can be found in the same relative position as in the section through the medulla oblongata. The MLF is an important coordinating pathway for eye movements and vestibulo-ocular reflexes (Chapter 11); it is closely related both to the vestibular and to the oculomotor nuclei.

The massive basal part of the pons contains a large number of transversely oriented fibers that originate in the pontine nuclei and enter the opposite half of the cerebellum via the middle cerebellar peduncle. The *pontocerebellar fibers* form the second link in an important pathway between the cerebral cortex and the cerebellum. The first link is represented by longitudinally oriented fiber bundles, the *corticopontine tracts*. Other components of the longitudinally oriented fiber bundles in the basal pons are the *corticobulbar* and *corticospinal tracts*.

Upper Pons (Fig. 10–3)

A cross-section through the upper part of the pons looks rather similar to the previous section except that the fourth ventricle is smaller and is bounded on its lateral sides by the superior cerebellar peduncles, which contain most of the cerebellofugal projection fibers. The *superior cerebellar peduncles* can be clearly identified as separate bundles as they proceed medially towards their area of decussation in the lower part of the midbrain (Fig. 10–4). As for the pontine tegmentum, the major difference

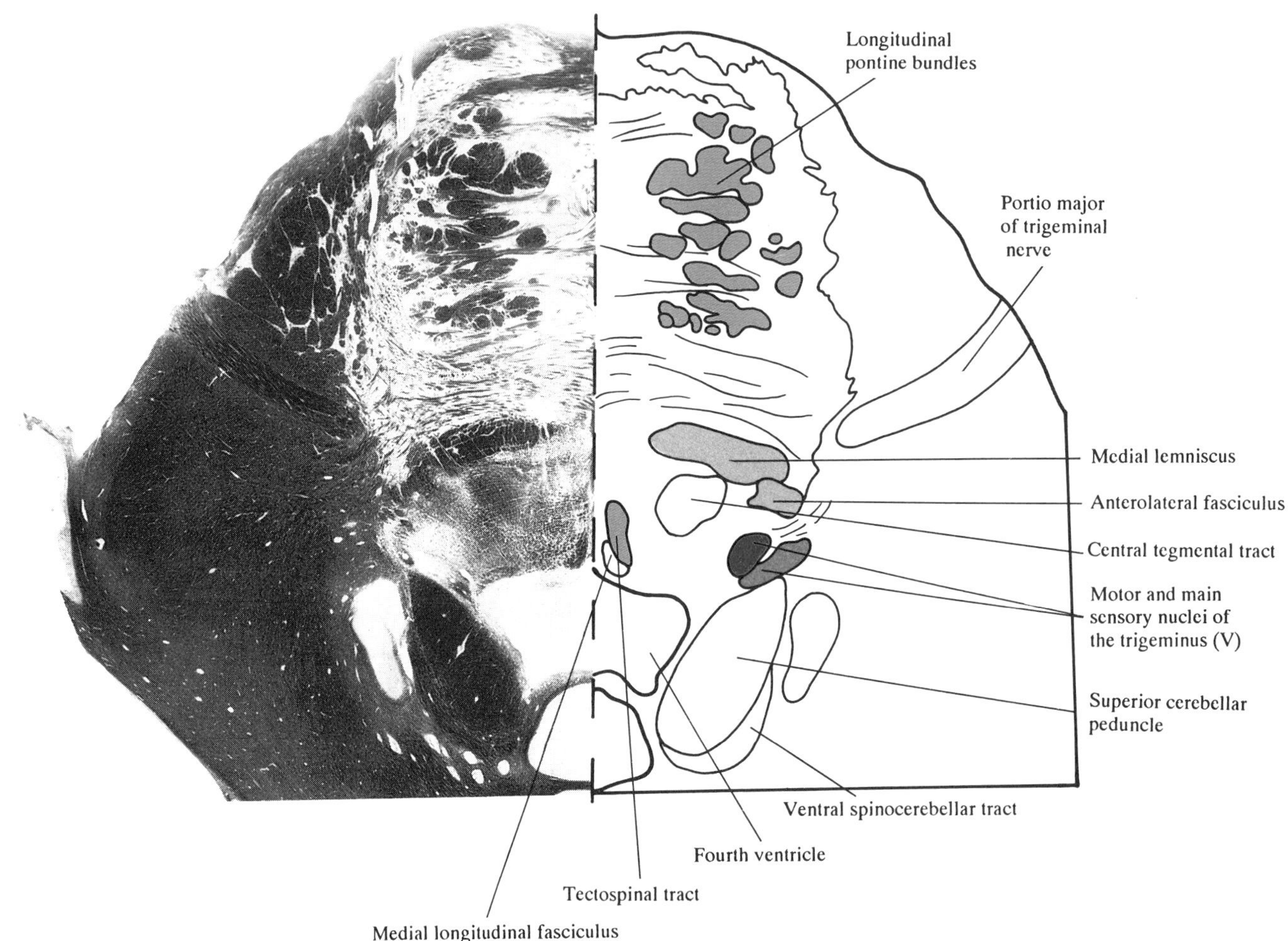

Fig. 10–3. Cross-Section Through Upper Pons (From: Fundamental Neuroanatomy by W.H. Nauta and M. Feirtag. Copyright 1986 by WH Freeman and Company. Reprinted by permission.)

from the previous section is that the spinal trigeminal nucleus has been replaced by the *motor* and *main sensory nuclei of the trigeminus* (V). The cisternal portion of the large trigeminal nerve, which exits the brain on the lateral side of the middle cerebellar peduncle, can be easily identified in axial MRI through the midportion of the pons (Fig. 11–5). The medial lemniscus has moved into a slightly more lateral position where it joins the anterolateral fasciculus in its projection towards the thalamus.

Lower Mesencephalon (Fig. 10–4)

The midbrain section looks totally different from the sections through the pons, primarily because the massive middle cerebellar peduncles are no longer present. The superior cerebellar peduncles, furthermore, have reached the midline, where they form the *decussation of the superior cerebellar peduncles*. Another notable difference compared with the pons sections is that the many longitudinal fiber bundles in the base of the pons, represented by *corticospinal*, *corticobulbar*, and *corticopontine fibers*, now form one massive bundle of descending fibers in the base of the mesencephalic peduncle. Another part of the base of the peduncle is occupied by the *substantia nigra*, an important component of the motor system (Chapter 16), which can also be clearly imaged, primarily because its largest part, the *pars reticulata*, is rich in iron (see Vignette). The other part of the substantia nigra, the *pars compacta*, has many dopamine-containing neurons whose axons form part of the nigrostriatal dopamine system (Fig. 10–7, see also "Important Side Loops" in Chapter 16).

Another significant change from the previous

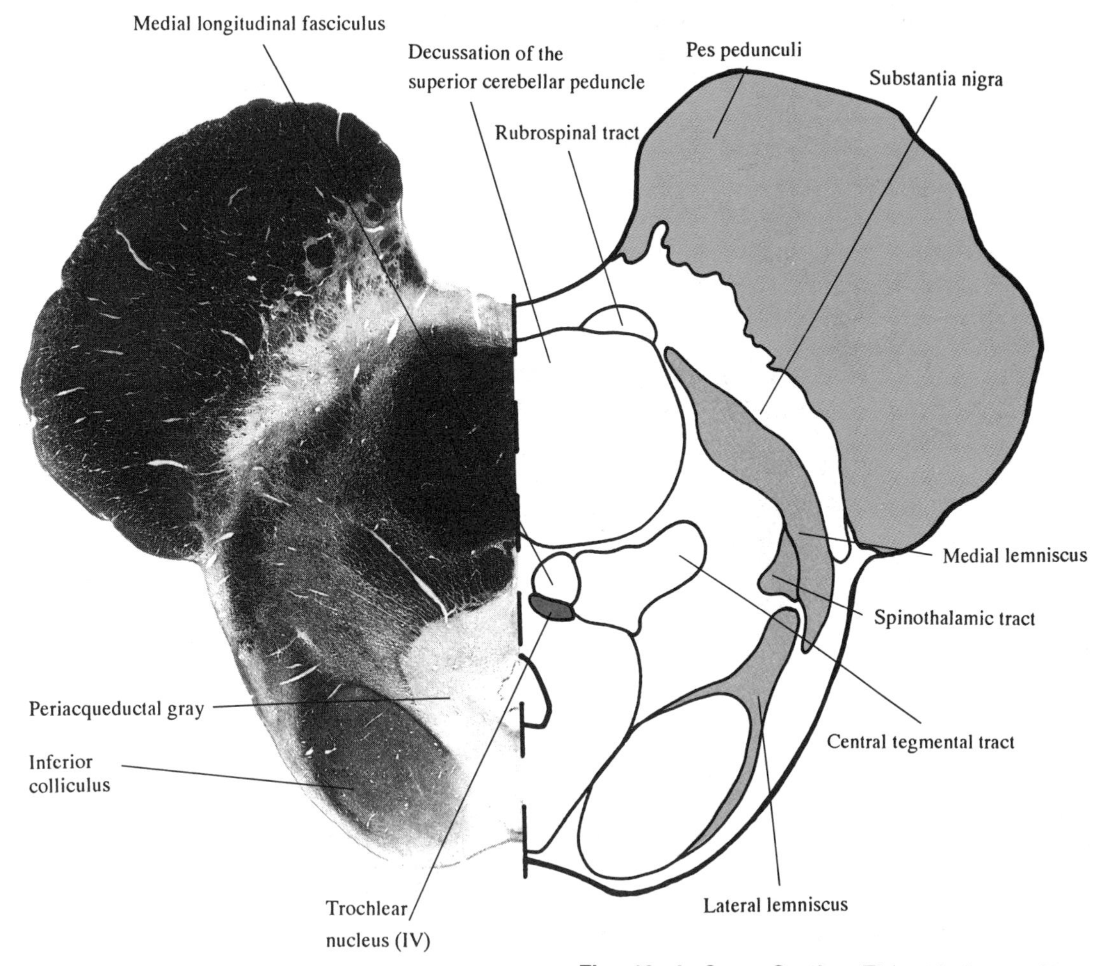

Fig. 10–4. Cross-Section Through Lower Mesencephalon
(From: Fundamental Neuroanatomy by W.H. Nauta and M. Feirtag. Copyright 1986 by WH Freeman and Company. Reprinted by permission.)

section is that the fourth ventricle has been replaced by the *cerebral aqueduct*, which in turn is surrounded by a thick layer of gray substance, the *periaqueductal gray*, sometimes referred to as the central gray. The periaqueductal gray is an important part of a pain-inhibiting system (Chapter 9), and as a component of the central autonomic network, it plays a special role in the integration of somatomotor and autonomic responses involved in various defense reactions (Chapter 19). The tectum is represented by the *inferior colliculi*, which serve as important relay centers in the auditory pathways from the cochlear nuclei to the medial geniculate bodies. A large part of the tegmental area lateral to the central tegmental tract is occupied by the *pedunculopontine tegmental nucleus* (Fig. 10–12), which is part of the anatomic substrate for the ascending activating system.

The cranial nerve nuclei represented at this level are the *trochlear nucleus* and the *mesencephalic trigeminal nucleus* which is embedded in the lateral part of the central gray. The MLF, however, which is closely related to the trochlear nucleus, forms a well-circumscribed bundle that can be identified in MRI. Although the *trochlear nerve*, which is the thinnest of the cranial nerves, is difficult to demonstrate even with MRI, it can be visualized in an appropriate parasagittal view as it curves around the cerebral peduncle (Fig. 11–2C).

Upper Mesencephalon (Fig. 10–5)

The appearance of the rostral midbrain section through the *superior colliculi* is reminiscent of the section through the caudal part of the midbrain. The main difference is that the decussation of the superior cerebellar peduncles has disappeared.

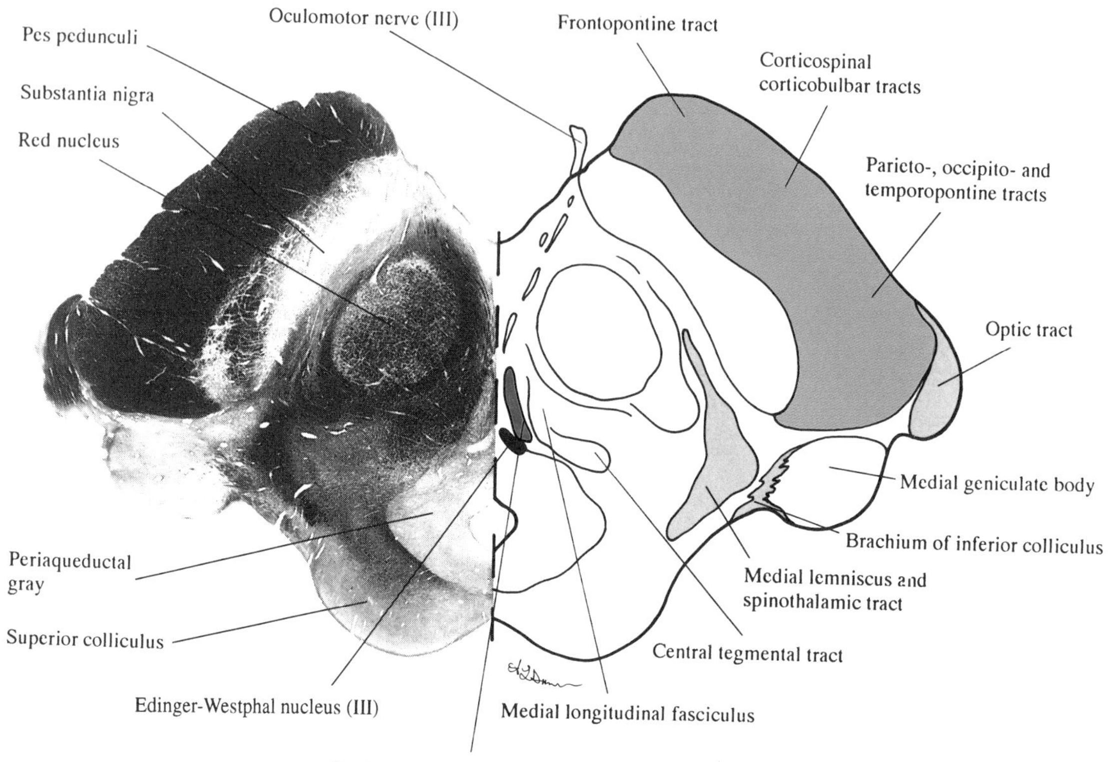

Fig. 10–5. Cross-Section Through Upper Mesencephalon
(From: Fundamental Neuroanatomy by W.H. Nauta and M. Feirtag. Copyright 1986 by WH Freeman and Company. Reprinted by permission.)

the compact MLF (Fig. 10–4), the considerably larger *oculomotor nucleus* (III), which is located in a corresponding position in a cross-section through the superior colliculus, leans against the whole extent of the MLF. The root fibers of the *oculomotor nerve* proceed in a nearly sagittal plane to their exit from the brain stem in the *interpeduncular fossa* (Fig. 10–5).

The *superior colliculi* are important reflex centers for orienting movements and postural adjustments of eyes, head, neck, and body (see "Brain Stem Centers for Control of Eye Movements", Chapter 11). They receive sensory impulses from the visual system, but also from the auditory and somatosensory systems, especially the trigeminus region. Instead, a large part of the tegmentum is occupied by the *red nucleus*, which, like the reticular part of the substantia nigra, contains iron and therefore can be easily recognized in MRI. Whereas the small *trochlear nucleus* is located in an indentation in

Monoaminergic Pathways

There are five major monoaminergic systems: the *5-hydroxytryptamine* (serotonin, 5-HT), *dopamine*, *noradrenaline*, *adrenaline*, and *histamine* neuron systems. Except for the histamine-containing neurons, which are located in the posterior hypothalamus, the monoaminergic cell bodies are located primarily in the brain stem, often within or in close relation to the reticular formation (see below). In spite of the fact that these cell groups are concentrated within a relatively small part of the brain, their axonal projections reach all major regions of the CNS.

Serotonin is synthesized from tryptophan, whereas the catecholamines (dopamine, noradrenaline, and adrenaline) are derived from tyrosine through the following steps: tyrosine → dopa (L-dihydroxyphenylalanine) → dopamine → noradrenaline → adrenaline (see Vignette, Chapter 23). Histamine is formed from the amino acid histidine.

The effect that a neurotransmitter exerts is dependent not only on its chemical composition, but also on the receptor with which it combines. Several types of receptors have been identified for all of the monoamine transmitters, and it is therefore possible for the same transmitter to have completely different effects, i.e., inhibitory or excitatory, on a target cell. For instance, the adrenergic receptors responding to peripheral sympathetic stimulation have classically been divided into α

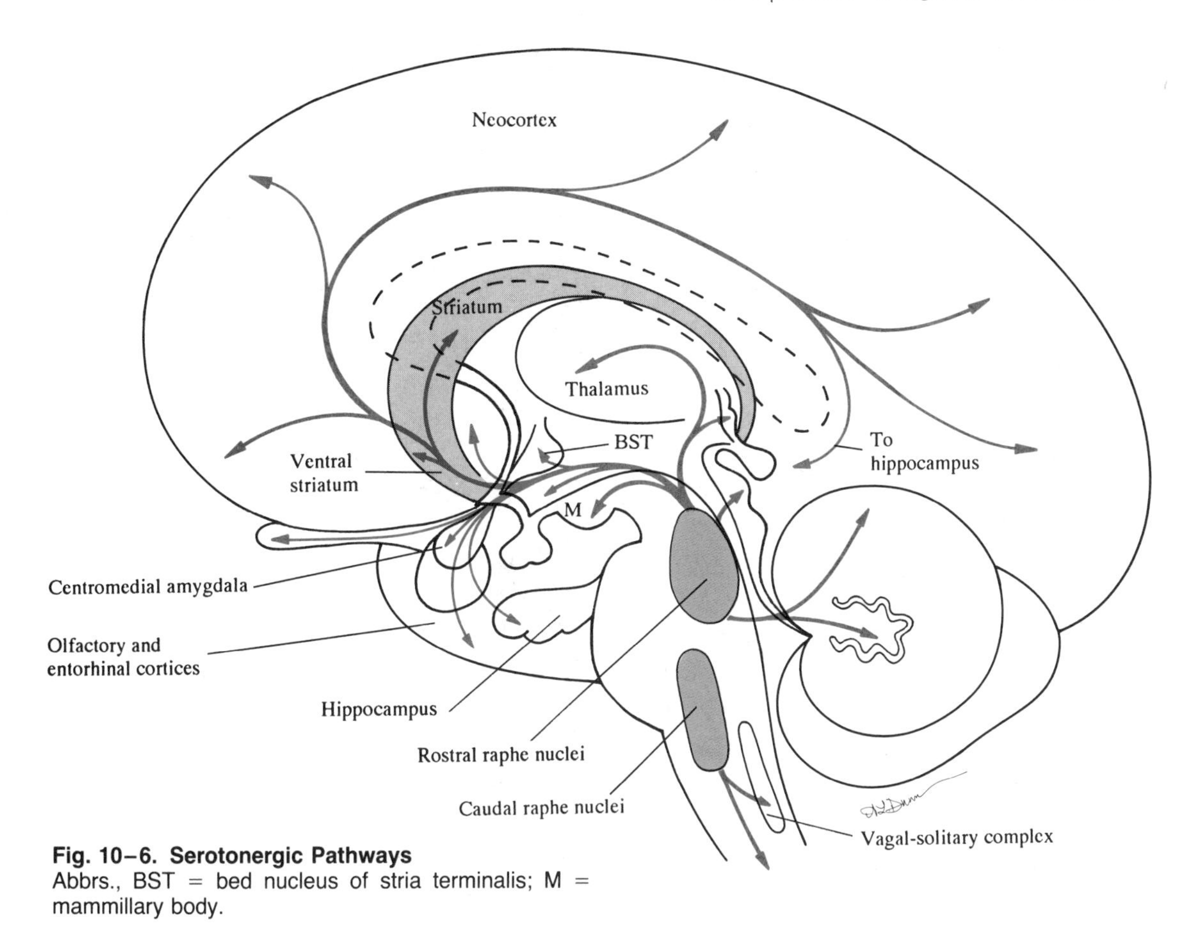

Fig. 10–6. Serotonergic Pathways
Abbrs., BST = bed nucleus of stria terminalis; M = mammillary body.

and β receptors, which have opposite effects on smooth muscle; α-adrenergic responses result in contraction of most smooth muscles, whereas β-adrenergic stimulation provides for smooth muscle relaxation. A similar situation exists in regard to the different serotonergic (5-HT) receptors; some have an excitatory effect and others an inhibitory effect on the postsynaptic neuron (Fig. 9–3).

Two major types of postsynaptic dopamine receptors (D_1 and D_2) are located in the distribution areas of the midbrain dopamine projection system. Recently, several subtypes of these main receptor types have also been identified, and some of these subtypes, characterized as D_3 and D_4, have a more restricted distribution than the D_1 and D_2 receptors in the sense that they are located primarily in the accumbens and some other parts of the basal forebrain rather than in all parts of the striatal complex. Several types of histamine (H) receptors have also been identified, and the role of histamine as a transmitter is now generally accepted, although its functional role is less clear.

The action of neuroactive drugs is in large part dependent on the effect that the drug has on a specific receptor. As additional receptor subtypes with specific functional attributes are identified, we will be in a better position to develop drugs that are restricted in their effect to a specific receptor subtype, thereby reducing the nonspecific side effects that may limit the usefulness of an otherwise effective drug. Progress in the field of neuropharmacology, therefore, which deals with neurotransmitters, receptors, and their interactions, is often of immediate clinical relevance.

Serotonergic Pathways

Many of the cells of origin of the serotonin system are located in the *raphe nuclei*, which form an extensive, more or less continuous collection of cell groups close to the midline throughout the brain stem (Fig. 10–6). Not all the cells in the raphe nuclei are serotonergic. As a matter of fact, serotonergic cells are in the minority in some of the raphe nuclei, whose cells contain many other

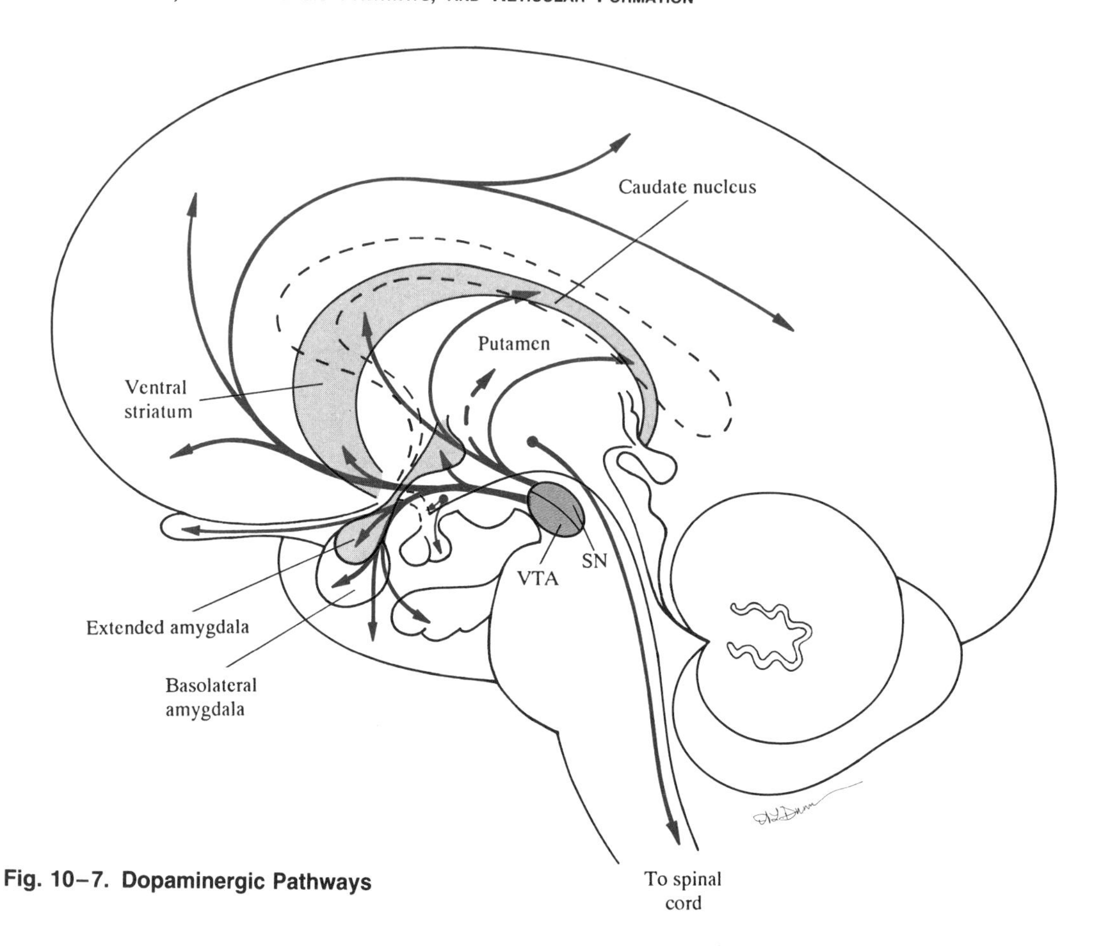

Fig. 10–7. Dopaminergic Pathways

neurotransmitters including dopamine, noradrenaline, GABA, and a number of neuropeptides. On the other hand, serotonergic cells are located also in adjacent parts of the reticular formation. Afferent fibers to the rostral group of raphe nuclei come from the prefrontal cortex and from a large number of basal forebrain regions including the extended amygdala, hypothalamus, diagonal band nuclei, and habenula. The caudal raphe group receives input from the prefrontal cortex, periaqueductal gray, several regions in the reticular formation, and the spinal cord.

The axons of the cells in the rostral group of raphe nuclei, which are located in the mesencephalon and the rostral part of the pons, are widely distributed in the forebrain and cerebellum, whereas the raphe nuclei in the lower pons and medulla oblongata send projections to the spinal cord. Through their widespread connections, the serotonergic pathways undoubtedly influence many functional–anatomic systems in the CNS, and they often contain another transmitter (e.g., substance P, enkephalin, or thyrotropin-releasing hormone) in addition to serotonin.

The descending serotonergic fibers subserve a variety of functions on the spinal cord level. Serotonergic fibers with terminations in the substantia gelatinosa form part of a central pain-control system (Chapter 9), whereas serotonergic projections to the anterior horn have an excitatory effect on motor neurons. Some of the descending serotonergic fibers that reach the intermediolateral cell column are involved in cardiovascular regulation.

The widespread ascending projections reach all parts of the forebrain including the cerebral cortex. Terminations are especially dense in the hypothalamus, basal ganglia, amygdaloid body, hippocampus, and cingulate gyrus. Serotonin has long been implicated in the sleep–wake cycle and in affective behavior, and there has also been much speculation about the influence of the serotonergic system on a number of other functions including food intake, hormone secretion, sexual behavior, and thermoregulation (Chapter 18). As for the role of serotonin in the sleep–wake cycle, it is interesting to note that the activity in the serotonergic system decreases during sleep, and it is at its lowest during rapid eye movement (REM) sleep (see below), when muscular relaxation is most pronounced. This is consistent with serotonin's excitatory effect on the motor neurons, as mentioned above.

The serotonergic system, finally, has been implicated in affective behavior, and the effect of hallucinogenic drugs, some of which resemble serotonin in chemical structure, has been specifically linked to their action on one or several of the subtypes of serotonin receptors.

Of great clinical importance is the fact that serotonergic fibers reach the cranial blood vessels, including those within the brain and in the pia mater, where they have a vasoconstrictor effect. Indeed, the potency of some of the new migraine drugs seems to be related to their ability to block serotonin transmission in cranial blood vessels.

Dopaminergic Pathways

A large number of dopaminergic cells are located in the midbrain, especially in the *substantia nigra, pars compacta* (Fig. 10–7) and in the *ventral tegmental area.* The dopaminergic mesencephalic cell groups receive input from several basal forebrain regions including both the basal ganglia (Chapter 16) and the extended amygdala (Chapter 20), as well as from the reticular formation in the mesopontine tegmentum and the raphe nuclei. The axons from the dopaminergic cells form a massive ascending projection system, the *mesotelencephalic dopaminergic* system.

Many of the fibers in the mesotelencephalic system traverse the internal capsule to reach all parts of the caudate nucleus and putamen; this is the famous *nigrostriatal dopamine pathway*, which has received much attention in relation to Parkinson's disease, in which there is a significant loss of dopamine cells in the mesencephalon with accompanying depletion of dopamine in the striatum (Chapter 16). Other fibers proceed forward through the lateral hypothalamus and underneath the lentiform nucleus toward the rostral parts of the basal forebrain, where they terminate primarily in ventral striatal areas, including the accumbens, and in the extended amygdala, but also in several other basal forebrain regions, including the basolateral amygdala, septal, and olfactory areas. Although most of the cerebral cortex receives dopaminergic innervation from the mesencephalic cell groups, the density of innervation is considerably less than in the striatum, and it becomes increasingly sparse in the caudal parts of the cerebral cortex.

Whereas a significant loss of dopaminergic neurons in the substantia nigra is the pathologic hallmark of Parkinson's disease (see Clinical Notes, Chapter 16), an increased activity in the ascending dopamine system, especially to the ventral striatum, has often been promoted as an explanation for schizophrenia. One of the strongest arguments that has kept the so-called dopamine hypothesis of schizophrenia alive for many years is the fact that most of the effective medicines for schizophrenia are antidopaminergic drugs, whose antipsychotic effect is closely correlated with their affinity for the D_2 receptors. The dopamine hypothesis of schizophrenia is discussed in Chapter 16.

Dopamine-containing cells are also located in the hypothalamus and several other basal forebrain regions, including the retina, olfactory bulb, cerebral cortex, and adjoining white matter. Dopaminergic cells of special importance for neuroendocrine functions are concentrated in the arcuate and periventricular nuclei. These give rise to the *tuberoinfundibular pathway*, which projects to the median eminence and the posterior lobe of the pituitary. Dopamine liberated from this system has been shown to have an inhibitory effect on the release of prolactin and melanocyte-stimulating hormone from the pituitary (Chapter 18).

The paraventricular and supraoptic nuclei also contain a significant number of dopamine cells, and there is evidence that some of the descending fibers from the paraventricular nucleus to the brain stem and spinal cord use dopamine as their transmitter.

Noradrenergic and Adrenergic Pathways

As indicated in Chapter 19, noradrenaline or norepinephrine is the transmitter substance that is released from all postganglionic sympathetic fibers except those to the sweat glands, which are cholinergic. Noradrenergic fibers in the CNS arise from special cell groups in the reticular formation, especially the locus ceruleus in the pons and the intermediate reticular zone in the medulla oblongata. Several of the cranial nerve nuclei, e.g., those in the vagal complex (dorsal nucleus of vagus, solitary and ambiguus nuclei) also contain moradrenaline-synthesizing cells. The highest level of noradrenaline in the medulla oblongata is present in the region of the area postrema, one of the circumventricular organs closely related to the vagal-solitary complex.

More than half of all noradrenergic neurons are located in a relatively small compact cell group referred to as the *locus ceruleus* (Fig. 10–8), which is located underneath the floor of the fourth ventricle at a slightly more rostral level than that shown in Fig. 10–3. The densest concentration of the large pigmented norepinephrine-producing cells occurs rostromedial to the motor nucleus of the trigeminus. The area of the locus ceruleus reveals itself in the rostral part of the rhomboid fossa by a

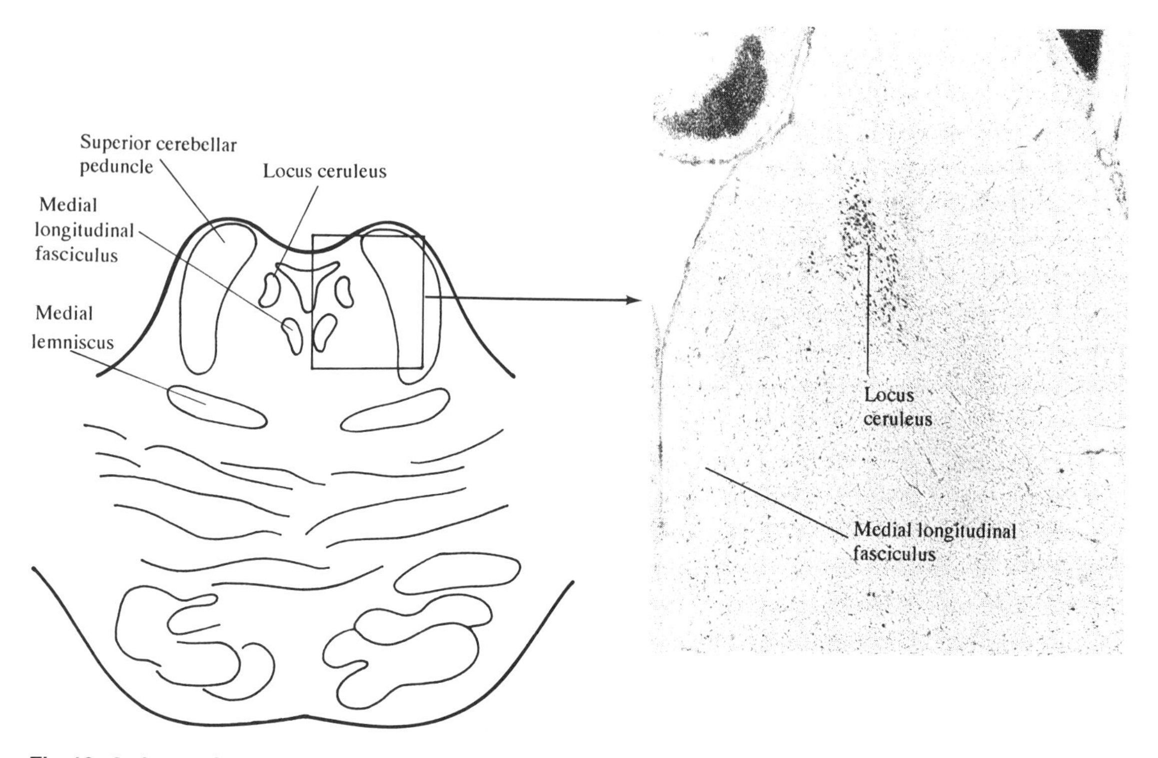

Fig. 10–8. Locus Ceruleus
The photomicrograph of the locus ceruleus (from a Nissl-stained cross-section of the human brainstem) was kindly provided by Drs. George Paxinos and Xu-Feng Huang. Note that the sections are presented in the traditional fashion, i.e. with the posterior part of the brainstem facing upwards.

bluish-gray tint [hence the name blue (ceruleus) place (locus)] in the region of the sulcus limitans. The pigment, neuromelanin, which gives rise to the blue color, is a polymer of dihydroxyphenylalanine (DOPA). The projections from the locus ceruleus are characterized by a widespread distribution (Fig. 10–9). Indeed, the nucleus affects every major region in the brain and spinal cord in spite of the fact that the number of neurons in the nucleus is rather limited, a little more than 10,000 on each side of the brain. To accomplish this, many of the axons branch profusely, but not at the expense of specificity. On the contrary, different regions of the CNS are characterized by specific patterns and densities of noradrenergic terminals. The afferent connections of the locus ceruleus are not yet well known; they may not be as diversified as previously thought. In fact, there is little evidence for the presence of forebrain projections directly to this nucleus. However, significant inputs have been discovered from areas in the rostral part of the medulla oblongata.

Like the ascending raphe system, the locus ceruleus seems to play an important role in the sleep–wake cycle. The noradrenergic neurons are almost silent during REM sleep, and their level of activity becomes more intense with increased wakefulness. The release of noradrenaline is especially prominent during situations of extreme vigilance or attention, and it is widely assumed that the noradrenergic system enhances the ability of its target neurons to respond to whatever input, e.g., sensory or otherwise, they have to deal with. Loss of noradrenergic neurons in the locus ceruleus is a prominent feature in both Parkinson's and Alzheimer's disease.

The intermediate reticular zone in the medulla oblongata and caudal pons, including the area postrema and the vagal group of cranial nerve nuclei, contains both noradrenergic and adrenergic neurons, which are involved in neuronal circuits related primarily to visceral, cardiovascular, and respiratory functions (Chapter 19). Whereas the ascending noradrenergic projection fibers from the intermediate reticular zone reach a number of structures both in the brain stem and the basal forebrain including septum, extended amygdala, and many hypothalamic nuclei, the ascending medullary adrenergic pathway has a more restricted projection to primarily the paraventricular nucleus, central

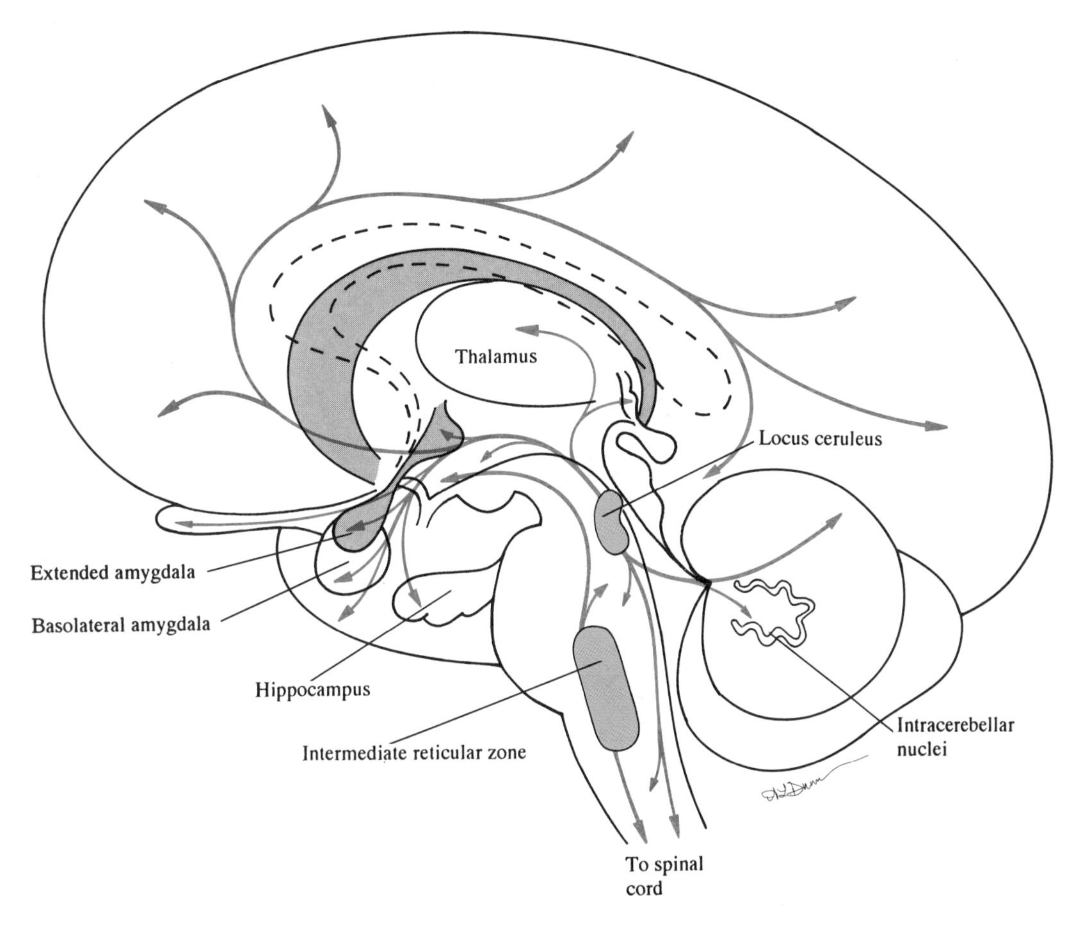

Fig. 10–9. Noradrenergic Pathways

gray, and locus ceruleus. The reticulospinal pathways from the intermediate reticular zone contain both adrenergic and noradrenergic fibers, which are headed primarily for the intermediolateral cell column containing the preganglionic autonomic motor neurons (Chapter 19).

Histaminergic Pathways

Unlike the monoamine systems discussed above, whose cells of origin are located primarily in the brain stem, the cells giving rise to the histaminergic projection system are located in the *tuberomammillary nucleus* in the posterolateral hypothalamus. Nonetheless, considering the many similarities that exist between the histamine system and the other monoamine systems, it makes sense to discuss the histaminergic pathways at this point, especially since the lateral hypothalamus for all practical purposes can be considered a rostral extension of the brain stem reticular core. The histaminergic projections reach all major regions of the forebrain and several structures in the brain stem and spinal cord (Fig. 10–10). Although the histaminergic projections are widespread, it appears that the input to the tuberomammillary nucleus is restricted primarily to fibers coming from the hypothalamus, extended amygdala, septum, and some allocortical and periallocortical areas.

There is considerable evidence, both from the effects obtained by neuropharmacologic manipulations (e.g., central administration of histamine or histamine antagonists) and on the basis of morphologic data, that the tuberomammillary histaminergic system is an important regulator of autonomic and neuroendocrine activities including, feeding, and drinking, and temperature regulation. Behaviors such as locomotion, and arousal may also be affected by histamine.

Since the histaminergic fibers have many varicosities but few synapses, it appears that the histaminergic projection system may use a "parasynaptic" or "paracrine" mode of communication, in which the

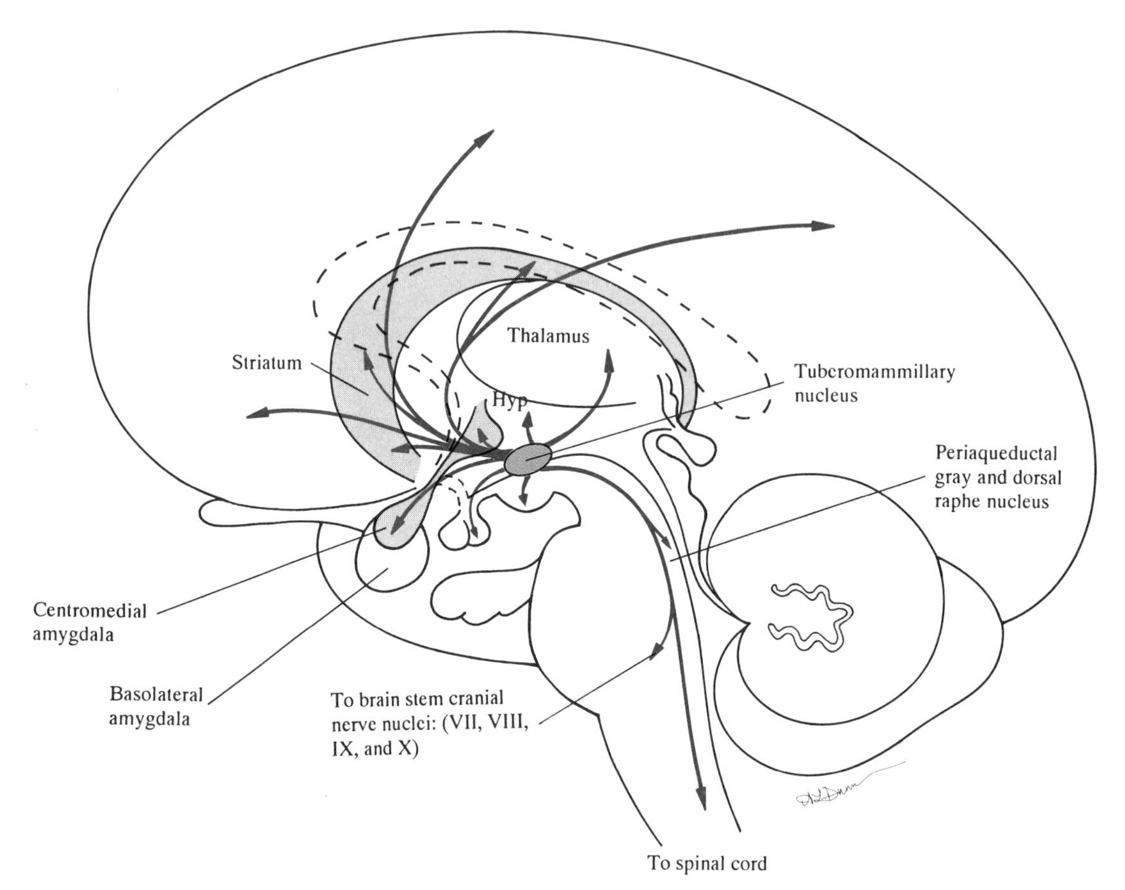

Fig. 10–10. Histaminergic Pathways
The tuberomammillary nucleus, which is located in the posterolateral hypothalamus, projects to all major regions of the forebrain and many structures in the brain stem and spinal cord.

transmitter would spread through the extracellular space and exert its actions at places where the appropriate receptors are located (Chapter 6). In this context, it should be pointed out that the varicosities mentioned above are in close proximity not only to neurons, but also to glia and blood vessels, both of which have been shown to contain histamine receptors.

Reticular Formation

The brain stem reticular core is continuous with the intermediate gray of the spinal cord caudally, and with the lateral hypothalamic and subthalamic regions rostrally. It is located in the central parts of the brain stem (Fig. 10–11) and is characterized by loosely aggregated cells of different types and sizes intermingled with a wealth of fibers of different orientation. The reticular (net-like) appearance is easily appreciated in histologic sections, where the reticular formation can be easily distinguished from surrounding long pathways and specific cell groups such as the red nucleus and the cranial nerve nuclei. The reticular formation, however, is not a diffuse, undifferentiated structure. Cytoarchitectonic and histochemical differences exist between various parts of the reticular formation, and different subdivisions are characterized by specific connections.

Although anatomists had long recognized the reticular formation as a special part of the brain stem, it was the notion that the reticular formation represents the anatomic substrate for an "ascending activating system" (see below) that triggered an intense interest in this part of the brain. In spite of much research, however, it is still not possible to give a meaningful characterization of all the different cell groups in the reticular core of the brain stem. Based on cytoarchitectonic features revealed in Nissl-stained material, anatomists have been able to identify more than 40 nuclei, although their borders are often poorly defined. However, as in

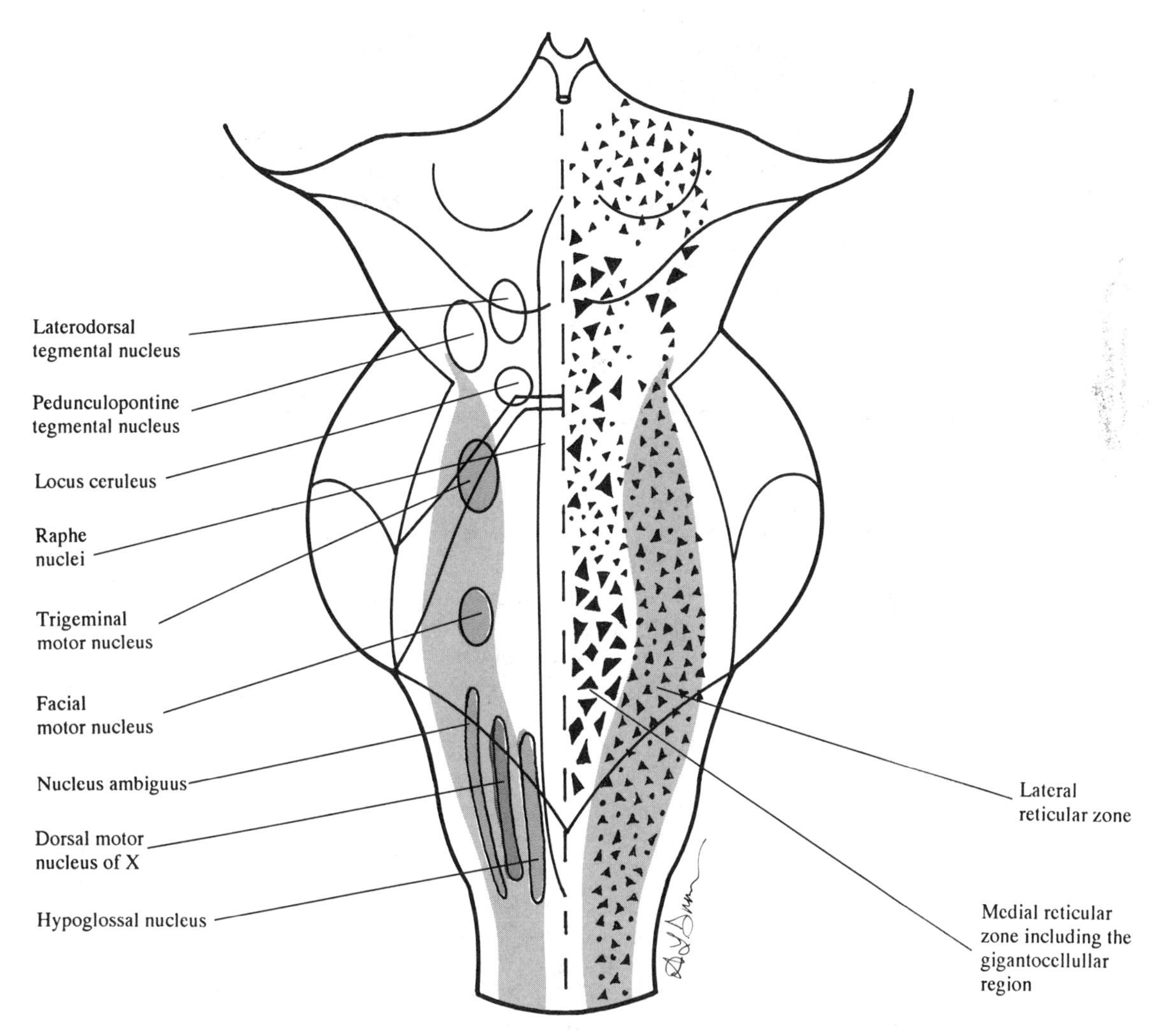

Fig. 10–11. The Reticular Formation
Diagram of the reticular formation and some related brain stem nuclei in a dorsal view of the brain stem. The reticular formation can be roughly subdivided into a medial longitudinal zone containing many large cells and a lateral zone with generally smaller cells. The raphe nuclei close to the midline form a third, paramedian, longitudinal zone. (Modified from Nieuwenhuys, R., J. Voogd, and C. van Huijzen, 1988. The Human Central Nervous System, 3rd ed. Springer-Verlag, and from Holstege, G., 1991. Descending motor pathways and the spinal motor system: Limbic and non-limbic components. In: Holstege, G. (ed.): Role of the Forebrain in Sensation and Behavior. Progress in Brain Research, Vol. 87, Elsevier, Amsterdam.)

many other parts of the brain, including the basal forebrain (Chapter 16), modern tracer methods and transmitter-specific techniques have revealed the existence of chemically specific cell groups and anatomic systems that do not always abide by the boundaries that have been identified on purely cytoarchitectonic grounds. For instance, when the monoaminergic cell groups were discovered in the 1960s (Chapter 7), it became clear that most of them were located in the brain stem reticular core, but it was also obvious that many did not conform to conventional neuroanatomic subdivisions. Therefore, an alphanumeric nomenclature was introduced which labeled the catecholaminergic cell groups A1–A15; noradrenergic groups in the medulla oblongata and pons are labeled A1–A7 from caudal to rostral, followed by the dopaminergic groups in the mesencephalon, hypothalamus, and olfactory bulb, which are classified as A8–A15. The serotonergic neurons, many of which are located in the raphe nuclei close to the midline, are labeled B1–B9 from the caudal to the rostral, and the adrenaline-containing neurons in the medulla oblongata are designated C1–C3.

The numerically labeled cell groups do in some instances conform to the boundaries of traditional anatomic subdivisions. Cell group A6, for instance, corresponds to the locus ceruleus, and the dopaminergic neurons in the substantia nigra, pars compacta, are equivalent to cell group A9. The alphanumeric nomenclature is exclusive, in the sense that it refers only to those neurons within a given area that contain the specific neurotransmitter in question. The conventional nomenclature, on the other hand, is all-inclusive in the sense that it covers all the neurons within a specific region.

The numerical nomenclature for monoamines was introduced at a time when it was difficult to imagine the explosive development in chemical neuroanatomy that was just around the corner, and which has led to the identification of a large number of chemospecific neuronal systems with their parent cell bodies located to specific regions in the CNS. Although a numerical nomenclature has recently been proposed also for the cholinergic cell groups (Chapter 16), there are no corresponding nomenclatures for the other transmitter-specific cell groups and systems that have been identified during the last 20 years (Chapter 23). Instead, these neuronal systems are described with reference to traditional cell groups and boundaries, and we will in general do the same in regard to the monoaminergic cell groups. However, we do not need to concern ourselves with all the nuclei that have been identified in the brain stem reticular core on the basis of cytoarchitectonic criteria, especially since their boundaries are often poorly defined and their functional significance largely unknown. Instead, we will focus attention on some general principles of anatomic organization.

Intrinsic Organization

Throughout most of the brain stem, the reticular core is characterized foremost by loosely arranged small and medium-sized neurons. The main exception to this general rule occurs in the central parts of the upper medulla oblongata and lower pons (Fig. 10–11), where a significant number of large cells intermingle with small and medium-sized cells. This region is referred to as the *gigantocellular reticular region*.

Many of the cells in the reticular formation are characterized by long radiating dendrites that spread out in a plane perpendicular to the long axis of the brain stem. The dendrites of different neurons overlap extensively, and the neurons seem to be eminently suited to sample information from different sources. This is consistent with physiologic studies that have shown a widespread convergence of afferent impulses on units in the reticular formation. Somatosensory and viscerosensory information reaches the reticular neurons via spinoreticular fibers and sensory cranial nerves. The reticular formation also receives information from many parts of the CNS, including the motor cranial nerve nuclei, cerebellum, hypothalamus, basal ganglia, extended amygdala, and cerebral cortex, especially the premotor cortex.

The axons of many of the reticular neurons remain within the reticular formation, where they take part in multisynaptic pathways and reflex circuits. Other axons form a widespread system of pathways that reach practically every part of the CNS, including all the major forebrain regions, cerebellum, and all segments of the spinal cord. Many of the axons are characterized by a large number of axon collaterals, and some of the large cells have axons that dichotomize in a long ascending and descending branch. However, most effector neurons in the reticular formation influence either the spinal cord, the forebrain, or the cerebellum. Nonetheless, there is a high degree of integration of ascending and descending activities. For instance, a change in the level of consciousness from a drowsy state to an extreme degree of attention (ascending activity) is usually accompanied by changes in the somatomotor and autonomic spheres (descending activity).

The raphe nuclei adjacent to the midline are also part of the reticular formation. They extend throughout most of the brain stem except in the basal part of the pons or in other areas where large fiber bundles cross the midline (e.g., the pyramidal decussation, the lemniscal decussation, and the decussation of the superior cerebellar peduncle). The raphe nuclei are best known for their content of serotonin-containing cells, whose highly branching axons reach practically every part of the CNS (see Fig. 10–6).

It is difficult to present a picture of the brain stem reticular formation that does justice to the many and often complicated anatomic relations, some of which are still poorly understood. The highly schematic illustration in Fig. 10–11 is based on a classic subdivision of the reticular formation into a lateral reticular zone, which is continuous caudally with the spinal intermediate gray matter, and a medial reticular zone, which contains the gigantocellular region in the central part of the medulla oblongata and the caudal part of the pons, and which continuous rostrally into the subthalamic and lateral hypothalamic regions. The lateral reticular zone is confined primarily to the pons and medulla oblongata, and a distinct boundary between the lateral and medial reticular zone can only be recognized at the level of the gigantocellular reticular region. The lateral reticular zone, like the spinal intermediate zone, contains interneurons for the primary motor neurons, which in the case of the brain stem are represented by the motor cranial nerve nuclei. The medial reticular zone gives rise to most of the long descending pathways involved in motor control. It also contains the oculomotor nuclei and related structures involved in coordinated eye–head movements (Chapter 11). Serotonergic, noradrenergic, and histaminergic pathways, as well as the ascending cholinergic brain stem pathways related to the ascending activating system, are also part of the medial reticular zone.

Functional–Anatomic Systems

The reticular core of the brain stem is involved in a multitude of functions, which are described in various chapters of this book. A specific part of the reticular core can usually be identified as a substrate for a certain function or a group of interrelated functions, even if it may be difficult to identify the exact boundaries for such regions. Nonetheless, the fact that the reticular formation can be subdivided into specific functional-anatomic systems should discourage the use of the term "reticular formation" as a synonym for the "ascending (reticular) activating system."

Many cell groups in the medial zone of the pontine and medullary reticular formation give rise to descending reticulospinal systems related to motor functions (Chapter 15). Other descending pathways originating in the periaqueductal gray and caudal raphe nuclei form part of a central pain-inhibiting system (Chapter 9). Interneurons in the paramedian pontine reticular formation are closely related to the superior colliculus and to the cranial nerve nuclei (III, IV, and VI) mediating eye movements. These structures, which lie adjacent to the medial longitudinal fasciculus, form a system of great importance for conjugated eye movements and for the coordination of eye and head movements (Chapter 11). Other motor cranial nerve nuclei (V, VII, X, and XII) are embedded in the lateral parts of the pontine reticular formation and in the medullary reticular formation (Fig. 10–11). The reticular interneurons in these areas typically participate in a multitude of brain stem reflexes implemented by these motor nuclei, or in the mediation of cortical input to the motor neuronal cell groups (Chapter 15). Autonomic regulatory centers in the medulla oblongata, including neuronal circuits for cardiovascular and respiratory functions, are closely related to the intermediate reticular zone, which in turn is characterized by its content of noradrenergic and adrenergic neurons (Chapter 19). Finally, it is widely believed that one of the most conspicuous collections of cholinergic neurons, which is located in and around the pedunculopontine tegmental nucleus in the upper part of the brain stem (Fig. 10–12), forms a crucial component of the anatomic substrate for the ascending activating system.

The Ascending Activating System

The ascending activating system is a physiologic concept that is based on the fundamental findings by Giuseppe Morruzi from Italy and Horace Magoun from the United States in the late 1940s of an arousal zone in the brain stem. They found that electrical stimulation of especially the upper part of the brain stem reticular core, or, for that matter, natural stimulation of spinal or cranial sensory nerves, produces widespread cortical activation or desynchronization, i.e., the high-voltage–slow-wave electroencephalogram (EEG) of quiet relaxation is replaced by the low-voltage–high-frequency EEG of intense wakefulness, also referred to as the arousal or alerting response.

The anatomic substrate for cortical activation or arousal is complicated and not fully understood, but it is generally agreed that many systems are involved, including not only ascending reticular pathways, but also the hypothalamic histaminergic system as well as cholinergic and noncholinergic corticopetal fibers from the basal nucleus of Meynert and related areas in the basal forebrain (Chapter 16). Some of the ascending pathways exert their effect by way of the thalamus, whereas others, including the basal nucleus of Meynert and histaminergic systems, reach the cerebral cortex without relay in the thalamus.

Of the ascending reticular pathways, it is widely believed that the cholinergic system that originates in a prominent collection of cholinergic neurons at the level of the pontomesencephalic transition area (Fig. 10–12) is of special significance. The two main parts of this cholinergic system are the pedunculopontine tegmental nucleus along the

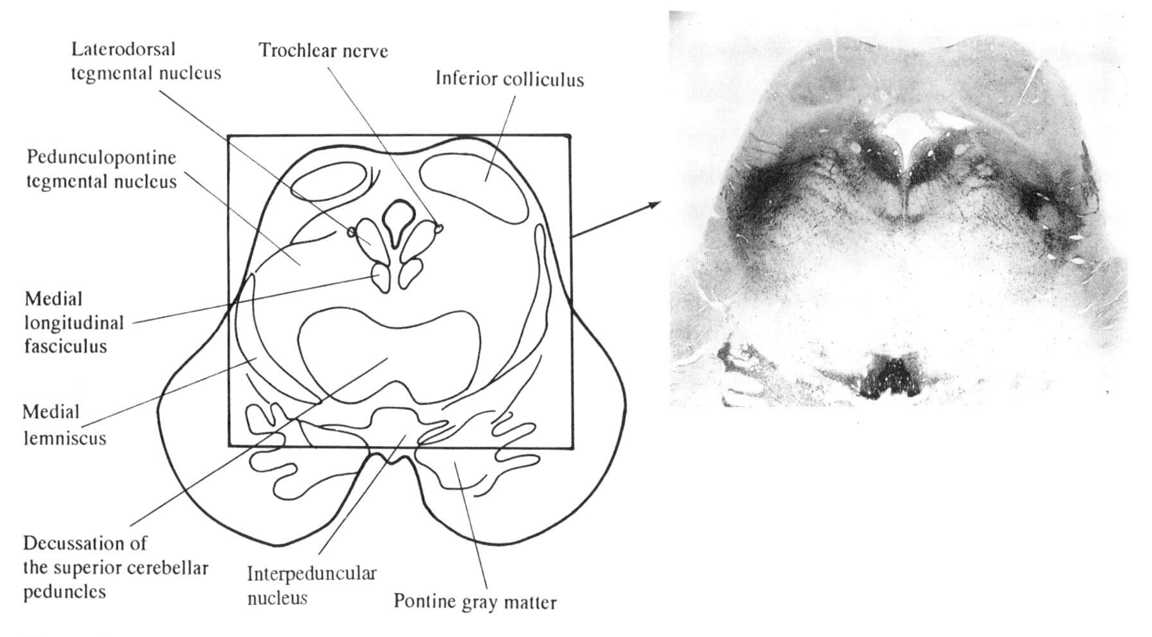

Fig. 10–12. Pedunculopontine-laterodorsal Tegmental Complex
Section through the pontomesencephalic transition area stained for acetylcholinesterase. The darkly stained pedunculopontine tegmental nucleus and the laterodorsal tegmental nucleus form an almost continuous complex of cholinergic neurons. (Photograph courtesy of Dr. Clifford Saper. From Saper, C. B., 1990. Cholinergic System. *In*: Paxinos, G. (Ed.). The Human Nervous System. With permission from Academic Press, San Diego.)

dorsolateral aspect of the superior cerebellar peduncle and the laterodorsal tegmental nucleus in the ventral part of the periaqueductal gray. As illustrated in Fig. 10–12, however, a number of cholinergic neurons bridge the gap between these two nuclear groups. The section illustrating the pedunculopontine–laterodorsal tegmental complex in Fig. 10–12 is at a slightly more rostral level than that demonstrating the locus ceruleus (Fig. 10–8), which explains the presence of the inferior colliculi in this figure. A large part of the brain stem cholinergic system reaches the intralaminar thalamic nuclei as well as many of the specific thalamic nuclei (Fig. 22–6), which in turn project to the cerebral cortex. Another major part of the brain stem cholinergic system projects to the lateral hypothalamus and the basal forebrain, and some fibers even reach the cerebral cortex directly.

Whereas the content of consciousness is a reflection of cortical activity, it is the ascending activating system, including reticulocortical and reticulothalamocortical pathways, that is primarily responsible for the conscious state, which is a prerequisite for perception. Physiologic studies and clinical observations have convincingly shown that the reticular formation is of crucial importance for maintaining a state of wakefulness. When an individual passes from sleep to wakefulness, the EEG changes from a synchronized slow-frequency pattern to a high-frequency desynchronized pattern similar to that seen after stimulation of the brain stem reticular core. If the ascending cortical activation is eliminated by a lesion, e.g., tumor or hemorrhage, in the upper part of the brain stem reticular formation, the patient loses consciousness and cannot be aroused to interact, a state referred to as coma (see Clinical Notes).

Sleep and Wakefulness

Sleep is a cyclical phenomenon, in which 90- to 100-minute periods of slow-wave sleep, characterized by synchronized EEG, are interrupted by 10- to 15-minute periods with desynchronized EEG and bursts of saccadic eye movements, i.e., rapid eye movement (REM) sleep. Since the EEG pattern of REM sleep is similar to that of the awake state, REM sleep is also referred to as active or paradoxical sleep. This active sleep state, however, is characterized by a profound loss of muscle tone (except in the eye muscles) and an increased threshold for awakening stimuli. REM sleep is also associated with dreaming.

Sleep is not merely a passive process of reduced activity in the ascending activating system. It is an active process, in the sense that it derives from neuronal activity in various brain regions, many of which are located in the brain stem reticular core. Like many other behavioral functions, the sleep–wake cycle is controlled by the circadian (24-hour) rhythm generator represented by the suprachiasmatic nucleus, which receives information about the dark–light cycle through its afferents from the retina. The medial preoptic–anterior hypothalamic area is also involved in promoting the sleep cycle, but it is not known how these hypothalamic areas interact with the various brain stem regions that are ultimately responsible for slow-wave and REM sleep and their cyclical pattern. Nor do we know the details of the brain stem circuits involved, but it is clear that the mechanisms for sleep engage a number of brain stem and basal forebrain regions, including serotonergic, catecholaminergic, and histaminergic cell groups, as well as several cranial nerve nuclei, e.g., the oculomotor and vestibular nuclei and associated reticular cell groups.

Clinical Notes

Many brain stem lesions are caused by cerebrovascular disorders, and they often give rise to typical brain stem syndromes, which are discussed in Chapter 24.

Disorders of Consciousness

Mental functions are often disturbed in patients with neurologic disorder, and it is important to evaluate a patient's mental state as it is expressed in such categories as level of consciousness, emotional reactivity, intellectual ability, and language.

Although it is difficult to define the term consciousness, a person is considered to be in a state of normal consciousness if he or she is fully awake and aware of self and environment. Further, the state of consciousness can be said to exist on a continuum with maximal alertness at one extreme and coma at the other. From a medical point of view it is convenient to distinguish between the following stages:

Confusion. A confused person cannot think clearly and is disoriented in time and place.

Somnolence. This is a condition of semiconsciousness; the patient can be aroused by various stimuli, but drifts back to sleep when the stimulus ceases.

Stupor. The patient can be aroused only by repeated and persistent stimulation and may be able to respond briefly to questions or commands.

Coma. The patient appears to be asleep and is in general unable to respond to external or internal stimuli. Reflex responses may or may not be absent depending on the degree of coma.

Consciousness can be disturbed for a variety of reasons, ranging from metabolic disturbances and drug intoxication to head injuries and large space-occupying lesions. Many brain regions and biochemical processes are important for consciousness, and it is usually impossible to identify all the pathophysiologic mechanisms by which consciousness is disturbed in a given case. It is appropriate, however, to make a few generalizations.

Metabolic diseases are often accompanied by a reduction in cerebral metabolism or blood flow that affects all neurons, and disturbance in consciousness at the cellular level generally is the result of altered function of both cortical and subcortical neurons. Hypoxia and hypoglycemia are examples of diffuse disease conditions that affect nerve cells both in the cerebral hemispheres and in the brain stem. Intracranial bleedings, tumors, and abscesses are examples of conditions that cross-compress the other hemisphere. Alternatively, the space-occupying lesions may produce a transtentorial herniation (Chapter 3; see also Clinical Example in Chapter 11) with subsequent compression of the upper brain stem reticular formation. Such herniations often compress the oculomotor nerve.

A major step in the clinical evaluation of an unconscious person is to determine whether the disease is in the cerebral hemispheres or primarily in the brain stem itself.

Suggested Reading

1. Brodal, A., 1981. Neurological Anatomy in Relation to Clinical Medicine, 3rd ed. Oxford University Press: New York, pp. 394–447.
2. Klemm, W. R. and R. P. Vertes, 1989. Brainstem Mechanisms of Behavior. John Wiley & Sons: New York.
3. Nauta, W. J. H. and M. Feirtag, 1986. Fundamental Neuroanatomy. W. H. Freeman and Co.: New York, pp. 163–219.
4. Nieuwenhuys, R., 1985. Chemoarchitecture of the Brain. Springer-Verlag: Berlin.
5. Nieuwenhuys, R., J. Voogd, and C. van Huijzen, 1988. The Human Central Nervous System. Springer-Verlag: Berlin, pp. 197–220.
6. Paxinos, G., I. Tork, G. Halliday, and W. R. Mehler, 1990. Human homologs to brainstem nuclei identified in other animals as revealed by acetylcholinesterase activity. In G. Paxinos (ed.): The Human Nervous system. Academic Press: San Diego, pp. 149–202.
7. Steriade, M. and R. W. McCarley, 1990. Brainstem Control of Wakefulness and Sleep. Plenum Press: New York.
8. Yamatodani, A., N. Inagaki, P. Panula, N. Itowi, T. Watanabe, and H. Wada, 1991. Structure and functions of the histaminergic neurone system. In B. Uvnäs (ed.): Handbook of Experimental Pharmacology, vol. 97. Springer-Verlag: Berlin, pp. 243–283.

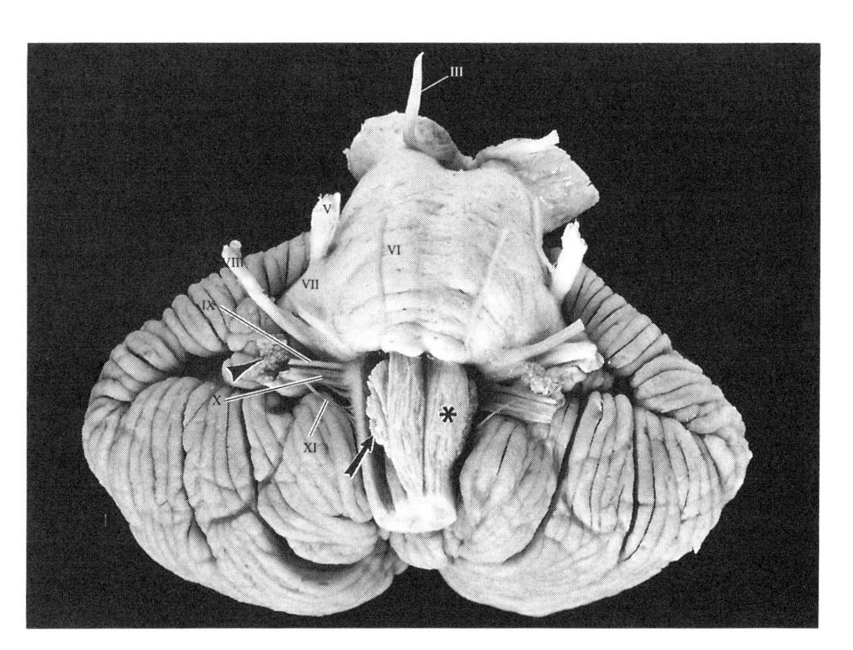

Basal aspect of the brain stem to demonstrate the exits of cranial nerves III and V–XI from the brain. The corticospinal tracts (*star*) and the inferior olivary nucleus on the left side (*arrow*) have been exposed by the dissection. Note how part of the choroid plexus of the fourth ventricle is protruding through the foramen of Luschka (*arrowhead*). (From Gluhbegovic, N. and T. H. Williams, 1980. The Human Brain: A Photographic Guide, JB Lippincott Co., Philadelphia. With permission of the publisher.)

11

Cranial Nerves

Except for the olfactory (I) and optic (II) nerves, the nuclei of the cranial nerves are located in the brain stem. Three of the cranial nerves are purely sensory (I, II, and VIII), five are motor (III, IV, VI, XI, and XII) and four are mixed (V, VII, IX, and X). See Table 11.1 at the end of chapter.

The cranial nerve nuclei, many of which have a columnar appearance, are arranged in an orderly fashion in the brain stem, in the sense that motor nuclei are located medial, and sensory nuclei lateral to the sulcus limitans.

As the name indicates, the cranial nerves emerge from the cranium, and they innervate structures in the head and neck. The IXth and Xth cranial nerves, the, glossopharyngeal and vagus nerves, in addition innervate visceral organs in the thorax and abdomen. The cranial nerves serve many functions related to the special sense organs; they also control oculomotor activities, mastication, vocalization, facial expression, respiration, heart rate, and digestion. The testing of the cranial nerves is an important part of the neurologic examination. One or several of the cranial nerves are often involved in lesions of the brain stem, and the location of such lesions can usually be determined if the topographic anatomy of the cranial nerves and their nuclei is known.

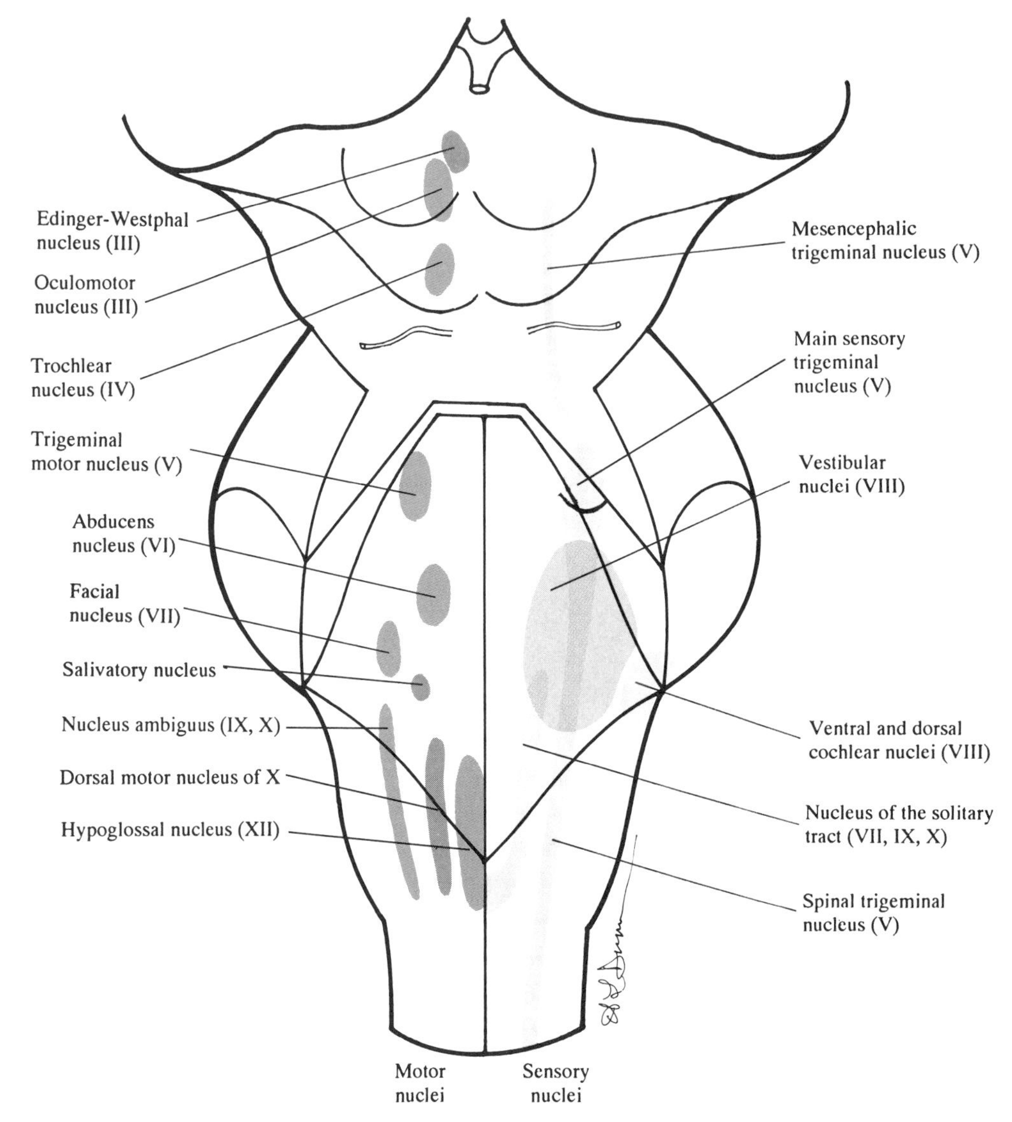

Fig. 11-1. Cranial Nerve Nuclei
The location of the cranial nerve nuclei in the brain stem as viewed from the dorsal side. Motor nuclei (somatomotor, *red*; visceromotor, *gray*) are shown to the *left* and sensory nuclei (*blue*) to the *right*. (Modified after Nauta, W. J. H. and M. Feirtag. In: Fundamental Neuroanatomy, 1986. W. H. Freeman and Company, New York.)

Olfactory (I) and Optic (II) Nerves

The Ist and IInd cranial nerves are not nerves in the usual sense. The olfactory nerve is represented by the central processes of neuroepithelial cells in the olfactory mucosa, and the optic nerve is in reality a central pathway. The olfactory and visual systems are discussed in Chapters 12 and 13.

Oculomotor (III), Trochlear (IV), and Abducens (VI) Nerves

Somatic Motor Parts

The nuclei of the oculomotor, trochlear, and abducens nerves are located close to the midline (Fig. 11-1). Their exits from the brain can be easily visualized in MRI (Fig. 11-2). The nuclear

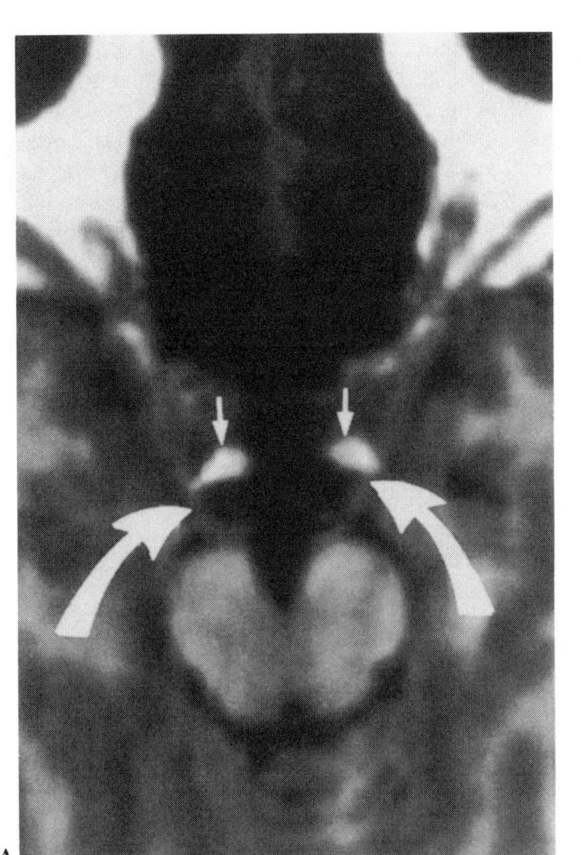

Fig. 11–2A.
Axial MRI showing the oculomotor nerves (white curved arrows) in the sellar region; posterior clinoid processes (tiny white arrows).

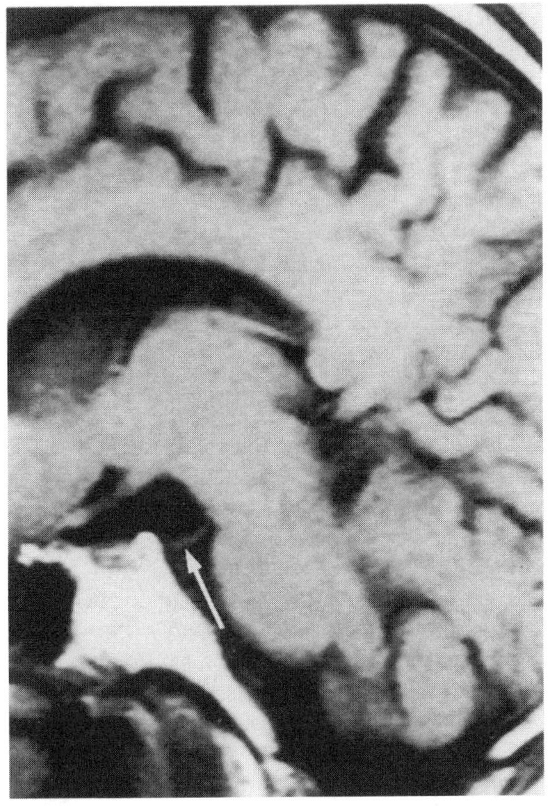

Fig. 11–2B.
Cisternal portion of the oculomotor nerve (arrow) in a sagittal MRI.

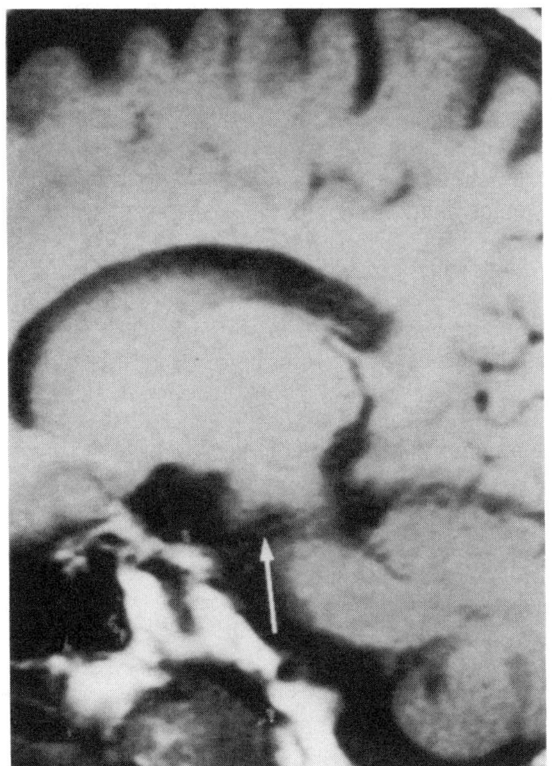

Fig. 11–2C.
Circum-mesencephalic portion of the trochlear nerve (arrow) in a sagittal MRI.

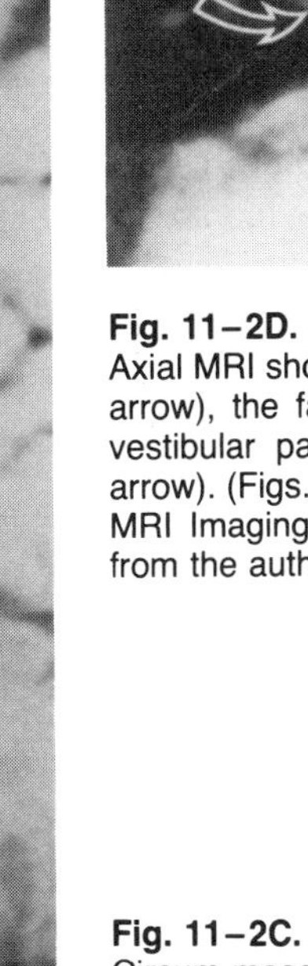

Fig. 11–2D.
Axial MRI showing exits of the abducens nerve (straight arrow), the facial nerve (solid curved arrow), and the vestibular part of eighth cranial nerve (open curved arrow). (Figs. 11–2A, B, C and D; from K. Sartor, 1992. MRI Imaging of the Skull and Brain. With permission from the author and Springer-Verlag, Berlin.)

complex of the IIIrd cranial nerve is located in the periaqueductal gray matter, where it extends from the diencephalic–mesencephalic junction area to the lower level of the superior colliculus (Fig. 10–5). It innervates the levator palpebrae superioris and all extrinsic eye muscles except the superior oblique and lateral rectus muscles. The contralateral superior oblique muscle receives innervation from the trochlear nucleus, which is located at the level of the inferior colliculus, and the lateral rectus muscle is innervated by the abducens nucleus in the lower pons.

The trochlear nerve is exceptional in two respects. It is the only cranial nerve that emerges on the dorsal aspect of the brain stem, and it crosses the midline in the anterior medullary velum before its exit.

Parasympathetic Part (III)

A group of preganglionic parasympathetic neurons, the *Edinger–Westphal nucleus*, is part of the oculomotor complex. The cells are situated close to the midline at the level of the superior colliculus (Fig. 10–5) and their axons proceed in the oculomotor nerve to the *ciliary ganglion*, where they establish synaptic contacts with postganglionic neurons innervating the sphincter pupillae and ciliary muscles. The Edinger–Westphal nucleus is an important component of the light reflex and the accommodation reflex (Chapter 13).

Eye Movements

Eye movements are highly coordinated and finely tuned. The motor units for the eye muscles, accordingly, are very small, on the order of three to six muscle fibers for each motor neuron. The extraocular muscles, furthermore, contain different types of muscle fibers, including some of the fastest recorded in mammals. Eye movements are in general conjugated, i.e., the two eyes move together in an almost identical manner in order for the image to always fall on corresponding parts of the two retinae. Double vision or diplopia (see Clinical Notes) may occur when the eye movements are not conjugated, which happens if one of the eye muscles is paralyzed. Eye movements, finally, can be either voluntary or reflexive in nature, and they occur in response to a variety of stimuli including visual impulses as well as messages from the vestibular nuclei, cerebellum, and cerebral cortex.

The motor neurons in the oculomotor, trochlear, and abducens nuclei represent the final common path for the control of the extraocular muscles. These nuclei are closely related to neighboring parts of the brain stem reticular formation and to the superior colliculus (Fig. 11–3). This system of interconnected brain stem areas, which is responsible for most of the neuronal processing and coordinating activities involved in conjugated eye movements, is closely related to the vestibular nuclear complex and to the cerebellum. The pretectal region, further, is noted especially for its involvement in the pupillary light reflex and the accommodation reflex (Chapter 13). Forebrain structures in the cerebral cortex and basal ganglia, finally, are also involved in visually guided and voluntary eye movements.

It is useful to distinguish between the following types of eye movements:

Saccades, or gaze shifts, are fast conjugated eye movements to reset the eye position so that the image of an object in the periphery falls on the fovea. Saccades are "ballistic" in the sense that they occur at very high speed and can hardly be modified once they have been initiated.

Smooth pursuit movements are used to follow a moving abject in order to keep the image on the fovea.

Vestibulo-ocular reflexes provide adjustment of eye position in response to head movement. Prolonged stimulation in one direction leads to *nystagmus*, i.e., rhythmic oscillation of the eyeballs with a slow (compensatory) phase alternating with a fast (reset) phase. The direction of the nystagmus is named according to the fast phase. The vestibulo-ocular reflexes are discussed further in relation to the vestibular nerve later in this chapter.

Optokinetic responses occur as a compensation for movement of the visual surround, and they can take the form of slow eye movements or saccades. Extended stimulation in one direction produces optokinetic nystagmus.

Vergence refers to binocular disjugate movements, i.e., each eye moves differently in order to keep the image in the fovea in both eyes when looking at objects at different distances from the viewer. If the object moves toward the viewer, the eyes converge; if it moves away, the eyes diverge. The *accommodation reflex*, i.e., adaptation of the eyes for near vision, involves convergence of the eyes as well as pupillary constriction and contraction of the ciliary muscle, which changes the radius of curvature of the lens.

Brain Stem Centers for Control of Eye Movements

Saccade Generating Centers

Neurons in the abducens nucleus and in the nearby paramedian pontine reticular formation (PPRF) are

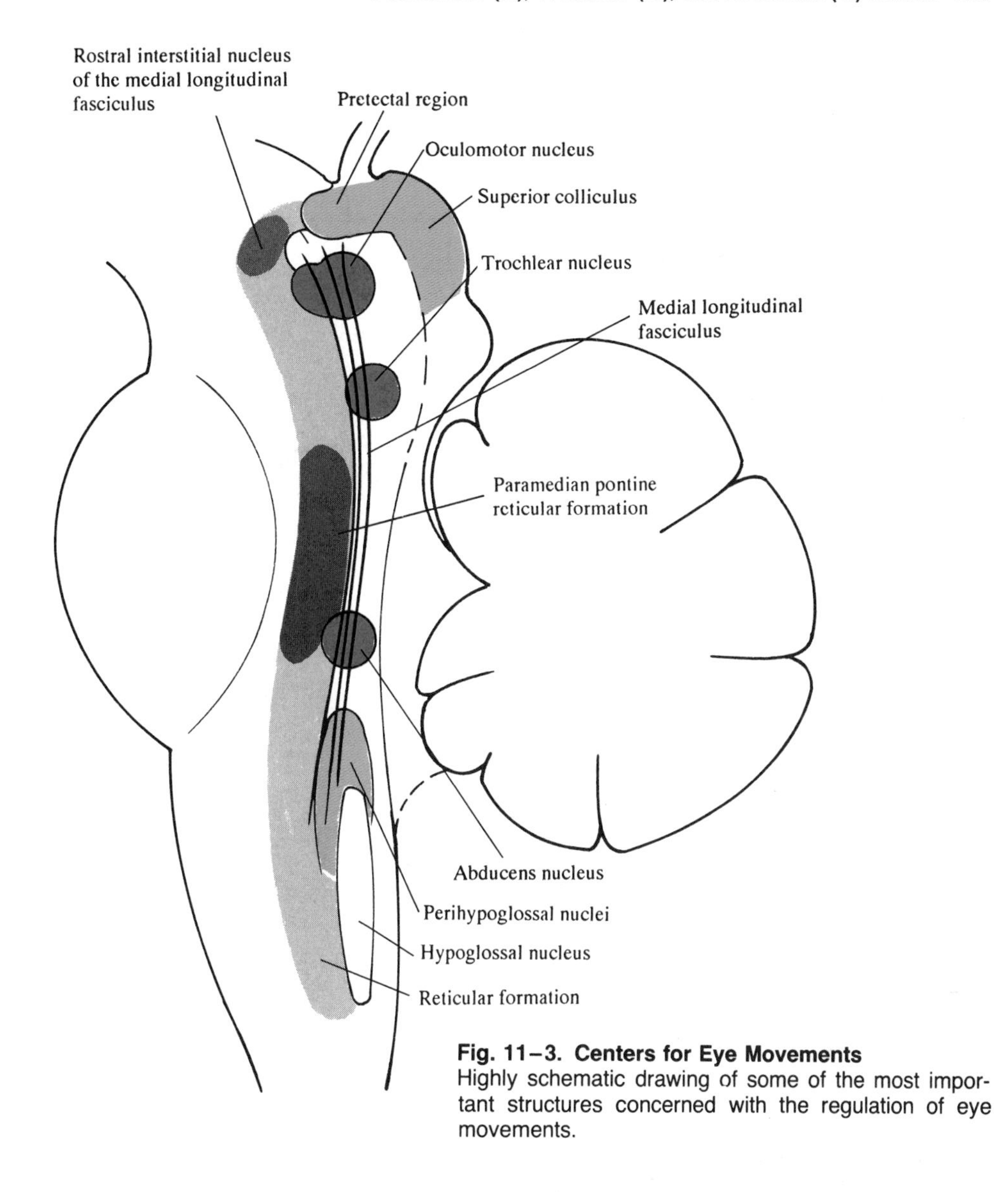

Fig. 11–3. Centers for Eye Movements
Highly schematic drawing of some of the most important structures concerned with the regulation of eye movements.

of special significance for horizontal saccades. These neurons, which are referred to as the "gaze center" or "saccade generating center" for horizontal eye movements, secure the coordination between the lateral rectus muscle on one side and the medial rectus muscle on the opposite side. They do so via a direct pathway from the abducens nucleus on one side to the group of oculomotor neurons innervating the medial rectus muscle on the opposite side (Fig. 11–4). The projection is incorporated in the *medial longitudinal fasciculus*, which is the most important coordinating pathway for conjugated eye movements.

Although the center for vertical saccades is often said to be located in the rostral interstitial nucleus of the medial longitudinal fasciculus at the rostral end of the mesencephalon, this nucleus receives important input from the paramedian pontine reticular formation, which therefore in effect also controls vertical gaze. The premotor neurons in the rostral interstitial nucleus project to the trochlear nucleus and to most of the oculomotor nuclear complex.

The perihypoglossal complex, which consists of several cell groups in the vicinity of the hypoglossal nucleus, is closely related to the oculomotor nuclei on one hand, and to the vestibular system and cerebellum on the other.

Superior Colliculi

The superior colliculi are multisensory in the sense that they receive input not only from the visual

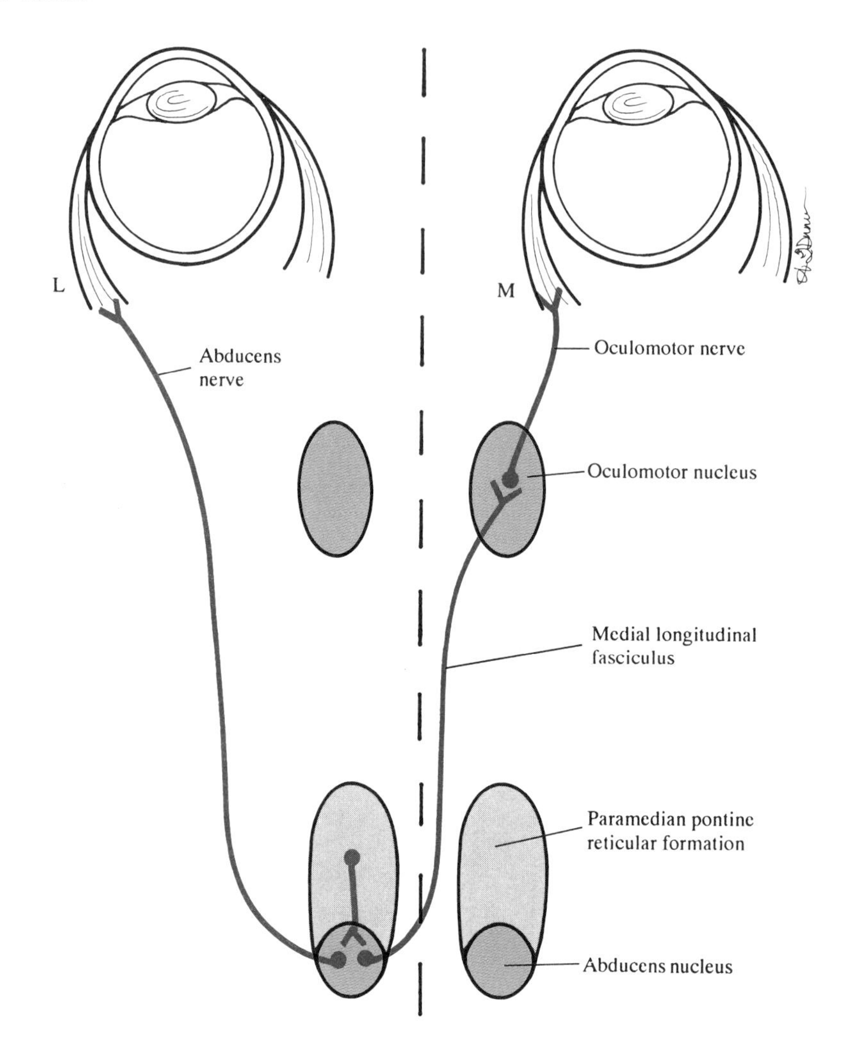

Fig. 11–4. Horizontal Conjugate Eye Movements
Highly simplified scheme illustrating how horizontal eye movements can be accomplished by the simultaneous activation of the lateral rectus muscle (L) on one side and the medial rectus muscle (M) on the opposite side. A unilateral lesion of the medial longitudinal fasciculus destroys fibers from the abducens nucleus to part of the oculomotor nucleus, which innervates the medial rectus muscle on the ipsilateral side. This produces a paralysis of medial gaze referred to as internuclear ophthalmoplegia. Only excitatory connections are illustrated. (Modified from Brodal, P., 1992. The Central Nervous System. Oxford University Press, New York.)

system but also from auditory, vestibular, and somatosensory systems. However, the visual system is especially well represented with afferents from the retina, visual cortex, and frontal eye field (FEF). The superior colliculi, accordingly, are involved in many aspects of visual functions, including reflexive eye and head movements in response to visual stimuli, the "visual grasp reflex."

Projections from the superior colliculi are also multifarious. The tectoreticulospinal tract (Chapter 15), for instance, reaches a number of areas in the brain stem reticular formation, including the paramedian pontine reticular formation and other regions related to the oculomotor nuclei. These connections provide the anatomic substrate for the involvement of the superior colliculi in saccadic eye movements. The superior colliculi, however, are not indispensable for the generation of visually guided saccades, which reflects the importance of some of the forebrain structures to be mentioned next.

Forebrain Structures Involved in the Control of Eye Movements

Several forebrain regions in both the cerebral cortex and the basal ganglia are directly involved in voluntary eye movements including visually guided saccades. Best known is the FEF in area 8, which is located in the posterior part of the middle frontal gyrus just in front of the premotor cortex (Fig. 22–2). The FEF receives input from several other cortical areas, but especially from the posterior parietal eye field, which is located in area 7 in the posterior parietal lobe.

There are distinct differences between the FEF and the posterior parietal eye field. The FEF is closely related to saccades in the sense that it serves as a "triggering area" for the brain stem saccade-generating centers, which can be affected either by direct projections from the FEF, or indirectly via frontal eye field projections to the superior colliculus, which in turn projects to the saccade-generating brain stem centers. A third possibility is provided by the motor circuit through the basal ganglia. Activation of this circuit by excitatory impulses from the FEF would serve to disinhibit the superior colliculus (Chapter 16). By removing the tonic inhibition of the superior colliculus, this structure would be in a position to initiate the saccades via its projections to the saccade-generating centers.

The posterior parietal eye field, which is more concerned with visual attention than with the motor aspects of saccades, is of importance for saccades only to the extent that visual attention is a necessary prerequisite for visually guided saccades.

Clinical Notes

Strabismus and Diplopia

Misalignment of the eyes may be noted by an examiner (as strabismus or squint) or may be apparent to the patient himself (as diplopia or double vision). It may be due to paralysis of one or several of the extraocular muscles. A *IIIrd cranial nerve* lesion results in a deviation of the ipsilateral eye to the lateral side, i.e., lateral or divergent *strabismus*, because of the unopposed action of the lateral rectus. *Ptosis* (drooping of the upper lid) and *mydriasis* (dilation of the pupil) are other characteristic signs. The ptosis is due to paralysis of the levator palpebrae muscle, whereas the dilation of the pupil and paralysis of accommodation is due to the interruption of the parasympathetic fibers in the IIIrd cranial nerve. *Trochlear nerve* lesions are uncommon. They cause vertical diplopia with a compensatory tilt of the head toward the opposite shoulder. Lesions of the *abducens nerve* result in paralysis of lateral movements of the eye, which deviates in a medial direction, i.e., medial or convergent strabismus, because of the unopposed action of the medial rectus muscle.

Ocular palsies may result from lesions of the eye muscle nuclei or from interruption of their respective nerves, two of which (III and VI) extend a considerable distance within the brain stem. Ocular palsies secondary to disorders in the brain stem, e.g., vascular lesions or tumors, are often accompanied by signs of other involvement of parts of cranial nerves or long ascending or descending pathways. The peripheral parts of the nerves are also vulnerable during their extended course through the cavernous sinus to the superior orbital fissure. Peripheral lesions are often attributable to aneurysms in the circle of Willis or tumors on the base of the brain. The oculomotor nerve, closely related to the posterior cerebral artery, is typically involved in a transtentorial herniation (Fig. 3–7).

Paralysis of Conjugated Eye Movements

Considering the extensive and complicated mechanisms involved in eye movements, it is not surprising that disturbances in ocular functions may occur as a result of lesions in many parts of the brain. Paralysis of conjugate eye movements, for instance, is a frequent symptom. An acute lesion in the frontal lobe usually causes paralysis of contralateral gaze, and the eyes may deviate toward the side of the lesion. Paralysis of ipsilateral conjugate gaze is seen following lesions of the saccade-generating center for horizontal eye movements in the abducens nucleus or paramedian pontine reticular formation. A lesion of the medial longitudinal fasciculus between the abducens and oculomotor nuclei, i.e., internuclear ophthalmoplegia (Fig. 11–4) destroys the ability to adduct the eye on the side of the lesion in conjugated horizontal eye movements. However, since the motor neurons to the medial rectus muscle are still intact, vergence movements are not affected. A mesencephalic lesion, e.g., a pineal tumor, may exert pressure on the saccade-generating center for vertical eye movements, which results in paralysis of upward, or sometimes downward, gaze.

Pupillary Light Reflex

Testing the direct and the consensual (indirect) light reflexes of both eyes is a routine part of the neurologic examination. In a IIIrd cranial nerve lesion, both the direct and the consensual light reflexes (Chapter 13) are absent on the side of the lesion as the result of paralysis of the pupillary sphincter, but both reflexes are intact in the opposite eye. A blind eye (e.g., optic nerve lesion), on the other hand, does not respond directly to light, but it responds consensually if light is focused on the opposite intact eye.

Trigeminal Nerve (V)

The Vth cranial nerve is a mixed nerve that supplies the skin and mucous membranes of the face with sensory fibers and the masticatory muscles with motor fibers. The sensory part of the nerve, the *portio major*, is much larger than the motor part, the *portio minor*. MRI of Vth cranial nerve is shown in Fig. 11–5.

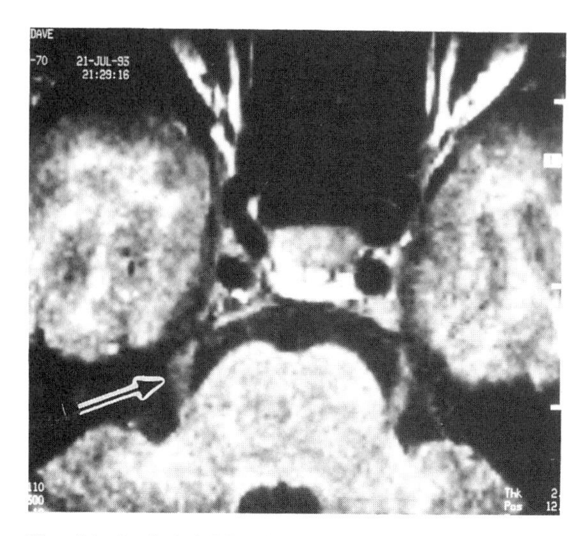

Fig. 11–5. Axial MRI Showing Exit of Trigeminal Nerves (*arrow*) from Lateral Side of Pons
(Courtesy of Dr. Jim Berlino.)

Somatosensory Part

Sensory fibers from pseudounipolar cells in the large *semilunar* or *trigeminal* (*Gasserian*) *ganglion* on the anterior surface of the petrous bone distribute peripherally in the three main divisions: the *ophthalamic*, *maxillary*, and *mandibular nerves* (Fig. 11–6). The central processes of the ganglion cells terminate in the three sensory trigeminal nuclei: the spinal, the main sensory, and the mesencephalic trigeminal nuclei (Fig. 11–1). Although it is generally held that the *spinal trigeminal nucleus* is concerned with pain and temperature impulses and the *main sensory nucleus* with tactile impulses, the situation is considerably more complicated and not fully understood. It is known, however, that many of the fibers that enter the brain through the main sensory portion give off one branch to the spinal trigeminal nucleus and another to the main sensory nucleus. The spinal nucleus, further, contains several different cell groups characterized by individual specific connections and functional attributes. The caudal part of the spinal trigeminal nucleus, which is continuous with the substantia gelatinosa of the spinal cord, seems to be of special significance for the mediation of pain impulses.

The *mesencephalic trigeminal nucleus* is unique in many respects. It is a long slender nucleus that reaches far rostrally to the level of the superior colliculus, and it receives proprioceptive impulses from both the extraocular and the masticatory muscles. Contrary to the general rule, the cell bodies for these sensory fibers are located within the mesencephalic nucleus rather than in the semilunar ganglion. The mesencephalic nucleus, therefore, is in part comparable to a sensory ganglion. Some of the sensory fibers in the mesencephalic root give off collaterals to the trigeminal motor nucleus, thereby providing the anatomic basis for the monosynaptic *jaw* (*masseter*) *reflex*, i.e., contraction of the masseters when the examiner's finger, placed on the patient's chin, is tapped with a percussion hammer. This reflex cannot usually be obtained in healthy individuals. If present, it may be a sign of a general hyperreflexia.

The central connections of the trigeminal nuclei are discussed in Chapter 9.

Special Visceral Motor Part

The motor part of the Vth cranial nerve, the *portio minor*, which originates in the *trigeminal motor nucleus* in the pons, is incorporated in the *mandibular nerve*. The fibers innervate the masticatory muscles (masseter, temporal, pterygoids), anterior belly of the digastric, tensor tympani, and tensor veli palatini. The trigeminal motor nucleus receives both crossed and uncrossed fibers from the cerebral cortex.

The masticatory muscles develop from the mesoderm of the first branchial arch and are related to the digestive tract. The trigeminal motor nucleus, therefore, is sometimes referred to as a special visceral motor nucleus. The facial nucleus and nucleus ambiguus (IX and X) can likewise be referred to as special visceral motor nuclei. However, since the neurons in these three nuclei (trigeminal motor, facial, and ambiguus) resemble other somatic motor neurons, and since they innervate striated rather than smooth muscles, it seems reasonable to refer to them as somatic motor.

Clinical Notes

The nuclei and nerves of the trigeminus have a wide distribution; partial affections of the trigeminal system are therefore quite common. Many disorders, including skull fractures, tumors, and aneurysms, e.g., of the internal carotid artery, may affect the trigeminus. The spinal trigeminal system is typically involved in the lateral medullary (Wallenberg) syndrome (see Clinical Examples).

Corneal Reflex

This reflex, i.e., bilateral blinking elicited by touching the edge of the cornea with a wisp of cotton, is a multisynaptic reflex with the afferent limb in the trigeminal nerve and efferent in the facial nerve. Interruption of the ophthalmic division of the Vth cranial nerve leads to loss of the blink reflex on the affected side. If the cornea is touched on the unaffected side, however, both eyes will close.

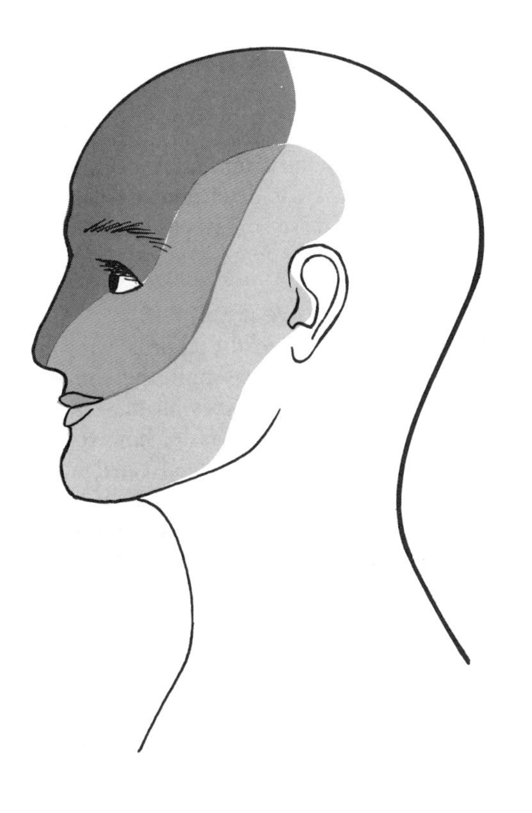

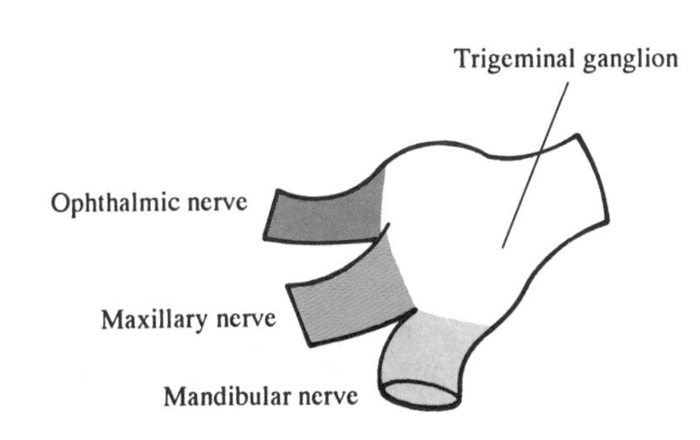

Fig. 11–6. Cutaneous Distribution of the Trigeminal Nerve
(Modified after Löfström, B., 1969. Block of the Gasserian ganglion. In: Illustrated Handbook in Local Anaesthesia (ed. Ejnar Eriksson). Year Book Medical Publishers, Inc., Chicago.)

Trigeminal Neuralgia (Tic Douloureux)

This is a common disorder of unknown etiology. It occurs in middle or late life and is characterized by paroxysms of sharp, excruciating pain in the distribution of one or more branches of the trigeminal nerve. The slightest tactile stimulus may trigger an attack. Many patients obtain relief from drug therapy, whereas others can be helped by some form of surgical interruption of the trigeminal pathways.

Ophthalmic Zoster

The virus *Herpes zoster* (see Clinical Notes, Chapter 8) may involve any part of the semilunar ganglion. The ophthalmic division is most often affected, and corneal scars can result that may interfere with vision.

Facial Nerve (VII)

The exit of the VIIth cranial nerve from the brain stem is easily visualized in MRI (Fig. 11–2D). The VIIth cranial nerve is a mixed nerve with several important functions. It is above all the motor nerve of facial expression, but it also innervates the submaxillary and sublingual glands as well as the lacrimal gland and glands of the nasal cavity, and it conveys taste impulses from the anterior two-thirds of the tongue.

Special Visceral Motor Part

The highly differentiated facial motor nucleus is located in the caudal part of the pons, ventromedial to the spinal trigeminal complex (Fig. 10–2). The axons detour in a dorsomedial direction around the abducens nucleus before turning ventrally and emerging on the ventrolateral aspect of the brain stem at the lower end of pons. The nerve enters the *internal auditory meatus* together with the vestibulocochlear nerve, continuous in the *facial canal*, and leaves the skull through the *stylomastoid foramen*. It branches in a fan-like manner in the region of the parotid gland and innervates the mimetic musculature, the stylomastoid, the posterior belly of the digastric muscle, and the platysma. During its course through the facial canal it gives off a branch to the stapedius muscle.

Parasympathetic Part

Preganglionic parasympathetic fibers from the *salivatory nucleus* join the *intermediate nerve*. Since this nerve also contains taste fibers and some cutaneous fibers from the external ear, it is often referred to as the sensory root of the facial nerve.

The parasympathetic fibers, which leave the main nerve during its course in the facial canal, form two major branches: (1) *the greater petrosal nerve*, which innervates the lacrimal and nasopalatine glands (postganglionic neurons in the sphenopalatine ganglion), and (2) the *chorda tympani* (postganglionic neurons in the submandibular ganglion), which innervates all salivary glands except the parotid, which receives secretory fibers through the glossopharyngeal nerve.

Visceral Sensory Part: Taste Pathways

Receptors

The sensory cells that receive the stimuli for the four basic taste sensations (sweet, sour, salty, and bitter) are located in *taste buds* on the surface of the *fungiform* and *vallate papillae*. The sensory cells, like the cells of the surrounding epithelium, have a life span of only a few days; they are continuously replaced by new sensory cells, which are derived from the basal cells through mitosis.

Peripheral Pathways

The cell bodies of the fibers that transmit taste impulses from the taste buds of the anterior two-thirds of the tongue are located in the *geniculate ganglion* in the facial canal. The peripheral processes proceed for a short distance in the facial nerve before they enter the *chorda tympani*, which joins the lingual branch of the mandibular nerve (Fig. 11–7).

Another important taste nerve, the *glossopharyngeal (IX) nerve*, carries impulses from the posterior third of the tongue, where a large number of taste buds are located on the sides of the vallate papillae. The *vagus (X)*, finally, carries some taste impulses from the extreme posterior part of the tongue and the epiglottis.

Central Pathways

The taste impulses transmitted in the above-mentioned three nerves reach the rostral part of the *nucleus of the solitary tract*, the main visceral sensory brain stem nucleus (Chapter 19). The further central connections for taste are complicated and not fully known. However, it appears that the majority of the taste-related fibers from the rostral third of the solitary nucleus reach the ipsilateral thalamic gustatory nucleus in the medial aspect of the *ventral posteromedial nucleus (VPM)* (somatosensory relay nucleus for the face and oral cavity). From here the impulses reach the *cortical taste area* in the opercular part of the inferior frontal gyrus and adjoining parts of the rostral insula. The cortical taste area, which is closely related to the somatosensory area for the tongue, subserves conscious sensation of taste. Brain stem taste fibers also reach the hypothalamus, where taste impulses presumably affect neuronal mechanisms involved in feeding behavior (Chapter 18).

Disturbances of taste functions are of minor significance compared with many other symptoms that may result from a lesion of the facial nerve. Such lesions give rise to various syndromes depending upon where the lesion is located (see below.)

Clinical Notes

Disorders of the facial nerve are common, and the topographic relationships of the facial canal take on a special significance in the diagnosis of facial nerve lesions. It is often possible to determine the location of the lesion on the basis of existing signs and symptoms.

Lesions of the Facial Nerve (Fig. 11–8)

A peripheral lesion distal to the chorda tympani (lesion A in Fig. 11–8) affects only the somatic motor component, i.e., it paralyzes all muscles of facial expression. The sense of taste, however, is intact. A lesion situated in the facial canal peripheral to the geniculate ganglion (lesion B) is likely to include the chorda tympani, in which case taste sensitivity is lost on the anterior two-thirds of the tongue on the ipsilateral side. Since the branch to the stapedius muscle leaves the facial nerve between the geniculate ganglion and the chorda tympani, hyperacusis (painful sensitivity to loud sound) may also be part of the syndrome. A lesion of the nerve in the proximal part of the facial canal (lesion C) is likely to involve the parasympathetic fibers to the lacrimal gland and usually results in loss of tear secretion in addition to all other symptoms. If the facial nerve is damaged in the internal auditory meatus, the VIIIth cranial nerve is likely to be involved, in which case tinnitus or deafness may be prominent.

The facial nerve can be damaged by many types of lesions including fractures of the petrous bone, middle ear infections, tumors of various kinds (e.g., cerebellopontine angle tumor, see Clinical Examples), and pathologic processes in the pons. The most common facial nerve disorder is *Bell's palsy*, which apparently is due to an inflammatory reaction in the facial canal, often in the course of an upper respiratory infection. If the patient loses the blinking reflex on the affected side, the eye can easily dry out and must be properly protected, especially during sleep. Recovery from Bell's palsy is usually spontaneous (see Clinical Examples).

Supranuclear Facial Paralysis

A supranuclear facial paralysis, e.g., as part of a capsular hemiplegia, is limited to the *lower part of the face*. This fact, which is of great diagnostic value, is explained by referring to the anatomic organization of the corticobulbar fibers to the facial nucleus (Fig. 11–9). The part of the facial nucleus that innervates the upper half of the face receives corticofugal fibers from the motor cortex of both sides, whereas the component for the lower part of

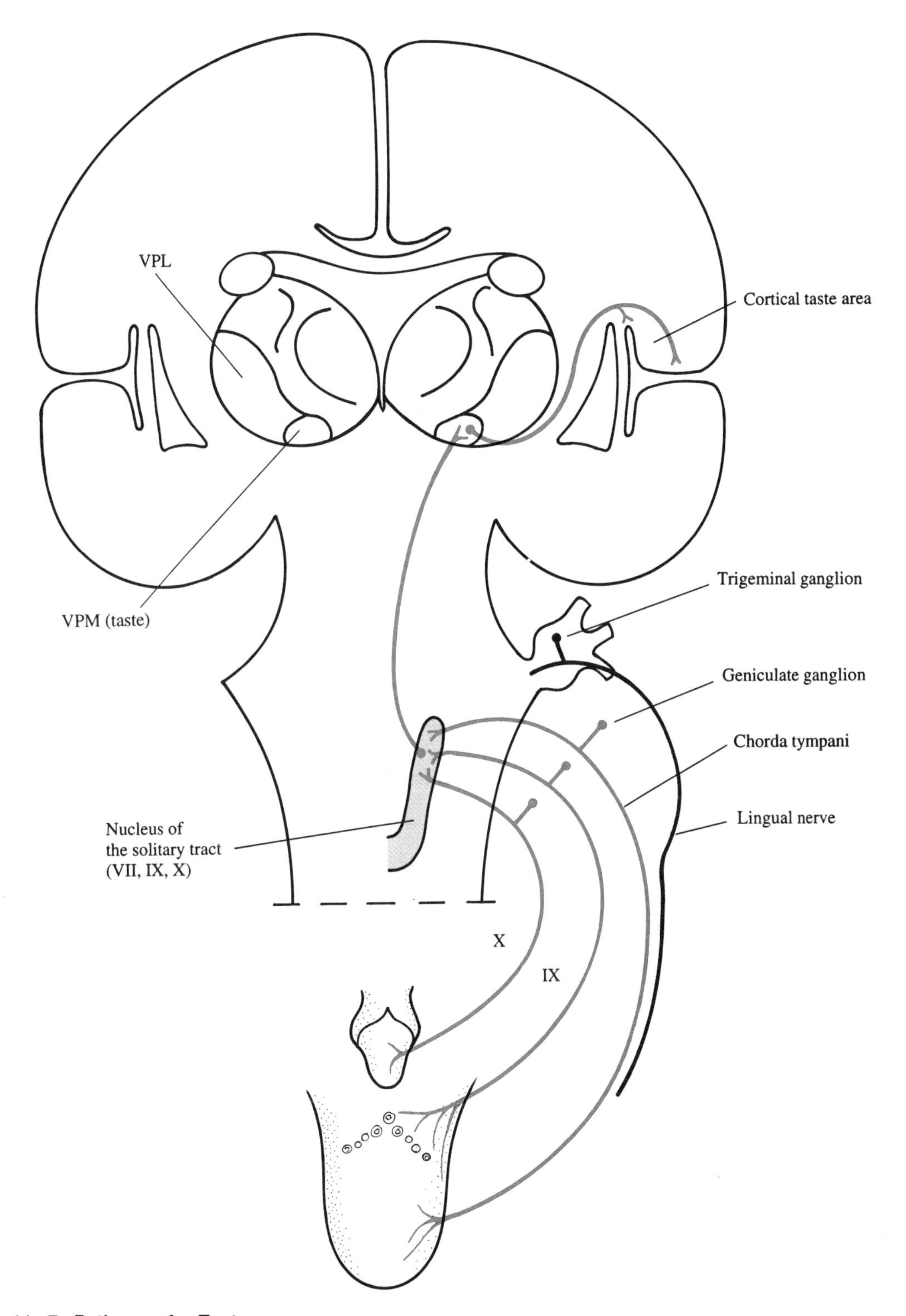

Fig. 11–7. Pathways for Taste
The rostral part of the nucleus of the solitary tract is concerned with the transmission of taste impulses travelling in the facial (VII), glossopharyngeal (IX), and vagal (X) cranial nerves. Impulses from the nucleus of the solitary tract reach the ventral posteromedial nucleus (VPM) in the thalamus before being transmitted to the gustatory cortex in the opercular part of the inferior frontal gyrus on the ipsilateral side. The caudal part of the nucleus of the solitary tract receives viscerosensory impulses through the glossopharyngeal and vagal nerves. These impulses are important for cardiovascular, respiratory, and other visceral reflexes.

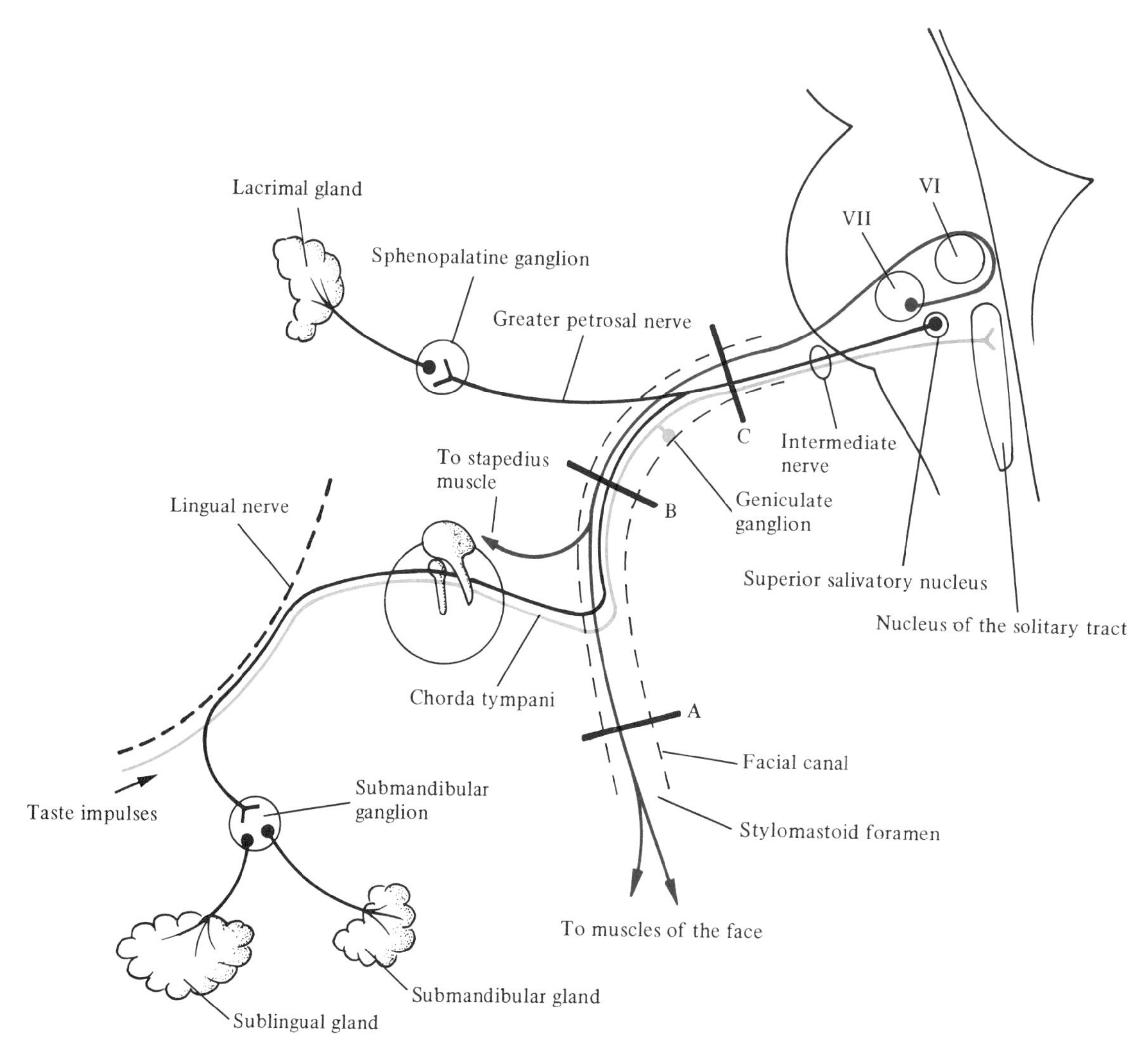

Fig. 11–8. Lesions of the Facial Nerve
Facial nerve lesions are common, and the symptoms will vary according to the site of the lesion. If the lesion is situated at **A**, i.e., distal to the departure of the chorda tympani, there is no loss of taste sensation on the anterior two-thirds of the tongue. A lesion at **B** involves the taste pathway from the anterior two-thirds of the tongue but leaves tear secretion intact. A lesion at **C** involves all components of the intermediate nerve in addition to the somatomotor component.

the face receives only a contralateral innervation. A patient with a central facial palsy can close both eyes voluntarily, but in efforts to retract the angles of the mouth (e.g., "social smile") the weakness is revealed on the side contralateral to the lesion. Curiously enough, a hemiplegia patient is able to smile normally when he enjoys a joke. This dissociation between voluntary and emotional innervation indicates that at least part of the forebrain projection system for control of emotional facial expression escapes destruction following a lesion in the internal capsule. These fibers, which in all likelihood come from the extended amygdala (Chapter 20) proceed through the lateral hypothalamus to the brain stem premotor neurons.

Vestibulocochlear Nerve (VIII)

The VIIIth cranial nerve consists of two divisions, the *cochlear* and *vestibular nerves*, both of which carry somatic sensory impulses from specialized receptors in the inner ear (Fig. 14–1). The two divisions,

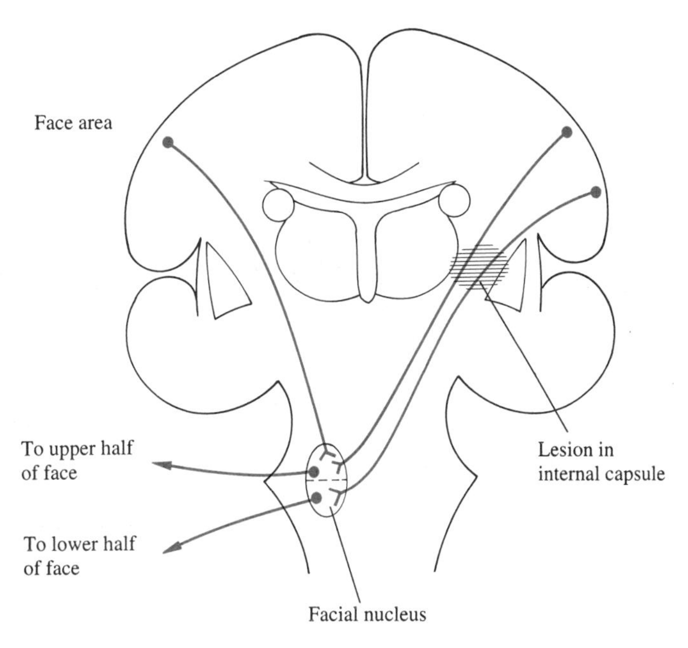

Fig. 11–9. Supranuclear Facial Paralysis
A supranuclear facial paralysis typically occurs in a capsular lesion. Since the peripheral motor neurons to the upper half of the face receive corticobulbar fibers from both hemispheres, a central facial paralysis is evident only in the lower half of the face on the opposite side of the lesion.

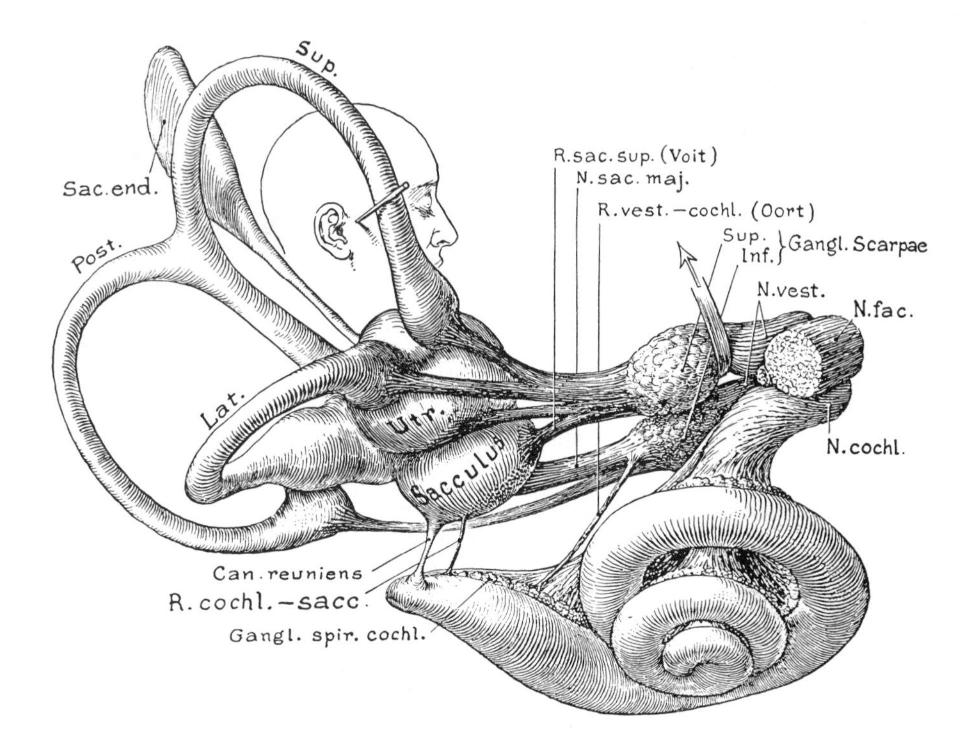

Fig. 11–10. *Membranous Labyrinth*
Sketch by M. Brödel (1946) showing the position of the membranous labyrinth within the head and its relation to the vestibulocochlear nerve.

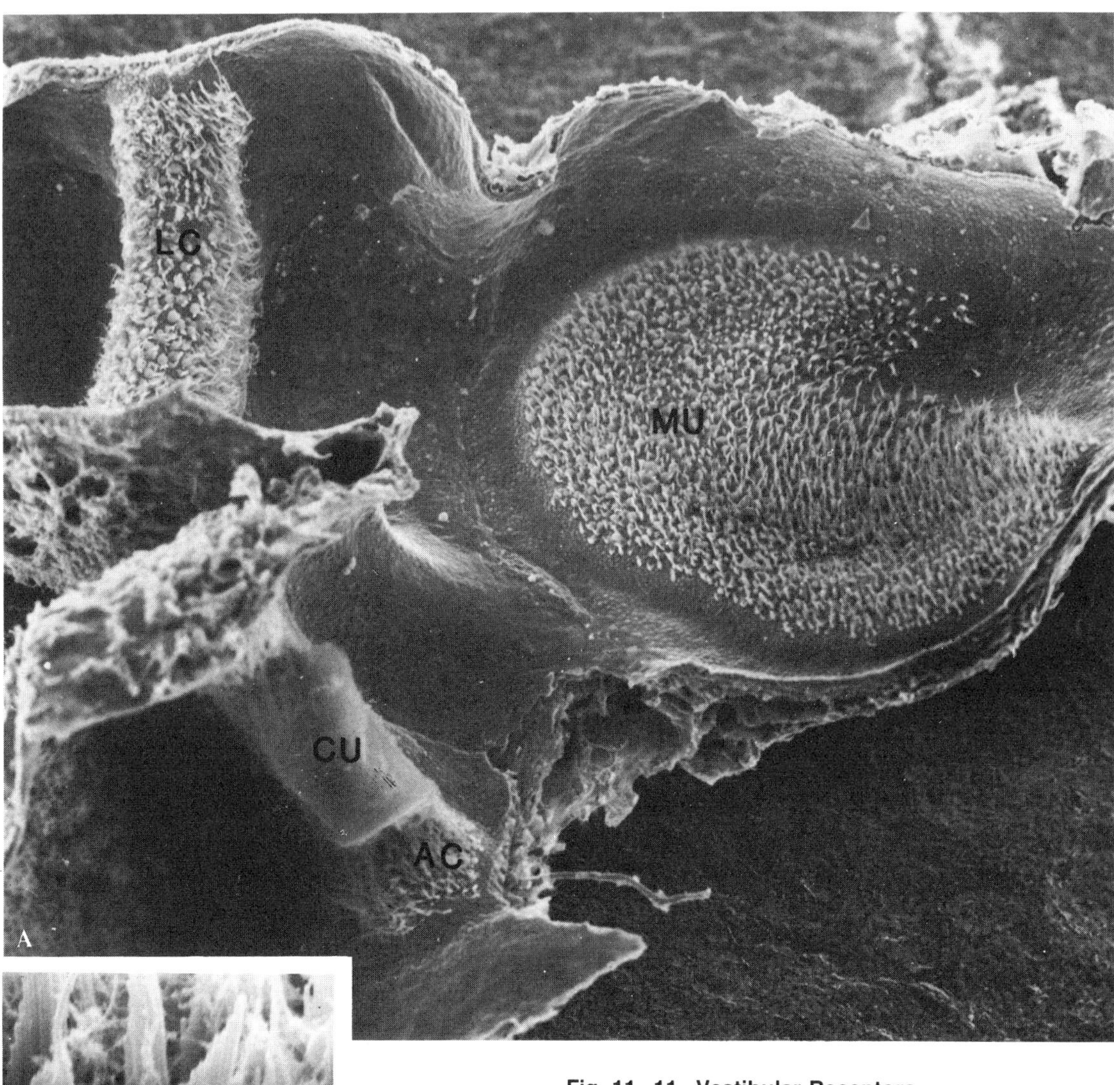

Fig. 11–11. Vestibular Receptors

A. Scanning electron micrograph showing relative orientation of the macula utriculi (MU), lateral crista (LC), and anterior crista (AC). The greatly shrunken cupula (CU) remains on the anterior crista.

B. Flask-shaped hair cell from macula utriculi. Note the bundle of stereocilia which is normally covered by the otolith membrane.

C. Scanning electron micrograph of the crista ampullaris (Cr) from one of the semicircular canals. The cupula (CU) appears shrunken following fixation and preparation for scanning electron microscopy.

(The three micrographs were kindly provided by Dr. Ivan Hunter-Duvar, Hospital for Sick Children, University of Toronto, Canada.)

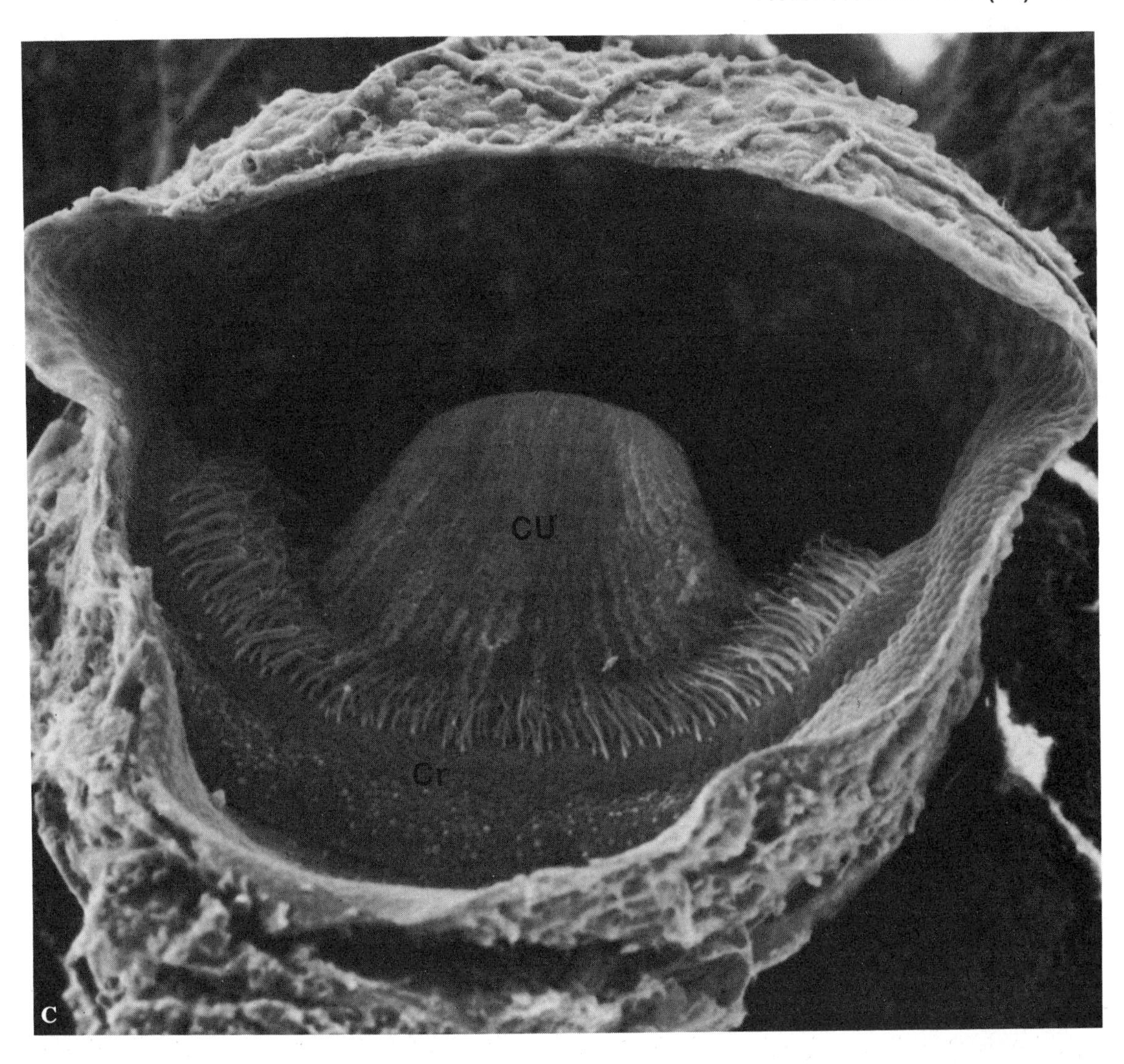

however, have different functions. The cochlear division carries auditory information from the organ of Corti in the cochlea, whereas the vestibular division transmits propioceptive impulses from the vestibular organ, i.e., the three semicircular canals, the utricle, and the saccule. The auditory system is discussed in Chapter 14.

The vestibular portion of the VIIIth cranial nerve is shown in the MRI in Fig. 11–2D (see also Fig. 4–7).

Vestibular Receptors

The vestibular receptors are represented by hair cells in the three *semicircular ducts* and in the vestibular portion of the labyrinth, which consists of two divisions, the *utricle* and the *saccule* (Fig. 11–10). The utricle and the saccule provide information about changes in the position of the head ("position recorders"). The semicircular canals, which are connected to the utricle, react to changes in angular movements of the head. The three canals are oriented in three planes, which are perpendicular to each other. The hair cells are located in special areas (Fig. 11–11A) of the membranous labyrinth, which is filled with endolymph. The membranous labyrinth, in turn, is surrounded by a space containing perilymph.

The vestibular organs, the maculae, in the saccule and utricle consist of hair cells (Fig. 11–11B) whose cilia project into a gelatinous mass, the *otolith membrane*. This membrane contains fine crystals of calcium carbonate, the *otoliths*. If the head is tilted, the gravitational force displaces the otolith membrane, and the otoliths stimulate the hair cells. The adequate stimulus is *linear acceleration* (or deceleration).

The hair cells in the semicircular canals, which look similar to the ones in the maculae, are located in dilatations of the canals called ampullae. Each ampulla contains a transversely oriented crista with hair cells (Fig. 11–11C). The crest is covered by a gelatinous mass, the *cupula*. Rotational movements of the head tend to deflect the cupula because of the inertia of the endolymph. This, in turn, stimulates the hair cells. The adequate stimulus is a change in speed rather than movement itself, i.e., *rotational acceleration* (or deceleration). The orientation of the semicircular canals, i.e., in three planes perpendicular to each other, guarantees stimulation of at least some hair cells regardless of the plane of rotation.

Vestibular Nerve and Nuclei

The bipolar ganglion cells of the vestibular nerve form the *vestibular ganglion* in the bottom of the internal auditory meatus. Their peripheral processes innervate the sensory cells in the vestibular apparatus, whereas the central processes form the vestibular nerve. Together with the cochlear nerve, the vestibular nerve enters the brain stem at the lower border of the pons, i.e., in the cerebellopontine angle (Fig. 4–6), and terminates in the four main vestibular nuclei, the *lateral (Deiters'), medial, superior*, and *inferior nuclei*. The vestibular complex occupies a rather large area in the lateral part of the brain stem just below the floor of the fourth ventricle (Fig. 11–1).

The vestibular apparatus as a whole is of importance for *maintenance of equilibrium* and *orientation in space*, and the different vestibular nuclei have specific connections and functional characteristics. For instance, the impulses from the utricular macula reach primarily the lateral vestibular nucleus, which gives rise to the lateral vestibulospinal tract. Most of the fibers from the semicircular ducts end in the superior and medial vestibular nuclei, which send significant projections into the medial longitudinal fasciculus. This pathway is of special importance for the vestibulo-ocular reflexes (see below). The inferior vestibular nucleus receives input from all parts of the vestibular labyrinth.

Central Connections

The information from the vestibular apparatus ultimately influences the motor system by means of connections between the vestibular nuclei and the spinal cord, the cerebellum, and the eye muscle nuclei (Fig. 11–12). There is also evidence that some ascending vestibular projection fibers reach the thalamus, which in turn would make it possible for vestibular stimuli to reach the cerebral cortex. The detailed anatomy of these ascending pathways, responsible for the conscious appreciation of vestibular stimuli, are not well known.

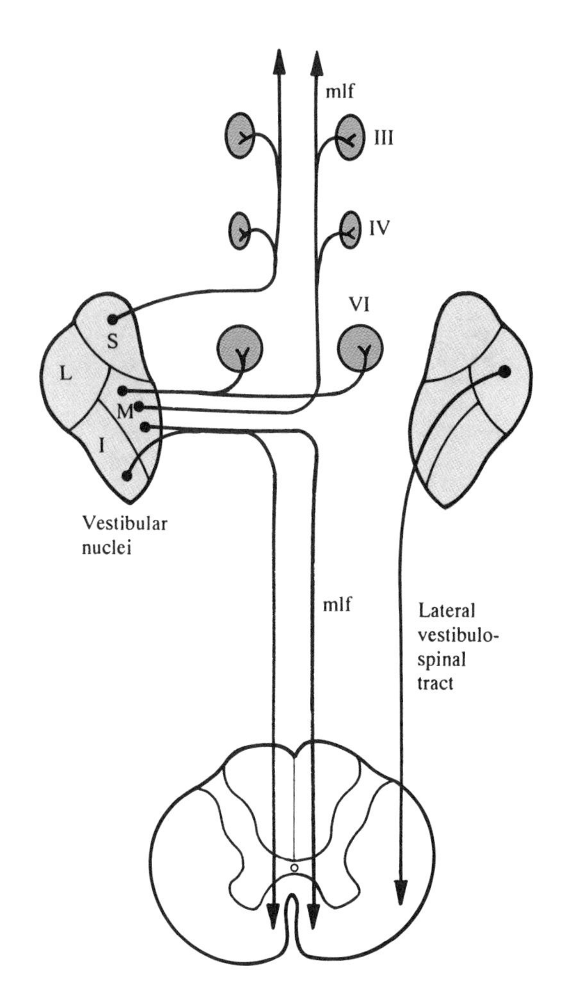

Fig. 11–12. The Vestibular Nuclei and the Motor System
Highly schematic drawing of the major efferent connections of the vestibular nuclei. The vestibulocerebellar connections have been deleted for didactic purpose. I, L, M, and S = inferior, lateral, medial, and superior vestibular nuclei; mlf = medial longitudinal fasciculus; III, IV, and VI = oculomotor, trochlear, and abducens nuclei.

Spinal Cord

Although the relationships between the vestibular nuclei and the spinal cord are reciprocal, the vestibulospinal tracts are more prominent than the spinovestibular tracts. The most important descending vestibular pathway, the *lateral vestibulospinal tract*, originates in Deiters' nucleus. The input from the utricular macula to the lateral vestibular nucleus signals the position of the head (and body) in space, and this information is critical for keeping balance and erect posture. The lateral vestibulospinal tract is an important component of this mechanism; it exerts a facilitatory action on the extensor motor neurons with concomitant inhibition of flexor motor neurons (Chapter 15).

Fibers from the medial and inferior vestibular nuclei contribute to the medial vestibulospinal tract, which is incorporated in the descending part of the medial longitudinal fasciculus. This tract is present only in the upper part of the spinal cord and is concerned primarily with head and neck movements in response to vestibular stimuli.

Cerebellum

Both direct fibers from the vestibular nerve and indirect fibers from the vestibular nuclei, primarily the superior, medial, and inferior nuclei, reach the flocculonodular lobe (vestibulocerebellum). Both the direct and the indirect fibers terminate as mossy fibers in the cerebellum. The vestibular nuclei can also influence cerebellar activity through the climbing fiber system, but this is accomplished indirectly via projections from the vestibular nuclei to the inferior olive.

The vestibulocerebellar fibers are reciprocated by fibers from the vestibulocerebellum to the vestibular nuclei. A prominent cerebellar projection to the vestibular complex also originates in the spinocerebellum and terminates in the lateral vestibular nucleus, which projects to the spinal cord through the lateral vestibulospinal tract.

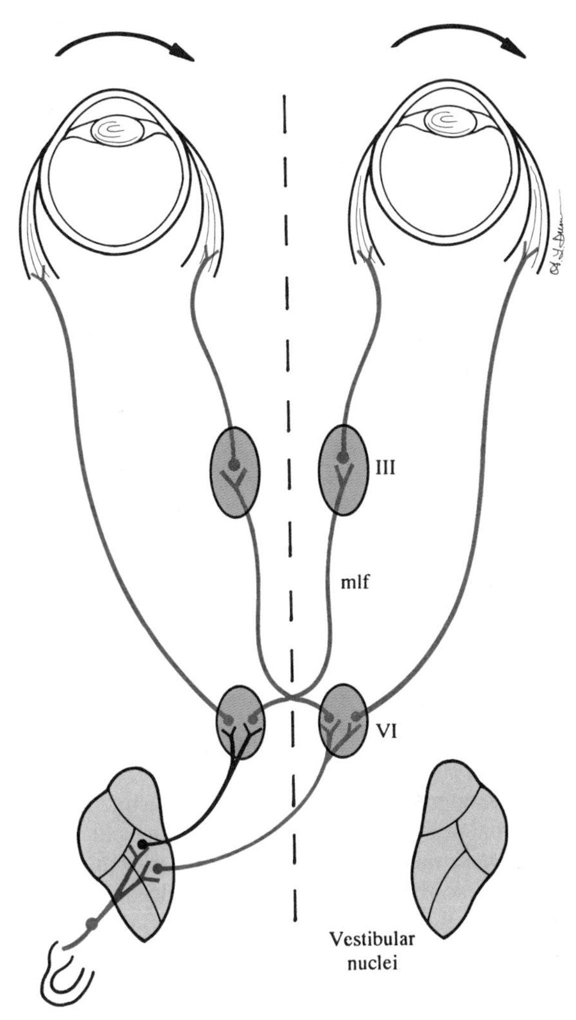

Fig. 11–13. The Vestibulo-Ocular Reflex
Simplified scheme for the horizontal vestibulo-ocular reflex showing only the pathways originating in the left horizontal semicircular canal. Excitatory neurons = red; inhibitory neuron = black. (Modified from Goldberg, M. E. et al. 1991. The Ocular Motor System. *In*: Kandel, E. R. et al., Principles of Neural Science, 3rd ed., Elsevier.)

Eye Muscle Nuclei and Vestibulo-Ocular Reflexes

The vestibulo-ocular reflexes are of great importance for the adjustment of the eyes in response to head movement. For instance, when the head is turned to the left, the eyes are turned to the right in order to keep the fields of vision unchanged. The scheme presented in Fig. 11–13 is highly simplified, but gives an idea of the basic structures involved in the vestibulo-ocular reflexes, i.e., the semicircular canals, vestibular nerve and nuclei, medial longitudinal fasciculus, and the oculomotor nuclei. Several other neuronal circuits and structures participate, including the vestibulocerebellum, which is especially important for fine tuning of the reflexes.

The scheme of the anatomic substrate for the horizontal vestibulo-ocular reflex that compensates for head-turning to the left (Fig. 11–13) includes only the pathways originating in the left horizontal semicircular canal. The impulses from the hair cells in the ampulla reach both excitatory (red) and inhibitory neurons (black) in the medial vestibular nucleus. The projections from the excitatory neurons terminate in the contralateral abducens nucleus, which effectuates contraction of the right lateral rectus muscle via a direct projection through the abducens nerve. The left medial rectus muscle is activated via a disynaptic pathway through the left medial longitudinal fasciculus to the oculomotor nucleus, which in turn innervates the muscle. The projections from the inhibitory neuron, in contrast, terminate in the ipsilateral abducens nucleus, thereby reducing the activity in corresponding pathways to the left lateral rectus muscle and the right medial rectus muscle.

Clinical Notes

Nystagmus, i.e., oscillating movements of the eye, is a frequent manifestation of neurologic disease, and spontaneous nystagmus is always pathologic. It usually signifies the presence of a pathologic process in the vestibular system, the cerebellum, or the brain stem. Nystagmus can also be induced as part of a neurologic examination to test the vestibular system and its connections to the oculomotor system.

Postrotatory nystagmus can be induced by turning the patient (with the head tilted slightly forward) in a revolving chair about 10 times. If the movement is suddenly stopped, the endolymph in the lateral, horizontal semicircular canals continues to flow in the direction in which the head turned. The induced nystagmus, with the fast component opposite to the direction of the rotation, lasts for about 30 seconds.

Nystagmus can also be induced by irrigating the external auditory meatus with warm or cold water, i.e., *caloric nystagmus*. In this test the patient's head is tilted so that the lateral semicircular canal is in the vertical plane. The water changes the temperature of the endolymph, which produces a current in the endolymph. This in turn stimulates the hair cells. One of the advantages of the caloric test is that each side of the vestibular pathways can be tested separately.

Vertigo, i.e., a sense of rotation, either of the individual or of the environment, is a common symptom in disorders of the vestibular system, especially the labyrinth. The most typical example of labyrinthine vertigo is caused by *Meniere's disease*, which is characterized by abrupt and severe attacks of vertigo, often combined with tinnitus and increasing deafness. Nystagmus is present during the attack, which can last for minutes or hours. The disease is caused by an overproduction of endolymph.

Vertigo can also indicate a lesion of the VIIIth cranial nerve, such as vestibular schwannoma (see Clinical Examples, Chapter 14: Cerebellopontine angle tumor), in which it is often preceded by tinnitus and deafness. A lesion of the vestibular nuclei, on the other hand, usually produces vertigo without loss of hearing.

Glossopharyngeal (IX) and Vagus (X) Nerves

The IXth and Xth cranial nerves (Fig. 11–14) are closely related both anatomically and physiologically, and they innervate in part the same structures. The vagus, further, innervates the thoracic and abdominal viscera, and is the most important parasympathetic nerve of the body. The nerves emerge on the lateral side of the medulla dorsal to the olive and leave the skull through the jugular foramen together with the accessory nerve (XI).

Functional Anatomy

Special Visual Motor Part

Both the glossopharyngeal and the vagus nerves contain fibers that originate in the *nucleus ambiguus* and that innervate the striated muscles of the pharynx. The vagus, in addition, innervates the larynx. The stylopharyngeal muscle is innervated by the glossopharyngeal nerve.

Parasympathetic Part

Fibers from the *salivatory nucleus* join the glossopharyngeal nerve and supply the parotid gland and pharyngeal glands with secretory fibers. The *dorsal motor nucleus of the vagus*, which is located underneath the vagal trigone in the floor of the rhomboid fossa, supplies the thoracic and abdominal viscera with visceromotor and secretory fibers (Fig. 19–3). The majority of the preganglionic parasympathetic cardioinhibitory fibers, however, originate in the nucleus ambiguus (Chapter 19).

Sensory Part

Most of the afferent fibers in the glossopharyngeal and vagus nerves convey visceral impulses from the pharynx, soft palate, posterior third of the tongue, tonsils, and tympanic cavity (glossopharyngeal nerve with cell bodies in the petrous ganglion) and from the thorax, abdomen, pharynx, and larynx (vagus nerve with cell bodies in the nodose ganglion). The fibers terminate in the *nucleus of the solitary tract*, which also receives taste impulses from the posterior third of the tongue. Both the IXth and the Xth cranial nerves contain somatic sensory fibers from the *external auditory meatus*.

Reflexes

The glossopharyngeal and vagus nerves participate in many important reflexes, e.g., the *respiratory and*

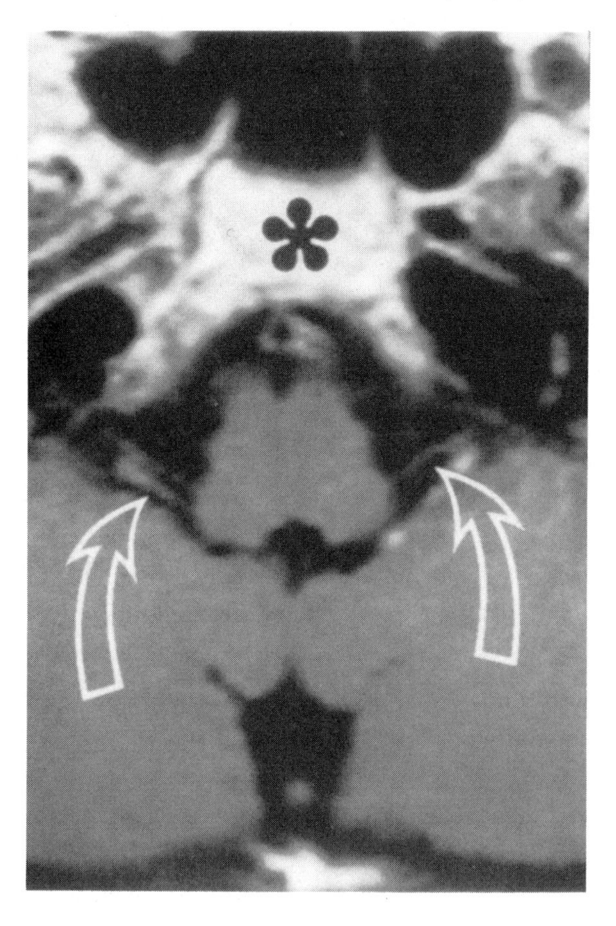

Fig. 11–14. IXth, Xth, and XIth Cranial Nerves
Axial MRI at the rostral level of medulla oblongata illustrating the cisternal portions of the IXth, Xth and XIth cranial nerves (large arrows). Asterisk = clivus. (From Sartor, K., 1992. MRI Imaging of the Skull and Brain. With permission of Springer-Verlag.)

cardiovascular reflexes (Chapter 19). Swallowing, vomiting, and coughing are other important reflex mechanisms that in part involve the IXth and Xth cranial nerves.

Swallowing

Stimulation of the pharynx or the back of the tongue initiates swallowing movements, i.e., a series of complicated reflexes that serve to introduce food and drink into the digestive tract. The most important afferent impulses are carried in the glossopharyngeal and vagus nerves to the nucleus of the solitary tract and nearby regions in the reticular formation, which coordinate the mechanisms involved in swallowing. The food is moved into the esophagus by the action of the tongue (hypoglossal nerve) and the palatal and pharyngeal muscles (glossopharyngeal and vagus nerves).

Vomiting

This is a protective mechanism whereby the gastric content is rapidly ejected through contraction of the abdominal muscle coupled with relaxation of the cardiac sphincter and esophagus. The concomitant closure of the glottis prevents aspiration of vomitus into the trachea. Irritation of the mucosa of the upper gastrointestinal tract gives rise to impulses that reach the nucleus of the solitary tract through the glossopharyngeal and vagus nerves. The vomiting response, which consists of a series of complicated somatic and visceral mechanisms, is coordinated by a "vomiting center" in the medulla oblongata. Efferent impulses to motor neurons of the spinal cord are responsible for contraction of the diaphragm and the abdominal muscles, whereas the autonomic innervation through one of the splanchnic nerves contracts the pylorus.

Vomiting can also be triggered by chemical substances in the bloodstream, e.g., apomorphine. The chemosensitive zone is apparently located in or near the *area postrema*, which is outside the blood–brain barrier and therefore more permeable to many substances in the blood than the rest of the CNS. The area postrema, which belongs to the circumventricular organs (Chapters 3 and 18) is closely related to the vomiting center.

Coughing

Irritation of the mucosa in the respiratory pathways elicits a cough response. The afferent impulses travel in the glossopharyngeal and vagus nerves to the nucleus of the solitary tract. The response is mediated through many channels. Motor impulses are mediated through the glossopharyngeal, vagus, and hypoglossal nerves to the pharynx, larynx, and tongue, and other impulses reach the spinal motor neurons that innervate the diaphragm, intercostal, muscles, and abdominal muscles.

Clinical Notes

Since both the glossopharyngeal and the vagus nerves receive innervation from the cerebral cortex on both sides, supranuclear lesions, e.g., tumors and bleeding in the cerebral hemispheres, do not affect the functions of these nerves if they occur on one side only. Nuclear and intranuclear lesions, however, do give rise to specific symptoms, and since the glossopharyngeal and vagus nerves are closely related both in their intramedullary and in the first part of their extramedullary course, they are often affected together by the same lesion.

The main symptoms of a glossopharyngeal nerve lesion are loss of sensibility and taste sensation in the posterior third of the tongue, whereas interruption of the vagus nerve has more widespread, but not necessarily alarming, consequences consisting primarily of transitory difficulties in articulation and swallowing. A bilateral vagus paralysis, on the other hand, is devastating.

Recurrent Laryngeal Nerve Lesion

The recurrent laryngeal nerve (X), which supplies all intrinsic muscles of the larynx except the cricothyroid, is easily damaged by tumors in the neck (e.g., thyroid tumor, metastatic lesions) or by aneurysms of the aortic arch. The voice is initially like a whisper or hoarse. The intact vocal cord, however, usually adapts to the condition with return of fairly normal voice, and the vocal cord paralysis can then be revealed only by a *laryngoscopic examination*.

Accessory Nerve (XI)

The XIth cranial nerve, which is purely motor, innervates the sternocleidomastoid and trapezius muscles. It is customary to divide the XIth cranial nerve into a spinal and a cranial part. The spinal part, which originates in anterior horn cells of the first five cervical segments of the spinal cord, enters the skull through the foramen magnum and joins the cranial part, which comes from the nucleus ambiguus and emerges from the brain together with the vagus nerve. The two parts leave the skull through the jugular foramen. However, since the cranial part soon rejoins the vagus nerve, it is more appropriate to refer to the cranial part as an aberrant vagus fascicle.

Hypoglossal Nerve (XII)

Functional Anatomy

The hypoglossal nucleus, which is located underneath the hypoglossal trigone in the floor of the fourth ventricle, innervates the muscles of the tongue. The fibers emerge from the medulla between the pyramid and olive as a series of 10–12 rootlets, which quickly come together as the hypoglossal nerve. The nerve leaves the skull through the hypoglossal canal.

Fibers from the Ist and IInd cervical nerves join the hypoglossal nerve and form the *ansa hypoglossi* together with the third cervical nerve. Branches from the ansa innervate the infrahyoid muscles.

Clinical Notes

Damage to the hypoglossal nerve results in paresis or paralysis of the ipsilateral half of the tongue. The paralysis can be easily demonstrated by asking the patient to protrude the tongue, which deviates toward the side of the lesion.

Although a hemiparalysis of the tongue can also result from a lesion of the hypoglossal nucleus, *nuclear lesions* are usually bilateral because of the close proximity of the right and left nucleus. *Chronic bulbar palsy* is a serious motor disorder that results from progressive degeneration of the motor nuclei in the medulla oblongata. A gradually worsening dysarthria and difficulties in swallowing are typical symptoms.

Since the hypoglossal nucleus receives corticobulbar fibers only from the contralateral hemisphere, a patient with *capsular hemiplegia* also suffers from a hypoglossal paralysis on the hemiplegic side.

Clinical Examples

Bell's Palsy

A 25-year-old healthy female was recovering from a mild upper respiratory infection when she awoke one morning and noticed that the right side of her face felt "pulled." While washing her hair that morning, she was bothered by the soap that kept coming into her right eye. Many sounds in her environment seemed excessively loud in her right ear.

Examination the next day showed moderate weakness of all facial muscles on the right side of her face. She could not taste well on the anterior surface of the right half of the tongue.

On the assumption that this patient suffered from Bell's palsy, i.e., acute peripheral facial paralysis, she was treated with steroids and her condition improved slowly.

Six weeks later a routine examination revealed that when she depressed her right eyelid the right corner of her mouth elevated slightly. In addition, she reported that tasting certain spices caused tearing from her right eye. The sense of taste on the right anterior surface of her tongue remained impaired.

1. This patient did indeed suffer from an acute neuropathy of the facial nerve. What part of the nerve was damaged?

2. When a damaged peripheral nerve regenerates, frequently the reinnervation does not totally follow the original pathways (aberrant reinnervation). What is the evidence for aberrant reinnervation in this case, and which branches of the facial nerve were involved in the process?

Discussion

Bell's palsy, or peripheral facial paralysis of unknown etiology, accounts for about 80% of all cases of facial palsy. It frequently follows exposure to cold on one side of the face or a mild viral respiratory infection.

Facial paralysis and taste deficit indicate that both the motor and sensory roots of VII are involved, and the hyperacusis reveals that the nerve was damaged proximal to the branch innervating the stapedius muscle (see Fig. 11–8).

As the damaged nerve recovered and reinnervation took place, the patient developed *synkinesis* or abnormal associated movements. In addition, there must have been damage at the level of the geniculate ganglion where parasympathetic fibers from the nervus intermedius leave the ganglion and travel in the greater superficial petrosal nerve. This is witnessed by the fact that the patient developed the "crocodile tear phenomenon" in which strong taste sensations produce involuntary lacrimation on the same side as the Bell's palsy. The phenomenon is believed to arise from regenerating autonomic fibers that should have reinnervated the salivary glands, but instead reinnervated the lacrimal glands.

Mediobasal Mesencephalic Syndrome of Weber

A 40-year-old man was suffering from headache and visual disturbance. One month earlier he had developed both horizontal and vertical diplopia and a drooping right eyelid (ptosis). He experienced progressive difficulty focusing his right eye at the same time as the left side of his body became weaker.

Examination confirmed that his right pupil was widely dilated. The direct light reflex was abolished in the right eye but a normal consensual response was present in the left eye. His right eye, further, was externally rotated and could not be brought across the midline or moved vertically. Ptosis was also evident on the right side. His left eye had full excursions in all directions, and the remainder of his cranial nerve functions were normal.

Motor examination showed a moderate left hemiparesis with increased stretch reflexes and Babinski's sign on the left.

A CT brain scan demonstrated an enhancing spherical lesion along the right ventral surface of the mesencephalon, and a subsequent angiography showed a large aneurysm arising from the right posterior communicating artery near its origin from the right posterior cerebral artery.

1. What cranial nerve is damaged in this patient and does the location of the aneurysm explain the involvement of that nerve?

2. Why does he have a left hemiparesis?

3. You receive a phone call from the patient's family physician, who tells you that 6 months ago the patient's right pupil was 3 mm larger than his left pupil, but his right extraocular functions were all normal and he did not have any hemiparesis. Does this make sense?

Discussion

The combination of ipsilateral IIIrd nerve paresis and contralateral hemiparesis is known as *Weber's syndrome* and is indicative of unilateral dysfunction of the mediobasal mesencephalon (Fig. 10–5). Although this syndrome was originally described in cases of midbrain infarction, the most common cause of Weber's syndrome is a mass lesion at the base of the midbrain, usually an enlarging posterior communicating artery aneurysm.

As the mass lesion expands, the IIIrd nerve is compressed, resulting initially in pupillary dilation and then ocular motor paresis. If the mass continues to expand, the adjacent cerebral peduncle is compressed, leading to a contralateral hemiparesis.

The pupillomotor fibers from the Edinger–Westphal nucleus are often the first fibers to be affected in a compression of the IIIrd nerve. The observation of pupillary dysfunction in the absence of extraocular dysfunction, therefore, makes good sense.

Brain Stem Hemorrhage

A 68-year-old woman with poorly controlled hypertension suddenly slumped to the ground and lost consciousness while shopping. She required assistance with respiration in the ambulance.

On examination in the Emergency Room, her blood pressure was 200/120. She appeared comatose, with her eyes closed, and had no spontaneous movements. Her respiratory rate was depressed, and she was intubated and artificially ventilated. The neurologic examination showed pinpoint-size pupils, and her corneal reflexes were absent bilaterally. Irrigation of either external ear canal with ice water caused no deviation of her eyes, i.e., her vestibulo-ocular reflex was absent. Her gag (pharyngeal) reflex was very weak. She responded with bilateral and symmetric extensor (decerebrate) posturing to painful stimuli, and Babinski's sign was present bilaterally.

An emergency enhanced CT scan was performed (see Fig. 11–15)

1. Where is the lesion?
2. Why were her pupils pinpoint in size?
3. Irrigation of the external ear canals with ice water (ice water calorics) is a commonly utilized tool to assess the connections between the vestibular and oculomotor nuclei. Why did her eyes not respond to this maneuver?
4. Occasionally, patients will survive this type of vascular catastrophe for varying periods of time, and they may even be able to respond to some written or spoken commands. Typically, the only response they are capable of making is elevating the eyelids or looking upwards or downwards, but not laterally. This condition is sometimes known as the "locked-in syndrome." Why should such patients be able to raise their eyelids voluntarily or look in a vertical direction?

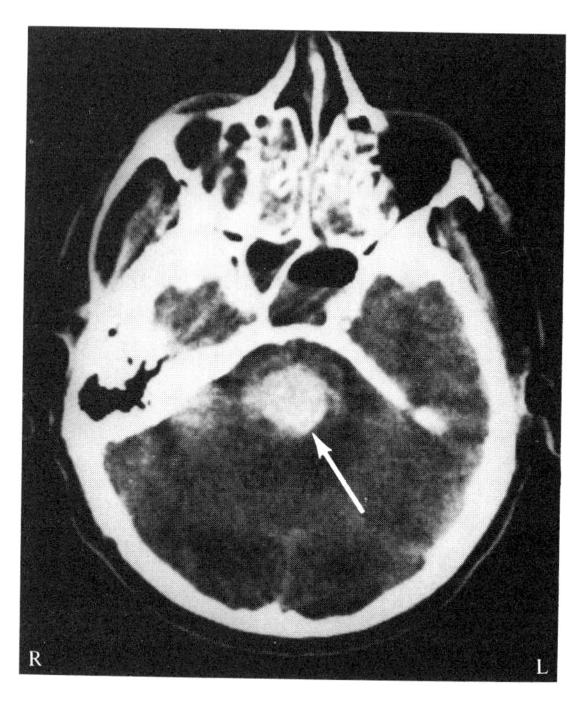

Fig. 11–15. Brain Stem Hemorrhage
The CT scan was kindly provided by Dr. V. Haughton.

Discussion

The lesion is a large, fairly symmetric hemorrhage in the pons. Intracranial hemorrhages typically occur in persons with chronic, poorly controlled hypertension in four characteristic locations: the lenticular nucleus, thalamus, pons, or cerebellum. In large pontine hemorrhages, the deficit is devastating because all long ascending and descending pathways including corticobulbar and corticospinal tracts are interrupted.

Patients with pontine hemorrhage have pinpoint pupils because the descending sympathetic fibers to the ciliospinal center are damaged with concomitant paralysis of the dilator pupillae muscles. However, the parasympathetic innervation (Edinger–Westphal nuclei and the IIIrd cranial nerves in the mesencephalon) to the pupillary constrictor muscles are not involved.

Damage to the vestibular nuclei or their connections with the oculomotor nuclei, i.e., the medial longitudinal fasciculus (Fig. 11–12), eliminates ice water caloric responses.

In the locked-in syndrome, the mesencephalic centers for the levator palpebrae superioris (IIIrd cranial nerve nuclei) and for vertical gaze are preserved. The supranuclear pathways innervating these structures are also intact. Thus, such patients are able to raise their eyelids and look up and down voluntarily, if consciousness is preserved. The pontine centers for lateral gaze, however, are damaged (see Figs. 11–3 and 11–4).

Lateral Medullary Syndrome of Wallenberg

A 63-year-old woman came to the Emergency Room with trouble swallowing, garbled speech, and numbness of the right arm that had lasted several hours. She also noted that she tended to fall toward her left side when walking. Six days previously she had had a similar episode of dysphagia lasting 2 hours. Other than arterial hypertension and diabetes mellitus, she had been well. There was no known history of heart disease or stroke.

Her examination was significant for mild hypertension and a systolic heart murmur heard at the base. The left pupil was smaller than the right with a slight droop of the left upper eyelid. A mild rotatory nystagmus was seen when the patient looked straight forward. Her left corneal reflex could not be elicited by stroking the left cornea, but blinking occurred bilaterally when the right cornea was stimulated. The uvula deviated to the right and the gag reflex was decreased on the left. Her voice was hoarse, and laryngoscopy revealed a paralyzed left vocal cord. Her stance and gait were mildly wide-based with a tendency to fall to the left. Sensory examination revealed diminished pin, touch, and temperature perception on the left side of the face and over the right arm and body. Position and vibration sensations were intact. Finger-to-nose and heel-to-shin tests were performed normally. Tendon reflexes and plantar responses were normal.

Initial CT scan of the head revealed no abnormalities. A barium swallow test showed good tongue movement but impaired soft palate and epiglottis function. MRI scan revealed a left lateral medullary infarct (see Fig. 11–16).

In the hospital, her hypertension and diabetes were well controlled with medications. Her gait and swallowing improved to the extent that she could walk unassisted and eat some foods without choking, but she remained hoarse at the time of discharge, one week after admission.

1. Why were the pupils of unequal size?
2. Explain the problems with swallowing and hoarseness.
3. What caused the uvula to deviate to the right?
4. Can a single lesion produce this diverse, bilateral syndrome, or is a "shower" of emboli from the heart more likely?

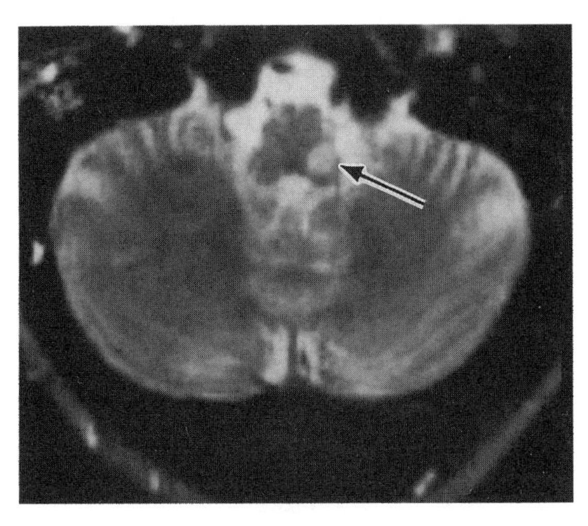

Fig. 11–16. MRI of Lateral Medullary Infarct (*arrow*) (Photograph courtesy of Dr. Maurice Lipper.)

Discussion

This patient has suffered an infarction involving the dorsolateral quadrant of her rostral medulla. In this location, a relatively small lesion can affect a wide variety of neurologic functions. Neural structures and their corresponding clinical deficits include vestibular nuclei and vestibulocerebellar connections (nystagmus, widened stance, falling to the left), descending sympathetic fibers (Horner's syndrome), the spinal trigeminal tract and nucleus (ipsilateral facial sensory loss), spinothalamic tract (sensory loss in the contralateral arm), and nucleus ambiguus (vocal cord paralysis, impaired palate and epiglottic movements, diminished gag reflex). The small left pupil and drooping left upper lid were part of the Horner's syndrome. Impaired pharyngeal and laryngeal muscles produced the hoarseness and dysphagia (trouble swallowing). This weakness caused failure of the larynx and epiglottis to close and sagging of the left soft palate. Under such circumstances, one cannot protect the upper trachea from accidental aspiration of liquids or foods, as was demonstrated on fluoroscopic viewing of swallowing efforts (barium swallow). The uvula is pitched into an asymmetric position by the fact that one side of the palate is normally elevated while the other sags. The dependent uvula appears to deviate toward the right (normal) side.

The corneal reflex was abolished on the left side because of damage to the spinal trigeminal nucleus on the left side. Stimulation of the right cornea, however, produced bilateral blinking and eye closure because the efferent limb of the reflex (facial nerve) was still intact on both sides.

Although Wallenberg's syndrome has often been said to be caused by occlusion of the posterior inferior cerebellar artery (PICA), recent investigations have shown that the syndrome is almost always caused by occlusion of the vertebral artery. Depending on the site of occlusion, the deficits may vary considerably from case to case. Most patients with this syndrome have a good prognosis and recover with little deficit.

Wallenberg is a household name to every neurologist. The present clinical case is almost identical to the one described originally by the German physician Adolph Wallenberg (1862–1949) in 1895 (see Wolf, 1971, p. 119).

Transtentorial Herniation

An 8-year-old boy was admitted to the hospital because of coma. He was one of many children in a family of itinerant farm workers. He had a history of frequent middle ear infections (otitis media), for which he received irregular medical care. One month before admission he developed a repeat attack of acute otitis media in his right ear, which resulted in localized right ear pain and fever. A physician prescribed an antibiotic which the patient took for only 2 days, after which his fever and pain lessened. He was well for the next week, but then began to complain of right-sided headaches that became progressively worse. On the morning of admission he was found by his parents to be very sleepy, could not be easily aroused, and was brought to the hospital.

On examination in the Emergency Room he was found to be comatose. He did not respond to visual stimuli (threats). His right pupil was dilated to 5 mm and did not react to light, and papilledema was present in both eyes. His gag (pharyngeal) reflex was present, but reduced in intensity. He moved all extremities in response to painful stimulation. While observed during the next 15 minutes, he developed bilateral extensor posturing (decerebrate posturing), and it was noted that the right extremities were as weak as the left. He showed no spontaneous movements. Babinski's sign was present bilaterally. Near the end of the examination it was noted that the left pupil had become dilated and fixed, and both eyes showed lateral deviation.

An emergency CT brain scan demonstrated a large hypodense hemisphere lesion which caused a 10-mm shift of the midline structures from right to left (see Fig. 11–17A). The picture is consistent with a very large brain abscess. On a lower CT scan (Fig. 11–17B) part of the right temporal lobe (long arrow) was seen to press against the mesencephalon. A linear enhancement, probably representing the tentorial margin, is seen along the left lateral border of the mesencephalon (short arrow). An area with decreased density, which could be consistent with an infarct, was also seen within the central substance of the mesencephalon just below the inferior colliculi.

1. What part of the right temporal lobe is pressing against the mesencephalon, and what membrane structure did it have to pass over to do so?
2. Which cranial nerve is related to the ocular abnormalities first seen in his right eye, and how did this cranial nerve become damaged?
3. Why did the patient have papilledema?
4. Why was the patient paretic on the right side of his body?

Discussion

This unfortunate boy demonstrates a well-known clinical syndrome that results from a rapidly expanding mass lesion in the supratentorial compartment. In this case, a brain abscess, which apparently arose from the middle ear infection, developed in the right hemisphere. The abscess grew so rapidly that the surrounding brain tissue was unable to accommodate the increased volume, resulting in a rapid rise in intracranial pressure. Once a critical pressure had been reached, the medial part of the temporal lobe, including the uncus and parahippocampal gyrus, herniated through the tentorial notch and pressed against the lateral aspect of the mesencephalon on the right side (see Fig. 3–7).

Damage to the IIIrd cranial nerve on the side of the herniation caused ipsilateral pupillary dilation and later complete oculomotor paralysis. The nerve, apparently, is either compressed by the herniating tissue or trapped between the posterior cerebral artery and the superior cerebellar artery as the brain stem is displaced downward (see Fig. 4–1).

Bilateral papilledema is caused by increase in intracranial pressure.

If the herniation is particularly severe, the mesencephalon is pressed laterally against the contralateral edge of the tentorium, which may result in a hemorrhagic infarct. The bleeding and the pressure on the base of the left cerebral peduncle may cause damage to the corticospinal tract, which could explain the paresis ipsilateral to the side of the herniation (false localizing sign). Ultimately, both oculomotor nerve become paralyzed.

Transtentorial herniation is always a very serious matter regardless of the cause. Hemorrhagic infarction of the brain stem due to occlusion or rupture of perforating arteries in the mesencephalon is rarely, if ever, compatible with survival.

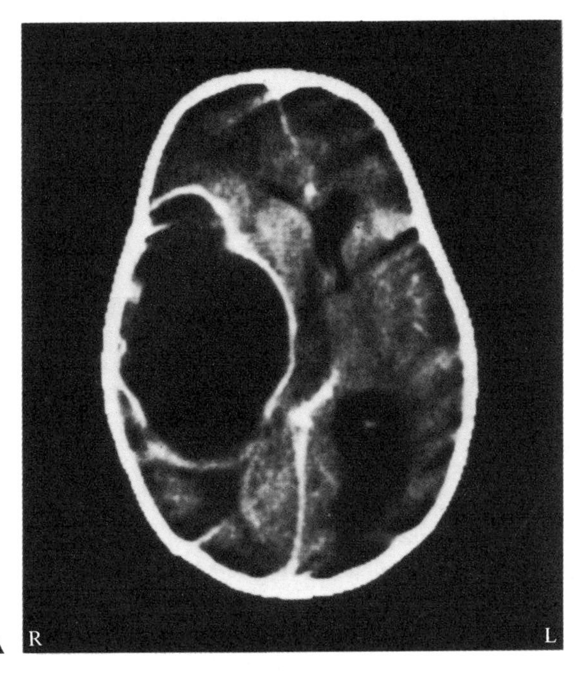

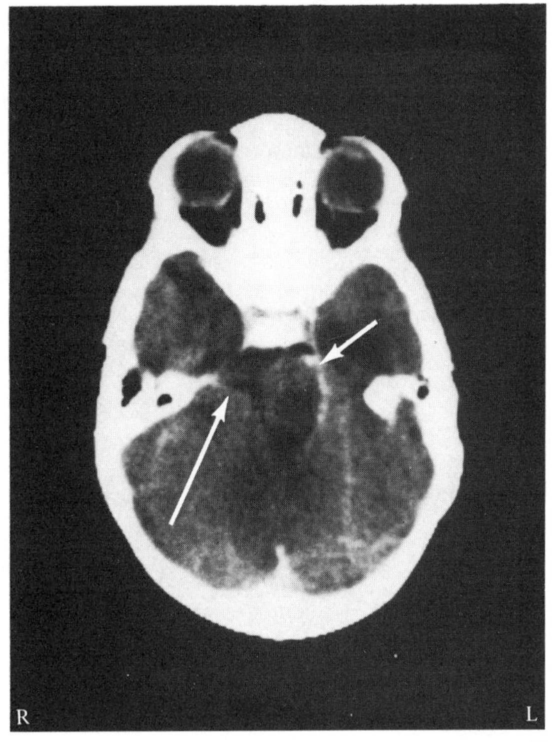

Fig. 11–17A&B. Transtentorial Herniation
The CT scans were kindly provided by Dr. V. Haughton.

Suggested Reading

1. Baloh, R. W. and V. Honrubia, 1989. Clinical Neurophysiology of the Vestibular System. 2nd ed. F. A. Davis Co., Philadelphia.
2. Brodal, A., 1981. The Cranial Nerves. In: Neurological Anatomy, 3rd ed. Oxford University Press, New York, pp. 448–577.
3. Büttner-Ennever, J. A. (ed.), 1988. Neuroanatomy of the Oculomotor System. Elsevier: Amsterdam.
4. Craigmyle, M. B. L., 1985. The Mixed Cranial Nerves. John Wiley and Sons: Chichester, UK.
5. Getchell, T. V., R. L. Doty, L. M. Bartoshuk and J. B. Snow, Jr. (eds.), 1991. Smell and Taste in Health and Disease. Raven Press, New York.
6. Hikosaka, O., 1991. Role of forebrain in oculomotor function. In G. Holstege (ed.): Role of the Forebrain in Sensation and Behavior. Progress in Brain Research, Vol. 87. Elsevier: Amsterdam, pp. 101–107.
7. Kellen, R. J. and R. M. Burde, 1990. Eye Movements and Vestibular System. In Pearlman, A. L. and R. C. Collins (eds.): Neurobiology of Disease. Oxford University Press, New York, pp. 149–167.
8. Kruger, L. and P. W. Mantyh, 1989. Gustatory and Related Chemosensory Systems. In Björklund et al. (eds.): Handbook of Chemical Neuroanatomy, Vol. 7. Elsevier: Amsterdam.
9. Leblanc, A., 1992. Anatomy and Imaging of the Cranial Nerves. A Neuroanatomic Method of Investigation Using Magnetic Resonance Imaging (MRI) and Computed Tomography (CT). Springer-Verlag: Berlin.
10. Leigh, R. J. and D. S. Zee, 1990. The Neurology of Eye Movements. 2nd ed. F. A. Davis Co., Philadelphia.
11. Wilson-Pauwels, L., E. J. Akesson, and P. A. Stewart, 1988. Cranial Nerves. Anatomy and Clinical Comments. B. C. Decker: Toronto.
12. Wolf, J. K., 1971. The Classical Brain Stem Syndromes, Charles C. Thomas, Springfield, USA.

Table 11–1. Nerve Components of the Cranial Nerves

Nerve	Component	Origin	Termination	Function
Oculomotor (III)	Somatic efferent	Oculomotor nucleus	All extraocular muscles except lateral rectus & superior oblique	Motor
	General visceral efferent	Edinger-Westphal (accessory oculomotor) nucleus	Ciliary and/or episcleral ganglia	Preganglionic parasympathetic neurons for ciliary muscle and constrictor muscle of iris
Trochlear (IV)	Somatic efferent	Trochlear nucleus	Superior oblique muscle	Motor
Trigeminal nerve (V)	Special visc. efferent	Motor nucleus of V	Chewing muscles	Motor
	General somatic afferent	Trigeminal ganglion	Spinal and pontine nuclei of V	Pain, touch & temperature from face & oral cavity
		Mesencephalic nucleus of V	Motor nucleus of V	Propriocep. from chewing musc. & temporomand. joint
Abducens nerve (VI)	Somatic efferent	Abducens nucleus	Lateral rectus muscle	Abducts the eye
Facial nerve (VII)	General visceral efferent	Lacrimal nucleus	Pterygopalatine ganglion	Parasymp. for lacrimal gland
		Superior salivatory nucleus		Parasymp. for nasal & palatine glands
			Submandibular ganglion	Parasymp. for sublingual gland
			Langley's ganglion	Parasymp. for submand. gland
			Ganglia near glands supplied	Parasymp. for glands ant. ⅔ tongue
	Special visc. efferent	Facial nucleus	Facial muscles	Motor
	General somatic afferent	Geniculate ganglion	Spinal nucleus of V	Pain, touch, temp. part of external ear
			Facial nucleus	Proprioception from facial muscles
	General visceral afferent		Nucleus parasolitarius	Impulses from mastoid air cells and middle ear
	Special visc. afferent		Dorsal visceral gray	Taste from anterior ⅔ tongue

Table 11–1. (cont.)

Nerve	Component	Origin	Termination	Function
Vestibulocochlear (VIII)	Special somatic afferent	Spiral ganglion	Dors/vent. coch. nuc.	Hearing
		Vestibular ganglion	Vestibular nuclei, cerebellum	Equilibrium
Glossopharyngeal (IX)	General visceral efferent	Inferior salivatory nucleus	Otic ganglion	Preganglionic parasympathetic for parotid gland
	Special visc. efferent	Nucleus ambiguus	Stylopharyngeus muscle	Motor
	General somatic afferent	Superior ganglion or IX	Spinal nucleus of V	Cutaneous sensibility from palate, post. ⅓ tongue, tonsils, ext. ear, pharynx
	General visc. afferent	Inferior ganglion of IX	Nucleus parasolitarius	Impulses from carotid body and carotid sinus
	Special visc. afferent		Dorsal visceral gray	Taste from the posterior ⅓ of the tongue
Vagus (X)	General visceral efferent	Dorsal efferent nucleus	Terminal ganglia in or near organs innervated	Pregang. parasympathetic for thoracic & most abdominal viscera
	Special visceral efferent	Nucleus ambiguus	Muscles of palate, larynx, pharynx, upper esophagus	Motor: swallowing, phonation, closure of nasal orifice
	General somatic afferent	Superior ganglion of X	Spinal ganglion of V	Cutaneous sensibility from part of external ear
	General visceral afferent	Inferior ganglion of X	Nucleus parasolitarius	Impulses from thoracic & abdominal viscera, aortic sinus & body
	Special visc. afferent		Dorsal visceral gray	Taste from epiglottis
Bulbar acc. (XI)	Visceral efferent	Supplements vagus nerve		
Spinal accessory (XI)	Special visceral efferent	Accessory nucleus (sp. cord)	Trapezius & sternocleidomastoid muscles	Motor
Hypoglossal (XII)	Somatic efferent	Hypoglossal nucleus	Muscles of tongue, (all but palatoglossus)	Motor

(By permission of Lockard, *Desk Reference for Neuroscience* 2nd ed, Springer-Verlag New York, 1992, pp 61–63.)

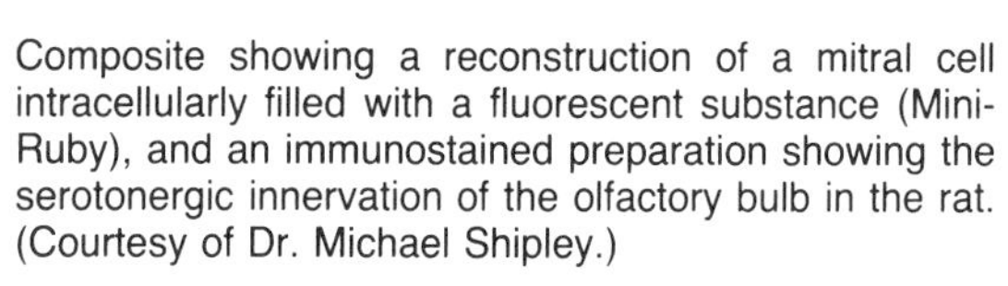

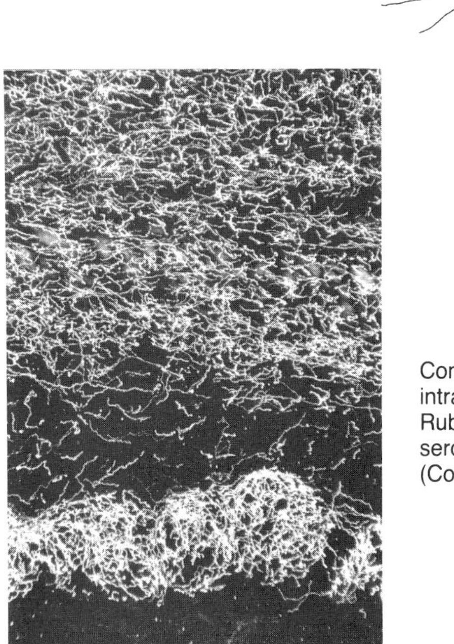

Composite showing a reconstruction of a mitral cell intracellularly filled with a fluorescent substance (Mini-Ruby), and an immunostained preparation showing the serotonergic innervation of the olfactory bulb in the rat. (Courtesy of Dr. Michael Shipley.)

12

Olfactory System

The central olfactory pathways constitute a rather small part of the human brain, and olfaction is undoubtedly of less clinical significance than vision and hearing. Nevertheless, the sense of smell is of considerable importance in our everyday life; it plays a significant role in food intake, in reproductive and endocrine functions, and in certain social behaviors of many mammalian species including man. Repugnant odors and sweet perfumes have a powerful effect on mood and behavior.

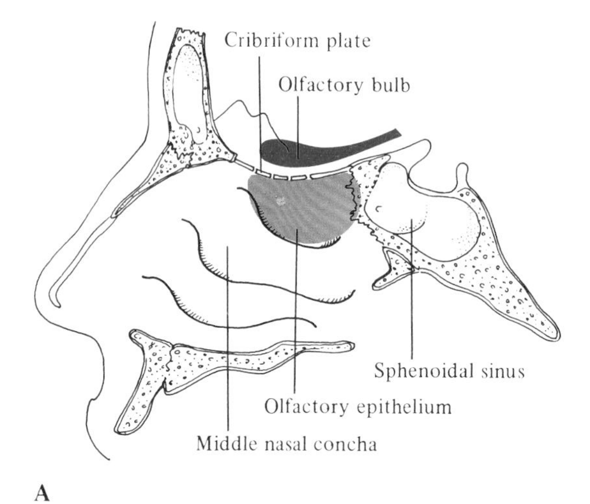

Fig. 12–1. The Olfactory Epithelium
A. The olfactory epithelium (*light red*) is located in the upper part of the nasal cavity. The olfactory bulb (*dark red*), which is located in the anterior cranial fossa, is separated from the olfactory epithelium by the cribriform plate of the ethmoid bone.
B. Scanning electron micrograph of the olfactory epithelium in the human. 15–18 cilia radiate into the overlying mucus from each olfactory vesicle. (From Fujita, T., K. Tanaka and J. Tokunaga, 1981. SEM Atlas of Cells and Tissue. Igaku-Shoin, Tokyo–New York. With permission of the authors and the publisher.)

Olfactory Epithelium

The olfactory epithelium occupies about a 2-cm^2 area in each nasal cavity, where it covers the upper concha and adjacent part of the nasal cavity (Fig. 12–1A). Because the epithelium is located in the upper part of the nasal cavity, only part of the air reaches the olfactory receptor cells in normal inspiration. Sniffing, which produces extra air currents in the nasal cavity, may therefore be necessary to bring enough odorous molecules in contact with the olfactory epithelium, especially if the odor is weak. The epithelium consists of olfactory receptor cells that are surrounded and supported by epithelial secretory cells called supporting cells (Fig. 12–2). The olfactory epithelium is covered by a secretion produced by the olfactory glands of Bowman and the supporting cells.

Olfactory Receptor Cells

These are represented by several million bipolar sensory neurons whose axons project to the olfactory bulb. The peripheral process or dendrite of the receptor cell contains a knob-like enlargement, the *olfactory vesicle*, from which several 100- to 200-µm-long hair-like *cilia* project into the mucus covering the epithelium (Fig. 12–2B). To reach the cilia, the chemicals in the air must be dissolved in the fluid that surrounds the cilia. Although the actual transduction mechanism is unknown, it is believed that the olfactory receptor cells are stimulated when odorants bind to specific receptor proteins located on their cilia. How different odors are coded and olfactory signals processed in the olfactory system is poorly understood.

There are strong indications that the olfactory receptor cells are renewed continuously throughout life. This constant turnover may reflect an adaptation on the part of the olfactory sensory neurons, which seem especially vulnerable to injuries from environmental toxic substances, since they are located in the surface epithelium. The reconstitution is accomplished through a process in which immature sensory cells (basal cells) in the deep part of the olfactory epithelium are transformed into mature receptor cells. This neurogenetic process, which is unique in the mature nervous system, raises some important questions related to the coding of olfactory stimuli. Because the phenomenon implies a continuous development, not only of receptor cell bodies in the olfactory epithelium but also of their axons projecting to the olfactory bulb, one wonders if the newly formed axons are related to a specific site in the glomerular layer or if their central connections are randomly made? If a certain group of receptor cells is responsive to a specific odor, one would expect that the axons of

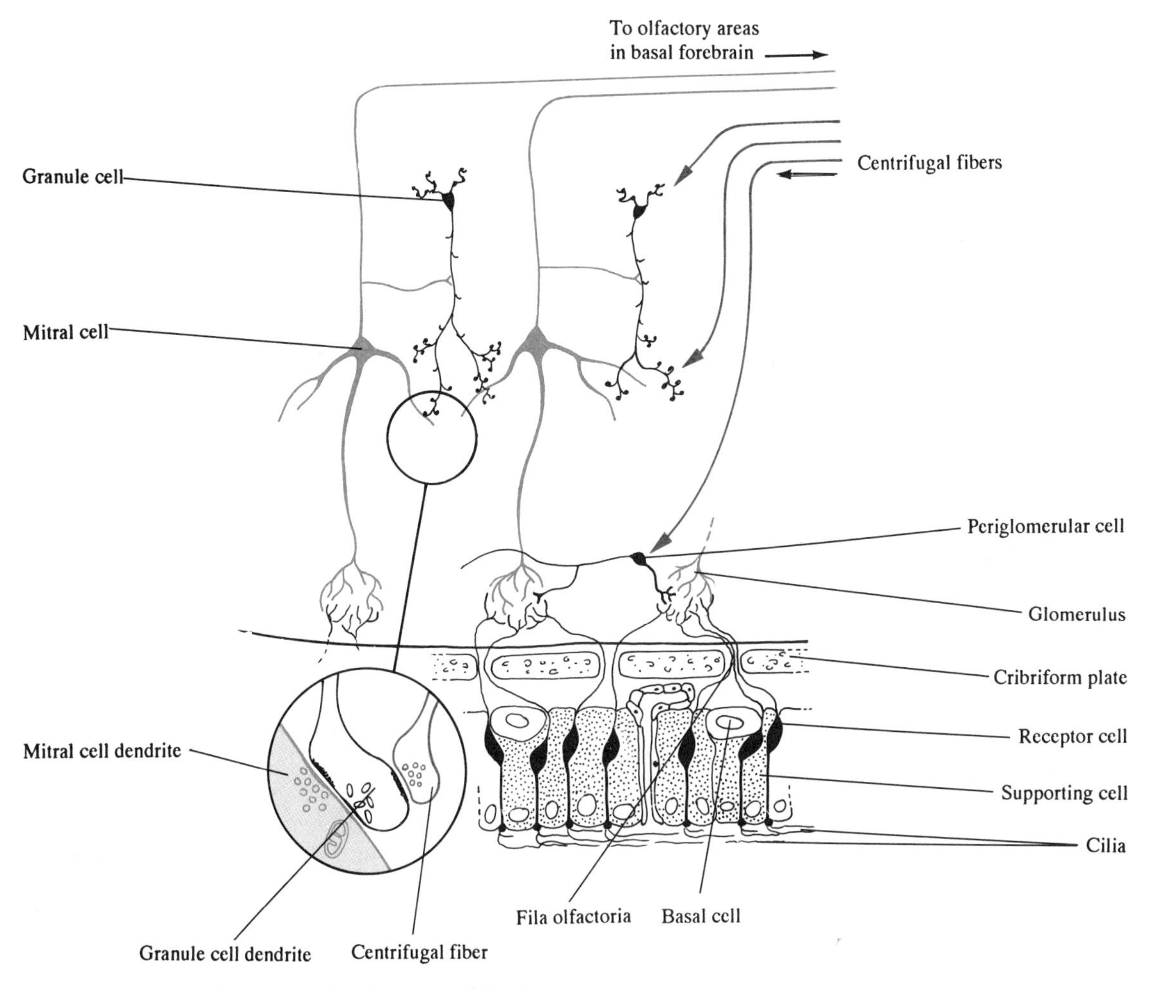

Fig. 12–2. The Olfactory Epithelium and Olfactory Bulb
Schematic drawing of the main neuronal elements of the olfactory bulb and their relationship to the olfactory receptor cells and the olfactory tract. The *inset* in the *lower left corner* shows a dendrodendritic synapse connection between a mitral cell dendrite (*blue*) and a granule cell dendrite (*black*). Note that the centrifugal fiber (*red*) from the basal forebrain forms an axodendritic synapse with the spine of the granule cell dendrite.

newly generated receptor cells should be able to find their way to a specific set of glomeruli. It must be emphasized, however, that our knowledge in this field is still rudimentary. For instance, we do not know if different odors are represented by spatially distinct groups of receptor cells.

Olfactory Pathways

Olfactory Nerve

The thin unmyelinated axons, *fila olfactoria*, of the receptor cells form about 20 slender bundles that pass through tiny openings in the cribriform plate of the ethmoid bone to reach the olfactory bulb (Fig. 12–2). These bundles represent the 1st cranial nerve, the *olfactory nerve*. As the bundles pass through the cribriform plate, they are surrounded by tubular sheaths of the meninges. The subarachnoid space, therefore, is in more or less direct continuity with the lymphatic system in the nasal cavity, which makes it relatively easy for a nasal infection to spread to the meninges and the brain.

Animal experiments, furthermore, have shown that viruses and other environmental toxins can be incorporated into the receptor cells and then reach other parts of the brain through the mechanism of axoplasmic transport. This has been suggested as a possible mechanism for the development of Alzheimer's disease (see Clinical Notes).

Olfactory Bulb

The olfactory bulb is the rostral expanded part of the olfactory stalk, which extends forward underneath the frontal lobe from its attachment to the brain in front of the anterior perforated substance. The large *mitral cells* (blue in Fig. 12–2) are the major relay neurons in the olfactory bulb, and their main function is to transmit olfactory stimuli from receptor cells to centrally located brain regions. The synaptic contacts between the incoming olfactory nerve fibers and the dendrites of the mitral cells take place in specialized structures called *glomeruli*. A glomerulus is a more or less spherical structure, partially or wholly encapsulated by glial processes, and characterized by a wealth of synaptic contacts. Many thousands of receptor cells synapse with the dendritic branches of one or only a few mitral cells within a glomerulus; this means that there is a high degree of convergence of olfactory receptors onto mitral cells.

Interneurons and Reciprocal Synapses

The transmission of olfactory impulses from receptor cells to mitral cells is only one aspect of the complicated interactions that take place in the olfactory bulb. Physiologic studies indicate that a considerable amount of information processing occurs in the bulb. A large number of inhibitory interneurons are present in the bulb, the most common being the *granule* and *periglomerular cells*. The interneurons (black in Fig. 12–2), which modulate and control the transmission of olfactory impulses at several points within the bulb, are in turn influenced by collateral branches from mitral cells as well as by centrifugal pathways from a number of regions in the basal forebrain and brain stem, including serotonergic and noradrenergic fibers.

The reciprocal *dendrodendritic synapse* for recurrent inhibition in the CNS was first discovered in the olfactory bulb, between the dendrites of mitral and granule cells (see inset in Fig. 12–2). The mitral cell excites the granule cell through one portion of the reciprocal synapse, whereupon the granule cell inhibits the same mitral cell through the other portion. This type of self-inhibition has now been described in many other parts of the nervous system. In view of the fact that one granule cell is related to several mitral cells, the inhibitory action released by a mitral cell can also affect nearby mitral cells; this is an example of *lateral inhibition*.

The layers of the olfactory bulb. Although a layering of the cellular elements is evident in the human olfactory bulb, it is less pronounced than in most other mammals. The following layers are recognized, beginning with the surface (Fig. 12–2).

1. *The fiber layer* is the most superficial layer; it consists of the incoming olfactory nerve fibers.
2. *The glomerular layer* consists of several rows of glomeruli that are surrounded by periglomerular cells.
3. *The plexiform layer* is characterized primarily by the dendritic branches of mitral cells and granule cells, and the numerous synaptic contacts that occur between them.
4. *The mitral layer*, which contains large mitral cells, is not well-defined in the human.
5. *The granular layer* contains a large number of small interneurons, the granule cells. The granule cell resembles the amacrine cell of the retina in that it lacks a morphologically distinct axon. The proximal dendrites of the granule cell branch locally in the granular layer, and a superficial spiny dendrite usually extends into the plexiform layer.

Olfactory Tract and Its Projection Areas

The axons of the mitral cells form the olfactory tract, which courses through the olfactory stalk to olfactory areas in the basal forebrain (Fig. 12–3). The main olfactory areas include the *anterior olfactory nucleus* at the base of the olfactory stalk and the *primary olfactory cortex* (piriform cortex), which forms a continuous field in part of the frontal and temporal lobes. The rostral part of the olfactory cortex is located on the orbital surface adjacent to the olfactory tract (often referred to as the lateral olfactory tract) as it proceeds towards the limen insulae. Some olfactory bulb fibers also reach the ventral striatum, which extends to the ventral surface of the brain in the region of the anterior perforated space. At the limen insulae, the olfactory tract turns sharply in a medial direction, and its fibers fan out over the dorsomedial surface of the temporal lobe in the region of the *uncus* of the parahippocampal gyrus. This is the temporal part of the olfactory cortex, which also includes the cortical amygdaloid nucleus and periamygdaloid cortex. Some olfactory bulb fibers even reach the entorhinal area. Most of the temporal olfactory cortex is located on the dorsal surface of the uncus and cannot be seen from the ventral side (Fig. 12–3).

In comparison with other sensory systems, a review of the olfactory pathways reveals some unique features. *First*, there is no indication that the olfactory pathways are topographic in the sense of a point-to-point relationship between the bulb and the olfactory cortex. *Second*, whereas there are at least three neurons that link the periphery to the cerebral cortex in other sensory systems, the olfactory pathway is characterized by a two-neuron chain. *Third*, the first-order neurons in the olfactory path are represented by the bipolar receptor

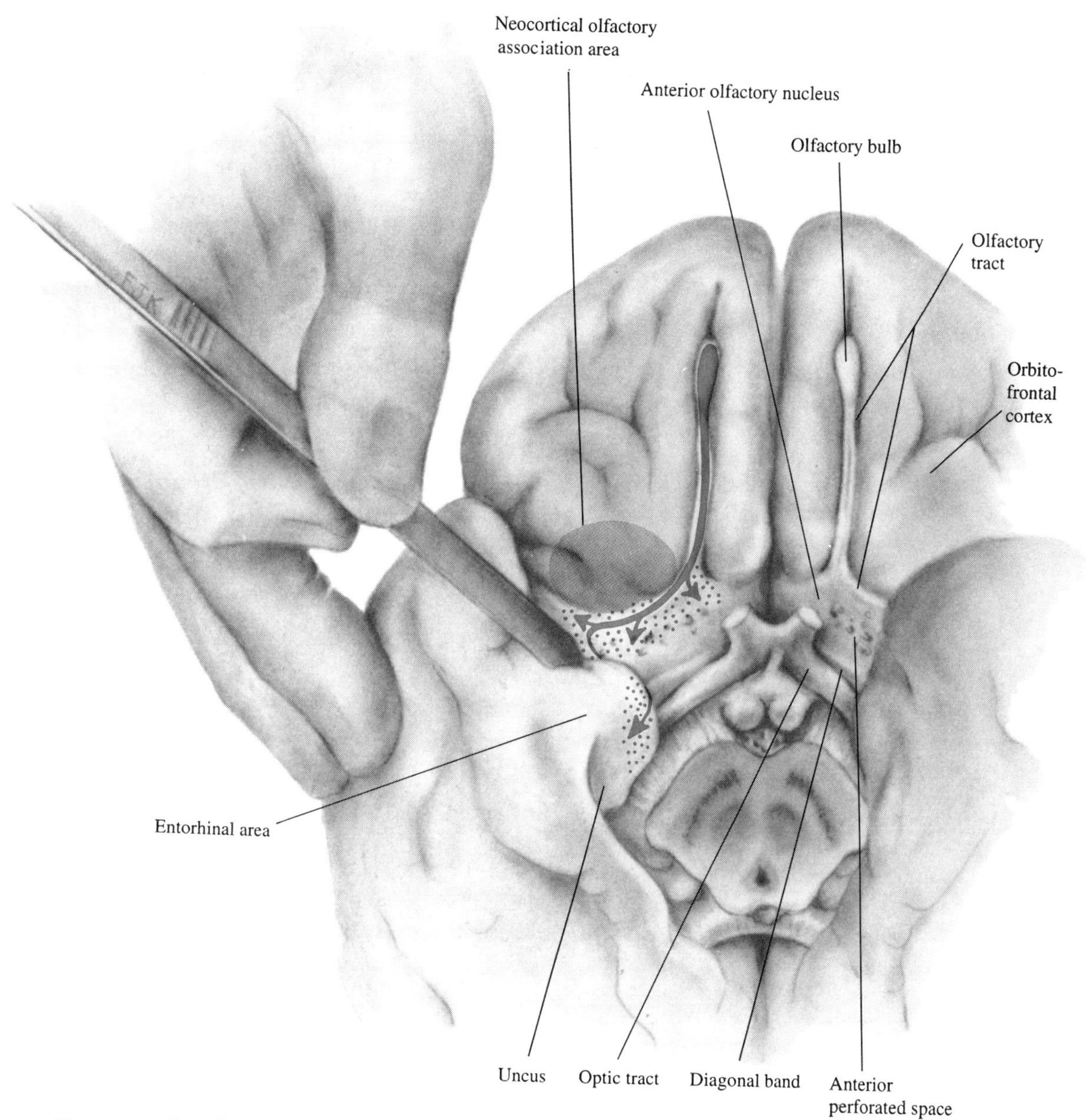

Fig. 12–3. The Olfactory Tract and Its Projection Areas
The projection areas of the olfactory bulb (*blue stippled area*) are easily identified on the basal surface of the brain. The temporal lobe has been deviated laterally to show the lateral extension of the olfactory tract, which makes a sharp bend in the region of the limen insulae to reach the uncus of the parahippocampal gyrus. The *light red* area indicates the approximate location of the neocortical olfactory association area in the orbitofrontal region.

cells, whose central processes form the Ist cranial nerve. *Fourth*, the second-order neurons, the mitral cells, project directly to the cerebral cortex rather than to the thalamus. In fact, the olfactory bulb can be considered a cortical structure since it derives embryologically from the same tissue as the cerebral cortex.

Olfactory Projections Beyond the Primary Olfactory Cortex

The olfactory cortex projects directly to several other parts of the brain including the ventral striatum, mediodorsal thalamus, insula, and orbitofrontal cortex (Fig. 12–3).

Although direct olfactohypothalamic connections are often promoted as the link whereby smell affects neuroendocrine mechanisms, there is little experimental evidence for a direct projection from the olfactory cortex to the hypothalamus. On the other hand, the *medial amygdala*, which is closely related to the hypothalamus, does receive input from both the cortical amygdaloid nucleus and the primary olfactory cortex.

Functional Aspects

Microsmatic Man

On the basis of the development of the olfactory system, vertebrates can be divided into *macrosmatic* (= those with a keen sense of smell), and *microsmatic* (= those with a poorly developed sense of smell). Although it is common practice to refer to man as a microsmatic mammal, one must nevertheless remember that the olfactory sense organ is capable of discriminating among thousands of different substances on the basis of their smell, and it can detect odors of which there are only traces in the atmosphere. As shown by perfumers and wine tasters, the efficiency of the olfactory sense can be greatly improved by training. Therefore, even if man is not critically dependent on the olfactory system for his survival, it is questionable if his sense of smell should be considered poorly developed.

The olfactory receptors are similar to taste receptors in that they are stimulated by chemicals dissolved in liquids, and both types of receptors are important for food selection. Much of what we perceive as the flavor of food is in fact related to the sense of smell. This becomes evident when a person suffers from a head cold, in which case excessive mucus prevents odors from reaching the olfactory receptors, and the food seems to have lost most of its "taste."

Functional Significance of Higher Order Olfactory Connections

The olfactory bulb and primary olfactory cortex are closely related to each other by reciprocal connections, and both structures are likely to be of importance for the perception of smell, but we know very little about the functional significance of the different primary olfactory areas. Recent studies indicate that the *orbitofrontal cortex* and part of the *insular cortex* may be of special significance for the discrimination of odors, and for the integration of olfactory impulses with gustatory and visceral, and maybe even visual, sensations. Possibilities for interaction between olfactory cortex and orbitofrontal cortex exist both via cortical association fibers and by way of the mediodorsal thalamic nucleus.

The sense of smell is essential for sexual behavior in many mammals, and as indicated by the success story of the perfume industry, olfaction is likely to play a role also in human sexuality. The olfactory part of the amygdaloid body, including the medial amygdala, which receives input from the olfactory cortex, is an integral part of the *extended amygdala* (Chapter 20), and it has been shown, in animal experiments that the medial extended amygdala, in conjunction with the medial preoptic hypothalamus, is crucially involved in male sexual behavior.

Centrifugal Pathways: Central Modulation of Olfactory Stimuli

Some of the most interesting discoveries in recent years are related to the centrifugal fiber systems to the olfactory bulb (red arrows in Fig. 12–2). Such pathways originate in widespread regions of the primary olfactory cortex, basal forebrain, lateral hypothalamus, and brain stem. The activities in these pathways undoubtedly serve to control and modulate the transmission of incoming olfactory stimuli, and they may play a role in the reduction of perceptual intensity that occurs during prolonged exposure to an odorous substance.

Clinical Notes

Anosmia

Lesions of the olfactory pathways may result in the loss of the sense of smell, i.e., *anosmia*. Unilateral anosmia is usually not noticed by the patient, whereas bilateral anosmia is readily apparent because food loses much of its flavor. As a matter of fact, the patient usually complains of loss of taste rather than of smell. The most common cause of anosmia is a nasal infection. However, more serious conditions, such as fractures of the ethmoid bone or orbitofrontal tumors, may also result in anosmia, and it is therefore important to know how to test the sense of smell.

Testing the Sense of Smell

Commonly known nonirritating odoriferous substances, such as chocolate, coffee, or tobacco can be used to test the sense of smell. The odor is applied to one nostril at a time by letting the patient sniff the test substance at the same time as he closes the other nostril. Even if the patient may not be able to identify the test substance, the olfactory nerves are usually intact as long as an odor can be detected. It is thus important that the test substance used be nonirritating since a positive response can occur due to irritation of the nasal mucosa, innervated by an intact trigeminal nerve.

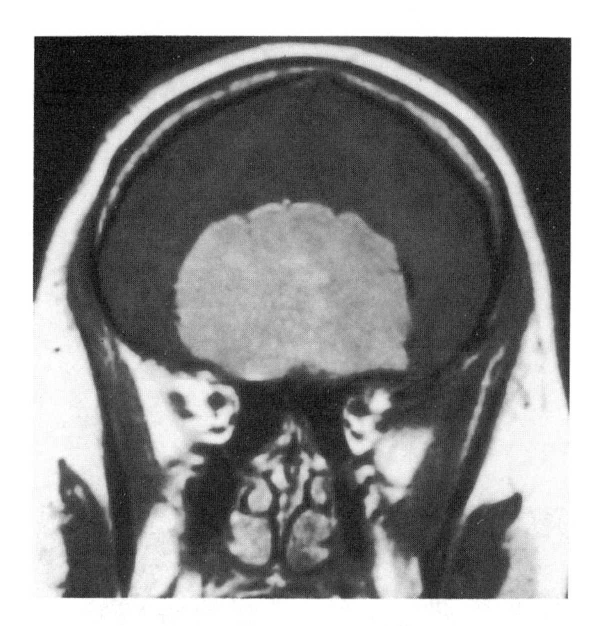

Fig. 12–4. Olfactory Groove Meningioma T1-weighted coronal MRI with gadolinium enhancement. (The MRI was kindly provided by Dr. Maurice Lipper, University of Virginia.)

Alzheimer's Disease

As mentioned earlier, it has been suggested that foreign substances, e.g., aluminosilicates, might enter the brain through the olfactory pathways and thereby possibly contribute to the development of Alzheimer's disease, and maybe even other neurodegenerative disorders. Although this is only an hypothesis, it is interesting to note that pathologic studies have shown that parts of the olfactory system can be as heavily inundated with neurofibrillary tangles as the hippocampus formation (see Clinical Notes, Chapter 21).

Uncinate Seizures

Lesions that involve the primary olfactory cortex can give rise to olfactory hallucinations, usually in the form of a disagreeable odor. This is occasionally the first sign of a focal epileptic discharge in the temporal lobe. The hallucination, which is often accompanied by chewing or smacking movements and a feeling of unreality, indicates that the irritation is located in the uncus of the parahippocampal gyrus or the adjoining regions. The syndrome is referred to as an uncinate seizure, which is one of the group of convulsive disorders known as complex partial epilepsy (Chapter 22).

Clinical Example

Olfactory Groove Meningioma (Subfrontal Meningioma)

A 70-year-old female suffering from senile dementia and headache was admitted for new onset of seizures. An MRI (Fig. 12–4) was obtained.

1. Where is the lesion?
2. Which cranial nerve functions might be affected by this lesion?

Discussion

The MRI demonstrates a lesion that extends from the inner table of the skull deep into the substance of the frontal lobe. The appearance is typical of a meningioma. These are usually benign tumors that arise from arachnoid cells and become symptomatic by exerting pressure on surrounding brain substance and/or cranial nerves. This tumor most likely arose from arachnoid cells along the cribriform plate, hence the name olfactory groove meningioma. One of the earliest and most important symptoms resulting from such a meningioma is unilateral or bilateral anosmia due to compression of the olfactory nerves or tracts (Fig. 12–1). Visual disturbances occur if the optic nerves become involved. Large bilateral frontal lesions also tend to obliterate the anterior part of the ventricular system. The patient was operated on through a bifrontal craniotomy and total removal of the tumor was achieved.

Suggested Reading

1. Eslinger, P. J., A. R. Damasio, and G. W. Van Hoesen, 1982. Olfactory dysfunction in man: A review of anatomical and behavioral aspects. Brain and Cognition. 1:259–285.
2. Getchell, T. V., R. L. Doty, L. M. Bartoshuk and J. B. Snow, Jr. (eds.), 1991. Smell and Taste in Health and Disease. Raven Press, New York.
3. Price, J. L., 1990. Olfactory system. In G. Paxinos (ed.): The Human Nervous System. Academic Press: San Diego, pp. 979–998.
4. Serby, M. J. and K. L. Chobor, 1991. Science of Olfaction. Springer-Verlag: New York.

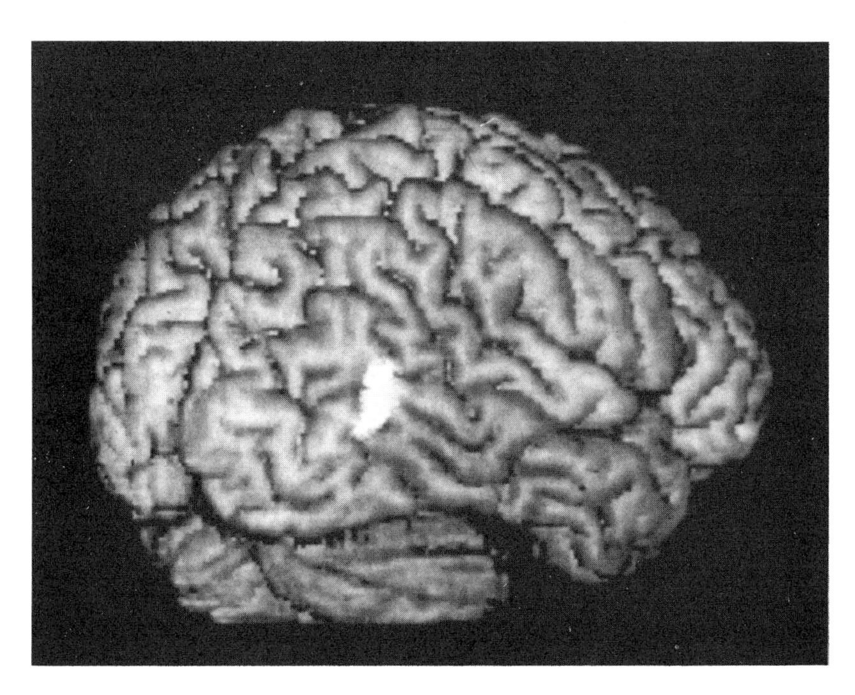

The visual motion area (V5) identified in combined PET–MRI study. (Courtesy of Dr. S. Zeki. From Watson, J. D. G. et al., 1993. Cerebral Cortex, Vol. 3, No. 2, pp. 79–94. With permission of Oxford University Press.)

13

Visual System

The phenomenal development that has taken place in our understanding of the nervous system during the last few decades is due in large part to new and sensitive methods for experimental investigations. This is especially well demonstrated in regard to the visual system, which is a favored topic in basic and clinical research. The biochemical processes of photoreception have been elucidated, and the intrinsic retinal circuits and central visual connections have been traced in great detail. The demonstration of how single cells and groups of neurons at successive stations along the visual pathways respond to different features in the visual image, and how various subdivisions of the visual system process specific features such as form, color, and movement, represent great success stories in modern neuroscience. The question of where and how these different components are finally brought together into a unified percept remains a formidable challenge.

The topography of the visual pathways is of great significance in clinical neurology. Since they extend from the frontal end of the brain to the occipital pole, they are often involved in brain lesions. The pathways are highly organized, and lesions in different parts of the visual system produce characteristic symptoms, which usually provide some clues about the location of the pathologic process.

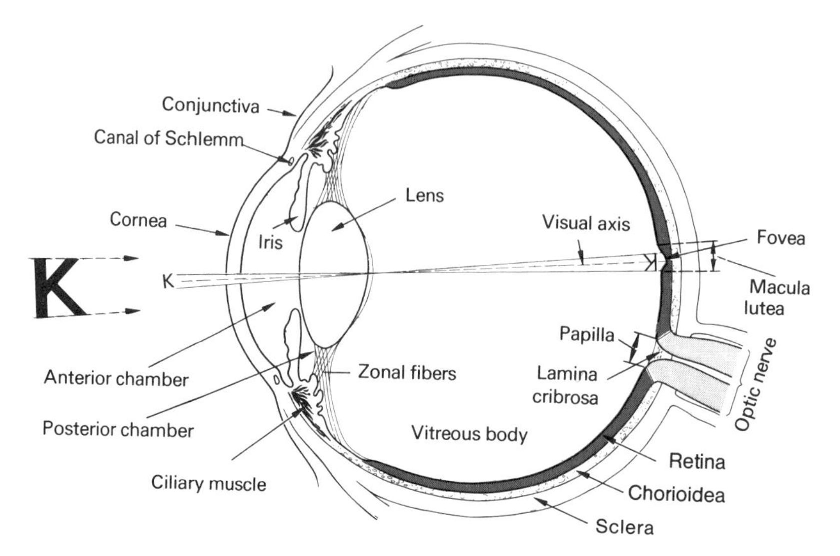

Fig. 13–1. The Eye
Horizontal section through the right eye. Retina is indicated with *red*. (From Grüsser, O. J. and U. Grüsser-Cornehls. Physiology of Vision. In: Fundamentals of Sensory Physiology (ed. R. F. Schmidt). Springer-Verlag, New York–Heidelberg–Berlin.)

The eye is a hollow, transparent structure covered by a thick, fibrous coat, the *sclera* (Fig. 13–1). The rostral part of the coat, the *cornea*, is transparent and serves as the "window" of the eye. Together with the *lens*, the cornea forms part of the refractive system, which focuses light rays on the sensitive part of the eye, the *retina*. The line of vision, i.e., the optic (visual) axis, extends from the object, seen through the center of the pupil, to the fovea centralis in the retina. Between the retina and the sclera is the highly vascularized *choroid layer*.

The *iris* controls the amount of incoming light by regulating the size of the *pupil*; it is analogous to the "f-stop" setting of a camera. Although the optic system of the eye is similar to that of a camera, the functions of the eye are considerably more intricate, and the comparison is especially weak in regard to the retina, which is infinitely more complicated than a film plate.

Retina

Photoreceptors

The sensory part of the eye, the retina, is an evaginated part of the forebrain, and it is connected to the rest of the brain by the optic nerve. The retina has a typical laminar organization (Fig. 13–2A) with three cellular layers in addition to two plexiform layers, where the synaptic interactions take place.

The visual receptors, the *rods* and *cones* (Fig. 13–2B), are located in the deep portion of the retina in close contact with the *pigmented epithelium*, which adjoins the *choroidea*. The light entering the eye, therefore, passes through the transparent retina in order to reach the photoreceptors. The pigment cells regulate the content of visual pigment in the outer segments of the receptor cells, where the receptor response is triggered by the radiant

Fig. 13–2. The Retina

→

A. Phase-contrast photomicrograph of a vertical section through the human retina close to the fovea. Note that the rods are thinner than the cones. The fiber layer between the outer nuclear layer and the outer plexiform layer consists of the axons of the receptor cells. The fact that the axons have a lateral orientation reflects the displacement of the inner retinal layers in the region of the fovea (compare **C**). (Modified from Boycott, R.B. and Dowing, L.E., 1969. Phil. Trans. R. Soc. Lond. (Biol.) 255: 109–184. With permission of The Royal Society, London.)

B. Scanning electron micrograph from a rhesus monkey showing the retinal surface with photoreceptors. (Micrograph courtesy of Dr. Masayaki Miyoshi, Japan.)

C. Schematic drawing of a microscopic section through the retina in the region of the fovea centralis. Note how the inner layers of the retina are pushed aside to reduce the dispersion and absorption of the light reaching the cones in the fovea. (From Lehrbuch der Histologie und der Mikroskopischen Anatomie des Menschen mit Einschluss der mikroskopischen Technik, 23. Auflage, 1933. With permission of Gustar Fischer Verlag, Stuttgart.)

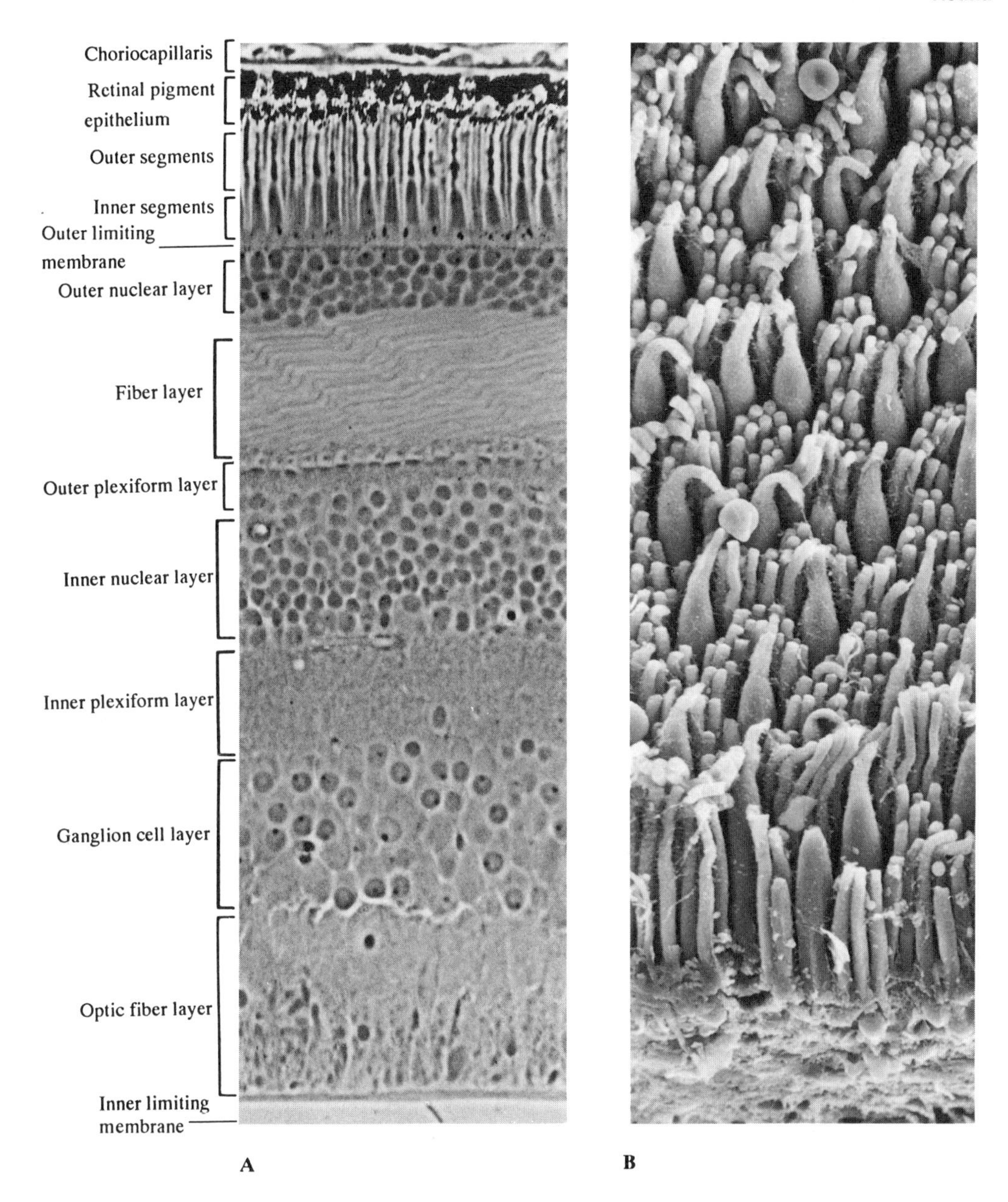
Choriocapillaris
Retinal pigment epithelium
Outer segments
Inner segments
Outer limiting membrane
Outer nuclear layer
Fiber layer
Outer plexiform layer
Inner nuclear layer
Inner plexiform layer
Ganglion cell layer
Optic fiber layer
Inner limiting membrane
A
B

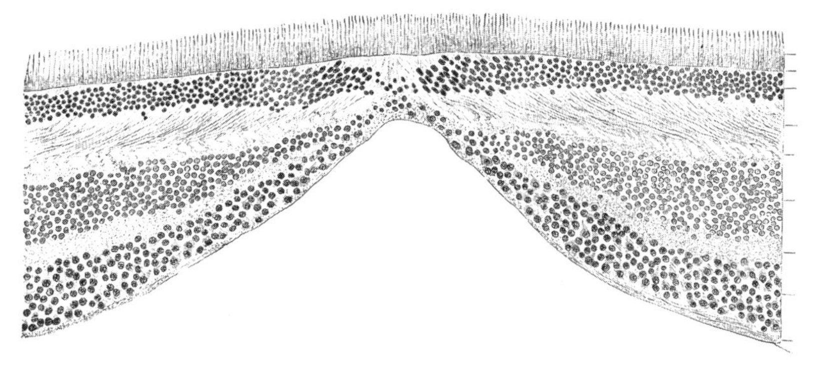
C

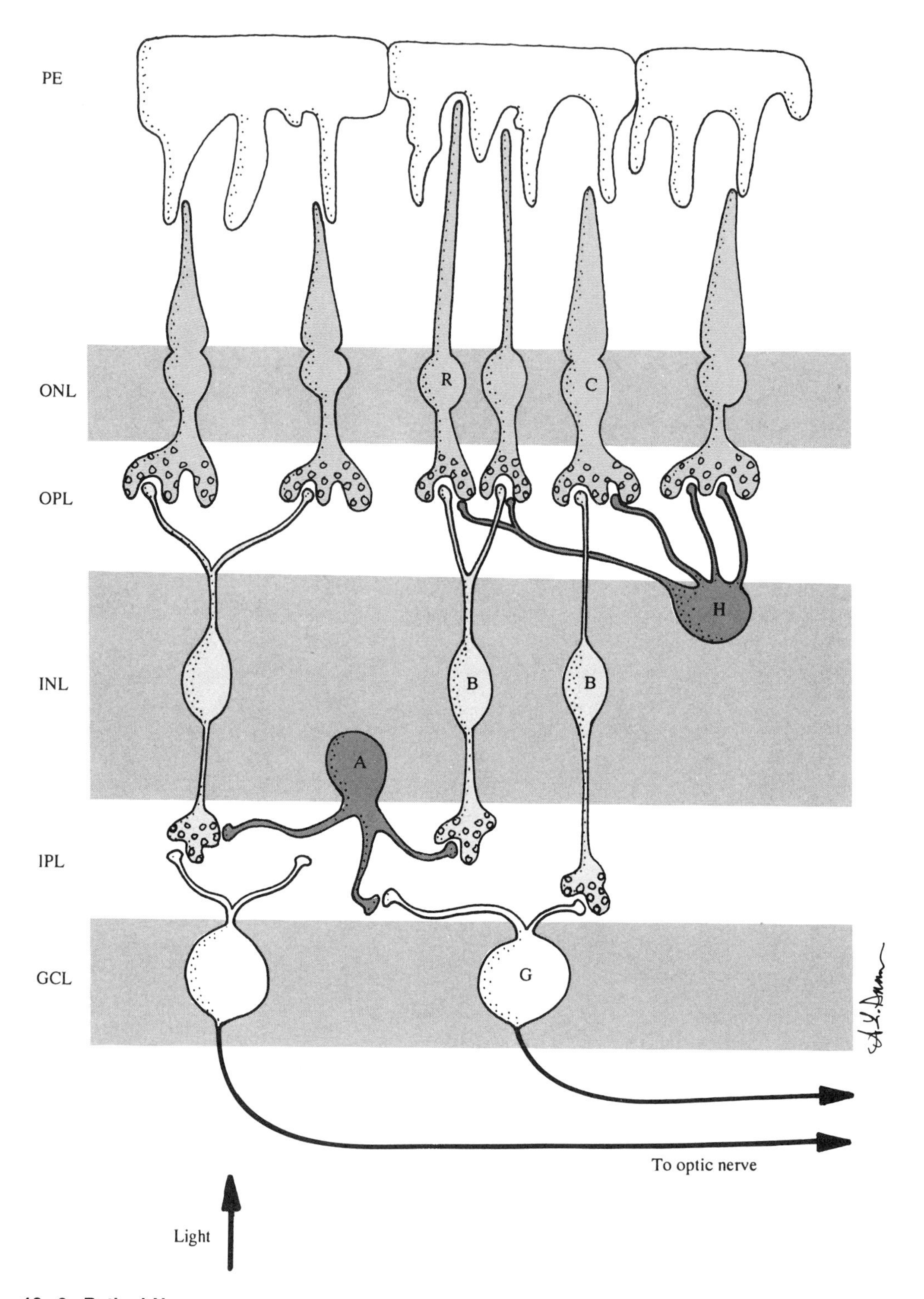

Fig. 13-3. Retinal Neurons
Highly schematic figure of the organization of the retina. Note that only the cone bipolar cells (B) synapse directly with the ganglion cells (G), whereas the rod bipolar cells use the amacrine interneurons (A) as intermediaries. The photoreceptors are *red*, and the interneurons, i.e., the horizontal and amacrine cells, are shown in *dark gray*. Bipolar cells, *blue*; ganglion cells, *white*. Abbreviations: A = amacrine cell; B = bipolar cell; C = cone; G = ganglion cell; GCL = ganglion cell layer; H = horizontal cell; INL = inner nuclear layer, IPL = inner plexiform layer; ONL = outer nuclear layer; OPL = outer plexiform layer; PE = pigment epithelium; R = rod. (Modified from Dowling, J. E., 1979. Information processing by local circuits: The vertebrate retina as a model system. In: F. O. Schmitt and F. G. Worden (eds.): The Neurosciences: Fourth Study Program. Cambridge, MA, MIT Press.)

energy of visible light (wavelength between 350 and 800 nm) through the mediation of the visual pigment.

The two types of photoreceptors, both of which apparently use glutamate as their transmitter, serve different functions. The rods are much more sensitive to light than the cones and they enable us to see in dim light. The visual pigment in rods is rhodopsin[1]. The cones mediate color vision in brighter light. There are three types of cone cells, each of which contains a different visual pigment for maximal absorption of light in the blue, red, and green parts, respectively, of the visible spectrum. The cones, furthermore, enable us to see sharp images, and the *fovea centralis*, the area of highest visual acuity, is populated by cones only (Fig. 13–2C). The concentration of cones decreases, while the concentration of rods increases, toward the periphery of the retina.

Retinal Neurons

Bipolar Cells. As a specialized part of the CNS, the retina contains several major classes of neurons besides the approximately 100 million receptor cells. Bipolar neurons transmit the impulses from the receptor cells to the ganglion cells (Fig. 13–3). Whereas the bipolar cells related to the cones form direct contact with ganglion cells, the rod bipolar cells use amacrine cells as intermediaries to the ganglion cells.

Amacrine and Horizontal Cells. Amacrine (axonless) cells, together with another type of interneurons, the horizontal cells, serve to modulate the visual information as it is transmitted from the receptor cells to the ganglion cells. The horizontal cells, some of which are GABA-ergic, receive input from receptor cells and direct their output to other neighboring receptor cells and to bipolar cell dendrites in the outer plexiform layer. The horizontal cells are primarily responsible for the phenomenon of *lateral inhibition*, thereby increasing the contrast between the illuminated and nonilluminated parts of the retina. The amacrine cells, like the granular cells in the olfactory bulb, lack a typical axon. Instead, their heavily branched processes have both axonal and dendritic characteristics. The amacrine cells constitute a morphologically heterogeneous group of neurons, which use a number of different transmitters (e.g., GABA, glycine, ACh, dopamine, and several peptides) in their interactions with bipolar cells, other amacrine cells, and ganglion cells. These synaptic interactions take place in the inner plexiform layer. Considering the large variety of amacrine cells, they are likely to be involved in a number of different interactions designed to shape the receptive field characteristics of the ganglion cells. As mentioned above, a special function is provided by those amacrine cells that form a direct link between rod bipolar cells and ganglion cells.

Ganglion Cells. The retinal ganglion cells, which form the ganglion cell layer next to the inner plexiform layer, are the output cells of the retina. Their axons run across the inner surface of the retina toward the *optic disc*, where they form the *optic nerve* as they leave the eye. The axons remain unmyelinated until they reach the optic disc. There are two main types of retinal ganglion cells, the *M cells* and the *P cells*, with different anatomic and physiologic characteristics. The M cells, which receive their designation from the fact that they project to the magnocellular layers of the lateral geniculate nucleus, are larger than the P cells, which project to the parvocellular layers of the lateral geniculate nucleus. The M cells also have a significantly wider dendritic field than the P cells (Fig. 13–4). As discussed in more detail below, the anatomic differences are matched by physiologic differences; movement and high-contrast images, for instance, are primarily the concern of the M cells and their central connections, whereas the more numerous P cells and their related central pathways are more involved with fine detail and color coding.

The capacity for spatial resolution, or the ability to see sharp images, is in part related to the extent of convergence in the pathway from receptor to ganglion cell. Since 100 million photoreceptors converge on one million ganglion cells, the average convergence is about 100:1. The amount of convergence, however, varies between cones and rods and between different parts of the retina. For instance, in the region of the fovea centralis, one or two cones may establish synaptic contact with one bipolar cell, which in turn is related to one ganglion cell. In regard to the rods, on the other hand, there is considerable convergence, on the order of more than 1,000 rods to around 100 bipolar cells, which in turn may converge on one ganglion cell.

Retinal Topography

The retina around the posterior pole of the eyeball, i.e., the *fundus oculi* (see Clinical Notes and Fig. 13–15), can be observed with the aid of an ophthalmoscope. The *optic disc* or papilla is a round or slightly oval, pale area of the retina where the optic nerve fibers leave the eye. The retinal vessels, which reach the eye through the optic nerve, radiate from the center of the disc. Since the region of the optic disc does not contain any receptor cells, it is referred to as the *blind spot* of the retina.

1. Rhodopsin is synthesized in the presence of vitamin A. Vitamin A deficiency, therefore, can cause night blindness.

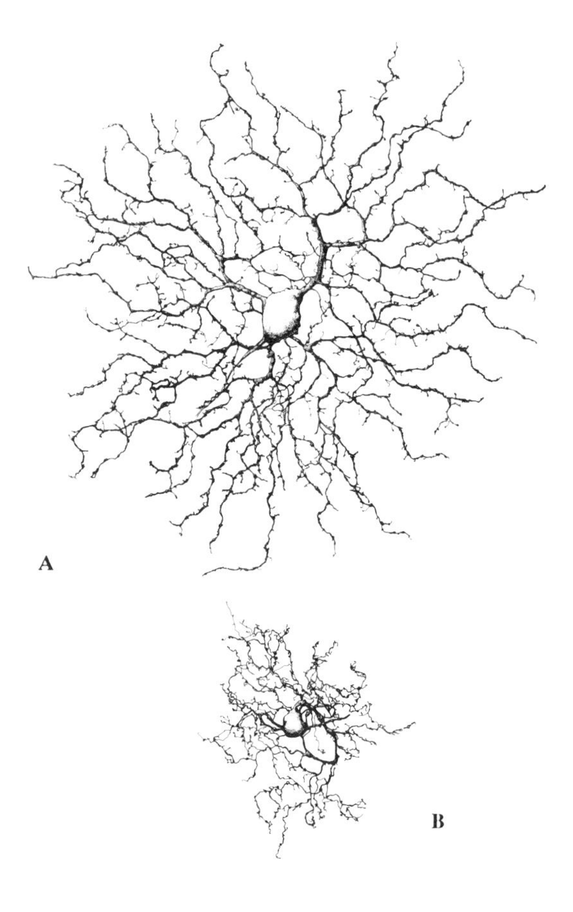

Fig. 13–4. Retinal Ganglion Cells
M cell (**A**) and P cell (**B**) from macaque retina. Note the significantly wider dendritic field of the M cell compared to the P cell. The cells were labeled by intracellular injections of HRP. Scale bar = 50 μm. (Courtesy of Dr. Robert Rodieck, from Watanabe, M. and R. W. Rodieck, 1989. Parasol and midget ganglion cells of the primate retina. J. Comp. Neurol. 289:434–454. Reprinted by permission of Wiley-Liss, a division of John Wiley and Sons, Inc.)

The blind spot in the right eye can be demonstrated by closing the left eye and fixating the right eye at a distant object. Hold a pencil at arm's length in front of the right eye and move it laterally from the line of fixation. The pencil disappears when its image passes across the blind spot.

The *fovea centralis*, the area of highest visual acuity, is a small pit in the center of the *macula lutea* (the yellow spot), which is slightly deeper in tint than the surrounding fundus. The fovea usually appears as a bright reflex about three disc diameters on the temporal side of the optic disc (Fig. 13–15A).

Note that the light must pass through the various layers of the retina before it reaches the photoreceptors, which are situated adjacent to the pigment epithelium (Fig. 13–2A). In the region of the fovea centralis, however, all blood vessels and most neuronal elements except the photoreceptors are pushed aside to reduce the dispersion and absorption of the light reaching the cones in the fovea (Fig. 13–2C).

Visual Centers and Pathways

Optic Nerve, Optic Chiasm, and Optic Tract

The optic nerve perforates the sclera and proceeds in a caudal direction to the *optic chiasm*, where the fibers from the nasal halves of each retina cross over to the optic tract on the opposite side (Fig. 13–5). The optic nerve is an integral part of the CNS, rather than a peripheral nerve, and, like the rest of the CNS, it is surrounded by dura, arachnoid, and pia mater. The subdural and subarachnoid spaces of the brain, therefore, are continuous along the optic nerve, and an increase in intracranial pressure is readily transmitted along the optic nerve to the optic disc, where it may manifest itself as papilledema (see Clinical Notes).

Since the fibers from the nasal half of each retina decussate in the optic chiasm, each optic tract contains the fibers from the temporal retina of the ipsilateral eye and the nasal retina of the contralateral eye. In other words, each optic tract (as well as the lateral geniculate nucleus and geniculocalcarine tract, see below) contains information from the contralateral visual "hemifield."

The optic tract sweeps around the cerebral peduncle (Fig. 13–6) to the *lateral geniculate nucleus (LGN)*, where most of the fibers terminate in a precise retinotopic pattern. Other fibers terminate in the superior colliculus and the pretectal region.

Lateral Geniculate Nucleus (LGN)

This nucleus in the posterior, ventrolateral part of the thalamus is a six-layered cone-shaped structure (Fig. 13–7). The axons of the type M ganglion cells terminate in the two ventral magnocellular layers (the layers with large cells), whereas the axons of the type P cells project to the four dorsal parvocellular layers (the layers with small cells). Axons from the two eyes, furthermore, terminate in different layers in the LGN. In other words, three of the six layers receive input from one eye and the three remaining layers receive input from the other eye, which means that impulses from the two eyes are kept separate in the LGN. This is the case also in the geniculocalcarine tract, and only when the visual information reaches the visual cortex do fibers carrying information from the two eyes eventually converge on single cells.

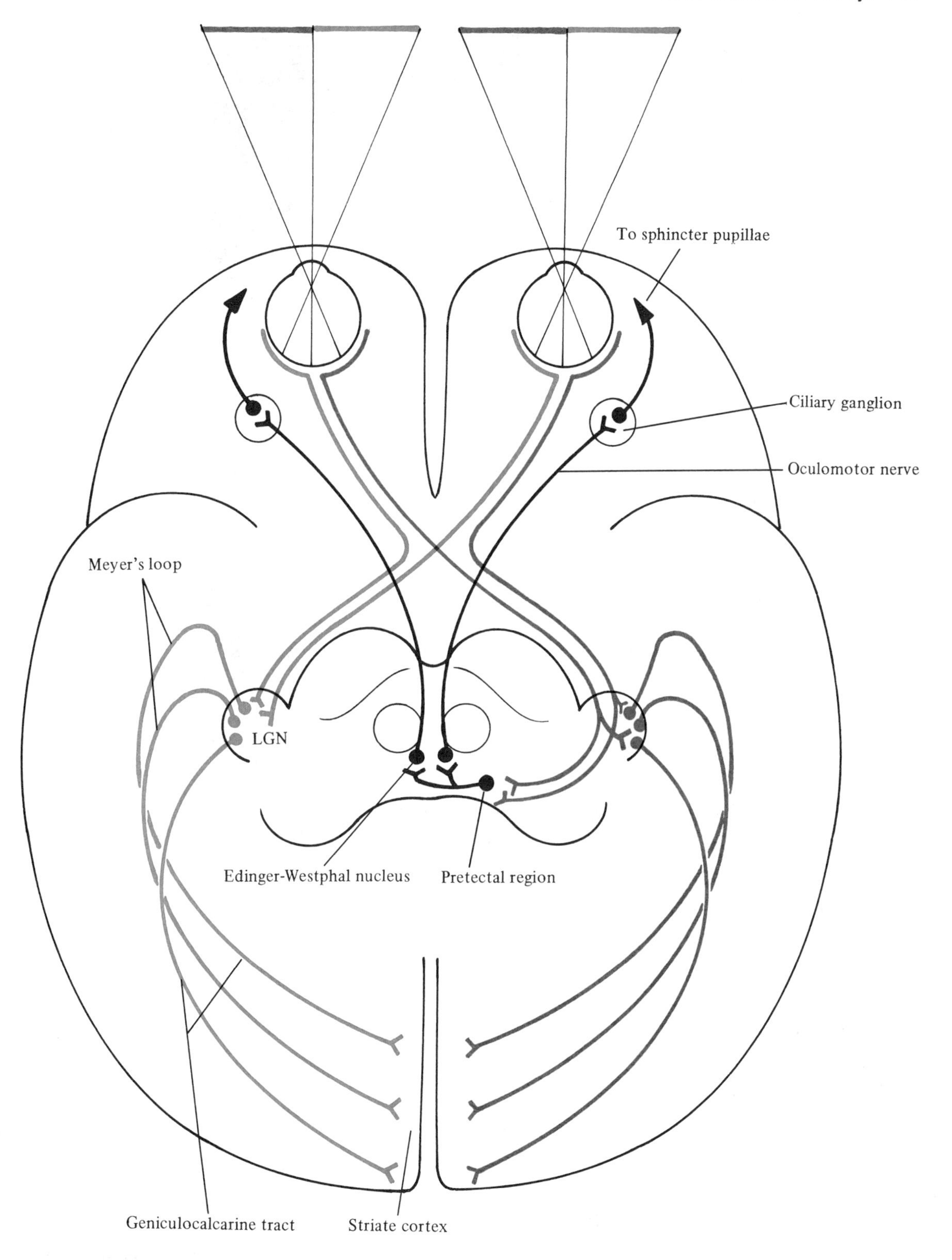

Fig. 13–5. Visual Pathways
The pathway for the pupillary light reflex is also illustrated. LGN = lateral geniculate nucleus.

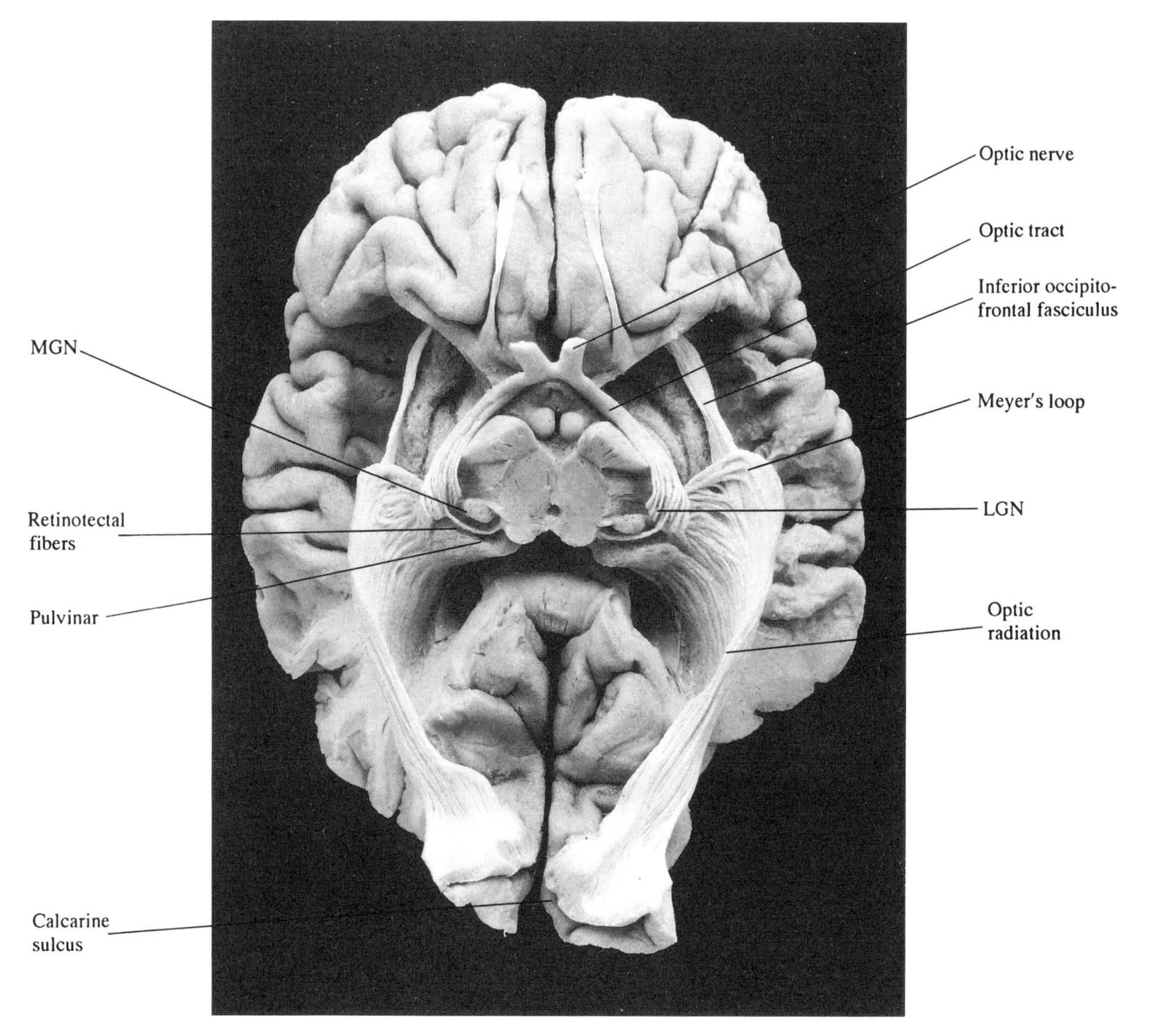

Fig. 13–6. Dissection of the Visual Pathways in Ventral View

The optic tract, which sweeps around the lateral side of the cerebral peduncle, terminates in the laminated lateral geniculate nucleus (LGN). The narrow bundle of intensely white fibers behind the medial geniculate nucleus (MGN) represents part of the retinotectal fiber system, which bypasses the LGN on the way to the superior colliculus in the mesencephalon. Note that only part of the massive fiber system that proceeds in a characteristically curved fashion from the diencephalon to the occipital lobe represents the geniculocalcarine tract. As clearly demonstrated in this dissection, other components of this thick sagittal stratum of white substance include fibers between the pulvinar and the occipital cortex as well as the inferior occipitofrontal association bundle. Nonetheless, the whole system is often, but incorrectly, referred to as the optic radiation. (The dissection was performed by the Dutch anatomist Dr. Nedzad Glutbegovic, who kindly provided the picture.)

In addition to the retinal afferents, the LGN receives input especially from the visual cortex but also from several brain stem centers. Monoaminergic fibers reach the LGN from the locus ceruleus and dorsal raphe nucleus, and the pedunculopontine tegmental nucleus provides a significant cholinergic input, which modulates the activity of the LGN neurons in response to attention and arousal (Chapter 10). For example, whereas the LGN neurons are relatively unresponsive during sleep, they are highly responsive to visual stimuli during wakefulness. Although the primary function of the LGN is to transmit information from the eyes to the visual cortex, the LGN is more than just a simple relay center. In other words, the retinal input seems to be significantly modified before it

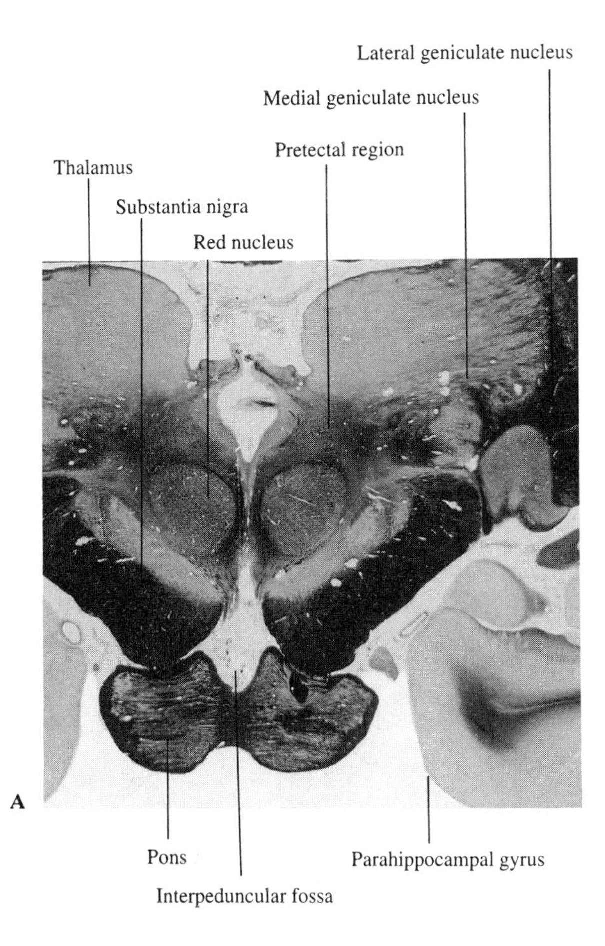

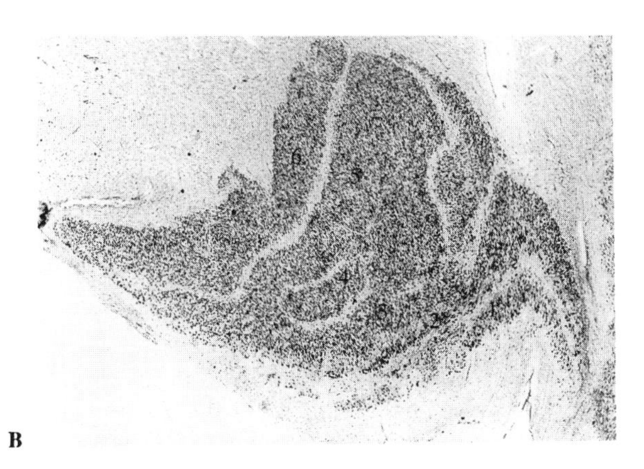

Fig. 13–7. Lateral Geniculate Nucleus
A. Myelin-stained section through the upper brain stem of the human brain. The section cuts through the posterior part of the diencephalon, the mesencephalon, and the most rostral region of the basal part of the pons. The lateral geniculate nucleus on the right side of the brain is located on the lateral side of the cerebral peduncle.
B. Nissl-stained section through the lateral geniculate nucleus of a human brain showing the laminar arrangement of neurons. (Courtesy of Dr. Edward Jones, from E. G. Jones, 1985. The Thalamus. Plenum Press, New York. With permission of the author and the publisher.)

is forwarded to the visual cortex, but how it is modified has not been clarified.

Geniculocalcarine Tract and Visual Cortex

Neurons in the LGN give rise to the *geniculocalcarine tract* or *optic radiation*, which projects to the primary visual cortex in the region of the calcarine sulcus in the occipital lobe. Since the fibers from the nasal halves of each retina cross in the optic chiasm, the visual cortex in each hemisphere receives information from the opposite visual hemifield (Fig. 13–5).

The optic radiation[2] proceeds through the retro- and sublenticular parts of the internal capsule, and the fan-like arrangement of its fibers can usually be appreciated during the blunt dissection of the brain (Fig. 13–6; see also Fig. 4–17 in the Dissection Guide). The fibers that carry impulses from the inferior retinal quadrants sweep in a lateral and slightly rostral direction around the temporal horn of the lateral ventricle (Fig. 13–8) before they proceed in a caudal direction to the inferior lip of the calcarine sulcus. The part of the optic radiation that makes a rostrolateral detour into the temporal lobe is referred to as *Meyer's loop*. The fibers carrying information from the upper retinal quadrants proceed more directly through the deep parts of the parietal and temporal lobes to the region above the calcarine sulcus. Although the optic radiation is spread out in a fan-like manner, its fibers are orderly arranged, and localized lesions tend to give rise to circumscribed homonymous defects, *scotomas*, in the opposite visual field (see Clinical Notes).

The *primary visual cortex* corresponds to Brodmann's area 17 (Fig. 22–2), most of which is located on the medial aspect of the occipital lobe, primarily in the depth of the calcarine sulcus (Fig. 13–9A). The primary visual cortex, also known as visual area 1 (VI), contains a horizontal stripe in layer 4, the *line of Gennari*, which can be seen even by the naked eye (see Fig. 5–20A). This stria, which is composed primarily of heavily myelinated intrinsic association fibers, gave rise to the term *striate cortex*.

The striate cortex is closely associated with surrounding cortical areas, the *extrastriate visual areas*, some of which have been labeled V2–V5. V2 and V3, which surround V1 (Fig. 13–9B), correspond

2. Although the optic radiation is a synonym for the geniculocalcarine tract, it is also generally applied to the sagittal stratum of white substance shown in Fig. 13–6. However as clearly illustrated in that figure, and as described in more detail in the figure text, the massive system of sagittally running fibers between diencephalon and the occipital lobe contains more than just the geniculocalcarine tract.

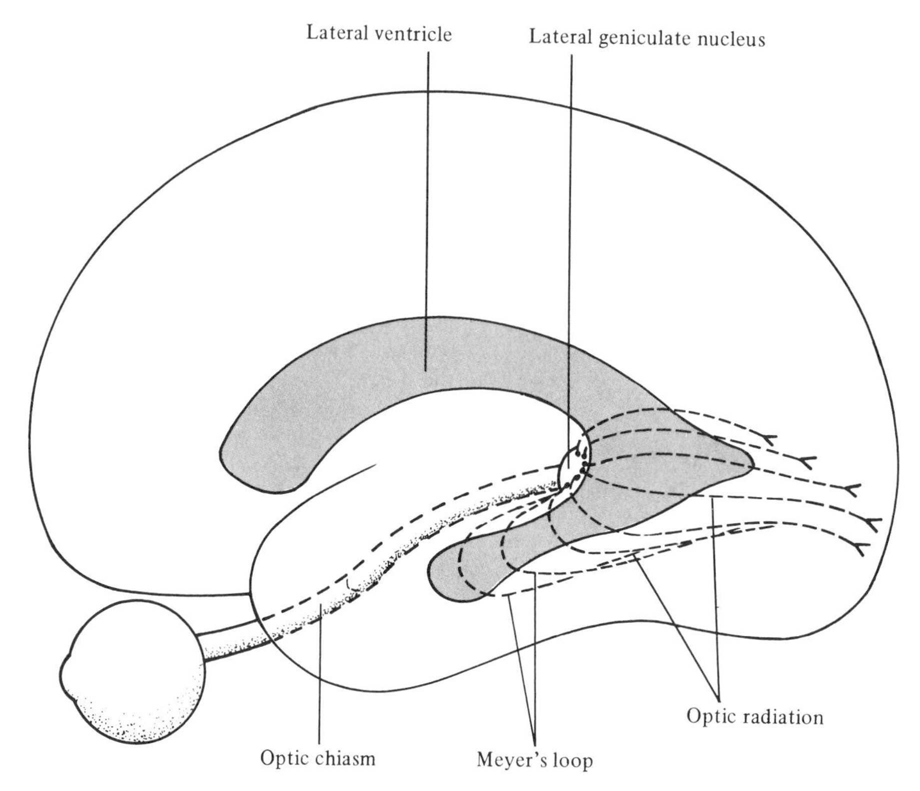

Fig. 13–8. Optic Radiation
Diagram illustrating the sweep of the geniculocalcarine tract on the outside of the lateral ventricle. The ventral part of the geniculocalcarine tract, Meyer's loop, makes a detour into the temporal lobe before it proceeds in a caudal direction to the calcarine sulcus. Note how the visual pathways extend from the eyeballs to the occipital lobes in the back of the head. (Modified from Sanford, H. S. and H. L. Bair, 1939. Visual disturbances associated with tumors of the temporal lobe. Arch. Neurol. Psychiatry (Chicago) 42:21–43.)

in large part to Brodmann's area 18. The main parts of V2 are located on the medial surface of the occipital lobe, where the part representing the upper retina (lower visual field) occupies part of the cuneus, whereas the part representing the lower retina (upper visual field) is located in the lingual gyrus (see Fig. 4–12). V4, and perhaps V5, are included in area 19, which apparently contains additional but little known visual areas. Whereas the location of V5 has been identified in the human brain (Fig. 13–10 and vignette), the location of V4 is more problematic, although there is some indication that it is located in the inferior occipitotemporal area in the region of the lingual or fusiform gyrus.

More than 20 extrastriate or prestriate visual areas have been identified in the monkey, and it is reasonable to expect that an equally large number of visual areas will eventually be identified also in the human. Both V1 and V2 are retinotopically organized, and different sets of neurons in these two visual areas are activated by different kinds of visual stimuli. The visual areas beyond V1 and V2 are associated with different visual functions. For instance, V3 is closely associated with form, V4 with color, and V5 with motion. Although most areas with predominantly visual functions are located in the occipital lobe, several regions with visual functions are located rostral to the occipital lobe in both the parietal and the temporal lobes. These areas are labeled remote association areas in Fig. 13–10, which represents a tentative map of the visual areas obtained in PET studies.

Topography of the Visual Pathway

The visual pathway from the retina to the primary visual cortex is highly organized in the sense that there is a point-to-point localization of each part of the retina throughout the system. Thus, the striate cortex is characterized by a precise retinotopic organization. However, the topographic projection of the retina is nonlinear in the sense that the central part of the retina, which is the region of highest discrimination, is represented by a relatively much larger cortical area than the peripheral parts of the retina (Fig. 13–9C). This reflects an important principle of cortical organization. The amount of cortical tissue devoted to a specific region of the body, including the retina, is related

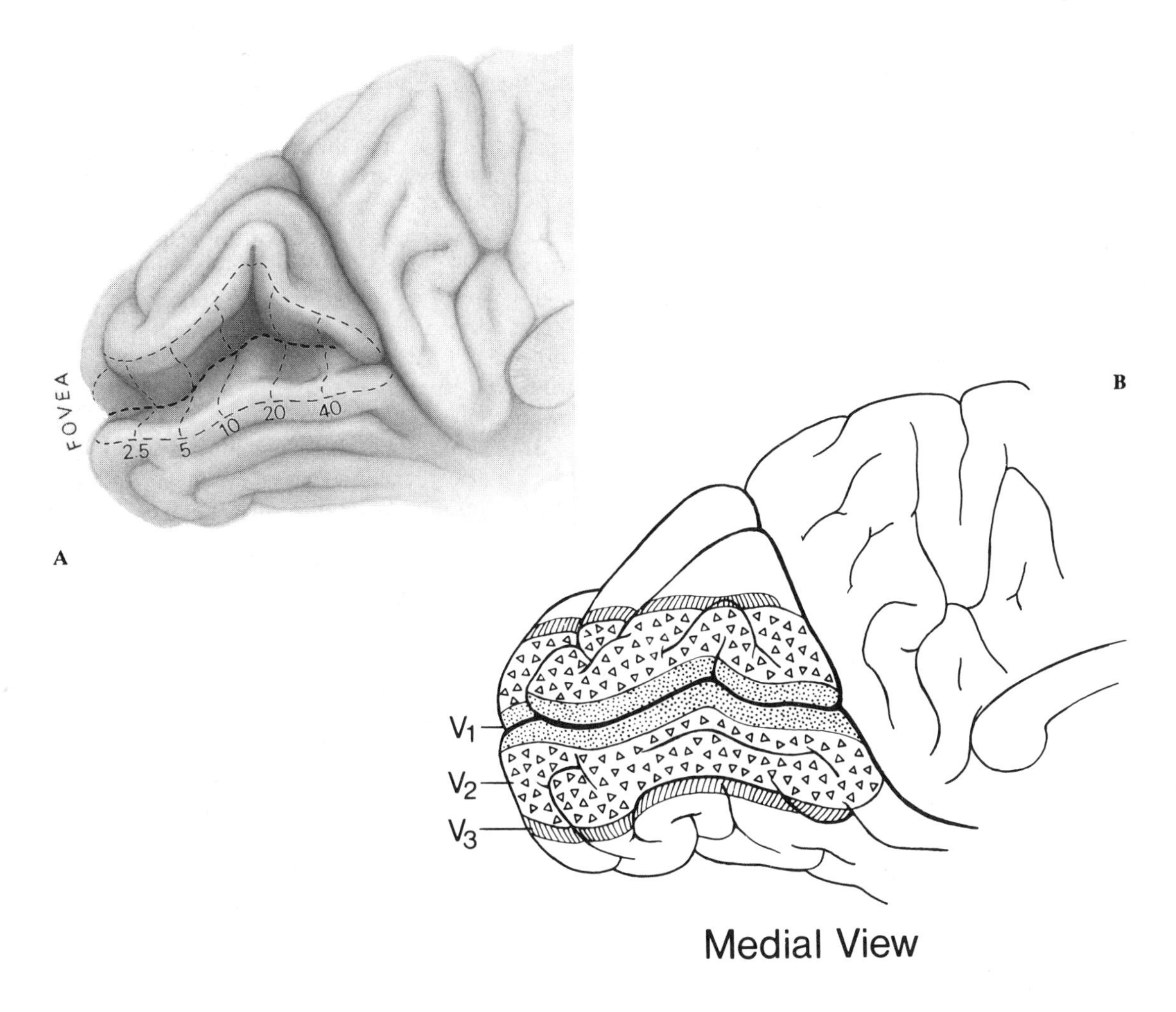

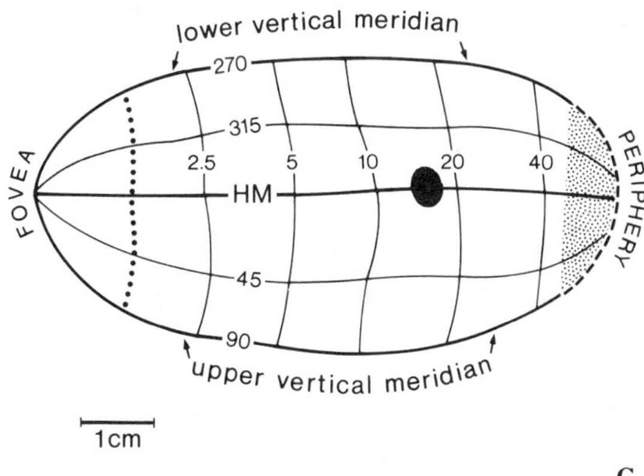

Fig. 13–9. Striate Cortex

A. Medial view of the left occipital lobe with the calcarine sulcus opened to expose the striate cortex (V1), most of which is buried within the sulcus. There are considerable variations among individuals in regard to the exact size and location of the striate cortex. All of the striate cortex is located on the medial surface of the brain in many individuals, whereas in others the striate cortex wraps around the occipital pole to extend for a short distance on the lateral surface.

B. Schematic diagram to show the arrangement of striate cortex (V_1) and neighboring visual areas (V_2 and V_3) on the medial surface of the brain.

C. Projection of the right visual hemifield on the left visual cortex by transposing the map illustrated in **A** onto a flat surface. The visual cortex has the shape of an ellipse, with the fovea centralis in this case represented on the lateral brain surface (the row of *dots* indicates where the striate cortex folds around the occipital pole). The *black oval* shows the location of the blind spot.

(**A** and **C**, from Horton, J. C. and W. F. Hoyt, 1991. Arch. Ophthalmol. 109:816–824. With permission from American Medical Association, Chicago.)

(**B**, from Horton, J. C. and W. F. Hoyt, 1991. Brain 114:1703–1718. With permission of Oxford University Press, New York.)

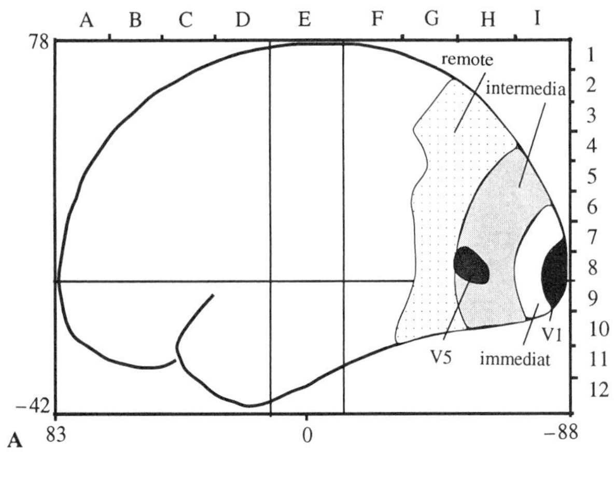

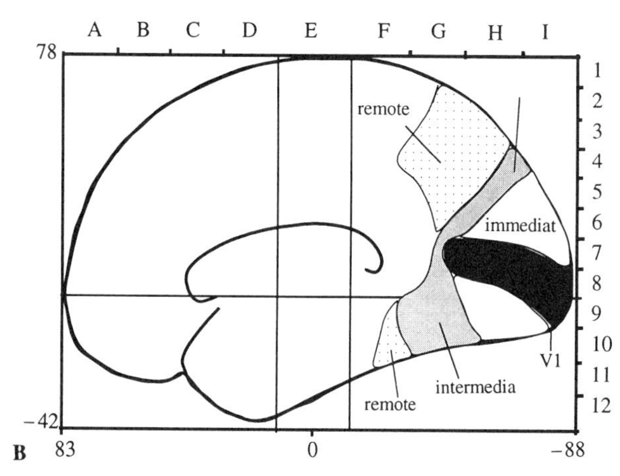

Fig. 13–10. Tentative Delineation of Visual Cortical Areas with PET

Regional cerebral blood flow was measured with PET in normal volunteers exposed to different visual patterns in order to determine the visual cortical areas on the lateral (**A**) and medial (**B**) surfaces of the cerebral hemisphere. (Courtesy of Dr. Per Roland, from Roland, P., 1993. Brain Activation. Wiley-Liss, New York. With permission of the author and the publisher.)

to the functional importance of the region rather than to its size.

Figure 13–9C illustrates some additional important points. First, the macula projects to the posterior part of the striate cortex, whereas the peripheral parts of the retina project to its anterior part. Second, the upper part of the visual field is represented below the calcarine sulcus, and the lower part above the sulcus.

Superior Colliculus and Pulvinar Thalami

A significant number of fibers in the optic tract terminate in the *superior colliculus*, especially in its superficial layers. Some of these fibers are collaterals of retinogeniculate axons, whereas others are dedicated solely to the retinotectal pathway. The superior colliculus also receives visual input both from the striate cortex and from extrastriate visual areas, and it is closely related not only to the LGN, but also to the *pulvinar*, another important visual thalamic structure. The functional significance of the massive visual input to the superficial parts of the superior colliculus is not well understood. However, as for the cortical cells projecting to the colliculus, it has been shown that they are highly directionally selective, and consequently can provide the colliculus with information about the direction of movement of objects in space. This would be appropriate for the collicular role in tracking objects in space.

Although the superior colliculus is widely known as a reflex center for orienting eye, head, and neck movements in response to visual stimuli (see "Brain Stem Centers for Control of Eye Movements," Chapter 11), it is questionable to what extent the superficial layers in the superior colliculus are involved in such functions. Orienting movements and postural adjustments appear to be more closely related to the deep part of the superior colliculus, which is multisensory in the sense that it is closely related not only to the visual system but also to the somatosensory and auditory systems. As one might expect from a structure devoted to sensorimotor integration, the deep part of the superior colliculus also has a significant motor representation, with prominent projections to brain stem premotor and motor structures including the motor eye nuclei. The frontal eye fields and the basal ganglia including the substantia nigra, pars reticulata, are also involved in the mechanisms that produce visually guided eye movements (see Forbrain Structures involved in the Control of Eye Movements, Chapter 11).

The pulvinar, which is the largest nuclear complex in the thalamus (Fig. 22–6), is an integral part of the visual system. Unlike the LGN, the pulvinar does not receive a significant input from the retina, but a substantial projection reaches it from the superficial parts of the superior colliculus, and it is reciprocally connected primarily to a number of extrastriate visual cortical areas in the occipital lobe and in the parietal and temporal territories related to visual functions. There is little doubt that the pulvinar has an important role to play in visual functions, but the specificity of this function has yet to be elucidated. A role in visual attention or salience has recently been suggested.

The Pretectum and the Pupillary Light Reflex

The size of the pupil is determined by the activity in the parasympathetic innervation of the pupillary sphincter, and to a lesser degree by the sympathetic innervation of the vessels of the iris and the dilator pupillae. The size variations are induced by different factors, e.g., by change in illumination or during accommodation, and by stress.

The well-known phenomenon of pupillary constriction in response to bright light is called the *direct light reflex*. The afferent link in the reflex arc is represented by fibers that reach the pretectal region through the optic nerve and the optic tract (Fig. 13–5). Neurons in the pretectal region connect with the parasympathetic Edinger–Westphal nucleus, which innervates the pupillary sphincter via the ciliary ganglion. Since fibers cross over from one side to the other both in the optic chiasm and in the pretectal area, the unexposed pupil constricts as well, which is referred to as a *crossed* or *consensual light reflex*. The evaluation of the pupillary light reflex is an important part of the neurologic examination.

Functional Considerations

Receptive Fields

The *receptive field* of a cell in the visual system is the area of the retina in which a particular stimulus can influence the electrical activity of the cell. The size of the receptive field varies with the location in the retina. The smallest receptive fields are found in the region of the fovea centralis, where the visual acuity is highest.

As discovered in pioneering physiologic experiments by the American Stephen Kuffler and the Englishman Horace Barlow, the retinal ganglion cells have fairly simple receptive fields, which are organized in a concentric manner with a circular central area surrounded by a ring-shaped outer zone (Fig. 13–11). Illumination of the central area results in either excitation (ON cell) or inhibition (OFF cell), whereas illumination of the surrounding zone (surround stimulation) has the opposite effect. The most effective stimulus for many of the retinal ganglion cells, therefore, is a circular spot of light of a particular size in a specific part of the visual field (the area within which an object is seen by the eye in a fixed position). An evenly illuminated area has very little effect on the ganglion cells, since simultaneous illumination of both center and surround tends to cancel each other out. The primary role of the center-surround arrangement is to signal the presence of contrast in the visual environment rather than the absolute amount of light reflected by objects in the environment. Together the ON and OFF center cells play an important role in form vision by indicating where contrasting boundaries are located.

The cells of the LGN have receptive fields similar to those of the retinal ganglion cells, i.e., with the center-surround configuration, and like the retinal ganglion cells they respond much better to regions of illumination change, i.e., contrast, than to areas with uniform illumination. Since the receptive fields of the LGN cells, like those of the ganglion cells, are circular symmetric, orientation becomes an important feature only when the information reaches the cerebral cortex. Apparently, the ON cells and the OFF cells and their central projections form systems or channels that in large part remain segregated until they reach the striate cortex, where the two systems converge on single cells.

Functional Segregation in the Visual Cortex

The physiologic studies of the retinal ganglion cells and their receptive fields in the middle of this century set the stage for the widely acclaimed investigations of the visual cortex by Drs. David Hubel and Torsten Wiesel in the 1960s and 1970s.

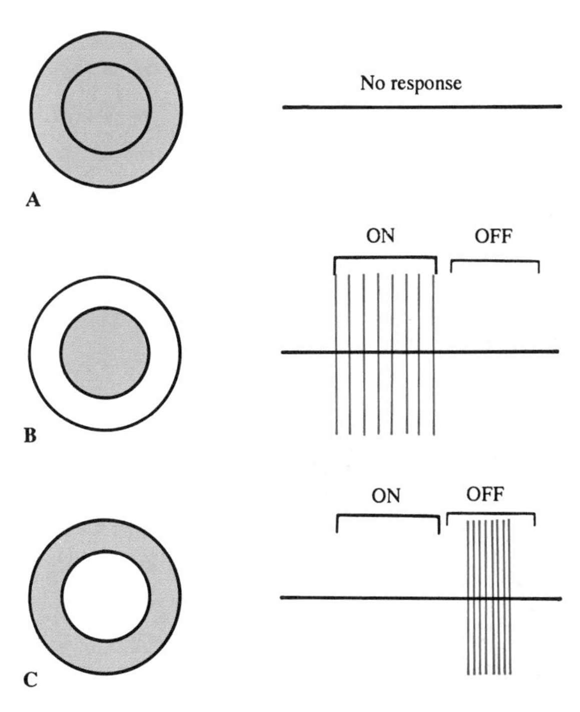

Fig. 13–11. Ganglion Cell Receptive Fields
Ganglion cells have circular receptive fields with a center and an antagonistic surround. The figure demonstrates the responses of an ON cell stimulated by diffuse light (**A**) and by a spot of light in the center of the receptive field (**B**). Light falling in the surround inhibits the cell, which responds only when the light is switched OFF (**C**). (From Zeki, S., 1993. A Vision of the Brain. With permission of Blackwell Scientific Publications, Oxford.)

Fig. 13–12. Ocular "Dominance" Columns
A. Dark-field autoradiograph of a cross-section through the primary visual cortex (area 17) of macaque monkey following an injection of tritiated proline into one eye. The label was taken up by cell bodies in the retina and transported to the LGN. Following transneuronal transport into LGN cells, the label reached the visual cortex via the optic radiation. The "ocular dominance" columns are confirmed by the presence of periodic bright patches. The banded projections are limited to layer IV, where most of the geniculocalcarine fibers terminate. (Micrograph courtesy of Dr. Torsten Wiesel. From Hubel, D. H. and T. N. Wiesel. Brain Mechanisms of Vision. Scientific American, September 1979, Vol. 241, No. 3.)
B. "Ocular dominance" columns demonstrated with the 2-deoxyglucose method in a rhesus monkey with the right eye occluded. The "ocular dominance" columns, i.e., the alternate dark and light striations, extend throughout the entire thickness of the cortex rather than being limited to layer IV, as in **A**. This is related to the fact that the 2-deoxyglucose method reveals the entire column of cells activated by one eye regardless of the number of synapses involved. (Micrograph courtesy of Dr. Louis Sokoloff. From Sokoloff. Localization of functional activity in the central nervous system by measurement of glucose utilization with radioactive deoxyglucose. J. Cereb. Blood Flow Metab. 1981, Vol. 1, 7–36. With permission of Raven Press, New York.)
C. The ocular dominance columns demonstrated by the autoradiographic method as in **A**, but the histologic section was cut parallel to the cortical surface instead of perpendicular as in **A**. (Courtesy of Dr. D. Hubel, 1977. From Proc. R. Soc. Lond. B 198:1–59. With permission of the Royal Society, London.)

Hubel and Wiesel discovered, fortuitously,[3] that cells in the primary visual cortex (area 17) respond to lines oriented in a specific direction. In other words, cells in the visual cortex are orientation-selective, and Hubel and Wiesel also demonstrated that different cells have different orientation preferences (see below). These dramatic findings had an energizing effect, and the visual cortex quickly became a favorite research object. It soon became evident that several other features in the visual image had the ability to activate a specific set of cortical cells. For instance, some of the cells in V1 are wavelength- or color-selective. Other cells are direction-selective, i.e., they respond to lines moving in a specific direction. Many cells are characterized by *binocular convergence*, i.e., they respond to stimulation of both eyes rather than to stimulation of one or the other eye, which is the case for the LGN neurons. Some cells even respond to the horizontal disparity between the input from the two eyes.

The above-mentioned functions of the primary visual cortex, e.g., orientation selectivity, direction selectivity, disparity selectivity, and binocular convergence, most likely represent early stages in the brain's analysis of form and movement, and perception of depths (stereoscopic vision). These discoveries were indeed revolutionary in the sense that they dispelled the old notion of a visual image being imprinted upon the retina and then transmitted to the striate cortex. Instead, it turns out that the striate cortex is subdivided into several different maps or zones, each devoted to a specific type of visual information. As we shall see below, the functional–anatomic segregation within the visual pathways begins already in the retina, and it continues beyond the striate cortex, from which several parallel pathways project to different extrastriate visual cortical areas associated with different submodalities of vision, e.g., color, motion, and form.

Cortical Columns and Blobs

The physiologic properties of the visual cortex are reflected in its architecture, in the sense that cells with common functional properties tend to be grouped together in repeating units referred to as *columns*. This important discovery was also made by Hubel and Wiesel, who identified both the *ocular dominance columns* and the *orientation columns*.

The ocular dominance columns are related to the eye preference of the cortical neurons. We mentioned earlier that the impulses from the two eyes are segregated in their passage through the central visual pathways to the striate cortex. This segregation of input from the two eyes is still evident in layer 4 of the striate cortex (Fig. 13–12A), where most of the fibers in the geniculocalcarine tract terminate (see also Fig. 22–5). The label in this autoradiographic darkfield photomicrograph is tritiated proline, which was injected into one eye of an adult macaque monkey and transported via the ganglion cell axons to the LGN. Following transneuronal transport of the label into LGN cells, the label was then transported via the optic radiation, i.e., LGN axons, to the striate cortex, where the bright patches represent labeled

3. Having discovered, by chance, that a crack in a slide had a powerfully stimulating effect on a cortical cell, Hubel and Wiesel settled on lines along a specific axis as the appropriate stimulus in their studies of the response properties of visual cortical cells. Chance, incidentally, has been a significant part of many important scientific discoveries. Alexander Fleming's fortuitous discovery of penicillin, i.e., through the accidental contamination of a culture dish with the mold penicillium, is a well-known example. However, having drawn attention to accidental findings in science, one should remember Louis Pasteur's famous quote: "Where observation is concerned, chance favors only the prepared mind."

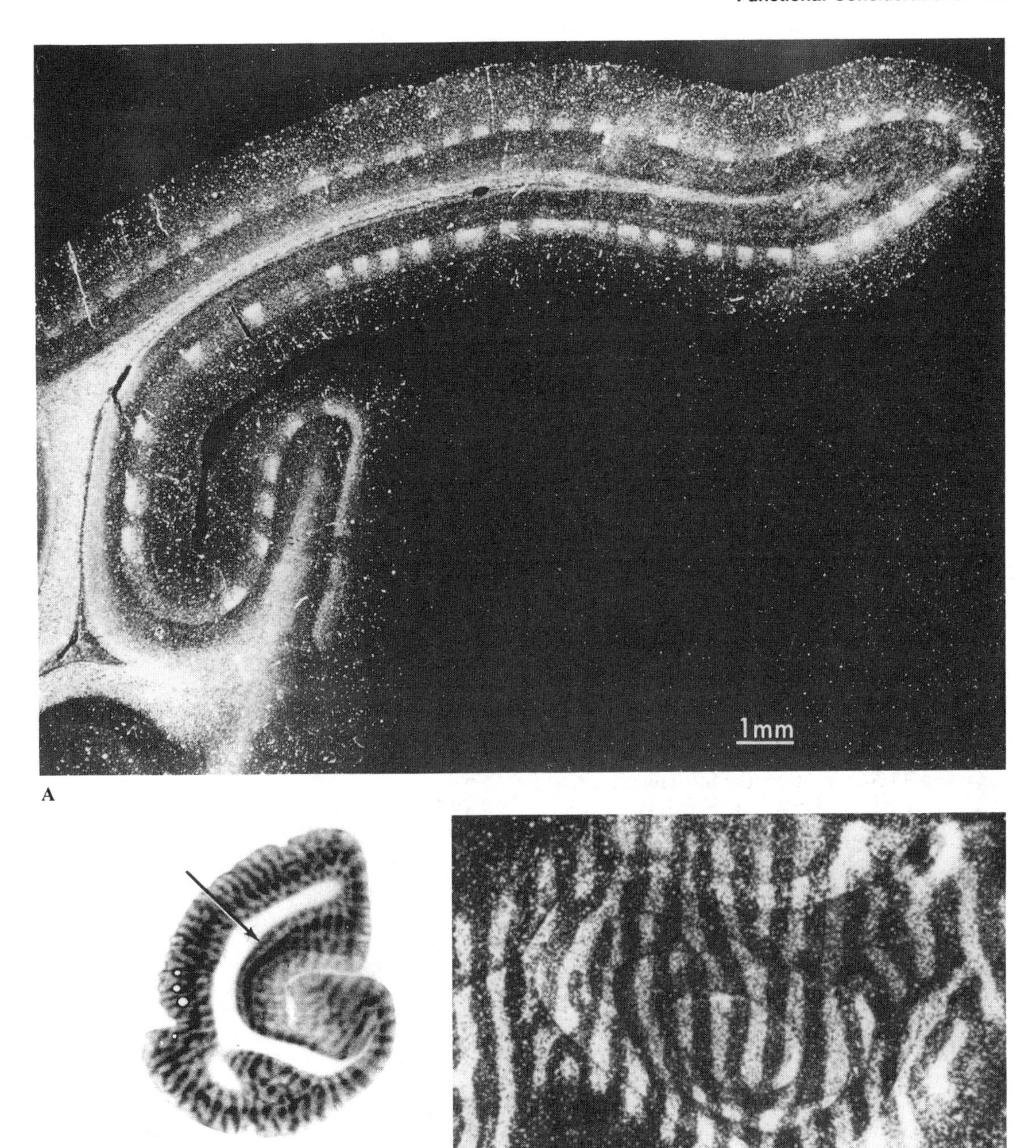
1mm
A
5.0mm
B
1mm
C

terminal arborizations in layer 4. The labeled patches are separated by unlabeled areas, which represent the ocular dominance columns of the other, noninjected eye.

In Fig. 13–12B, the ocular dominance columns are visualized by the aid of the 2-deoxyglucose method, which demonstrates the cell assemblies that are active following monocular visual stimulation (see Transneuronal Tracer Methods in Chapter 7). In this case, the ocular dominance columns (the alternating dark and light stripes) extend throughout the thickness of the cortex, which indicates that monocular stimulation tends to activate vertically oriented columns or slabs of cells. The main lines of communication within the cortex, therefore, seem to be in a direction perpendicular to the surface of the cortex, i.e., within the vertically oriented columns. On the other hand, there are also significant interactions between neighboring columns, which in part explains the fact that many cortical cells are binocular, i.e., they are influenced by both eyes.

If a section is cut tangentially, i.e., parallel to the surface of the cortex (Fig. 13–12C), the ocular dominance columns appear as more or less parallel bands. This demonstrates that the columns are in reality slabs of tissue.

Orientation columns were the first columns to be described in the visual cortex by Hubel and Wiesel, who discovered that a sequence of cells responded to a line with a specific orientation as they inserted a microelectrode perpendicular to the surface of the visual cortex. A complete set of orientation columns, representing 360°, forms an orientation hypercolumn, which covers about 750 μm in the primate. This is roughly equivalent to the dimension of a pair of ocular dominance columns.

Cytochrome oxidase-stained blobs represent aggregations of cells with apparent chromatic properties. The fact that they show a strong reaction with the cytochrome oxidase stain indicates that they are metabolically more active than their neighbors. The blobs are most pronounced in layers 2 and 3. A tangential cut through the superficial layers of the striate cortex, therefore, evokes the picture of regularly placed patches (Fig. 13–13).

Since the majority of blob cells are color-coded and do not show orientation selectivity, the blob system is apparently of special significance for color vision. However, there is uncertainty as to how the blobs are involved in the perception of color. Most of the cells in the regions between the blobs, i.e., the interblob cells, are orientation-selective but are not color-coded. The interblob system, therefore, appears to be concerned with form rather than color (Fig. 13–14).

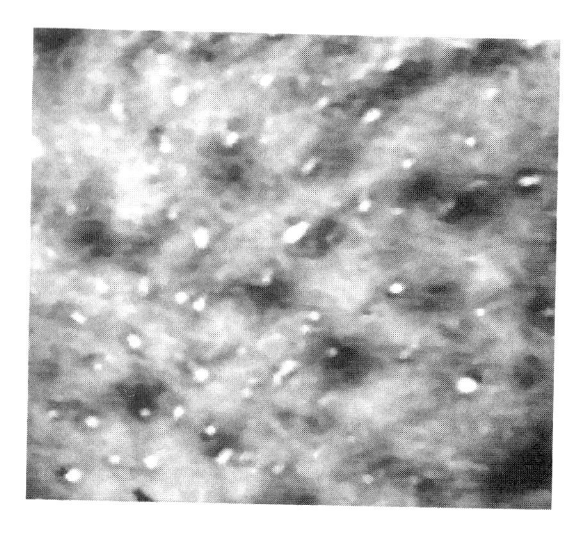

Fig. 13–13. Color Blobs
A cytochrome oxidase-stained tangential section of striate cortex (V1) in man showing the blobs in layers 2 and 3 in the region of the fovea centralis. (Courtesy of Dr. Roger Tootell, Department of Neurobiology, Harvard Medical School.)

The Magnocellular and Parvocellular Systems

One of the most exciting developments in visual research is the realization that various aspects of the visual image are analyzed by different subdivisions of the visual pathways. At least four parallel pathways have been identified so far for the processing of color, form, and movement. As indicated above, the functional-anatomic segregation of the visual pathways begins already in the retina (Fig. 13–14), in the sense that the axons of the two types of ganglion cells, the M cells (blue circles) and the P cells (red circles) project to different layers in the LGN. The M cells project to the two ventrally located magnocellular layers and the P cells to the four dorsal parvocellular laminae. As indicated in Fig. 13–14, the segregation continues in the optic radiation into the striate cortex and even beyond into the extrastriate visual areas (see drawing in lower left-hand corner in Fig. 13–14).

The *magnocellular pathway* (blue in Fig. 13–14) projects to layer 4Ca in the striate cortex, and from there to layer 4B. This layer contains both orientation-and direction-selective cells, which after further processing give rise to two different channels. The M channel, related to the orientation-selective

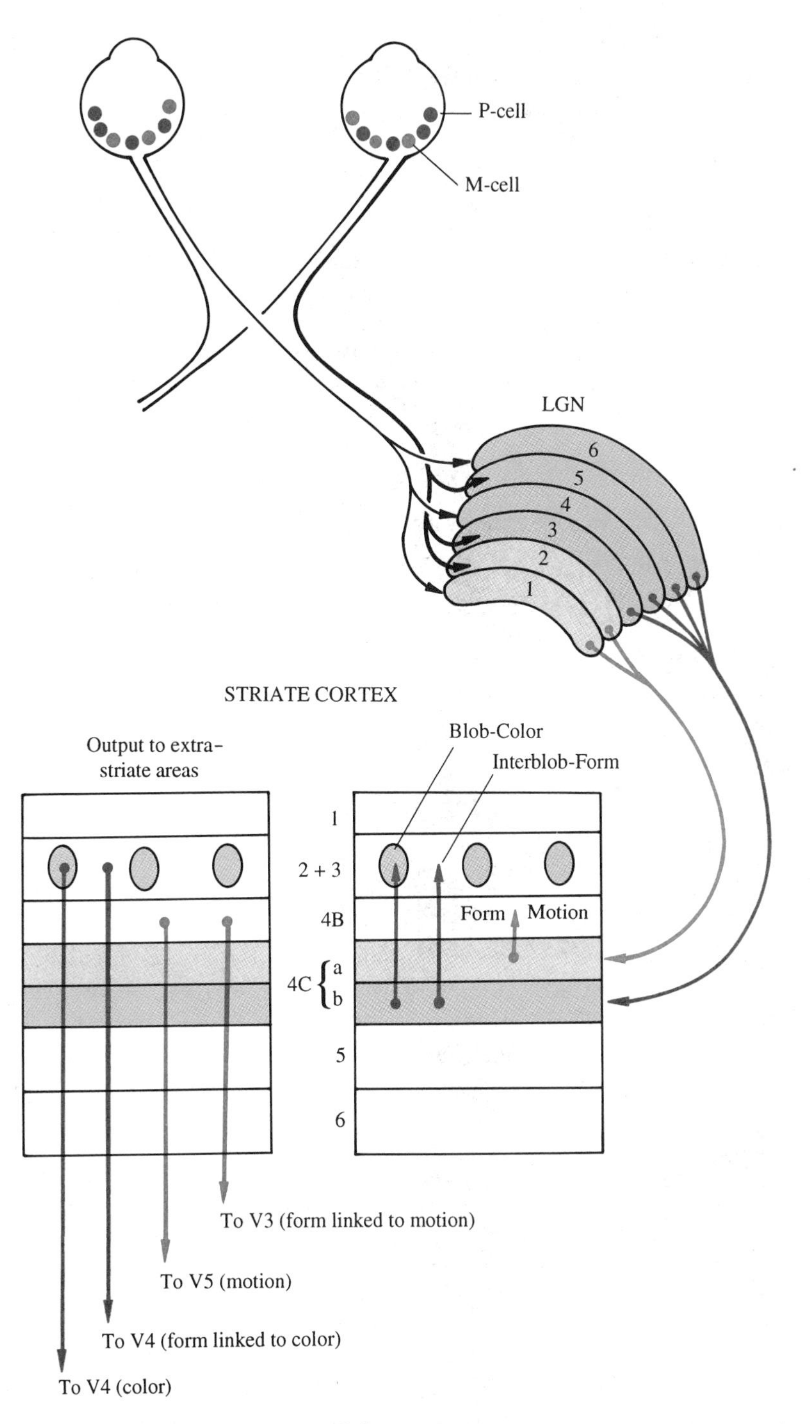

Fig. 13–14. The Magnocellular and Parvocellular Systems

Highly schematic drawing showing the parallel channels in the visual system. The M-type ganglion cells terminate in the two ventrally located magnocellular layers of the lateral geniculate nucleus (LGN), whereas the P cells terminate in the four dorsal parvocellular laminae. The two pathways, indicated by *blue* (magnocellular pathway) and *red* (parvocellular pathway), continue beyond the LGN to the striate cortex where the two original pathways give rise to four parallel channels to the extrastriate visual areas. But the segregation is not complete. There are significant interactions between the functional compartments at all stages along the visual pathways. It should also be noted that the four output channels from the striate cortex proceed directly to the extrastriate areas (as indicated in the lower left-hand corner of the figure) as well as via V2.
(Modified figure based on original drawings in Livingstone, M. and Hubel, D., 1988. Science 240:740–749; Zeki, S., 1993. A Vision of the Brain. Blackwell Scientific Publications, Cambridge, Massachusetts.)

cells, which are associated with form vision, projects to V3, both directly and via V2. The M-derived form channel is likely to be linked to motion, i.e., dynamic form. The M channel associated with the direction-selective cells, i.e., the motion pathway, reaches the motion area of the extrastriate cortex, area V5, either directly or via V2 (Brodmann's area 18). V5 was the first extrastriate visual area to be positively associated with a specific visual function on the basis of experimental physiologic studies in the monkey. The English neuroscientist Semir Zeki discovered that most of the cells in V5 are direction-selective. Dr. Zeki and his collaborators have also identified a visual motion area in the human brain. Aided by a combination of PET and MRI, they have been able to pinpoint its location to an architectonically distinct area in the rostral part of the occipital lobe where it borders the temporal lobe (Fig. 13–14; see also the vignette for this chapter)[4]. In contrast to the surrounding areas, V5 is already myelinated at birth. One of the major projections from V5 reaches the posterior parietal cortex, which is of special importance for visuospatial perception and visuomotor activities.

The *parvocellular pathway* (red in Fig. 13–14) reaches layer 4Cb, which in turn projects to layers 2 and 3, where the blob and interblob cells are located. In the further processing within the P system, therefore, at least two different parvocellular channels can be recognized: the parvocellular blob system for the processing of color, and the parvocellular interblob system for the detection of form. These two channels proceed to area V2 and from there to V4, which is especially concerned with form in association with color. As indicated in Clinical Notes, clinical observations are consistent with this type of segregation of visual functions in the extrastriate cortical areas.

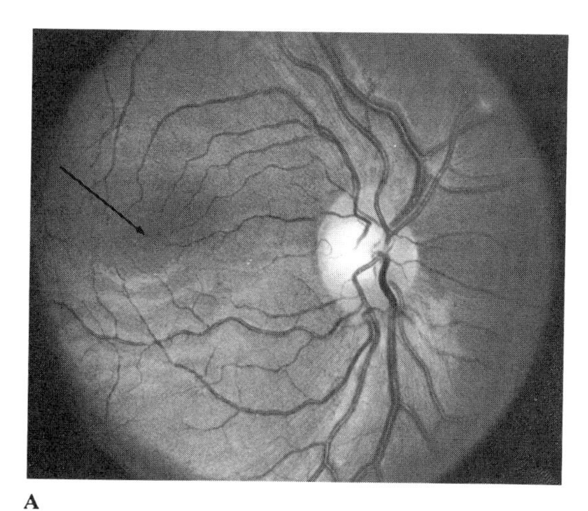

A

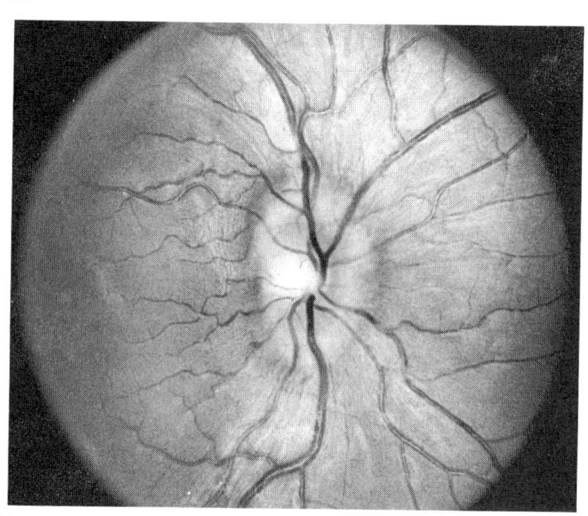

B

Fig. 13–15. Fundus Oculi
A. Normal fundus. Note the sharp border of the optic disc. The *arrow* points to the fovea centralis.
B. Papilledema with congestion of the disc and blurring of its margins. Note also that the optic cup (the central whitish area) is reduced in size compared with the normal optic cup. (The photographs were kindly provided by Dr. S. Newman.)

Integration in the Visual System: A Diffuse Multistage Process?

With segregation of function being such a prominent feature in the visual cortex, including the extrastriate areas, one wonders how the brain solves the problem of bringing together the different visual submodalities into an integrated visual image. Apparently, there is no single area to which the various extrastriate visual areas project, and where impulses from the various channels could come together to produce a unified visual percept. However, there is a significant degree of interaction among these functional compartments at all stages along the visual pathways. For instance, there are local interactions among the compartments within individual visual areas, and there are also direct connections among various extrastriate visual areas related to the different channels. Therefore, even if parallel processing is an important feature in the visual system, there is opportunity for considerable

4. Although the pictures obtained by modern imaging methods are truly magnificent, the results obtained in PET studies should be viewed with some caution. Technical difficulties, methodologic differences, and anatomic variations conspire against the scientists looking for clear-cut results. For instance, although the vignette shown at the beginning of this chapter indicates that there is one specific motion area in the cerebral cortex, other studies show multiple areas (e.g., Gulyás et al., 1993).

interaction between the channels, thereby providing the opportunity for convergence of different submodalities of vision.

Another crucially important aspect in the context of integration is the existence of reciprocal or feedback connections. In other words, an area can modify the activity in the area from which it receives its input. Some of the re-entrant circuits in the visual channels have been studied in great detail. For instance, it appears that the re-entrant circuits that reach the striate cortex from some of the specialized extrastriate areas may influence not only the cells that project to the specific region giving rise to the re-entrant circuit, but also cells related to different visual submodalities. This arrangement would provide another opportunity for convergence of signals subserving different submodalities.

The picture that emerges is highly complex, and it has been suggested that integration within the visual system is a multistage process (Zeki, 1993) involving the simultaneous activation of a number of visual channels and visual areas including both the specialized extrastriate areas and the striate cortex. One should also keep in mind that there are significant reciprocal connections between the extrastriate visual areas and the thalamus, whose functions, although little known, might be important in this context.

Clinical Notes

Most visual disturbances are caused by refractive errors or diseases of the eye, and an ophthalmologic disorder, therefore, should be excluded before a neurologic deficit is considered.

Cataract and glaucoma are two of the most common non-neurologic causes of blindness. *Cataract* is a loss of the transparency of the lens or its capsule. Since the cornea is the main refracting medium in the eye, the lens can be surgically removed and the loss of refractive power can usually be compensated for by eyeglasses or by an artificial lens. *Glaucoma* is characterized by increased intraocular pressure caused by restricted re-entry of the aqueous humor into the bloodstream. Continuous high pressure in the eye fluid causes destruction of the optic nerve fibers, resulting in visual field defects, and it is therefore important to make a correct diagnosis at an early stage.

Ophthalmoscopy: The Fundus Oculi

The ophthalamoscopic examination of the fundus is an important part of the neurologic examination; indeed, every physician should be able to examine the fundus since many systemic diseases, e.g., diabetes mellitus, hypertension, and arteriosclerosis, produce characteristic changes in the retina.

The *normal fundus* is shown in Fig. 13–15A. The optic disc stands out sharply because of its pale appearance and well-defined margin. The retinal vessels radiate from the center of the disc and divide repeatedly as they spread over the fundus.

A disease in the optic nerve may cause *optic atrophy*, which is characterized by a very pale or even chalk-white disc. Changes in the retinal vessels are seen primarily in arteriosclerosis and hypertension. The picture of *hypertensive retinopathy*, which is especially dramatic, contains arterial constriction, obliteration of the veins at the arteriovenous crossings, and retinal exudates. In more advanced stages there may even be pronounced swelling of the disc, i.e., *papilledema*, which is indistinguishable from that seen in a patient with increased intracranial pressure.

Papilledema ("Choked Disc")

Swelling of the optic papilla (Fig. 13–15B) is one of the most important signs of increased intracranial pressure, which is a common sequela of expanding intracranial processes, including brain tumor, bleeding, and infection. Although increased intracranial pressure is not the only cause of papilledema, it is by far the most important. Pronounced venous congestion, bleeding, and exudates appear in later stages of papilledema.

The optic nerve, like the brain, is surrounded by both dura mater and pia-arachnoid, which are continuous with the meningeal spaces surrounding the brain. Increased CSF pressure in the skull, therefore, is immediately transmitted to CSF along the optic nerve, which interferes with venous outflow and probably also lymphatic drainage of the retina.

Lesions of Visual Pathways

Since the visual pathways extend all the way from the eyeball in front to the occipital lobe in the posterior part of the hemisphere, they can be interrupted by lesions in many parts of the brain. Further, the visual pathways are highly organized, and lesions in different parts of the visual pathways produce characteristic visual field defects, which make it possible in many cases to determine the location of the lesion. Countless disorders can affect the visual pathways, but some should be noted in particular. Pituitary tumors, meningiomas, and aneurysms of the carotid artery or the circle of Willis often affect the optic chiasm or nearby regions of the optic nerve or tract. Intracerebral tumors and bleedings can easily encroach on the optic radiation.

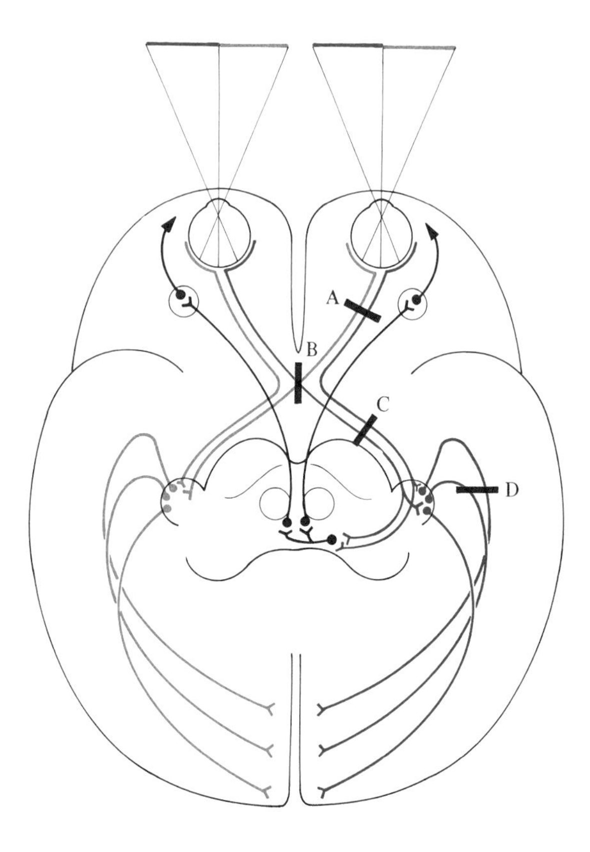

Visual field defects
L R

A. Monocular blindness

B. Bitemporal hemianopsia

C. Homonymous hemianopsia

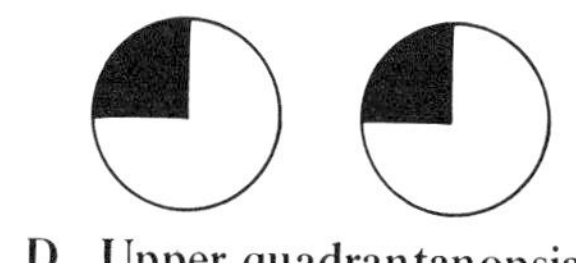

D. Upper quadrantanopsia

Fig. 13–16. Lesions of the Visual Pathways
Diagram illustrating some of the most typical lesions of the visual pathways. Whereas complete lesions at **A**, **B**, and **C** give rise to field defects as indicated by the figure, lesions of the optic radiation produce homonymous field defects that are often incomplete and asymmetric. A homonymous superior quadrantanopsia may occur in temporal lobe lesions involving Meyer's loop (**D**).

Since lesions interrupting various parts of the visual pathways cause specific visual field defects, examination of the visual fields constitutes a very important part of the neurologic examination. The fields of vision can be accurately delineated by *perimetry*, but they can be estimated in a matter of seconds by the *confrontation method*. With one eye covered, the patient is instructed to fix his gaze on one of the examiner's eyes and report when he first sees a small object, e.g., a pencil, which the examiner moves into the patient's field of vision. By bringing the pencil into the visual field from many directions it is often possible to detect a significant visual field defect. Partial defects within the field of vision are called *scotomas*. Visual defects, further, are always named according to the visual field loss and not according to the corresponding part of the nonfunctional retina.

Lesions of the Optic Nerve

Prechiasmal lesions cause monocular abnormalities, which can vary from *complete blindness* (lesion A in Fig. 13–16) to various types of restricted areas of blindness, i.e., *scotomas*. If the lesion of the optic nerve is complete, the afferent fibers for the pupillary light reflex will be interrupted, and the pupil will not react to light stimulation. However, the pupillary response can be elicited by stimulating the opposite eye (consensual reflex).

Lesions of the Optic Chiasm

A chiasmal lesion, which interrupts the midline decussating fibers, produces a typical *bitemporal hemianopsia*[5] (lesion B in Fig. 13–16). If such a

5. Hemianopsia (hemianopia) means blindness in one-half of the visual field of one or both eyes.

lesion is caused by a pituitary tumor, the defect usually starts as an *upper bitemporal quadrantanopsia*, since the tumor first exerts pressure on the inferior part of the chiasm, where the fibers from the lower regions of the nasal halves of the two retinae cross. An aneurysm in the anterior cerebral or the anterior communicating artery is likely to produce a *lower bitemporal quadrantanopsia.*

Lesions in the Optic Tract and Geniculocalcarine Tract

Like the chiasmal lesion, the postchiasmal lesion produces bilateral field defects, but the typical symptom is a *homonymous* defect, i.e., a defect that affects the corresponding halves of the two visual fields. A complete lesion of the optic tract results in a *contralateral homonymous hemianopsia* (lesion C in Fig. 13–16). The same defect occurs if the whole geniculocalcarine tract or the striate cortex is destroyed. Lesions of the geniculocalcarine tract, however, are more apt to be partial due to the widespread extent of the geniculocalcarine fiber system and to produce *incomplete homonymous hemianopsia.* For instance, a temporal lobe lesion may interrupt Meyer's loop, the inferior portion of the optic radiation, and thereby produce a *contralateral upper quadrantanopsia* (lesion D in Fig. 13–16). Such lesions are hardly ever as regular as depicted in D.

Since the retina, and hence the visual field, is mapped in a topographic fashion in the striate cortex (with the upper contralateral visual field represented in the lower bank of the calcarine sulcus, and the lower contralateral visual field represented in the upper bank of the calcarine sulcus), the location of a focal lesion in the striate cortex can be correctly determined on the basis of the visual field defect.

Vision is sometimes preserved in the area of the macula lutea, i.e., *macular sparing.* This phenomenon is usually seen in vascular lesions of the occipital lobe. It is usually explained on the basis of overlapping blood supply (from middle and posterior cerebral arteries) in the occipital pole, where the macula is represented.

Lesions in Extrastriate Visual Areas

Focal brain lesions can be localized with great accuracy using modern imaging methods, and it has therefore been possible to correlate rather closely the location of cortical lesions with specific dysfunction. Since the occipital lobe and large regions in the parietal and temporal lobe are related to visual functions, defects in the visual domain without significant involvement of the striate cortex can occur with a variety of cortical lesions. Often, cortical lesions are large, resulting in a number of different deficits that may be difficult to sort out, especially since such patients are often in poor condition.

Visual agnosias[6], or visual recognition defects, are relatively common following large cortical lesions, often involving the lateral and ventral parts of the occipitotemporal visual regions. They are in essence a breakdown of the integrative process; for instance, the patient cannot recognize an object or indicate what it is used for (visual object agnosia), in spite of the fact that the patient's mind is clear and vision intact, or so it appears.

Visual agnosia is a complicated syndrome, which appears in many variations, and which may be accompanied by scotomas, achromatopsia, and prosopagnosia (see below). Considering the previous discussion of integration as a multistage process involving both the striate cortex and the extrastriate visual areas, one would hardly expect a clear-cut distinction between two separate cortical domains, one for "seeing" and one for "understanding." Clinical experience would seem to support this notion; total loss of form vision, for instance, is hardly ever encountered in clinical practice.

On the other hand, there are several examples in the clinical literature where localized cortical lesions have produced specific forms of visual syndromes including achromatopsia (inability to recognize colors), prosopagnosia (inability to recognize familiar faces), and akinetopsia (inability to see motion). Such disorders testify to the remarkable degree of functional specialization in the visual cortex. Alexia (inability to read), is discussed in Chapter 22 (under "Language Disorders").

Achromatopsia, complete color blindness[7] is often the result of a lesion in the inferior occipitotemporal cortex in the region of the lingual or fusiform gyri (V4). The loss of color vision is related to the whole visual field (bilateral lesions) or to a hemifield (unilateral lesions), and the patient experiences only black and white, or a "wash-out" of the colors.

Prosopagnosia, face agnosia, is a rare form of

6. The Greek word for knowledge is *gnosis*, and *gnosia* refers to the ability to recognize the form and nature of persons and things. This faculty depends on a comparison of present sensory input with past experience. The lack of understanding the significance of sensory stimuli is referred to as *agnosia*.

7. A much more common form of "color-blindness" results from a lack of retinal photoreceptors, usually those responsible for green or red color. This abnormality is related to the absence of the appropriate color genes in the X chromosomes. Since this represents a recessive trait, men are more commonly affected than women.

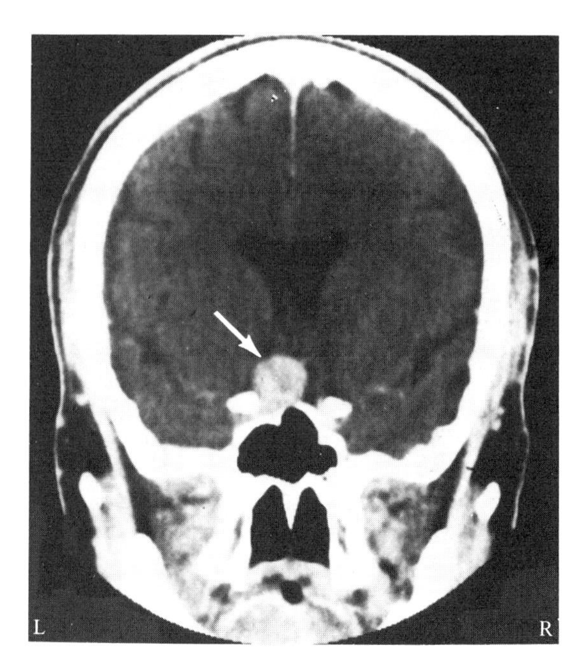

Fig. 13–17A. Internal Carotid Artery Aneurysm
The CT scan was kindly provided by Dr. V. Haughton.

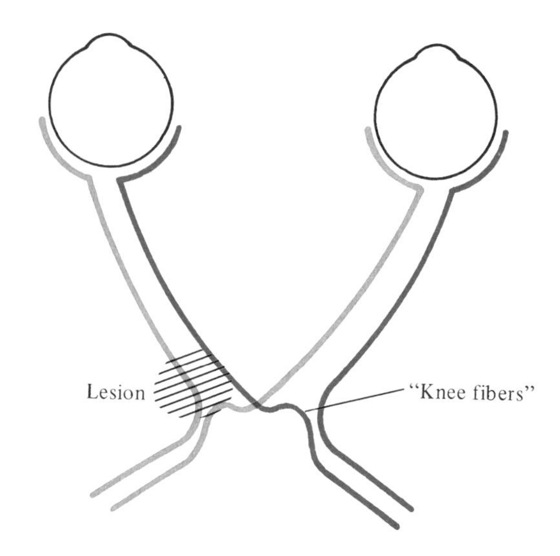

Fig. 13–17B. Supra-sellar Mass Lesion
Diagram of the arrangement of visual fibers in the chiasm. The ventrally located fibers crossing in the chiasm make a loop into the terminal portion of the opposite optic nerve. This explains why a lesion that primarily involves the optic nerve on one side may cause superior temporal quadrantanopsia in the opposite eye. Further, since the "knee fibers" come close to the lateral surface of the chiasm after the crossing, lateral pressure on the chiasm usually causes visual field defects of both eyes.

visual memory defect in which the patients are unable to recognize familiar faces, although they have not lost the concept of a face. In other words, they know a face when they see one, but they cannot recognize it as a particular face belonging to a specific individual. Prosopagnosia results from bilateral lesions, and it is usually associated with achromatopsia, since the areas responsible for face recognition and color vision are apparently located side-by-side in the inferior occipitotemporal region.

Akinetopsia has occasionally been described in patients with bilateral lesions involving the V5 region (see Vignette). This is a specific defect in visual motion perception; the patient cannot see objects in motion, although the appreciation of movement elicited by nonvisual stimuli is still intact.

Clinical Example

Internal Carotid Artery Aneurysm

A 49-year-old woman had been bothered by mild headache for half a year and she had lately started to complain about pain in and around her left eye. She sought medical attention when she experienced a gradual deterioration of vision in the left eye.

When seen by the neurologist, she had almost complete blindness in her left eye and a superior temporal quadrantanopsia in the right eye. The pupillary light reflex could not be elicited by light stimulation of her left eye, but a consensual reflex was obtained by stimulating the opposite eye.

A CT scan in a coronal plane (Fig. 13–17A) was obtained after intravenous infusion of contrast. The enhancing mass lesion (arrow) is resting on the base of the skull immediately above the sphenoid sinus. Note the asymmetric lateral ventricles.

1. The spherical shape and generally smooth border suggest that this is an aneurysm. Assuming that this is the case, what artery did it arise from?
2. At what point in the optic pathway can a single lesion cause monocular blindness and contralateral quadrantanopsia? Is this lesion in the appropriate place to cause such a defect?
3. If this lesion were to extend in the posterior direction, what other important structures located at the base of the brain could be damaged?

Discussion

This lesion represents a "giant" aneurysm of the internal carotid artery. Most intracranial aneurysms arise in arteries that are related to the anterior portion of the circle of Willis and they are usually around 1 cm in diameter when they start to produce symptoms. This aneurysm, which has a diameter of about 2 cm, probably arises from the internal carotid artery near the origin of the ophthalmic artery (supraclinoid aneurysm).

A lesion in this region is likely to involve the optic nerve and chiasm. Compression of the rostral part of the chiasm, from the lateral and ventral side, apparently involves the "knee-fibers," which would explain the superior temporal quadrantanopsia in her right eye (see Fig. 13–17B). Compression of the optic nerve on the left side causes blindness in her left eye. A lesion in this location could conceivably involve the oculomotor nerve, but this was not the case in this patient.

The sella turcica, pituitary gland, infundibular stalk, and hypothalamus all lie in the immediate neighborhood of the lesion. Therefore, if the aneurysm were to expand further, signs of hypothalamic or hypothalamohypophysial dysfunction would be likely to appear. This aneurysm is occupying and distending a portion of the subarachnoid space and, if ruptured, would give rise to a subarachnoid bleeding, which is a very serious condition (see Chapter 4, First Dissection).

Suggested Reading

1. Dreher, B. and S. R. Robinson (eds.), 1990. Neuroanatomy of the Visual Pathways and their Development. Vision and Visual Dysfunction, Vol. 3. CRC Press: Boca Raton.
2. Garey, L. J., 1990. Visual system. In G. Paxinos (ed.): The Human Nervous System. Academic Press: San Diego, pp. 945–977.
3. Goodale. D. H. and A. D. Milner, 1992. Separate visual pathways for perception and action. Trends Neurosci. 15(1):20–25.
4. Gulyas, B., D. Ottoson, and P. E. Roland, 1993. The Functional Organization of the Human Visual Cortex. Pergamon Press. Oxford.
5. Hubel, D. H. and T. N. Wiesel, 1977. Ferrier Lecture: Functional architecture of macaque monkey visual cortex. Proc. R. Soc. Lond. (Biol.) 198:1–59.
6. Leventhal, A. G., 1990. The Neural Basis of Visual Function. Vision and Visual Dysfunction, Vol. 4. CRC Press: Boca Raton.
7. Livingstone, M. and D. Hubel, 1988. Segregation of form, color, movement, and depth: Anatomy, physiology, and perception. Science 240:740–749.
8. Parnavelas, J. G., A. Dinopoulus and S. W. Davies, 1989. The Central Visual Pathways. In A. Björklund et al. (eds.): Handbook of Chemical Neuroanatomy, Vol. 7. Elsevier: Amsterdam.
9. Zeki, S., 1993. A Vision of the Brain. Blackwell Scientific Publications: Cambridge, Massachusetts.

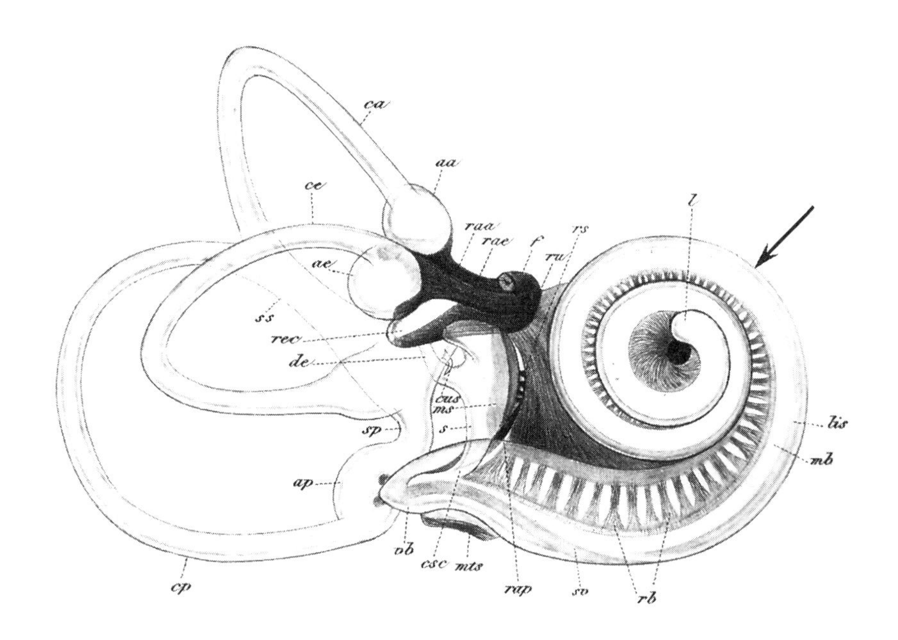

The Inner Ear
Drawing of the inner ear in a 5-month human embryo (From Gustav Retzius, 1884, Das Gehörorgan der Wirbelthiere, Vol. II, Samson & Wallin, Stockholm.)

14

Auditory System

The auditory system has a great analytical capacity. The pitch and intensity of sound can be determined with great accuracy, and the ear can differentiate between sounds of different quality, i.e., different voices and musical instruments. The anatomic substrate for these amazing functions consists of an intricate receptor apparatus, the cochlea, in the inner ear, and a series of complicated pathways and CNS structures that have been extensively studied in recent years.

Hearing and speech constitute some of the most important means of communication, and hearing loss is a great handicap in social behavior. Ten percent of the population suffers from measurable hearing loss.

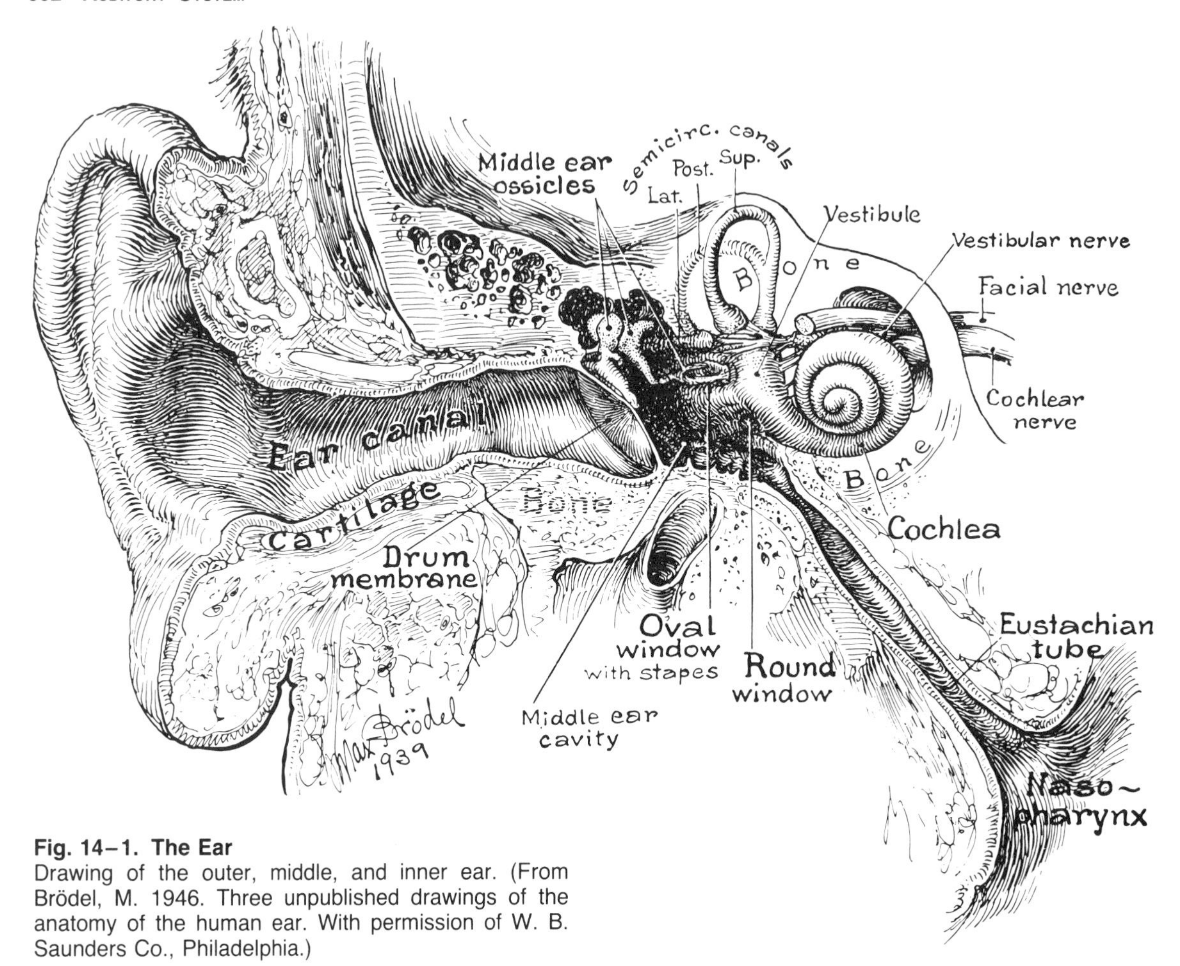

Fig. 14–1. The Ear
Drawing of the outer, middle, and inner ear. (From Brödel, M. 1946. Three unpublished drawings of the anatomy of the human ear. With permission of W. B. Saunders Co., Philadelphia.)

The internal ear is related to hearing and equilibrium, and its sensory organs, the cochlea and the vestibular apparatus, are innervated by the vestibulocochlear nerve (VIII). The vestibular apparatus and the vestibular division of the VIIIth cranial nerve are described in relation to the other cranial nerves in Chapter 11.

The cochlear division is part of the auditory system, which informs us about sounds in the environment. Sound is propagated as pressure waves, and the *pitch* of the sound is determined by the frequency of the sound waves, expressed in cycles per second (cps).[1] The *intensity* of the sound is related to the amplitude of the waves, and it is measured in decibels (db).[2]

External and Middle Ear

The external ear, the *auricle* (pinna) and the *external auditory canal* (meatus), direct the sound waves to the *tympanic membrane*, or eardrum, which serves as a boundary between the external and the middle ear (Fig. 14–1). The air-filled cavity of the middle ear, the *tympanic cavity*, which contains the three ossicles, is connected to the pharynx through the

1. The human ear can detect sound waves with frequencies between 20 and 16,000 cps or Hz (Hertz). The human ear has the greatest sensitivity for sounds around 1,000 Hz. The greater the frequency, the higher the pitch.

2. Normal conversation at 3 feet is usually carried on with an intensity of 60–70 db. The noise from a jet engine at close range has an intensity of 120–140 db, which is close to the destructive limit of the human ear. Since the intensity scale is logarithmic, the jet engine noise has an intensity that is many thousand times higher than the normal conversation voice.

eustachian tube. The tube is normally closed, but it opens during swallowing, thereby permitting equilibration of air pressure on the two sides of the eardrum.

Middle Ear Ossicles: A Mechanical Transformer

The three ossicles, the *malleus* (hammer), *incus* (anvil), and *stapes* (stirrup), form a bridge between the eardrum and the oval window of the inner ear (Fig. 14–2), and they are joined in such a way that they transmit the oscillations of the eardrum to the footplate of the stapes, which is attached to the oval window by an annular ligament. By channeling the energy of vibration of the large eardrum to the much smaller area of the oval window, the ossicular chain produces the additional force required to set the fluid of the cochlea in motion.

Middle Ear Muscles and Middle Ear Reflex

The position and tension of the auditory ossicles can be regulated by two muscles, the *tensor tympani* and *stapedius*, which attach to the malleus and the stapes, respectively. The tensor tympani is innervated by the motor trigeminal nucleus and the stapedius by the facial nucleus. The muscles

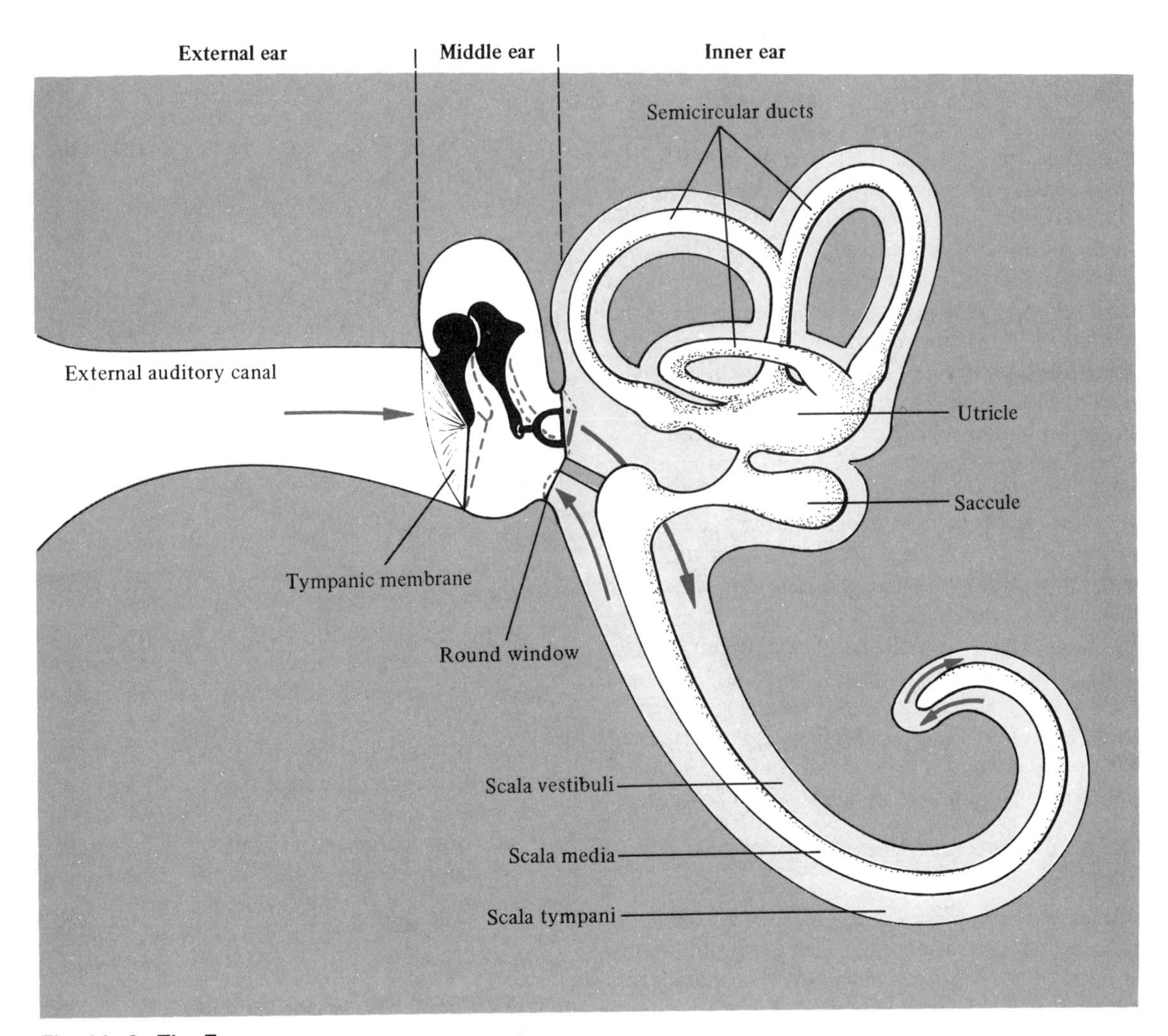

Fig. 14–2. The Ear
Diagram of the external, middle, and inner ear. The middle ear ossicles, the malleus, incus, and stapes, are *black*. The extreme positions to which the ossicles can be moved by oscillations of the eardrum as a result of incoming sound waves are indicated by *red*. The cochlea has been partly uncoiled to show the fluid movement in the perilymph (*red arrows*). The fluid motions that originate at the oval window by movement of the stapes are dissipated at the round window. (Modified from Klinke, R. Physiology of Hearing. In: Fundamentals of Sensory Physiology (ed. R. F. Schmidt), 1978. Springer-Verlag, New York, Heidelberg, Berlin.)

respond in a reflexive manner to various stimuli, especially to loud noise.

The reflexive contraction of the middle ear muscles, known as the *acoustic* or *middle ear reflex*, decreases the sound transmission through the middle ear by making the ossicular chain more rigid. It is a protective mechanism against excessive stimulation, which can easily cause damage to the sensitive receptors or hair cells in the inner ear. The reflex may, in addition, have more subtle functions designed to improve speech discrimination in noisy environments. For instance, the reflex is more efficient in the low-frequency range, and it may therefore reduce the masking of high frequencies by low-frequency sounds.

Inner Ear

The Cochlea

The auditory part of the inner ear, the *cochlea*, is shaped like a snail shell (see Vignette at the beginning of chapter), and it contains perilymph like the other spaces of the *bony labyrinth*, i.e., the vestibule and the semicircular canals. The *osseus cochlea* is coiled around a bony core, the *modiolus* (Fig. 14–3A), which contains the ganglion cells of the cochlear portion of the VIIIth cranial nerve (Fig. 14–3B). Inside the osseus cochlea is the *membranous cochlea* or *cochlear duct*, which contains endolymph, and which is attached to a thin bony shelf, the *spiral lamina*, which winds around the modiolus like the threads of a screw. Since the cochlear duct is broadly attached to the lateral wall of the osseus cochlea as well, it divides the bony canal into an upper canal, the *scala vestibuli*, and a lower canal, the *scala tympani* (Figs. 14–3A and B). The cochlear duct, which lies between the two bony compartments, is referred to as the *scala media*. The cochlear duct ends as a closed sac but there is an opening, the *helicotrema*, at the apex of the cochlea, where the fluids of the scala tympani and scala vestibuli communicate with each other.

Organ of Corti and Mechanism of Transduction

The organ of Corti (Fig. 14–3C), which contains the auditory receptors, is located on the upper surface of the *basilar membrane*. This membrane, which forms the floor of the cochlear duct, is a crucial component in the mechanism that finally transduces the mechanical energy of the sound waves to chemoelectrical potentials. The sound vibrations that enter the scala vestibuli and the perilymph at the oval window produce displacement of the basilar membrane before they finally dissipate back to the middle ear by movements of the membrane covering the round window. Oscillations of the basilar membrane produce a shearing force on the stereocilia of the receptor cells, which are in firm contact with the non-oscillating gelatinous *tectorial membrane*. The tilting of the stiff cilia, is the adequate stimulus for the auditory receptor cells.

Tonotopic Organization: Pitch Discrimination

The Hungarian scientist Georg von Békésy demonstrated more than 50 years ago that each sound initiates a traveling wave along the whole length of the cochlea. Different frequencies produce waves that affect large parts of the basilar membrane but with peak amplitudes at different points along the membrane. A very important feature of the displacement mechanism of the basilar membrane, therefore, is the fact that different parts of the membrane respond with maximal efficiency to sounds of different frequency. For instance, high-frequency sounds produce maximal displacement of the basal part of the membrane, whereas progressively lower frequencies involve more and more apical parts of the membrane. The reason for this is primarily related to the fact that the elastic fibers that form the membrane are relatively short at the base of the cochlea, but become longer toward the apex.

The basilar membrane is lined with two groups of hair cells along the length of the cochlear duct: a single row of *inner hair cells* and three to five parallel rows of *outer hair cells*. The two types of hair cells can be distinguished morphologically, and show striking differences in patterns of innervation (see below). Recently, it has been discovered that the tonotopic localization is dependent not only on the properties of the basilar membrane, but also on different types of hair cells, which are tuned to different frequencies. Since different parts of the basilar membrane stimulate specific sets of hair cells, the spatial representation of frequency, i.e., *tonotopicity*, is preserved in the transduction mechanism. Indeed, the principle of tonotopic organization characterizes most of the central auditory system.

Innervation of the Receptor Cells

The hair cells are innervated by the dendritic terminals of large bipolar neurons whose cell bodies are located in the modiolus. They are also innervated by efferent fibers of the olivocochlear bundle (see below).

The *cochlear nerve*, which is formed by the central

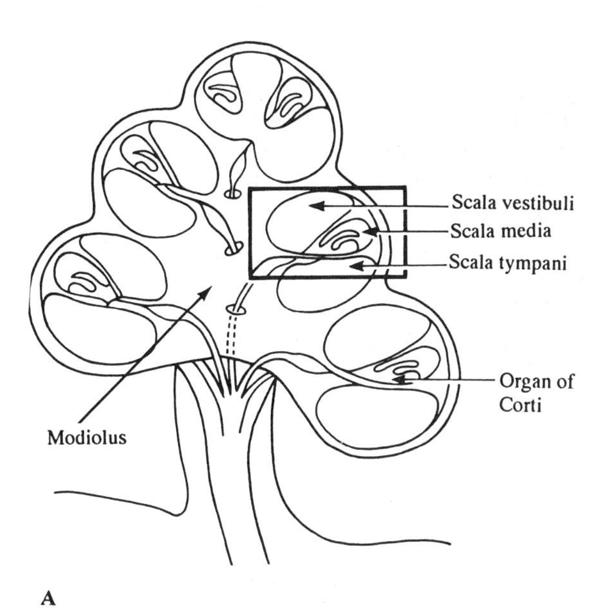

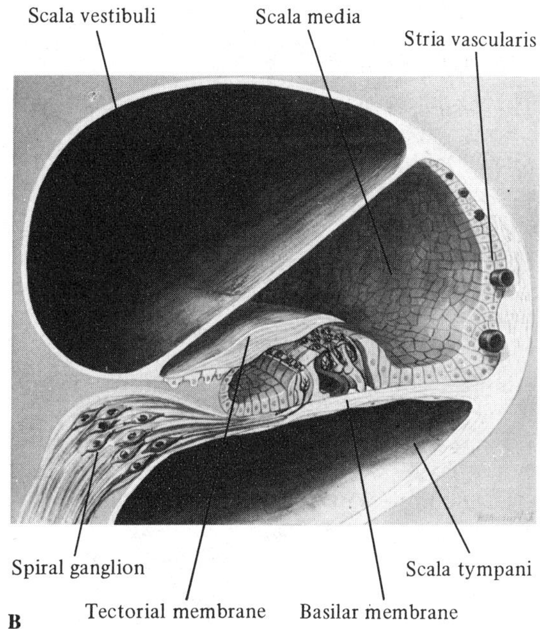

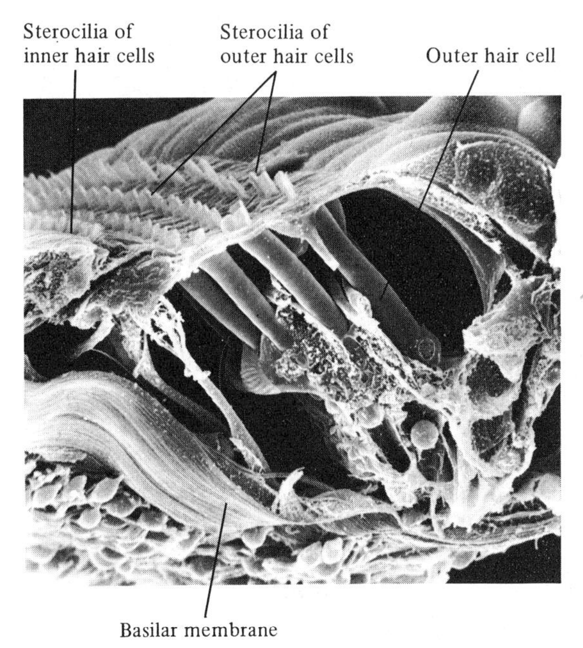

Fig. 14–3. The Organ of Corti
A. Section through the cochlea in the direction indicated by the *arrow* in the vignette at the beginning of the chapter. (Courtesy of Dr. J. Ballantyne, London. From Friedmann, I. and Ballantyne, J. (eds.), 1984. Ultrastructural Atlas of the Inner Ear. With permission of Butterworth and Co.)
B. Drawing of a cross-section through one turn of the cochlea. (Courtesy of J. Kálmánfi.)
C. Scanning electron micrograph of the organ of Corti of a guinea pig. The tectorial membrane has been removed in order to illustrate the stereocilia, which are normally imbedded in the tectorial membrane. (Micrograph courtesy of G. Bredberg. From Nolte, J., 1981. The Human Brain. With permission of C. V. Mosby Company, St. Louis, Toronto-London.)

processes of the ganglion cells in the spiral ganglion, passes through the internal auditory canal together with the vestibular and facial nerves and enters the brain stem at the pontomedullary junction (Fig. 4–6). About 95% of the spiral ganglion cells, whose peripheral processes proceed to the organ of Corti, innervate the inner hair cells, which are the main transducers in the auditory system. The remaining 5% of the neuronal population in the cochlea innervate the outer hair cells.

Central Auditory Pathways

The Cochlear Nuclei

The cochlear nerve enters the ventral cochlear nucleus on the ventrolateral side of the inferior cerebellar peduncle at the level of the pontomedullary junction. The nerve fibers bifurcate within the nucleus; a couple of branches terminate within the ventral nucleus, whereas another branch proceeds to the dorsal cochlear nucleus on the dorsolateral aspect of the inferior cerebellar peduncle (Fig. 14–4).

The cochlear nuclei contain many different cell types, which can be distinguished on the basis of their appearance, electrophysiologic properties, and specific connections. The most important principle of organization, however, is the fact that the cochlear nerve fibers terminate in a strict cochleotopic or tonotopic pattern within the various subdivisions of the cochlear nucleus. This tonotopicity, which is based on a spatial organization of the neurons involved, is maintained throughout the central auditory pathways.

The Trapezoid Body, Lateral Lemniscus, and Auditory Radiation

The central auditory connections from the cochlear nuclei to the auditory cortex in the superior temporal gyrus are complex, and they are illustrated in a very schematic fashion in Fig. 14–4. Nonetheless, the figure emphasizes several important features of the central auditory pathways, including the fact that fibers from one ear reach auditory centers on both sides of the brain and that several nuclei are intercalated in the central auditory pathways.

The cochlear nuclei and their various subdivisions are characterized by a multitude of different excitatory and inhibitory pathways; only the major ascending projections are indicated in Fig. 14–4. First, axonal projections from the ventral cochlear nucleus (the ventral acoustic stria) form the *trapezoid body*, which proceeds in a medial direction across the tegmentum and midline and terminates in the *superior olivary complex* on both sides. Other fibers join the lateral lemniscus on the opposite side to ascend to the inferior colliculus in the midbrain. Second, fibers from the dorsal cochlear nucleus (dorsal acoustic stria) cross the midline and then join the *lateral lemniscus* on the opposite side. Fibers from both the ventral and the dorsal cochlear nuclei also terminate in the nuclei of the lateral lemniscus on the opposite side on their way to the midbrain (not indicated in the figure). Third, fibers from the superior olive ascend in the lateral lemniscus on both sides and terminate in the nuclei of the lateral lemniscus and in the *inferior colliculus*.

The fibers that ascend in the lateral lemniscus, regardless of their origin, terminate in the inferior colliculus, establishing synaptic contacts with inferior collicular cells, which in turn project through the brachium of the inferior colliculus to the medial geniculate body of the thalamus. A few of the collicular cells project to the *medial geniculate body* on the opposite side, but the largest decussations in the midbrain are the commissure of the inferior colliculus interconnecting the left and right colliculi and the commissure of Probst interconnecting the left and right nuclei of the lateral lemniscus. The *auditory radiation*, i.e., the fibers from the medial geniculate body to the auditory cortex on the upper bank of the superior temporal gyrus, pass through the sublenticular part of the internal capsule.

It is worth re-emphasizing that the central auditory pathways have several features that set them apart from other sensory pathways. Many cell groups are integral parts of the central auditory pathways. These include obligatory synapses in the cochlear nuclei, inferior colliculus, and medial geniculate body, but also nonobligatory synapses in the superior olivary complex and the nuclei of the lateral lemniscus, which are embedded in the lateral lemniscus. Although little is known about many of these intercalated cell groups, it is evident that a considerable amount of information processing occurs at the subcortical level as exemplified below in regard to the superior olivary complex. Another important aspect of the central auditory pathway relates to the crossing and recrossing at several levels, which explains why damage to central auditory structures seldom produces deficits related to only one ear. In fact, restricted unilateral lesions of the ascending auditory pathways are usually of little consequence to hearing.

Superior Olivary Complex: Localization of Sound in Space

Of the many brain stem nuclei that are intercalated in the auditory pathways, the superior olivary complex has attracted special attention. A medial subdivision of the complex is populated by transversely oriented bipolar neurons with a lateral and a medial dendrite. The lateral dendrite receives auditory input from one ear and the medial dendrite from the other ear. These neurons are sensitive to the interaural time difference of sounds reaching the two ears, which is one of the two chief cues used to localize sound in space. A lateral subdivision of the superior olivary complex is concerned with

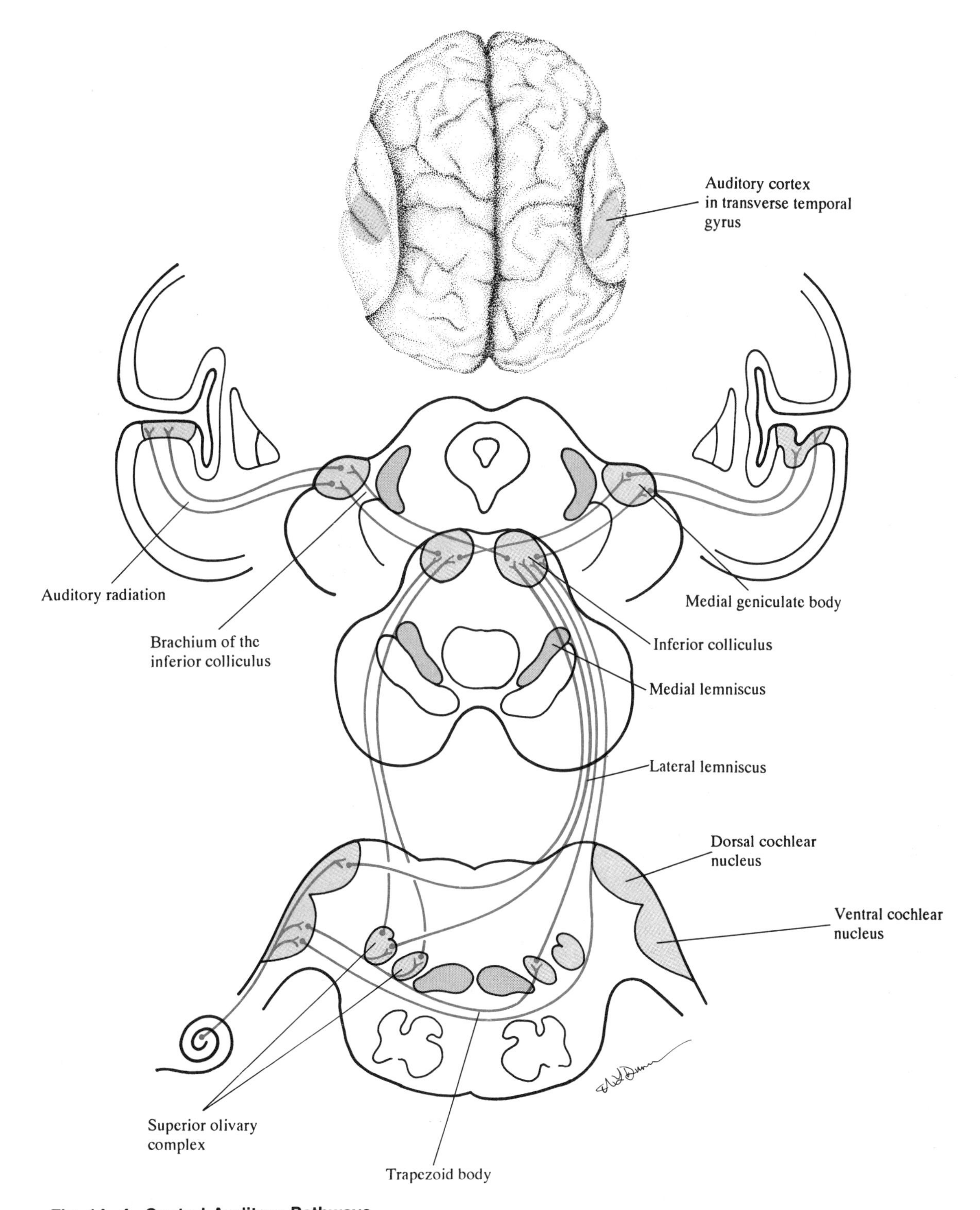

Fig. 14–4. Central Auditory Pathways
Highly schematic diagram of the central auditory pathways. Note that fibers from one ear reach auditory centers on both sides of the brain. The auditory cortex in Heschl's gyrus is hidden in the depth of the lateral sulcus.

the other main cue for sound localization based on interaural intensity differences.

Auditory Cortex

The auditory cortex is located in the *transverse temporal gyrus* (*Heschl's gyrus*) on the upper bank of the superior temporal gyrus. It is represented by Brodmann's cortical area 41 and part of 42, and is hidden in the depth of the lateral sulcus. The tonal frequencies, represented by functional frequency columns, are organized in a tonotopic pattern with the highest frequencies represented in the posterior and medial parts of Heschl's gyrus. Other sound variables, e.g., sound location, may be represented orthogonally to the frequency map in the auditory cortex.

A wealth of association fibers connect the auditory cortex with surrounding cortical regions, i.e., auditory association areas, which are located both on the dorsal and lateral aspect of the superior temporal gyrus, and which include parts of area 42 and area 22 (Wernicke's area; see Chapter 22). These areas are of importance for the interpretation of auditory information, especially speech sounds. The area behind Heschl's gyrus, i.e., the *temporal plane*, is usually larger on the left side of the brain (Fig. 4–10). Although this size difference may be related to the linguistic dominance of the left hemisphere, the proportion of individuals with larger temporal plane on the left side (about 65%) is considerably less than that for left hemisphere language dominance (90–95%). As the most celebrated asymmetry in the human brain, the temporal plane asymmetry has been the focus of much interest, but its significance is still poorly understood.

Since the auditory cortex on each side receives impulses from both ears, unilateral cortical lesions have only a slight effect on sound sensitivity. If the auditory cortex on both sides is destroyed, e.g., in the case of bilateral temporal infarcts, which is extremely rare, the patient seems to be nearly deaf, ignoring all but the loudest sounds (cortical deafness). A somewhat different syndrome, referred to as *auditory agnosia* (word deafness) can result from a bilateral lesion of auditory areas, but also from unilateral lesions, usually on the left side. Such patients are able to hear and recognize sounds, but they cannot understand a spoken word. Since these patients can still read, write, and often even speak, they are not considered to suffer from aphasia (Chapter 22). However, a patient's inability to understand clearly his own speech often results in an unwillingness to speak that can resemble true *sensory aphasia* (inability to understand speech).

The Olivocochlear Bundle: Centrifugal Control of the Cochlea

Descending pathways from the auditory cortex to the medial geniculate body, the inferior colliculus, and the other auditory brain stem nuclei parallel the ascending auditory pathways. The final link in this centrifugal fiber system is the *olivocochlear bundle*, which originates in the superior olivary complex in the pons (Figs. 14–4 and 14–5). Some of these centrifugal fibers, which originate in the ipsilateral superior olivary complex, terminate in the inner hair cell region, where they contact the afferent fibers. Other fibers, primarily from the contralateral medial part of the superior olive, terminate on the base of the outer hair cells.

The outer hair cells seem to serve as effectors or modulators of the cochlear environment rather than as neural transducers. They contain actin, and one of their functions appears to be to provide mechanical modulation of the cochlear mechanism, in part by changing their length in response to efferent stimulation. This contraction affects the position of the tectorial membrane, which in turn influences the sensitivity of the inner hair cells.

Functional Aspects

The detection and analysis of sound involve very complicated mechanisms, which to some extent are reflected in the complicated anatomy of the auditory system. Many of the functional aspects of hearing are directly related to our ability to recognize and appreciate speech, and they are therefore of vital importance in our everyday life. The abilities to discriminate between different frequencies and intensities of sound, or to localize the source of a sound, are important aspects of hearing.

Pitch Discrimination

A tonotopic organization, i.e., a spatial representation of tonal frequencies, is one of the most striking characteristics of the auditory pathway. The spatial separation of tones with different frequencies occurs in the cochlea, where the basal part of the organ of Corti is stimulated by high-frequency tones, whereas progressively lower frequencies involve more and more apical parts of the organ. A strict cochleotopic pattern is then preserved throughout the different relays in the auditory path to the cerebral cortex, where high-pitch tones are represented in the medial part of the auditory cortex close to the insula, whereas low-frequency sounds are represented in the lateral part, close to the lateral surface of the brain.

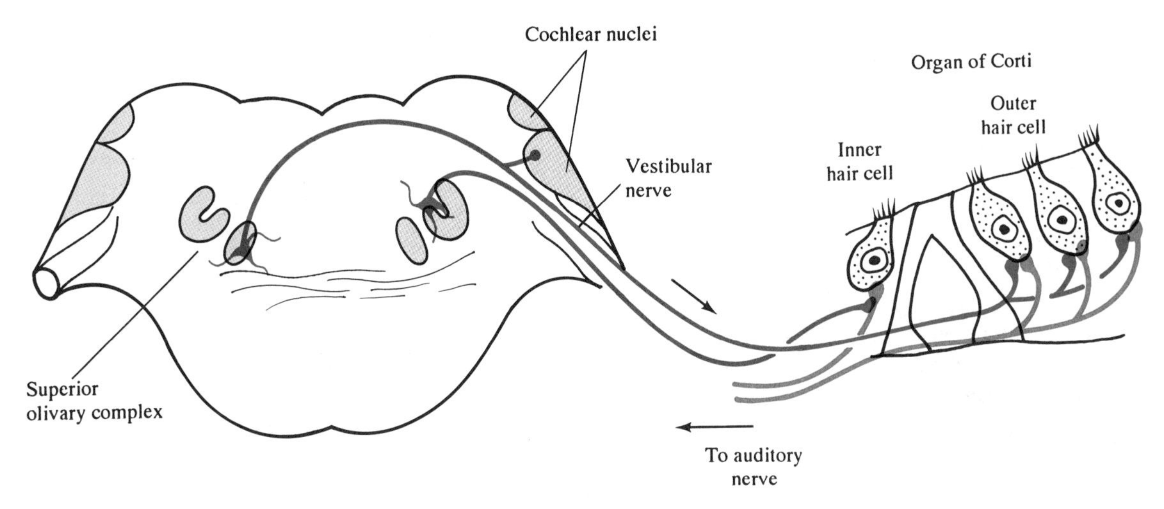

Fig. 14–5. Olivocochlear Bundle
The olivocochlear bundle exits the brain stem through the vestibular nerve and reaches the organ of Corti via a vestibulocochlear anastomosis. (Modified from Schuknecht, H. F., 1993. Pathology of the Ear. Lea & Febiger, Philadelphia.)

Although the auditory cortex is tonotopically organized, it does not seem to be absolutely essential for frequency discrimination. The analytic capacity of the auditory system, however, extends beyond the pure discrimination of tonal frequencies; we can effectively distinguish between different voices and different musical instruments and it is likely that the auditory cortex and its related association areas become increasingly involved in more subtle aspects of sound discrimination.

Intensity Discrimination

This type of discrimination, apparently, is dependent both on the number of auditory receptors stimulated and then on the discharge rate in the individual nerve fibers. The cerebral cortex is not essential for intensity discrimination.

Sound Localization

This important function is dependent foremost on binaural representation in some of the auditory structures of the brain. A sound that comes from a specific direction will usually reach the two ears with a slight time difference and with somewhat different intensity, and these two factors are important cues in sound localization. Consequently, deafness in one ear sharply reduces the ability to localize sound.

The central neural mechanisms responsible for localization of sound in space are not completely known; nor is it known to what extent the cerebral cortex is involved. It has been shown, however, that two of the cell groups in the *superior olivary complex* are populated by bipolar cells, with a medial and lateral dendrite. The two dendrites receive input from opposite ears, which makes it possible for the neurons to register the interaural time or intensity differences. The superior olivary complex and other structures with similar features would seem to be likely candidates for functions related to sound localization.

Modulation of Auditory Input

The olivocochlear bundle is the final link in an extensive set of feedback pathways in the auditory system. The functions of the centrifugal pathways have been the focus of intense research; they appear to be important both for tuning the cochlear mechanism to sounds of particular interest and for increasing the "signal-to-noise" ratio by suppressing unwanted auditory signals.

Clinical Notes

Hearing Loss

Hearing loss is a problem of gigantic proportions. Ten percent of the population have hearing defects serious enough to interfere with the understanding of speech. Two types of hearing loss can be distinguished: (1) *conductive* hearing loss and (2) *sensorineural* hearing loss. The first type is related to defects in conductive mechanisms in the middle ear resulting from conditions such as otitis media and otosclerosis. Sensorineural hearing loss is caused by disease in the cochlea or its central connection, the auditory nerve (VIIIth cranial nerve). Hearing loss of cochlear origin is common and can result from a variety of conditions including tumors, infections, and temporal bone fractures, or from exposure to excessive noise or ototoxic drugs. Senile hearing loss (*presbycusis*) is an especially common form of sensorineural deafness, but also is a result of conductive hearing loss.

Tinnitus, i.e., noise (ringing, humming, whistling, etc.) in the ears, is a common symptom in disorders of the inner ear, but it can also occur in disease of the VIIIth cranial nerve, e.g., acoustic neurinoma (see below). Even wax in the ears can cause tinnitus.

Tests of Hearing

Many simple procedures can be used to test hearing. A rough estimate can be obtained by recording the distance at which the paitent is able to hear a whisper or a watch tick with one ear while the other ear is covered.

The *Rinne* test can be used to differentiate between conductive and sensorineural hearing loss. The stem of a vibrating tuning fork of 256 or 512 Hz frequency is placed on the mastoid process (bone conduction) until it is no longer heard by the patient, at which time the fork is held at the external auditory canal (air conduction). Since air conduction should be greater than bone conduction, the sound can be heard longer if the air conduction is intact (normal Rinne test). However, the sound cannot be heard if the air conduction is severely impaired by a middle ear disease (abnormal Rinne test). In cases of sensorineural hearing loss the Rinne test is normal but both bone and air conduction is quantitatively decreased.

More elaborate tests can be performed with sophisticated *audiometers*, which can sequentially present all combinations of frequencies and intensities through the whole range of hearing.

Cochlear Implants

During the last 20 years, considerable effort has been devoted to the development of cochlear prostheses capable of restoring at least part of the hearing in patients suffering from profound sensorineural hearing loss as a result of depletion of hair cells. The device consists of several pairs of electrodes implanted in the cochlea in order to stimulate directly the auditory nerve fibers, thereby bypassing the defective hair cells. In addition, there are components which are worn externally consisting of a microphone, speech processor, and transmitter.

Considering the difficulties of mimicking the very complicated way speech normally activates the auditory nerve fibers via the cochlear mechanism, it should come as no surprise that cochlear implants have yet to provide full recovery of hearing. However, since loss of hair cells is a common reason for sensorineural hearing loss, it is exciting that recent studies have shown that sensory epithelia in the inner ear do have the potential for regenerating new hair cells. This may provide an opportunity to facilitate the production of new hair cells to replace the missing or damaged.

Otosclerosis

Otosclerosis can cause a conductive hearing loss. The disease usually starts in the second or third decade and is characterized by progressive ossification of the annular ligament around the base of the stapes, which eventually becomes fixed in its position in the oval window. Surgical procedures to replace the stapes usually result in a significant improvement in hearing.

Vestibular Schwannoma (Acoustic Neurinoma)

The most common form of cerebellopontine angle tumor (see Clinical Examples) is the acoustic neurinoma, which originates from Schwann cells of the sheath surrounding the VIIIth cranial nerve in the internal auditory canal. The first symptoms is usually tinnitus followed by hearing defects, loss of corneal reflex (Vth cranial nerve), and cerebellar signs.

Hereditary Hearing Loss

Genetic abnormalities can cause a number of syndromes in which hearing loss is a prominent feature. Other defects associated with hereditary hearing loss include malformations of the external ear; albinism; ocular abnormalities; mental deficiencies; or neurologic abnormalities including progressive ophthalmoplegia, polyneuropathy, and cerebellar ataxia.

Clinical Example

Cerebellopontine Angle Tumor

A 65-year-old woman was seen by her local physician for evaluation of hearing loss on the right side. Her physician found fluid behind the right tympanic membrane and prescribed a decongestant, but her hearing did not improve. She ignored the hearing loss for several months, but finally came back to her doctor, because of increasing dysequilibrium and some incoordination in her right arm; she continued to note hearing loss and tinnitus (ringing noise in the ear).

On examination her vision, pupils, optic fundi, and extraocular movements were normal. Stimulation of her right cornea with a wisp of cotton did not elicit blinking or closure of the eyes, whereas stimulation of the left cornea produced blinking not only of the left eye but also of the right (consensual response). She did not taste salt or sugar well on the anterior surface of her tongue on the right side, but there was no significant weakness of her right facial muscles. She had a sensorineural type of hearing loss on the right side and a diminished response to caloric stimulation on the right. The remainder of her cranial nerve examination was normal.

Generalized weakness with some difficulty in maintaining muscle contractions in the right arm was noted, and the muscle tone was reduced in the right arm and leg. The finger-to-nose test and the heel-to-shin test (placing the right heel on the left knee) revealed ataxia in her right arm and leg. She had difficulty with tandem gait and tended to veer off to the right side.

An axial MRI with gadolinium enhancement was obtained, revealing a posterior fossa tumor (Fig. 14–6A).

1. Which cranial nerves are involved?
2. Why was the sense of taste on the right side of the tongue affected?
3. What type of lesion is most likely responsible for the symptoms?

Discussion

The patient's neurologic deficits involve the right Vth and VIIIth cranial nerves, as well as the intermediate nerve (VII) including especially the taste fibers to the anterior part of the tongue. Cerebellar functions are also prominently affected by the lesion.

The examination suggests an intracranial extra-axial lesion where the above-mentioned cranial nerves are close to one another and to the cerebellum (see Figs. 14–6 and 14–6B). This places the lesion in the posterior fossa at the base of the skull, or more precisely in the angle between pons and cerebellum, i.e., the cerebellopontine angle. There are two types of tumors that commonly arise in this location, *acoustic neurinomas* (or better, *vestibular schwannomas*, since the tumor usually originates in the vestibular portion of the VIIIth nerve) and *meningiomas*. This patient had a vestibular schwan-

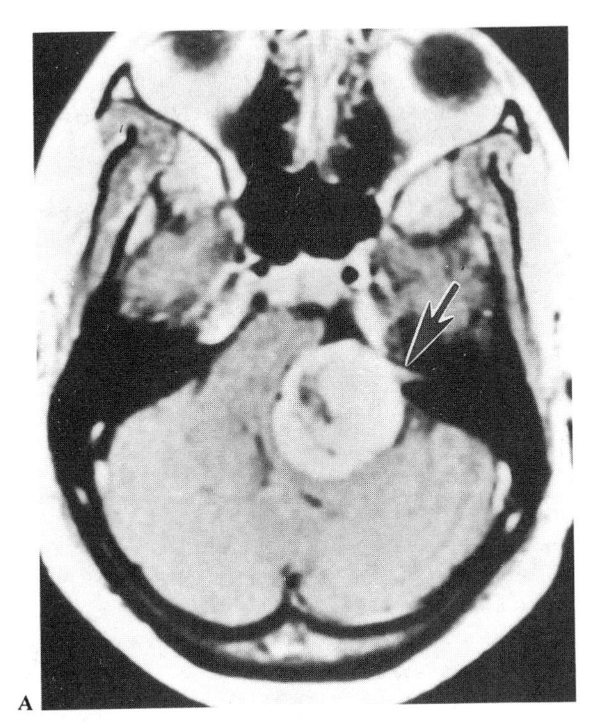

Fig. 14–6A. Cerebellopontine Angle Tumor
Arrow indicates the intracanalicular portion of the tumor. (Axial MRI with gadolinium enhancement was provided by Dr. Maurice Lipper.)

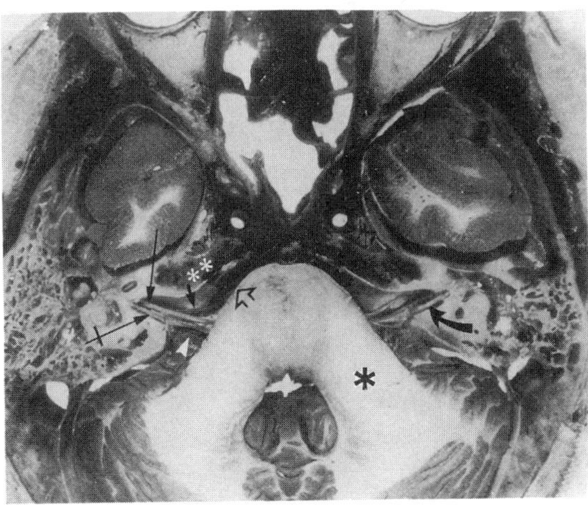

Fig. 14–6B. Cerebellopontine Angle
Horizontal cryotomic section at the level of the internal auditory canal showing the cerebellopontine angle and its contents. Cochlear nerve (*arrow*); inferior vestibular nerve (*crossed arrow*); facial nerve (*curved arrow*); anterior inferior cerebellar artery (*short arrow*); flocculus (*white arrowhead*); petrous bone (*white asterisks*); lateral surface of pons (*open arrow*); middle cerebellar peduncle (*black asterisk*). (Photograph by Prof. S. Kubik, Department of Anatomy, University of Zürich. From Valavanis, A. et al., 1986. Clinical Imaging of the Cerebellopontine Angle. Springer-Verlag.)

noma, which typically begins in the internal auditory canal and then grows medially, involving the VIIIth, Vth, and VIIth cranial nerves, usually in that order. The cerebellum is also often affected at a relatively early stage as in this case.

Although the disappearance of the corneal reflex on the side of the lesion is often said to be due to compression of the Vth cranial nerve, a tumor in this location may also interfere with the corneal reflex by compressing the spinal nucleus and tract of the trigeminus in the medulla oblongata (Fig. 10–1), which may well be the case in this patient. Note that the exit of the Vth cranial nerve is located in front of the cerebellopontine angle (Fig. 4–6, see also Vignette, Chapter 11) at some distance from the exits of the VIIth and VIIIth cranial nerves.

The tumor shows intense contrast enhancement; also note the intracanalicular portion of the tumor (arrow in Fig. 14–6A). This large tumor has displaced the brain stem and has also affected the cerebellum and some of the cerebellar pathways to a significant degree, which explains the cerebellar symptoms. Although there is distortion of the fourth ventricle, there is no hydrocephalus and no obvious signs of increased intracranial pressure. Most patients with a cerebellopontine angle tumor seek medical help before the tumor reaches this size. Nonetheless, following a successful operation via a suboccipital craniectomy, the patient's symptoms resolved, except for the profound hearing loss, which was permanent.

Suggested Reading

1. Aitkin, L. M., 1989. The Auditory System. In A. Björklund et al. (eds.): Handbook of Chemical Neuroanatomy, Vol. 7. Elsevier: Amsterdam.
2. Aitken, L., 1990. The Auditory Cortex. Chapman and Hall: London.
3. Altschuler, R. A., R. P. Bobbin, B. M. Clopton, and D. W. Hoffman (eds.), 1991. Neurobiology of Hearing. The Central Auditory System. Raven Press: New York.
4. Brodal, A., 1981. Neurological Anatomy in Relation to Clinical Medicine, 3rd ed. Oxford University Press: New York, pp. 602–639.
5. Corwin, J. T. and M. E. Warchol, 1991. Auditory hair cells: Structure, function, development, and regeneration. Annu. Rev. Neurosci. 14:301–333.
6. Masterton, R. B., 1992. Role of the central auditory system in hearing: The new direction. Trends Neurosci. 15(8):280–285.
7. Schuknecht, H. F., 1993. Pathology of the Ear, Lee and Febiger: Philadelphia.
8. Webster, W. R. and L. J. Garey, 1990. Auditory system. In G. Paxinos (ed.): The Human Nervous System. Academic Press: San Diego, pp. 889–944.

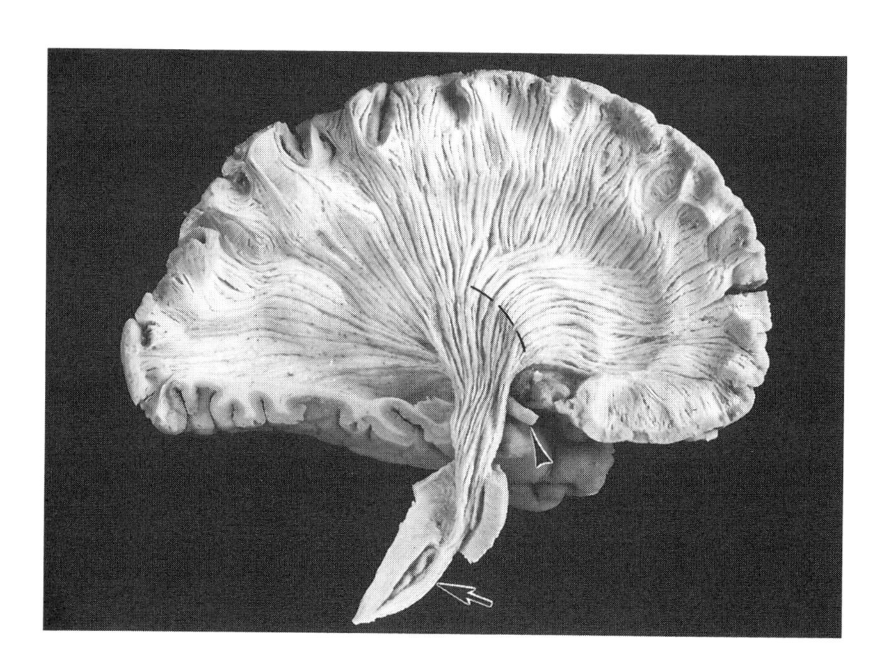

Brain dissection showing the corona radiata, ther internal capsule (genu indicated by a broken line), and the corticospinal tract forming the pyramid in the medulla oblongata (arrow). The optic tract (*arrowhead*) is located lateral to the base of the cerebral peduncle. (From N. Gluhbegovic and T.N. Williams. 1980. The Human Brain: A Photographic Guide. JB Lippincott Co., Philadelphia. With permission of the publisher.)

15

Spinal Cord Motor Structures and the Descending Supraspinal Pathways

The motor neurons in the spinal cord and the cranial nerve nuclei innervate the skeletal muscles. They constitute the "final common path" to the striated muscles for all the impulses that influence motor behavior. All movements are brought about by influences that ultimately converge on these neurons from many sources, including peripheral sensory receptors, as well as regions in the brain stem and cerebral cortex. These latter structures exert their influence through the descending supraspinal pathways.

One of the main functions of the CNS is to produce purposeful movements, and since many parts of the brain and spinal cord are concerned with motor functions, motor disorders are very common following lesions in the nervous system. The type of motor disturbance inflicted on a patient is to a significant degree dependent on the part of the motor system that has been damaged, and a good understanding of its various components is usually the first step to a right diagnosis.

Synopsis of the Motor System

Coordinated movements result from harmonious interactions between a variety of different structures throughout the nervous system. To be able to perform in a harmonious fashion, the various parts of the motor system need continuous feedback from the special and general sensory organs including information from proprioceptors within the muscles themselves. A dramatic example of the importance of proprioceptive feedback is presented by patients with dorsal column lesions (see Clinical Notes, Chapter 9).

By subdividing the subject of the motor system into a few major topics (Fig. 15–1), it will be easier to appreciate its organization and to recognize the main clinical syndromes resulting from lesions of its various parts. The spinal cord contains motor neurons and premotor interneurons that form the basis for spinal reflexes and basic motor patterns, which in turn are modulated by descending supraspinal pathways, the most famous of which is the *corticospinal* or *pyramidal tract.* This pathway, which carries impulses from the cortical motor fields to the primary motor neurons and their related interneurons, is especially important for the control of finely tuned voluntary movements, as exemplified by delicate finger movements.

The functions of the corticospinal tract are superimposed on the control exerted by the descending brain stem pathways, which originate in many parts of the reticular formation, the vestibular nuclei, and some midbrain areas. All movements are influenced by these pathways, which are also of special importance for the regulation of muscle tone and the maintenance of erect posture.

Two other brain structures, the *cerebellum* and the *basal ganglia*, are crucially important for motor functions. The cerebellum, which plays a critical role in the coordination of movements and in postural adjustments, is discussed in Chapter 17, and the basal ganglia, whose modulatory role is based on a double inhibitory circuit, are examined in the next chapter. Part of the activity generated in the cerebellum and basal ganglia is channeled through the brain stem descending pathways. In addition, both structures form "key elements in two parallel, re-entrant systems, which return their influences to the cortex through discrete and separate portions of the ventrolateral thalamus" (Alexander and DeLong, 1991).

That emotions influence motor behavior is to say the obvious. Only during the last 10 to 15 years, however, has it become evident that several telencephalic structures known for their close relation to emotional behavior have direct access to brain stem areas controlling motor neurons. These structures, including some parts of the amygdaloid body, the ventral components of the basal ganglia, and some hypothalamic areas, form a highly integrated system, which is especially concerned with emotional and motivational states, including their influence on motor behavior. The structures and pathways that are of special importance for motor aspects of emotion, drive and motivation (sometimes referred to as the emotional motor system) are reviewed in Chapters 19 and 20.

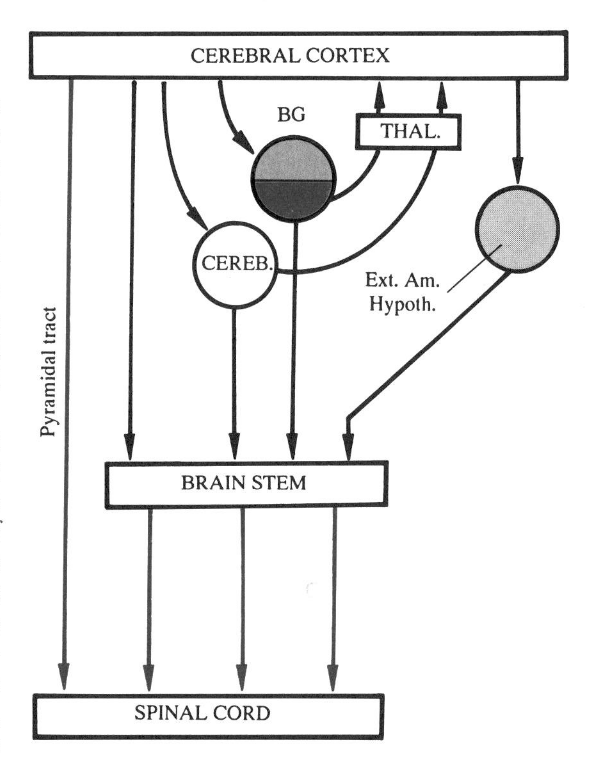

Fig. 15–1. Components of the Motor System
Highly schematic diagram showing the relations among the major components of the motor system. Note the parallel projections from the cerebral cortex to different levels of the motor system. The basal ganglia (BG) and the cerebellum are best known for their role in two parallel corticosubcorticocortical re-entrant circuits, but they also influence motor functions directly via the brain stem descending pathways. Thal. = thalamus; Ext. Am = extended amygdala; Cereb. = cerebellum; Hypoth. = hypothalamus. (Modified from Alexander, G. E. and M. R. DeLong, 1992. Central mechanisms of initiation and control of movement. In: A. K. Asbury et al. (eds.): Diseases of the Nervous System. W. B. Saunders, Philadelphia, pp. 285–308.)

Motor Neurons, Muscle Receptors, and Spinal Reflexes

Motor Neurons and the Motor Unit

The motor neurons whose axons project to the striated muscles are located in the ventral horns of the spinal cord and in some of the cranial nerve nuclei, and they represent the "final common path" for all impulses to the striated muscles.

The large motor neurons that innervate the striated muscle fibers are referred to as *α motor neurons*. Scattered among the large motor neurons are many small neurons called *γ motor neurons*; they send their axons to *intrafusal muscle fibers*, which are located within special receptors in the muscles called *muscle spindles* (see below). The *γ motor neurons* control the sensitivity of the muscle spindles.

Topography of Motor Neuronal Cell Groups

The spinal motor neuronal cell groups, which together form Rexed's lamina IX, are located according to a general somatotopic pattern (Fig. 8–4). The neurons that supply the axial musculature, including the neck muscles, are located ventromedially, whereas the neurons for the musculature of the limbs are situated in the lateral part of the ventral horn. The ventromedial cell column is present throughout the spinal cord. The lateral cell groups, however, are present only in the lumbar and cervical enlargements, and they can in turn be divided into three divisions: the central, ventrolateral, and dorsolateral columns. The *central column* innervates the girdle muscles, the *ventrolateral column* innervates the proximal limb muscles, and the *dorsolateral column* supplies the distal limb muscles. Furthermore, the motor neurons related to the distal musculature, i.e., those in the dorsolateral column, are located not only dorsal but also caudal to those innervating the proximal part of the limbs.

The somatotopic localization of the motor neurons is important for understanding the relationship among the descending supraspinal pathways and various motor neuron pools (see below). It also explains why a localized lesion in the ventral horn may result in segmentally distributed paralysis.

The Motor Unit and the Neuromuscular Junction

The axon of a motor neuron arborizes extensively to supply many muscle fibers. The motor neuron, together with the muscle fibers it supplies, is called a *motor unit*. Muscles used for delicate movements, like the hand muscles, have small motor units in the sense that one motor neuron may supply less than 100 muscle fibers; the very small eye muscles have even smaller motor units, on the order of 10 muscle fibers per unit. In some of the larger muscles in the body, e.g., leg muscles, one motor unit may contain several hundreds or even thousands of muscle fibers.

The junction between the terminal branches of the axon and the muscle fiber, the *neuromuscular junction*, is shown in Fig. 15–2. The axonal

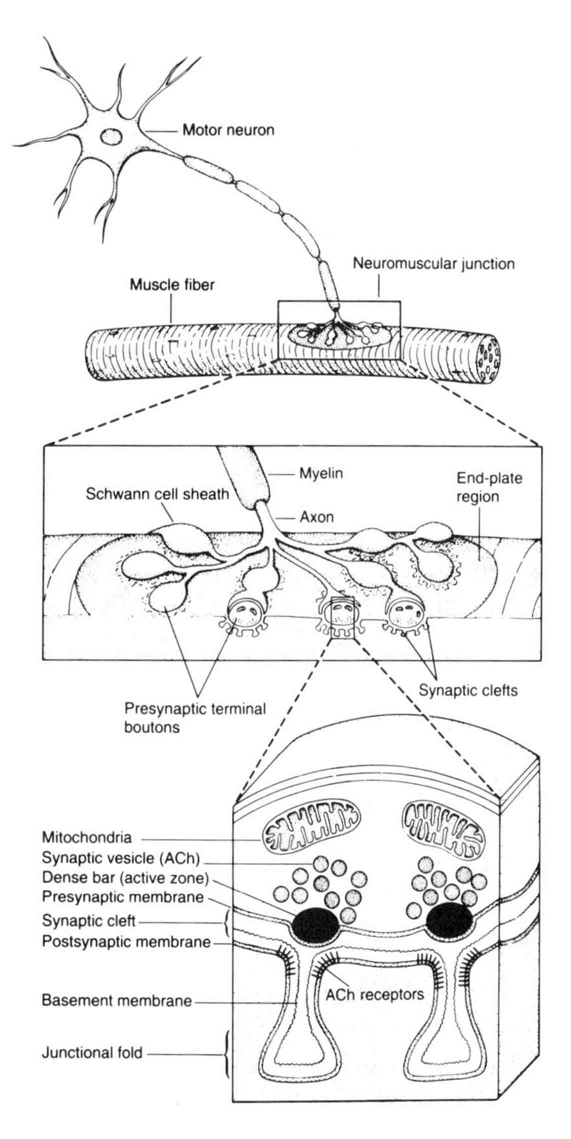

Fig. 15–2. The Neuromuscular Junction
The drawings from top to bottom show progressive enlargements of the neuromuscular junction, where the terminals of the motor nerve fibers establish synaptic contacts with a specialized region of the muscle fiber membrane called the motor end-plate. (From Kandel, E.R. and S.A. Siegelbaum, 1991. Directly gated transmision at the nerve-muscle synapse. In: E.R. Kandel, J.H. Schwartz, and T.M. Jessell (eds.): Principles of Neural Science, pp. 135-152. Appleton & Lange, Norwalk, CT. With permission of the publisher.)

branches contain a number of boutons or terminals, which contain synaptic vesicles and mitochondria. The boutons dip into grooves or troughs in the *motor end-plate*, which is a specialized region of the muscle fiber membrane. The trough, in turn, contains a number of junctional folds, which greatly enlarge the surface area of the muscle membrane. When the action potential reaches the end-plate, the neurotransmitter acetylcholine (ACh) is released into the synaptic cleft. The transmitter binds to specific receptors (nicotinic ACh receptors) on the muscle cell membrane, thereby producing an *excitatory postsynaptic potential*, i.e., the *end-plate potential*. This triggers an *action potential* in the muscle fiber, which results in contraction of its myofibrils. The muscle fiber action potential is recorded in the examination known as *electromyography* (*EMG*) (see Clinical Notes).

Premotor Interneurons and Propriospinal Neurons

Most of the boutons that impinge on the α and γ motor neurons belong to *interneurons* that are distributed in lamina VII (intermediate gray) and lamina VIII. The importance of the interneuronal system can hardly be overestimated; the interneurons participate in polysynaptic spinal reflexes, and many of the motor commands that reach the motor neurons through the descending supraspinal pathways are mediated by interneurons. Circuits of interneurons also form specialized *pattern generators* for rhythmic movements in activities like locomotion (see below). In other words, the interneuronal system is in large part responsible for the great versatility that characterizes our motor behavior, and the flexibility that the motor system shows in response to various stimuli.

Propriospinal neurons project through the *fasciculi proprii* to other segments of the spinal cord. They form the anatomic substrate for intersegmental coordination of various muscles in movement synergies. Long propriospinal pathways with cells of origin located primarily in the medial part of the intermediate gray are responsible for the coordination of axial body movements and for the coordination of arm and leg movements.

The Renshaw Cell

A special type of inhibitory interneuron is represented by the *Renshaw cells*, which are activated by axon collaterals from motor neurons. The Renshaw cells, in turn, inhibit the same neurons from which they receive the collaterals, as well as motor neurons innervating synergistic muscles. At the same time they disinhibit antagonistic motor neurons (Fig. 15–3) by an action on interneurons. The

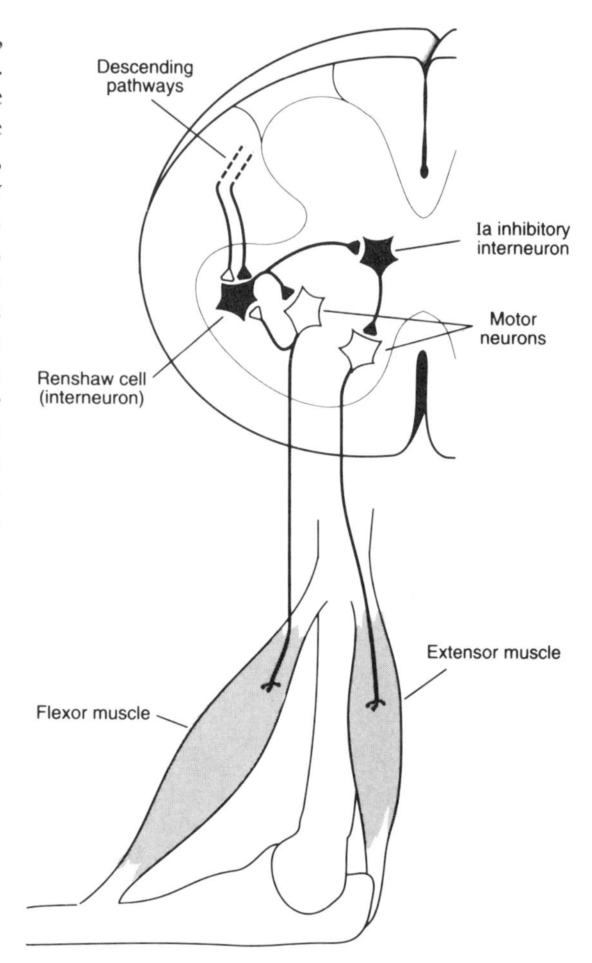

Fig. 15–3. The Renshaw Cell
The Renshaw cell, which is activated by collaterals from motor neurons, produces recurrent inhibition of motor neurons. (From Gordon, J., 1991. Spinal mechanisms of motor coordination. In: E. R. Kandel, J. H. Schwartz, and T. M. Jessell (eds.): Principles of Neural Science. Elsevier, New York, pp. 581–595.)

inhibition produced by the Renshaw cells is also known as *recurrent inhibition*, since their effects are elicited from the axon collaterals of motor neurons.

Central Pattern Generators

Based on mutual interrelations and specific connections with motor neurons, certain populations of interneurons take on a special significance in the sense that they form highly sophisticated circuits, which are connected in a specific pattern to flexor and extensor motor neurons, and which are capable of facilitating and sustaining coordinated locomotor activities and rhythmic limb movements if properly activated. This is the essence of the so-called *central pattern generators*, whose activity can be easily appreciated by anyone who has seen a decapitated hen run across a barnyard. The central pattern generator can be regarded as "a central framework for the stereotyped locomotor movements" to quote the Swedish neurophysiologist Sten Grillner, who has successfully promoted this concept. Each limb, apparently, has its own central pattern generator, and the coordination between the upper and lower limbs is provided by propriospinal neurons.

In the human, such spinal locomotor circuits cannot generate rhythmic movements, e.g., walking, in the absence of supraspinal control. However, in response to commands and modulatory influences reaching the spinal cord from the periphery and through the supraspinal descending pathways, the locomotor generators are able to carry out many of the detailed and complicated neuronal interactions that are a prerequisite for the highly coordinated rhythmic movements that characterize locomotion. A special area capable of inducing rhythmic locomotor movements, the *mesencephalic locomotor region*, has been identified in the mesencephalic tegmentum (see below). Neurons in this region appear capable of activating neurons in the lower brain stem, which project to the spinal cord and activate the spinal pattern generators.

Muscle Receptors

The Muscle Spindle

Intrafusal Muscle Fibers. The *muscle spindle* is an intricate sense organ consisting of an encapsulated bundle of specialized muscle fibers, the *intrafusal fibers*. The intrafusal fibers attach to the extremities of the *capsule*, which ultimately attaches to the tendon just as the extrafusal muscle fibers do. The intrafusal fibers, therefore, are arranged in parallel with the extrafusal fibers. This means that the tension in the intrafusal muscle fibers increases if the muscle is stretched and decreases when the muscle contracts. The muscle spindles, therefore, can be referred to as "length recorders," since they record the length of the muscle. This is known as the *static response*. The muscle spindle can also mediate a *dynamic response*, which provides information about the speed of stretching. The static response is mainly related to thin intrafusal fibers called *nuclear chain fibers*, whereas the dynamic response originates in thicker intrafusal fibers called *nuclear bag fibers* (Fig. 15–4).

Primary and Secondary Endings. There are two types of sensory nerve endings in the muscle spindles, referred to as *primary* and *secondary endings*. Both types of intrafusal fibers have primary endings wrapped around their central portion, whereas secondary endings are related primarily to nuclear chain fibers. The impulses from the endings, which are located in the central parts of the intrafusal fibers, are transmitted to the CNS either by large, fast-conducting group Ia afferent fibers (from primary endings) or by thinner and more slowly conducting group II fibers (from secondary endings).

Fusiform Fibers. The contractile polar parts of the intrafusal fibers are innervated by *fusimotor fibers*, i.e., the axons of the γ motor neurons, which are intermingled with the α motor neurons of the same muscle. The spindle, in other words, has its own motor innervation and its sensitivity is subjected to central control. It is important to

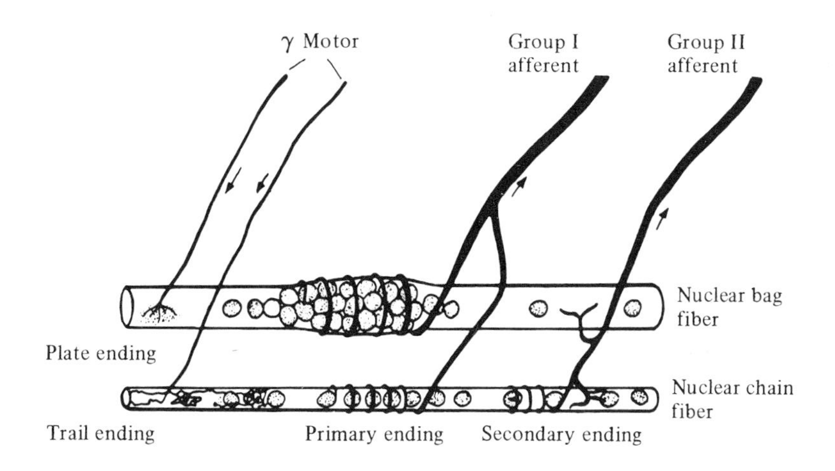

Fig. 15–4. The Muscle Spindle
Simplified diagram of the central region of a muscle spindle showing the two types of intrafusal muscle fibers and their sensory and motor innervation. (From Brodal, A., 1981. Neurological Anatomy, 3rd ed. With permission of the author and Oxford University Press.)

recognize this fact, because it means that the sensory part of the spindle can be kept under tension and continue to send information about the length of the muscle during the full range of a contraction. If the intrafusal fibers could not be contracted, the spindle would quickly be unloaded or silenced at the very beginning of the muscle contraction. In general, the descending pathways activate the α and γ neurons simultaneously, i.e., there is an $\alpha-\gamma$ *coactivation* or linkage. However, it is not a rigid linkage. The contractions needed for complicated movements call for a high degree of flexibility in the $\alpha-\gamma$ linkage and for the possibility of activating the α and γ systems independently.

The Golgi Tendon Organ

This is another important muscle receptor, which is also innervated by fast-conducting afferent fibers (group 1b) similar to the primary muscle spindle endings. The *Golgi tendon organ* consists of a small capsule that surrounds a number of collagen fiber bundles. Although some tendon organs are located within the tendon proper, most line the *aponeuroses*, i.e., the fibrous sheets that provide attachment to the muscle fibers.

Five to fifteen muscle fibers insert themselves into one tendon organ. Therefore, a tendon organ is in series with only a few muscle fibers, whereas it is arranged in parallel with the rest of the muscle fibers. Since the tendon organ is extremely sensitive to contraction of the muscle fibers inserted into the organ, it is an effective monitor of muscle contraction and of the overall tension of the contracting muscle fibers. It is not a "stretch-tension recorder" for safeguarding the muscle from excessive contraction, as previously believed. Rather, the tendon organs work closely together with the muscle spindles in the modulation of movements through various reflexes including long-loop reflexes that involve even the cerebral cortex.

Spinal Reflexes and Muscle Tone

Reflexes are preprogrammed stereotyped reactions that occur in response to stimuli, which in the case of spinal reflexes arise in receptors in the muscles, joints, or skin. Spinal reflexes are important elements of all types of movements that involve the axial musculature or the extremities, and they form the basis for muscle tone. The testing of spinal reflexes is an important part of the neurologic examination; the absence of a reflex, or an exaggerated reflex response, gives the physician valuable information.

A reflex consists of five elements: (1) A *receptor* receiving the stimulus; (2) an *afferent fiber* carrying the information to the CNS; (3) a *reflex center* in the CNS, where the stimulus can be influenced in various ways; (4) an *efferent fiber* carrying the impulse to (5) the *effector organ*. The nature of the reflex center determines to a large extent the complexity of the reflex. The simplest form of reflex is composed of only two neurons, in which case it is called a *monosynaptic reflex*. Examples of monosynaptic reflexes are the stretch or myotatic reflexes. If the reflex center consists of one or more interneurons, the reflex is referred to as a *polysynaptic reflex*. Examples of polysynaptic reflexes are the flexor withdrawal reflex and the crossed extensor reflex. The reflexes mentioned below are only some of the best-known spinal reflexes, which have attracted the detailed attention of a large number of neurophysiologists since the turn of the century. Some of these scientists, i.e., Sir Charles Sherrington, Sir John Eccles, and Ragnar Granit, have been rewarded with the Nobel Prize for their discoveries. The many functions of the spinal cord are dependent on a multitude of reflexes.

The Stretch Reflex and Muscle Tone

The best-known stretch reflex is the *quadriceps (patellar) reflex* or *knee jerk* (Fig. 15–5), produced by tapping the patellar tendon. This reflex is initiated by the muscle spindles, which are sensitive to stretch. The impulses generated by the receptors are transmitted through primary afferent fibers (blue in Fig. 15–5) to the spinal cord, where the fibers establish synaptic contact with motor neurons (red), which in turn produce contraction of the quadriceps and extension of the leg at the knee. At the same time as the quadriceps contracts, the antagonistic muscles, i.e., the flexors of the knee, are inhibited. The inhibition of the flexors is mediated by polysynaptic reflex arcs, and since the motor neurons for the flexors are located in more caudal segments than the motor neurons innervating the quadriceps, the inhibitory reflex is intersegmental.

Some stretch reflexes that are routinely tested in neurologic examination have the following segmental reflex centers:

1. *Biceps reflex* (flexion of the elbow by tapping the biceps tendon): C5–6
2. *Brachioradial reflex* (flexion of the elbow and supination of the forearm by tapping the styloid process of the radius): C5–6
3. *Triceps reflex* (extension of the elbow by tapping the triceps tendon): C7–8
4. *Patellar reflex* or knee jerk (extension of the knee by tapping the ligamentum patellae): L3–4

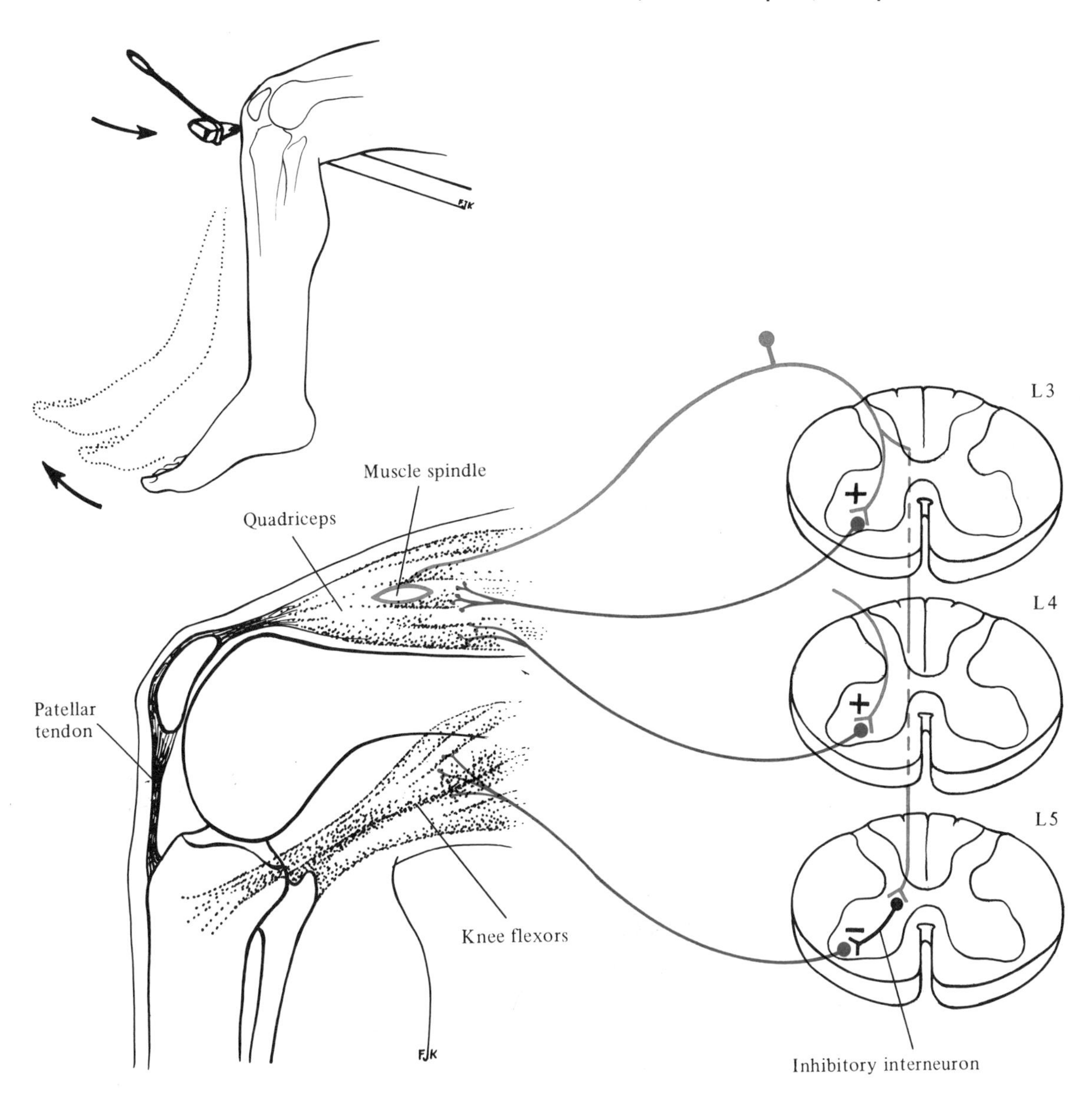

Fig. 15–5. The Quadriceps Reflex
Diagram of the pathways involved in a well-known stretch reflex, the quadriceps or knee jerk reflex. If the quadriceps is suddenly stretched, usually by tapping the patellar tendon with a reflex-hammer (*inset*), the muscle spindles generate impulses that are transmitted in primary afferent fibers (*blue*) to the spinal cord, where they establish contact directly with motor neurons (*red*) innervating the quadriceps. This monosynaptic reflex results in extension of the leg at the knee. For this to occur, there must be a reduction of tension in the antagonists, the knee flexors, which is accomplished by the aid of a polysynaptic intersegmental reflex using an inhibitory interneuron (*black*).

5. *Achilles reflex* (plantar flexion of the foot by tapping the achilles tendon): S1

Muscle Tone. The tension in a muscle during the resting condition or slow imposed movement is referred to as *muscle tone*. It is in large part a reflex phenomenon dependent on the activity in the muscle spindles. Muscle tone can be tested by passively manipulating the limbs of the patient. The resistance to the movements indicates the muscle tone, and experience tells the clinician when the muscle tone is abnormal. Abnormal muscle tone can appear as either *hypotonia* or *hypertonia* (see Clinical Notes).

Flexor Reflex and Crossed Extensor Reflex

The polysynaptic *flexor reflex* serves important protective functions, including the rapid withdrawal of a limb in response to painful or unexpected cutaneous stimuli, i.e., the *withdrawal reflex*. To maintain balance, a flexor withdrawal reflex of one leg is accompanied by increased contraction in the muscles of the opposite limb, which carries the body weight through the action of the *crossed extensor* reflex.

Descending Supraspinal Pathways

As indicated above, the spinal cord contains a multitude of spinal reflexes and highly organized interneuronal systems capable of performing many of the detailed interactions required for coordinated motor activities. However, the spinal cord motor apparatus is dependent on energizing and modulatory influences from the brain. To expose the various brain structures and pathways involved in the control of motor functions in the same picture, we will use an idealized drawing in which the medial surface of the cerebral hemisphere is shown together with a parasagittal section through the basal ganglia, thalamus, and brain stem (Fig. 15–6). The pre-eminent control systems represented by the motor structures in the brain exert their influence through the *descending supraspinal pathways* that originate in the cerebral cortex and in different parts of the brain stem (Fig. 15–7A).

Following a series of detailed tract-tracing experiments with the aid of the Nauta–Gygax silver method (Chapter 7) in the 1960s, it became apparent that the supraspinal tracts that descend in the lateral funiculus terminate in relation to the motor neuronal cell groups (and their interneurons) innervating the limb musculature (Fig. 8–4), whereas the pathways in the ventromedial funiculus distribute their terminals among the motor neuronal pools related to axial and girdle musculature. Based

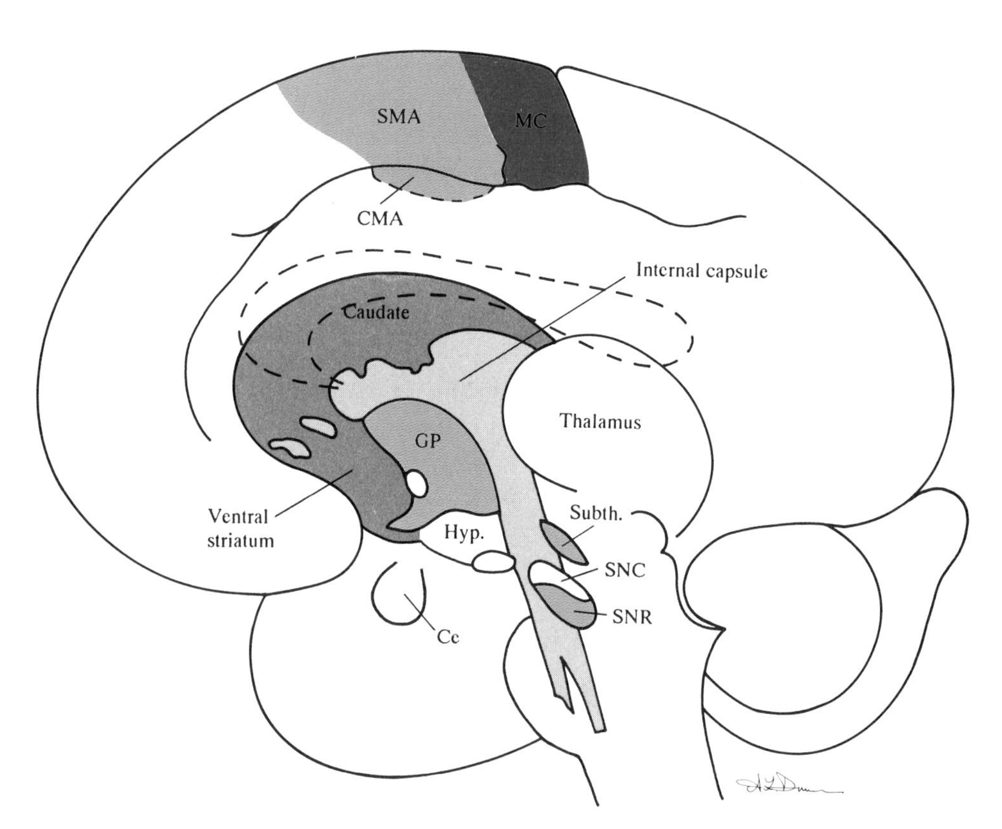

Fig. 15–6. Overview of the Motor Structures
An idealized drawing in which the medial surface of the cerebral hemisphere is combined with a parasagittal section through the basal ganglia and brain stem to show the topographic relations between the cortical motor areas, the basal ganglia, and some of the brain stem motor structures. Abbreviations: Ce = central amygdaloid nucleus; CMA = cingulate motor area; GP = globus pallidus; Hyp = Hypothalamus; MC = motor cortex; SMA = supplementary motor area; SNC = substantia nigra, pars compacta; SNR = substantia nigra, pars reticulata; Subth. = subthalamic nucleus.

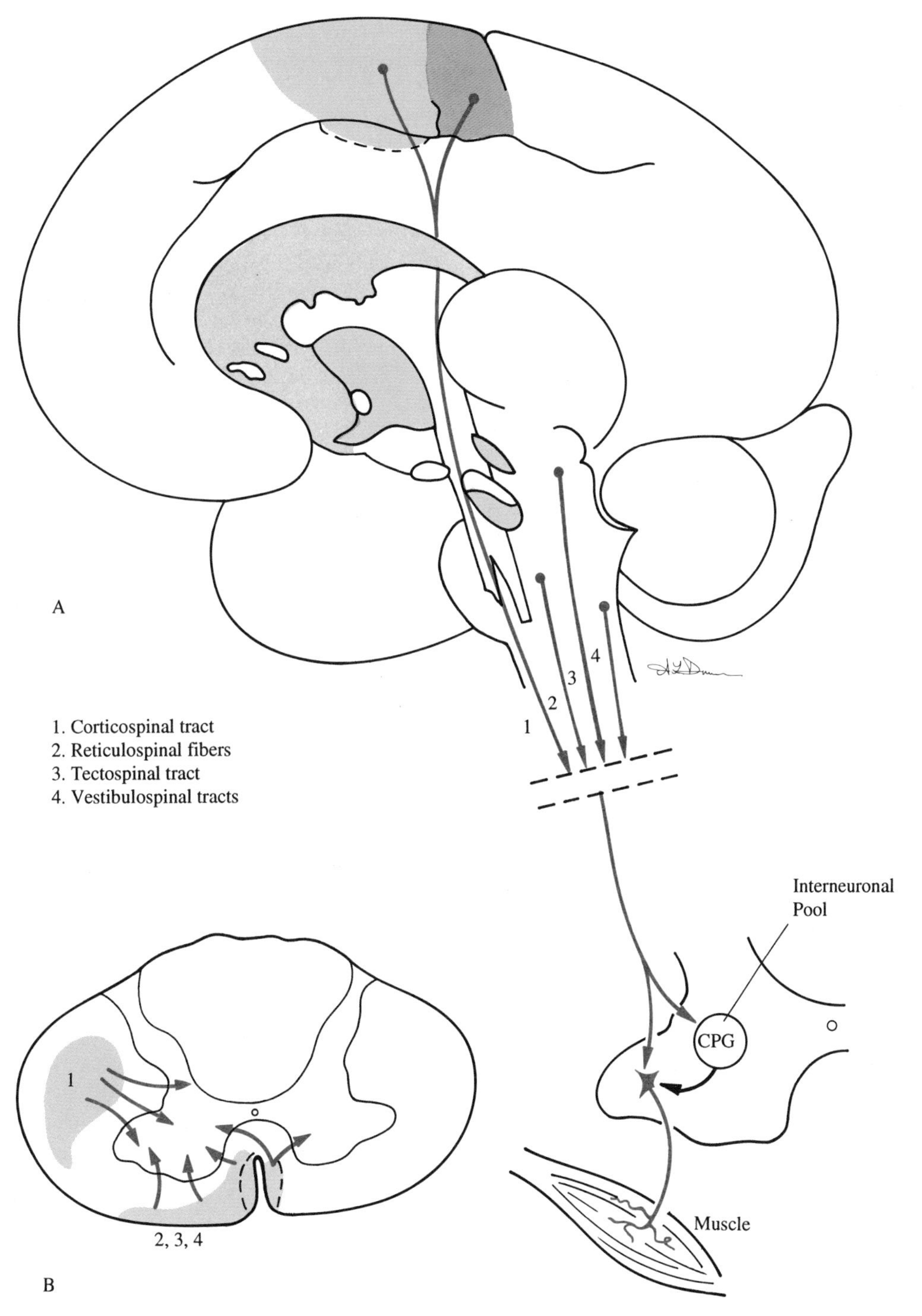

Fig. 15–7. Descending Supraspinal Pathways
A. The most significant descending pathways for the control of the spinal motor neurons originate in the cerebral cortex and in different brain stem structures including the reticular formation, the tectum mesencephali, and some of the vestibular nuclei. The termination of the descending pathways in relation to the motor neurons and the central pattern generators (CPG) in the spinal cord is shown in the lower right-hand corner. Note that only corticospinal tract fibers terminate directly on the spinal motor neurons.
B. Cross-section of the spinal cord showing the general position of the descending supraspinal pathways (see Fig. 15–6 for labeling of structures).

on these experiments, especially those by two Norwegian neuroanatomists, Alf Brodal and Rolf Nyberg-Hansen, and on functional–anatomic studies in the monkey by Donald Lawrence and Hans Kuypers, it became fashionable (for good reasons, one might add) to subdivide the supraspinal motor pathways into a lateral and a ventromedial group. The *lateral group*, including the lateral corticospinal tract, is related primarily to the independent control of limb movements, with special importance for skilled hand movements. The *ventromedial group* of descending supraspinal pathways, including many of the brain stem or "bulbospinal" pathways (vestibulospinal, tectospinal, and many reticulospinal fibers), is more directly concerned with gross movements of the axial musculature for the control of balance and erect posture, and for intralimb coordination. Although this attractively simple functional–anatomic scheme of very complex control mechanisms does have some merit, especially from a didactic point of view, it is no doubt an oversimplification, which tends to hide the fact that we are still far from a complete understanding of many of the descending pathways.

The *rubrospinal tract* deserves special mention. This tract, which in experimental animals has been shown to share many anatomic and physiologic features with the corticospinal tract, is usually included in the lateral group of pathways. However, since the tract does not seem to extend beyond the first cervical segments in humans, it may not have much of an effect on the spinal cord motor structures. Nonetheless, the red nucleus is an important player in a multineuronal circuit that transmits information from many cortical areas, via the red nucleus and the inferior olivary nucleus, to the cerebellum and back to the motor cortex (Chapter 17).

In spite of the fact that we are still far from understanding the detailed functions of many of the descending supraspinal pathways, considerable progress has been made during the last 20 years, with regard to both the functional significance of the various cortical motor areas and the influence of the descending pathways on the spinal motor apparatus. Our appreciation of the neural control of locomotion has benefitted greatly from the notion of the central pattern generators discussed earlier, and from the discovery of a mesencephalic locomotor region by the Russian scientists Mark Shik and Grigori Orlovski in the 1960s.

The Corticospinal Tract

Origin of the Corticospinal Tract; The Cortical Motor Fields

The *cortical motor fields* contain the *pyramidal neurons*, whose axons form the *corticospinal* or *pyramidal tract*[1]. It is estimated that about 80% of the corticospinal tract fibers originate in the motor cortex (area 4) and in the premotor area (area 6), especially in the part that is located on the medial side of the hemisphere, and which is referred to as the *supplementary motor area* (SMA) (Fig. 15–8A). The contribution is especially heavy from the arm representation in these fields. Recently, an additional premotor area has been discovered in the monkey in the anterior part of the cingulate gyrus (*CMA*) (Fig. 15–6), which has traditionally been considered part of the limbic system (Chapter 21). To what extent this area contributes to the corticospinal tract in the human is not known. It is interesting to note that even the postcentral gyrus contributes fibers to the corticospinal tract. Many of these fibers, however, originate in area 3a, which is closely related to the adjoining motor cortex, and which is activated especially by muscle spindles and other deep receptors. A sparse contribution to the corticospinal tract, however, comes from the other somatosensory areas in the postcentral gyrus. These fibers apparently terminate in relation to the dorsal horn of the spinal cord and are concerned with somatosensory functions, although these may well be related to the ongoing movements.

Motor Cortex. Much of the *motor cortex* (*MC*) (area 4), or *primary motor area* as it is sometimes called, is hidden in the central sulcus, and it is only in the more dorsal part of the hemisphere and on its medial side that a significant part of the MC spreads over the free cortical surface (Fig. 15–8A). The MC receives input from other cortical areas, especially the premotor areas and the somatosensory cortex in the postcentral gyrus, and from the ventrolateral thalamus, which in turn is related both to the cerebellum and the basal ganglia (Chapters 16 and 17). The major efferent pathway is the corticospinal tract and it is estimated that about 40–50% of the fibers in the medullary pyramids originate in the MC.

1. Since each corticospinal tract contains about 1 million fibers and there are only 20 to 30 thousand giant pyramidal cells of Betz in the motor cortex, only 2–3% of the pyramidal tract fibers come from the giant cells. The pyramidal tract received its name from the fact that its fibers form the pyramid of the medulla oblongata, and not because its fibers originate in pyramidal cells in the cerebral cortex.

The motor functions of the cerebral cortex were demonstrated more than 100 years ago in a famous experiment by two German scientists, Eduard Hitzig and Gustav Fritsch, who obtained muscular contraction on the opposite side of the body following electrical stimulation of the cerebral cortex. Since then, neurophysiologists have used increasingly refined methods to study the finer details of cortical motor control. Characteristically, discrete movements can be obtained by the application of weak electrical stimulation of the MC. Such studies have shown that the MC has a somatotopic organization referred to as the motor map of the body (Fig. 15–8B).

The homunculus conveys beautifully the idea that different parts of the body are represented in the cortex according to their functional importance. Based on recent PET studies, however, it appears that the motor mouth area is located higher up on the lateral brain surface than shown in this figure (see Fig. 22–8).

It has been shown that activity in the pyramidal neurons in the primary motor cortex produces contraction of single muscles on the opposite side of the body. However, following weak stimulation there is usually some movement also in other muscles, which is consistent with the fact that individual pyramidal tract axons diverge to several different motor neuronal cell groups in the spinal cord. The branching of individual corticospinal fibers, however, is more pronounced for the cell groups innervating the proximal muscles than for the finger muscles.

Premotor Cortical Areas. These cortical fields include the *premotor cortex (PMC)* on the lateral aspect of the frontal lobe and the *supplementary motor area (SMA)* on its medial aspect, both in area 6, as well as a newly discovered premotor area (*CMA* in Fig. 15–6) in the anterior cingulate gyrus (parts of areas 23 and 24). The premotor areas send fibers to the primary motor cortex, and the SMA in particular contributes fibers to the corticospinal tract. Like the MC, the premotor areas are somatotopically organized, but in contrast to the MC, the premotor areas are bilaterally organized.

Although both the PMC and the SMA are characterized by direct projections to the MC, there are also significant differences between the two premotor areas. The PMC on the dorsolateral side of the frontal lobe, while contributing only slightly to the corticospinal tract, is instead characterized by especially prominent projections to the

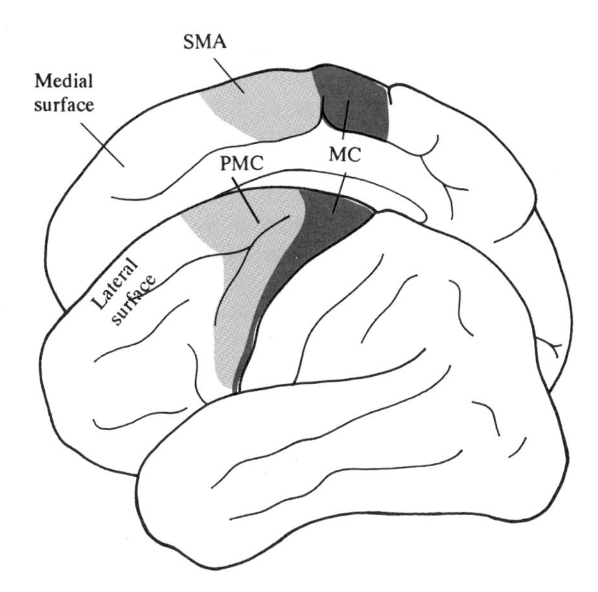

A. Origin of corticospinal tract

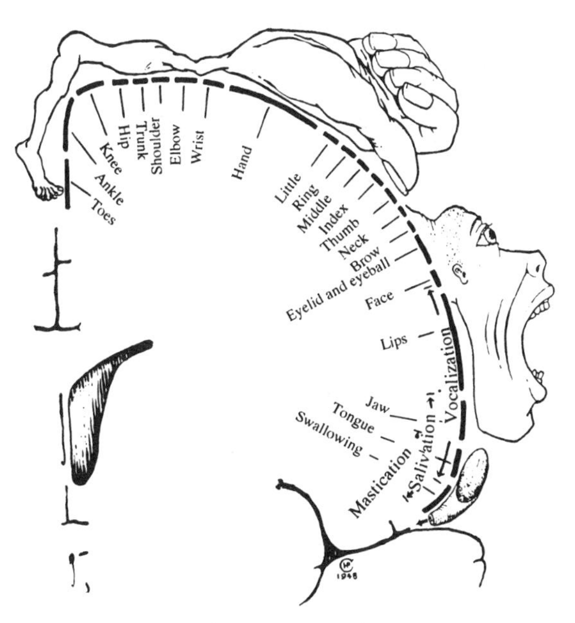

B. Motor homunculus

Fig. 15–8. Cortical Motor Fields

A. Location of the motor and premotor fields in the human brain. The medial aspect of the hemisphere has been projected as a mirror image. (Modified from Alexander, G.E. and M.R. DeLong, 1992. Central mechanisms of initation and control of movement . In: A.K. Asbury et al. (eds.): Diseases of the Nervous System, pp. 285-308. WB Saunders, Philadelphia. With permission.)

B. Diagram showing the motor representation of the various parts of the body in the motor cortex. Note the large areas devoted to the fingers, thumb, and face. (From Penfield, W. and Rasmussen. T. 1955. The Cerebral Cortex of Man. With permission of Macmillan, New York.)

medial part of the brain stem reticular formation, which in turn projects to the spinal cord as part of the ventromedial descending system, i.e., through fibers in the ventromedial funiculus. Via this pathway the PMC plays a major role in postural stabilization and in the control of proximal limb musculature and intralimb coordination. Areas 5 and 7 in the superior parietal lobule send projections to the PMC, and it has been shown in cerebral blood flow studies that the PMC is activated especially in relation to voluntary movements guided by tactile (area 5) and visual (area 7) information. Most skilled movements, such as eating with knife and fork or tying a shoestring, require sophisticated sensorimotor integration. The PMC, like the MC, receives thalamic input from the ventral anterior–ventral lateral (VA–VL) thalamic complex.

The SMA is characterized by a multitude of inputs from other cortical areas in the frontal, parietal, and temporal lobes, and from subcortical regions including the amygdaloid body and several thalamic nuclei. The SMA receives thalamic input from the VA–VL thalamic complex which is related to the basal ganglia and cerebellum (Chapters 16 and 17). The SMA seems to be a focal point for a variety of different pathways from cortical and subcortical regions, which therefore are in an excellent position to influence motor activities directly through the SMA.

The output from the SMA is channeled both directly through the corticospinal tract, and indirectly via the MC. Most of the effect on the distal limb muscles is relayed in the MC, whereas the direct corticospinal system from the SMA seems to be more concerned with the proximal limb musculature. In short, the SMA is an important motor field for voluntary movements of all kinds, and it plays an especially significant role in the planning and initiation of movements.

Course and Termination of the Corticospinal Tract

The corticospinal tract fibers converge toward the posterior limb of the internal capsule, where they occupy a rather small area. As the tract descends through the internal capsule, its position vis-à-vis the genu of the capsule changes. At dorsal levels, the corticospinal tract is located in the anterior half of the posterior limb close to the genu, but it gradually moves into the posterior part of the posterior limb in increasingly more ventral sections (Figs. 15–9A and B). This change in position is due to the fact that the axis of the bend in the capsule is at a slight angle to the axis of the fibers in the capsule (see vignette), which in effect increases the distance between the genu and the corticospinal tract as one examines more ventral sections of the brain. As the relative position of the fibers in the internal capsule changes, many of the fibers that originally were located in the anterior limb at dorsal levels will be accommodated in the anterior part of the posterior limb in more ventral sections. These include the frontopontine fibers, which further down occupy the medial part of the pes pedunculi (Fig. 15–9C). It is useful for the clinician or radiologist to appreciate this shift in position of the corticospinal tract (see Clinical Example), since it is now possible to locate even very small lesions with modern MRI techniques.

As the corticospinal tract descends through the capsule, its fibers intermingle with those from other fiber systems. Therefore, a capsular lesion that affects the corticospinal tract will of necessity influence other also fiber systems, including corticorubral, corticoreticular, and corticopontine fibers (see Pyramidal Tract Syndrome in Clinical Notes).

After having passed through the middle part of the base of the cerebral peduncle (Fig. 15–9C), the pyramidal fibers break up into several bundles in the basilar part of the pons before they come together again to form the *pyramid of the medulla oblongata* (Fig. 15–9D). At the lower level of the medulla, most of the fibers (70–90%) cross to the opposite side in the *pyramidal decussation* and proceed as the *lateral corticospinal* tract in the lateral funiculus to all levels of the spinal cord. Although many of the corticospinal fibers terminate on interneurons in the intermediate gray, many others synapse directly on the α and γ motor neurons, particularly those controlling hand and finger movements. The number of monosynaptic connections between corticospinal fibers and primary motor neurons is especially high in the human, where the voluntary control of finely tuned finger movements, especially, has reached its highest development.

The fibers that do not cross in the pyramidal decussation form the *ventral corticospinal tract*, which terminates bilaterally in the medial parts of the intermediate gray and lamina VIII and related ventromedial motor columns innervating the axial musculature (Fig. 15–9).

Corticobulbar Tract. The corticofugal fibers to the cranial nerve motor nuclei are collectively referred to as the *corticobulbar tract*, and since the function of the corticobulbar fibers is similar to that of the rest of the pyramidal tract fibers, the corticobulbar tract is usually conceived of as an integral

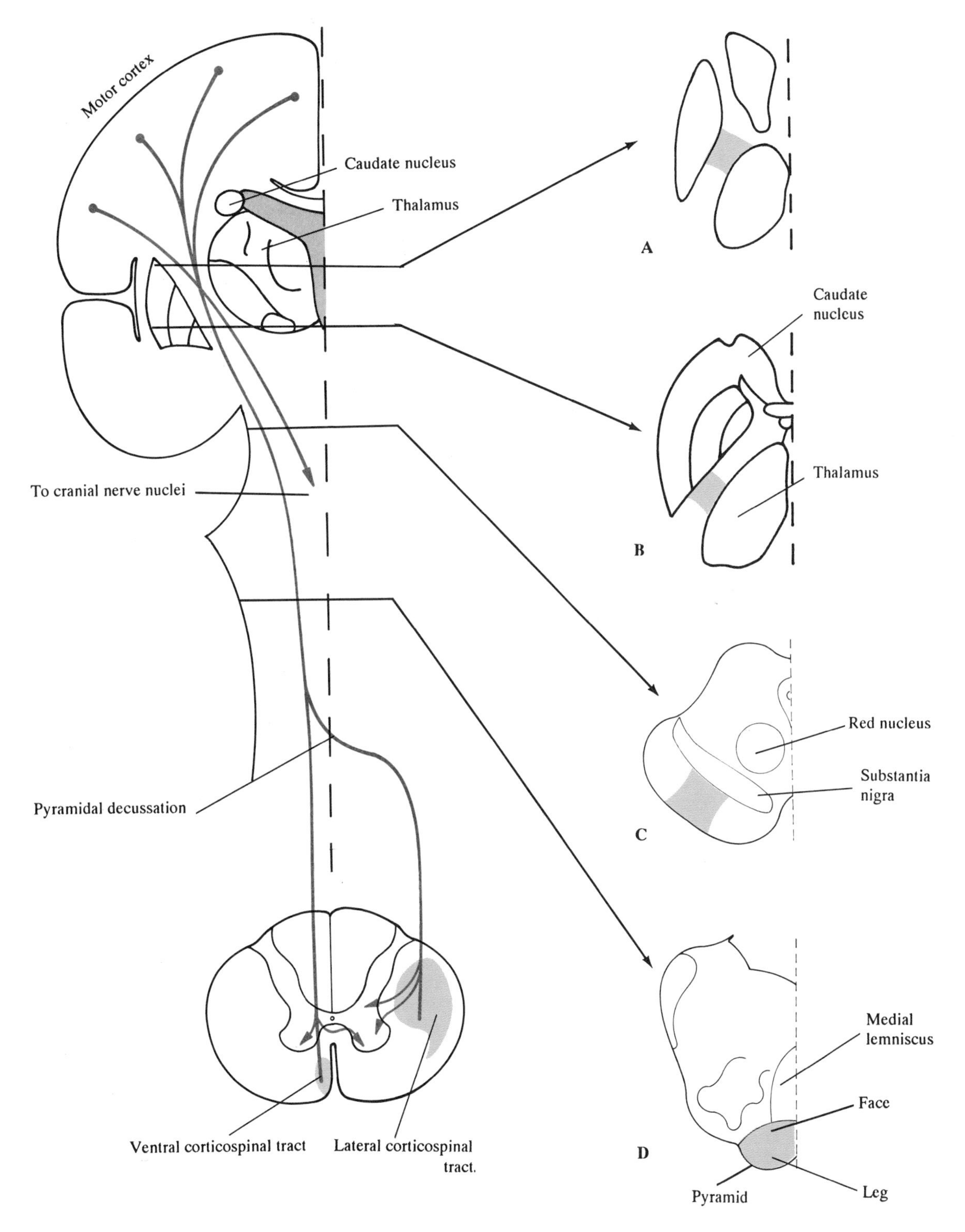

Fig. 15–9. Corticospinal Tract
Diagram of the pyramidal or corticospinal tract. The corticofugal fibers to the cranial nerve nuclei are referred to as the corticobulbar tract.

A and **B.** Internal Capsule
C. Mesencephalon
D. Medulla Oblongata

part of the pyramidal tract. The corticobulbar fibers originate in the face areas of the cortical motor fields, especially in the MC and SMA. Cranial nerve motor nuclei receiving corticobulbar fibers are those of V (trigeminus), VII (facial), IX (glossopharyngeal), X (vagus), XI (accessory), and XII (hypoglossal). In addition, corticofugal fibers from the frontal eye fields in area 8 and from area 46 in the middle frontal gyrus are of special significance for voluntary eye movements. These fibers do not project directly to the eye muscle nuclei, but rather to the superior colliculus and to centers in the brain stem reticular formation, which influence the motor nuclei of cranial nerves III (oculomotor), IV (trochlear), and VI (abducens). Further details regarding the innervation of the cranial nerve nuclei are presented in Chapter 11.

Descending Pathways from the Brain Stem

The dramatic expansion of the cerebral cortex and the concomitant development of the corticospinal tract in primates and particularly humans is a reflection of the important role that the cerebral cortex plays in all aspects of human life including motor behavior. Nonetheless, there are many subcortical structures that also have a significant role to play, and their activity is to a large extent conveyed through brain stem areas with direct lines of communication to the motor neurons in the spinal cord. Continuous modulation of muscle tone and posture is crucial for an effective performance of almost any kind of movement, not least in locomotion, and these adjustments and the many postural reflexes involved are in large part executed by the descending pathways from the brain stem.

The Reticulospinal Fiber Systems

Fibers from the brain stem reticular formation descend in all parts of the spinal cord white matter except in the dorsal columns, and they terminate in all parts of the gray matter to innervate neurons related to sensory functions (Chapter 9), autonomic activities (Chapter 19), and movements. The reticulospinal fibers controlling somatomotor functions originate primarily in the medial parts of the pontomedullary reticular formation, and they are sometimes divided into an uncrossed *medial reticulospinal tract* and a *lateral reticulospinal tract*, which originates on both sides in the reticular formation. Descending fibers from the monoaminergic cell groups in the brain stem are also included in this large group of reticulospinal pathways (Chapter 10).

Considering the large number and wide distribution of reticulospinal fibers, it is a fair guess that the reticulospinal systems are involved in all types of synergistic movements; they work closely together with the vestibulospinal pathways in the control of movements and maintenance of posture. The reticulospinal system, furthermore, seems capable of producing a wide range of highly coordinated movements without the support or control superimposed by the corticospinal tract (see Pyramidal Syndrome in Clinical Notes). As mentioned earlier, the medial reticular formation receives prominent projections from the cerebral cortex, especially the PMC. This projection constitutes the first link in a corticoreticulospinal system which, in view of its considerable size, is likely to exert a powerful effect on the spinal cord motor apparatus. But the input to the reticulospinal neurons in the pontomedullary reticular formation comes from a variety of different sources, including sensory areas in the spinal cord, cerebellum, vestibular nuclei, and mesencephalic structures, and the functions of many of these inputs are still unclear.

The Lateral and Medial Vestibulospinal Tracts

The vestibulospinal fibers that originate in the large cells of the lateral vestibular nucleus of Deiters form the prominent *lateral vestibulospinal tract*, which descends in the ipsilateral ventral funiculus and gradually moves medially as it descends through the spinal cord. The vestibulospinal fibers terminate mostly on interneurons and to some extent on primary motor neurons related to both trunk and limb muscles. The main physiologic effects of stimulation of the lateral vestibulospinal tract are excitation of extensor motor neurons with concomitant inhibition of flexor motor neurons. These effects are in large part responsible for maintaining erect posture and for the various reflexes needed to maintain balance in response to changing conditions. Many of these postural reflexes are modified by an often powerful cerebellar input to the vestibular nuclei (Chapter 11).

Fibers from the medial and inferior vestibular nuclei descend bilaterally in the medial longitudinal fasciculus only to the midthoracic level. This tract, therefore, appears to be concerned primarily with head and neck movements in response to vestibular input.

Mesencephalic Projections to the Spinal Cord

Although the pontomedullary reticular formation and the vestibular nuclei provide the origin of the large majority of the descending motor pathways from the brain stem, several midbrain areas are also involved in the control of motor functions. Some of this control, however, is effectuated via synaptic relays in the pontomedullary reticular formation. This appears to be the case especially for the *periventricular gray* and the nearby *mesencephalic locomotor region*, two regions that have received considerable attention recently. The periaqueductal gray was presented as an important component in a central pain control system in Chapter 9, and it is also known as a key structure for integrating autonomic and somatomotor activities in emotional behavior (Chapter 19). Stimulation of the mesencephalic locomotor region in the nearby mesopontine tegmental area elicits locomotor behavior in experimental animals including primates, and although the significance of this region in human locomotor activities is not known, it is reasonable to expect that the circuits for such basic rhythmic activities as walking and running are rather similar in different mammals including the human. The *mesopontine tegmental area* and its relevance for rhythmic activities in general were discussed in Chapter 10.

In addition to these important indirect motor pathways from the midbrain, there are also projections directly to the spinal cord motor neuronal pools, especially from the superior colliculus and closely related oculomotor centers in the midbrain. Descending fibers from the deep layers of the superior colliculus form the *tectospinal tract*, which crosses the midline in the dorsal tegmental decussation. On its way to the cervical part of the spinal cord, the tectospinal fibers establish contact with groups of interneurons, so-called gaze centers, that connect in a highly organized fashion with the motor nuclei innervating the external eye muscles. It is through the mediation of these gaze centers that conjugated eye movements are made possible (Chapter 11). The final destination of the tectospinal tract is the cervical spinal cord, where its fibers reach the motor neuronal pools innervating the neck muscles. Since the superior colliculus receives direct visual input from the retina, the tectospinal system emerges as an important link through which visual stimuli can directly influence eye and head movements. The fast saccadic eye movements we use as we look from one point to another are elicited from the superior colliculus. It appears that not only visual impulses, but also stimuli from the auditory and somatosensory systems, can influence such movements via the deep layers of the superior colliculus and their projection through the tectospinal tract. The orientation of the eyes and the head and neck toward a novel stimulus is known as the *orienting reflex*, which functions as an important tracking mechanism.

The tectospinal tract is only a part of the prominent *medial longitudinal fasciculus*, which contains both descending and ascending fibers, which are related not only to the oculomotor nuclei and the gaze centers, but to many other regions including the vestibular and reticular nuclei. The medial longitudinal fasciculus extends throughout the brain stem underneath the cerebral aqueduct and the fourth ventricle, and it proceeds downward in the ventromedial funiculus to the midthoracic level, but some fibers continue throughout the spinal cord (Fig. 8–6). The bundle contains many different components, not only from motor regions in the midbrain, but also from other areas of the brain stem, including the medial vestibulospinal tract, and serves as a conduit for axons carrying a large variety of signals to the spinal cord.

The Brain Stem Descending Pathways as a Focus for Converging Impulses from Other Brain Regions

As indicated above, the brain stem areas that are involved in motor control are in turn influenced by many other brain regions and systems including especially the cerebral cortex, extended amygdala and hypothalamus, basal ganglia, cerebellum, and the various sensory systems (Fig. 15–10). Indeed, all major subdivisions of the brain and spinal cord have made it their business to communicate with the brain stem regions that ultimately give rise to the descending pathways to the motor apparatus of the spinal cord. Cortical input to the motor areas in the brain stem reticular formation comes primarily from the PMC and the SMA, and there is considerable evidence that the corticoreticulospinal system, which contains many different components, is directly involved in all kinds of both voluntary and involuntary movements.

The black arrow indicating projections from the ventral part of the basal ganglia, amygdaloid body, and hypothalamus to the brain stem in Fig. 15–10 reflects the important influence that emotion and motivation exert on our motor behavior (Chapter. 19). This single arrow represents a number of different pathways, many of which terminate in mesencephalic structures, including the mesencephalic locomotor region and the periaqueductal gray. The term "emotional motor system" (Holstege, 1992) has been applied to this group of pathways. However, this should not be taken as an indication that the activity in the corticofugal pathways cannot be influenced by emotions.

The basal ganglia, i.e., the striatum, globus pallidus, and substantia nigra, have long been a favorite

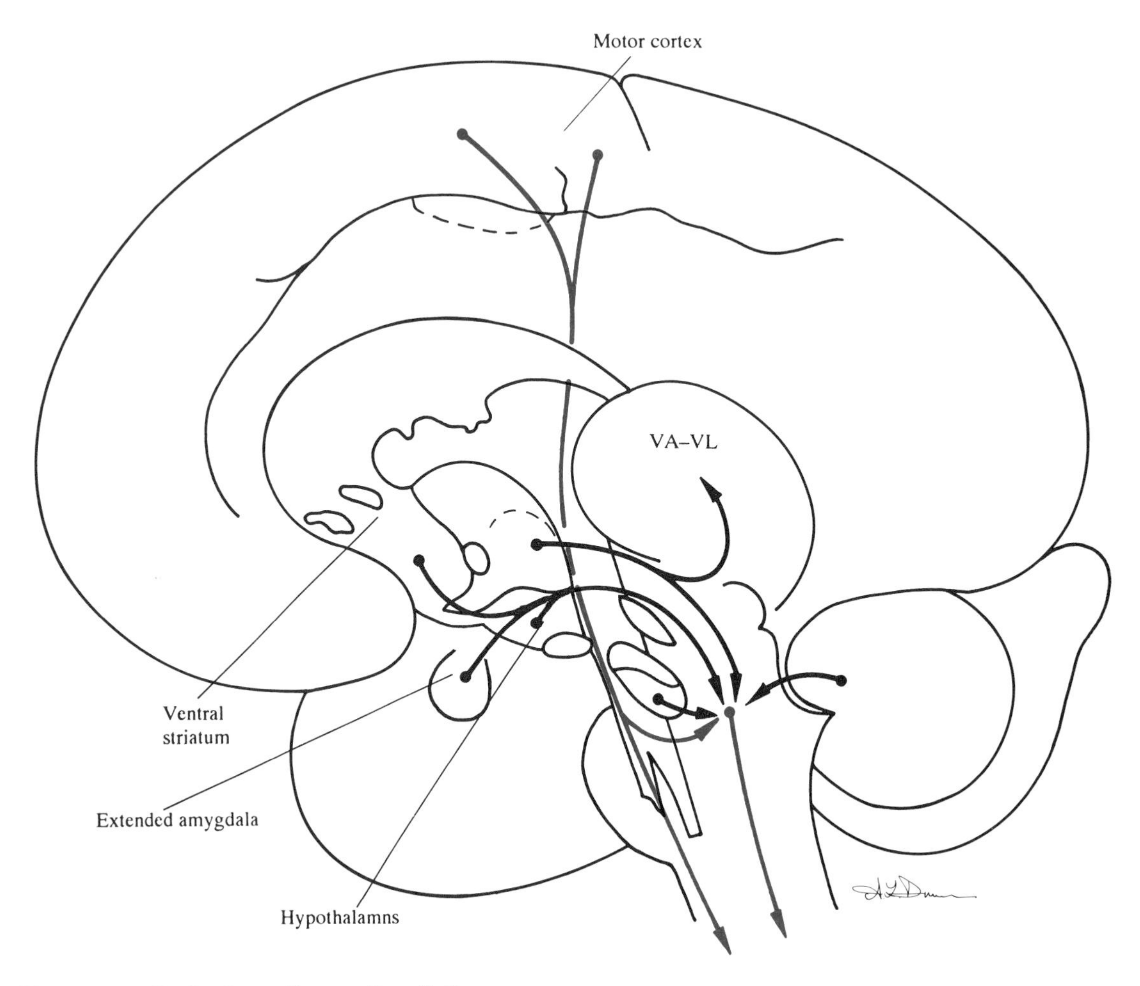

Fig. 15–10. Brain Stem Descending Pathways as a Focus for Converging Inputs from other Brain Regions

subject for those interested in the pathophysiology of motor disorders, especially since lesions in different parts of the basal ganglia produce characteristic motor abnormalities. Although the prominent basal ganglia–thalamocortical re-entrant circuit has attracted most of the attention (see side arrow to the thalamus in Fig. 15–10; see also Chapter 16), the downstream projections to the mesencephalic locomotor region and other brain stem regions (represented in Fig. 15–10 by the black arrows from the globus pallidus and substantia nigra, the two output stations of the basal ganglia) are by no means insignificant. The basal ganglia are discussed in the next chapter.

The cerebellar influence on brain stem motor mechanisms, finally, is prominent and crucially important for balance and posture (Chapter 17).

Clinical Notes

Motor Neuron Paralysis

A lesion that destroys the peripheral motor neurons in the ventral horn or interrupts their axons produces a loss of voluntary movements in the affected muscles. All reflexes, including the stretch reflex, are abolished, and the muscles lose their normal tonus. This produces a *flaccid paralysis*, which together with a *pronounced atrophy* constitutes a characteristic lower motor neuron (LMN) paralysis. This form of paralysis is typical for poliomyelitis, a viral infection of the LMNs. In areas where vaccinations are common, poliomyelitis is seldom seen.

If only part of the motor innervation to a muscle is lost, there will be a partial paralysis, known as *paresis*, and the stretch reflex will be weakened, rather than abolished. The atrophy, likewise, will be less pronounced. In the clinic, the word "paralysis" is sometimes used for both complete and partial loss of motor function.

If sensory deficits accompany the flaccid paralysis, either a mixed nerve is involved or the lesion affects both ventral and dorsal roots.

Electromyography, Fibrillations, and Fasciculations

The electrical activities in a muscle can be tested by inserting needle electrodes into the muscle. The electrodes pick up the electrical activity of the muscle fibers as they contract. The potentials are transmitted, amplified, and displayed on a cathode ray oscilloscope. The technique, which is referred to as electromyography (EMG), is a valuable aid in the diagnosis of motor neuron disorders and in primary muscle disorders.

There should be no electrical activity in a relaxed muscle. Spontaneous activity is abnormal. The spontaneous discharge of a single muscle fiber is known as *fibrillation*, which is characterized by small potentials of short duration. Although fibrillation usually appears in denervated muscles, it can also occur in primary diseases of the muscle. A fibrillation cannot be seen through the intact skin, since the contraction of a single muscle fiber is too small. Fibrillation is an EMG observation.

The spontaneous contraction of the muscle fibers comprising a motor unit or a group of motor units is known as *fasciculation*, and it may be noticed as a twitching under the skin. Fasciculation is a sign of irritation rather than denervation, and it is a common phenomenon in chronic affections of the motor neurons, such as amyotrophic lateral sclerosis, a progressive and ultimately fatal disease of unknown cause. It is important to remember, however, that fasciculations do occur in healthy persons. Therefore, they can usually be ignored, unless accompanied by other signs of disease, e.g., muscle atrophy or weakness.

Paralysis Due to Lesion in CNS

The descending supraspinal pathways can be interrupted[2] by lesions at many levels from the cerebral cortex to the spinal cord, and the distribution of the paralysis is to some extent dependent on the location of the lesion. A lesion involving only the MC may give rise to only a circumscribed weakness or a paralysis of a limb (monoplegia) or part of a limb, whereas a lesion in the internal capsule (capsular lesion) or the brain stem usually involves both the arm and the leg on one side of the body (hemiplegia). Such a lesion involves the impulse traffic from the cortex to both the brain stem centers and the spinal cord.

A chronic paralysis is characterized by *spasticity*,[3] i.e., weakness with increased muscle tone and exaggerated stretch reflexes. There is no significant atrophy of the affected muscles. As a result of the increased reactivity to stretch stimuli, a "clasp-knife phenomenon" may be elicited. That is, if an affected muscle is stretched rapidly it increases muscular resistance to a certain point; then all of a sudden, the resistance melts away, which is due to inhibitory effects from muscle receptors. *Clonus* is another sign of spasticity worthy of attention. It is a series of rhythmic contractions elicited by a sudden stretch of a muscle.

Spastic Hemiplegia

The most common form of paralysis in a human is spastic hemiplegia. A patient suffering from a spastic hemiplegia secondary to a cerebral lesion presents the following signs and symptoms: (1) weakness or paralysis in arm, leg, and often lower half of face contralateral to the cerebral lesion, (2) hyperreflexia, i.e., exaggerated stretch reflexes, and (3) Babinski sign (dorsiflexion of the great toe instead of the normal plantar flexion to plantar stimulation by stroking the sole of the foot with a blunt point).

Such a paralysis primarily affects limb muscles. It seldom involves all the muscles on one entire side of the body. Although most cranial motor neurons and spinal motor neurons innervating neck and axial musculature receive corticofugal innervation from both hemispheres, the part of the facial nucleus that innervates the lower half of the face receives only crossed innervation. The lower half of the face, therefore, is usually affected in a hemiparesis, but there is sparing of muscles in the upper part of the face, because their cranial motor neurons receive bilateral innervation from the motor cortex. The hypoglossal nucleus and the nucleus of the accessory nerve are two other motor

2. Paralysis due to lesions of the descending supraspinal pathways (i.e., the corticospinal, tectospinal, reticulospinal, and vestibulospinal tracts) is sometimes referred to as *upper motor neuron paralysis* to distinguish it from paralysis due to lesions the spinal motor neurons, *lower motor neuron paralysis*.

3. Often the acute paralysis following an intracerebral catastrophe such as intracranial bleeding (cerebral stroke) is flaccid (shock stage). The stretch reflexes subsequently slowly reappear and the paralysis becomes increasingly more spastic as the patient recovers.

nuclei that receive only crossed innervation from the cerebral cortex.

It must be emphasized that a hemiplegic patient is often suffering from other symptoms as well, and the site of the lesion can often be determined on the basis of the additional signs and symptoms. The most common cause of hemiplegia is a vascular lesion in the internal capsule, and a capsular lesion is likely to involve sensory and visual fibers with sensory impairments and homonymous hemianopsia in addition to the hemiplegia. An extensive lesion in the cerebral cortex and the subcortical white matter, especially in the left hemisphere (see Clinical Examples, Chapter 24) may result in speech defects (aphasia) in addition to the hemiplegia.

A brain stem lesion can also cause a spastic hemiplegia. Depending on the level of the damage, one or several of the cranial nerve nuclei are usually involved. Typically, the hemiplegia is contralateral to the lesion, whereas the cranial nerve disorder appears on the side of the lesion. This is referred to as a "crossed (alternating) syndrome" (e.g., Wallenberg's syndrome, see Clinical Examples, Chapter 11).

The So-Called Pyramidal Syndrome

Although the motor symptoms that characterize an upper motor neuron (UMN) paralysis following a capsular lesion are sometimes referred to as a "pyramidal tract syndrome," it must be emphasized that such a lesion involves more than the pyramidal tract. Even if a lesion is restricted to the posterior limb of the internal capsule, it always involves other fiber systems besides the corticospinal tract. Corticostriatal, corticothalamic, corticopontine, and corticoreticular fibers, as well as ascending fiber systems from the thalamus, are also likely to be involved to some extent. The use of the term "pyramidal tract syndrome" for a spastic hemiplegia, therefore, is not correct and should be discontinued. The only place where the corticospinal tract can be damaged selectively is in the pyramids of the medulla, and such pathologic processes are extremely rare. If it occurs, the patient initially becomes hemiplegic, but gradually regains all, or practically all, ability to perform useful and well-coordinated movements, including finger movements. The only residual effects are slightly exaggerated stretch reflexes and a Babinski sign. This is not to say that the corticospinol tract is of no use under normal circumstances. However, it does indicate that other corticofugal fiber systems with relay in the brain stem can fairly well compensate for the lack of corticospinal function.

Comparison Between Motor Neuron and CNS Lesions

Motor neuron lesion	*CNS lesion*
Flaccid paralysis	Spastic paralysis
Pronounced atrophy	No significant atrophy
Hyporeflexia	Hyperreflexia; clonus
No Babinski sign	Babinski sign usually present
Fibrillations and fasciculations	No fibrillations or fasciculations

Monoplegia, Paraplegia, and Tetraplegia

Paralysis of one limb, i.e., monoplegia, is often the result of a motor neuron disorder, in which case there might be a significant degree of muscle atrophy as well as fasciculations and fibrillations. Monoplegia without muscular atrophy may be seen in UMN disorders. A lesion in the paracentral lobulus, for instance, may cause a spastic paralysis of the leg and foot.

Paralysis of the two lower or upper limbs, *paraplegia*, or of all four limbs, *tetraplegia*, occurs

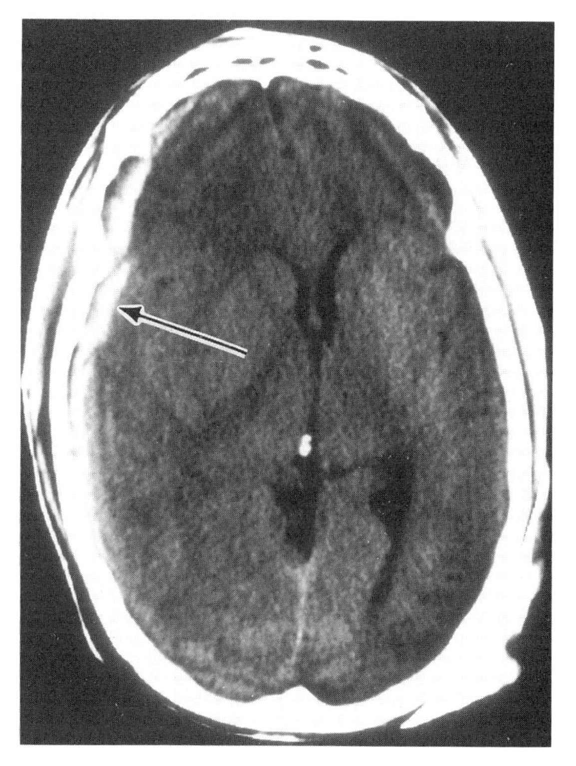

Fig. 15–11. Subdural Hematoma
The CT scan was kindly provided by Dr. Maurice Lipper.

especially in spinal cord disorders. A lesion confined to the thoracic or lumbar part of the spinal cord results in a spastic paraplegia, whereas a lesion in the cervical part of the spinal cord may give rise to a spastic tetraplegia.

Clinical Examples

Subdural Hematoma

A 45-year-old man suffering from chronic alcoholism was brought to the Emergency Room after having been found lying on the street in a daze. He had been hospitalized 2 months earlier, when, in a drunken condition, he had fallen against a lamp post. At that time he was observed for a couple of days, his sensorium cleared without any neurologic deficits, and he was released.

The man was stuporous when he was brought to the hospital, but he could be aroused to answer simple questions. His speech was slow but not noticeably slurred. The only abnormality in regard to his cranial nerves was some difficulty with conjugate lateral gaze to the left. Motor examination showed a mild left hemiparesis, with increased reflexes and a Babinski sign on the left side. His serum alcohol level was not significantly raised.

In view of the focal neurologic signs, a CT scan was obtained (Fig. 15–11).

1. Where is the abnormality located?
2. Why does he have a left hemiparesis?
3. Why does he have difficulty with conjugate lateral gaze to the left side?
4. What is the elliptically shaped white structure near the midline, and why is the space around this structure dark?

Discussion

The CT scan demonstrates a crescent-shaped dense area that overlies the right frontal and parietal lobes, and is causing compression of the right hemisphere with a shift of the midline brain structures from right to left.

The clinical history and radiologic findings are typical for a subdural hematoma, which is particularly common in persons who drink large amounts of alcohol or other persons subjected to repeated head injury. The may also occur without a history of trauma in elderly individuals. The neurologic deficits are often appropriate to the side of the lesion, but this is not always the case.

Because the hematoma is pressing primarily on the right frontal lobe, it is reasonable to expect that the patient may have deficits referable to the motor areas in the right frontal lobe, i.e., left-sided hemiparesis. His difficulty with conjugate eye movements to the left is consistent with the fact that supranuclear control of conjugate lateral gaze originates in the frontal eye field (area 8) contralateral to the direction of intended gaze (signals to look rapidly to the left originate in the right frontal lobe).

The hematoma was evacuated and the patient recovered fully.

The structure near the midline is the pineal gland, which is calcified in about 50% of individuals over the age of 25. A calcified pineal gland is helpful in delineating the midline on plain skull X-rays. The space around the pineal gland, the superior cistern, is dark because it is filled with CSF, which has a low density compared with surrounding brain structures. The speckled structure on either side of the superior cistern corresponds to the choroid plexus, which is also commonly calcified.

Capsular Infarct

A 70-year-old male was admitted for evaluation of acute onset of weakness and sensory loss. He had at least a 20-year history of hypertension. Erratic compliance with medication, however, resulted in poor control of his hypertension. On the morning of admission, while having his breakfast as usual, he noticed a feeling of numbness in his right leg and arm. When he tried to leave the table he realized that his right leg was too weak to support him.

When seen in the Emergency Room 2 hours later, he was alert and oriented, and his blood pressure was 180/110. Neurologic examination revealed a mild weakness in the lower right side of the face but no other cranial nerve deficits. Muscle strength and tonus were significantly reduced in his right leg and arm. There was a moderate sensory deficit involving pain, temperature, vibration, and proprioception in his right leg and arm. The patient was admitted for observation, and a CT scan obtained shortly thereafter was normal. During the following days his blood pressure decreased slowly with medication. Whereas the somatosensory deficits became less and less pronounced, the weakness in his arm and leg did not improve significantly. Muscle tone and stretch reflexes slowly increased in the right arm and leg associated with the development of a Babinski sign on that side. Two weeks after admission another CT scan was obtained. This time a lesion was clearly seen in the brain substance (see arrow in Fig. 15–12).

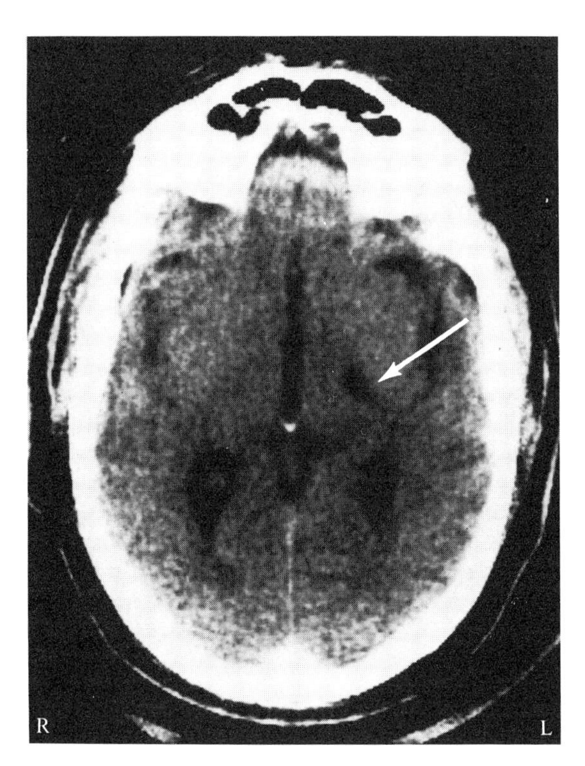

Fig. 15–12. Capsular Infarct
The CT scan was kindly provided by Dr. V. Haughton.

1. Where is the lesion located?
2. Why does he have weakness in his right arm and leg but only in the lower half of the face on the right side?
3. Why did he initially have sensory loss?

Discussion

The lesion, an ischemic infarct, is located in the posterior limb of the left internal capsule. This type of lesion, involving the small penetrating branches of the cerebral arteries, is commonly found in persons with long-standing hypertension and atherosclerosis and is designated a "lacunar infarct." When the softened tissue in the infarct has been removed it leaves a cavity, which can be as large as 1–1½ cm in diameter.

The lesion involves the posterior limb of the internal capsule, apparently including the corticospinal tract (see Fig. 15–9B). The corticospinal fibers innervating the face may be less involved because they are located at the rostral border of the infarct. Since the part of the facial nucleus that innervates the upper half of the face receives corticofugal fibers from both hemispheres, there is no weakness in the upper part of his face on the right side.

The transient hemisensory deficit was most likely due to some involvement of the ventral posterolateral nucleus (VPL) or the very first part of the somatosensory radiation before its fibers start to diverge to different parts of the cerebral cortex. The optic radiation, which is located primarily in the retro- and sublenticular parts of the internal capsule, was not involved.

The lacunar infarct described above is somewhat unusual in the sense that it involves both motor and sensory pathways in the internal capsule. The lacunae are usually small enough and strategically located in the basal ganglia, thalamus, or internal capsule to produce either pure motor strokes when the lesion is located in the internal capsule or the base of the pons, or pure sensory strokes, especially if the lacuna is located in the VPL. Lacunar infarcts have become more rare after the introduction of effective antihypertensive therapy.

Suggested Reading

1. Adams, R. D. and M. Victor, 1993. Principles of Neurology, 5th ed. McGraw-Hill: New York, pp. 39–55.
2. Alexander, G. E. and M. R. DeLong, 1991. Central mechanisms of initiation and control of movement. In A. K. Asbury, G. M. McKhann, and W. J. McDonald (eds.): Diseases of the Nervous System, pp. 285–308. W. B. Saunders Philadelphia.
3. Asanuma, H., 1988. The Motor Cortex. Raven Press: New York.
4. Bucy, P. C., J. E. Keplinger, and E. B. Siqueira, 1964. Destruction of the "pyramidal tract" in man. J. Neurosurg. 21:385–398.
5. Freund, H.-J., 1984. Premotor Areas in Man. Trends in Neurosciences, Vol. 7, pp. 481–483. Elsevier: Amsterdam.
6. Grillner, S. and R. Dubuc, 1988. Control of locomotion in vertebrates: Spinal and supraspinal mechanisms. In S. G. Waxman (ed.): Advances in Neurology, Vol. 47, pp. 425–453.
7. Holstege, G., 1992. The emotional motor system. Eur. J. Morphol. 30:67–79.
8. Kuypers, H. G. J. M., 1981. Anatomy of the descending pathways. In Handbook of Physiology, Sec. 1, The Nervous System, Vol. 2: Motor Control Part 2. American Physiological Society, pp. 597–666.

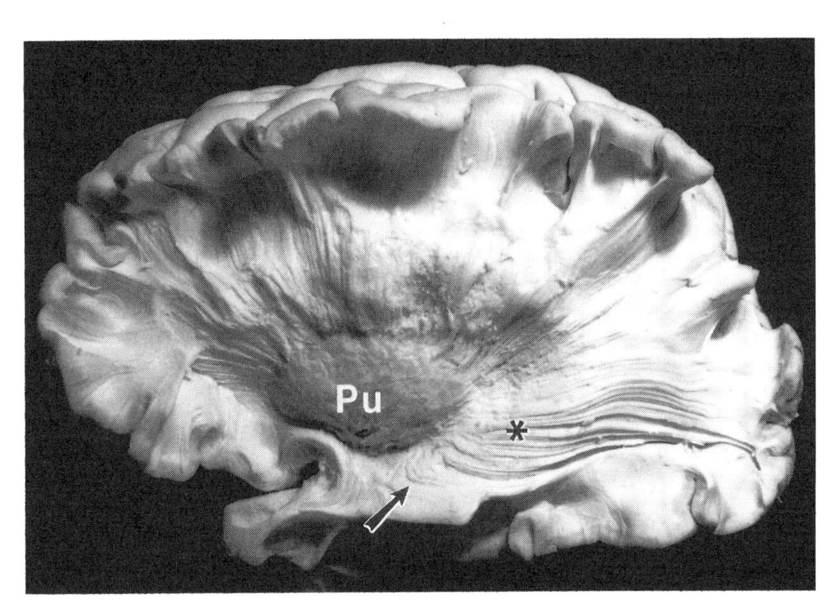

Putamen in Lateral View
Blunt dissection of the left cerebral hemisphere from the lateral side has exposed the putamen (Pu), the most lateral part of the basal ganglia, all of which are located in the ventral half of the hemisphere. Note how the geniculocalcarine tract (*star*) is coming into view underneath the caudal half of the putamen. Some of its fibers make a rostrally directed detour referred to as Meyer's loop (*arrow*) outside the temporal horn of the lateral ventricle before turning caudally towards the occipital lobe. The U-shaped bundle between the orbital part of the frontal cortex and the temporal lobe is the uncinate fasciculus.

16

Basal Ganglia and Related Basal Forebrain Structures

The term basal ganglia refers to some large nuclear masses in the basal telencephalon, including the caudate nucleus, putamen, and globus pallidus. Two closely related structures, the substantia nigra and the subthalamic nucleus, are also included. Through the use of modern tracer methods, the connections of the basal ganglia have been largely clarified during the last two decades. The most prominent pathways are represented by the many corticosubcortical loops that connect all parts of the cerebral cortex, via the basal ganglia and thalamus, with motor, premotor, and prefrontal cortical areas in the frontal lobe. Important circuits also connect the basal ganglia with motor and premotor areas in the brain stem.

The basal ganglia are best known for their motor functions, and they seem to be involved in both the intiation and control of movements. Diseases of the basal ganglia, accordingly, produce characteristic motor symptoms. The classic syndromes of akinesia, rigidity, and tremor seen in Parkinson's disease and chorea–athetosis of Huntington's disease represent the two extremes of basal ganglia dysfunction.

The basal forebrain has recently come into sharper focus, and it has become apparent that the ventral parts of the basal ganglia, especially, are intimately related to two other basal forebrain systems, i.e., the basal nucleus of Meynert and the extended amygdala. These interdigitating and partly overlapping anatomic systems appear to be involved, together with other forebrain regions, in some of the most devastating neuropsychiatric disorders, such as schizophrenia and Alzheimer's disease.

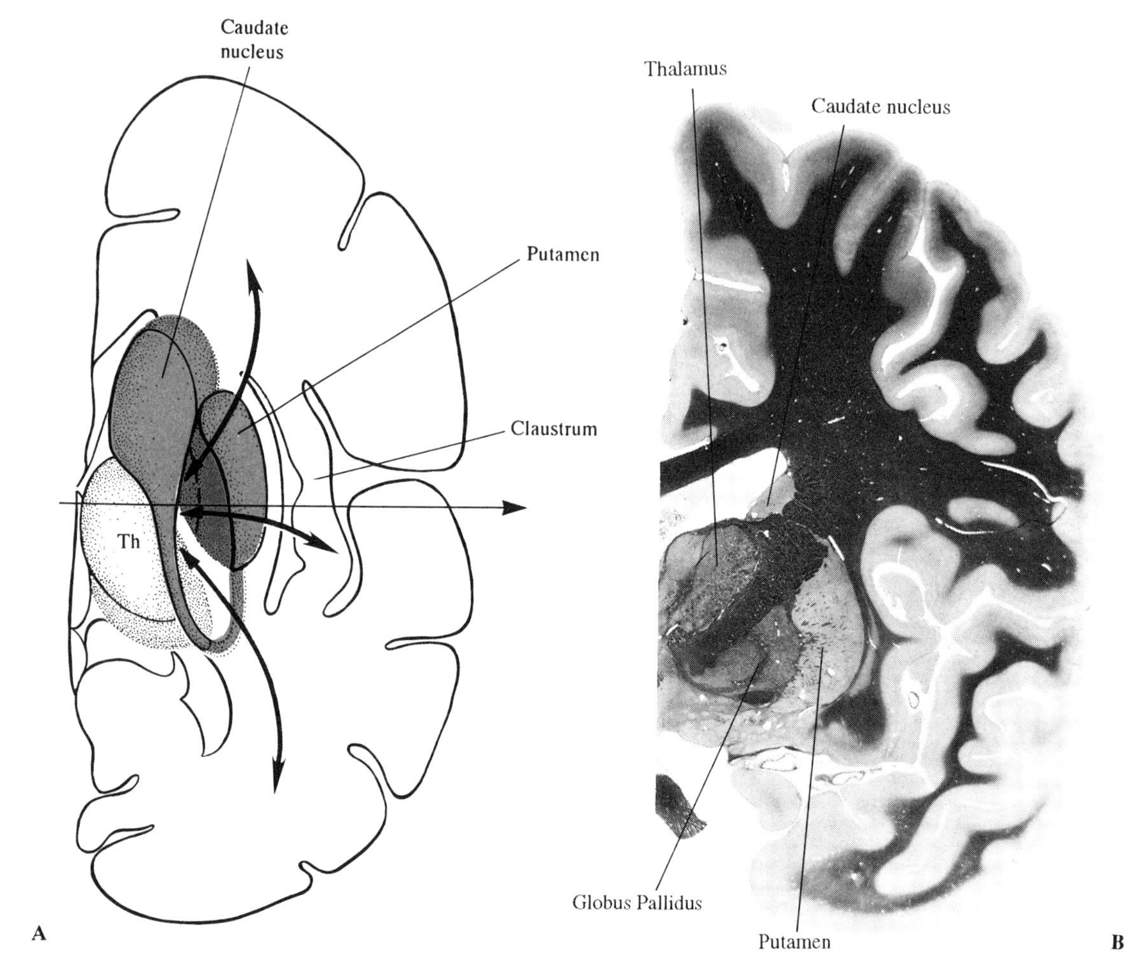

Fig. 16–1. Topography of Basal Ganglia and Thalamus
A. Idealized drawing showing the basal ganglia and thalamus (Th) in three-dimensional rendition within a horizontal section through the hemisphere. Caudate and putamen (*blue*), globus pallidus (*red*).
B. Myelin-stained frontal section through the right hemisphere at the approximate level indicated in **A**.

Components of the Basal Ganglia

There is no generally accepted definition of the term basal ganglia. Early anatomists included most of the forebrain subcortical gray matter in the term, i.e., the caudate nucleus, lentiform nucleus, claustrum, thalamus, and amygdaloid body. Clinicians, however, often use the term basal ganglia in the context of the motor system, and since the claustrum, the amygdaloid body, and a large part of the thalamus are conceived of as primarily nonmotor in function, these structures are generally excluded. The *substantia nigra* and the *subthalamic nucleus*, on the other hand, are widely known for their important motor functions, and are therefore often included in the basal ganglia. Perhaps it is somewhat ironic, considering this past history, that one of the new trends in this field is to think of direct basal ganglia involvement in both motor and nonmotor functions.

The main components of the basal ganglia, the *caudate* and *lentiform nuclei*, are large subcortical structures, which are easily recognized in macroscopic dissection of the brain (Fig. 16–1A). The lentiform nucleus consists of two parts, the *putamen* and the *globus pallidus* (*pallidum*), which are characterized by great differences in histologic structure and anatomic relationships. The putamen, which is the lateral part of the lentiform nucleus, is similar to the caudate nucleus, and the two structures together are called the *striatum*.

Caudate Nucleus and Putamen

Although the caudate nucleus and the putamen are partly separated by the fiber bundles of the internal capsule, bridges of cells connect the two nuclei in many places. Such striatal cell bridges, which indicate the common origin of the caudate and the putamen, are especially pronounced in the anterior limb of the internal capsule (Fig. 5–6).

The striatum is the main target for the afferents that reach the basal ganglia from cortical and subcortical structures. The input impinges on the medium-sized striatal neurons, whose dendritic branches are densely covered with spines (Fig. 6–4B). These spiny striatal neurons also serve as projection neurons, and their axons reach either the globus pallidus or the substantia nigra, which are the two output structures of the basal ganglia. The profusely branching axon collaterals of the medium-sized striatal neurons, furthermore, contribute to most of the intrinsic striatal circuitry, which is augmented by several types of interneurons, including large cholinergic neurons.

The Striatal Mosaic. A compartmental nature of the striatum can be demonstrated by a number of markers, including various peptides and receptors, and it is strikingly demonstrated in histochemical preparations stained for acetylcholinesterase or tyrosine hydroxylase (Fig. 16–2B). Whereas the matrix of the striatal tissue is strongly stained with these two markers, patches of lightly stained areas are scattered throughout the striatum. These more or less irregularly shaped areas were labeled *striosomes*[1] by the American neuroscientist Ann Graybiel.

The work by Graybiel and her American compatriot Charles Gerfen and their colleagues has contributed significantly to the intense interest surrounding the striatal mosaic organization, especially since it turned out that the striatal compartments are in large part correlated with the striatal connections. Although the "striosomal" organization represents a striking morphological feature, which has been compared with the columnar organization of the cerebral cortex, its functional significance is still far from clear. This is a wide open field for scientific inquiry, and the many notions surrounding this subject are beyond the scope of the present text.

Globus Pallidus

The globus pallidus, which forms the medial cone-shaped portion of the lentiform nucleus, consists of an external (lateral) and an internal (medial) division (Fig. 16–1B), the latter of which is closely related to the more caudally located substantia nigra, pars reticulata. The globus pallidus can easily be distinguished from the outer part of the lentiform nucleus, i.e., the putamen, in a freshly cut brain because of its pale color, which reflects the presence of a large number of myelinated fibers. The pallidal neurons are generally larger than the typical striatal cells, and their long, thick dendrites are characterized foremost by the large number of boutons, which often cover the entire dendritic surface.

Ventral Striatum and Ventral Pallidum

The rostral parts of the basal ganglia are covered on their ventral side by the orbitofrontal cortex (Fig. 16–2A). The broad continuity that exists between the caudate nucleus and the putamen at these levels is referred to as the fundus striati, which includes the nucleus accumbens. Further back, at the level of the optic chiasm, it becomes more difficult to appreciate the fact that part of the striatum reaches all the way to the ventral surface of the brain (Fig. 16–3). Nonetheless, the ventromedial extension of the putamen can be followed in continuity towards the ventral surface of the brain in regular Nissl sections (Fig. 16–4). Although this fact was not lost on some of the earlier anatomists, it is only recently, using contemporary histochemical methods that visualize striatal and pallidal markers, that a more precise delineation of the ventral striatal and pallidal territories has been obtained. Such studies leave little doubt that most of the region that has traditionally been referred to as the rostral substantia innominata (between the surface of the anterior perforated space ventrally, and the posterior limb of the anterior commissure dorsally) is in reality occupied by ventral striatal and pallidal territories.

1. Although the striosomes ("striatal bodies") appear as more or less irregularly shaped areas in cross-section, they form in reality branched three-dimensional labyrinths within the more extensive matrix. Recently, a further compartmentation has been discovered within the matrix itself, where cells with specific projections to either pallidum or substantia nigra tend to form clusters referred to as *matrisomes*.

Fig. 16–2. Rostral Parts of Basal Ganglia
A. Myelin-stained frontal section through the rostral part of the brain at the level of the fundus striati or nucleus accumbens. The rectangle indicates the approximate location of the area shown in **B**.
B. Frontal section of the human basal ganglia (at the approximate level shown in **A**) stained for acetylcholinesterase. Note the distinction between the lightly stained striosomes (*arrow*) and the strong immunoreactivity of the matrix. Note the striatal cell bridges in the internal capsule (ic). C = caudate nucleus; GP = globus pallidus; Pu = putamen.
C. Using in situ hybridization, specific nucleic acid sequences can be detected within individual neurons. This picture shows the location of mRNA coding for enkephalin (ENK) in striatal cells in the human. Note the patchy appearance. (Photograph courtesy of Dr. Suzanne Haber, University of Rochester.)

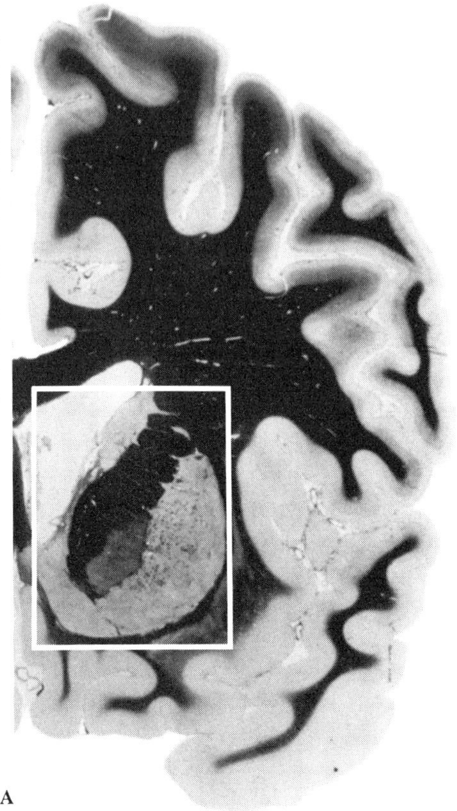

A

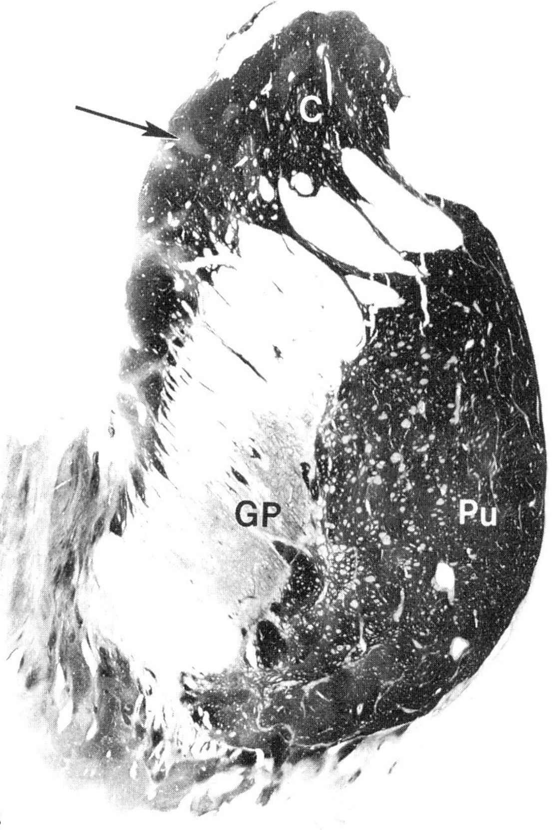

B

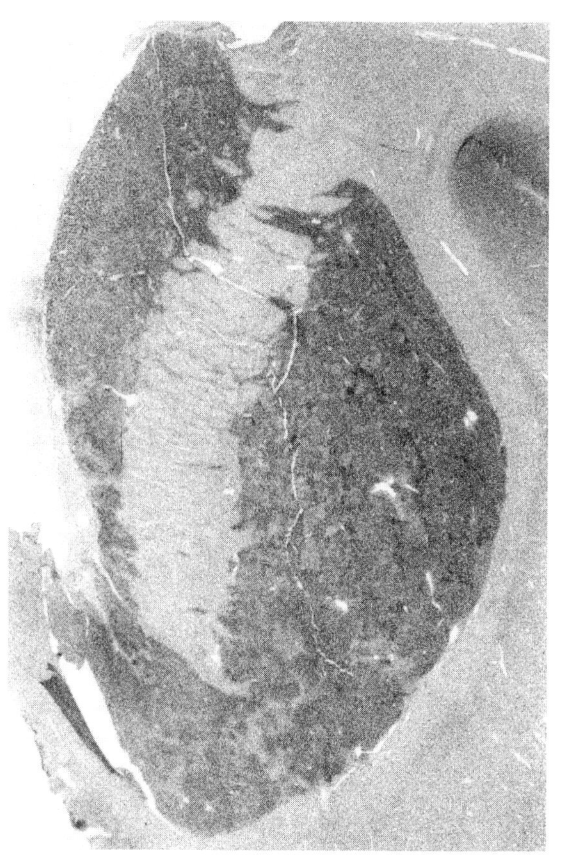

C

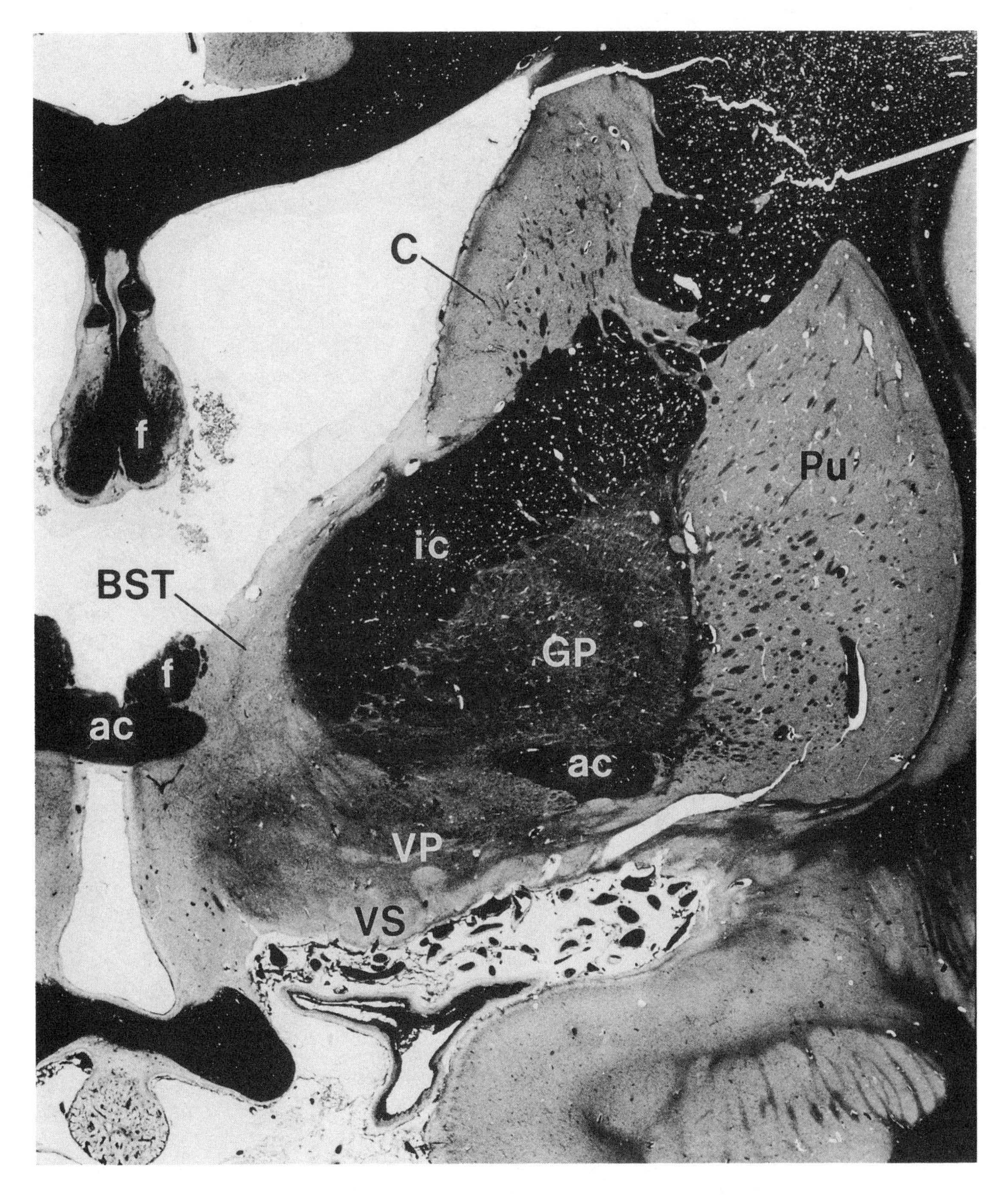

Fig. 16–3. Basal Ganglia at the Level of the Anterior Commissure

Myelin-stained frontal section of the human basal forebrain at the level of the crossing of the anterior commissure (ac) shows the extensions of the basal ganglia to the ventral surface of the brain, and the relation between the basal ganglia and bed nucleus of the stria terminalis (BST). Other abbreviations: C = caudate nucleus; f = fornix; GP = globus pallidus; ic = internal capsule; Pu = putamen; VP = ventral pallidum; VS = ventral striatum. (From Alheid, G. F. et al., 1990. Basal ganglia. In: G. Paxinos (ed.): The Human Nervous System. With permission of Academic Press.)

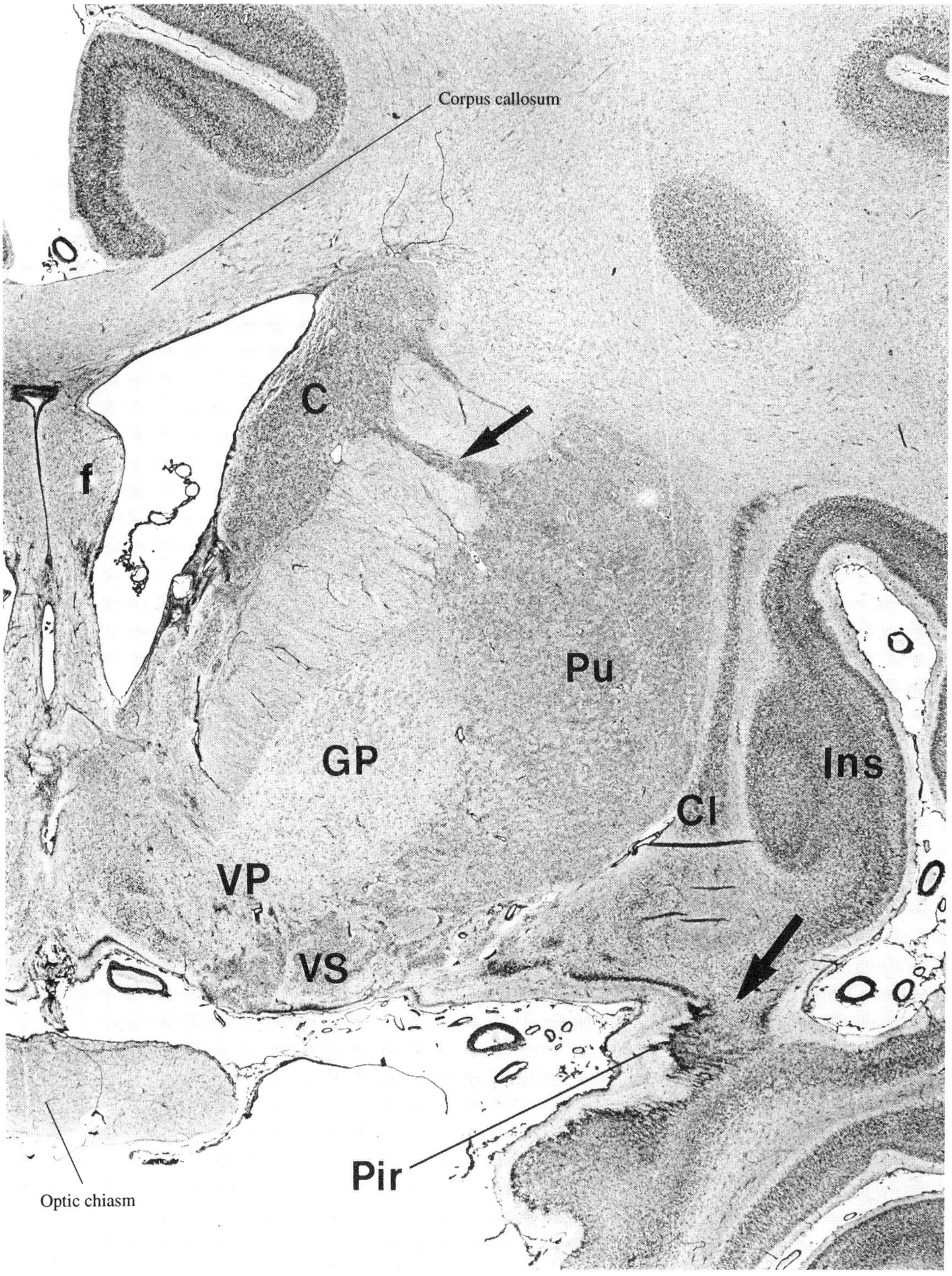

Fig. 16–4. Basal Ganglia

Nissl-stained frontal section at a slightly more rostral level than the one shown in Fig. 16–3. Note the continuity between the caudate nucleus (C) and the putamen (Pu) represented by the cell bridges connecting these two striatal territories (*small arrow*). The ventral striatal tissue extends medially and ventrally from the putamen towards the surface of the brain. The *large arrow* indicates the limen insulae. Cl = claustrum, Ins = insula, Pir = piriform (olfactory) cortex. Other abbreviations as in Figure 16–3.

Substantia Nigra and Ventral Tegmental Area

The *substantia nigra* is a flattened oval structure that rests on the dorsal side of the basis pedunculi or crus cerebri and that extends throughout the mesencephalon from the rostral border of the pons into the subthalamic area (Figs. 10–5 and 5–22A). It can be easily recognized with the naked eye (Fig. 16–5A) because the neurons in its cell-rich part, the *pars compacta*, contain *neuromelanin*[2]. The neuromelanin-containing cells (Fig. 16–5C), which project to the large dorsal expanse of the striatum, synthesize and use *dopamine* as their transmitter. Dopaminergic cells closer to the midline, in the ventral tegmental area, are continuous with the dopaminergic cells in the substantia nigra and project to the ventral striatum, amygdala, and cerebral cortex.

Degeneration of the dopaminergic cells, primarily in the substantia nigra, is the cardinal pathologic sign of *Parkinson's disease* (Fig. 16–5B; see also Clinical Notes). Alzheimer's disease (Clinical Notes, Chapter 21) is also characterized by degeneration of pigmented brain stem cells, but in this case the locus ceruleus and ventral tegmental area are usually more heavily affected than the substantia nigra.

The cell-sparse part of the substantia nigra, the *pars reticulata*, is located ventral and adjacent to the dopaminergic cell groups of the pars compacta. The pars reticulata and the medial pallidum share many (but not all) of the same histologic characteristics, and since both structures represent the output side of the basal ganglia, they are often considered together as part of the same entity.

Subthalamic Nucleus

This is a lens-shaped nucleus located on the medial side of the internal capsule at the border between the diencephalon and the mesencephalon, where it partly overlies the substantia nigra (Fig. 5–11A). The subthalamic nucleus is a key structure in the basal ganglia circuitry, and like the striatum, it influences motor activities primarily through its prominent projections to the two output structures of the basal ganglia, the globus pallidus and the substantia nigra, pars reticulata. In addition to its intimate relations with the other basal ganglia structures (see below), the subthalamic nucleus receives direct input from motor regions in the cerebral cortex. A lesion in the subthalamic nucleus produces an unusual disorder called *hemiballismus* (see Clinical Notes).

Connections

Afferent Connections

Corticostriatal Projections

The most massive input to the basal ganglia comes from the cerebral cortex, and the whole cortical mantle is involved in this highly organized projection system to the striatum (Fig. 16–6). The somatosensory and motor cortices project somatotopically to the putamen, and this input is "funneled" through the basal ganglia (including the substantia nigra, pars reticulata) and the "motor" thalamus (ventral anterior–ventral lateral or VA–VL complex)[3] back to motor and premotor cortex including the supplementary motor area (SMA). The putamen and related parts of the globus pallidus, therefore, have the most obvious influence on motor functions. The cortical association regions in the frontal, temporal, and parietal lobes project instead to the caudate nucleus, where the axonal projections from various cortical regions terminate in elongated longitudinal territories, which often involve the head, body, and tail of the caudate. The hippocampal formation, particularly the subiculum and the entorhinal area, the olfactory cortex, the insula, and the orbitofrontal regions, as well as the basolateral amygdaloid complex, project primarily to the ventral striatum.

Other important afferents reach the striatum from the intralaminar and midline thalamic nuclei (Fig. 16–6), and from the dopaminergic cell groups in the substantia nigra and ventral tegmental area.

Striatopallidal Projections

The medium-sized spiny striatal neurons, which are the recipients of most of the input to the

2. Neuromelanin, which is a polymerized waste product of dopamine and norepinephrine metabolism, is also located in the cells of some other brain stem nuclei, including the locus ceruleus.

3. The terminology in this part of the thalamus is a source of much confusion. The lateral part of the VA nucleus, which receives input from the internal pallidal segment, is often included in the VL thalamic complex and referred to as VLo (or sometimes VLa). The term VLo (ventral lateralis oralis) was coined by the Polish neuropathologist Jerzy Olszewski, who is well known especially for his cytoarchitectonic studies of the brain stem. In order to avoid this terminologic quagmire, we will use the more general term "VA–VL complex" or "motor" thalamus, and simply acknowledge the fact that the various inputs from the globus pallidus, substantia nigra, and cerebellum terminate in different parts of the VA–VL complex. These various inputs to the "motor" thalamus, however, are apparently integrated at the cortical level.

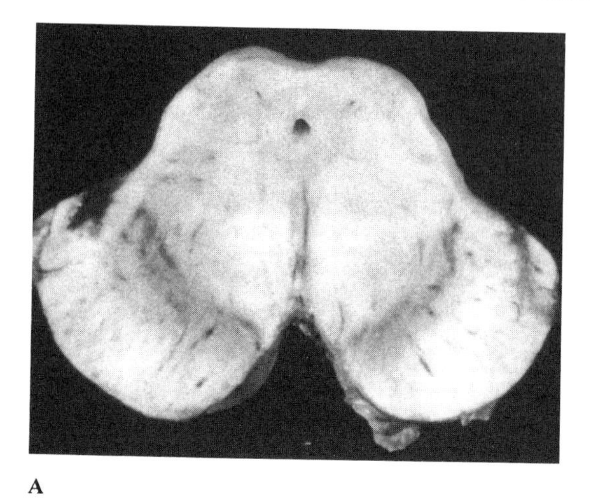

A

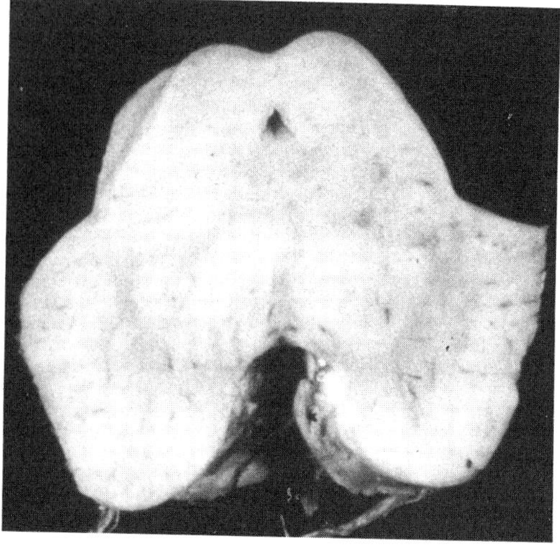

B

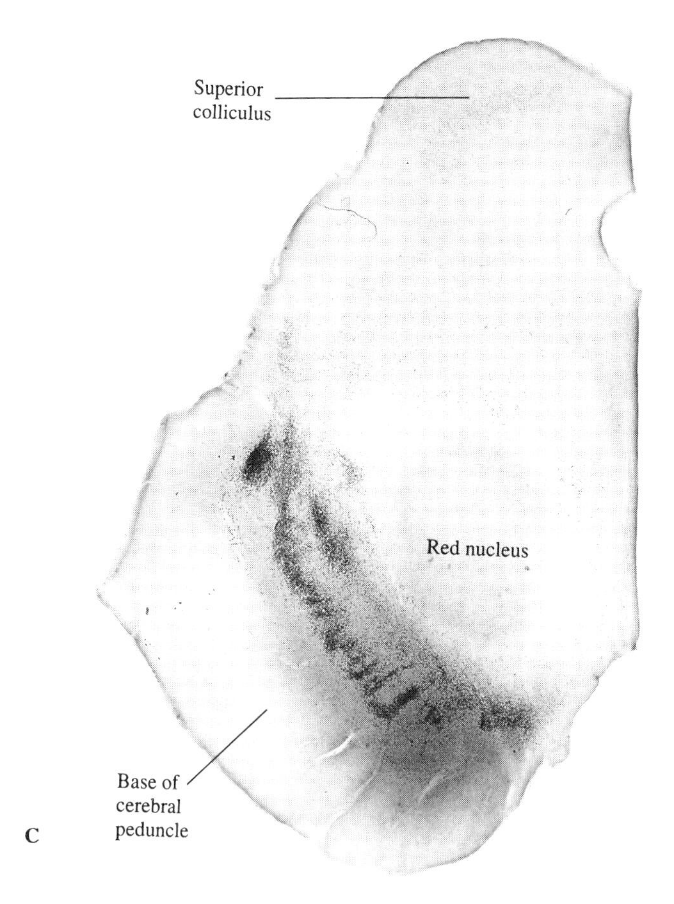

C

Fig. 16–5. Substantia Nigra

A. Unstained cross-section through the human mesencephalon. The dark layer, representing the melanin-containing dopaminergic cells in the substantia nigra, can be easily appreciated.

B. Depigmentation of the substantia nigra in a patient who suffered from Parkinson's disease. (**A** and **B** courtesy of Dr. Lewis Sudarsky. From Sudarsky, L., 1990. Pathophysiology of the Nervous System. With permission of Little, Brown and Company, Boston.)

C. Cross-section through the human mesencephalon stained for the pigment neuromelanin, which is a good indicator of the dopaminergic cells in the pars compacta. (Courtesy of Dr. Hendrik Jan ten Donkelaar, Holland. From van Domburg, P. H. M. F. van, and H. J. ten Donkelaar, 1991. The Human Substantia Nigra and Ventral Tegmental Area. Advances in Anatomy, Embryology and Cell Biology, Vol. 121, pp. 1–132. With permission of Springer-Verlag, Berlin.)

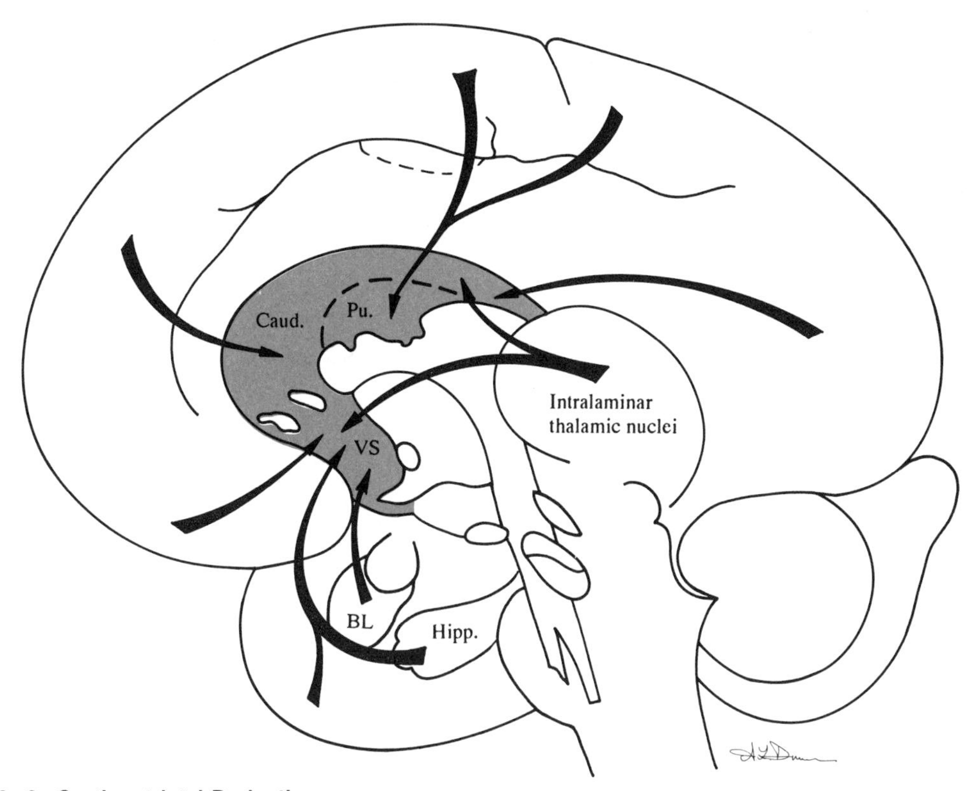

Fig. 16–6. Corticostriatal Projections
The motor and somatosensory cortices project primarily to the putamen (Pu), whereas the frontal, parietal, and occipitotemporal association areas project to the caudate nucleus, and the orbitofrontal cortex, hippocampal formation, basolateral amygdala, and related temporal lobe areas primarily to the ventral striatum (VS). BL = basolateral amygdaloid complex.

striatum, also represent the projection neurons of the striatum, since only their axons reach the two output structures of the basal ganglia, the globus pallidus and the substantia nigra (Fig. 16–7). The projection neurons use GABA as their transmitter, and they can be divided into different subpopulations depending on which target they reach and which peptide they contain in addition to GABA. For instance, the striatal GABA neurons projecting to the external pallidal segment contain primarily enkephalin, whereas GABA and substance P coexist in the neurons that reach the internal pallidal segment and the substantia nigra. Although the caudate projects primarily to the substantia nigra and the putamen primarily to the globus pallidus, there is not an absolute dichotomy in this regard.

Huntington's disease, a well-known but fortunately not very common hereditary disorder, which ultimately affects all striatal neurons, appears to be characterized in its early stage by a selective involvement of the projection neurons to the external pallidal segment (see Clinical Notes).

Efferent Connections

The two output structures of the basal ganglia, the medial division of the globus pallidus and the substantia nigra, pars reticulata, project to the thalamus and the brain stem (Fig. 16–8). The pathways to the thalamus are topographically organized, with the globus pallidus projecting to the VA–VL complex, the ventral pallidum primarily to the mediodorsal thalamic nucleus (MD), and the substantia nigra to the VA–VL complex and MD nuclei, but to somewhat different parts than those receiving the pallidal output. The VA–VL complex and MD, in turn, project to motor–premotor and prefrontal cortical areas.

The pallidal efferents proceed via two different routes. Fibers from the dorsal part of the medial division of the globus pallidus traverse the internal capsule as the *fasciculus lenticularis* (Fig. 16–9), whereas axons of more ventrally located cells in the pallidal complex sweep around the medial edge of the internal capsule as the *ansa lenticularis*. After having either traversed or looped around the in-

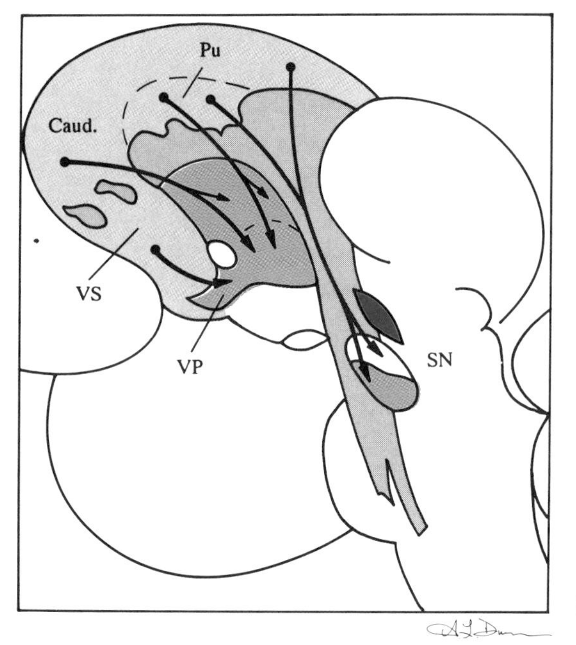

Fig. 16–7. **Striatopallidal Projections**
There is a tendency for the caudate to project primarily to the substantia nigra and for the putamen to reach primarily the globus pallidus, but there is not an absolute dichotomy in this regard. Note that the striatal efferents terminate in both the external and internal divisions of the globus pallidus. Pu = putamen; SN = substantia nigra; VP = ventral pallidum; VS = ventral striatum.

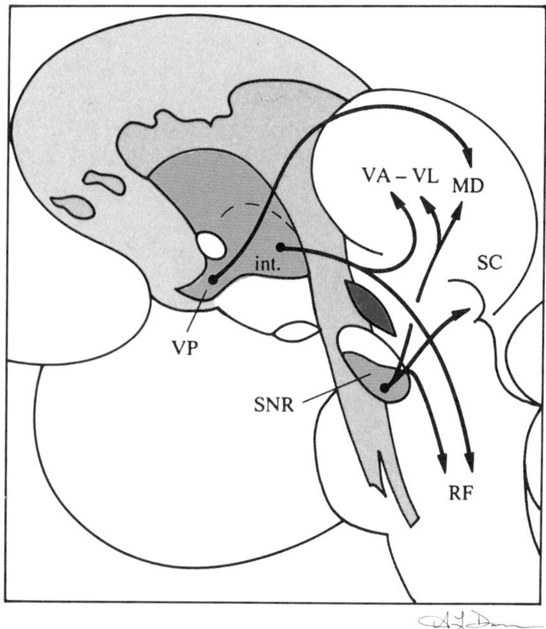

Fig. 16–8. **Pallidofugal Projections**
The output structures of the basal ganglia, i.e., the internal segment of the globus pallidus (int) and the substantia nigra, pars reticulata (SNR), project to the thalamus (VA-VL and MD), the superior colliculus (SC), and the brain stem reticular formation (RF). The output from the ventral pallidum (VP) is less well known, but it appears that a major projection reaches the mediodorsal thalamic nucleus (MD).

ernal capsule, the fibers combine to form a massive fiber system (field H2 of Forel), which courses medially and caudally to the prerubral area (field H of Forel) just rostral to the red nucleus. At this point, the fibers make a sharp U-turn and proceed to the VA via the thalamic fasciculus (field H1 of Forel).

Although the basal ganglia–thalamocortical re-entrant circuits (see below) have attracted most of the attention, the output channels from the basal ganglia, especially from the substantia nigra, pars reticulata, also reach the superior colliculus and the reticular formation in the mesopontine tegmentum. The projection to the mesopontine tegmentum is likely to involve motor circuits including the mesencephalic locomotor region (see "Descending Pathways from the Brain Stem in Chapter 15), whereas the projection to the superior colliculus influences oculomotor functions which are mediated by pathways from the superior colliculus to brain stem regions that control eye and head movements (Chapter 11).

Recently, another projection from the basal ganglia to the reticular thalamic nucleus appears to have been discovered. This nucleus, which surrounds the thalamus like a shield, is in a strategic position to integrate thalamocortical and corticothalamic activities (Chapter 22), and a projection to this part of the thalamus would provide the basal ganglia with another important avenue to influence thalamocortical functions.

Corticosubcortical Re-entrant Circuits

Studies of lesions in humans and animals, and recordings from basal ganglia neurons in experimental animals, tend to show that the basal ganglia are involved not only in motor functions, but also in cognitive functions that may not be primarily related to the control of movements. This opinion has recently been bolstered by the notion of multiple basal ganglia-thalamocortical re-entrant loops (Alexander et al., 1989). For instance, the impulses that come from the sensory-motor cortical areas are channeled through the striatum

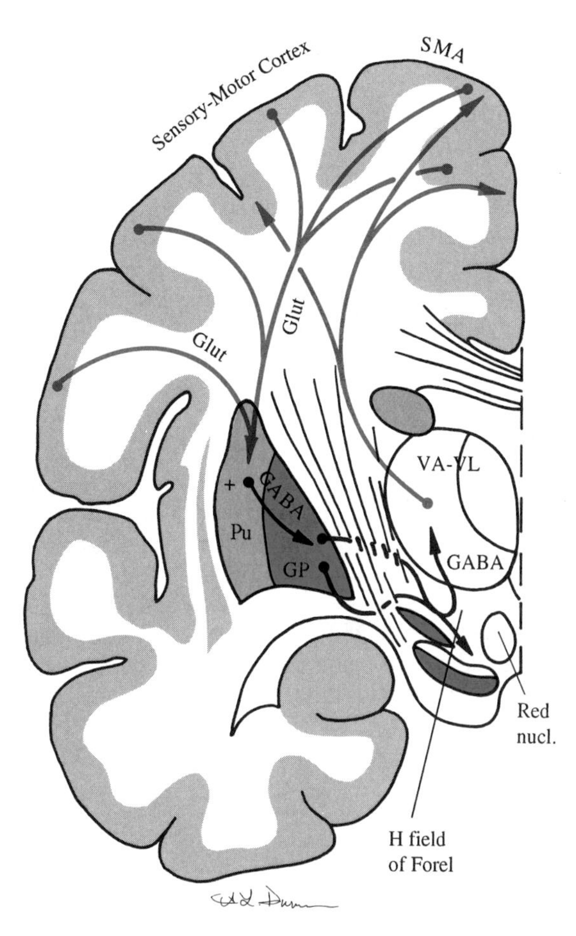

Fig. 16–9. The "Motor" Circuit
Impulses from sensorimotor cortex are channeled through the putamen and globus pallidus to VA-VL and back to the motor–premotor cortex including especially the supplementary motor area (SMA). The corticostriatal projections are glutaminergic (Glut) and excitatory (*red*), whereas the striatopallidal and pallidothalamic projections are GABA-ergic and inhibitory (*black*).

(putamen), pallidum, and thalamus back to cortical motor regions. This is the "motor" circuit (Fig. 16–9). Another re-entrant loop, which reaches the frontal eye field (area 8), is referred to as the "oculomotor" circuit. Several other circuits have been recognized that ultimately influence prefrontal areas rather than motor and premotor cortical areas. A ventral striatopallidal circuit, for instance, reaches the anterior cingulate area via the mediodorsal thalamic nucleus. The functional significance of many of these circuits is still a matter of speculation, nor do we know how the activity in these basal ganglia-thalamoprefrontal circuits ultimately affect behavior.

Important Side Loops

The Nigrostriatal Pathway

The nigrostriatal dopaminergic pathway, which is part of a more widespread mesotelencephalic dopamine projection system (Fig. 10–7), is one of the most studied pathways in the brain. In fact, it was the rapidly increasing knowledge of the nigrostriatal pathway in the late 1950s and 1960s that led to the development of L-DOPA, the drug that revolutionized the treatment of Parkinsonism (see Clinical Notes).

The dopamine-synthesizing cells are located not only in the pars compacta of the substantia nigra, but also in the more medially located ventral tegmental area. This continuous system of dopaminergic cells projects in a more or less topographic fashion to all parts of the striatum, in the sense that the main dorsal part of the striatum (caudate nucleus and putamen) receives input from the pars compacta of the substantia nigra, whereas the ventral striatum, including the accumbens and the olfactory tubercle, receives input primarily from the ventral tegmental area.

Side Loops Involving the Subthalamic Nucleus

The subthalamic nucleus has a pivotal position in the basal ganglia circuitry. It receives major inputs from cortical motor areas and from the external segment of the globus pallidus, and it can influence directly the two output structures of the basal ganglia, the internal pallidal segment and the substantia nigra, pars reticulata (Fig. 16–10).

Functional Considerations

Considering the relatively large size of the various basal ganglia–thalamocortical re-entrant systems, including the "motor" circuit, it is difficult to argue with those who contend that one of the main roles of the basal ganglia is to "regulate cortical functions via their influence on thalamocortical projections" (Albin et al., 1989). Nonetheless, it is important to be aware of the fact that the internal segment of the globus pallidus and the substantia nigra, pars reticulata, can directly influence motor areas in the brain stem reticular formation (Fig. 16–8) including the mesencephalic locomotor region (see "Descending Pathways from the Brain Stem in Chapter 15).

Only the "motor" circuit, which proceeds primarily through the putamen rather than the caudate nucleus (Fig. 16–9), has been investigated in detail by scientists, including the American neurophysiologists Edward Evarts and Mahlon DeLong and their colleagues, who have shown that

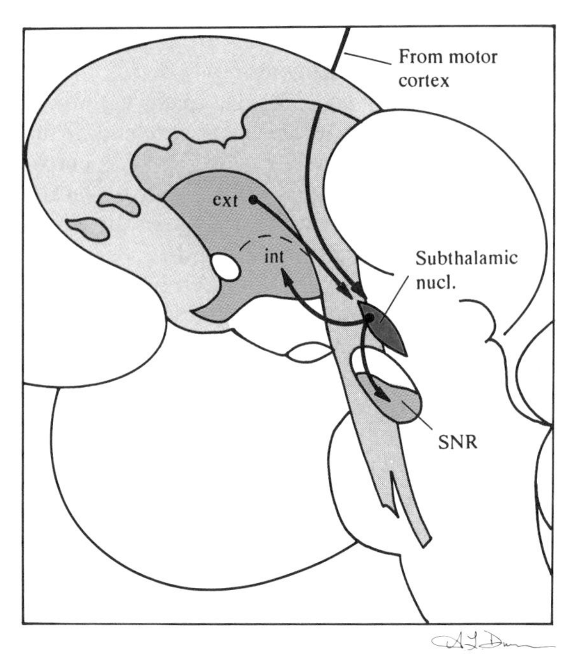

Fig. 16–10. Connections of Subthalamic Nucleus
Abbrs: ext = external segment of globus pallidus; int = internal segment of globus pallidus; SNR = substantia nigra, pars reticulata.

individual neurons in the circuit are tuned to different aspects of motor processing. A striking somatotopic organization and functional specificity, furthermore, seem to be maintained throughout the circuit. Some of the prominent motor disturbances that occur in basal ganglia disorders, including Parkinson's disease, can now be reasonably well explained on the basis of our current understanding of the functional anatomy of the basal ganglia.

The "Direct" and "Indirect" Projections Through the Basal Ganglia

The neurotransmitters employed by the various connecting links in the corticosubcortical loops through the basal ganglia and thalamus are indicated in the diagram of the "motor" circuit in Fig. 16–9. The corticostriatal projections are glutaminergic and excitatory, whereas the two other intermediaries, i.e., the striatopallidal and pallidofugal connections, are GABA-ergic and inhibitory. This means that activation of the corticostriatal system disinhibits the thalamic regions, as well as the brain stem areas, which are innervated by the output structures of the basal ganglia. Many of the theories of basal ganglia movement disorders have focused on this double inhibitory action in the "motor" circuit.

According to the scheme in Fig. 16–11, which relates to the "motor" circuit, there are two major projections through the basal ganglia, i.e., from the input structure (putamen in the case of the "motor" circuit) to the output structures (represented by the internal segment of the globus pallidus). The "direct" pathway projects directly from a certain subpopulation of striatal neurons in the putamen to the internal segment of the globus pallidus. The "indirect" pathway, which has its origin in a different population of striatal neurons, proceeds through the external segment of the globus pallidus and the subthalamic nucleus before reaching the internal segment of the globus pallidus. As indicated in Fig. 16–11, all intrinsic projections are inhibitory (black arrow) except the projection from the subthalamic nucleus to the internal segment of the globus pallidus, which is excitatory (red arrow).

The ascending dopamine fibers from the substantia nigra, pars compacta (left-hand side of Fig. 16–11), impinge directly on the striatal projection neurons. Overall evidence suggests that dopamine has an inhibitory action on the striatal GABA-ergic cells projecting to the external segment of the

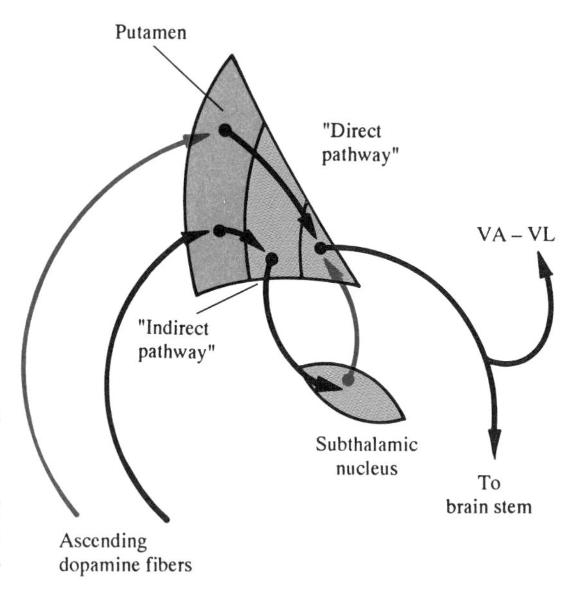

Fig. 16–11. The "Direct" and "Indirect" Projections Through the Basal Ganglia
The ascending dopamine fibers (*left-hand* side of figure), which impinge directly on the striatal neurons in the putamen, differentially affect the "direct" and "indirect" pathways, apparently by acting on different types of dopamine receptors (see text). Excitatory pathway (*red*). Inhibitory pathway (*black*). Decreased dopamine activity would increase the inhibitory activities of the basal ganglia on the thalamocortical and brain stem mechanisms.

globus pallidus, but an excitatory action on the neurons projecting directly to the internal pallidal segment (and the substantia nigra, pars reticulata). There is some evidence that these contrasting effects of dopamine are mediated by different types of dopamine receptors, in the sense that D_2 receptors are located on the striatal cells projecting to the external pallidal segment, and D_1 receptors on neurons projecting to the internal pallidal segment (and the substantia nigra, pars reticulata). One important effect of increased dopamine activity, therefore, would seem to be to decrease the inhibitory effect that the internal pallidal segment has on the thalamocortical and brain stem systems. This, in turn, would have a facilitatory effect on movements generated by cortical or brain stem activity. Decreased dopamine activity, as in Parkinson's disease, would have the opposite effect, i.e., an increase of the inhibitory actions of the basal ganglia on thalamocortical and brain stem mechanisms. This may in part explain the slowness of movements that is one of the cardinal symptoms of Parkinson's disease (see Clinical Notes).

The neurons in the subthalamic nucleus, presumably glutaminergic, have an excitatory effect on the two basal ganglia output structures, i.e., the internal pallidal segment and the substantia nigra, pars reticula, which in turn have an inhibitory effect on their thalamic and brain stem targets. A loss of subthalamic input to the basal ganglia output structures, therefore, would result in a disinhibition of their thalamic and brain stem targets, which might explain the hyperkinetic movements in hemiballismus following destruction of the subthalamic nucleus, and maybe also to some extent the involuntary movements in Huntington's chorea (see Clinical Notes). In Parkinson's disease, on the other hand, the tonic inhibition by basal ganglia output structures may be exacerbated by the action of the subthalamic nucleus. As expected, reducing the excitatory impact from this structure has been shown to reduce akinesia in primates with experimental Parkinson's disease.

Despite the intense research efforts generated by the involvement of the basal ganglia in motor disturbances, many aspects of the basal ganglia movement disorders are still poorly understood. Nor do we know to what extent the basal ganglia are involved in cognitive and other nonmotor functions. Striatal and pallidal lesions in animals, for instance, can produce a variety of cognitive symptoms and deficits in more "complex" behaviors. It is also the case that patients with basal ganglia disorders are usually suffering from other problems besides motor defects, e.g., cognitive and emotional symptoms, and this is sometimes taken as a sign that the basal ganglia are concerned with other functions besides movements. Although this may well be the case, it is also true that the pathology in most cases of basal ganglia disorders is not restricted to the basal ganglia. Another complicating factor, which has only recently come into sharper focus, is the fact that the ventral parts of the basal ganglia do intermingle to a considerable extent with other major functional–anatomic systems in the basal forebrain (see Emerging Concepts of Basal Forebrain Organization, below).

The basal ganglia will remain for the foreseeable future a prime area of neuroscience research, where the results of molecular and genetic studies will be increasingly integrated with the insights gained into the systematic organization of these complex structures based on anatomic, physiologic, and pharmacologic studies in animals and humans. These basic science gains will be an essential foundation for understanding the complex data emerging from the growing research field of human and animal in vivo functional imaging.

Clinical Notes

Clinicians often refer to basal ganglia and cerebellar disorders as *extrapyramidal*[4]. Such disorders are characterized by involuntary movements, difficulties in movement initiation and control, and change in muscle tone, as opposed to other disorders of central origin (Chapter 15), which are characterized by a loss of voluntary movements and spasticity. This classification is useful because it makes a distinction between two major supraspinal motor syndromes, one characterized by alteration in the execution of voluntary movements, i.e., extrapyramidal disorders, and the other by paresis or paralysis. It should be kept in mind, however,

4. Many clinicians use the term extrapyramidal disorders as a synonym for basal ganglia disorders only. Considering the origin of the term extrapyramidal, however, it seems more appropriate to include both basal ganglia and cerebellar disorders. In the past, the motor system was divided into pyramidal and extrapyramidal parts, which implied that every motor structure, except the pyramidal or corticospinal tract, belonged to the extrapyramidal system. This distinction, however, is highly artificial. For instance, the cortical areas that give rise to the corticospinal fibers also give rise to other descending (extrapyramidal) fibers. Furthermore, the two main components of the extrapyramidal system, i.e., the basal ganglia and the cerebellum, use the corticospinal tract to a considerable extent to influence the spinal motor neuron.

that all motor activities of supraspinal origin are conveyed through the same descending pathways.

Parkinson's Disease (Paralysis Agitans or Shaking Palsy)

That the basal ganglia have an important role to play in movements is painfully obvious to everyone who has observed the gradually declining motor functions in persons suffering from Parkinson's disease, a common degenerative disorder that was first described almost 200 years ago by the English physician James Parkinson. The disease is probably of environmental origin, and it often begins insidiously in middle or old age. The four cardinal manifestations are muscular rigidity, slowness of movements (bradykinesia), tremor at rest, and postural instability. The typical Parkinson patient can be recognized at a glance by the stooped posture (Fig. 16–12), the slow movements and muscle stiffness with decreased facial expression (mask-like face), and a particular resting tremor that involves the hand (pill-roller's tremor) more than other parts of the body. The patient, in addition, may develop a shuffling gait with short, increasingly rapid steps (festination), micrographia, and a low-volume voice difficult to hear.

Although motor symptoms are usually the most prominent features of Parkinson's disease, most Parkinson patients also suffer from cognitive impairments and various behavioral signs, e.g., indecision and depression. Some patients even develop a severe dementia reminiscent of an Alzheimer dementia (Chapter 21). This indicates that the pathophysiologic mechanisms in Parkinson's disease often involve more than just motor structures. As a matter of fact, cell loss has been described in other brain stem regions, such as the locus ceruleus and the pedunculopontine tegmental nucleus, as well as in the basal forebrain (basal nucleus of Meynert). Furthermore, degeneration of dopaminergic cells in the ventral tegmental area does occur in a significant number of patients, even if it is not as common as involvement of the substantia nigra. When it occurs, however, it has the potential to play havoc with many important forebrain systems, including not only the striatal system, but also the extended amygdala, the basal nucleus of Meynert, and the frontal cortex, all of which receive dopaminergic input from the midbrain. This is likely to lead to disruption of a number of functions besides motor behavior.

Serious efforts to unravel the biochemistry of Parkinson's disease began in the late 1950s, especially by two pharmacologic research teams, one headed by Oleh Hornykiewicz from Switzerland and the other by Arvid Carlsson from Sweden. The discovery that Parkinson's disease was related to selective reduction of dopamine caused by degeneration of the nigrostriatal dopamine system indicated for the first time that a disorder in the CNS could be caused by a deficiency of a specific neurotransmitter. As indicated above, decreased activity in the nigrostriatal dopamine system is likely to increase the inhibitory action of the basal ganglia on thalamocortical and brain stem motor mechanisms, which would explain at least one of the most prominent symptoms of Parkinson's disease, the slowness of movements.

Since heredity does not seem to play a major role in Parkinson's disease, interest has focused on environmental toxins, especially following a tragic episode in northern California. Several drug abusers fell seriously ill with typical parkinsonian motor symptoms following the use of a potent synthetic narcotic. One of the byproducts of this "designer" drug was identified as MPTP (1-methyl-4-phenyl-1,2,3,6,-tetrahydropyridine), which is

Fig. 16–12. Typical Posture of Patient with Parkinson's Disease

(Drawing from Gowers, W. R. 1893. A Manual of Diseases of the Nervous System, 2nd ed. Vol. 2. Blakiston, Philadelphia.)

now known to selectively destroy the dopaminergic cells of the substantia nigra. Other toxins known to produce Parkinson-like extrapyramidal symptoms include carbon monoxide, manganese, and cyanide. In the early part of this century, furthermore, a large number of patients suffering from encephalitis developed a postencephalitic parkinsonian syndrome.

Another challenge is presented by the apparent relationship between aging and parkinsonian symptoms. It remains to be seen if the accelerated process of neuronal degeneration in old age, which appears to be especially prominent in the substantia nigra as well as in the basal nucleus of Meynert, is the result of a lifetime of exposure to environmental substances, or the result of endogenous toxic products, e.g., free radicals, some of which are formed by the auto-oxidation of catecholamines.

Since the Parkinson's patient suffers from a reduction of dopamine in the striatum, a rational replacement therapy would be to administer dopamine. Dopamine, however, is rapidly metabolized and does not readily cross the blood–brain barrier; the patients are therefore treated with a precursor of dopamine, the amino acid L-DOPA, which is converted to dopamine in the brain, probably in the striatum. Although L-DOPA therapy has a dramatic effect, especially on the bradykinesia, it does not prevent or retard the degenerative process. Combined with this ongoing process, L-DOPA eventually results in disturbing side effects, e.g., irregular athetoid (wormlike) movements, autonomic dysfunctions, changes in mood, or even psychosis. These latter changes are hardly surprising considering the fact that dopamine is involved as a transmitter, not only in the striatum, but in several other brain regions including the hypothalamus, amygdaloid body, and frontal cortex. Therefore, the search continues for more effective medications, which would have an effect, not only on the symptoms of the disease, but on arresting or reversing the disease itself. Recently, promising results have been obtained by monoamine oxidase inhibitors, which reduce the amount of potentially toxic by-products, i.e., free radicals, which are formed by the oxidation of catecholamines.

The transplantation of embryonic dopamine cells into the patient's striatum represents another exciting and potentially effective form of therapy. Following some discouraging transplantation experiments in humans using brain or adrenal tissue in the 1980s, a large number of systematic animal experiments have examined the various problems and variables confronting brain tissue transplantations. It now appears that carefully planned transplantations of human embryonic dopamine cells into the striatum of diseased patients can result in significant and reliable improvements, especially in patients with selective damage of the dopaminergic neuronal system.

Huntington's Disease

Huntington's disease is a rare hereditary disorder[5], which usually starts in the fourth decade of life, and which is caused by degeneration of neurons primarily in the striatum and the cerebral cortex. It is characterized by gradually worsening, involuntary choreiform (dancelike) movements, accompanied by personality changes and progressive dementia.

In some respects, Huntington's disease represents the opposite of Parkinson's disease. In Parkinson's disease, the major alteration is progressive loss of dopaminergic neurons in the substantia nigra, which results in decreased dopamine content in the striatum. In Huntington's disease, on the other hand, the major basal ganglia deficits are due to the loss of GABA-ergic neurons in the striatum. If Parkinson's disease is viewed as a case of functional dopamine deficiency, Huntington's disease may, to some extent at least, be characterized by relative dopamine hyperactivity in the striatum. The explanation for this is that the loss of inhibitory GABA-ergic striatonigral neurons projecting to the substantia nigra, pars compacta, results in hyperactivity of the nigrostriatal dopamine neurons. This is consistent with the experience that antidopaminergic therapies tend to reduce the choreiform movements.

Although most striatal neurons are eventually affected in Huntington's disease, it appears that in the early stages of the disease, the cell loss involves primarily the GABA-ergic neurons projecting to the external pallidal segment. This loss of inhibitory input to the external pallidal segment increases the tonic inhibitory pallidal influence on the subthalamic nucleus. The reduced activity of the subthalamic nucleus, in turn, reduces inhibition of the thalamic and brain stem targets by the two basal

5. Huntington's disease is inherited in an autosomal dominant pattern, and the gene responsible for the disease has recently been isolated. It is a series of repeated sequences located on chromosome 4. This important discovery provides the opportunity to test if a person carries the Huntington gene long before the symptoms have appeared. It is also likely to give scientists a better understanding of the pathophysiologic mechanisms involved, which in turn may lead to an effective therapy or even an eventual cure.

ganglia output structures, i.e., the internal pallidal segment and the substantia nigra, pars reticulata. Disinhibition of thalamic and brain stem targets results in increased activity, which may explain the spasmodic involuntary movements characteristic of the disease.

Although the genetic basis of Huntington's disease is known, it may yet be some time before we understand the phenotypic expression of this defect. It is widely believed that the selective destruction of some neuronal populations with sparing of others is related to a neurotoxic process involving the excitatory amino acid receptors, especially the NMDA (*N*-methyl-D-aspartate) glutamate receptors. Quinolinic acid, a metabolite of tryptophan, may be one of the endogenous excitotoxins. The hard-won progress in clinical research, based on advances in understanding the anatomy of the basal ganglia, as well as their physiology and pharmacology, has been combined with diligent work in genetic epidemiology and molecular biology to provide a bright beacon of hope for the patients afflicted with this devastating disease.

Hemiballismus

The basic pathophysiologic mechanism suggested for the generation of involuntary movements in Huntington's disease (see above) may also be involved in a rare disorder, known as *hemiballismus*, which is usually caused by a vascular lesion in the subthalamic nucleus. In this instance, the loss of excitatory subthalamic input to basal ganglia output structures results in disinhibition of their targets, which results in suddenly developing symptoms of violent flinging movements of the extremities contralateral to the lesion.

Extrapyramidal Syndromes and Antipsychotic Drugs

The discovery of the antipsychotic drugs in the 1950s revolutionized the treatment of psychotic patients, especially those suffering from schizophrenia (see Clinical Notes, Chapter 20). Most antischizophrenic drugs, however, have undesirable side effects, including the appearance of parkinsonian rigidity and bradykinesia. It is believed that these extrapyramidal symptoms appear because the drugs block dopamine transmission.

A relatively common and often irreversible motor disorder, *tardive dyskinesia*, has been associated with chronic treatment with antipsychotic drugs. It is characterized by choreiform movements of the face, mouth, and tongue, and occasionally the extremities. Typically, tardive (late) dyskinesia occurs after many years of drug treatment, and it is thought that one reason for its occurrence is that the dopamine receptors have become hypersensitive as a result of the prolonged blockade. Fortunately, effective antipsychotic medicines that do not give rise to extrapyramidal side effects, the so-called atypical antipsychotic drugs, have now become more generally available. Clozaril (clozapine) is one such medicine, which has received considerable attention.

Emerging Concepts of Basal Forebrain Organization

Three major functional–anatomic systems, i.e., the ventral striatopallidal system, the extended amygdala, and the basal nucleus of Meynert, occupy the region often referred to as the *substantia innominata*. The anatomic relations of these systems have been the subject of intense interest over the last decade. As a result, it is now possible to discern some fundamental principles that underlie the organization of this part of the brain, which in the past deserved its status as an unidentifiable substance, i.e., substantia innominata or unnamed substance[6]. The existence of the functional–anatomic systems mentioned above was barely known fifteen years ago, and yet they have come to be the focus of a large body of modern research in neuropsychiatric disorders.

The Ventral Striatopallidal System

That the basal ganglia extend to the surface of the human brain in the region of the anterior perforated space can be demonstrated with the aid of a frontal section at the level of the crossing of the anterior commissure (Fig. 16–13A). This situation is illustrated in a schematic drawing in Fig. 16–13B. The general area occupied by the ventral striatum (VS) and ventral pallidum (VP) in this figure is often referred to as the substantia innominata.

As indicated in the section on basal ganglia connections, the ventral striatopallidal system receives its telencephalic input primarily from the hippocampus formation, the basolateral amygdaloid complex, and some temporal and prefrontal cortical areas.

6. The mediobasal forebrain region between the temporal limb of the anterior commissure and the ventral surface of the brain is often referred to as the substantia innominata of Reil. Johann Christian Reil was a German anatomist who in 1809 referred to the area as "die ungenannte Mark-Substanz," since he was unable to understand its organization.

The Bed Nucleus–Amygdala Continuum; The Extended Amygdala

In the early part of this century, the American comparative neuroanatomist J. B. Johnston pointed out that the bed nucleus of the stria terminalis, which is located in the ventral part of the forebrain close to the midline (Fig. 16–3), is continuous with the amygdaloid body in the temporal lobe through cell columns along the stria terminalis (Fig. 20–1). The notion of a continuum between the amygdala and the bed nucleus of the stria terminalis has recently been developed further by the Argentinean neuroanatomist José de Olmos, whose studies of the amygdala using silver stains identified another substantial ventral component of this continuum between the bed nucleus of the stria terminalis and the centromedial part of the amygdaloid body. This continuum, which traverses the substantia innominata immediately behind the ventral striatopallidal system (Fig. 16–14A), is described also in Chapter 20. With the extension of the centromedial amygdala both alongside the stria terminalis and through ventrally located cell columns in the substantia innominata, the whole continuum, which has been labeled the *extended amygdala*, does indeed appear as a ring formation around the internal capsule (Fig. 16–14B).

It should be noted that the extended amygdala does not include the large basolateral part of the amygdaloid complex, which differs in many ways from the centromedial part of the amygdala and its extension into the bed nucleus of the stria terminalis. It is also important to realize that although the continuity between the bed nucleus of the stria terminalis and the centromedial amygdaloid body was originally discovered in comparative studies of the embryonic human brain by Johnston (1923), the greatest amount of research has been based on the rodent, and it is only recently that this system has been re-examined in the primate and human brain (Alheid and Heimer, 1988; Martin et al., 1991).

The Basal Nucleus of Meynert

The third major component of the poorly defined region referred to as the substantia innominata is the basal nucleus of Meynert, which contains large cholinergic neurons that project to the cerebral cortex, basolateral amygdala, basal ganglia, and thalamus, especially the reticular nucleus (Figs. 16–13 and 16–14). The basal nucleus is part of a widely dispersed, more or less continuous collection of cholinergic and noncholinergic (e.g., GABA, peptides) cells, which in part intermingle with other functional–anatomic systems. This extensive collection of large and medium-sized cells is also referred to as the magnocellular basal forebrain complex. Besides the basal nucleus of Meynert, two other basal forebrain regions contain a large number of corticopetal cells, i.e., the septum and the nucleus of the diagonal band of Broca. Both these areas contain a mixture of cholinergic and noncholinergic cells, many of which project to the hippocampus.

The basal nucleus of Meynert is so conspicuous in Nissl-stained sections (Fig. 16–13A) as to completely overshadow the other major basal forebrain systems, which have often been ignored to the point where the basal nucleus of Meynert is sometimes called the "nucleus of the substantia innominata," or simply referred to as the "substantia innominata." But this is misleading, since the basal nucleus is only one of several neuronal systems in this area. Furthermore, and as indicated above, neurons with all the characteristics of cells in the basal nucleus of Meynert are located in many basal forebrain regions outside the area commonly referred to as the substantia innominata. In fact, many of these neurons intermingle to a significant degree with the ventral striatopallidal system, the dorsal pallidum, and even the extended amygdala (Fig. 16–14C).

Although the basal forebrain corticopetal complex is conspicuous, it has been difficult to elucidate its exact functions. Nonetheless, it is generally believed that the large cholinergic cells are an integral part of the *ascending activating system* (Chapter 10), and as such of great significance for cortical arousal and related mechanisms of learning and memory. Interest in the basal nucleus of Meynert, which had long been known to be especially vulnerable in the aging process, has intensified greatly during the last decade as a result of the focus on Alzheimer's disease. Like many other areas of the brain (and especially the temporal lobe, Chapter 21), the basal nucleus of Meynert is often involved in Alzheimer's disease, and particular significance has been attached to the neuropathologic observation that severe decrements in the cholinergic innervation of the cortex are a common feature in Alzheimer's patients.

Connections

The three functional–anatomic systems described above have their own characteristic histologic, histochemical, and connectional features, which set them apart from each other. As indicated above, the ventral striatopallidal system receives input pri-

Fig. 16–13. Ventral Striatopallidal System as part of the Basal Ganglia
A. Nissl-stained frontal section through the human basal forebrain at the level of the optic chiasm and the crossing of the anterior commissure (ac). Note that the putamen extends towards the base of the brain, thereby forming part of the ventral striatum (VS). The basal nucleus of Meynert (B) is a conspicuous cell group of this level. BST = bed nucleus of stria terminalis, GP = globus pallidus, VP = ventral pallidum. One of the striatal cell islands is indicated by a *small straight arrow*. The *large curved arrow* points to a group of large hyperchromic cells, which are part of the magnocellular corticopetal system.
B. Schematic drawing of a frontal section at approximately the same level as in **A**. The region between the anterior commissure and the ventral surface of the brain is often referred to as the substantia innominata. However, as indicated in this drawing and in the Nissl-stained section in **A**, it is occupied in large part by the ventral striatum (VS), ventral pallidum (VP), and basal nucleus of Meynert (B). The basal nucleus of Meynert is a prominent collection of large cholinergic neurons (*black triangles*) that project to the cerebral cortex. The *open triangles* represent cholinergic interneurons in the striatum. Other abbreviations: ac = anterior commissure, BST = bed nucleus of stria terminalis, f = fornix, GPe = external segment of globus pallidus, GPi = internal segment of globus pallidus, ic = internal capsule, POA = preoptic area.

marily from the basolateral amygdala, the hippocampal formation, and some prefrontal and temporal association areas, in the same way as the dorsal parts of the corpus striatum receive projections from the rest of the neocortex (Fig. 16–6). The output from the ventral striatopallidal system is channeled via the mediodorsal thalamus to prefrontal cortical areas and the anterior part of the cingulate gyrus. The subthalamic nucleus, substantia nigra, and midbrain tegmentum also receive projections from the ventral striatopallidal system.

The cortical projections to the extended amygdala mirror in large part those to the ventral striatopallidal system. The efferent connections of the extended amygdala, however, are considerably more diversified than those of the basal ganglia output structures, in the sense that they reach a variety of neuroendocrine, autonomic, and somatomotor areas in the hypothalamus and brain stem (Chapter 19). Contrary to the situation in the striatopallidal complex, which is characterized by the presence of parallel channels, various parts of the extended amygdala are interconnected via association pathways, which suggest a certain functional homogeneity along the mediolateral axis of the extended amygdala. Another outstanding feature of the extended amygdala is its rich content of neuropeptides, many of which, however, also exist in the nearby striatal complex.

The basal nucleus of Meynert is part of the magnocellular corticopetal neuronal system, whose cells project not only to the cerebral cortex, but also to the basolateral amygdala and the reticular nucleus of the thalamus (Fig. 16.15). Most of the cells projecting to the neocortex are cholinergic, whereas a mixture of cholinergic, GABA-ergic, and peptidergic cells project to the hippocampal formation, olfactory cortex, basolateral amygdala, and some of the closely related cortical regions in the temporal lobe and orbitofrontal cortex. These non-neocortical areas, incidentally, receive the densest innervation from the basal forebrain corticopetal system, and as mentioned previously they are also the areas that project to the ventral striatopallidal system and the extended amygdala. Moreover, some of the corticopetal cells in the ventral pallidum may be integral parts of the ventral striatopallidal system in the sense that they receive some of their input from the ventral striatum. A similar situation may exist in regard to the intermingling of the extended amygdala and basal nucleus of Meynert cells, but the details of such potential interactions are not yet well defined.

An even more intriguing situation is encountered at the border region between the striatal complex and the extended amygdala, where the two structures appear to intermingle to a significant degree, especially alongside the temporal limb of the anterior commissure (Fig. 16–14A). Such transitional areas take on an identity of their own, as an increasing number of histochemical studies make them conspicuous owing to their unusual content of certain neuropeptides or receptors. Some of these have received special attention in discussions of the pathophysiology of schizophrenia (e.g., neurotensin and cholecystokinin) or as culprits in neuronal degeneration (e.g., excitotoxic amino acid receptors).

In this context, it is interesting to reflect for a soment on the well-known fact that many Huntington's disease patients suffer from a variety of emotional disturbances (including schizophrenia-like symptoms), which may precede the motor

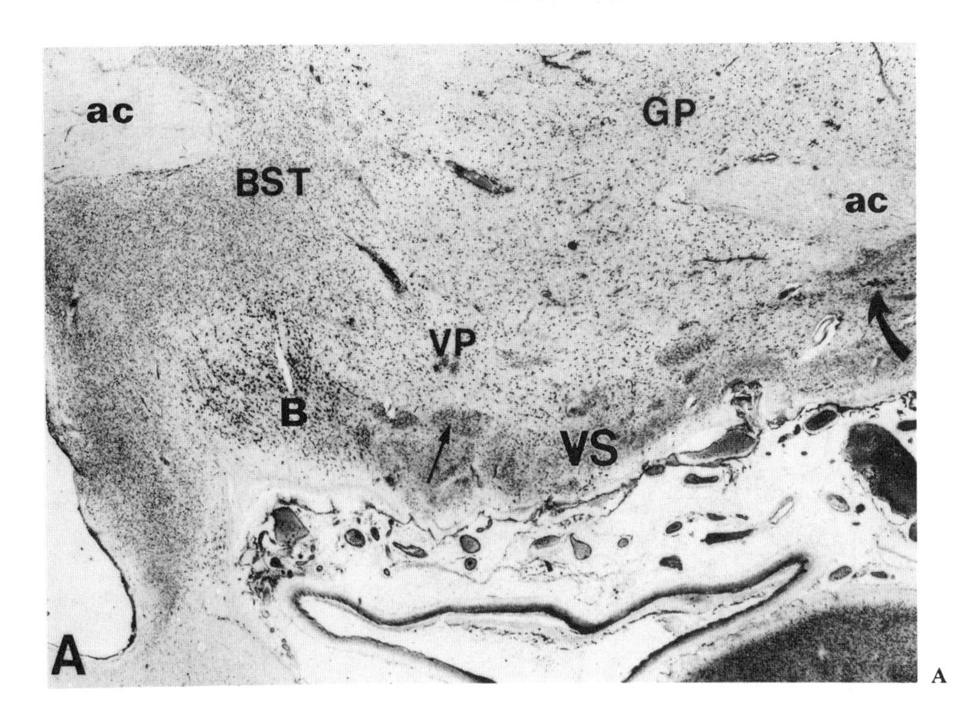
ac
GP
BST
ac
VP
B
VS
A

A

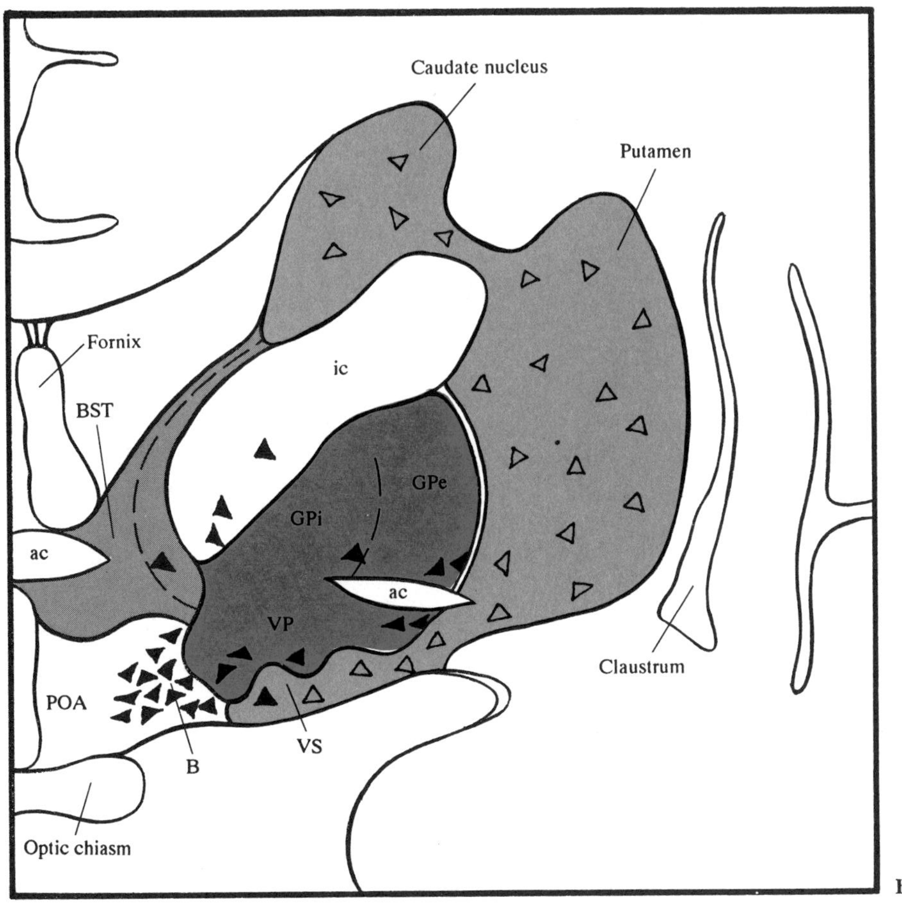
Caudate nucleus
Putamen
Fornix
ic
BST
GPe
GPi
ac
ac
VP
POA
B
VS
Claustrum
Optic chiasm

B

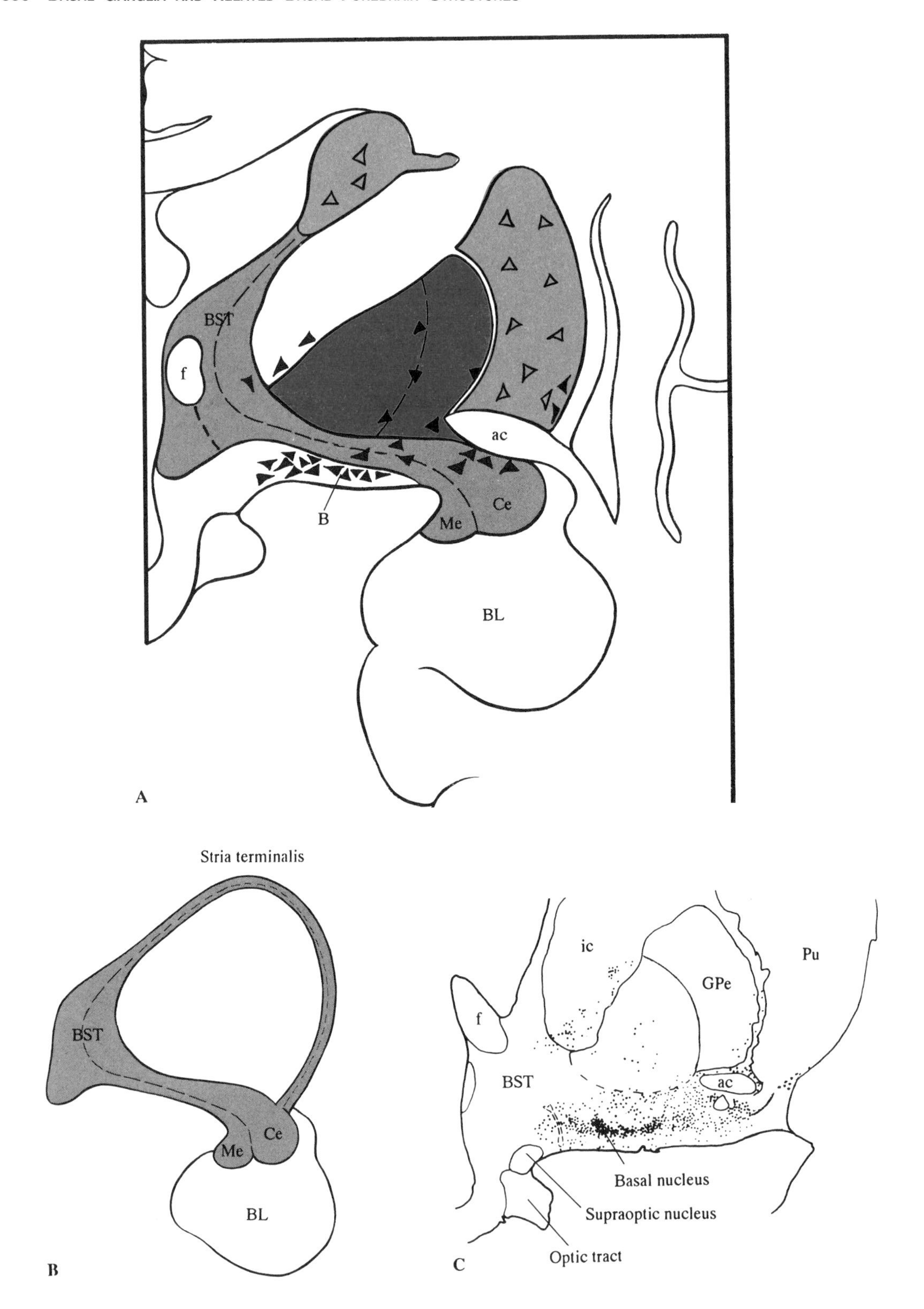
BST
f
ac
B
Me
Ce
BL
A
Stria terminalis
BST
Ce
Me
BL
B
ic
GPe
Pu
f
BST
ac
Basal nucleus
Supraoptic nucleus
Optic tract
C

Fig. 16–14. Extended Amygdala and the Basal Nucleus of Meynert

A. Schematic drawing of a frontal section behind that shown in Fig. 16–13. At this slightly more caudal level, the ventral striatopallidal system has been replaced by an extension of the central and medial amygdaloid nuclei. This extension reaches far medially to include the bed nucleus of the stria terminalis (BST). Note that the extended amygdala does not include the large basolateral complex (BL) of the amygdala. The basal nucleus of Meynert (B) constitutes another important functional–anatomic system at this level. The black triangles represent cholinergic corticopetal neurons, whereas the open triangles represent cholinergic striatal interneurons.

B. The extended amygdala shown in isolation from the rest of the brain. with the extensions of the central (Ce) and medial (Me) amygdaloid nuclei alongside the stria terminalis and through the substantia innominata to the bed nucleus of the stria terminalis (BST), the extended amygdala appears as a ring formation, which in effect surrounds the internal capsule.

C. Schematic drawing showing the cholinergic basal forebrain complex at approximately the same level as in **A.** The histologic section from which this drawing was obtained was stained with a method showing acetylcholinesterase-containing neurons, which at this level signify cholinergic neurons. Although the basal nucleus of Meynert appears as a concentrated group of neurons, it is obvious that the large cholinergic neurons form a widely dispersed system, whose cells intermingle to a significant degree with other functional–anatomic systems including the basal ganglia and the extended amygdala. Same abbreviations as in Fig. 16–13B. (Courtesy of Dr. C. B. Saper. Modified from Saper, C. B., 1990. Cholinergic system. In: G. Paxinos (ed.): The Human Nervous System. Academic Press.)

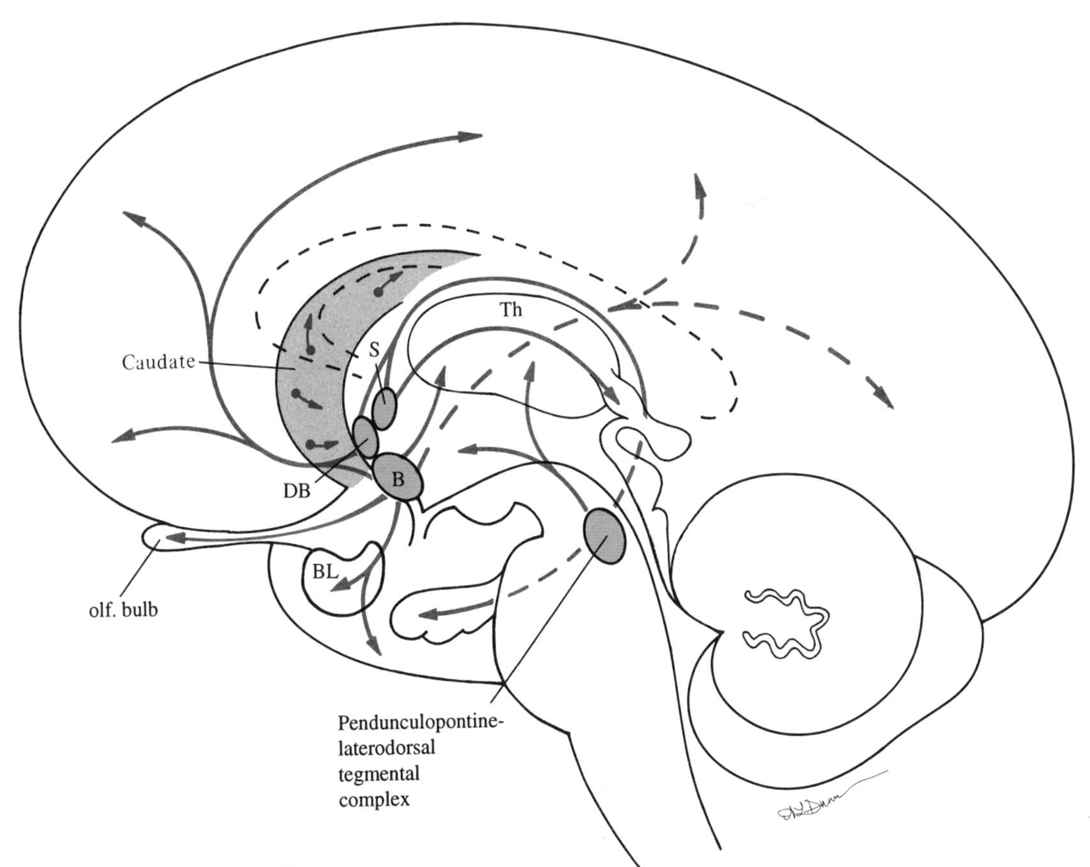

Fig. 16–15. Cholinergic Pathways

Diagram of the cholinergic pathways from the brain stem and basal forebrain. The *small arrows* in the caudate nucleus represent cholinergic interneurons. (B = basal nucleus of Meynert, BL = basolateral amygdaloid complex, DB = diagonal band of Broca, S = septal nuclei, Th = thalamus.)

symptoms by several years. Since these symptoms tend to suggest the involvement of an amygdaloid-related system, it is especially intriguing to acknowledge a number of recent reports which show that the first brain regions to be compromised in Huntington's disease are the dorsal parts of the striatum. This may involve the dorsal arch of the extended amygdala alongside the stria terminalis (Fig. 20–1) and the corresponding border regions between the dorsal striatal territories and this part of the extended amygdala.

Clinical–Anatomic Correlations

Without a good understanding of the functional–anatomic systems in the brain, it is difficult to make sense of clinical signs and symptoms. For instance, it is generally recognized that pathologic changes in the basal forebrain, often loosely defined as the basal nucleus of Meynert or substantia innominata, are a prominent feature in chronically disabling brain disorders like Alzheimer's disease and schizophrenia. If the notion prevails that the symptoms of patients with lesions in this part of the brain are accounted for by involvement of the large basal nucleus of Meynert cells, because their disappearance is easily observed in routine histologic sections, it may never occur to the casual observer that derangement in the ventral parts of the basal ganglia or the extended amygdala may provide a more coherent explanation of the patient's symptoms.

Consider, for example, *Alzheimer's disease* and *schizophrenia*, two of the most common and arguably among the worst diseases to affect human beings. Although many different brain regions besides the basal forebrain are eventually involved in these two disorders, recent studies using histologic techniques and modern imaging methods tend to show that medial temporal lobe structures, including the hippocampal formation and amygdaloid body, are especially vulnerable and are often involved at an early stage (see Clinical Notes, Chapters 20 and 21). As indicated above, the subcortical projections of both the hippocampal formation and the basolateral amygdaloid body, and for that matter most of the cortical areas in and around the parahippocampal gyrus, involve significantly the ventral striatopallidal system and the extended amygdala. Therefore, if Alzheimer's disease and schizophrenia are characterized in their early stages by conspicuous pathologic changes in medial temporal lobe structures, it is only reasonable to expect that these changes will have a direct impact on those basal forebrain structures with which these temporal lobe structures are directly connected. In other words, what happens in the medial aspect of the temporal lobe is likely to disturb thc physiology, and maybe even in due time change the anatomic circuits related to the basal nucleus of Meynert, the striatopallidal system, and the extended amygdala. Considering the significance of these basal forebrain systems in various neuroendocrine, autonomic, somatomotor, and even cognitive functions, as mentioned above, it should come as no surprise that patients suffering from these disorders may encounter problems in a number of areas ranging from motor functions to basic drives like eating, drinking, and sexual behavior to personality changes involving stress, mood, and higher cognitive functions.

One of the central themes in this discussion of clinical–anatomic correlations is the proposition that the connections between the medial temporal lobe structures and the mediobasal forebrain may be crucially important for understanding some of the pathophysiological aspects of Alzheimer's disease and schizophrenia. Considering the magnitude of these connections, it is difficult to escape the conclusion that the basal forebrain systems discussed in this section represent likely targets for biochemical, physiologic, and anatomic changes that may result from primary lesions in the medial part of the temporal lobe. Such secondary changes can be referred to as *transsynaptic effects*. It is in this context that the anatomic details of the ventral striatopallidal system and the extended amygdala, and their relation to the basal nucleus of Meynert become important, since such knowledge is likely to provide new insights into the confusingly heterogeneous symptomatology of neuropsychiatric disorders.

The Dopamine and Glutamate Hypotheses of Schizophrenia

Although schizophrenia is discussed briefly in Chapter 20, its relation to the dopamine system can best be appreciated in the context of the basal forebrain functional–anatomic systems described in this chapter. It should be emphasized that the following discussion, like the previous one concerning the clinical–anatomic correlations, is speculative. However, in areas where hard scientific data are difficult to obtain, speculations are not necessarily out of place, and may in fact be the ember that sparks the scientific fire.

It was discovered by chance in the early 1950s

that chlorpromazine had a significant antipsychotic effect, and when Arvid Carlsson and his colleague Margit Lindqvist 10 years later suggested that the clinically relevant action of antipsychotic drugs like chlorpromazine and haloperidol is to block dopamine receptors, the dopamine hypothesis of schizophrenia was born. The fact that dopamine agonists, such as *d*-amphetamine and L-DOPA, can produce a picture of paranoid schizophrenia in healthy persons or increase the symptoms in schizophrenic patients, only served to solidify the dopamine hypothesis, which has now held center stage for over 30 years. In short, the dopamine hypothesis of schizophrenia postulates that the main etiologic factor in schizophrenia relates to dysfunction in dopamine neurotransmission, and there has been considerable speculation that the dysfunction involves primarily the dopamine system related to the ventral striatum, the extended amygdala, and the frontal lobe. This part of the dopamine system is often referred to as the "mesolimbic" dopamine system.

Although there is little direct evidence that schizophrenia is caused by a selective dysfunction of dopaminergic transmission, dopamine nevertheless does seem to play a significant role in the generation of some of the psychotic symptoms, as indicated by the fact that most, if not all, antipsychotic agents block dopaminergic transmission. This is to some extent also the case for the so-called atypical antipsychotic medicines like Clozaril (clozapine). Although Clozaril is not an especially effective antidopaminergic agent, it does seem to have a preferential effect on the dopaminergic system that originates in the ventral tegmental area and terminates prominently in the basal forebrain, including especially the ventral striatum and the extended amygdala (see below), as well as the prefrontal cortex.

As mentioned above (see also Clinical Notes, Chapter 20), a significant number of schizophrenic patients have pathologic changes in the medial temporal lobe structures, which are characterized by prominent projections especially to the above-mentioned mediobasal forebrain systems that receive the major portion of the dopaminergic input from the ventral tegmental area. It is likely that substantial pathologic derangement in the medial temporal lobe will lead to changes in the mediobasal forebrain regions to which the affected parts of the temporal lobe project. Since all of these regions, both in the mediobasal forebrain and the temporal lobe, receive dopaminergic innervation from the midbrain, there is reason to expect that dopamine antagonists would be effective in schizophrenia as they are in disorders such as Huntington's disease, where the persisting dopamine innervation is acting on a deteriorating target structure.

It is important to realize that the basal forebrain contains many other transmitters besides dopamine. One such transmitter that has received special attention in recent years is *glutamate*, especially since it was discovered that PCP (phencyclidine or "angel dust") can produce symptoms that mimic schizophrenia. PCP is an NMDA antagonist, and since the cortical projections, including those from medial temporal lobe structures to the ventral striatum and extended amygdala, use excitatory amino-acids (e.g., glutamate or aspartate) as their transmitter, their effect is likely to be mediated in large part by NMDA receptors.

One important link between the basal forebrain (and its glutaminergic input) and the dopamine system is represented by the striatonigral fibers from the ventral parts of the striatal complex and projections from the extended amygdala to the dopamine cells in the substantia nigra. These fibers terminate not only in the ventral tegmental area but also in the substantia nigra, which gives rise to the ascending nigrostriatal dopamine system controlling the "motor" circuit. In other words, besides their potentially damaging effects on various physiologic functions and transmitter systems in the extended amygdala and the ventral striatopallidal system, lesions in the medial temporal lobe structures, with concomitant reduction of glutaminergic input to these forebrain structures, are likely to involve secondarily the entire nigrostriatal dopamine system, which has repercussions for the whole range of corticosubcortical re-entrant circuits, including the "motor" circuit discussed in relation to the basal ganglia earlier in this chapter. This may create a situation reminiscent of that in Huntington's disease, i.e., unbalanced exaggeration of dopaminergic activity with concomitant hyperactivity in the thalamic and brain stem targets of the basal ganglia (see discussion of Huntington's disease in Clinical Notes). Incidentally, motor abnormalities, including orofacial dyskinesias, are integral parts of the schizophrenic symptom complex.

Needless to say, there are no easy answers to these complicated problems, and the apparent complexities of the mediobasal forebrain have in the past been a formidable obstacle to progress in this field of clinical–anatomic correlations. However, as reflected in this short overview of the emerging concepts of basal forebrain organization, the more prominent functional–anatomic systems in this part of the brain have recently come into sharper focus. Therefore, if the basal forebrain is significantly compromised in some forms of neuropsychiatric disorders, as implied in the previous discussion, one important avenue for progress is to develop reasonable hypotheses based on our understanding of the functional–anatomic organization of the basal forebrain as it is now emerging.

It has often been proposed that schizophrenic patients suffer from a defect in the "filtering" of incoming sensory impulses, which would lead to over-arousal. Carlsson and Carlsson (1990), in particular, postulate that an increased dopaminergic or a reduced glutamatergic tone leads to an increased transmission of sensory impulses through the thalamus to the cortex, resulting in increased arousal and psychomotor activity. If these changes surpass the integrative capacity of the cerebral cortex, they may provoke a schizophrenic psychosis including disorganized thinking. The forebrain circuits related to the basal ganglia, extended amygdala and the basal nucleus of Meynert would all be potential players in such a phenomenon.

Suggested Reading

1. Albin, R. L., A. B. Young, and J. B. Penney, 1989. The functional anatomy of basal ganglia disorders. Trends Neurosci. 12:366–375.
2. Alexander, G. E., M. R. DeLong, and P. Strick, 1989. Parallel organization of functionally segregated circuits linking basal ganglia and cortex. Annu. Rev. Neurosci. 9:357–381.
3. Alheid, G. F. and L. Heimer, 1988. New perspectives in basal forebrain organization of special relevance for neuropsychiatric disorders; the striatopallidal, amygdaloid and corticopetal components of substantia innominata. Neuroscience 27:1–39.
4. Alheid, G. F., L. Heimer, and R. C. Switzer, 1990. Basal ganglia. In G. Paxinos (ed.): The Human Nervous System. Academic Press: San Diego, pp. 483–582.
5. Basal Ganglia Research (Special Issue), 1990. Trends Neurosci. 13(7).
6. Carlsson, M. and A. Carlsson, 1990. Interactions between glutamatergic and monoaminergic systems within the basal ganglia—implications for schizophrenia and Parkinson's disease. Trends Neurosci. 13(7):272–276.
7. DeLong, M. R. and T. Wichmann, 1993. Basal ganglia-thalamocortical circuits in parkinsonian signs. Clin. Neurosci. 1:18–26.
8. Gerfen, C. R., 1993. Relations between cortical and basal ganglia compartments. In A.-M. Thierry et al. (eds.): Motor and Cognitive Functions of the Prefrontal Cortex, Springer Verlag: Berlin, pp. 78–92.
9. Graybiel, A. M., 1991. Basal ganglia—input, neural activity, and relation to the cortex. Curr. Opinion Neurobiol. 1:644–651.
10. Heimer, L., J. de Olmos, G. F. Alheid, and L. Zaborszky, 1991. "Perestroika" in the basal forebrain: Opening the border between neurology and psychiatry. In G. Holstege (ed.): Progress in Brain Research, Vol. 87. Elsevier: Amsterdam, pp. 109–165.
11. Hoover, J. E. and P. L. Strick, 1993. Multiple output channels in the basal ganglia. Science 259:819–821.
12. Johnston, J. B., 1923. Further contributions to the study of the evolution of the forebrain. J. Comp. Neurol. 35:337–481.
13. Kim, J. S., H. H. Kornhuber, J. Kornhuber, and M. E. Kornhuber, 1986. Glutamic acid and the dopamine hypothesis of schizophrenia. In Ch. Chagass, et al. (eds.): Biological Psychiatry 1985. Elsevier, Amsterdam, pp. 1109–1111.
14. Lipska, B. K. and D. R. Weinberger, 1993. Cortical regulation of the mesolimbic dopamine system. In P. W. Kalivas and C. D. Barnes (eds.): Limbic Motor Circuits and Neuropsychiatry. CRC Press: Boca Raton, Florida, pp. 329–349.
15. Markham, C. H., 1993. Parkinson's disease. Clin. Neurosci. 1(1):1–67.
16. Martin, L. J., R. E. Powers, T. L. Dellovade, and D. L. Price, 1991. The bed nucleus-amygdala continuum in human and monkey. J. Comp. Neurol. 309:445–485.
17. Mesulam, M.-M., D. Mash, L. Herch, M. Bothwell, and C. Geula, 1992. Cholinergic innervation of the human striatum, globus pallidus, subthalamic nucleus, substantia nigra and red nucleus. J. Comp. Neurol. 323:252–268.
18. Percheron, G., J. S. McKenzie, and J. Feger (eds.), 1993. Basal ganglia IV. New data and concepts on the structure and function of the basal ganglia. Plenum Press: New York.

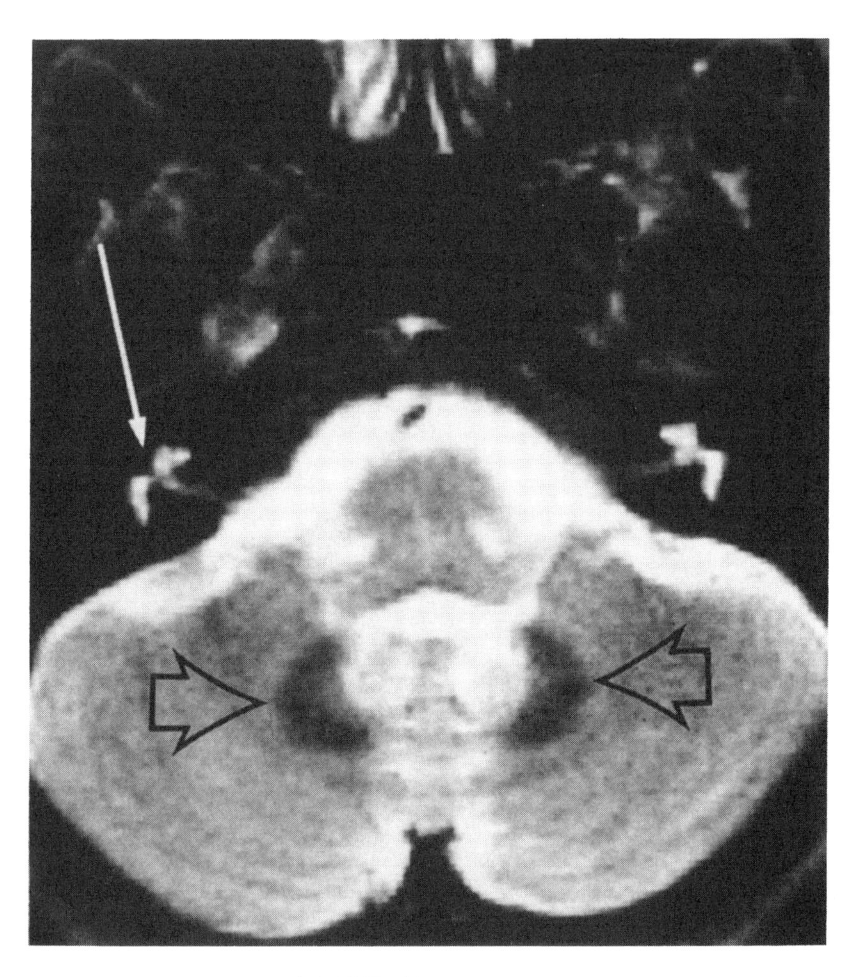

Axial MRI (T2-weighted image) of normal cerebellum. *Open arrows* = dentate nuclei; *long white arrow* = inner ear structures. (From Sartor K., 1992. MR Imaging of the Skull and Brain. Springer-Verlag Berlin. With permission of the publisher.)

17

Cerebellum

The overall structure of the cerebellum is similar to that of the cerebrum in the sense that superficial cellular layers form a cortex, which covers several deep nuclei embedded in the white substance. The cerebellar cortex is heavily folded, but its histologic structure is stereotyped and relatively simple compared with that of the cerebral cortex.

By observing patients with cerebellar disease, it becomes evident that the cerebellum is an important center for control of movements and postural adjustments. To regulate these functions, the cerebellum receives information from all parts of the body: proprioceptive and interoceptive impulses from muscles and joints and from visceral organs; messages from the skin and from the visual, the auditory, and the vestibular systems; as well as a variety of impulses from other parts of the CNS.

Cerebellar Cortex

Cortical Layers

One of the most characteristic features in the cerebellum is the stereotyped histology of its cortex. The cerebellar cortex is in this respect different from the cerebral cortex, in which various regions can be identified on the basis of their histologic characteristics. Although the cerebellar cortex contains several billions of neurons, most of them granule cells, its design is nevertheless relatively simple compared with that of the cerebral cortex. The five major cell types of the cerebellar cortex, *granular cells*, *Golgi cells*, *Purkinje cells*, *basket cells*, and *stellate cells* are arranged in a trilaminar pattern. From within, moving outward, these layers are the granular layer, the Purkinje layer, and the molecular layer (Fig. 17–1).

1. *The granular layer* contains an abundance of small relay neurons, the granular cells (red in Fig. 17–1), as well as a more limited number of larger interneurons, the Golgi cells (in white), whose large dendritic tree extends in all directions in the molecular layer.
2. *The Purkinje cell layer* is formed by a sheet of large Purkinje cells, whose richly branching, characteristically flattened dendritic tree extends throughout the thickness of the molecular layer in a plane perpendicular to the longitudinal axis of the folium.
3. *The molecular layer* contains two types of interneurons, the stellate cells and the basket cells (in black), in addition to the dendritic arborizations of Purkinje and Golgi cells. It is also characterized by an abundance of tightly packed granular cell axons, i.e., *parallel fibers*, which run parallel to the direction of the folia.

Intrinsic Circuitry

Three different categories of afferent nerve fibers conduct nerve impulses to the cerebellum, the *climbing fibers* (CFs), the *mossy fibers* (MFs), and the *aminergic fibers* (AFs). Many of the incoming fibers send collateral axons to one of the cerebellar nuclei before they reach the cerebellar cortex to ultimately affect the Purkinje cells.

The arrangements through which the CFs and the MFs influence the Purkinje cells are very different. A CF, which typically originates in the inferior olive, makes multiple and powerful contacts directly with proximal dendritic branches of a limited number of Purkinje cells. Each one of the more numerous MFs, on the other hand, which come from a variety of different sources, influences a large number of Purkinje cells in an indirect fashion. To accomplish this, the MF branches extensively both in the white matter and in the granular layer, and establishes synaptic contacts with the dendrites of several granule cells in complex formations called *glomeruli*.[1] The glomeruli contain a central MF terminal and boutons of the Golgi cell axons, both of which synapse with the granule cell dendrites (inset in Figs. 17–1 and 17–2). The axons of the granule cells, in turn, project upward to the molecular layer, where they bifurcate in a typical T-shaped manner to form the *parallel fibers*, each of which runs for several millimeters along the longitudinal extent of a folium. In other words, the parallel fiber passes perpendicular to the flattened dendritic tree of the Purkinje cell and it establishes synaptic contacts with spiny distal dendritic branchlets of many hundreds of Purkinje cells. Whereas most neurons in the brain receive an average of 1,000 synapses, it is estimated that each Purkinje cell receives input from as many as 200,000 parallel fibers. Therefore, whereas the CF provides a powerful input to a few specific Purkinje cells, the MF input is greatly dispersed to a large number of Purkinje cells through glomerular relays in the granular layer. The Purkinje cells project to the deep cerebellar nuclei, which ultimately give rise to the cerebellar efferent pathways.

Afferent axons in the third category, the AFs, are characterized by widespread distribution in the cerebellar cortex. The aminergic fibers, which are inhibitory, are of two different types, serotonin axons originating in the raphe nuclei of the brain stem, and norepinephrine-containing axons originating in the locus ceruleus.

Three types of inhibitory interneurons, the Golgi cells (white in Fig. 17–1) and the basket and stellate cells (black), which are contacted either directly or indirectly by extrinsic afferent fibers of all categories, serve as modulators in the cerebellar cortex. A remarkable feature of the cerebellar cortex is that its only output neurons, the Purkinje cells, have an inhibitory effect on the intracerebellar nuclei and the lateral vestibular nucleus. As indicated earlier, however, the cerebellar nuclei do get excitatory input via axon collaterals of MFs and CFs. Like the three types of interneurons just mentioned, the Purkinje cells use γ-aminobutyric acid (GABA) as their neurotransmitter. One of the

1. The name glomerulus derives from the similarity this formation shows with the classic glomerulus in the olfactory bulb.

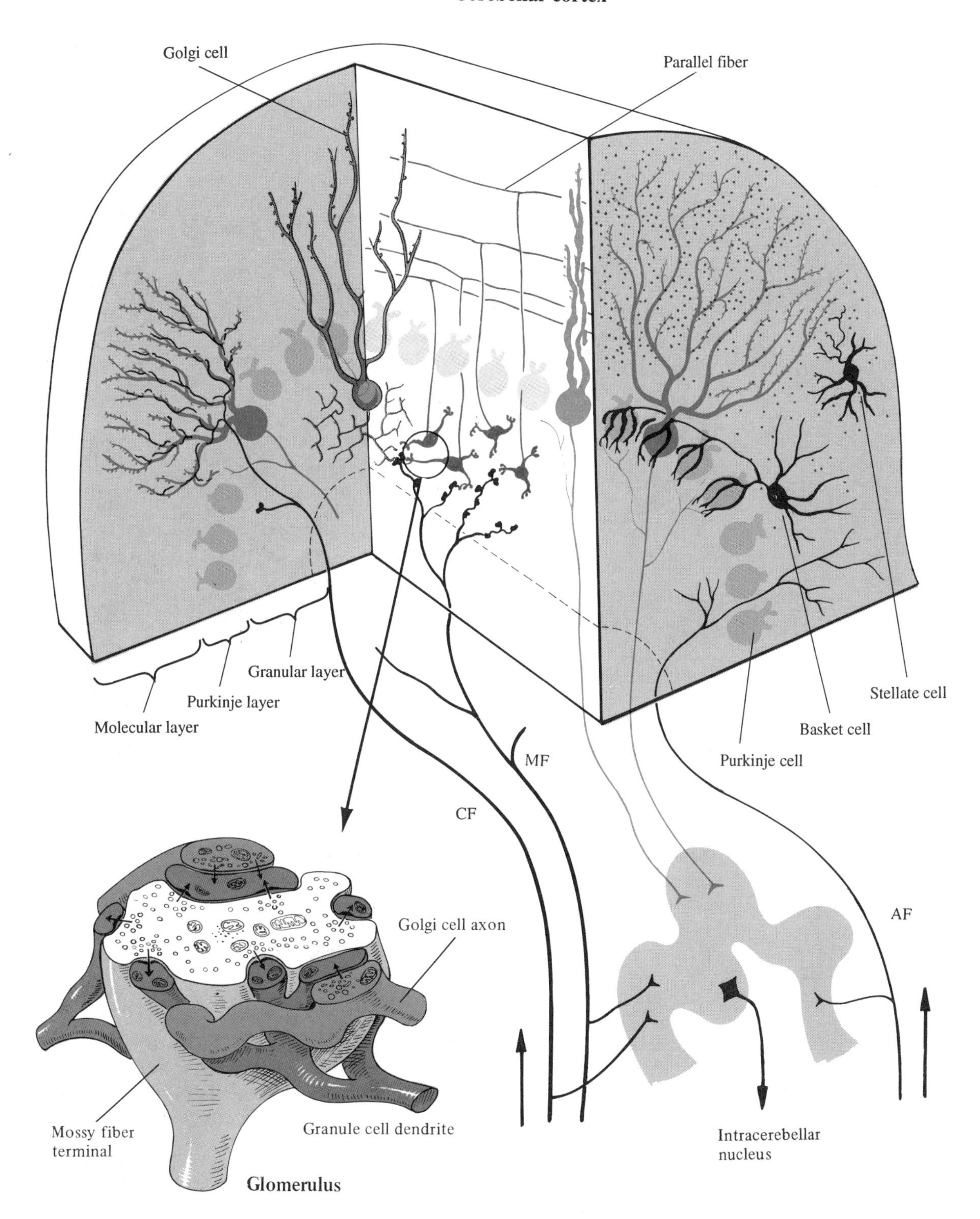

Fig. 17–1. Cerebellar Cortex
A single cerebellar folium has been sectioned vertically, both in longitudinal and transverse planes to illustrate the general organization of the cerebellar cortex. Purkinje cells, *blue*; granule cells, *red*; Golgi cell, *white*; basket and stellate cells, *black*. CF, climbing fiber; MF, mossy fiber; AF, aminergic fiber. The stereodiagram in the *lower left-hand corner* illustrates the structure of a cerebellar glomerulus.

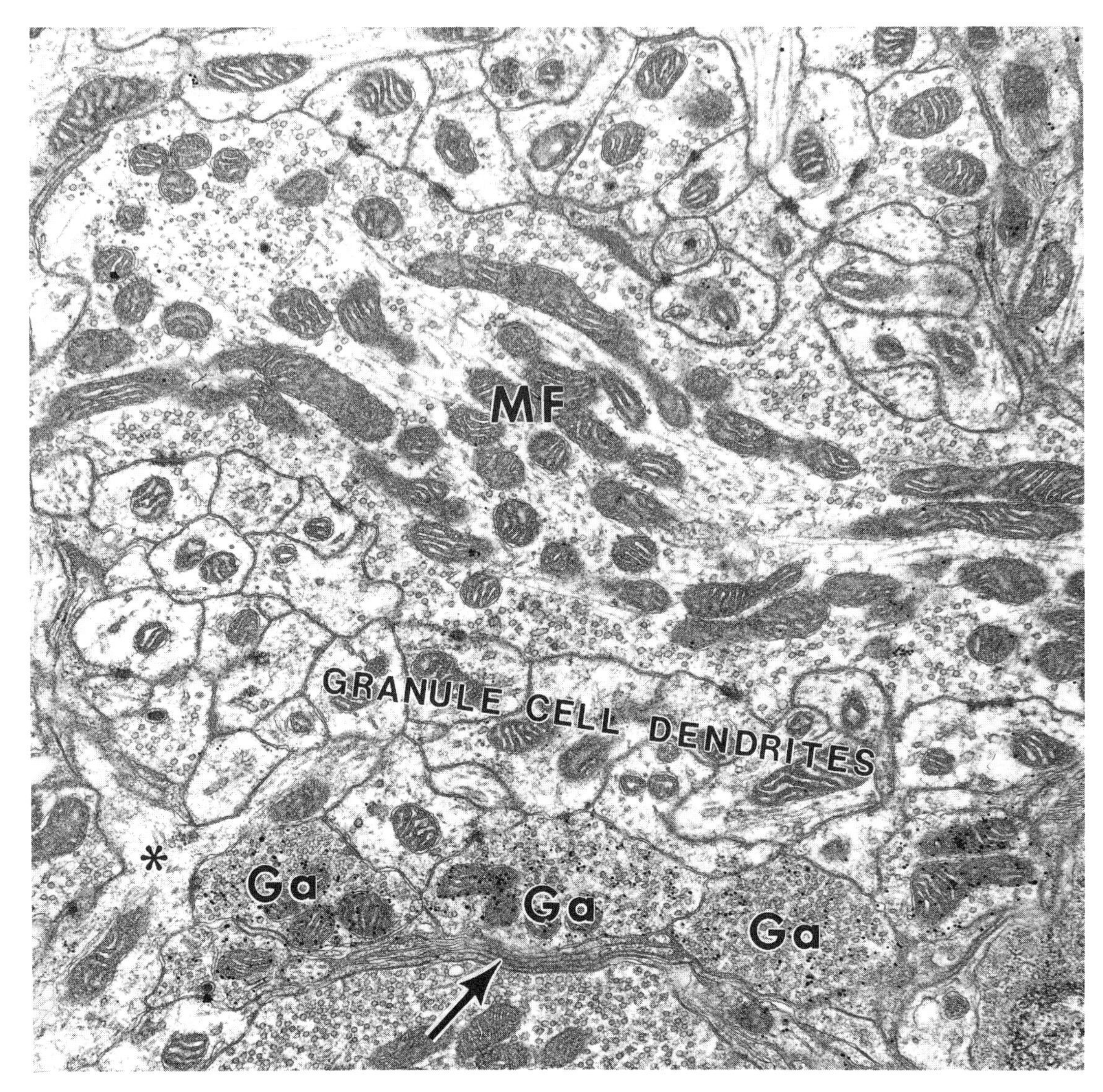

Fig. 17–2. Cerebellar Glomerulus: Ultrastructure
Mossy fiber terminal (MF) surrounded by granule cell dendrites, which are also postsynaptic to Golgi cell axon terminals (Ga). The glomerulus is surrounded by an astroglial sheath (*arrow*). The Golgi cell axons are immunolabeled by gold particles with antiserum to GABA. Asterisk = incoming granule cell dendrite. (Micrograph courtesy of Dr. Enrico Mugnaini.)

markers for GABA-ergic neurons is glutamic acid decarboxylase (GAD), the biosynthetic enzyme for GABA, and this enzyme has been localized by the aid of immunohistochemical methods in Purkinje, basket, stellate, and Golgi cells (Fig. 17–3A), as well as in synaptic terminals of these cells (Fig. 17–3B).

Without going into a detailed analysis of the synaptic relations, the flow of information through the cerebellum can be summarized with reference to two major circuits (Fig. 17–4). The major inputs, reaching the cerebellum via MFs and CFS, activate the cerebellar nuclei through their axon collaterals. This excitatory "main circuit" is in turn modulated by an inhibitory cortical "side loop," which is activated by the same input that reaches the cerebellar nuclei.

Functional–Anatomic Organization

The nervous elements of the cerebellar cortex and their afferent and efferent connections are organized in two perpendicularly oriented planes (Fig. 17–5). The Purkinje cells and their axons, which constitute the only efferent system of the

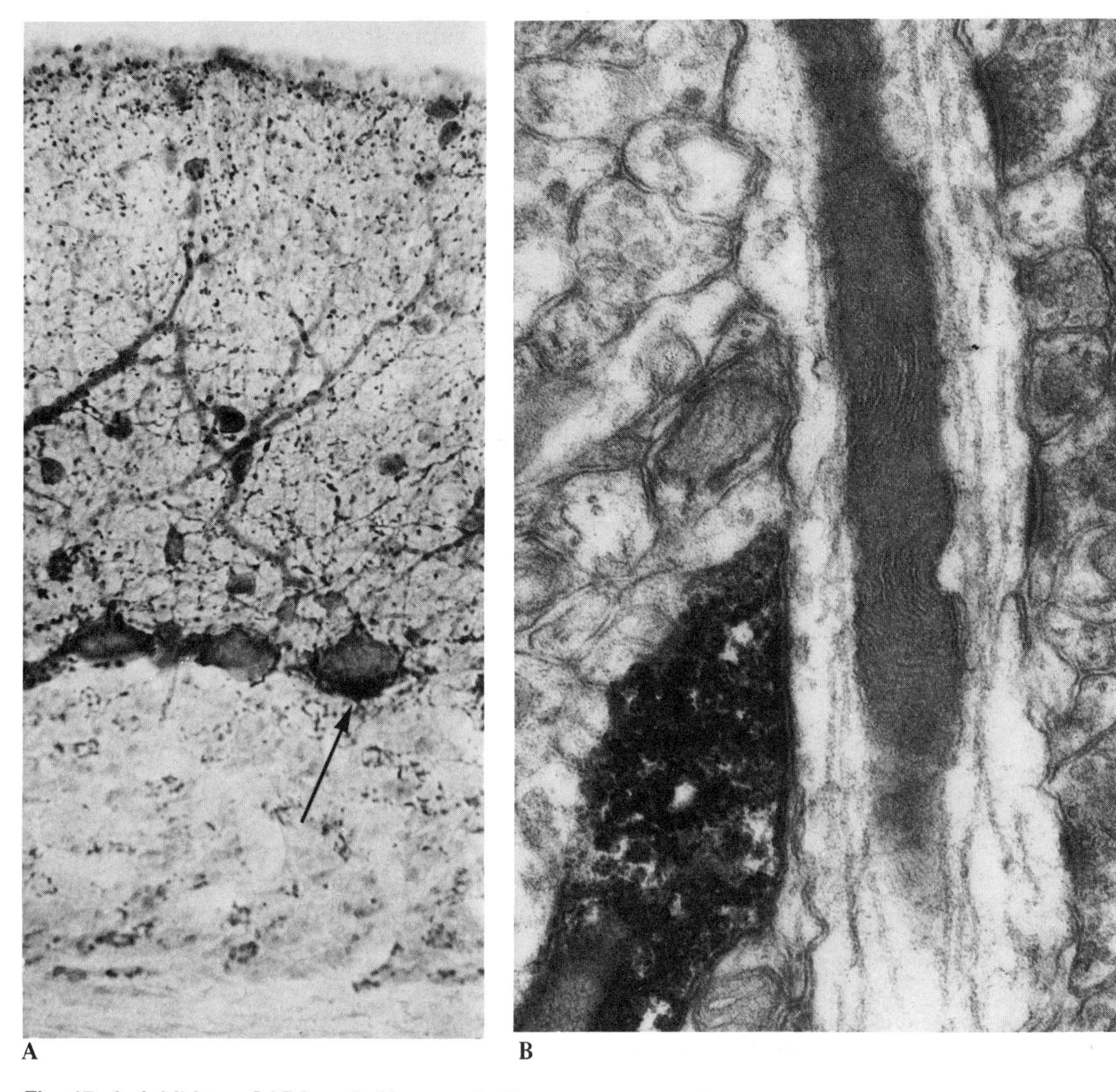

Fig. 17–3. Inhibitory GABA-ergic Neurons in Cerebellar Cortex

A. Light micrograph of a histologic section of the cerebellar cortex stained with an immunohistochemical technique for the localization of glutamic acid decarboxylase (GAD). Three large and moderately stained Purkinje cells are surrounded by heavily stained fibers and terminals. Note the heavy stain around the initial axon segment of one of the Purkinje cells (*arrow*). This is the place where basket fibers and terminals are known to congregate. Purkinje cell dendrites as well as many smaller neurons (basket and stellate cells) are stained in the molecular layer (*upper half of figure*). Most of the black granules in the molecular layer probably represent GABA-ergic inhibitory boutons.

B. Electron micrograph from the molecular layer showing a heavily stained terminal forming a symmetric synapse with a dendrite. The two boutons forming asymmetric contacts on the right side of the dendrite are unstained. (A and B, Micrograph courtesy of Dr. Enrico Mugnaini.)

cortex, are oriented in the parasagittal plane. Purkinje cells, whose axons terminate in a certain cerebellar nucleus, constitute narrow sagittal zones, which are distributed across the cerebellar cortex, parallel to the median line. These longitudinal zones cut across the transversely organized cerebellar fissures. With the exception of the border between vermis and hemisphere, the borders between these zones are not visible at gross inspection. A similar longitudinal zonal organization is present in the afferent system of climbing fibers that originates in the inferior olive and terminates directly on the Purkinje cell dendrites.

The mossy fiber-parallel fiber system exhibits a

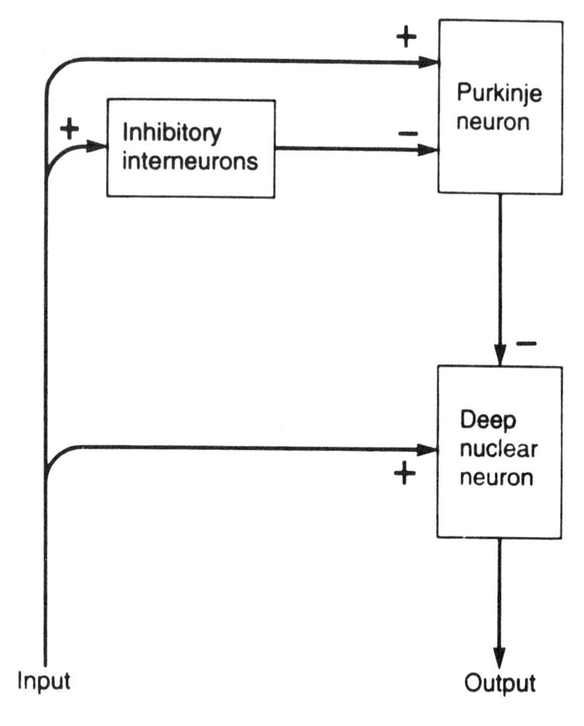

Fig. 17–4. Diagram of the Main Cerebellar Circuits The excitatory input is modulated by inhibitory interneurons (From Ghez, C., 1991. The cerebellum. In: E.R. Kandel, J.H. Schwartz, and T.M. Jessell. Principles of Neural Science, 3rd ed. Appleton and Lange, Norwalk, CT. With permission of the publisher.)

mainly transverse orientation. Functionally different mossy fiber systems tend to terminate in different cerebellar lobules. In this way the cerebellum can be subdivided in a somewhat imprecise manner into three different parts that receive mossy fibers from the vestibular apparatus, the spinal cord, and the cerebral cortex.

The vestibulocerebellum. (archicerebellum: obsolete term) consists primarily of the flocculonodular lobe. Apart from vestibulocerebellar mossy fibers, it receives pontocerebellar fibers concerned with the oculomotor system. Purkinje cell axons from the vestibulocerebellum terminate mainly in the vestibular nuclei, which are comparable to the cerebellar nuclei in their relations to the cerebellar cortex.

The spinocerebellum. (palaeocerebellum: obsolete term) consists of the anterior lobe and the region immediately caudal to the primary fissure as well as the pyramis. Apart from the direct and indirect mossy fiber pathways from the spinal cord and secondary trigeminocerebellar connections, the spinocerebellum receives a corticopontocerebellar projection from the sensorimotor cortex.

The pontocerebellum. (neocerebellum: obsolete term) consists of the portions of the vermis and hemispheres between the two spinal and the vestibular region of the cerebellum. The pontocerebellum, therefore, consists of two largely separate regions (Fig. 17–5). The pontocerebellum receives mossy fibers belonging to corticopontocerebellar projections primarily from sensorimotor, visual and cingulate areas of the cerebral cortex. It comprises the major part of the cerebellar hemisphere, and its development parallels that of the cerebral cortex.

Afferent Cerebellar Connections

The fiber tracts, which carry information to the cerebellum from different sources in the spinal cord, the brain stem, and the cerebral cortex, converge primarily on the inferior and the middle cerebellar peduncles (Fig. 17–6). The major afferent systems can be subdivided as follows:

Mossy fibers:
1. From spinal cord
2. From vestibular system
3. From reticular formation
4. From pontine nuclei

Climbing fibers:
5. From inferior olive

Aminergic fibers:
6. From raphe nuclei and locus ceruleus

Some of the fibers arising from the spinal cord and the vestibular system are ipsilateral, whereas others have a bilateral distribution. The fibers originating in the inferior olive and the pontine nuclei cross the midline to terminate on the opposite side of the cerebellum. The reticulocerebellar fibers, as well as the aminergic fibers, i.e., serotonin and norepinephrine fibers from the raphe nuclei and the locus ceruleus, are both crossed and uncrossed.

Spinocerebellar Tracts

Through the spinocerebellar tracts (Fig. 8–6B) the cerebellum receives information from both the periphery and motor reflex centers in the spinal cord. According to the incoming information, the cerebellum can then elicit the appropriate regulatory

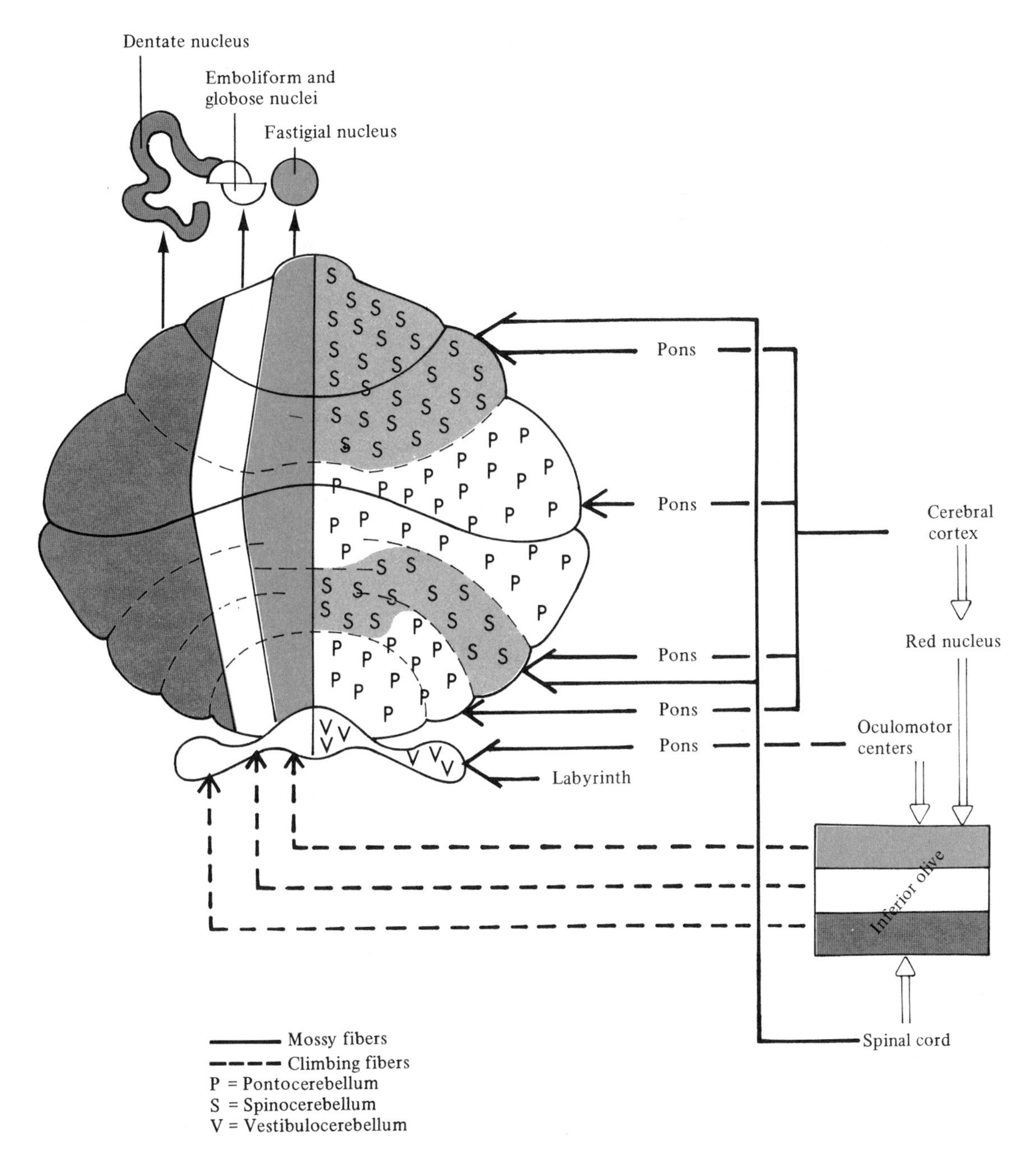

Fig. 17–5. Connections of the Cerebellar Cortex
Summary diagram of the afferent and efferent connections of the cerebellar cortex. The *left half of the figure* illustrates the climbing fiber input from the inferior olive to the Purkinje cells as well as the projections from the Purkinje cells to the intracerebellar nuclei. Both systems show a longitudinal zonal organization. The *right half of the figure* illustrates the mossy fiber systems from the spinal cord, vestibular apparatus, and cerebral cortex (via pons), all of which exhibit a mainly transverse orientation. Compare with Fig. 4–37, which illustrates the main subdivisions of the cerebellum. P, Pontocerebellum; S, spinocerebellum; V, vestibulocerebellum. (Based on an original drawing by J. Voogd.)

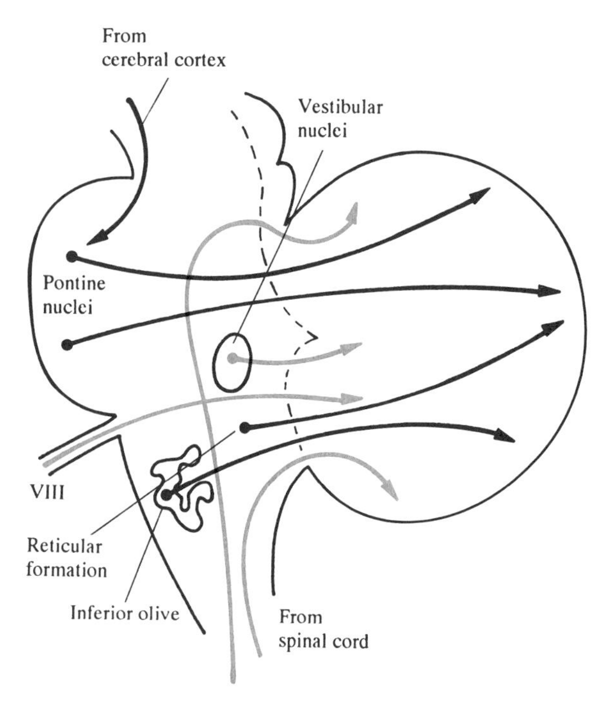

Fig. 17–6. Afferent Cerebellar Connections

response. The cerebellar response is transmitted to the motor cortex and to different brain stem centers for further delivery to motor centers in the spinal cord. All spinocerebellar pathways, except the ventral spinocerebellar tract, which enters through the superior cerebellar peduncle, reach the cerebellum through the inferior cerebellar peduncle.

Dorsal Spinocerebellar Tract

The *dorsal spinocerebellar tract* (DSCT) and the *cuneocerebellar tract* (CCT) transmit highly specific and discriminating proprioceptive and exteroceptive impulses from lower and upper parts of the body, respectively. The DSCT originates in the *nucleus dorsalis of Clarke* (T1–L2), whereas the CCT originates in the *external* (accessory) *cuneate nucleus* (proprioceptive impulses) and in the *internal cuneate* and *gracile nuclei* (exteroceptive impulses). The cells in Clarke's column receive dorsal root afferents from the lower part of the body (up to about midthoracic level) and the cells in the external cuneate nucleus from the upper part of the body.

Ventral Spinocerebellar Tract

The *ventral spinocerebellar tract* (VSCT) seems to transmit information about activity in motor reflex centers in the spinal cord. It originates in cell bodies located in the intermediate and ventral gray matter in the lower half of the spinal cord. The equivalent pathway from the cervical part of the spinal cord is called the *rostral spinocerebellar tract*. The VSCT enters the cerebellum through the superior cerebellar peduncle. Most of its axons cross twice, both in the spinal cord and in the cerebellar white matter.

Trigeminocerebellar Fibers

The *trigeminocerebellar fibers* in the inferior cerebellar peduncle transmit proprioceptive and exteroceptive impulses from areas of the face to the cerebellum.

Somatotopic Organization in the Cerebellar Cortex. The spinocerebellar fibers terminate in the "spinocerebellum," represented by the anterior lobe, and by pyramis and adjoining parts of the hemisphere in the posterior lobe. The fibers terminate in a somatotopic fashion, in the sense that different regions of the body are represented separately in the anterior lobe and in the posterior lobe (Fig. 17–7). There are, in other words, two sets of topographically organized spinocerebellar receiving areas in the cerebellar cortex.

Vestibulocerebellar Fibers

The vestibular impulses, which arrive through the inferior cerebellar peduncle, provide the necessary information for proper reflex control of posture and eye position during head movements and changes in position of the head.

Primary Fibers

The *primary fibers*, with cell bodies located in the vestibular ganglion, carry information directly from the labyrinth receptors primarily to the ipsilateral cerebellum.

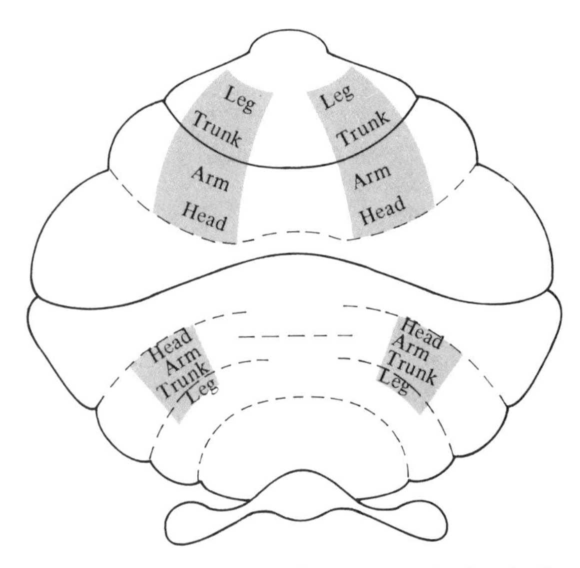

Fig. 17–7. Somatotopic Organization in Cerebellar Cortex

Secondary Fibers

The *secondary fibers* arise from cell bodies in the vestibular nuclei and have a symmetric, bilateral distribution.

"***Vestibulocerebellum.***" The vestibulocerebellar fibers terminate primarily in the flocculonodular lobe, which therefore is referred to as the "vestibulocerebellum."

Reticulocerebellar Fibers

Fibers from the reticular formation arise from several nuclear groups and project primarily through the inferior cerebellar peduncle to both the ipsilateral and the contralateral halves of the cerebellum. The reticular nuclear groups, such as the inferior olive, serve as relay and integration centers through which information from the spinal cord as well as other parts of the CNS can reach the cerebellum.

Pontocerebellar Fibers

The *pontocerebellar fibers*, which convey impulses from the cerebral cortex to the cerebellum, form the massive middle cerebellar peduncles. The first link in this important cerebrocerebellar projection system (Fig. 17–8A) is represented by the *corticopontine fibers*, which converge toward the internal capsule primarily from the motor and premotor areas and the somatosensory cortices, but also from the visual cortex and the cingulate gyrus. The corticopontine fibers proceed through the cerebral peduncle to the base of the pons, where they terminate in the *pontine nuclei*. The axonal projections of the pontine nuclei, the pontocerebellar fibers, enter the cerebellum through the middle cerebellar peduncles. Although it is generally believed that the pontocerebellar fibers cross over in the pons before they enter the cerebellum, recent studies indicate that a significant number do not cross. The large number of pontocerebellar fibers reflects the importance of close cooperation between the cerebral cortex and the cerebellum in the regulation of movements. Although the majority of the pontocerebellar fibers terminate in the so-called pontocerebellum in the cerebellar hemispheres, there are a significant number of pontocerebellar projection fibers also to the more medial "spinocerebellar" parts of the cerebellar cortex, as well as to the "vestibulo-cerebellum" (Fig. 17–5).

Olivocerebellar Fibers; Sagittal Zonal Organization

The *inferior olive* is the only known source of the climbing fibers. It constitutes an important relay and integration center, through which information from many different sources in the spinal cord as well as in the brain stem and the forebrain, including the cerebral cortex, is processed in parallel with the corresponding mossy fiber systems. The fibers from the inferior olive cross the midline before they enter the cerebellum to terminate in the intracerebellar nuclei and, as climbing fibers, directly on the Purkinje cells throughout the cerebellar cortex.

Prominent among the projection areas to the inferior olive are the spinal cord, the dorsal column nuclei, oculomotor centers in the midbrain, including the superior colliculus, and the parvocellular part of the red nucleus, which mediates the projections from the cerebral cortex, including a somatotopically arranged projection from the sensorimotor cortex. Direct cortico-olivary fibers are probably few in number. The axons from different groups of neurons in the inferior olive project to narrow sagittal zones, which may extend over the entire rostrocaudal length of the cerebellum, and which contain the Purkinje cells projecting to a certain central cerebellar nucleus. These longitudinal zones cut through the transverse, spinal, vestibular, and pontine subdivisions of the cerebellum. This means that the information from vestibular, spinal, and cortical levels, which reaches the inferior olive, becomes fragmented over a number of longitudinal zones. Each of these zones contains the appropriate vestibular, spinal, and cortical information in its portion corresponding to the particular transverse subdivision of the cerebellum.

The input to the cerebellar cortex therefore shows a high degree of convergence of different pathways carrying information of the same modality. Spinocerebellar and corticopontocerebellar mossy fiber pathways and the different longitudinally organized olivocerebellar paths, which receive somatotopically organized information from the spinal cord and the sensorimotor cortex through the red nucleus, converge on the appropriate portions of the spinocerebellum. Similar examples can be cited for the afferent connections of the "vestibulocerebellum" and the pontocerebellum.

Aminergic Fibers from Raphe Nuclei and Locus Ceruleus

Serotonin-containing fibers from the raphe nuclei and norepinephrine-containing fibers from the locus ceruleus reach the intracerebellar nuclei and cerebellar cortex through the superior and inferior cerebellar peduncles. The functional significance of biogenic amine-containing fibers is not clear. Their anatomic arrangement suggests a modu-

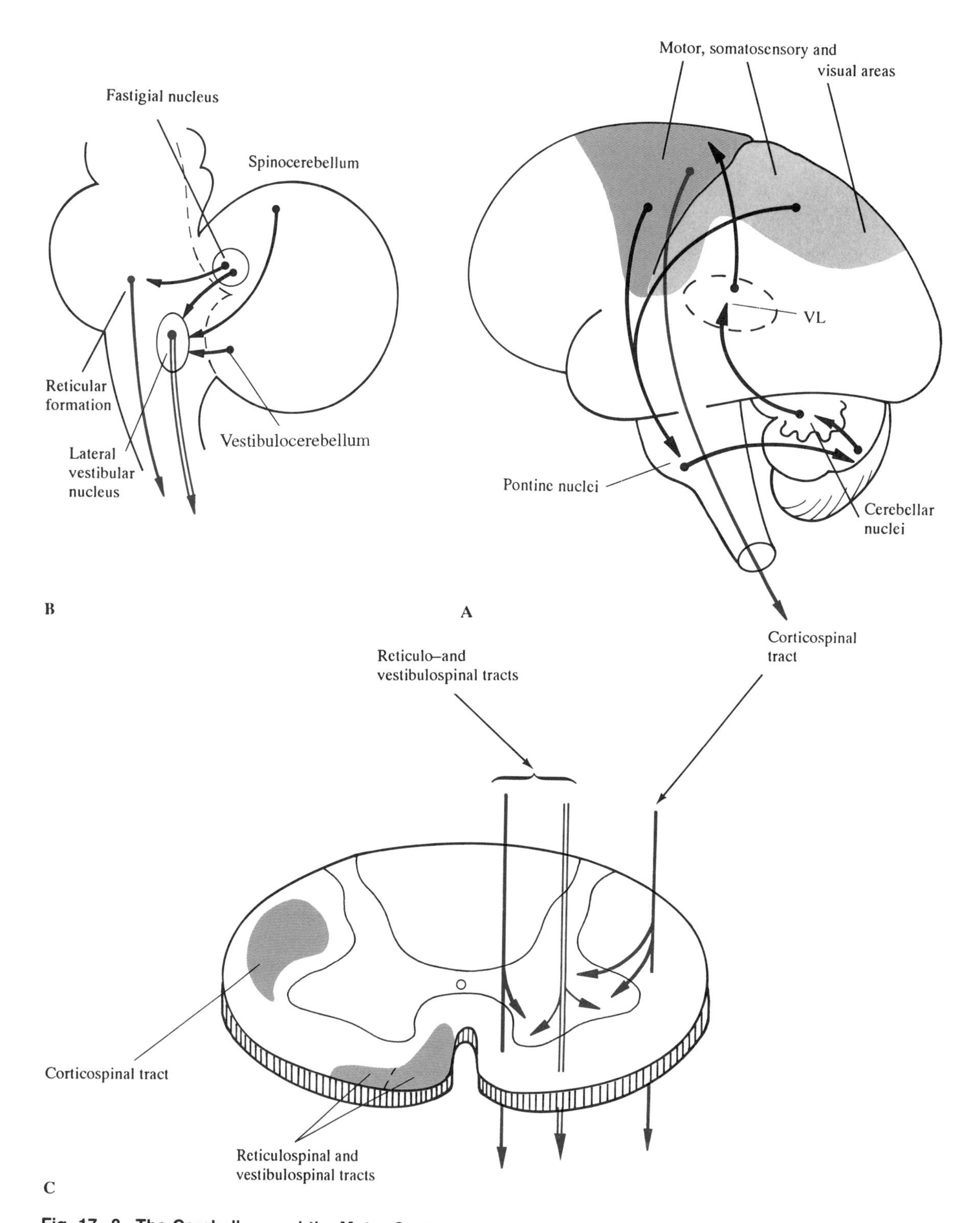

Fig. 17–8. The Cerebellum and the Motor System
A. Highly schematic drawing of the corticopontocerebellar projection system and the efferent cerebellar projections to the red nucleus and the VA–VL complex of the thalamus. The VA–VL complex in turn, projects to the motor cortex, which provide access to the lateral descending system in the spinal cord.
B. Efferent cerebellar connections to the lateral vestibular nucleus and the reticular formation provide access to vestibulospinal and reticulospinal pathways.
C. Descending pathways in the spinal cord.

lating effect on neuronal activities in general. The norepinephrine input from the locus ceruleus may be of special importance for cerebellar mechanisms related to the regulation of autonomic responses.

Cerebellar Nuclei

The cerebellar nuclei within the white matter are, from lateral to medial (Fig. 4–40B in dissection guide):

1. *Dentate nucleus*
2. *Emboliform nucleus*
3. *Globose nuclei*
4. *Fastigial nucleus*

The cerebellar nuclei receive excitatory input through collateral projections of the afferent cerebellar fibers that pass by en route to the cerebellar cortex. This input to the cerebellar nuclei forms part of the "main line" through the cerebellum. The same input that reaches the cerebellar nuclei also reaches the cerebellar cortex. The activity generated in this "side loop" through the cortex is ultimately channeled to the cerebellar nuclei and to the vestibular nuclei via the Purkinje cells.

Corticonuclear Relationships

The projections from the cerebellar cortex to the intracerebellar nuclei tend to be organized in a longitudinal or sagittal fashion, in the sense that the axons of the Purkinje cells extend to the nearby deep nucleus (Fig. 17–5). In other words, a medial zone represented by the vermis is primarily related to the fastigial nucleus and Deiters' nucleus, an intermediate zone consisting of the medial parts of the cerebellar hemisphere is related to the emboliform and globose nuclei, and a lateral zone including the rest of the hemisphere is related to the dentate nucleus. Although this subdivision of the cerebellum into three longitudinal zones is an oversimplification, the terms medial, intermediate, and lateral zones, which reflect the anatomic relationships between the cerebellar cortex, the intracerebellar nuclei, and the inferior olive, have physiologic and clinical significance (see below).

Efferent Cerebellar Connections

The distribution patterns of the cerebellofugal projections are very complicated and only the main features are summarized below. The efferent fibers, which leave the cerebellum in the inferior or superior cerebellar peduncle, terminate in brain stem and forebrain regions, many of which are directly or indirectly related to descending spinal pathways. These regions include the vestibular nuclei, the reticular formation, oculomotor centers in the brain stem, the red nucleus, and the ventrolateral thalamus.

Cerebellovestibular Fibers

The fibers to the vestibular nuclei project through the inferior cerebellar peduncle. Since many of the fibers arise directly from the cerebellar cortex, the vestibular nuclei can to some extent be compared with the intracerebellar nuclei. The two major contingents of cerebellovestibular fibers are: (1) fibers from the flocculonodular lobe ("vestibulocerebellum") and the fastigial nucleus to all four vestibular nuclei; and (2) fibers from the anterior vermis ("spinocerebellum") to the lateral vestibular nucleus of Deiters.

Cerebelloreticular Fibers

A considerable portion of the cerebelloreticular fibers originate in the fastigial nucleus and pass through the inferior cerebellar peduncle to the reticular formation in the pons and the medulla oblongata.

Cerebellorubral and Cerebellothalamic Fibers

Ascending cerebellar projections, which originate in dentate, emboliform, and globose nuclei, leave the cerebellum in the superior cerebellar peduncle and cross the midline in the decussation of the superior cerebellar peduncles. Important areas of termination include the red nucleus and the oculomotor centers in the midbrain and the ventral lateral (VL) nucleus of the thalamus. The ventral anterior–ventral lateral (VA–VL) complex of the thalamus also receives significant input from the basal ganglia, but the pathways from the cerebellum and the basal ganglia reach different parts of the VA–VL complex. The thalamic zones in the VL that receive input from the cerebellum project primarily to the motor cortex (MC) and to some extent the premotor cortex (PMC), while those in the VA–VL receiving afferents from the basal ganglia project to the supplementary motor area (SMA).

The projection to the VL, which in turn projects to the motor cortex, establishes a cerebellocerebral projection system, which in part reciprocates the corticopontocerebellar projection system. However, there is an important difference between the two systems. Whereas the cerebellum receives information from both motor and sensory regions of the cerebral cortex through the corticopontocerebellar system, the cerebellar activity mediated through the cerebellothalamocortical pathway affects cortical motor areas, especially the primary motor cortex.

Cerebellar Functions

A review of cerebellar anatomy reveals some striking features. Quite characteristic, for instance, is the stereotyped cortical histology. However, different parts of the cerebellum are related to different regions of the nervous system. Therefore, although the intrinsic neuronal mechanisms are similar throughout the cerebellum, the regulatory activities in different parts of the cerebellum are apparently related to different functions or to different parts of the body.

Another striking feature is the large number of pathways carrying information to the cerebellum from a variety of different sources. The large number of fibers in the corticopontocerebellar system is especially conspicuous, and since this massive input to the cerebellum comes from sensorimotor areas around the central sulcus, it can only mean that the cerebellum is kept abreast of voluntary motor activities to be able to take the appropriate corrective measures.

A survey of the efferent cerebellar pathways indicates that a large part of the cerebellar output ultimately affects the motor system (Fig. 17–8). How the cerebellum is related functionally to the rest of the motor system is not altogether clear. What is obvious, however, is that cerebellar lesions do affect the control of both voluntary and involuntary motor functions (see Clinical Notes).

The Cerebellum and the Motor System

In order to appreciate the clinical significance of cerebellar lesions, it is useful to consider the arrangement in longitudinal zones depicted in Fig. 17–8.

Fibers from the emboliform and globose nuclei and from the dentate nucleus project to the red nucleus, and via the VL of the thalamus, to the motor cortex. Through these connections the intermediate and lateral parts of the cerebellum have direct access to the lateral descending system in the spinal cord (see Chapter 17–8), i.e., the corticospinal tracts (Fig. 17–8A and C), which are primarily related to limb musculature and fine manipulative movements. There are some indications that the intermediate parts of the hemispheres, together with the emboliform and globose nuclei, are directly involved with the execution and adjustments of movements, whereas the lateral parts, including the dentate nuclei, are primarily concerned with mechanisms related to planning and preparation for motor actions. In clinical practice, however, these different functions cannot be correlated with a specific localization of a hemispheric cerebellar lesion.

The fastigial nucleus and the vermis represent major areas of origin for fibers to the reticular formation and the vestibular nuclei. The lateral vestibular nucleus of Deiters is influenced primarily by the vermis in the anterior lobe. The reticular formation and the vestibular nuclei give rise to the reticulospinal and vestibulospinal tracts that control both postural activity and locomotion (Figs. 17–8B and C).

Relationships between the Cerebellum and the Vestibular System and Oculomotor Centers

The vestibular system and the cerebellum cooperate closely in motor functions related to the maintenance of equilibrium. They are also closely related to various oculomotor centers in the brain stem, and the "vestibulocerebellum" plays an important regulating role in the so-called vestibulo-ocular reflex, i.e., adjustment of eye position in response to head movement. The importance of cerebellar control of vestibular and vestibulo-ocular reflexes is reflected in the massive reciprocal connections between the vestibular system and the cerebellum. Cerebellar relationships with the oculomotor centers are established via the inferior olive and its projections to the cerebellum, as well as by efferent cerebellar connections to the oculomotor centers. Impulses between the vestibular system and the eye movement centers are mediated by the medial longitudinal fasciculus (Fig. 11–12).

Through its relationships to both the vestibular system and the oculomotor centers, as well as its projections to the red nucleus and the motor cortex, the cerebellum emerges as a crucial component in the neuron circuitry responsible for the coordination of voluntary eye, head, and hand movements that characterize most human activities.

Clinical Notes

Cerebellar Ataxia

Ataxia, loss of muscular coordination, is most strikingly seen in cerebellar disease. It is best detected in the patient's gait and complicated limb movements. The gait is broad-based, irregular, and staggering, like that of an intoxicated person. When the patient is asked to put his finger on his nose (finger-to-nose test), the finger starts on its course normally but gradually begins to oscillate more and more violently as it approaches the nose. These tremulous movements are referred to as "cerebellar tremor" or "intention tremor." Cerebellar ataxia is often evident also in conjugate eye movements, which become jerky, and in speech, which has a tendency to become irregular and explosive with a slurring of words (cerebellar speech).

Although a cerebellar lesion is a common cause of ataxia, incoordination of movements can also be caused by loss of afferent proprioceptive input, e.g., by a lesion in the somatosensory pathways, in which case it is referred to as *sensory ataxia*.

Localization of Cerebellar Disease

Although there seems to be some validity in the functional subdivisions of the cerebellum into "vestibulocerebellum," "spinocerebellum," and "pontocerebellum," considerable overlap exists among the three parts, and it is often difficult to determine the position of a focal cerebellar lesion, e.g., a tumor, on the basis of the patient's symptoms. The situation is further complicated by the fact that there is little room for expansion in the infratentorial part of the cranial cavity. A space-occupying lesion in the posterior fossa, therefore, may cause symptoms from pressure on surrounding brain stem and cerebellar structures.

Of importance, however, is the fact that each cerebellar hemisphere exerts its influence primarily on the ipsilateral half of the body. In other words, if cerebellar ataxia is evident in limb movements on only one side of the body, the lesion is likely to be present in the ipsilateral cerebellar hemisphere or its extrinsic pathways. The explanation for this is quite simple. Most of the cerebellofugal fibers from the hemisphere depart through the superior cerebellar peduncle and cross in its decussation before terminating in the red nucleus and the motor cortex (via the VL thalamic nucleus) on the opposite side. The motor cortex in turn gives rise to the corticospinal tract, which primarily affects the motor centers in the contralateral half of the spinal cord, i.e., that same side from which the cerebellofugal part of the circuit originated.

Whereas cerebellar hemisphere lesions primarily cause ipsilateral incoordination (ataxia and tremor), midline or vermis cerebellar disease may be manifest by impairment of equilibrium. Such deficits in upright posture cause titubation (a staggering and stumbling gait with shaking of the trunk and head), wide-based stance, and Romberg's sign (difficulty in standing erect with feet approximated and eyes closed).

In summary, one may say that the most common objective signs of a cerebellar tumor are disturbances of gait and posture, regardless of whether the tumor is located in the area of the vermis or in the hemisphere. Although ataxic limb movements can be seen in patients with both lateral and medial disease, unilateral limb movement disturbances indicate the presence of a lesion in the ipsilateral cerebellar hemisphere. It is important to remember, however, that a space-occupying lesion in the posterior fossa quickly produces increased intracranial pressure. Therefore, subjective symptoms, such as headache and vomiting with or without disturbances in balance, may be the first symptoms of a cerebellar tumor.

Clinical Example

Metastatic Cerebellar Hemisphere Lesion

A 65-year-old male with a long history of cigarette smoking was found on routine chest X-ray to have a lesion in the upper lobe of his right lung, and bronchoscopy and biopsies revealed a lung cancer (bronchogenic squamous cell carcinoma). Examination of his scalene lymph nodes showed evidence of metastatic tumor, and he was treated with palliative radiotherapy.

Three months later, the man developed progressive gait difficulties. Examination at that time showed that he was alert and oriented, his cranial nerves were normal, but he had decreased muscle tone in his right arm and leg. The stretch reflexes were reduced on the right side compared with the left. He also had "intention tremor" in his right arm and leg while performing the finger-to-nose test or when trying to put his right heel on his knee when lying in bed (heel-to-shin test). Although his balance was relatively good, he did have difficulty with heel-to-toe walking, and tended to veer toward the right side.

An MRI (Fig. 17–9) was obtained after infusion of intravenous contrast.

1. Where is the lesion in the CNS?
2. Given the clinical history, what do you think the lesion most likely represents?
3. The lesion is on the right side of his brain. Why then are his deficits also on the right side?
4. If the lesion were to expand, what adjacent brain structure would be compressed?

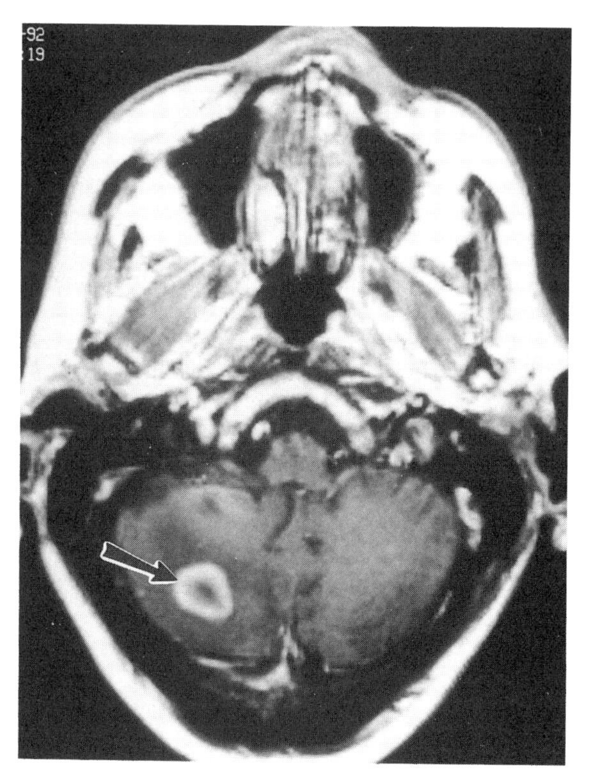

Fig. 17–9. Metastatic Cerebellar Hemispheric Lesion T1-weighted axial MRI with gadolinium enhancement. (Photograph courtesy of Dr. Maurice Lipper.)

Discussion

The CT scan, which is taken through the posterior fossa, shows an elliptically shaped lesion in the right cerebellar hemisphere (*arrow*). With a history of lung cancer that had already spread to regional lymph nodes, the most likely etiology is further metastatic spread of the lung tumor. Brain metastasis can involve virtually any part of the brain, but typically occurs at the junction of gray and white matter.

The patient's symptoms and signs are typical for a hemispheric cerebellar lesion in the sense that the cerebellar deficits (decreased muscle tone, hyporeflexia, and ataxia) involve the arm and the leg on the same side of the body as the lesion. This reflects the fact that the cerebellar hemisphere is involved primarily in the control of the ipsilateral limbs. Spinocerebellar pathways to the cerebellar hemispheres are mainly uncrossed, and the cerebellofugal pathways from the cerebellar hemispheres via the motor cortex to the spinal cord are "doubly crossed" (see Fig. 17–8A).

If the lesion were to expand, the adjacent brain stem would gradually be compressed. In addition, the fourth ventricle would probably become obliterated thereby causing an acute obstructive hydrocephalus. Because of their accessibility in the posterior fossa, cerebellar metastases are commonly removed surgically if they are the only metastasis found in the brain. An alternative mode of therapy would be radiation treatments to the posterior fossa.

Suggested Reading

1. Adams, R. D. and M. Victor, 1993. Principles of Neurology, 5th ed. McGraw-Hill: New York, pp. 74–82.
2. Brodal, A., 1981. Neurological Anatomy in Relation to Clinical Medicine, 3rd ed. Oxford University Press: New York, pp. 294–393.
3. Ghez, C., 1991. The cerebellum. In E. R. Kandel, J. H. Schwartz, and T. M. Jessell (eds.): Principles of Neural Science, 3rd ed. Elsevier: New York, pp. 626–646.
4. Gilman, S., 1992. Cerebellum and motor dysfunction. In A. K. Asbury, G. M. McKhann, and W. J. McDonald (eds.): Diseases of the Nervous System, 2nd ed. Saunders: Philadelphia, pp. 319–341.
5. Gilman, S., J. R. Bloedel, and R. Lechtenberg, 1981. Disorders of the Cerebellum. F. A. Davis: Philadelphia.
6. Palay, S. L. and V. Chan-Palay, 1974. Cerebellar Cortex: Cytology and Organization. Springer-Verlag: Heidelberg.
7. Thach, W. T. Jr. and E. B. Montgomery, 1992. Motor system. In A. L. Pearlman and R. C. Collins (eds.): Neurobiology of Disease. Oxford University Press: New York, pp. 168–196.

In situ hybridization autoradiography for oxytocin mRNA in a frontal section through the human hypothalamus. Note the labeling of cells in the paraventricular (Pa) and supraortic (SO) nuclei. (Courtesy of Dr. Stephan Guldenaar, Amsterdam).

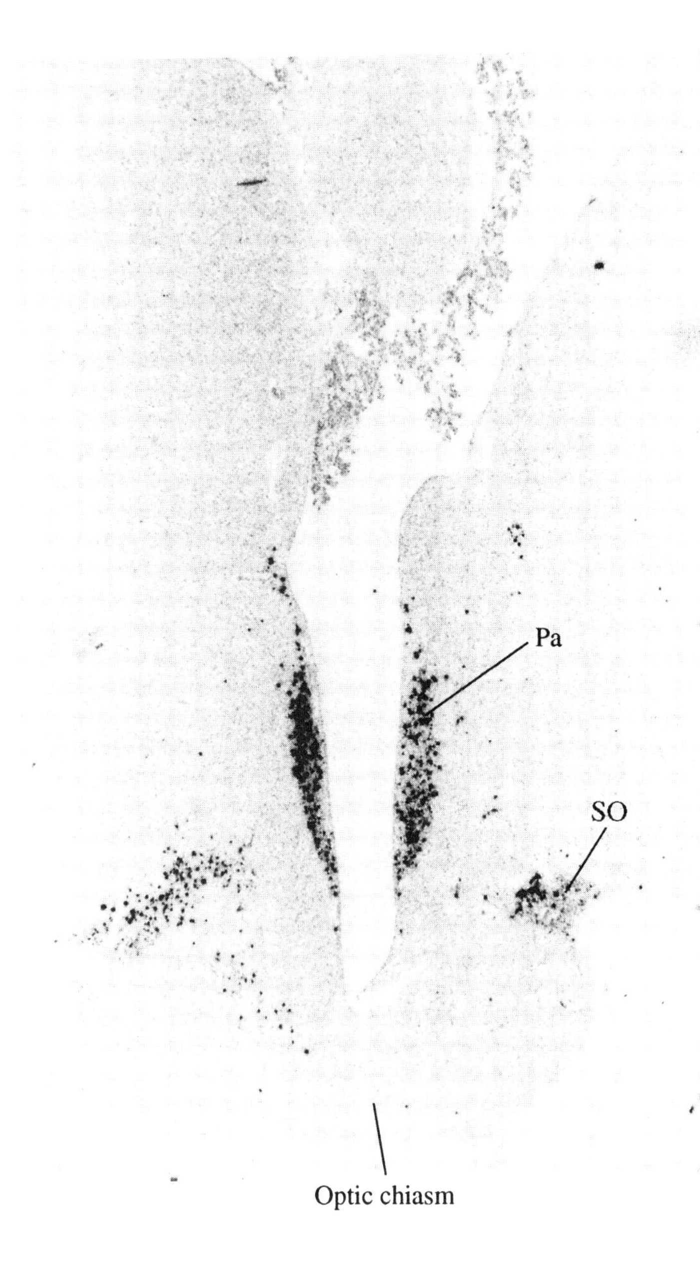

18

The Hypothalamus and the Hypothalamohypophysial Systems

Although the hypothalamus weighs only about 4 g, it contains a large number of more or less well defined cell groups that are critically important for preserving the individual and the species. It is characterized by numerous connections with practically every major part of the CNS including the cerebral cortex, hippocampus and amygdaloid body, as well as the brainstem and the spinal cord.

Through its intimate neuronal and vascular relationships with the pituitary, the hypothalamus controls the release of the pituitary hormones, which in effect means that the entire endocrine system is brought under the control of the CNS. Neuroendocrinology, the study of the anatomic and functional relationship between the brain and the endocrine system, has developed into a large scientific discipline that is concerned not only with the control of endocrine secretion, but also with the effect of hormones on the brain and through it, on behavior. With the isolation and synthesis of various hypothalamic hormones and their clinical application in diagnosis and treatment of endocrine disorders, clinical neuroendocrinology has become one of the most rapidly expanding fields in clinical medicine.

The hypothalamus is concerned with generalized response patterns that often involve autonomic, somatomotor, and hormonal systems, and which serve a variety of functions subserving homeostasis[1] (such as fluid and electrolyte balance, temperature regulation, and energy metabolism), as well as sexual behavior and a variety of other social behaviors and emotional responses.

1. Homeostasis refers to the many processes through which the body maintains a state of equilibrium in its internal environment as reflected in, for instance, temperature, heart rate, blood pressure, and the chemistry of tissues and body fluids.

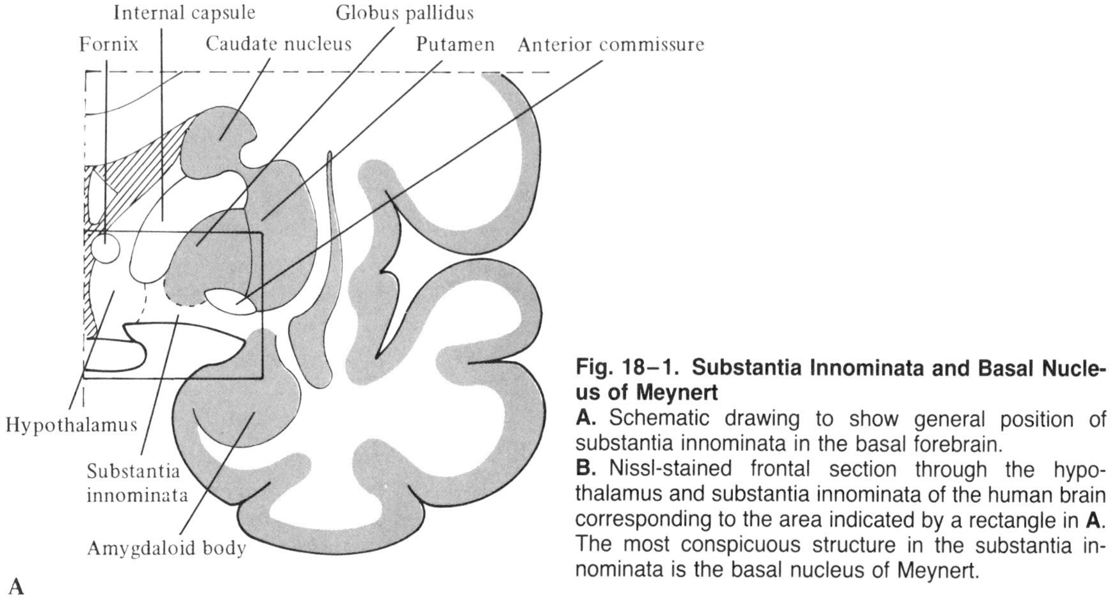

Fig. 18–1. Substantia Innominata and Basal Nucleus of Meynert

A. Schematic drawing to show general position of substantia innominata in the basal forebrain.

B. Nissl-stained frontal section through the hypothalamus and substantia innominata of the human brain corresponding to the area indicated by a rectangle in **A**. The most conspicuous structure in the substantia innominata is the basal nucleus of Meynert.

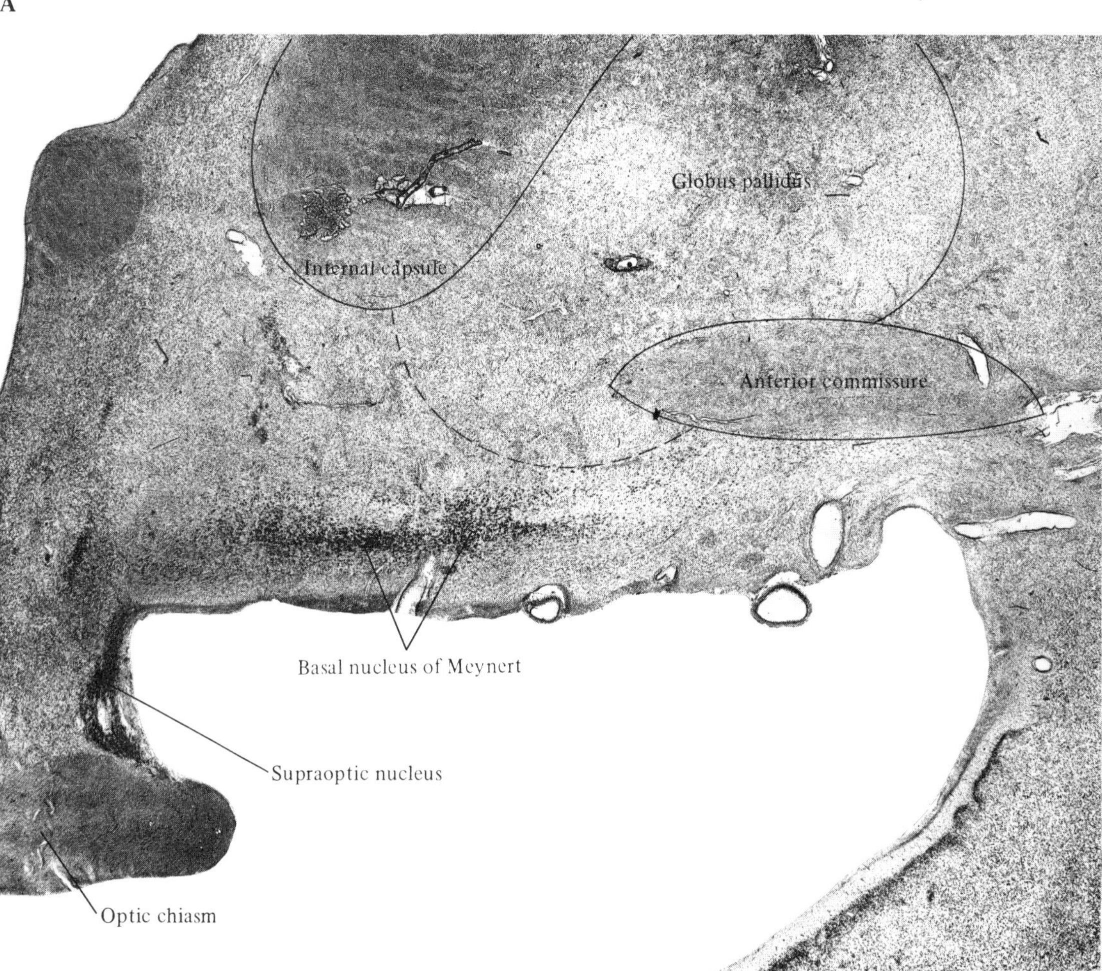

Anatomic Organization

Boundaries and Subdivisions

The hypothalamus is located at the base of the diencephalon, where it surrounds the lower part of the third ventricle underneath the thalamus. The boundary between the thalamus and the hypothalamus is indicated on a midsagittal section by a distinct groove, the *hypothalamic sulcus* (Fig. 4–8). The *lamina terminalis* indicates the rostral boundary of the hypothalamus, and an imaginary line that extends from the posterior commissure to the caudal limit of the mammillary body represents the caudal boundary (Fig. 4–7). Dorsolaterally, the hypothalamus extends to the medial edge of the internal capsule (Fig. 5–9). Although the hypothalamic boundaries are thus easily described, it is important to realize that the hypothalamus does not form a sharply circumscribed region of the CNS. On the contrary, it is continuous with the surrounding parts of the brain. For instance, the hypothalamus is continuous rostrally with the preoptic area and the septal area in the mediobasal parts of the telencephalon and with various components of the anterior perforated substance. Its anterior part is continuous laterally with the sublenticular part of the substantia innominata (Figs. 18–1 and 18–2; see also Figs. 16–13 and 16–14), whereas the caudal part is continuous with the central gray matter and the tegmentum of the mesencephalon.

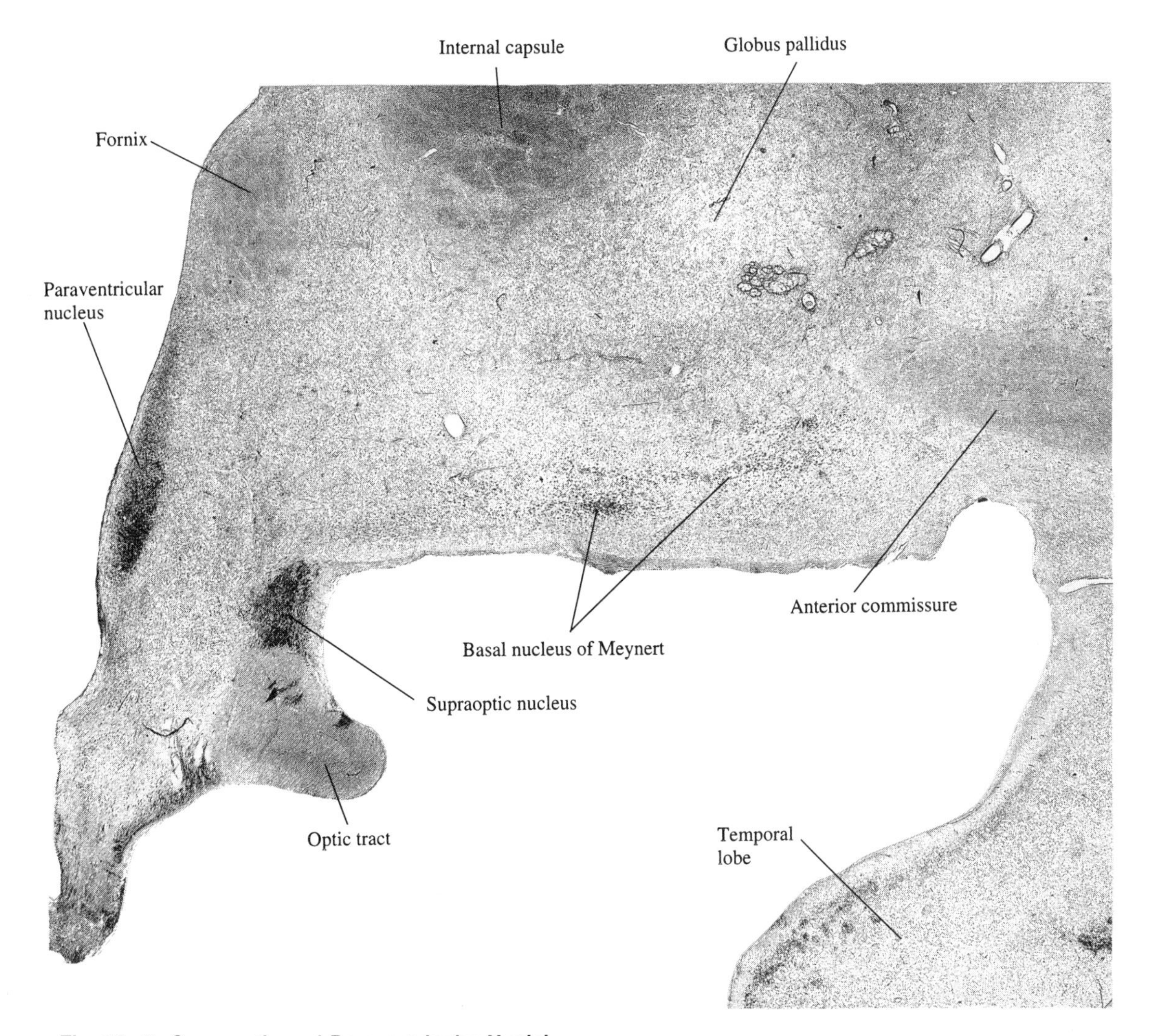

Fig. 18–2. Supraoptic and Paraventricular Nuclei
Nissl-stained cross-section through the anterior hypothalamus of a human brain showing the circumscribed supraoptic and paraventricular nuclei.

The basal surface of the hypothalamus is characterized by the *mammillary bodies* caudally and the *optic chiasm* rostrally (Fig. 4–6). Between these two structures is a gray swelling, the *tuber cinereum*, which is hidden behind the hypophysis in Fig. 4–6. The tuber cinereum tapers ventrally into a funnel-shaped structure, the *infundibulum*, which represents the most proximal part of the neurohypophysis (Fig. 18–3A). The infundibulum and the infundibular part of the adenohypophysis together form the *hypophysial stalk*. The infundibular region, often referred to as the *median eminence*, constitutes an important link between the hypothalamus and the pituitary (see below). The *lateral eminence* in the lateral part of the tuber cinereum is a rounded prominence produced by a group of superficially located nuclei, the *lateral tuberal nuclei* (Fig. 18–3C).

The conspicuous landmarks on the ventral surface of the brain can be used to divide the hypothalamus into three parts (Fig. 18–3): the *chiasmatic* or *supraoptic* (anterior), the *tuberal* (middle), and the *mammillary* (posterior) part.

Each half of the hypothalamus can also be subdivided into a periventricular, a medial, and a lateral zone. The *periventricular zone*, which borders the wall of the third ventricle, is a thin layer of cells that contains most of the neurosecretory neurons regulating the anterior pituitary functions. The more voluminous ventral part of the periventricular zone in the middle part of the hypothalamus is referred to as the arcuate (infundibular) nucleus.

The *medial zone* contains some of the most well-demarcated nuclei in the hypothalamus, including the paraventricular and supraoptic nuclei in its anterior part. As components of the magnocellular neuroendocrine system, these two nuclei are well known for their intimate relation to the neurohypophysis; the paraventricular nucleus, in addition, projects to the median eminence as well as to the brain stem and spinal cord autonomic cell groups. The ventromedial and dorsomedial hypothalamic nuclei and the mammillary bodies are also located in the medial longitudinal zone.

The *fornix* forms an easily recognizable boundary between the medial and the lateral hypothalamus in the middle and posterior regions (Fig. 18–3). Although there are some prominent nuclei also in the basolateral part of the hypothalamus, e.g., the lateral tuberal nuclei, most of the lateral zone cannot be easily subdivided into nuclei. Rather, it forms a continuum, the lateral hypothalamic area, of loosely scattered cells with long overlapping dendrites similar to the cells of the reticular formation. The cells have widespread connections with many areas in the forebrain, including the basal ganglia and the cerebral cortex, and in the brain stem and spinal cord. A number of neuroactive substances, including a melanin-concentrating hormone and histamine, have been identified in lateral hypothalamic neurons.

Hypothalamic Nuclei

The best-known hypothalamic nuclei are the discrete cell groups of the medial zone. They are illustrated as if projected on the wall of the third ventricle in Fig. 18–3A, and they are shown in cross-sections through the anterior, middle, and posterior parts of the hypothalamus in Figs. 18–B, C, and D. Although some of the cell groups, e.g., the supraoptic and paraventricular nuclei, are easily identified as separate units (Fig. 18–2), others cannot be sharply delimited from the surrounding substance.

Anterior Group (Figs. 18–3A and B)

Important nuclei in this group are the *preoptic* and *suprachiasmatic nuclei*. Animal experiments have shown that the preoptic nucleus and the adjoining regions of the anterior hypothalamus are involved in sexual and maternal behavior. The reproductive functions are consistent with the sexual dimorphism (structural difference between male and female) that characterizes the preoptic area; one of its cell groups is larger in males. The preoptic–anterior hypothalamic region is also implicated in the integration of several homeostatic functions including temperature regulation and the sleep–wake cycle.

The suprachiasmatic nucleus, which is located in the periventricular zone at the base of the third ventricle just dorsal to the optic chiasm, receives visual input from both the retina and the lateral geniculate body, and is closely related to the preoptic region and several other hypothalamic areas. It serves as a circadian (24-hour) rhythm generator for many functions, including the sleep–wakefulness rhythm. The suprachiasmatic nucleus also has a special relation to the pineal gland, which will be discussed in more detail in relation to the circumventricular organs below.

Included in the anterior group are the prominent *supraoptic* and *paraventricular nuclei*. The large cells that characterize these two magnocellular nuclei send their axons to the posterior lobe of the pituitary. The paraventricular nucleus, furthermore, contains a large number of parvocellular (small-celled) cell groups, some of which project to the median eminence, whereas others project to autonomic neurons and reticular neurons in the brain

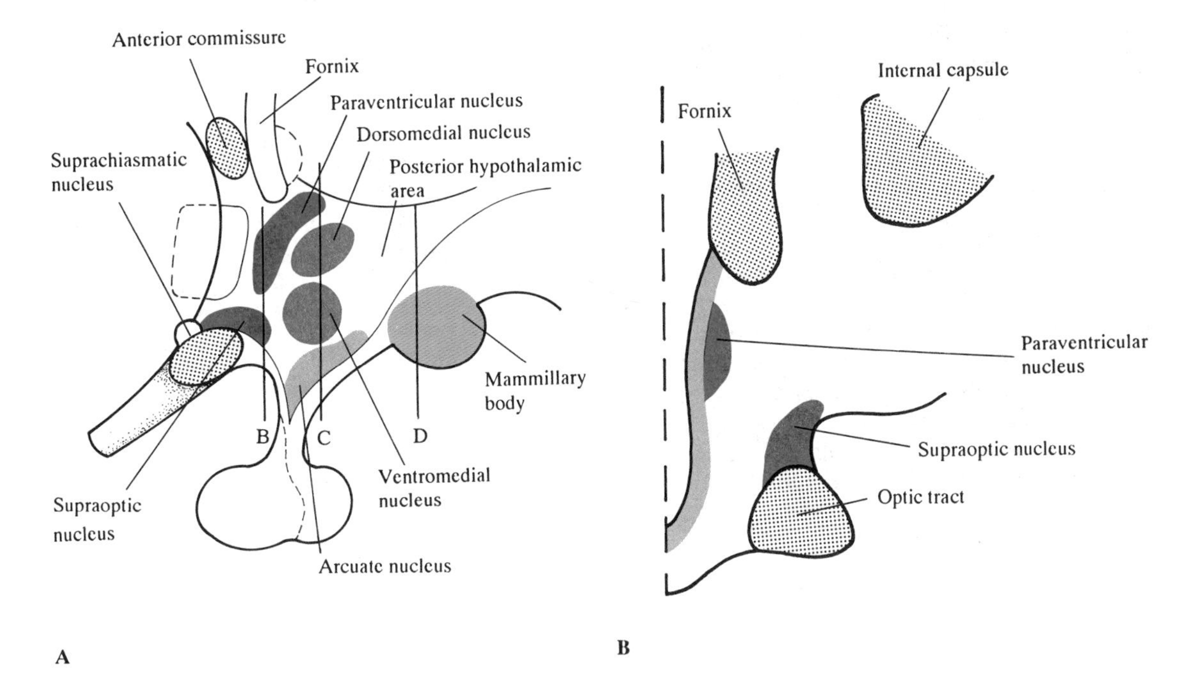

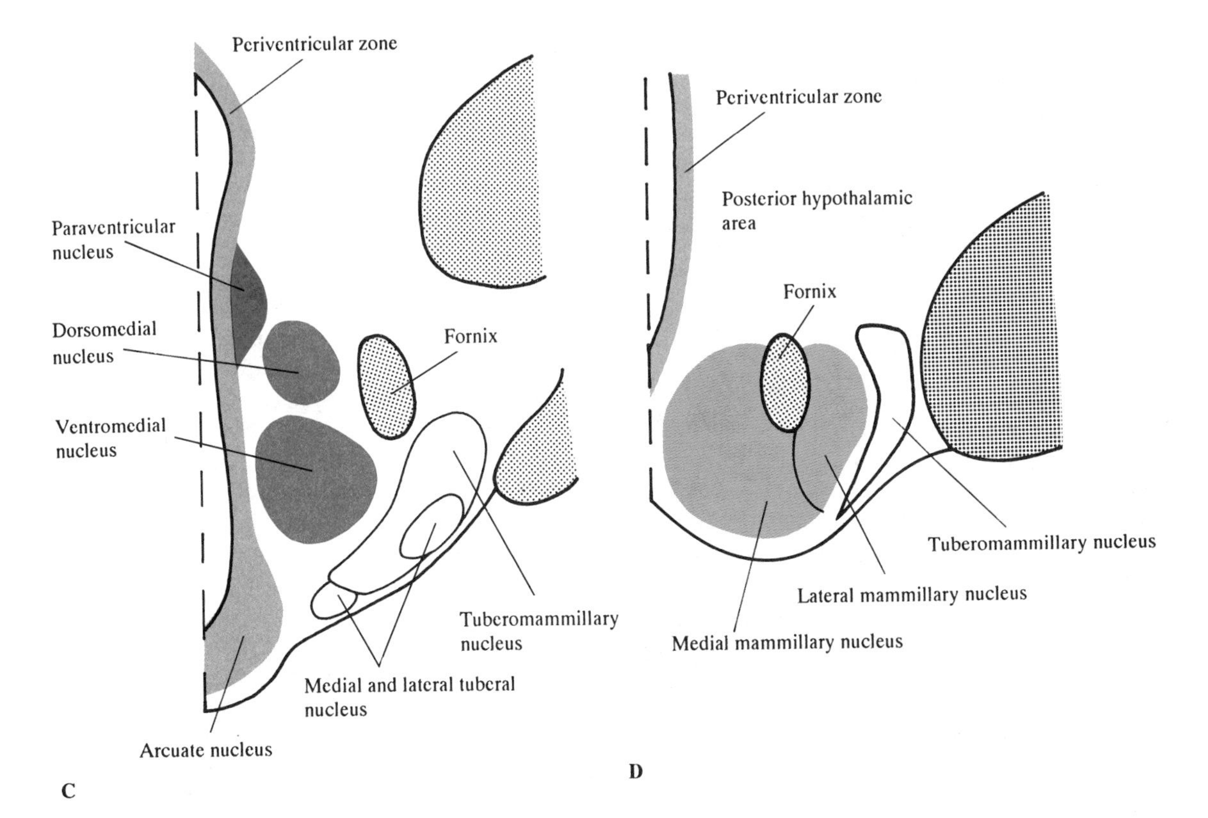

Fig. 18–3. Hypothalamic Nuclei
A. Diagram of the ventricular surface of the hypothalamus showing the approximate location of the most important hypothalamic nuclei (Modified from W. E. Le Gros Clark, 1936, J. Anat. (Lond.)).

B, C, and D. Cross-sections through the anterior (**B**), middle (**C**), and posterior (**D**) hypothalamus.

stem and spinal cord. Based on these projections, the nucleus occupies a pivotal position in the hypothalamus in the sense that it can directly control the endocrine activities of the pituitary, the preganglionic cell groups in the autonomic nervous system, and somatomotor areas in the brain stem reticular formation.

Middle Group (Fig. 18–3C)

The middle hypothalamus contains the *dorsomedial* and *ventromedial nuclei* in the medial zone and the *arcuate (infundibular) nuclei* in the periventricular zone surrounding the ventral part of the third ventricle. The ventromedial nucleus occupies a strategic position in the hypothalamus. It has widespread afferent and efferent connections with many other regions of the CNS, including the amygdaloid body and many brain stem areas, including the reticular formation and the periaqueductal gray.

Axons from cells in the arcuate nucleus and surrounding parts of the basomedial and periventricular hypothalamus form a diffuse projection system that terminates on the hypophysial portal vessels in the median eminence. This provides for a neurovascular link between the hypothalamus and the anterior pituitary.

The *medial* and *lateral tuberal nuclei* are more or less embedded in the large *tuberomammillary nucleus*, a neuronal complex that occupies a large part of the basolateral hypothalamus. The large tuberomammillary cells, which contain a number of different neurotransmitters including the monoamine histamine, are noted for their widespread conections to many parts of the CNS including the cerebral cortex. The tuberomammillary nucleus and the tuberal nuclei give rise to a prominent bulge on the ventral surface, the *lateral eminence*.

Posterior Group

The posterior part of the hypothalamus consists of the *posterior hypothalamic area* and the *mammillary body*. The posterior hypothalamic nucleus is a large but poorly defined cell group that is continuous with the mesencephalic reticular formation. The mammillary body, which is a focal point for several prominent fiber bundles, consists of a large medial and a smaller lateral nucleus. It is surrounded by a capsule of heavily myelinated fibers. Bordering the lateral part of the mammillary body is the prominent *tuberomammillary nucleus*.

Circumventricular Organs

As the name indicates, the *circumventricular organs* (CVOs) are located around or in relation to the ventricular system (Fig. 3–4), and several of them are also closely related to the hypothalamus. The CVOs are highly vascularized structures, and since they lack a blood–brain barrier, they provide for the exchange of substances between the blood and the brain. On the other hand, tight junctions between the ependymal cells of the CVOs form a CVO–CSF barrier. Some of the CVOs, i.e., the *subfornical organ* and the *vascular organ of the lamina terminalis*, contain receptors for angiotensin II and atrial natriuretic factor (ANF), and there is evidence that neurons in these organs can influence water balance and blood pressure in response to blood-borne angiotensin II. Neurons in the CVOs are also influenced by osmotic changes in the blood, substances in the CSF, and afferent fibers from other parts of the nervous system. To be able to affect various mechanisms related to water balance, drinking behavior, and blood pressure, the neurons in the subfornical organ and the vascular organ of the lamina terminalis are connected with several nearby basal forebrain and hypothalamic regions, including the paraventricular and supraoptic nuclei, which contain neurons that secrete the antidiuretic hormone, also known as vasopressin.

The *pineal gland* or *epiphysis* was identified as an anatomic structure more than two thousand years ago, but its functions are still to some extent shrouded in mystery. It contains astrocytes and a special type of secretory cells, *pinealocytes*, which have many processes, all of which remain in the gland. The pinealocytes secrete melatonin and polypeptides into both the blood and the CSF. Although the functions of the pineal gland are not well understood, its hormones are believed to play a regulatory role in the functions of several of the endocrine organs including the pituitary, adrenal glands, and gonads. The major input to the pineal gland comes from the suprachiasmatic nucleus, which provides information about the day–light cycle. This input reaches the pineal gland via descending pathways from the suprachiasmatic nucleus to the intermediolateral nucleus in the upper part of the thoracic cord, from which it is transmitted by way of the sympathetic nervous system to the gland.

The *median eminence* and the *neurohypophysis* are two important members of the CVOs, which will be discussed in more detail below. The *area postrema*, which is located in the lower brain stem, receives input from some hypothalamic regions including the paraventricular nucleus, but is otherwise related mostly to several neighboring brain stem regions. The functional role of the area postrema is largely unknown, although it has long been considered to be a trigger zone for vomiting.

Connections

Some of the myelinated hypothalamic fiber tracts, e.g., the fornix, mammillothalamic tract, stria terminalis, and stria medullaris, were dissected or depicted in the dissection of the brain (Chapter 4). Needless to say, macroscopic dissections can only give an idea of the general course of a myelinated fiber bundle; knowledge about the precise origin and termination of axonal projections has been obtained primarily from light and electron microscopic tracer studies in experimental animals. Such studies have revealed that the hypothalamic connections are considerably more complicated and widespread than the following list of myelinated fiber bundles might suggest. The hypothalamus is reciprocally related to a large number of forebrain areas including the extended amygdala, ventral striatum, septum, hippocampus, and many other cortical regions, as well as to regions in the brain stem and spinal cord (Figs. 18–4 and 18–5).

Fornix

The fornix is a large fiber bundle that connects the hippocampal formation with the septal area, anterior thalamus, and hypothalamus. Judged by the gross anatomic dissection of the fornix, one gets the impression that the whole bundle proceeds to the mammillary body. Microscopic examination, however, indicates that many fibers leave the main bundle along its course through the hypothalamus, where they terminate in both its medial and lateral parts.

Mammillothalamic Tract and Mammillary Peduncle

The mammillary body is different from the rest of the hypothalamus in several regards. It is sur ounded by a capsule of heavily myelinated fibers, and it is not as closely related to autonomic and endocrine functions as the rest of the hypothalamus. Its functions, incidentally, are not well known. Most of its efferent fibers leave the medial mammillary nucleus in a dorsal direction as the mammillothalamic tract, which proceeds in the internal medullary lamina of the thalamus to the anterior thalamic nuclei. Collaterals of the mammillothalamic fibers form the less conspicuous mammillotegmental tract (Fig. 18–4), which projects to dorsal tegmental cell groups in the mesencephalon. These cell groups in turn give rise to the mammillary peduncle, which projects back to the hypothalamus and terminates in the mammillary body.

Stria Medullaris

The stria medullaris, which can be easily recognized on the mediodorsal side of the thalamus (Fig. 4–31), connects the lateral preoptic–hypothalamic region with the habenular complex. However, like most other hypothalamic pathways, the stria medullaris is a complicated bundle that contains many different fiber components with various origins and terminations.

Stria Terminalis

The stria terminalis reciprocally connects the amygdaloid body and the medial hypothalamus (Fig. 18–5). Similar to the fornix, the stria terminalis makes a dorsally convex detour behind and above the thalamus. It can be identified in the floor of the lateral ventricle, where it accompanies the thalamostriate vein in the groove that separates the thalamus from the caudate nucleus (Fig. 4–31). In the region of the anterior commissure, the stria terminalis divides into different components, which distribute their fibers to the bed nucleus of the stria terminalis, the medial hypothalamus, and other areas in the basal parts of the forebrain. The stria terminalis is an important pathway for amygdaloid modulation of hypothalamic functions. The amygdaloid body, especially its centromedial part (Chapter 20), is also related in a reciprocal fashion to the hypothalamus through a diffuse ventral pathway that spreads out underneath the lentiform nucleus. This ventral amygdaloid fiber system also contains a significant component from the basolateral amygdaloid complex to the ventral striatum.

Dorsal Longitudinal Fasciculus

The dorsal longitudinal fasciculus is a component of an extensive periventricular system of descending and ascending fibers that connects the hypothalamus with the periaqueductal gray and with the pons and medulla oblongata, including preganglionic autonomic nuclei.

Medial Forebrain Bundle

Many of the fibers in the medial forebrain bundle, like those in the dorsal longitudinal fasciculus, connect the hypothalamus with the brain stem and spinal cord. Through descending fibers that originate primarily in the paraventricular nucleus and the posterolateral hypothalamus, the hypothalamus can directly influence preganglionic sympathetic and parasympathetic nuclei in the brain stem and spinal cord. Other fibers reach somatomotor areas in the brain stem reticular formation.

The medial forebrain bundle is one of the most famous hypothalamic fiber bundles, but also one of the most incomprehensible. Some of its components are long axons that arise from olfactory-related areas in the forebrain or from monoaminergic cell groups in the brain stem. Other

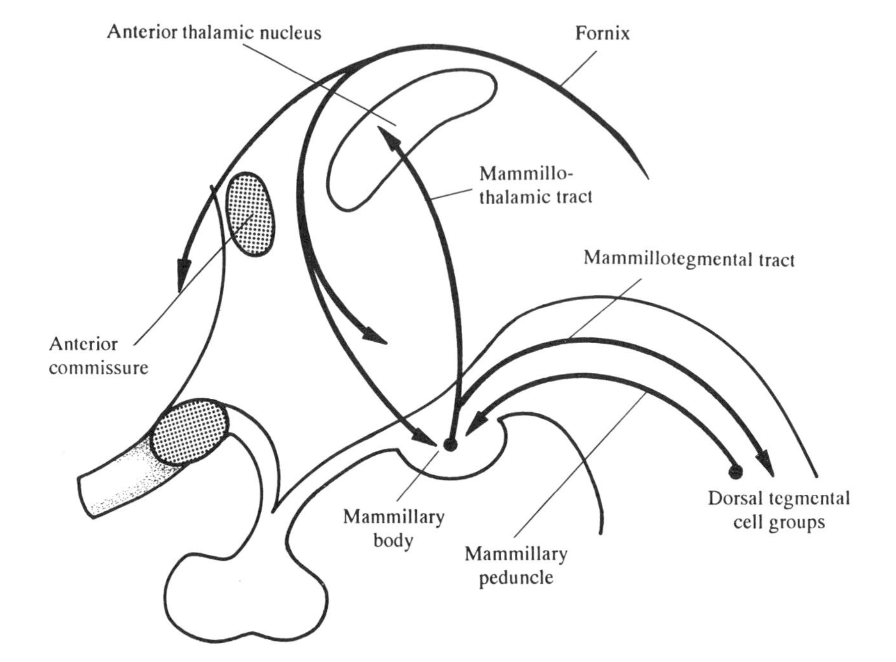

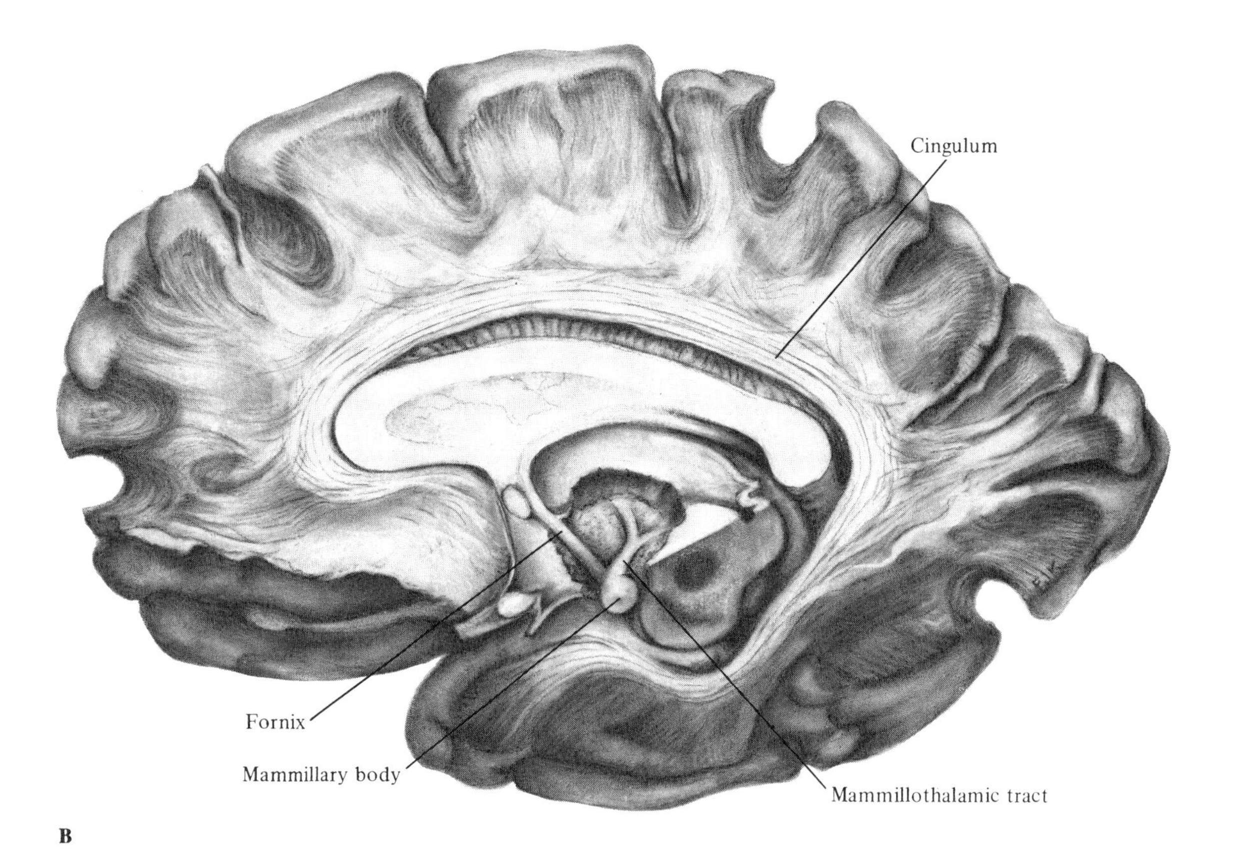

Fig. 18–4. Connections of the Mammillary Body

A. Highly schematic drawing of the main fiber connections of the mammillary body.

B. Blunt dissection of the human brain from the medial side. The gray substance of the hypothalamus has been removed to expose the fornix and the mammillothalamic tract. The medial parts of the hemisphere have also been removed to show the cingulum.

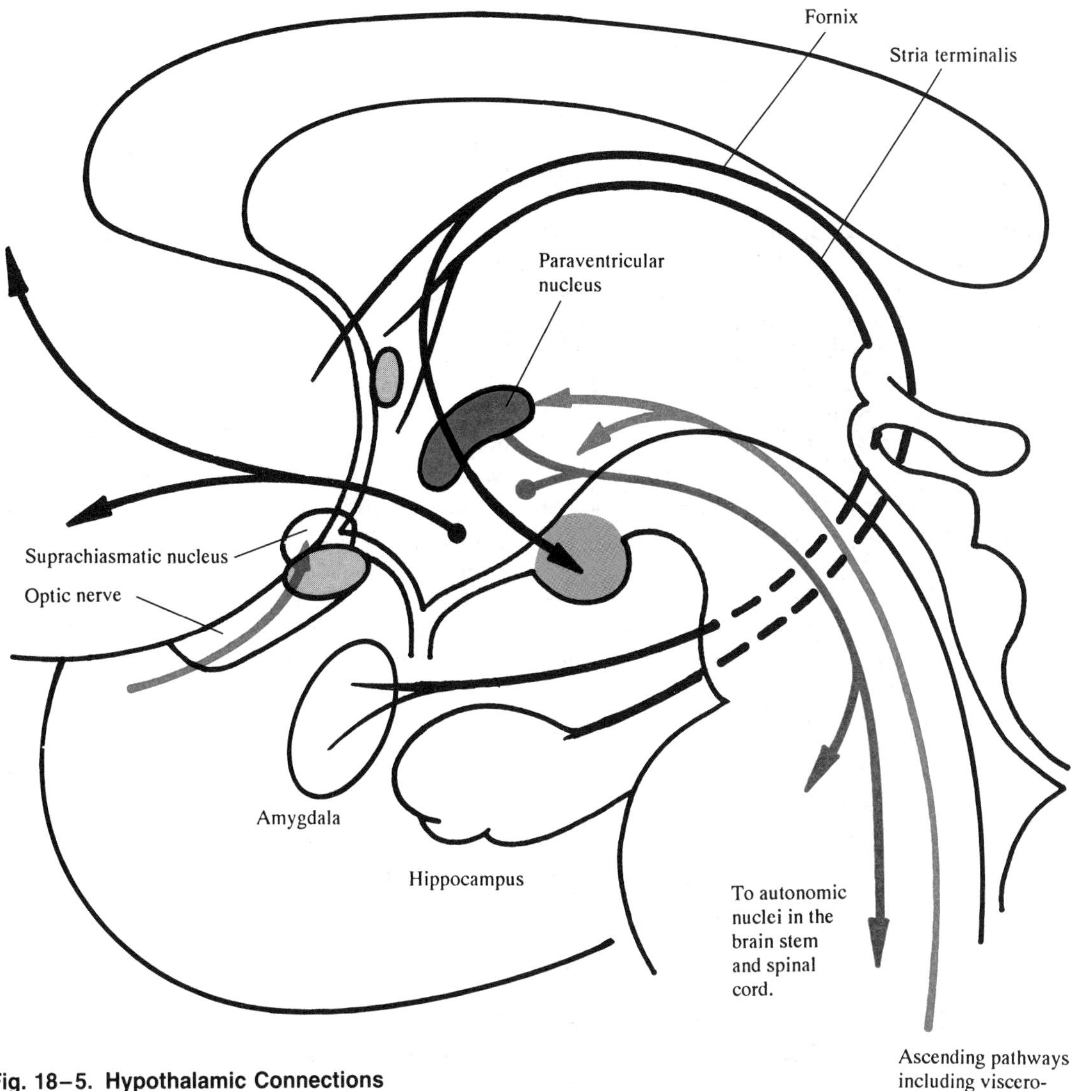

Fig. 18–5. Hypothalamic Connections
Some of the most prominent pathways have been indicated in this schematic drawing.

components are diffusely organized ascending and desending links similar to those found in the brain stem reticular formation. Some of the components are collaterals of other well-known pathways, such as the stria medullaris or the mammillary peduncle. Still others connect the amygdaloid body with the lateral hypothalamus and a variety of brain stem areas (see the "Central Autonomic Network" in Chapter 19). Although it may be convenient to continue to use the term medial forebrain bundle for this multitude of different and often loosely arranged fiber systems, one should remember that the fibers included do not form a well-defined bundle; the term is used to designate a heterogeneous group of axons that run through the lateral hypothalamus.

Hypothalamohypophysial Pathways

The hypothalamus is closely related to the *pituitary*, also known as the *hypophysis*. The *magnocellular neuroendocrine system* is represented by axons from large oxytocin- and vasopressin-containing neurons in the supraoptic and paraventricular nuclei that terminate in the posterior lobe of the pituitary. The *parvocellular neuroendocrine system* is formed by neurosecretory fibers from smaller neurons in the arcuate nucleus and other periventricular parts of the hypothalamus, including the paraventricular nucleus. These neurosecretory fibers project to the *median eminence*, where they terminate on the basal laminae that line the perivascular spaces surrounding the capillary loops of the *hypophysial portal system*. The capillaries in the median eminence, as in the other CVOs, are fenestrated, which allows for the passage of macromolecules between the brain and the blood. The parvocellular neuroendocrine system together with the portal vessels forms a neurovascular transport system for the control of the various anterior lobe hormones (see below).

The fiber tracts mentioned above reflect the diversity of the hypothalamic pathways, and some of the important connections between the hypothalamus and the cerebral cortex, amygdaloid body, brain stem, and spinal cord are summarized in Fig. 18–5 and discussed further in the next chapter.

In summarizing the hypothalamic connections, it should be emphasized that it is only during the last 20 years that the extraordinary complexity of the hypothalamic neuronal circuits and their relation to the rest of the brain have become apparent. Thanks to modern tracer techniques and immunohistochemical methods (Chapter 7), it has become possible to map in detail some of the intrinsic and extrinsic hypothalamic pathways related to specific functions and behaviors. The connections of the paraventricular nucleus and their relations to water balance and food intake are good examples (see below).

When considering the importance of the hypothalamus for brain functions in general, special attention should be paid to the large number of cells located in the lateral hypothalamus including the tuberomammillary nucleus. The cells in the tuberomammillary nucleus contain histamine, often colocalized with other transmitters or neuropeptides, whereas many of the other lateral hypothalamic neurons contain an MCH (melanin-concentrating hormone)-like peptide. Together, these neuronal systems reach practically every major region in the CNS (Fig. 10–10); often a single neuron projects to several places by the aid of long axon collaterals, and it is widely believed that these systems are important for cortical arousal and for sensorimotor integration in general. Incidentally, the lateral hypothalamic neurons that project to the cerebral cortex form as massive a corticopetal system as that represented by the better-known basal nucleus of Meynert (Chapter 16).

Chemical Neuroanatomy and Functional Aspects

Visionary ideas and technical breakthroughs often conspire to produce remarkable scientific discoveries, and the history of chemical neuroanatomy is a particularly striking example of this synergy. It started with the notion of neurosecretion, which was envisioned in 1928 by the German zoologist Ernst Scharrer, who together with his wife Berta Scharrer developed this concept further in a series of papers in which they showed that the magnocellular neurons in the supraoptic and paraventricular nuclei have glandular functions characterized by secretory activities. In the late 1940s, another German scientist, Wolfgang Bargmann, and his collaborators succeeded in staining selectively the neurosecretory material. It became clear that the neurosecretory material is manufactured by the large cells and transported down their axons to the axon terminals in the posterior lobe of the pituitary, where the material is released into the bloodstream to act on distant targets. Soon afterwards, the hormones of the magnocellular neurosecretory system, i.e., the neuropeptides oxytocin and vasopression, were characterized and synthesized.

Various aspects of the relationships between the hypothalamus and the adenohypophysis were also being investigated by a number of scientists during the middle part of this century. This work culminated with the discoveries by the English anatomist–physiologist Geoffrey Harris and his collaborators, who provided strong support for the now widely accepted notion of a neurovascular pathway linking the hypothalamus to the anterior lobe of the pituitary. Hypothalamic hormones are produced by parvocellular neurons located primarily in the periventricular zone of the hypothalamus. The hormones are carried via axoplasmic transport to the base of the brain, released into the blood of the portal veins, and carried across the pituitary stalk to the adenohypophysis, where the hypothalamic hormones regulate the release of the anterior lobe hormones.

It has been shown that the hypothalamic hormones oxytocin and vasopressin, as well as several of the releasing factors and even adenohypophysial hormones, e.g., adrenocorticotropic hormone (ACTH), are widely distributed in the brain including areas not primarily involved in endocrine functions. Well-known examples are provided by descending oxytocin or vasopressin fibers from the paraventricular nucleus to the brain stem and spinal cord autonomic centers (see below). At the same time as the role of the "hypophysiotrophic" peptides has expanded, new substances are continuously being added to the list of substances that may serve as transmitters or modulators, often in the hypothalamus and closely related areas, but also in other parts of the nervous system. With the help of the extremely powerful immunohistochemical techniques, the number of neuroactive substances or neuromediators may, according to the experts, reach into the hundreds, and the study of their distribution in the nervous system, "chemical neuroanatomy," is one of the most active disciplines in contemporary neuroscience.

Many of the neuropeptides are especially prominent in the hypothalamus and in other periventricular forebrain and brain stem structures closely related to the hypothalamus. These regions have recently been referred to as the "paracrine core" of the nervous system, and the suggestion has been made that within this core, which is devoted to a number of vital functions, interneuronal communication may be accomplished to a large extent by a "parasynaptic" or "paracrine" mode of interaction, contingent on the diffusion of the neurotransmitters through the extracellular space (see "The Synapse", Chapter 6).

Interactions between the brain and the immune system have attracted much interest in recent years. Classical immune mediators, like the cytokines—tumor necrosis factor and, for example, interleukins 1 and 6—are also released from brain cells and control the release of hypothalamic and pituitary hormones as well as neuronal function. In turn, "classical pituitary hormones" have also been shown to regulate the release of several cytokines. Hence, we are moving into an era of blurred traditional boundaries in which neuroendocrine-immunology represents a more global concept.

It is a generally held belief that neuronal plasticity, including modifications in the synapse, i.e., synaptic plasticity, is responsible for lasting changes in behavior and for long-term memory. A remarkable form of intercellular plasticity in response to physiologic stimuli has recently been discovered. Using the magnocellular neurosecretory system as a model, the American anatomist Glenn Hatton and his colleagues have demonstrated a rapid (within minutes) withdrawal of glial processes between the magnocellular neurons followed by the formation of new synapses in response to stimuli that call for increased hormone production. Similar results have been obtained in a few other anatomic systems in experimental animals, and the prospect that similar types of "rapid synaptogenesis" may exist in the adult human brain in response to rapidly changing demands is indeed fascinating.

The Hypothalamus and the Neurohypophysis: The Magnocellular Neurosecretory System

The neurohypophysis, which consists of the *infundibulum* and the *neural lobe*, is an outgrowth of the diencephalon, and it receives axonal projections from large neurosecretory neurons, most of which are located in the supraoptic and paraventricular nuclei of the hypothalamus. These axons form the supraoptico- and paraventriculohypophysial pathways (Fig. 18–6A), and they carry neurosecretory material containing vasopressin or antidiuretic hormone (ADH) and oxytocin to the neural lobe, where the axon terminals release the hormones into the perivascular spaces surrounding the fenestrated capillaries.

Both the supraoptic and the paraventricular nuclei contain vasopressin- as well as oxytocin-producing cells, which receive input from various intra- and extrahypothalamic sources. Whereas the output from the supraoptic nucleus is primarily restricted to the neurohypophysis, neurosecretory cells of the paraventricular nucleus project not only to the posterior lobe of the hypophysis, but also to the median eminence and to several extrahypothalamic regions including autonomic centers in the brain stem and spinal cord.

As the name indicates, ADH (sometimes referred to as AVP = arginine-vasopressin) controls water balance. In particular, it is responsible for the retention of water by the distal tubules of the kidneys. The release of vasopressin is controlled by osmotic and volemic changes in the circulating blood. The osmoreceptors that respond to changes in osmotic pressure are believed to be located in the anterior periventricular region of the hypothalamus rather than in the magnocellular nuclei. This region includes the organum vasculosum of the lamina terminalis (OVLT), which projects to the two magnocellular nuclei. Another circumventricular organ, the subfornical organ (SFO), is also involved in the regulation of vasopressin release. The stimulus in this case is angiotensin[1], a blood-borne substance that has a direct effect on the neurons in the SFO. These neurons, like those in the OVLT, are characerized by direct input to the magnocellular neurons in the supraoptic and paraventricular nuclei.

Angiotensin also regulates body fluid through its effect on drinking behavior (see below), and as a powerful vasoconstrictor, it is involved in blood pressure regulation (Chapter 19).

---

1. Angiotensin II is a product of the renin–angiotensin system. Renin is a proteolytic enzyme secreted by the kidney. Within the blood, renin cleaves plasma angiotensinogen into angiotensin I, which is then converted into the highly active angiotensin II. The secretion of renin, which is regulated by local mechanisms in the kidney, can be increased by a number of factors that stimulate the sympathetic nervous system, including exercise, stress, and hemorrhage.

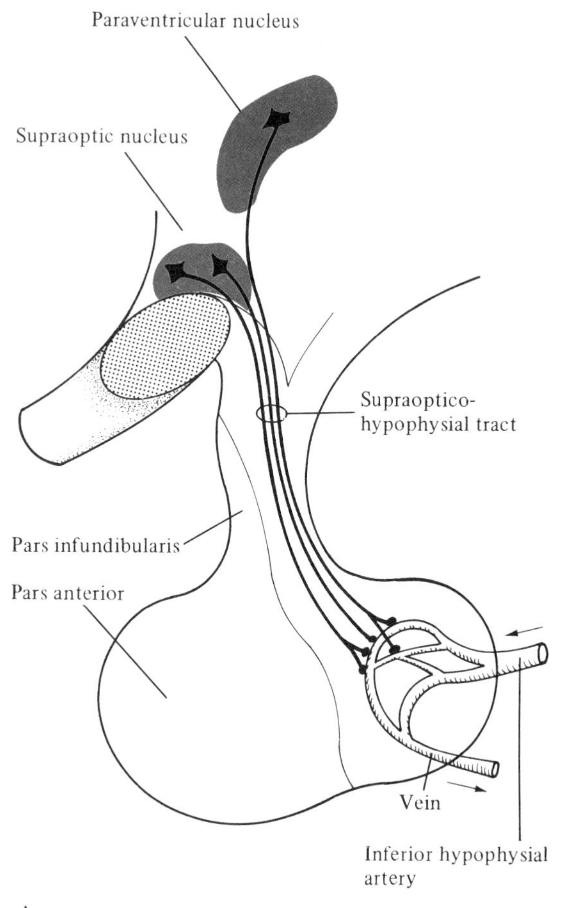

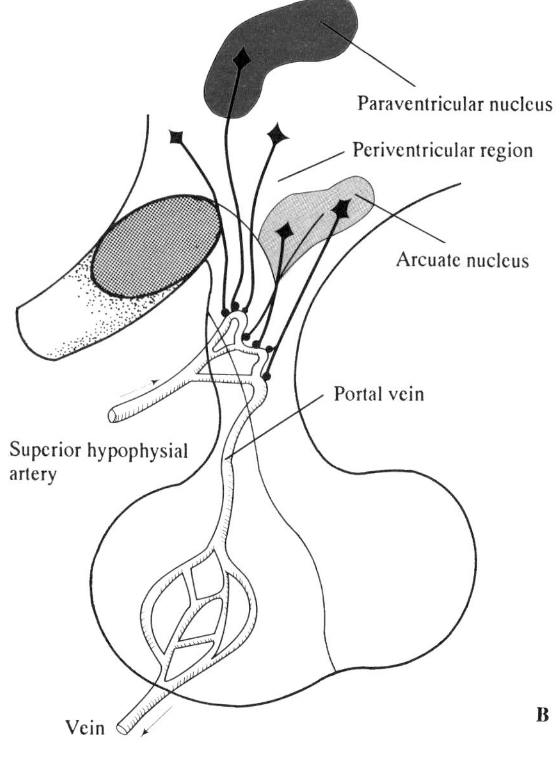

Fig. 18–6. Hypothalamohypophysial Pathways
A. Magnocellular neuroendocrine system.
B. Parvocellular neuroendocrine system.

Another important input to the magnocellular vasopressin neurons is represented by ascending noradrenergic fibers (part of the ascending lateral tegmental noradrenergic projection system, Fig. 10–9) from the ventrolateral medulla oblongata. Changes in blood pressure and blood volume, such as those that accompany hemorrhage, provide powerful stimuli for vasopressin release through this system[2]. The ventrolateral medulla oblongata, a critical part of the cardiovascular regulatory system (Chapter 19), is responsible for integrating various corrective measures in response to changes in blood volume and blood pressure.

Oxytocin, through its effect on the uterine smooth muscle and the myoepithelial cells of the mammary glands, promotes uterine contraction during birth and milk ejection after birth. Potent stimulatory input for uterine contraction reaches the brain via afferents from the vagina or cervix, whereas stimulation of the nipple by suckling promotes milk let-down.

2. Only in extreme conditions, e.g., dehydration or hemorrhage, does the vasopressin content in the blood reach a level at which it can act as a powerful vasoconstrictor in certain vascular beds.

The Hypothalamus and the Adenohypophysis: The Parvocellular Neurosecretory System

The neurosecretory cells of the parvocellular system are scattered throughout the periventricular zone and the preoptic area, and there is a distinct localization for each of the hypothalamic releasing hormones or factors that influence the adenohypophysis. The hormones are carried by axoplasmic transport to the median eminence where they are discharged into the perivascular spaces surrounding the portal capillaries, formed by the superior hypophysial arteries. The median eminence belongs to the circumventricular organs, which are characterized by a porous blood–brain barrier since they contain fenestrated capillaries. The portal capillaries join to form the portal veins through which the hormones are carried to the vascular sinusoids in the adenohypophysis. Here they influence the secretion of the pituitary hormones: TSH (thyroid-stimulating hormone or thyrotropin), ACTH (adrenocorticotropin), FSH (follicle-stimulating hormone), LH (luteinizing hormone), GH (growth hormone or somatotropin), and PRL (prolactin).

After years of intense research, the location of the various cell groups that produce the hypothalamic releasing hormones for transport to the median eminence have been fairly well established. For instance, the cells producing luteinizing hormone-releasing hormone (LHRH, also known as gonadotropin-releasing hormone, GnRH) are rather widely distributed in the periventricular zone including the medial preoptic area and the arcuate nucleus. Parvocellular neurons producing growth hormone-releasing hormone (GRH) are limited primarily to the arcuate nucleus. Somatostatin (SST) inhibits the release of GH, and the SST neurons responsible for this effect, i.e., those that project to the median eminence, are located primarily in the anterior periventricular hypothalamus. SST also inhibits the release of TSH and PRL. Thyrotropin-releasing hormone (TRH) neurons, as well as corticotropin-releasing hormone (CRH) neurons regulating ACTH secretion, are found primarily in the paraventricular nucleus, where the cells are distinct from the magnocellular neurosecretory neurons. As indicated before, several of the hypophysiotropic factors, i.e., those that serve as releasing and inhibiting hormones, have, like most other neuroactive substances, a wide distribution within the brain and are therefore likely to serve as modulators in many other pathways besides the hypothalamohypophysial system.

The various mechanisms controlling hormone secretion are complicated. Suffice it to say that the magno- and parvocellular cell groups producing the hypothalamic hormones receive a variety of stimuli from different parts of the brain, primarily within the hypothalamus, but also from extrahypothalamic regions including the extended amygdala, the hippocampus, and various brain stem areas. Furthermore, it is well known that monoamines and several neuropeptides serve as modulators of the neuroendocrine system, and both monoaminergic and peptidergic fibers, besides those carrying the specific hypothalamic hormones, can be traced to the periventricular zone, and even into the median eminence, where they have an opportunity to interact with the parvocellular neurosecretory system or even discharge neuroactive substances directly into the portal system. A well-known example of a nonpeptidergic inhibiting factor is dopamine, which is synthesized in neurons of the arcuate nucleus and released into the portal system through the *tuberoinfundibular dopamine pathway*. Dopamine has an inhibitory effect on prolactin secretion.

The subject of neuroendocrine control mechanisms is complicated further by the fact that many neurons in the nervous system, including the hypothalamic magnocellular and parvocellular neurosecretory neurons, contain two or even several neuroactive substances. A well-known example is provided by the parvocellular CRH neurons in the paraventricular nucleus. They also contain vasopressin and the two substances are released together into the portal vessels, through which they are likely to cooperate in the control of ACTH release from the adenohypophysis.

Hypothalamic neurons, including the neurosecretory neurons, are also subject to hormonal feedback control. Such feedback mechanisms are often quite complicated in the sense that they involve not only the neurosecretory hypothalamic neurons but also hormone-sensitive cells in other brain regions, which in turn are in a position to modulate hypothalamohypophysial functions.

Integration of Autonomic, Endocrine, and Somatomotor Aspects of Vital Functions

Homeostasis and the Notion of Set Point

Although the eminent English neurophysiologist Sir Charles Sherrington called the hypothalamus "the head ganglion of the autonomic nervous system," many of the orderly routine functions of the visceral organs are organized and controlled by reflex arcs on the spinal cord or brain stem level or even within the walls of the peripheral organs, as in the enteric nervous system (Chapter 19). Generally, hypothalamic functions are related not so much to isolated autonomic reactions, but rather to integrated patterns of autonomic, endocrine, and somatomotor responses necessary for homeostasis and for the survival of the individual (e.g., regulation of water balance, food intake, and temperature; defense reactions) and the species (e.g., reproduction).

In general, it appears that the hypothalamic regulation of homeostatic functions is based on the principle of set points within rather narrow limits, similar to the regulation of room temperature by a thermostat. For instance, if the body temperature increases, thermosensitive neurons in the hypothalamus initiate a series of responses designed to lower the body temperature (see below).

Although the hypothalamus is crucial for these functions, it is important to remember that the organization of the many response patterns in which the hypothalamus is involved is often dependent on a network of central neuronal circuits that includes not only the hypothalamus but also several other areas in both the brain stem and the forebrain (Chapter 19).

Eating

Experimental studies in the middle of this century showed that bilateral lesions in the region of the ventromedial hypothalamus resulted in overeating (*hyperphagia*) and obesity, whereas lesions in the lateral hypothalamus produced the opposite effect, i.e., *aphagia*. This led to the dual-center hypothesis of a "satiety center" located in the ventromedial hypothalamus and a "feeding center" in the lateral hypothalamus. Attractive as this notion may be, recent studies have shown that the central nervous control of food intake and metabolism is considerably more complicated. The hypothalamus, no doubt, plays a key role in food ingestion, but the dual-center hypothesis has been replaced by a more differentiated view. Destruction of several other hypothalamic areas, including in particular the paraventricular nucleus, also has a dramatic effect on eating, and a number of transmitters, including both monoamines and neuropeptides, have been shown to be important for the regulation of food intake. Noradrenaline, adrenaline, and neuropeptide Y (NPY), apparently acting on the paraventricular nucleus, have a stimulatory effect on eating, particularly with regard to the ingestion of carbohydrates. Opioid peptides, acting on different hypothalamic areas, also stimulate ingestion of food, especially fat and proteins. Some of the monoamines, i.e., dopamine and serotonin, and the gut peptide cholecystokinin (CCK), have been shown to have an inhibitory effect on eating behavior.

How the hypothalamus regulates feeding behavior in response to various sensory and metabolic activities, and how it interacts with other parts of the CNS and with the autonomic and endocrine systems, is the focus of much research. Therefore, it is likely that the network of neuronal circuits that form the anatomic substrate of feeding may soon be revealed in considerable detail. As indicated above, the paraventricular nucleus is uniquely equipped to control endocrine, autonomic, and somatomotor systems, and it is apparent that this nucleus plays a key role in many of the integrated responses related to food intake, as well as in many other activities necessary to maintain homeostasis. It is populated by both magno- and parvocellular neurosecretory neurons that project to the neurohypophysis and to the portal system in the median eminence. The regulation of the hypothalamo-pituitary-adrenal axis, which is controlled by CRH-secreting neurons in the paraventricular nucleus, may be of special importance in this context. Corticosterone is known to affect carbohydrate metabolism, and stress, which releases ACTH, is noted for its influence on eating behavior. Parvocellular neurons in the paraventricular nucleus also project to the brain stem reticular formation and to the preganglionic autonomic nuclei in the brain stem and spinal cord (Fig. 18–6). Through these pathways, the paraventricular nucleus can directly influence the secretion of pancreatic hormones (insulin and glucagon) and of catecholamines from the adrenal medulla. It may also control somatomotor activities that may be relevant in feeding behavior.

The paraventricular nucleus, furthermore, receives signals from many different sources that are likely to be of importance in feeding. Major inputs reach the nucleus from other hypothalamic regions (e.g., the medial preoptic area, the suprachiasmatic nucleus, and the subfornical organ) and from a variety of extrahypothalamic sources, including the extended amygdala (especially the bed nucleus of the stria terminalis) and several brain stem regions, including viscerosensory information from the nucleus of the solitary tract as well as impulses from noradrenergic cell groups in the medulla oblongata.

Drinking

Increased water intake is an important means of replenishing body fluids. Although we often drink spontaneously, as a matter of routine, drinking can also be activated by water deficit, i.e., deprivation-induced drinking. Deprivation-induced drinking is regulated primarily by osmotic change in the blood or change in blood volume, e.g., hemorrhage. The osmoreceptors for drinking, like the ones regulating vasopressin release, are located in the OVLT and neighboring medial preoptic area. Drinking in response to reduced blood volume is initiated by two different mechanisms. One type of input originates in mechanoreceptors in the pulmonary artery and the vena cava, and reaches hypothalamic integration centers for drinking via the nucleus of the solitary tract. Another important stimulus is blood-borne, i.e., angiotensin II, which activates the neuronal circuit for drinking behavior through its action on the subfornical organ. Animal experiments have shown that the subfornical organ is also an intermediary structure in another important mechanism for fluid homeostasis. It is the place through which blood-borne angiotensin II, released in response to reduced blood volume, can activate the vasopressin system to reduce water loss through the kidneys.

Temperature Regulation

In order to sense changes in body core temperature and in the ambient temperature, thermoreceptors are located largely in the brain and the skin, respectively. In the brain, thermosensitive neurons are found primarily in the preoptic–anterior hypothalamic area. These thermosensitive neurons sense the temperature of the blood that passes through this richly vascularized region. Based on this information and that from free nerve endings in the skin and other peripheral receptors, neuronal assemblies in both the anterior and posterior hypothalamus coordinate the activity needed to maintain the body temperature within fairly narrow limits. In response to increased body temperature, hypothalamic neurons initiate a series of processes that result in heat loss, including peripheral vasodilation and sweating. A drop in the surrounding temperature leads to a series of events including peripheral vasoconstriction, piloerection, increased metabolism, and shivering in order to preserve heat.

It should be emphasized that most species, including humans, thermoregulate through behavior, i.e., they seek a warmer or cooler environment depending on the circumstances. The physiologic mechanisms described above are in general auxiliary to the behavioral thermoregulatory mechanisms, and cannot keep the appropriate body temperature in the face of environmental extremes (see Clinical Notes).

Reproduction

The hypothalamus plays a major integrative role in the control of reproductive physiology and behavior, including sexual development and differentiation, sexual behavior, and maternal behavior. Important stimuli for the various reproductive functions come from a variety of exteroceptive and interoceptive sources including circulating gonadal steroids.

Differences between male and female are not limited to sexual organs and secondary sex characteristics; they are also evident within the CNS. This important fact was discovered in 1971 by two contemporary English anatomists, Geoffrey Raisman and Pauline Field, who demonstrated a significant difference in synaptic organization within the preoptic area between male and female rats. Later, other sexually dimorphic areas have been demonstrated also in the human brain and spinal cord. A special cell group, referred to as the sexually dimorphic nucleus of the preoptic area, is considerably larger in males, and the same is true for a cell group in the sacral spinal cord known as the nucleus bulbocavernosus, which controls the reflexes of the penis and the clitoris.

The sex hormones, i.e., androgens and estrogens, play important roles both in the development and differentiation of the male and female sex organs and in sexual behavior. LHRH-producing neurons, responding to sensory input and to circulating gonadal steroids, control the secretion of LH and FSH from the anterior pituitary. LHRH-secreting neurons with projections to the portal system in the median eminence are located especially in the preoptic–anterior hypothalamus periventricular zone, but also in the periventricular zone of the tuberal region of the hypothalamus in primates. Cells in other hypothalamic areas, including the preoptic area, the anterior hypothalamus, and the ventromedial and infundibular nuclei, are important for reproductive functions, as they modulate the secretion of LHRH, which is the final link between the CNS and the peripheral parts of the endocrine reproductive system.

Sexual behavior involves a number of general (e.g., respiratory, cardiovascular) and specific (e.g., penile erection and ejaculation, vaginal lubrication) responses mediated in large part by the autonomic nervous system (Chapter 19). Although several of these specific responses represent involuntary or reflex phenomena, descending pathways from the hypothalamus and other basal forebrain regions play a significant modulatory role. Generally, it is not useful or even possible to identify specific hypothalamic "centers" for various activities, including reproductive functions. Rather, functions in the brain are related to neuronal networks involving several brain regions, some of which may be more important than others. The region in and around the ventromedial hypothalamic nucleus, for instance, seems to be especially important for female sexual behavior. Several hypothalamic regions, including the medial preoptic–anterior hypothalamus, are undoubtedly critical for the integration of psychogenic sexual responses. The American neuroscientist Paul MacLean, well-known for his stimulating papers on the limbic system (Chapter 21) and provocative observations on the functional–anatomic organization of the brain in general, discovered that topical stimulation in medial parts of the diencephalon, especially the medial preoptic–anterior hypothalamus and regions along the medial forebrain bundle in monkeys, resulted in penile erection. Although the medial preoptic area seems to be a nodal point for facilitating effects on sexual behavior, especially in males, other forebrain structures, including the extended amygdala and prefrontal cortical regions, may also be important for the integration of the general arousal and executive mechanisms associated with sexual behavior, but the details of their involvement are less clear.

Clinical Notes

Hypothalamohypophysial Disorders: Diversity of Symptoms

The hypothalamus and the pituitary influence a large number of both endocrine and nonendocrine functions, and it is difficult to give a systematic account of the number and variety of possible symptoms in hypothalamohypophysial disorders. Damage to different parts of the hypothalamohypophysial system may result in various neuroendocrine disturbances. Autonomic dysfunctions in the respiratory, cardiovascular, and gastrointestinal systems are commonly seen, as are disturbances in temperature regulation, water balance, sexual behavior, and food intake. Hypothalamic lesions can also change the level of consciousness, the sleep–wake cycle, and emotional behavior.

Many pathologic processes can damage the hypothalamus but tumors are the most common. Pituitary tumors can become clinically apparent through problems caused by their enlargement, through problems due to oversecretion of hormones specific for the type of tumor cell, or through problems due to inadequate secretion of hormone from adjacent cells damaged by the tumor bulk. Primary pituitary tumors and the early signs of a pituitary tumor may be related to the involvement of the hypothalamus. To cause hypothalamic symptoms, the lesion must ordinarily involve both sides of the hypothalamus.

Since the optic pathways have a strategic position at the base of the hypothalamus, they are often affected by hypothalamohypophysial tumors. A pituitary tumor, for instance, that expands in a dorsal direction is likely to encroach on the optic chiasm or one of the optic tracts, causing visual field defects. The classic symptom is a bitemporal visual field defect (bitemporal hemianopsia) caused by compression of the medial fibers of the optic tracts just posterior to the chiasm (Fig. 13–16). These fibers originate in the nasal retina of each eye; thus, their destruction causes blindness in each temporal visual field. Equally possible is lateral tumor growth into the cavernous sinus with dysfunction of one or several of the cranial nerves, i.e., oculomotor, trochlear, ophthalmic division of the trigeminal and abducens, coursing through this region.

Hypothalamic Syndromes

Many of the symptoms related to hypothalamic diseases can also be seen in lesions of other brain regions, especially those regions closely related to the hypothalamus, e.g., the prefrontal cortex, the amygdaloid body and extended amygdala, and various structures in the brain stem. The number of truly hypothalamic syndromes, therefore, is rather limited. The best known are diabetes insipidus and Froehlich's syndrome. Disturbances in ingestion and temperature regulation are also typical of hypothalamic disorders.

Diabetes Insipidus

Lesions at the base of the brain, e.g., tumors (sarcoidosis, craniopharyngiomas, meningiomas) and skull fractures involving the supraoptic and paraventricular nuclei or the hypothalamohypophysial tract, are likely to interfere with the formation, transport, or release of ADH. This may result in a condition known as diabetes insipidus, which is characterized by an increased production of urine (polyuria), up to 10 liters/day. The loss of fluid, in turn, results in an excessive thirst (polydipsia). In many cases, no explanation can be found for this disease.

Dystrophia Adiposogenitalis (Froehlich's Syndrome)

Although obesity is a common occurrence in many cultures, it is only occasionally the result of a pathologic lesion, e.g., a tumor, that involves the ventromedial hypothalamus and the pituitary. The adiposity is then often associated with underdevelopment of the genitalia, in which case it is referred to as Froehlich's syndrome.

Anorexia Nervosa

This is a serious disorder in which the patient refuses to eat. Those affected are almost always young white women, 10% of whom may actually die of starvation. Endocrine disturbances are very common, but the cause of the disease is unknown; no structural lesion has been discovered in the hypothalamus. Psychological factors undoubtedly play a role, and the degree to which vital hypothalamic mechanisms are involved may in the end determine the outcome of this truly psychosomatic disease.

Disturbances in Temperature Regulation

By balancing heat production in the body with heat dissipation from the surface of the body, the body temperature is normally maintained at about 37°C (98°F). Inflammatory conditions are characterized by fever, which in effect means that the thermoregulatory set point has been raised, but only rarely above 41.5°C (106°F). Fever is caused by bloodborne substances called pyrogens, which are released into the bloodstream by leucocytes. The pyrogens act on the thermosensitive hypothalamic neurons, with the effect that they interpret the body temperature as being too low. This, in turn, evokes a coordinated response aimed at increasing the body temperature through peripheral vasoconstriction and shivering. The constriction of the skin vessels results in a decreased skin temperature, which produces a chilly sensation.

High environmental temperature, especially in combination with high relative humidity and strenuous exercise, can lead to heat exhaustion or even heat stroke, especially in children and old people. A decrease or disappearance of sweating (hot and dry skin) indicates a breakdown of the hypothalamic thermoregulatory mechanism. A rectal temperature above 41.5°C (106°F) is a poor sign. Hypothermia can be a serious problem in the elderly, especially if they are exposed to cold temperatures, drugs, or alcohol.

Hypothalamic lesions, e.g., tumors, vascular accidents, or congenital defects, can cause disturbances in temperature regulation, usually hypothermia but sometimes in the form of hyperthermia.

Psychosomatic Disorders

People who live in "modern" highly competitive and hectic societies are apparently afflicted by psychosomatic disturbances to a greater extent than people who live in stable social environments characterized by traditions and age-dependent rather than rapid change and competition. Environmental stress, e.g., broken families, unstable work situations (including fight for laboratory space at medical schools), and financial problems, results in emotional arousal and neuroendocrine changes that eventually may lead to pathologic changes in the organism. Such psychosomatic disturbances are often an important factor in the development of commonly known diseases including many forms of cardiovascular disorders, asthma, peptic ulcer, and eating disorders. Although the psychosomatic diseases are not referred to as hypothalamic disorders, it is by way of the hypothalamus that the psychosomatic disturbances are usually executed.

Clinical Example

Pituitary Adenoma (Fig. 18–7)

A 45-year-old male was seeking medical attention for a series of disturbing but seemingly unrelated symptoms. He had been suffering from increasingly severe headaches for a couple of years, and about a year ago he developed hypothyroidism and was treated with thyroid extract. Three months ago he realized that he was impotent, and he started to complain of visual deficits at about the same time.

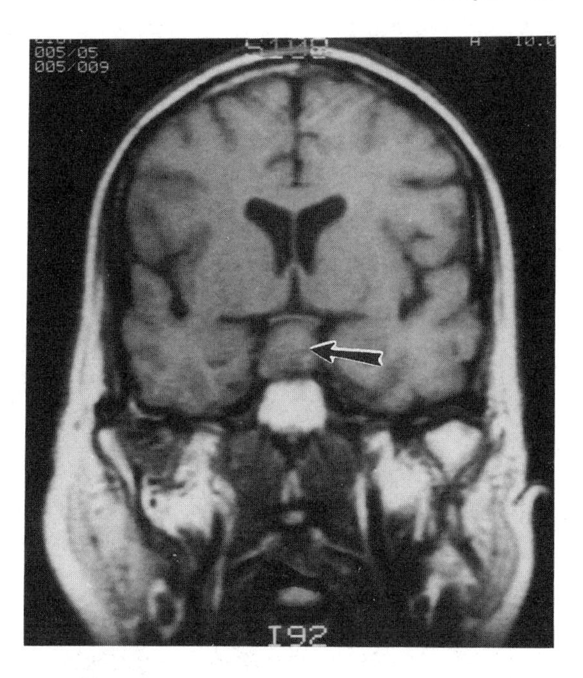

Fig. 18–7. Pituitary Adenoma
Pituitary adenoma (*arrow*) with compression of optic chiasm. T1-weighted coronal MRI. (Photograph courtesy of Dr. Maurice Lipper.)

His appearance was that of a chronically ill man, with a sallow complexion typical of panhypopituitarism. He had decreased pubic hair and some testicular atrophy.

Neurologic examination revealed a complete bitemporal hemianopsia, but he had no other cranial nerve deficits. However, he seemed to have a general mild weakness in his musculature.

Laboratory studies showed normal thyroid hormone levels, but he had abnormally low serum testosterone. Intravenous administration of thyrotropin-releasing hormone produced only a small elevation of thyroid-stimulating hormone in the serum.

An MRI (Fig. 18–7) was obtained.

1. Where is the lesion?
2. Does the lesion account for his visual deficits?
3. How did his endocrine abnormalities arise?
4. What is the solid white area in the middle lower half of the picture?

Discussion

The MRI demonstrates a mass lesion growing out of the sella turcica (arrow) and extending towards the hypothalamus. Considering the clinical findings, the most likely diagnosis is a pituitary adenoma (prolactinoma). The tumor, apparently, compresses the optic chiasm in the midline, thereby damaging the crossing fibers, i.e., those coming from the nasal part of the retina of each eye (Fig. 13–5). The solid white area represents fatty marrow in the sphenoid bone.

The endocrine deficiencies result from destruction of the anterior lobe of the pituitary. Pituitary adenomas are slowly growing tumors, and they may reach a large size before they interfere with the well-being of the patient. Indeed, visual failure, rather than endocrine symptoms, often brings the patient to the doctor.

The patient was operated upon through a transsphenoidal route and the tumor was removed. Complete resolution of the visual symptoms was achieved.

Suggested Reading

1. Ader, R., D. Felten, and N. Cohen, 1990. Interactions between the brain and the immune system. Annu. Rev. Pharmacol. Toxicol. 30:561–602.
2. Ganten, D. and D. Pfaff, 1986. Morphology of hypothalamus and its connections. Current Topics in Neuroendocrinology, Vol. 7. Springer Verlag: Berlin.
3. Martin, J. B., S. Reichlin, and G. M. Brown, 1987. Clinical Neuroendocrinology, 2nd ed. F. A. Davis: Philadelphia.
4. Nieuwenhuys, R., 1985. Chemoarchitecture of the Brain. Springer-Verlag: Berlin.
5. North, G., A. M. Moses and L. Share (eds.), 1993. The neurohypophysis: a window on brain function. Annals of New York Acad. of Sciences, Vol. 689.
6. Saper, C. B., 1990. Hypothalamus. In G. Paxinos (ed.): The Human Nervous System. Academic Press: San Diego, pp. 389–414.
7. Saper, C. B., 1990. Hypothalamus. In A. L. Pearlman and R. C. Collius (eds.): Neurobiology of Disease. Oxford University Press: New York, pp. 1095–1113.
8. Swaab, D. F., M. A. Hofman, M. Mirmiran, R. Ravid and F. W. van Leeuwen, 1992. The human hypothalamus in health and disease. Prog. Brain Res. Vol. 93. Elsevier: Amsterdam.
9. Swanson, L. W., 1987. The Hypothalamus, In A. Björklund et al. (eds.): Handbook of Chemical Neuroanatomy, Vol. 5. Elsevier: Amsterdam.

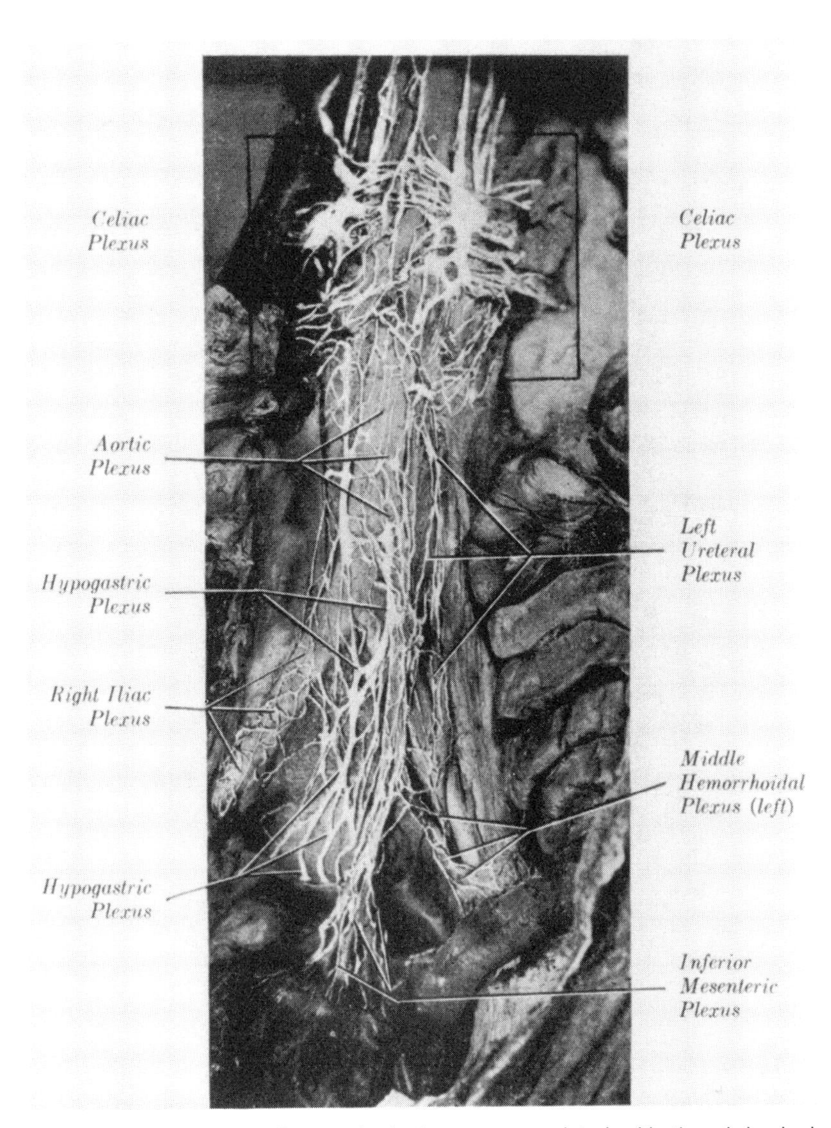

Prevertebral plexuses associated with the abdominal aorta. (Photograph of human dissection by Dr. J. D. Humber. By permission of A. Kuntz, The Autonomic Nervous System, 4th ed. Lea & Febiger, 1953, p. 35.)

19

The Autonomic Nervous System

Through its innervation of the visceral organs, the glands, and the blood vessels, the autonomic nervous system regulates the internal environment; it is largely responsible for maintaining normal bodily functions. The functions of the autonomic nervous system are to a considerable extent based on local activities that involve neurons in the viscera and/or peripheral ganglia. The local activities are regulated by spinal reflexes and by tonically active bulbar centers that control vital functions including blood pressure and respiration. The autonomic centers in the brain stem and spinal cord, in turn, are reciprocally related to several forebrain structures, including the hypothalamus. These structures and circuits, which are collectively referred to as the "central autonomic network," are essential for the integration of autonomic, endocrine, and somatomotor functions.

Many diseases affect the autonomic pathways, in the peripheral and the central nervous systems, and thus cause derangement of autonomic functions. Hypothalamic syndromes, in particular, are characterized by autonomic disturbances, and many emotional states, both normal and pathologic, are accompanied by autonomic nervous system activities. The autonomic system is the means for physical expression of emotion, as in the blush of embarrassment or the tachycardia of fright.

The autonomic ("vegetative" or "visceral") nervous system innervates smooth muscles, glands, and cardiac muscle. It consists of two major divisions, a *sympathetic* (or thoracolumbar) and a *parasympathetic* (or craniosacral) division, which usually have antagonistic effects. Neuronal plexuses in the gastrointestinal tract form the *enteric nervous system*, which is now commonly considered as a third division.

The sympathetic system responds to needs requiring a mobilization of all resources, e.g., in stressful situations. It prepares the body for an emergency: the pupils dilate and heart rate, blood pressure, and cardiac output increase, favoring blood flow to skeletal muscles, myocardium, and brain. At the same time, the peristaltic activity of the gastrointestinal (GI) tract is suppressed by adrenergic nerve inhibitory effects.

In contrast, the parasympathetic system reduces heart rate but stimulates the peristaltic and secretory activities of the GI tract. Autonomic activity, however, is seldom solely either sympathetic or parasympathetic; it is usually the product of a well-balanced interplay between the two parts, thereby providing an important mechanism for the maintenance of homeostasis, i.e., a condition of dynamic equilibrium in the internal environment.

The enteric nervous system, with neurons located in the walls of the stomach and intestines, and in the gallbladder and the pancreas, regulates motility and secretion in the GI tract with a great deal of autonomy.

Originally, the autonomic nervous system was considered to be a peripheral system with purely motor functions. However, it was soon discovered that the CNS also contains important autonomic centers, and it is now generally recognized that afferent fibers and pathways are as much part of the autonomic nervous system as they are of the somatic system. In fact, the majority of the nerve impulses in the "great visceral nerve," the vagus, are *afferents*, conveying important information about pressure, stretch, chemical environment, etc., from the visceral organs, e.g., the GI tract, the heart, the central arteries, and the veins. Further, the sympathetic nerves to the visceral organs also contain many "nociceptive" afferents, which convey visceral pain and distress.

Peripheral Parts

Pre- and Postganglionic Neuron

One of the most characteristic features of the autonomic nervous system is that a significant number of the efferent neurons are located outside the CNS in *autonomic ganglia*. The efferent peripheral pathway, therefore, consists of two neurons, a *preganglionic neuron*, whose cell body is located within the CNS, and a *postganglionic neuron*, whose cell body is located in a ganglion outside the CNS (Fig. 19–1). Because one preganglionic neuron usually contacts many postganglionic neurons in the peripheral ganglion, there is generally a considerable divergence of activity following stimulation of only one preganglionic neuron. This arrangement is the basis for the "mass effect" of the autonomic, especially the sympathetic, system. Nevertheless, this system of divergence-convergence is highly specific, as it serves to bind together into "functional units" such parts as the vascular bed or the GI system, which for central control require unification. This almost simulates military command systems. Principal orders are given to subordinate organizations with their own, local command systems, *without* any detailed "order" to the individual soldiers, or muscle-gland cells, which are instead integrated by local orders/mechanisms.

Anatomic Differences Between the Sympathetic and Parasympathetic Systems

Whereas the postganglionic sympathetic neurons are located in sympathetic ganglia at some distance from the effector organs (Fig. 19–2) the postganglionic parasympathetic neurons are located either in the wall (Fig. 19–2) of the organs they innervate (e.g., in the myenteric and mucosal plexuses of the GI tract) or in close proximity to the organs of innervation (e.g., in pulmonary, cardiac, or gastric plexuses). The parasympathetic activity, therefore, is generally more localized in its effect than the sympathetic activity.

Pharmacologic Differences Between the Sympathetic and Parasympathetic Systems

Both sympathetic and parasympathetic preganglionic neurons use *acetylcholine* (ACh) as their transmitter. ACh is also released at the axon terminals of the postganglionic parasympathetic fibers (cholinergic fibers), whereas *noradrenaline* (norepinephrine) is the primary transmitter substance that is released from most postganglionic sympathetic fibers (adrenergic fibers). However, the sympathetic postganglionic fibers to the sweat glands are cholinergic.

A neuropeptide is usually colocalized with the classical transmitter in autonomic postganglionic neurons. For instance, the cholinergic fibers to the salivary glands contain vasoactive intestinal polypeptide (VIP), which causes vasodilation, whereas ACh has a direct effect on the secretory cells.

Furthermore, it has recently been shown that adenosine triphosphate (ATP), which often serves

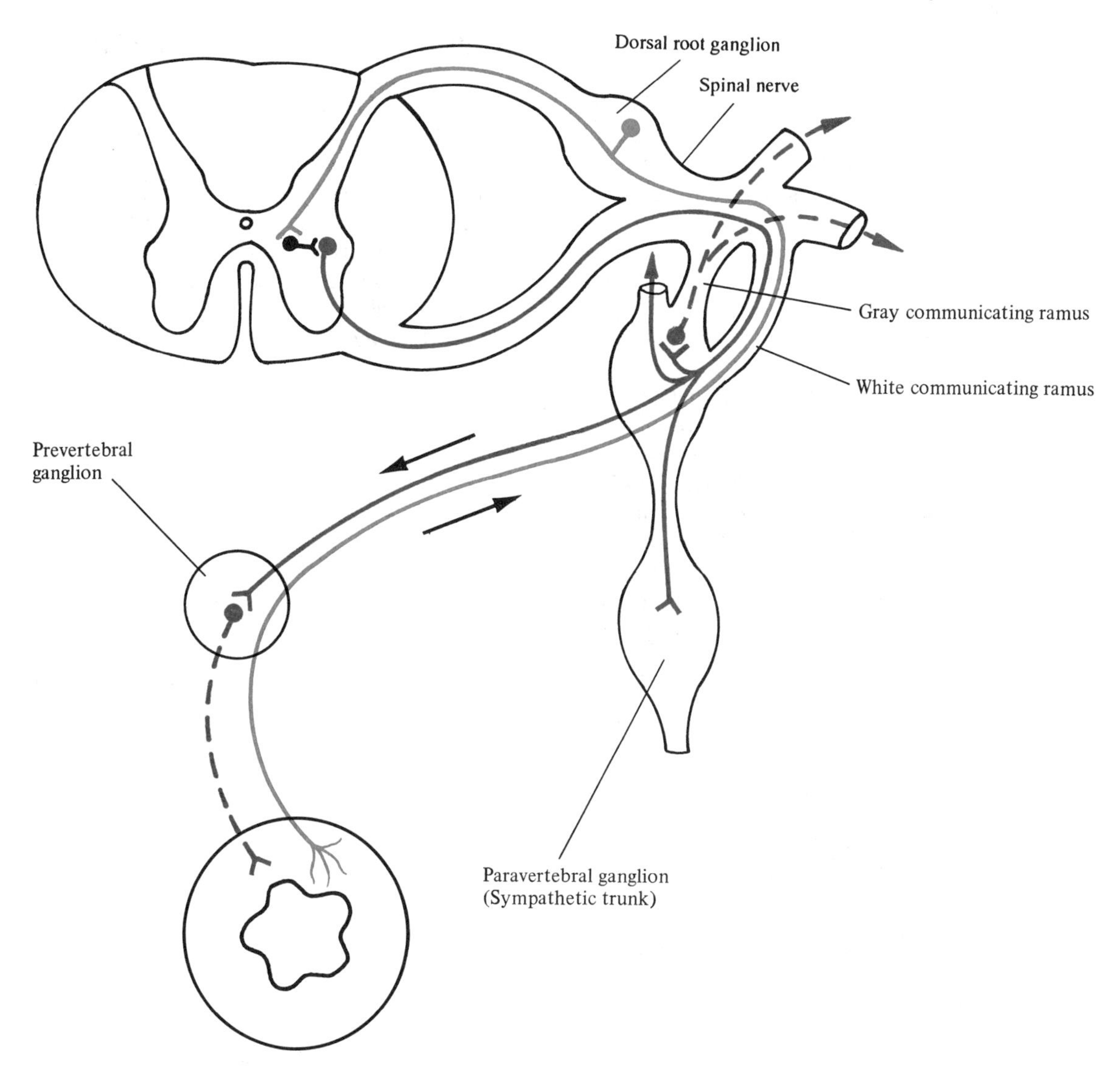

Fig. 19–1. Pre- and Postganglionic Neurons
Diagram showing the arrangement of pre- and postganglionic neurons in the sympathetic autonomic system. Preganglionic neuron, *solid red*; postganglionic neuron, *dashed red*; afferent neuron, *blue*; interneuron, *black*.

as a cotransmitter with, for instance, norepinephrine, is responsible for fast "phasic" effects and responses, while tonically maintained actions are dominated by the release of norepinephrine.

Thoracolumbar or "Sympathetic" Division

The cell bodies of the preganglionic motor neurons of the sympathetic division (red in Fgi. 19–3) are located in the *intermediolateral cell column* in the lateral horn of the spinal cord. This cell column extends from the first thoracic to the second or third lumbar segment. The preganglionic fibers exit via ventral roots and spinal nerves of T1–L2–3 and reach the *paravertebral ganglia* of the *sympathetic trunk*, which extends from the base of the skull to the coccyx on the ventrolateral side of the vertebral column. The connection between the spinal nerve and the sympathetic trunk is formed by the *white communicating ramus* (Fig. 19–1), which contains both preganglionic sympathetic fibers and visceral afferent fibers. The color of these rami is whitish because the preganglionic fibers are myelinated.

Having reached the sympathetic trunk, the preganglionic fibers can:

either pass through the paravertebral ganglia as *splanchnic nerves* and synapse in one of the *prevertebral ganglia*, such as the *celiac ganglion*, the *superior mesenteric ganglion*, or the *inferior mesenteric ganglion*,

or pass up or down in the sympathetic trunk before establishing contacts with postganglionic neurons at different levels of the sympathetic trunk ganglia,

or, finally, they may establish synaptic contacts with postganglionic neurons in the paravertebral ganglion at the level of entrance.

Many of the postganglionic fibers that emerge from the different paravertebral and prevertebral ganglia join arteries and follow these, as plexuses of nerve fibers, to the different internal organs. Other postganglionic fibers return to the spinal nerves from the different levels of the sympathetic trunk as the *gray communicating rami*, so named because the majority of the postganglionic fibers are unmyelinated.[1] In contrast with the white rami, which are present only in the thoracolumbar (T1–L2–3) part of the sympathetic trunk, the gray rami are found along the entire trunk. The postganglionic fibers that join the spinal nerves distribute to the sweat glands in the skin and to the blood vessels and the hair follicles (pili muscles) in the peripheral parts of the body. These structures, incidentally, do not receive a parasympathetic innervation.

The sensory fibers in the sympathetic division have their cell bodies in the dorsal root ganglia (Fig. 19–1), and many of these dorsal root ganglion cells contain one or several neuropeptides, including substance P, enkephalin (ENK), cholecystokinin

1. Although the terms white and gray communicating rami indicate a difference in color, it is not possible to identify the two types on the basis of their color during surgery.

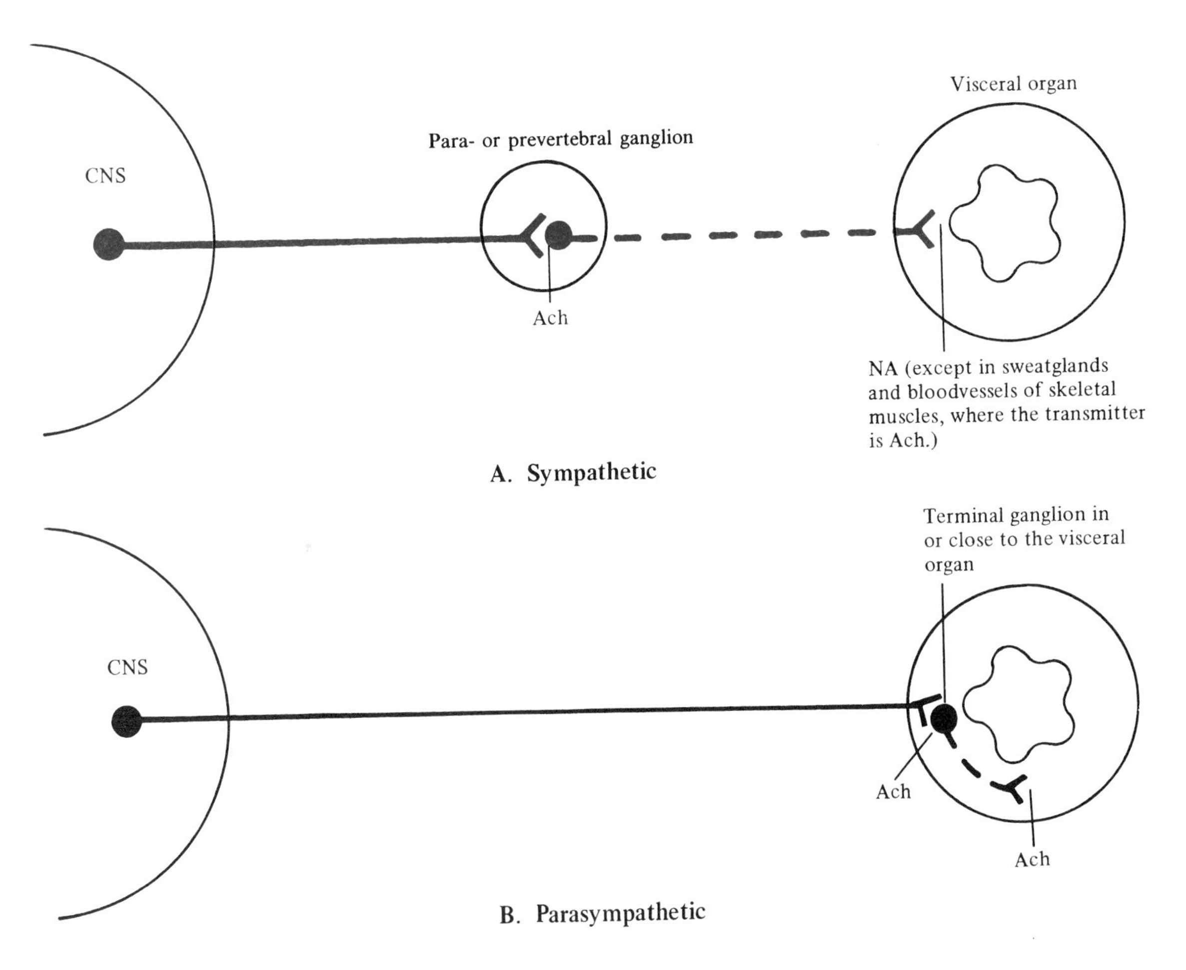

Fig. 19–2. Major Differences Between the Sympathetic and Parasympathetic Systems

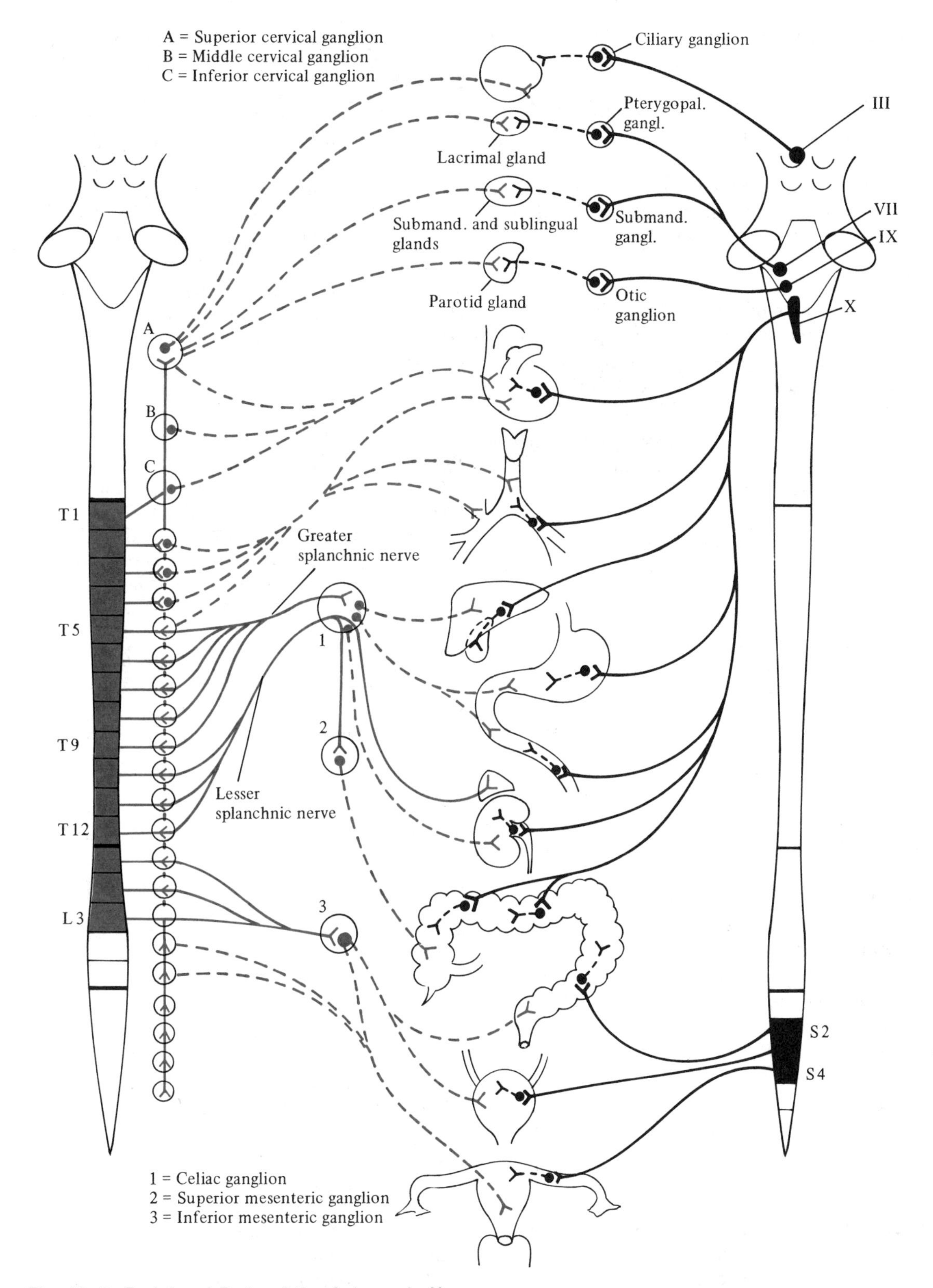

Fig. 19–3. Peripheral Parts of the Autonomic Nervous System
Scheme of the efferent pathways of the sympathetic thoracolumbar (*red*) and parasympathetic craniosacral (*black*) divisions of the peripheral autonomic system.

(CCK), vasoactive intestinal polypeptide (VIP), and others.

Craniosacral or "Parasympathetic" Division

The preganglionic efferent fibers in the parasympathetic division (black in Fig. 19–3) pass through the IIIrd, VIIth, IXth, and Xth cranial nerves (the cranial part) and through the second, third, and fourth sacral nerves (the sacral part).

Cranial Part

The cranial part of the parasympathetic division originates in the brain stem, and the preganglionic fibers emerge along with the following four cranial nerves: the oculomotor (III), facial (VII), glossopharyngeal (IX), and vagus (X) nerves.

Oculomotor Nerve. The parasympathetic oculomotor fibers originate in the parasympathetic nucleus (Edinger–Westphal) of the oculomotor nuclear complex. They course along in a branch of the IIIrd cranial nerve to the *ciliary ganglion*, where preganglionic fibers synapse with the postganglionic neurons that innervate the ciliary muscle and the sphincter pupillae.

Facial Nerve. Preganglionic fibers from cells in the *salivatory nucleus* emerge from the skull in the nervus intermedius. Some of the fibers reach the *pterygopalatine ganglion* via the great petrosal branch and the nerve of the pterygoid canal. Others of these preganglionic axons proceed through the chorda tympani and the lingual nerve to synapse in the *submandibular ganglion*. The postganglionic fibers originating in the pterygopalatine ganglion innervate the lacrimal gland and those from the submandibular ganglion terminate in the submandibular and sublingual glands.

Glossopharyngeal Nerve. Preganglionic secretomotor fibers originate in the *salivatory nucleus* and reach the *otic ganglion* through the tympanic branch of the glossopharyngeal nerve and the lesser petrosal nerve. The postganglionic fibers innervate the parotid gland.

Vagus Nerve. The vagus nerve is the principal nerve of the parasympathetic division. It contains preganglionic parasympathetic fibers that originate in the *dorsal motor nucleus of the vagus* and the *nucleus ambiguus*. The postganglionic neurons are located in ganglia that lie in the walls of the visceral organs of the thorax and the abdomen or in close proximity to the organs. Only a small number (10–15%) of the fibers in the vagus are efferent; the majority are afferent fibers from receptors in the different visceral organs. The visceral afferent fibers, like the autonomic afferent fibers in the facial and glossopharyngeal nerves, terminate in the *solitary nucleus*.

Sacral Part

The preganglionic fibers of the sacral part of the parasympathetic system originate in cells of the second, third, and fourth sacral segments of the spinal cord. The fibers emerge in the ventral root and later leave the sacral nerves to form the *pelvic nerves* (nervi erigentes) on each side of the rectum. Here they intermingle with sympathetic fibers. The parasympathetic ganglia containing the postganglionic neurons are located close to the peripheral organs, i.e., the descending and sigmoid colon, the rectum, the bladder, and the genitalia.

Afferent Visceral Fibers

Impulses from visceral receptors reach the CNS through afferent fibers, which follow the efferent autonomic fibers peripherally (Fig. 19–1). The afferent fibers, whose cell bodies are located in the dorsal root ganglia and in the cranial nerve ganglia of the facial (VII), the glossopharyngeal (IX), and the vagus (X) nerves, are equal in importance to the efferent fibers. The visceral sensory impulses play an important role in various reflexes for the coordination and adjustment of visceral functions. After synapse in the spinal cord or nucleus solitarius, input from the visceral organs reaches many areas in the brain stem and the forebrain (see below).

Vagal afferents from the cardiovascular system carry information about chemical composition of the blood and about wall tension and distension of the heart and large vessels, serving as "volume receptors" and "baroreceptors" with both "phasic" and "tonic" information conveyed via both myelinated and unmyelinated afferents. Vagal afferents from the GI organs carry information about filling, wall tension and distension, and chemical composition of local contents, serving as afferent limbs both for reflexes and for central information of, e.g., satiety, comfort, discomfort, etc. Far too little is known about the overall functional importance of GI afferents, which greatly outnumber the efferent fibers.

From a clinical point of view, the most prominent visceral sensation is pain, which is usually poorly localized and of aching quality. Visceral pain is also characterized by its capacity to induce strong autonomic responses and by its tendency to be displaced or reflected from the true source in the visceral organ to the body surface (see Referred Pain in Clinical Notes). The vast majority, perhaps all, of the pain fibers from the internal organs pass with the sympathetic innervation, whereas nearly all the other afferents controlling autonomic function run in the parasympathetic system.

Enteric Nervous System

The enteric nervous system, which innervates the GI tract as well as the pancreas and gallbladder, regulates gut motility and secretion. It is an intramural system, i.e., it is located within the walls of the various GI organs, and it consists of two interconnected sheets of neurons or plexuses. The *myenteric* (Auerbach's) *plexus* is located between the external longitudinal and circular smooth muscle layers, whereas the *submucosal* (Meissner's) *plexus* is located within the submucosa. Although many of the basic functions of the enteric nervous system occur without significant extrinsic interference, the rest of the autonomic nervous system, i.e., the sympathetic and parasympathetic divisions, does modulate its activities, and it does so by innervating an interneuronal system, which in turn controls the enteric system. In emergency situations, for example, the sympathetic nervous system can effectively override the activities of the enteric system by reducing secretion and peristalsis, and by increasing the sphincter tone. It does so, mainly or entirely, by suppressing the activity of the intramural neurons exciting gland cells or muscle cells, rather than directly inhibiting these cells.

The intrinsic circuitry of the enteric nervous system and its many specialized functions are complicated. Suffice it to say that the neurons representing the system are of many different types, including motor neurons, sensory neurons, and interneurons, and they use a number of different transmitters including many neuropeptides. Somewhere around 40 to 50 different transmitter or modulator "candidates" have so far been identified, and many of these have subsequently been identified in the hypothalamus and related basal forebrain regions.

Autonomic Innervation of Some Organs (Fig. 19–3)

Eye

Through its regulation of the diameter of the pupil, the autonomic innervation of the eye is crucial for accommodation and for visual acuity. It also controls the blood flow to the eye, and by doing so it indirectly controls the intraocular pressure.

The preganglionic sympathetic fibers to the eye, the cell bodies of which lie in the *intermediolateral cell column* in the first and second thoracic segments (ciliospinal center), proceed through the sympathetic trunk to the *superior cervical ganglion*, where they establish synaptic contacts with the postganglionic fibers. The postganglionic fibers follow the internal carotid and the ophthalmic arteries and pass uninterruptedly through the *ciliary ganglion* and the *short ciliary nerves* to end in the dilator pupillae and the superior tarsal muscle. A lesion of the sympathetic pathway to the eye results in Horner's syndrome (see Clinical Notes).

Preganglionic parasympathetic fibers, which originate primarily in the *Edinger–Westphal nucleus* of the oculomotor nuclear complex, reach the *ciliary ganglion* through the oculomotor nerve. The postganglionic fibers pass forward in the *short ciliary nerves* and innervate the constrictor pupillae and the ciliary muscle.

Through its regulation of the diameter of the pupil, the parasympathetic innervation of the eye is crucial for accommodation and for visual acuity. The sympathetic innervation controls the blood flow to the eye, and by doing so it influences the intraocular pressure.

Heart

The preganglionic sympathetic neurons are located in the upper thoracic part of the *intermediolateral cell column* (T1–T5). They end in the three *cervical ganglia* and the upper *thoracic ganglia*. The postganglionic fibers reach the heart by way of the *cardiac nerves*. Stimulation of the sympathetic fibers that innervate the pacemaker cells and cardiac muscle cells increases heart rate, transmission velocity, excitability, and the contractility of the cardiac muscle cells. This latter important effect increases both the speed and force of the cardiac systole.

Activation of the parasympathetic fibers, which originate both in the dorsal motor nucleus of the vagus and in the nucleus ambiguus, and which reach the heart through the vagus, results predominantly in a reduction of the heart rate, and delays the transmission of the atrial excitation to the ventricles. Important areas for the coordination of sympathetic- and parasympathetic-mediated cardiovascular reflexes are located in the medulla oblongata.

Lungs

The postganglionic sympathetic fibers, which arise from level T2–T5 of the sympathetic trunk, produce bronchodilation and vasoconstriction, whereas stimulation of the vagus results in constriction of the bronchioli and increased glandular secretion.

A large number of afferent fibers carry information about distension, mucosal irritation, and pain, and provide the opportunity for reflex responses to this information.

Adrenal Medulla

Cells of the adrenal medulla are derived from the neural crest and represent postganglionic sympathetic neurons, often referred to as *chromaffin cells*. Accordingly, they are innervated by preganglionic sympathetic fibers, which emerge from the *lesser splanchnic nerve* (Fig. 19–3). Stimulation of the adrenal medulla is followed by liberation of epinephrine and norepinephrine into the bloodstream. This provides for a relatively rapid and widespread reinforcement of sympathetic activities during emergency situations and conditions of stress. The hormonal component is particularly important for the mobilization of substances such as glucose and fatty acids, from nutritional depots, and in some vascular circuits, as in skeletal muscle, via β_2-adrenergic receptors, causing vasodilation while the direct nervous influence on the heart and vessels dominates the cardiovascular adjustments.

Urinary Bladder and Sexual Organs

The complex neural mechanisms regulating the urogenital functions, i.e., micturition and penile and clitoral erection, as well as various secretory functions and ejaculation, depend on the integration of a number of autonomic and somatic neural circuits involving both the lumbosacral spinal cord and different regions in the brain stem and the forebrain. The bladder and the sex organs are innervated by three sets of nerves: the parasympathetic *pelvic splanchnic nerves* with preganglionic neurons in spinal cord segments S2–S4, the sympathetic *hypogastric nerves* with preganglionic neurons in segments T11–L2, and the somatic *pudendal nerves* from segments S2–S4.

The control of micturition is primarily a parasympathetic function. When the volume of urine in the bladder reaches a certain level, impulses from stretch receptors in the wall of the bladder pass through the pelvic nerves to the sacral cord where they activate parasympathetic preganglionic neurons. Increased activity in their axons, which pass through the pelvic splanchnic nerves to the bladder, results in contraction of the detrusor muscle. There is concomitant inhibition of the somatomotor neurons that innervate the external sphincter. Furthermore, once the urethra becomes distended, stretch-sensitive afferents in the urethral wall cause a maintained reflex detrusor contraction, which ensures that the bladder contracts until entirely emptied. Cell groups in the dorsolateral pontine tegmentum are important for coordinating these activities. This explains why patients with spinal cord transections above the sacral level have difficulty emptying the bladder, since the reticulospinal fibers that carry impulses from the pontine "micturition center" are damaged. The emptying of the bladder is predominantly under reflex control in the newborn, but gradually becomes subjected to conscious regulation.

Although there are obvious anatomic differences between male and female sex organs, the physiologic responses during the sex act, e.g., penile and clitoral erection, secretory functions, and smooth and striated muscle contractions, are in many ways similar. The transmitter involved in penile erection appears to be nitric oxide (NO). The spinal reflex phenomena involve stimulatory input through afferent fibers in the pudendal nerves and efferent stimuli through the sacral parasympathetic fibers. The neural mechanisms for penile and clitoral erections are usually a combination of spinal cord reflexes and psychogenic stimuli involving supraspinal pathways.

Secretion from the various genital glands, including the seminal vesicles and prostate gland in the male, is controlled by sacral and thoracolumbar autonomic systems. Emission of seminal and prostate fluid into the urethra is primarily a sympathetic function, whereas ejaculation is accomplished by activation of somatomotor efferents to bulbocavernosus and ischiocavernosus muscles with reflexive closure of the internal urinary sphincter (sympathetic innervation) and the proximal part of the urethra to prevent reflux of the semen.

Central Parts

Although many of the visceral functions are regulated by reflex arcs within the peripheral organs, most prominently in the case of the enteric nervous system, or by spinal cord reflexes, several supraspinal regions are involved in the overall control of the autonomic functions. A prominent position is held by the nucleus of the solitary tract, which receives input from all visceral organs. Part of this input is transmitted to nearby brain stem regions involved in various reflexes, including those controlling blood pressure and respiratory functions, which do not depend on feedback loops or regulation from the forebrain for their basic operations.

The nucleus solitarius, in addition, is directly and reciprocally related to several structures in the upper part of the brain stem and in the forebrain, including parts of the hypothalamus, the extended amygdala, and the insular and prefrontal cortical regions. The neuronal circuits that unite these forebrain structures with the nucleus solitarius and with other autonomic nuclei in the brain stem and spinal cord are collectively referred to as the "*central autonomic network*" or the "*visceral forebrain system*." It forms the anatomic substrate for the adjustments that are needed in the basic visceral, cardiovascular, and respiratory functions as they relate to the many activities in which the body is involved, i.e., from regulation of temperature and control of food intake, thirst, sexual and other types of emotional behavior, to muscular exercise and mental activities.

Control of Blood Pressure and Respiratory Functions

The basic control mechanisms for blood pressure and respiratory functions are located in the medulla oblongata. It is no accident that the anatomic substrates regulating these two vital functions are closely related and to some extent overlapping. After all, the fundamental purpose of both cardiovascular and respiratory systems is to secure the supply of oxygen to the tissues, and the activities of the two systems must be coordinated in order to maintain "cardiorespiratory homeostasis."

The *nucleus solitarius*, which plays an important role in both cardiovascular and respiratory functions, is an extensive and highly differentiated nucleus with many subdivisions, each of which contains a large number of neuropeptides as signal substances. As the major visceral sensory nucleus in the brain, it integrates a number of important visceral reflexes, and it serves as a relay for visceral sensory impulses to other parts of the CNS. For the sake of simplicity, the nucleus can be divided into three main regions: rostral, intermediate, and caudal divisions. The *rostral* third is primarily devoted to the transmission of taste impulses (Fig. 11–7). The *intermediate* part is organized in a viscerotopic fashion in the sense that cardiovascular, pulmonary, and gastrointestinal impulses terminate in different, but partly overlapping, parts of the nucleus. This intermediate region is of particular importance for cardiovascular reflex adjustments and it also contains important respiratory neurons, which are collectively referred to as the dorsal respiratory group (see below). The *caudal* part also receives visceral afferents from various sources, and represents the main region for transmission of visceral sensory input to the "central autonomic network."

Other medullary regions involved in the regulation of autonomic functions are the *dorsal motor nucleus of the vagus*, which flanks the medial border of the solitary complex, and cell groups in the *intermediate reticular zone*, which extends obliquely across the middle part of the medulla oblongata (Fig. 10–1). The term *ventrolateral medulla* is often used for the lateral part of this region, which is poorly defined anatomically. It is delimited by the spinal trigeminal nucleus and tract dorsally, the inferior olivary complex ventrally, and the nucleus ambiguus dorsomedially.

Brain Stem Control of Cardiovascular Functions

The preganglionic and postganglionic sympathetic neurons that serve to constrict arterioles and venous "capacitance" vessels, and to increase heart rate and stroke volume, maintain a low-frequency tonic activity, perhaps 0.5–1–2 Hz during resting equilibrium. In addition, important brain stem reflexes are designed to keep the blood pressure reasonably well stabilized. Afferent impulses in the glossopharyngeal and vagus nerves reach the nucleus solitarius from many sensory receptors, including *baroreceptors* in the aortic arch and the carotid sinuses, and cardiac "*volume*" *receptors* in the atrial and ventricular walls. If the blood pressure increases and the vessel wall stretches, or if the heart is increasingly filled in diastole, the respective receptors increase the discharge of afferent impulses in both myelinated and unmyelinated fibers to the nucleus solitarius, thus delivering multifaceted information on filling, pressure, and wall tension. This nucleus, in turn, projects to the nucleus ambiguus, where parasympathetic cardioinhibitory neurons are activated. Other neurons in the nucleus solitarius project to inhibitory interneurons in the ventrolateral medulla, which reduce the activity in the reticulospinal vasomotor neurons, i.e., those that project to the sympathetic preganglionic neurons in the spinal cord. This results in reduced activity in the sympathetic fibers to the heart and the arterioles, thereby reducing heart rate and resistance to blood flow, as well as to the venous "capacitance" side, which favors increased peripheral pooling of blood, and hence a reduction of venous return to the heart. As a result of these adjustments, i.e., reduction of both cardiac output and vascular resistance, the blood pressure falls. However, during intense physical exercise or mental arousal, higher brain centers can more or less suppress these reflex inhibitory actions, allowing for intense tachycardia and heart activation despite a rise in pressure and filling.

The carotid body and aortic arch also contain *chemoreceptors* that respond to oxygen, carbon dioxide, and hydrogen ion levels in the blood. Impulses from these receptors are transmitted to the nucleus solitarius, which in turn activates control mechanisms that modify both respiration and cardiac output.

Brain Stem Control of Respiratory Functions

The medullary respiratory cell groups, which are closely related to the cardiovascular regions, are concentrated primarily in and around the nucleus solitarius and nucleus ambiguus. The respiratory neurons are divided broadly into two groups, a *dorsal respiratory group* (DRG) located in the middle third of the solitary nucleus at the level of the obex, and a larger *ventral respiratory group* (VRG) within the nucleus ambiguus and neighboring areas of the reticular formation.

Most neurons in the DRG are active during the inspiratory phase of the respiration, i.e., inspiratory neurons. These neurons, which send their axons through the reticulospinal tract to the spinal cord, transmit inspiratory monosynaptic drive to the phrenic and intercostal motor neurons. Additional inspiratory drive, especially to the intercostal motor neurons, comes from inhibitory bulbospinal neurons in the VRG. In quiet breathing, only the inspiratory movement is active, and the inspiratory neurons are driven by impulses from both central and peripheral chemoreceptors and various reflexes. When the inspiratory activity has reached a critical threshold, the inspiration is "switched off" by a neural control mechanism that apparently gets input from several sources, including impulses through the vagus nerve from *pulmonary stretch receptors* (Hering–Breuer reflex).

The VRG is represented by an extensive column of cells, which includes the nucleus ambiguus. Most of the column contains both inspiratory and expiratory neurons; some of these cells send their axons to the respiratory motor neurons in the spinal cord, whereas others form local circuits for interaction with other respiratory neurons or cardiovascular units. Glossopharyngeal and vagal motor neurons for the innervation of larynx and pharynx are also part of the VRG, and their activity is closely coordinated with the rest of the respiratory cycle.

Exactly how the respiratory rhythm is generated has long been a primary concern to respiratory physiologists, and it is only recently that a group of neurons with a spontaneous rhythmical activity has been identified in a region close to the rostral pole of the nucleus ambiguus. It remains to be seen if this group of pacemaker neurons, which is referred to by the peculiar term *pre-Botzinger complex*, has the intrinsic qualities and connections necessary for providing the oscillatory drive to the respiratory circuits.

The brain stem respiratory areas for rhythmic respiration do not function in isolation from the rest of the brain; instead, their activity is constantly being modified voluntarily or involuntarily as part of various activities, e.g., physical exercise, different emotional states, changes in sleep–wake states or temperature regulation, swallowing, coughing, and a number of other conditions. Stimuli for adjusting the respiratory pattern in various conditions come from many sources within the nervous system, including but not limited to the "central autonomic network" (see below), and they are transmitted through pathways that impinge either on the brain stem respiratory areas or on respiratory motor neurons in the spinal cord. Such stimuli also include modifying impulses generated by chemoreceptors in the large arteries of the heart and by mechanoreceptors of the lungs. In addition, there is a particularly sensitive set of mechanisms for direct chemical stimulation of the respiratory "center," which responds especially to an increase in the carbon dioxide content of the blood and to changes of pH in the extracellular fluid.

The "Central Autonomic Network"

Several structures in the forebrain and upper brain stem are directly related to the nucleus solitarius and to preganglionic autonomic nuclei in the lower brain stem and spinal cord. The "central autonomic network" (Fig. 19–4) is used as a collective term for these structures and the pathways that connect them with the autonomic nuclei. The major brain structures included in this network are the periaqueductal gray; the hypothalamus, especially the paraventricular nucleus and lateral hypothalamic areas; the extended amygdala; and some cortical regions, primarily in the insula and adjoining frontoparietal operculum and medial prefrontal cortex. However, it is important to realize that these structures are not only involved in autonomic functions; their importance in the context of autonomic functions is primarily related to the fact that they serve as areas for integration of autonomic, endocrine, and somatomotor activities that accompany specific behaviors. The term "central autonomic network," therefore, does not do full justice to the great importance of these structures, which are vital to survival, as they initiate and convey most of those overall responses to serious changes of both the "internal" and the "external" environments, which subserve the protection of the individual and the species.

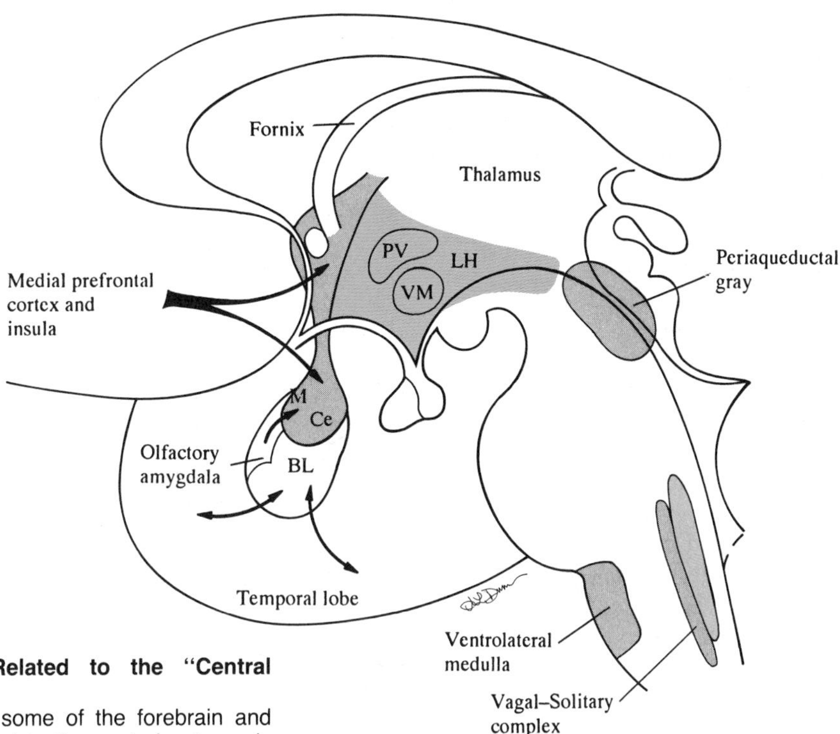

Fig. 19–4. Structures Related to the "Central Autonomic Network"
Schematic figure showing some of the forebrain and brain stem structures related to the central autonomic network. BL = basolateral amygdala; Ce = central amygdaloid nucleus; LH = lateral hypothalamus; Me = medial amygdaloid nucleus; PV = paraventricular hypothalamic nucleus; VM = ventromedial hypothalamic nucleus.

Insular and Medial Prefrontal Cortex

Physiologic studies have uncovered a crude viscerotopic organization within the insular cortex and adjoining frontoparietal operculum, which is recognized as the visceral sensorimotor region of the cerebral cortex. The cortical taste area, for instance, occupies part of the insular region, which spills over into the overlying opercular region and the somatosensory cortex (Chapter 9). In this context, it is interesting to note that although most of the olfactory bulb projection fibers terminate in the frontal olfactory cortex and in the region of the uncus (Chapter 12), some of them also reach the anteroventral part of the insula. The insular cortex is closely related to the nearby medial orbitofrontal cortex and to the caudal parts of the medial prefrontal cortex in front of the lamina terminalis and the genu of the corpus callosum (anterior cingulate area). Both the insula and the medial prefrontal cortex are closely and reciprocally related to the amygdaloid body including the extended amygdala, and to the lateral hypothalamus and autonomic nuclei in the brain stem.

The Extended Amygdala

The bed nucleus of the stria terminalis, which is a large structure located close to the midline in the human brain, is directly continuous with the centromedial amygdala in the temporal lobe (Fig. 20–1). The attenuated cell columns that form the continuity between these two basal forebrain structures are located underneath the lentiform nucleus in an area that is sometimes referred to as the substantia innominata. Together, these structures form the "extended amygdala." If the amygdaloid body is presented as an anatomic and functional unit, which is often the case both in gross anatomic dissections and in physiologic reviews, some of the most prominent and no doubt functionally relevant attributes of this intriguing structure are concealed.

As discussed in some detail in Chapter 20, there are fundamental differences between the basolateral and the centromedial part of the amygdaloid body. Whereas the direct cortical relationships of the centromedial part of the amygdaloid body are considerably more limited than those of the "cortical-like" basolateral part, the centromedial part receives significant input from the basolateral cell group, which in turn receives afferent pathways from modality-specific and multimodal cortical association areas. Even more striking differences are evident in regard to the subcortical relations of these two parts of the amygdaloid body. Whereas the basolateral part, like the rest of the cerebral cortex, projects to the striatum, the centromedial part is characterized foremost by its close relationship to the hypothalamus and to the brain stem reticular formation and autonomic nuclei. Furthermore the bed nucleus of the stria terminalis is not only continuous with the centromedial amygdala, it also shares the same connections, and it contains the same type of transmitters and neuropeptides as the centromedial amygdala. The striking similarities in histology, connectivity, and neurochemistry between the bed nucleus of the stria terminalis and the centromedial amygdala suggest functional similarities. In short, the dumbbell-shaped extended amygdala constitutes a functional–anatomic system that sets it apart from the basolateral part of the amygdaloid body. This situation, which is illustrated in a schematic fashion in Fig. 19–4, is not easily appreciated if attention is focused on the amygdaloid body as an anatomic and functional unit.

The extended amygdala has emerged as a large forebrain system that evaluates the significance of external and internal events, and effects and coordinates appropriate behavioral responses in close cooperation with the hypothalamus and some regions in the upper brain stem, e.g., the periaqueductal gray (see below). The extended amygdala receives cortical input from the insula, medial prefrontal cortex, and hippocampal formation. In addition, activity generated in most cortical areas can affect the extended amygdala by way of the cortical-like basolateral amygdaloid complex. Through its widespread and highly organized downstream projections, the extended amygdala can directly affect autonomic and neuroendocrine centers in the hypothalamus, as well as somatomotor and autonomic areas in the brain stem with subsequent effect on the spinal cord, in terms of important behavioral-autonomic-hormonal response patterns, like the defense reaction.

The notion of the extended amygdala as a key structure in motivational and adaptive behavior does not imply that this whole forebrain structure responds in its entirety to any stimulus. The extended amygdala is a highly differentiated structure, characterized by several subdivisions and neuronal circuits, whose specific functional relevance is not known. As a first approximation, it is useful to recall that the central amygdaloid nucleus and its extension into the lateral part of the bed nucleus of the stria terminalis have a direct effect on autonomic and somatomotor activities, whereas the medial amygdaloid nucleus and its extension into the medial part of the bed nucleus is more closely related to endocrine functions.

Hypothalamus

The functions of the hypothalamus and its close relation to the autonomic nervous system were beautifully demonstrated in a series of systematic electrical stimulation experiments in the 1930s and 1940s by the Swiss neurophysiologist N. R. Hess, who was awarded the Nobel Prize in 1949 for his pioneering studies. Suffice it to say that the hypothalamus, and especially the paraventricular nucleus, occupies a unique position in the "central autonomic network," since it seems to be the only area in the brain that can directly influence not only the endocrine system and somatomotor areas in the brain stem, but also the entire visceral organ system through its pathways to the sympathetic and parasympathetic cell groups in the brain stem and spinal cord. Thus, it contains neuron pools controlling hunger–satiety, thirst, sexual behavior, fight–flight reactions, submissive responses, etc., all placed very close together and all expressing themselves in the efferent directions as differentiated and appropriate behavioral-autonomic-hormonal adjustments affecting the whole organism.

The Periaqueductal Gray and the Defense Reaction

The midbrain periaqueductal (central) gray, which is strategically located at the transition between the brain stem and the forebrain, is reciprocally related to most major regions of the CNS. In spite of its seemingly uniform character, it is highly differentiated and known as a key area for the coordination of somatic and autonomic responses in affective behavior, including the defense reaction (fight–flight response). The defense reaction is typical for all mammals, including man, who are faced with a painful stimulus or a threatening situation. The periaqueductal gray, incidentally, is also an important part of an endogenous "pain-inhibiting system" (Chapter 9), and it is probably no accident that the system for pain modulation is in close proximity to the parts responsible for the fight–flight reaction.

The neuronal circuits related to the periaqueductal gray and their involvement in emotional behavior, such as the defense reaction, exemplifies the general idea and significance of the "central autonomic network" as an integral part of more widely distributed neuronal circuits that generally

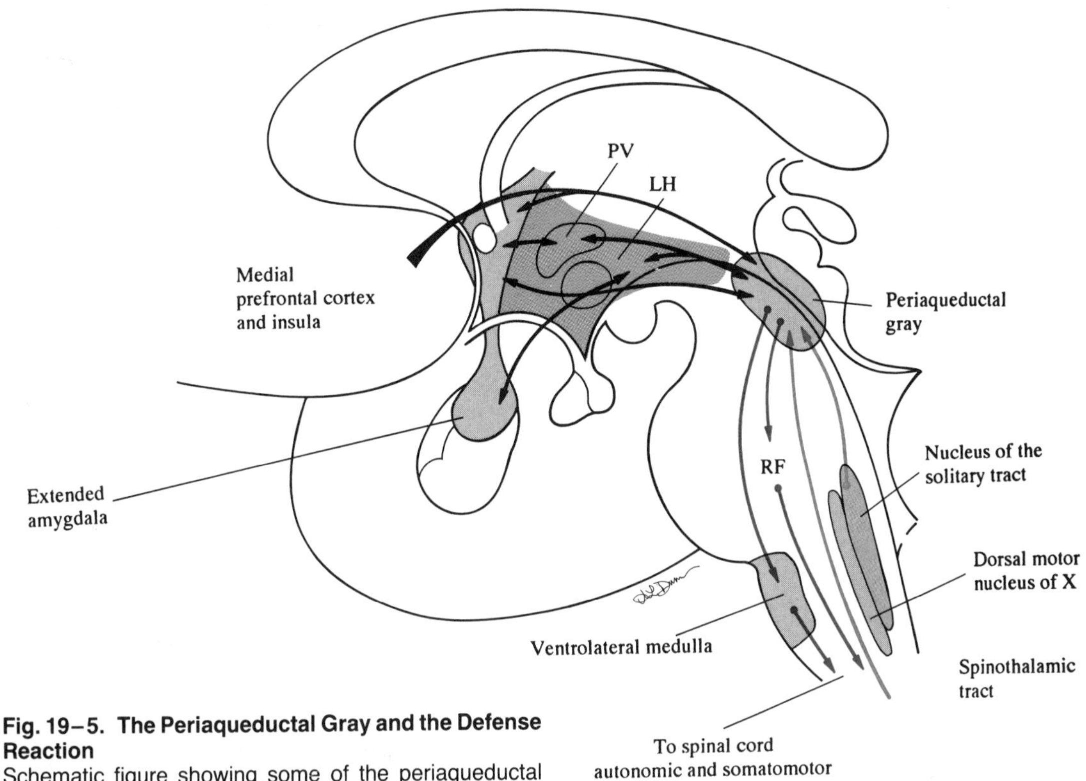

Fig. 19–5. The Periaqueductal Gray and the Defense Reaction
Schematic figure showing some of the periaqueductal gray pathways relevant to the defense reaction. The periaqueductal gray is a key area for the coordination of somatic and autonomic responses in affective behavior, including the defense reaction. RF = reticular formation.

involve sensorimotor and endocrine functions as well. The final response is determined by many factors, including the nature of the afferent input and the emotional significance attached to the stimulus. Some of the anatomic pathways through which these factors can influence the periaqueductal gray are illustrated in Fig. 19–5. Sensory input can reach the periaqueductal gray direct from the nucleus solitarius and through ascending spinomesencephalic and trigeminomesencephalic afferent pathways. The unfolding of the defense reaction is dependent on the emotional and mental state of the individual, and this is where the input from the cerebral cortex, extended amygdala, and hypothalamus becomes crucial. The extended amygdala plays a major role in evaluating the significance of external and internal events, and it projects prominently to the periaqueductal gray. Although the specific roles played by the hypothalamus and the periaqueductal gray in the integration of the defense reaction are not fully understood, it appears that the hypothalamus and the periaqueductal gray cooperate closely in many of the aspects that characterize affective behavior, and they are related to each other by means of prominent reciprocal connections. Of the many descending projections from the periaqueductal gray, some of the most conspicuous are the ones terminating in the ventrolateral medulla, which is one of the most prominent autonomic regulatory regions in the brain, and in brain stem areas related to motor neuronal activity. Through the reticulospinal pathways from these brain stem regions, the whole spectrum of cardiovascular, respiratory, pupillary, and somatomotor functions that characterize the defense reaction can be activated.

Emotional fainting is the antithesis of the defense reaction. It is characterized by a feeling of weakness and dizziness in response to an emotional situation characterized by a strong vasovagal depressor reaction. A similar response can be seen in some animals, e.g., rabbits and opossums, and it is known as the "playing dead" reaction or "playing possum," since it is especially common in the opossum, which suddenly drops as if dead when confronted with imminent danger.

The "*defeat reaction*" is a less well-known, but from a pathophysiologic perspective very important,

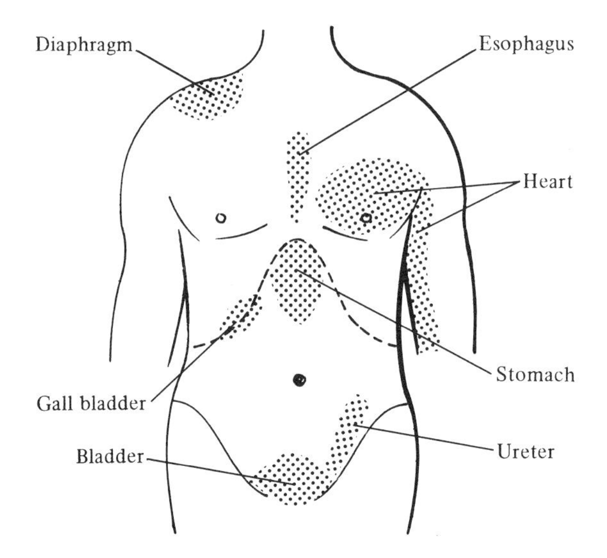

Fig. 19–6. Referred Pain
Diagram showing cutaneous sites of reference of visceral pain commonly encountered in medical practice. It is worth emphasizing that there are significant variations from case to case.

syndrome, which can result from severe grief or from protracted or repeated psychic stress situations. The physiology of the "defeat reaction" is characterized by a complicated pattern of sympathetic and parasympathetic responses with potentially harmful effects on blood vessels. There is also reduced secretion of sex hormones and increased secretion of ACTH-glucocorticoids, which, if longstanding, can lead to significant depression of the immune system with increased risk for infections and tumors.

Clinical Notes

Referred Pain

Referred pain is a clinically important phenomenon in which the pain is displaced or referred from its true source in a visceral organ to the surface of the body. The pain is usually referred, not to the skin overlying the affected organ, but to the skin that is innervated by the same spinal cord segment as the diseased organ (Fig. 19–6). For example, the pain of angina pectoris is usually felt in the upper thorax (T2–T4) or down the inside of the left arm (T1), whereas gallbladder pain is referred to thoracic dermatomes (T6–T8), usually in the epigastric region or dorsally underneath the right shoulder blade. Pain in the shoulder region (C3–C4) may be caused by irritation of the diaphragm.

Visceral pain is mediated by sympathetic afferents. Although there may be several explanations for referred pain, the most likely is that the visceral afferents in the sympathetic division, which have their cell bodies in the dorsal root ganglia, converge on the same spinothalamic cells that receive cutaneous pain afferents from the same segment. Since only a small percentage of the afferent fibers in the dorsal roots originate in the visceral organs, and because somatic pain is much more common than visceral pain, the brain becomes accustomed to interpreting the activity in a specific pool of spinal cord neurons as the result of pain stimuli in a particular skin region. When the same neural pool is stimulated by impulses from the visceral organ, the brain "misinterprets" the message as coming from the skin.

Horner's Syndrome

Interruption of the sympathetic fibers to the eye results in *miosis* (constriction of the pupil) and *ptosis* (drooping of the eyelid) because of paralysis of the dilator pupillae and the superior tarsal muscle. There is also *anhydrosis* (loss of sweating) on the affected side of the face and neck. The eyeball also appears to lie deeper in the orbit, i.e., *enophthalmos*, but whether this is real, as a result of paralysis of the smooth muscles in the inferior orbital fissure, or only apparent because of the ptosis, is unclear.

Horner's syndrome can result from damage either to peripheral sympathetic structures, or to autonomic fibers that descend from the hypothalamus through the lateral parts of the brain stem and cervical spinal cord to the intermediolateral cell group in the first and second thoracic segments (see medullary syndrome of Wallenberg in Clinical Examples, Chapter 11).

Autonomic Dysfunction in Spinal Cord Injuries

Trauma to the spinal cord is common in highly industrialized societies and in times of war. Complete transection of the spinal cord results in paralysis and anesthesia of all body segments below the lesion. The difficult and time-consuming treatment of such patients (called paraplegics, if lower limbs and body are paralyzed, and tetraplegics, if all four limbs are involved) is often carried out in special medical centers.

The paralysis of the parasympathetic innervation to the bladder and the rectum deserves special attention. Immediately following a spinal transection autonomic functions are completely suppressed. There is retention of urine and lack of activity in the bowel. Urination and defecation must be induced by catheter and enema. As the "spinal shock" wears off after a few weeks, reflex activities resume in the bladder and the rectum, providing the lesion is rostral to the sacral parasympathetic part (S2–S4) of the spinal cord. Urination becomes automatic and precipitous, a condition known as *hypertonic* or *hyperreflexive bladder.* If the sacral segments of the spinal cord are destroyed there is no reflex control of the bladder, which will fill to capacity and overflow. This is known as *flaccid, hypotonic,* or *hyporeflexive bladder.*

"Postural Hypotension"

This condition, which is often seen in elderly people, results from a failure of cardiac volume— and arterial baroreceptors to induce the appropriate reflex sympathetic activation and vagal inhibition. Thus, these patients show a blood pressure fall, even towards fainting when erect, with little or no trace of reflex heart rate increase or peripheral vasoconstriction. Most likely, there is some damage to the bulbar reflex centers.

Psychosomatic Disorders

Although many of the vegetative activities, e.g., secretions and motility of the GI tract, contraction of the heart, kidney function, etc., are basically automatic in nature, the autonomic control of visceral functions is closely correlated with the activity in the rest of the nervous system. This intimate relationship is clearly reflected in the organization of the central parts of the autonomic system.

The intimate relationships between mental activities and vegetative functions are familiar to everybody. Salivation can occur by the mere thought of delicious food, embarrassment causes blushing, and sudden psychic stress can cause a dramatic change in heart rate. This intricate relationship between mental activities and visceral functions may have far-reaching consequences for a person's well-being; continuous frustration, or emergency responses that are repeated too often, will eventually lead to neuroendocrine disturbances that may result in pathologic changes in visceral organs. This psychosomatic component may be a key factor in the development of many commonly known diseases such as peptic ulcer, asthma, hypertension, and coronary atherosclerosis to mention a few.

Suggested Reading

1. Appenzeller, D., 1990. The Autonomic Nervous System. An Introduction to Basic and Clinical Concepts, 4th ed. Elsevier BioMedical Press: Amsterdam.
2. Cervero, F. and J. F. B. Morrison, 1986. Visceral Sensation. Prog. Brain Res. Vol. 67.
3. Cervero, F. and K. A. Sharkey, 1985. More than just feelings about visceral sensation. Trends Neurosci. 8:188–190.
4. Ciriello, J., 1987. Organization of the autonomic nervous system: central and peripheral mechanisms. Liss, New York.
5. Depaulis, A. and R. Bandler, 1991. The Midbrain Periaqueductal Gray Matter. Plenum Press: New York.
6. Furness, J. B. and M. Costa, 1987. The Enteric Nervous System. Churchill-Livingstone, London/ Edinburgh.
7. Heimer, L., J. de Olmos, G. F. Alheid, and L. Záborszky, 1991. "Perestroika" in the basal forebrain: Opening the border between neurology and psychiatry. Prog. Brain Res. 87:109–165.
8. Holstege, G., 1992. The emotional motor system. Eur. J. Morphol. 30:67–79.
9. Loewy, A. D. and K. M. Spyer, 1990. Central Regulation of Autonomic Functions. Oxford University Press: New York.
10. Martin, Lee J., R. E. Powers, T. L. Dellovade, and D. L. Price, 1991. The bed nucleus-amygdala continuum in human and monkey. J. Comp. Neurol. 309:445–485.

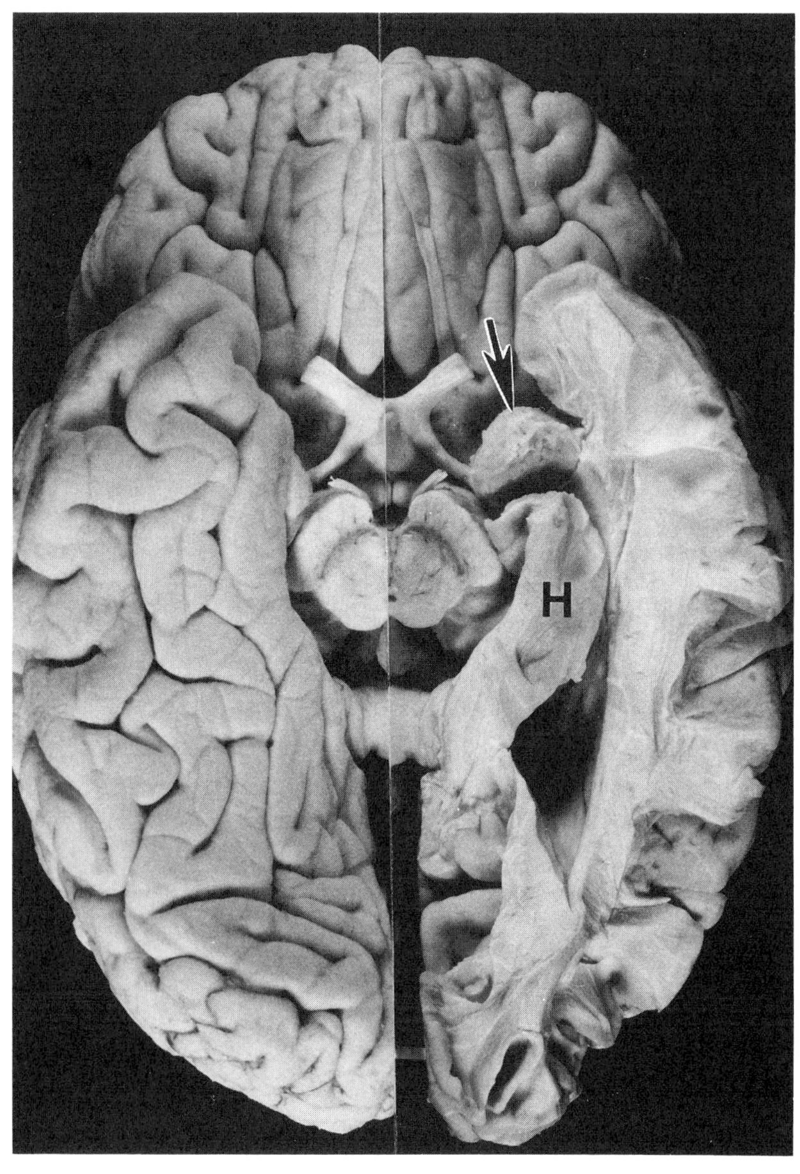

Blunt dissection of the amygdaloid body (*arrow*) from the ventral side to show its position in relation to the surface topography of the anterior part of the temporal lobe. H = Hippocampal formation.

20

Amygdaloid Body and Extended Amygdala

The amygdaloid body is a large subcortical structure located in the rostromedial part of the temporal lobe in front of the hippocampus. The multitude of pathways that connect the amygdaloid body with many other parts of the brain provide the opportunity for the amygdaloid body to be involved in a large number of functions ranging from motor activities, basic drives like eating and sexual behavior, cardiovascular and endocrine mechanisms, to memory and other higher cognitive processes. Some of the most prominent pathways to the amygdaloid body come from cortical areas in the temporal and frontal lobes. These pathways transmit highly processed sensory information to the amygdaloid body, where the information is evaluated in terms of its biological significance.

As a modulator of emotional behavior, part of the amygdaloid body, i.e., the centromedial part, is characterized by a highly organized system of pathways to many hypothalamic and brain stem areas responsible for the generation of various aspects of emotional expression. The centromedial part of the amygdala performs this function in conjunction with another structure close to the midline of the brain, the bed nucleus of the stria terminalis, with which it is directly continuous. This continuum, which extends beyond the classical boundaries of the amygdaloid body, is referred to as the extended amygdala.

The amygdaloid body and other medial temporal lobe structures are often the focus of significant atrophy and other pathologic changes in several common disorders including temporal lobe epilepsy, schizophrenia, and Alzheimer's disease.

Subdivisions

The amygdaloid body is located underneath the uncus of the parahippocampal gyrus (see vignette and Fig. 5–10A), and in front of the hippocampus and the temporal horn of the lateral ventricle (Fig. 20–1A). It is a heterogeneous structure containing many different cell groups with significant differences in cytoarchitecture, histochemical characteristics, and connections. In this sense it is reminiscent of the thalamus. A distinction can be made between a basolateral, cortical, and centromedial nuclear group.

Basolateral Amygdala

A large part of the amygdaloid complex is occupied by the basolateral nuclear group, which is closely related to the cerebral cortex. As a matter of fact, even if the basolateral part of the amygdala is not laminated, its histology, histochemistry, and connections are in many ways reminiscent of a cortical structure. Therefore, the "cortical-like" basolateral part of the amygdaloid body (BL) has been coded with the same grayish color as the rest of the cerebral cortex (Fig. 20–1B), in order to distinguish it from the centromedial part of the amygdaloid body (Ce-M), which has distinctly different characteristics.

Olfactory Amygdala

Of the three major nuclear groupings, the cortical nuclear group, i.e., the olfactory amygdala, is the smallest part of the human amygdala. It is located on the surface of the amygdaloid body adjacent to the temporal olfactory cortex and the medial amygdaloid nucleus. Since the primary input to the cortical nuclear group comes from the olfactory bulb and the olfactory cortex, the cortical nuclear group is referred to as the olfactory amygdala.

Centromedial Amygdala and Extended Amygdala

The central and medial nuclei of the amygdaloid body are directly continuous with another large nucleus close to the midline of the brain, i.e., the *bed nucleus of the stria terminalis* (BST), which therefore can be considered to be an integral part of the central and medial amygdaloid nuclei. The continuity between the centromedial amygdala and the bed nucleus of the stria terminalis is apparent both by cell columns across the ventral part of the brain (Fig. 20–1B) in a region sometimes referred to as the substantia innominata, and by the presence of cells alongside the stria terminalis. This centromedial amygdaloid–bed nucleus of stria terminalis continuum, which is referred to as the *extended amygdala*, appears as a ring formation around the internal capsule and the basal ganglia (Fig. 20–1B). The extended amygdala and its relation to other forebrain systems was discussed in Chapter 16.

The amygdaloid body contains a large number of neuropeptides, and most of the peptidergic axonal plexuses as well many peptidergic neurons are located in the centromedial nuclear group and the rest of the extended amygdala. The neuropeptides include corticotropin-releasing factor (CRF), neurotensin (NT), angiotensin, somatostatin (SST), enkephalin (ENK), vasoactive intestinal polypeptide (VIP), cholecystokinin (CCK), and substance P (SP), many of which are colocalized with GABA.

Connections

The amygdaloid body is closely related to the cerebral cortex, including the hippocampus, as well as to the striatum, thalamus, hypothalamus, and brain stem; the three main subdivisions of the amygdala, i.e., the basolateral, cortical, and centromedial divisions, have in general very different anatomic affiliations.

Basolateral Amygdala

This is the part of the amygdaloid body that is closely related to many cortical areas in the prefrontal, temporal, insular, and occipital regions. Characteristically, it receives highly processed and partly converging sensory information both from modality-specific sensory association areas and from regions that already contain polysensory information. In this way, all sensory modalities can influence the activity in the basolateral amygdala. The corticoamygdaloid pathways, which use glutamate and/or aspartate as their neurotransmitter, are reciprocated by amygdalocortical pathways (Fig. 20–1B), which indicates that the basolateral part of the amygdaloid body is in a position to influence the cortical processing of sensory information.

Like the rest of the cerebral cortex, the basolateral amygdala projects to the striatum, especially its ventral part, and it also has close relations to the thalamus in the sense that it projects to the mediodorsal thalamus, which in turn projects to the prefrontal cortex. This provides another route, albeit indirect, whereby the amygdaloid body can influence the activity primarily in the medial and orbital prefrontal cortex. Thalamoamygdaloid projections are also prominent, primarily from viscerosensory and auditory relay nuclei in the

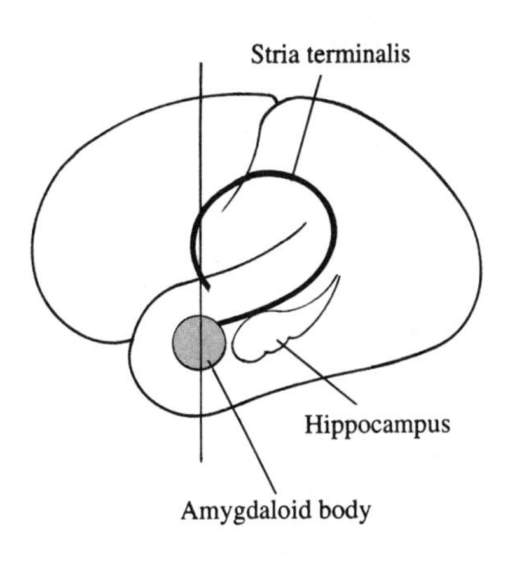

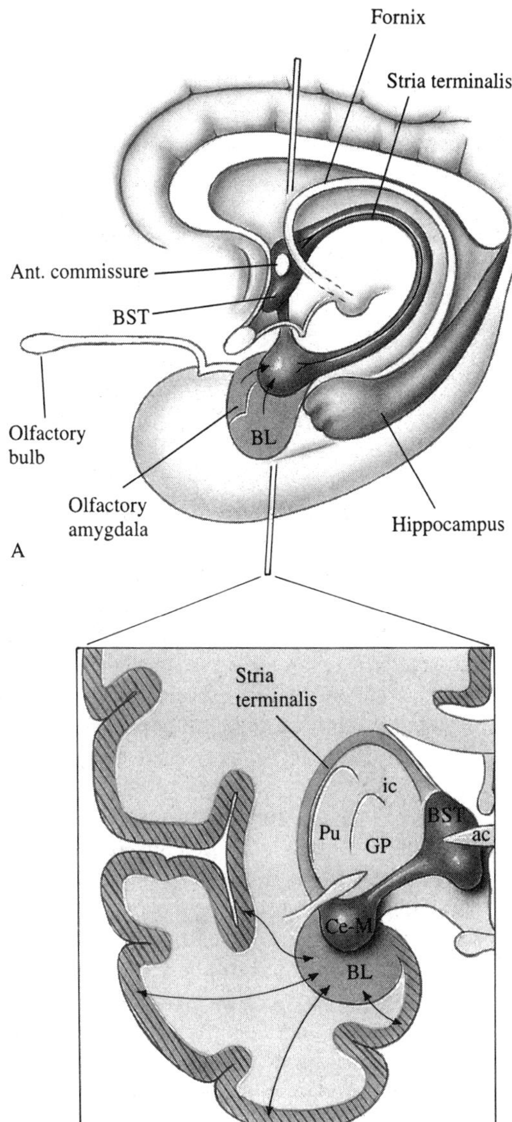

Fig. 20–1. The Amygdaloid Body and the Extended Amygdala
The *key figure* in the upper right corner indicates the location of the amygdaloid body in the rostral part of the temporal lobe, and the approximate location of the stria terminalis.
A. The topographic relations between the amygdaloid body, extended amygdala, hippocampus, and related pathways (stria terminalis and fornix) in a three-dimensional rendition of part of the right hemisphere.
B. Cross-section through the basal part of the hemisphere at the level of the anterior commissure (ac). Some of the prominent connections between the basolateral amygdala and the cerebral cortex are indicated by *arrows*. Abbreviations: BL = basolateral amygdaloid complex; BST = bed nucleus of stria terminalis; Ce-M = centromedial amygdaloid complex; GP = globus pallidus; ic = internal capsule; Pu = putamen. (Art by Medical and Scientific Illustration, Crozet, Virginia.)

thalamus. Another striking similarity between the basolateral amygdala and the rest of the cerebral cortex is that both structures receive direct cholinergic input from the basal nucleus of Meynert.

Olfactory Amygdala

Olfactory input reaches superficial parts of the amygdala, i.e., the cortical nuclear group, both directly from the olfactory bulb and indirectly from the olfactory cortex. The olfactory amygdala, in turn, projects to the centromedial amygdala (Fig. 20–1A) and to the hypothalamus (Chapter 12).

Centromedial Amygdala and Extended Amygdala

Although the centromedial amygdala and the rest of the extended amygdala also receive input direct from the cerebral cortex, it is largely restricted to fibers from the hippocampal formation, insula, and orbitofrontal cortex (Fig. 19–5). However, it is important to realize that the activities that are generated in widespread areas of the cerebral cortex can reach the centromedial amygdala via the basolateral part of the amygdala. A massive system of intra-amygdaloid association fibers ties together various parts of the amygdaloid body, and the majority of these fibers reach the centromedial amygdala and the rest of the extended amygdala from the olfactory amygdala and especially from the basolateral amygdala, which, as already indicated, receives a wealth of modality-specific and multimodal sensory input from the cerebral cortex. Other fibers in the stria terminalis tie together various parts of the amygdaloid body with the bed nucleus of the stria terminalis, which represents the medial expansion of the extended amygdala.

The centromedial amygdala does not project to the striatum; instead it projects, together with the rest of the extended amygdala, to many regions in the hypothalamus and brain stem (Fig. 19–5). It is through these pathways, which proceed both in the stria terminalis and in a massive ventral pathway, that the amygdaloid body can directly influence areas that generate various autonomic, endocrine, and somatomotor aspects of emotional behavior and motivational states or drives, e.g., eating, drinking, and sexual behavior (Chapter 19). Many of the descending projection systems are reciprocated. The number of peptidergic pathways is especially numerous between the extended amygdala and the hypothalamus and brain stem regions, especially the dorsal vagal complex including the nucleus of the solitary tract (see also Fig. 7–7), and the periaqueductal gray. Ascending monoaminergic projections, finally, terminate throughout the extended amygdala and to a lesser extent in the basolateral amygdala.

Functional Aspects

The Klüver–Bucy Syndrome; Role of the Amygdala in Sensory–Affective Associations

Whereas initially considerable attention focused on the hippocampal formation as a key structure in the "limbic system" theory of emotion (Chapter 21), it is now well established that the amygdaloid body is more prominently involved in emotional processing than the hippocampus. Bilateral destruction of the amygdaloid body and adjacent cortex in primates results in dramatic changes in emotional behavior referred to as the Klüver–Bucy syndrome. Following bilateral amygdalectomies, previously wild and fierce monkeys become tame and easy to handle, and they are willing to approach normally fear-inducing stimuli without display of anger or fear, i.e., affective blunting. They also show inappropriate sexual behavior and a compulsive tendency to put inedible objects in the mouth. If the amydalectomized monkeys are free-ranging rather than living in cages, the affective blunting is combined with complete social isolation rather than hyperorality, and the animals do not survive for long.

The various symptoms in the Klüver–Bucy syndrome seem to be related to a general defect that makes it difficult for the animal to relate sensory information to past experience, or to evaluate sensory stimuli in terms of their biologic significance. It is in this context of sensory–affective associations that the role of the amygdala in emotional and social behavior can best be appreciated. The sensory information that is channeled through the various cortical association areas is highly relevant in behavioral terms. In order to respond appropriately to this information, however, its affective relevance must be registered. Exactly how and where this takes place is not known, but it is clear that the amygdaloid body seems to be a crucial component in a functional system responsible for attaching significance to sensory information. Physiologic experiments have demonstrated many cells in the amygdala that respond to stimuli signifying impending positive or negative reinforcement. These cells do not respond if the stimulus has not been previously paired with these events. If this mechanism is destroyed by a lesion in or around the amygdaloid body, the response to sensory stimuli is likely to be inappropriate, as witnessed in the Klüver–Bucy syndrome, or for that matter in many cases of serious mental disorders (see Schizophrenia in Clinical Notes).

Significant changes in emotional behavior are also seen in humans with amygdaloid lesions, although humans do not usually become as emotionally blunted as monkeys. In the famous patient HM, for instance, the medial temporal lobes, including both amygdaloid bodies as well as major parts of the two hippocampal formations, were removed in an attempt to cure epilepsy (Chapter 21). Following the operation, the patient developed severe amnesia because of the destruction of hippocampus. In addition, and most likely related to the amygdaloid removal, he was also complacent, as reflected in the following evaluation: "He rarely complains about anything, even if he is unwell." However, he was not completely without emotion.

Results obtained in psychosurgery[1] directed at the amygdaloid body also reflect the involvement of the amygdala in emotions. In the 1950s, complete removal of the amygdaloid bodies, i.e., amygdalectomy, was attempted in efforts to treat schizophrenia. However, since this type of operation seldom relieved the schizophrenic symptoms, the procedure was soon abandoned. Following introduction of stereotaxic surgery in the 1960s, more restricted lesions of part of the amygdaloid body have been made in efforts to treat violent behavior or extreme forms of self-mutilation or psychoses. The results obtained have often included some loss of aggression or other affects, as reflected in the term "sedative surgery" for these types of operations.

Amygdaloid Body as Part of Distributed Anatomic Systems

Initially, the teaching of neuroanatomy tends to emphasize individual structures and their topographic relations. But cell groups and brain structures do not exist in isolation from the rest of the nervous system. Rather, they derive their importance from the fact that they represent essential parts of neuronal circuits that involve many other parts of the nervous system.

Amygdalocortical Circuits

As stated earlier, the amygdaloid body is closely related to many nearby cortical regions, especially in the temporal lobe, orbitofrontal cortex, and anterior cingulate gyrus. Assemblies of neurons in the amygdaloid body, therefore, are part of a variety of distributed networks or neuronal circuits involved in various functions, including aspects of human emotions. By virtue of its ability to attach significance or affective value to current events, the amygdaloid body undoubtedly is in a position to provide an emotional component to many learning experiences. Through its widespread cortical connections, the amygdaloid body can influence, not only the hippocampal mechanisms for establishing long-term memories (Chapter 21), but also the activities in cortical association areas, known to be of importance for higher cognitive functions including memory storage.

Amygdalohypothalamic and Amygdalo-Brain Stem Circuits

Although the amygdaloid body undoubtedly is a key structure for the mediation of various autonomic and endocrine responses accompanying emotional behavior, it is not the only structure involved. First, it is important to emphasize the fact that the centromedial amygdaloid nuclei are part of a continuum, i.e., the extended amygdala, that extends through the substantia innominata to the bed nucleus of the stria terminalis. It is reasonable to expect that the far-reaching anatomic similarities that exist between different parts of the extended amygdala (see above) imply functional similarities. In other words, if the centromedial amygdala is crucially involved in the modulation of autonomic and endocrine functions, these roles should apply also to the much larger continuum of which it is a part. Physiologic studies do provide support for this thesis.

There is a close and in large part reciprocal relationship between the extended amygdala and various autonomic and endocrine centers in the hypothalamus and brain stem. In general, the central amygdaloid nucleus and its extension into the bed nucleus of the stria terminalis are closely related to autonomic centers in the hypothalamus and brain stem; together, these areas form part of the central autonomic network (Chapter 19). The medial extended amygdala has especially close relations with the neuroendocrine parts of the hypothalamus (Chapter 18). Although the functional significance of many of these highly specific neuronal circuits is poorly understood, many studies have shown that a variety of autonomic and en-

1. The history of psychosurgery, i.e., the treatment of psychiatric symptoms by destruction of a portion of the patient's brain, is noted for its controversies. Nonetheless, Egas Moniz, who was also a pioneer in cerebral angiography, received the Nobel Prize in 1949 for the introduction of prefrontal lobotomy. The theoretical basis for prefrontal lobotomy was highly speculative, but the alleged purpose was to transect the projection fibers in the frontal lobe. The introduction of antipsychotic drugs, as well as the undesirable side effects following the operations, led to a rapid decline in psychosurgery during the second half of the 1950s. By that time, however, more than 40,000 patients had received some form of psychosurgical treatment in the United States alone.

Since some patients cannot be helped by pharmacologic agents or other forms of conventional treatments, psychosurgery is still used in a limited way. However, the field of psychosurgery has changed character with the introduction of stereotaxic operative procedures, and as a result of advances in our understanding of how different brain structures contribute to emotional behavior. The aim of modern psychosurgical methods is to sever specific fiber bundles, e.g., the cingulum (cingulotomy), or to destroy certain cell groups in, for instance, the amygdaloid body (amygdalotomy).

It is difficult to be completely objective when discussing psychosurgery. The prospect of destroying a part of a person's brain is frightening to most people, and some have adopted an uncompromisingly negative attitude in the sense that they condemn any type of psychosurgery. Occasionally, however, an emotional disorder may be resistant to any form of conventional therapy. As a noted neurosurgeon explains in reference to the patients for which he recommends psychosurgery: 'The brains of our patients are so disturbed that they only produce suffering."

The field of psychosurgery is loaded with difficult medical, ethical, and social considerations, but neither is conventional psychiatric pharmacologic therapy completely without hazards and deleterious side effects. One especially troubling problem relates to the development of a "late" and sometimes disabling movement disorder known as tardive dyskinesia following long-term treatment with some antipsychotic drugs.

docrine functions as well as emotional behavior and motivational drives can be modulated by the amygdaloid body and the extended amygdala.

Based on animal experiments, specific neuronal circuits including parts of the amygdaloid body, thalamus, and brain stem have also been identified for specific emotions, e.g., defensive and aggressive behavior and some forms of emotional conditioning, e.g., fear learning.

Clinical Notes

Epilepsy

Disturbances in medial temporal lobe structures, including the amygdaloid body, have long been known to play a pivotal role in *complex partial seizures* (psychomotor epilepsy or temporal lobe epilepsy), which is the most common form of epilepsy (see Clinical Notes, Chapter 22). Since the seizure activity sometimes originates in other lobes, including especially the frontal lobe, the term temporal lobe epilepsy is not always appropriate. Complex partial seizures are characterized by subjective phenomena such as auditory, visual, gustatory, or olfactory hallucinations, emotional experiences, and stereotyped movements, usually in the form of purposeless activities such as smacking the lips or fumbling with clothes. In the past, much attention focused on the hippocampus as the generator of seizures in this type of epilepsy (see Functional Aspects and Clinical Notes, Chapter 21). However, many of the most characteristic symptoms experienced by the patient, e.g., epigastric and other autonomic sensations as well as experiential phenomena ("like a real life experience") with affective qualities, often of fear, can most easily be explained with reference to the role of the amygdala in autonomic and emotional functions. In fact, an assessment of circumscribed pathologic lesions, e.g., tumors and vascular lesions causing complex partial seizures, suggests more frequent and more severe involvement of the amygdaloid body than of the hippocampus.

Although epilepsy can often be treated successfully with drugs, some patients suffer from intractable seizures that can only be managed with surgical therapy, i.e., excision of part of the temporal lobe. The question of whether the epileptogenic focus is located in the amygdala or the hippocampus becomes a matter of great significance for the neurosurgeon, who must decide the extent of the temporal lobe resection. If the focus is located in the amygdaloid body, it may be sufficient to resect only the anterior part of the temporal lobe and preserve most or all of the hippocampus, which is located just behind the amygdaloid body (Fig. 20–1). This would ensure minimal involvement of memory functions. The recent advances in neuroimaging, including MRI, have greatly improved the chances for a correct preoperative assessment of the location of the epileptogenic focus, which in turn is likely to lead to more accurate and successful surgical operations.

Schizophrenia

"Schizophrenia is arguably the worst disease affecting mankind, even AIDS not excepted" (editorial in the science magazine *Nature*, Vol. 336, Nov. 10, 1988, p. 95). It is a common and often catastrophic illness, which usually strikes a person in late adolescence or early adulthood, and which in many cases causes complete disability and lifelong suffering. It is characterized by a mixture of symptoms, including thought disorder, delusions (false beliefs or wrong judgments), hallucinations (false perceptions), blunted affect, lack of motivation, and social withdrawal. The etiology of schizophrenia is unknown, and it has been extraordinarily difficult to relate a specific brain lesion to this disease. However, evidence from recent imaging studies and postmortem examination of brains from schizophrenic patients indicate that pathologic processes in the medial temporal lobe, including the amygdaloid body and hippocampal formation, are of special importance in a significant number of schizophrenic patients. Recent evidence also indicates that the pathology may be consequent to developmental abnormalities, which is somewhat surprising considering the common onset of schizophrenia in late adolescence. It is generally agreed, however, that a genetic factor is involved.

As indicated above, the amygdaloid body is part of a system that is crucially involved in attaching emotional and biologic relevance to sensory stimuli. If this mechanism is destroyed or disturbed by a lesion in or around the amygdaloid body, the emotional response to external events and socially important stimuli may be diminished (blunted affect, emotional withdrawal) or distorted, e.g., in the form of excessive suspiciousness (paranoia), symptoms commonly seen in schizophrenia as well as in some cases of temporal lobe epilepsy. In either case, social interactions are likely to be severely disrupted.

When considering the pathophysiology of brain disorders, it is important to remember that lesions in one part of the brain tend to disrupt the activities also in other brain regions that are closely related to the damaged part. For instance, since the medial temporal lobe structures, which are implicated in a significant number of schizophrenic patients, are closely related to many other cortical areas in the temporal and parietal lobes, one should not be surprised that derangements of cognitive functions, e.g., distorted thinking and false perceptions, are prominent features in the schizophrenic syndrome. Furthermore, lesions in medial temporal lobe structures are likely to disrupt the activity in prefrontal and orbitofrontal cortices, which are characterized by prominent reciprocal connections with the medial temporal lobe. Prefrontal and orbitofrontal cortices, like the medial temporal lobe structures, have figured prominently in discussions of the pathogenesis of schizophrenia, especially since many of the symptoms in this disorder are reminiscent of symptoms following damage to the prefrontal-orbitofrontal areas (Chapter 22).

Lesions in the medial temporal lobe structures are likely to have a direct and significant impact, not only on nearby cortical regions including the prefrontal regions, but also on several subcortical structures. Foremost among these are the extended amygdala, the ventral striatopallidal system, and the basal nucleus of Meynert, which receive significant input from the hippocampal formation, basolateral amygdala, and related cortical areas in and around the parahippocampal gyrus. The involvement of these basal forebrain structures may have repercussions for many functions and involve a significant number of transmitter systems, including the dopamine system (see Emerging Concepts of Basal Forebrain Organization, Chapter 16). Finally, and although it appears that the most prominent pathology in schizophrenia is related to the medial temporal lobe, pathology in other interconnected structures, e.g., the prefrontal and orbitofrontal cortices, can possibly produce some of the symptoms consistent with medial temporal lobe pathology.

As a way of summarizing this difficult field where theories abound, it should be emphasized that schizophrenia has many faces; the symptoms vary from patient to patient, and even in the same patient over time. No single symptom is pathognomonic (distinctively characteristic) for schizophrenia; nor has a single brain structure, or set of structures, been consistently related to schizophrenia. Therefore, even if the primary pathology of this dreaded disease in many cases seems to be located in the medial temporal lobe and closely related areas, it is clear that the magnitude of the abnormality, as well as the structures affected by the lesion, can vary significantly from patient to patient.

Alzheimer's Disease

Although the temporal cortex, including the hippocampal formation, is generally the most severely affected region in Alzheimer's disease (see Clinical Notes, Chapter 21), cortical atrophy is usually widespread, especially in the late stages of the disease. The amygdaloid body is also a focus of considerable atrophy and other pathologic changes, which may well contribute to some of the emotional symptoms that characterize Alzheimer's disease.

Suggested Reading

1. Aggleton, J. P. (ed.), 1991. The Amygdala. Wiley-Liss: New-York.
2. Heimer, L., J. de Olmos, G. Alheid, and L. Zaborszky, 1991. "Perestroika" in the basal forebrain: Opening the border between neurology and psychiatry. Progress in Brain Research, Vol. 87, pp. 109–167. Elsevier: Amsterdam.
3. Martin, L. J., R. E. Powers, T. L. Dellovade, and D. L. Price, 1991. The bed nucleus–amygdala continuum in human and monkey. J. Comp. Neurol. 309:445–485.
4. Price, J. L., F. T. Russchen, and D. G. Amaral, 1987. The limbic region. II, The amygdaloid complex. In A. Björklund, T. Hökfelt, and L. W. Swanson (eds.): Handbook of Chemical Neuroanatomy. Vol. 5. Elsevier Science Publishers: Amsterdam, pp. 279–388.
5. Roberts, G. W., 1990. Schizophrenia: The cellular biology of a functional psychosis. Trends Neurosci. 13(6):207–211.
6. Valenstein, E. S., 1973. Brain Control. A Critical Examination of Brain Stimulation and Psychosurgery. John Wiley & Sons: New York.

Intraventricular (dorsal) aspect of the hippocampus. The temporal horn has been opened and the choroid plexus removed. (The photograph was kindly provided by Dr. Henri Duvernoy from his book, *The Human Hippocampus*, with permission of J.F. Bergmann Verlag, Munich.)

21

Hippocampal Formation

The hippocampus is a beautifully rounded cortical structure in the floor and medial wall of the temporal horn of the lateral ventricle. This by itself sets it apart from other cortical areas, which form the exterior surface of the brain rather than the ventricular surface. The hippocampus and neighboring regions of the parahippocampal gyrus form a larger functional–anatomic unit referred to as the hippocampal formation.

The hippocampus, with its relatively simple and orderly arranged histology, has long been a favorite research object for anatomists, physiologists, and molecular biologists. Its intrinsic circuits and extrinsic pathways, including a widely distributed and highly organized network of connections with other parts of the cerebral cortex, are known in great detail from animal experiments. Physiologic and behavioral experiments as well as careful case histories of humans suffering from lesions in the hippocampal formation have established its preeminent role in memory functions.

A Note on the Limbic System

Since the hippocampus is recognized as a key structure in the limbic system, a short review of the limbic system concept is appropriate.

In 1939, the American neuroanatomist James Papez proposed that "The hypothalamus, the anterior thalamic nucleus, the cingulate gyrus, the hippocampus and their interconnections, constitute a harmonious mechanism which may elaborate the functions of central emotion as well as participate in emotional expression." This was at the time a daring proposition, and it marks the beginning of the notion of the limbic system[1], which has become one of the most celebrated functional-anatomic systems in the brain. Papez emphasized a neuronal circuit that leads from the mammillary body to the anterior thalamic nucleus via the mammillothalamic tract, and then by way of the thalamocortical radiations to the cingulate gyrus. From the cingulate gyrus, impulses can reach the hippocampus formation via the cingulum bundle. Finally, the hippocampus, or rather the subiculum, projects to the mammillary body via the fornix. This multineuronal circuit, the *Papez circuit*, is by no means a closed loop, and most agree that it does not serve a specific functional purpose.

From a modest beginning, the limbic system has grown into a multifarious system that includes a large part of the basomedial telencephalon as well as diencephalic and mesencephalic structures believed to be of special significance for emotional and motivational aspects of behavior. Regular components of the limbic system include the cingulate and parahippocampal gyri, the hippocampus, the septum, and the amygdaloid body. These structures are to a considerable extent related to the hypothalamus, which is the main brain structure for the integration of various autonomic effects accompanying emotional expressions. The hypothalamus, therefore, is often included in the limbic system. Many other parts of the forebrain are more or less intimately associated with the above-mentioned structures. Such affiliated members of the limbic system include neocortical areas in the basal frontotemporal region, the olfactory cortex, the ventral parts of the striatal complex, the anterior and medial thalamic nuclei, and the habenula. Recently, there has been a tendency to widen the concept even further to include a number of brain stem areas that are directly related to the above-mentioned limbic forebrain structures. When and to what extent these various parts are included in the limbic system concept seems to be a matter of convenience and personal preference.

The notion of the limbic system has served as a powerful stimulus for a variety of scientific experiments, and it has especially attracted the attention of psychologists and psychiatrists, who see an opportunity to correlate emotional behavior and disorders with specific parts of the brain. Despite its popularity, however, the limbic system remains an enigma to many neuroscientists, and there is no generally accepted definition of the system. Often, the term "limbic" structure is used in a vague fashion to distinguish it from motor structures, or sometimes more specifically from basal ganglia, thalamic, or even cortical structures, although basal ganglia (e.g., accumbens), thalamic (e.g., medial and anterior thalamic nuclei), and cortical (e.g., hippocampus) structures are all usually considered integral parts of the "limbic system." Inherent in the concept of the "limbic system" or 'limbic circuits" is the notion that they subserve emotional

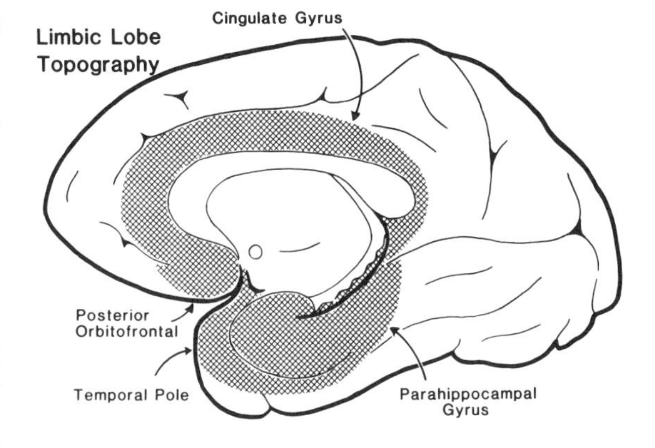

Fig. 21–1. The Limbic Lobe
Diagram of the medial surface of a primate brain depicting the topography of the limbic lobe (hatched area). (From Damasio, A. R. and G. van Hoesen, 1985. The limbic system and the localization of herpes simplex encephalitis. J. Neurol. Neurosurg. Psych. 48:297–301. With permission of the authors and J. Neurol. Neurosurg. Psych.)

1. The term "limbic" was used by Broca (1878), who noted that the cingulate and parahippocampal gyri, two of the main structures of the limbic lobe (Fig. 21–1), form a limbus or border around the diencephalon. The olfactory bulb and tract seemed to project to both the cingulate and the parahippocampal gyri, and he referred to all these structures as "le grand lobe limbique." Since Broca's limbic lobe was generally believed to subserve olfactory functions, it has also been referred to as rhinencephalon (smell brain). However, only a small part of the limbic lobe is directly related to the sense of smell (Chapter 12), and the olfactory system is not unique in this context; olfaction is only one of many sensory inputs to the limbic lobe. The term "rhinencephalon," therefore, is no longer used in this context.

functions. However, there is no evidence that the hippocampus is involved in emotional functions to any significant degree.

The "limbic system" may be a convenient term for many purposes, but the justifications for its existence are becoming increasingly contrived, and some experts contend that the notion of a "limbic system" has outlived its usefulness both as a scientific concept and as a catchword for clinical consumption. Others, who likewise do not find the concept very useful, especially since it has become rather obvious that the "limbic system" never functions as a unit, have suggested that the "limbic system" be limited to include only telencephalic "limbic" structures.

Clearly, the only thing that seems certain about the concept of the "limbic system" is that people will continue to use it in many different ways until we have learned more about the functional-anatomic systems in the basomedial forebrain. As reflected below, the anatomic and functional characteristics of the hippocampus are distinct enough to justify its presentation in a separate chapter, and as the main component of a memory system, rather than as a component of a "limbic system". As new basal forebrain systems, e.g., the ventral striatopallidal system and the extended amygdala (Chapter 16), as well as the central autonomic network (Chapter 19) have emerged, it will be more important to identify their intrinsic characteristics and their contribution to specific and functionally relevant neuronal circuits. In the meantime, it would seem appropriate to try to limit the use of the term "limbic system", especially since the notion of the "limbic system" tends to perpetuate an old-fashioned and simplistic anatomic viewpoint according to which the forebrain consists of two large territories represented by a neocortical–basal ganglia system and an allocortical–hypothalamic, "limbic"-related system.

Several of the so-called limbic structures display separate anatomic and functional identities, which makes it difficult to conceive of the limbic system as a functional unit. For instance, two of the key structures in the limbic system, the hippocampus and the amygdaloid body, have specific connections and functional significance, and in order to appreciate the unique qualities of these structures, it may be necessary to approach them individually rather than in the framework of the limbic system concept, which invites a preconceived view in regard to their anatomic affiliations and functional significance. As for the hippocampus, it can safely be stated that it is not devoted primarily to olfactory functions. Nor does it seem to be significantly involved in emotional mechanisms, as implied in the notion of the limbic system. The key structure in emotional behavior is the amygdaloid body (see next chapter), whereas the hippocampus and closely related cortical areas in the temporal lobe are increasingly recognized for the crucial role they play in memory functions. Experience from neurology, furthermore, indicates that the right hemisphere and maybe particularly the parietal lobe plays a role in emotions.

Hippocampal Formation

The hippocampus, dentate gyrus, and some closely related cortical areas in the parahippocampal gyrus, including the subiculum, are together referred to as the *hippocampal formation*.

The hippocampus (Ammon's horn) appears as a curved cortical structure in the floor and medial wall of the temporal horn of the lateral ventricle largely behind the amygdaloid body (see Vignette, Chapter 20). The surface facing the ventricle is the deepest part of a gyrus that has been folded inward in the region of the hippocampal fissure. The dentate gyrus, which is interlocked with the hippocampus, is a long narrow gyrus joined side-to-side to the hippocampus.

Histology

The hippocampus has a tortuous form anteriorly and posteriorly, and only in the midportion is it possible to get a good overview of the various hippocampal fields (Fig. 21–2). Three cortical layers are distinguished in the hippocampus. The structure can be divided into three fields, CA1, CA2, and CA3, on the basis of architectonic differences. There is basically only one cell layer, and the most conspicuous cells are the pyramidal cells, which have basal dendrites directed outward toward the ventricular surface, and apical dendrites extending in the other direction. The myelinated axons of the pyramidal cells pass to the ventricular surface, where they form a lamina of white matter, the *alveus*. CA1, which is the largest hippocampal field in the human brain, forms part of *Sommer's sector*. The other part of Sommer's sector is the subiculum. The pyramidal cells in Sommer's sector are especially sensitive to anoxia and other metabolic disturbances, and they are often affected in temporal lobe epilepsy (see Clinical Notes).

The dentate gyrus is typically crenated and it can be easily viewed during gross dissection of the human brain, where it is situated underneath the

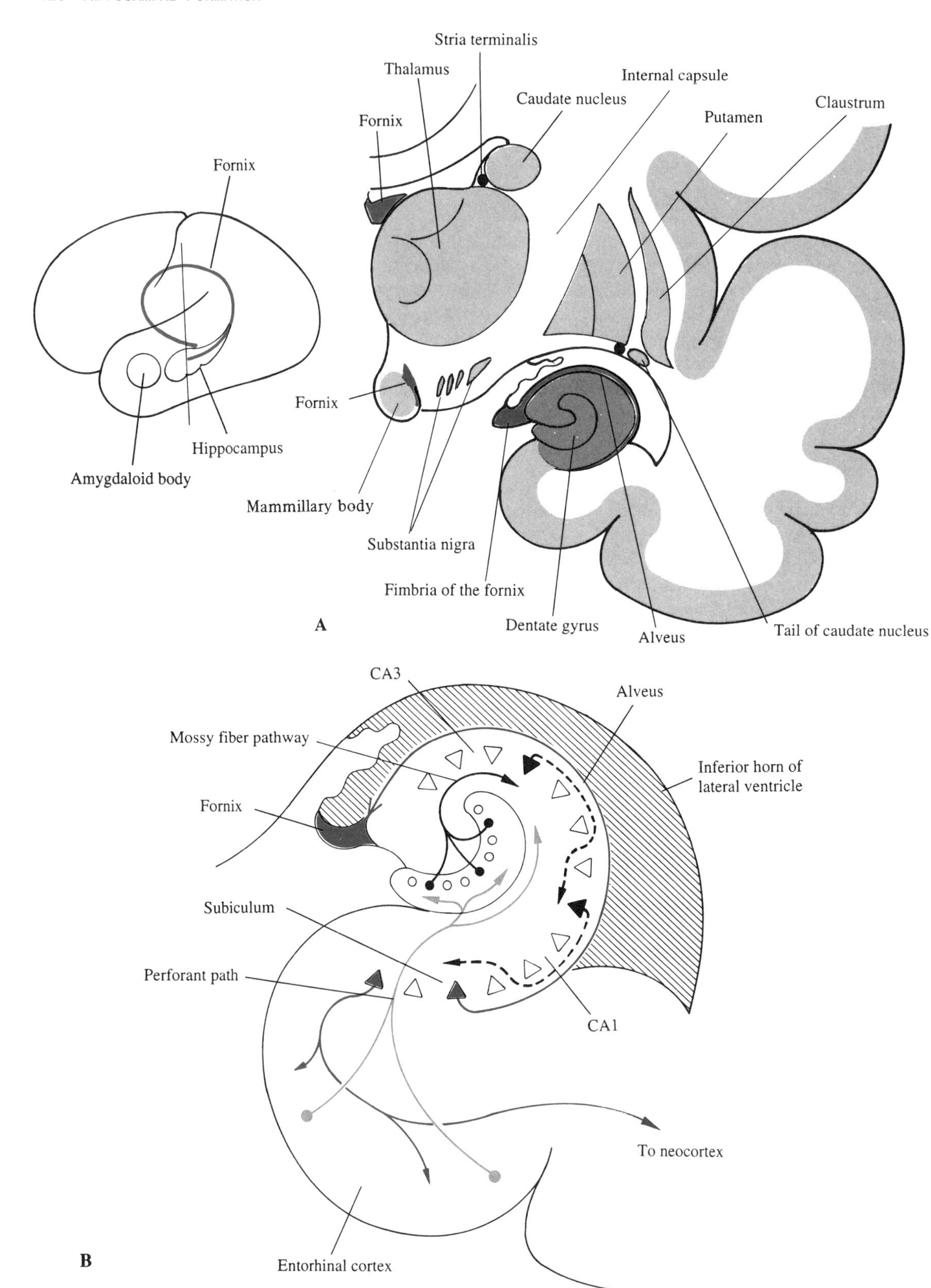
Stria terminalis
Thalamus
Internal capsule
Caudate nucleus
Claustrum
Putamen
Fornix
Fornix
Fornix
Hippocampus
Amygdaloid body
Mammillary body
Substantia nigra
Fimbria of the fornix
A
Dentate gyrus
Alveus
Tail of caudate nucleus
CA3
Alveus
Mossy fiber pathway
Inferior horn of
lateral ventricle
Fornix
Subiculum
Perforant path
CA1
To neocortex
B
Entorhinal cortex

←

Fig. 21–2. Hippocampus Formation
A. The key figure in the *upper left-hand corner* indicates the approximate level for the cross section in **A**. Note that the fornix (*red*) has been transected in three places at this level.
B. Hippocampus formation and its main connections. Efferent connections (*red*) from the subiculum reach both cortical and subcortical regions. Perforant path, *blue*; intrinsic hippocampal pathways, *black*. (Modified after an original drawing by G. van Hoesen.)

fimbria of the fornix on the medial aspect of the temporal lobe. Like the hippocampus proper, the dentate gyrus is a trilaminar structure in which the main layer is composed of small, round, densely packed granular cells. The granular cells give origin to an important association pathway, the *mossy fiber system*, through which the activity in the dentate gyrus is conveyed to the pyramidal cells in CA3, thus linking these two parts of the hippocampal formation together.

The hippocampus is directly continuous with the *entorhinal cortex* (Brodmann's area 28) of the parahippocampal gyrus through another extensive field called the *subiculum*. The subiculum gives rise to a significant part of the subcortical projections of the hippocampal formation, and both the subiculum and the entorhinal area are important way-stations for the transfer of information between the hippocampus and the rest of the cerebral cortex (see below). The entorhinal cortex can be easily identified in histologic preparations because its superficial cell layer is broken up into islands of large multipolar neurons. The islands are associated with wart-like bumps, which look like orange peel on the surface of the parahippocampal gyrus during gross anatomic dissection of the brain (Fig. 4–30). Although critical parts of the hippocampal formation are involved in Alzheimer's disease, the entorhinal area is usually the most heavily damaged cortical field in this devastating disorder (see below).

Connections

Extrinsic Connections

Cortical Connections. The hippocampus is intimately related to the rest of the cerebral cortex. Information from many cortical regions including the sensory cortical areas converges on the entorhinal cortex in the parahippocampal gyrus through a widespread system of anatomic pathways. The entorhinal cortex, in turn, gives origin to a massive association system, the *perforant path*, which terminates in the CA1 field of the hippocampus, subiculum, and dentate gyrus (Fig. 21–2B). The entorhinal area, therefore, serves as an important gateway for cortical input to the hippocampus. Most of the axons in the perforant path belong to large multipolar neurons in the superficial cell islands described above. The cortical afferents to the hippocampus are mirrored by projections from the hippocampus back to the rest of the cerebral cortex. This output route from the hippocampus proceeds either directly or through the subiculum to other parahippocampal areas, including the entorhinal area, and further on to association areas in all four lobes.

The intimate two-way relations that the hippocampus has with the rest of the cerebral cortex are consistent with its critical role in memory functions. Based on the input it receives from various cortical regions, the hippocampal formation performs a function that is essential for the formation of long-term memories, but the memories themselves are apparently stored more permanently in other cortical regions. In other words, both the processing and the storage of memories rely on an interchange between the hippocampal formation and the rest of the cerebral cortex, and the widespread connections between the hippocampal formation and the neocortical association areas provide the anatomic substrate for such interactions.

Subcortical Connections. The fornix is a well-known pathway containing both hippocampofugal and hippocampopetal fibers that connect the hippocampus with several subcortical structures. Whereas some of the axons in the fornix arise from pyramidal cells in the hippocampus, many come from the subiculum, and the axons proceed through the alveus to enter the fimbria of the fornix. At the level of the anterior commissure, many fibers leave the fornix bundle to reach the precommissural septum[2]. This part of the fornix is referred to as the *precommissural fornix*, and some of its fibers also reach the ventral striatum including the medial accumbens. The main part of the fornix descends

---

2. The septum is composed of two parts. The dorsal part, the *septum pellucidum*, is a thin lamina of glia and fibers that stretches between the anterior part of the corpus callosum and the fornix. This part is generally devoid of nerve cells. The ventral part of the septum, which is referred to as the precommissural septum or the "true" septum (*septum verum*), contains important cell groups located in the subcallosal area in front of the anterior commissure (Fig. 4–8).

behind the anterior commissure as the postcommissural fornix (Fig. 18–4), which terminates in the mammillary body. Other fibers reach the anterior thalamic nucleus.

As already indicated, a number of ascending fiber systems reach the hippocampus, and the most prominent of these systems, which originates in the region of the precommissural septum and the diagonal band of Broca, and which reaches the hippocampus through the fornix, contains both cholinergic and GABA-ergic fibers. This system is part of a widely distributed corticopetal neuronal system, which supplies all cortical regions with cholinergic, GABA-ergic, and other noncholinergic fibers (see Fig. 16–15). Additional subcortical input to the hippocampus formation comes from the monoaminergic cell groups in the ventral tegmental area (dopaminergic fibers), locus ceruleus (noradrenergic fibers), and raphe nuclei (serotonergic fibers). These various subcortical inputs are likely to have a modulatory effect on memory functions.

The amygdaloid body is another subcortical structure that is closely and reciprocally related to the hippocampal formation. Unlike the other subcortical connections, which form part of the fornix system, the amygdalohippocampal fibers take more direct routes through adjacent temporal white matter. The amygdalohippocampal connections are likely to provide part of the anatomic substrate for the well-known effect of emotion on memory funcions, and other interactions among these functions.

Intrinsic Connections

The mossy fiber pathway forms an important link in an intrinsic unidirectional neuron circuit that can be schematically outlined as follows (Fig. 21–2B). The perforant path fibers project to the dentate gyrus, which gives origin to the mossy fiber system. This system, in turn, projects to the pyramidal cells in CA3, which send collaterals to the CA1 pyramids. The axons of the CA1 pyramidal cells, finally, proceed to the subiculum. The latter, like the hippocampus itself, projects to the entorhinal cortex and related parahippocampal areas. As already indicated, axons of subicular neurons also form an important part of the postcommissural fornix projection to the mammillary body.

Another major intrinsic circuit excludes the dentate gyrus. As indicated in Fig. 21–2B, the perforant path (blue) projects not only to the dentate gyrus but also to the hippocampus proper (primarily CA1) and subiculum, thereby bypassing the dentate gyrus–mossy fiber system.

Functional Aspects and Clinical Notes

Memory Functions

Although a number of functions, including motivational and attentional mechanisms, aggressive behavior, inhibitory processes, as well as autonomic and endocrine activities, have been ascribed to the hippocampal formation through the years, there is little consensus and much speculation in regard to many of these functions. However, it is becoming increasingly evident that the main function of the human hippocampal formation is related to learning and memory[3] as reflected in the conscious recollection of specific events and facts ("knowing that"). This type of memory, i.e., *declarative or associative memory*, is impaired in amnesia, which is a memory disorder characterized by difficulties in learning new information and in recalling past experiences. *Procedural memory*, i.e., memory for skills and procedures ("knowing how"), is not dependent on the hippocampus but rather on corticostriatal circuits, and it is usually not affected in patients suffering from amnesia. Memory functions are also discussed in Chapter 22.

That the hippocampal formation is of special importance for memory was suggested at the turn of the century by the eminent Russian neuropathologist Vladimir Bekhterev, who at autopsy discovered bilateral softening of the hippocampal formation and adjoining temporal cortex in a patient who had suffered from profound memory disturbances (see Finger, 1994).

Fifty years later, the role of the hippocampus role in memory functions was clearly established (Scoville and Milner, 1957). HM, an epileptic patient, was suffering from intractable seizures, and in order to alleviate his symptoms, a neurosurgeon removed large parts of the medial temporal lobe including the amygdaloid body and most of the hippocampal formation on both sides. As a result of the operation, HM's epileptic condition improved considerably, but he became severely amnesic. Although his intellect was preserved, his inability to learn anything new or to remember and keep track of the events of daily life was so devastating that he needed constant supervision. On the basis of similar operations in other patients, it was also

3. Learning is the process whereby we obtain new knowledge, whereas memory refers to the process through which this knowledge is retained. There is considerable evidence that longterm memory involves structural changes in relevant neural circuits, especially alterations in the number and patterns of synaptic connections.

realized that the severity of the amnesia depended on the extent of the hippocampal removal. If the operations were restricted to the rostromedial parts of the temporal lobes, including primarily the uncus and the amygdaloid bodies, memory was not affected.

HM lost the ability to learn and form new memories (anterograde amnesia), and he also lost memory for events that occurred during the last few years before the operation (retrograde amnesia). However, his immediate or short-term memory was intact in the sense that he could retain a limited amount of information for less than a minute if not distracted, and he also retained old or remote memories. The hippocampal formation, therefore, does not seem to be essential for the initial information analysis taking place in various specialized brain systems, nor does it represent the final storage site for memories. Widely distributed networks in neocortical areas are likely to be involved in memory storage.

The many careful studies performed on HM and other patients with brain lesions, including some with lesions restricted to parts of the hippocampus, have convincingly shown that the hippocampal formation is an essential part of a neural system that participates in the formation of long-term memories. The parts of this functional–anatomic system for learning and memory include the entorhinal area and some other temporal cortical areas closely related to the hippocampus. Research with animals seems to indicate that the different structures included in the hippocampal formation (hippocampus, subiculum, entorhinal cortex) differ in how they contribute to learning and memory. Many clinical studies and a large number of animal experiments have also indicated the importance of several other brain structures in the context of memory, including the mediodorsal thalamic nucleus and the prefrontal cortex. Neuronal systems in the basal forebrain, including cholinergic and GABA-ergic cell groups that are closely related to the hippocampus, mediodorsal thalamus, and prefrontal cortex, have also been implicated in memory. Although it is still not possible to present a coherent picture of how these various structures fit into an overall memory system of the brain, it is reasonable to assume that each part makes a different contribution. For instance, as indicated in Chapter 22, the prefrontal cortex is important for awareness of our intents to order and produce sequences in our patterns of behavior. In other words, it is memory for the order of things the student is going to do after the class, but which is not retained on a long-term basis. This type of memory is referred to as working memory.

The amygdaloid body is another structure that is sometimes implicated in learning and memory, but it seems important to make a distinction between the amygdaloid body and the hippocampus formation in this context. There is no convincing evidence that damage limited to the amygdaloid bodies produces amnesia. As discussed in more detail in the next chapter, the amygdaloid body is a key structure in emotional behavior, and it seems to be involved in memory functions to the extent that the events have an emotional attribute. For instance, it has been shown in animal experiments that the amygdaloid body is part of the neuronal circuitry for emotional conditioning, e.g., fear learning. In the context of memory functions, therefore, the amygdaloid body, like many other parts of the brain, contributes to specific, nondeclarative memory functions, but does not appear to be dedicated to associative memory functions, i.e., the conscious recollection of facts and data.

Alzheimer's Disease

Alzheimer's disease, which has received so much attention during the last decade, is the most common dementing disorder. It is characterized foremost by cognitive impairments, often accompanied by agitation and emotional outbursts. Memory defects are especially common, and in the early stages of the disease, they present themselves as confusion and difficulties in remembering events in daily life. Considering the importance of the hippocampus in memory function, it is not surprising to learn that pathologic changes appear early in the hippocampal formation, and they are especially prominent in the posterior part of the hippocampal formation. In the end stages of Alzheimer's disease, when the patient is severely demented, there is widespread cortical atrophy, especially in the temporal lobe, parietal lobe, and prefrontal cortex (Fig. 21–3A).

Alzheimer's disease is a neurodegenerative disorder, and the most characteristic pathologic features are cell loss, intraneuronal neurofibrillary changes, i.e., neurofibrillary tangles (Fig. 21–3B), and extracellular amyloid-containing clusters of degenerating terminals, i.e., senile plaques. The pathologic changes appear most pronounced in the entorhinal area and in Sommer's sector (CA1 and subiculum), with the result that the hippocampus is effectively disconnected from the rest of the brain including the cerebral cortex (Fig. 21–2).

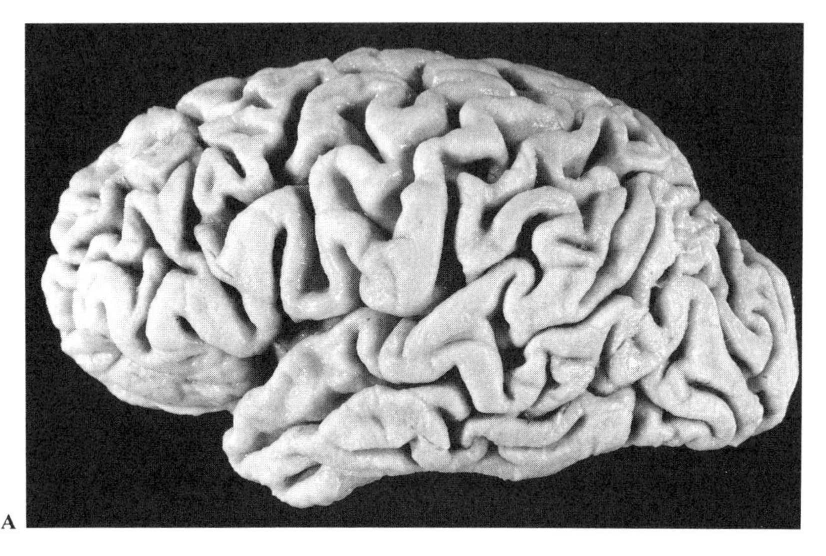

A

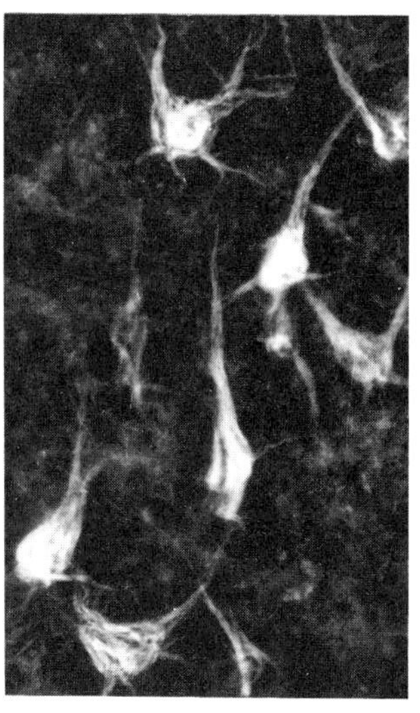

B

Fig. 21–3. Alzheimer's Disease
A. The brain of a 71-year-old Alzheimer's disease patient who was severely demented but had relatively well-preserved sensory, motor, and language functions at death. Note that whereas the central parts of the hemisphere (including pre- and postcentral gyri as well as Broca's area in the inferior frontal gyrus) are relatively well preserved, the prefrontal cortex and the parieto-temporal association areas are very atrophic, as is the limbic lobe.

B. Neurofibrillary tangles from layer 4 of the entorhinal cortex from the brain shown in **A**. The abnormal filamentous changes in the cell bodies and proximal dendrites, stained with thioflavin-S, mark the place where viable pyramidal neurons used to be.
(**A and B**, courtesy of Dr. Gary van Hoesen, Iowa University.)

Although it has been suggested that Alzheimer's disease may result from toxic substances, e.g., aluminum silicate, entering the brain through the olfactory receptors (Chapter 12), clear evidence for such a mechanism is still lacking.

Schizophrenia

Together with the prefrontal cortex, the medial temporal lobe structures, including the hippocampal formation and amygdaloid body, are currently the focus of much interest in schizophrenia (see Clinical Notes, Chapter 20).

Hippocampal Sclerosis and Epilepsy

Severe depletion of neurons with pronounced shrinkage of the tissue, especially in the CA1 and subicular fields (Sommer's sector), is referred to as *hippocampal* or *Ammon's horn sclerosis*. Although hippocampal sclerosis may be a focus in producing seizures, it is as likely to be a consequence of other abnormalities. Hippocampal sclerosis is often associated with onset of epilepsy at an early age, when the hippocampal neurons seem to be especially sensitive and likely to be damaged by metabolic hyperactivity occurring in epileptic seizures. Furthermore, sclerosis of the amygdaloid body and surrounding areas in the parahippocampal gyrus is also prominent in epileptic patients, and since many of the symptoms in complex partial seizures (see Clinical Notes in Chapter 22) reflect disturbances in autonomic and emotional functions related to the amygdaloid body, the seizure-inducing pathology can as well be located in the amygdaloid body (see Clinical Notes in Chapter 20). This is a classic "chicken or the egg" question whose details remain an enigma. Suffice it to say that dysregulation of excitatory amino acids may create conditions of excitotoxicity and lead to neuron death and dysfunction, resulting clinically in seizure activity and epilepsy.

Suggested Reading

1. Amaral, D. G. and R. Insausti, 1990. Hippocampal formation. In G. Paxinos (ed.): The Human Nervous System. Academic Press: San Diego, pp. 711–755.
2. Duvernoy, H. M., 1988. The Human Hippocampus: An Atlas of Applied Anatomy. J. F. Bergmann Verlag: Munich.
3. Finger, S., 1994. Origins of Neuroscience: A History of Explorations into Brain Function. Oxford University Press: New York, pp. 349–368.
4. Scoville, W. B. and B. Milner, 1957. Loss of recent memory after bilateral hippocampal lesions. J. Neurol. Neurosurg. Psychiat. 20:11–21.
5. Squire, L. R., 1987. Memory and Brain. Oxford University Press: New York.
6. van Hoesen, G. W., 1982. The parahippocampal gyrus: New observations regarding its cortical connections in the monkey. Trends Neurosci. 5(10): 345–350.
7. Zola-Morgan, S., L. R. Squire, and D. G. Amaral, 1986. Human amnesia and the medial temporal region: Enduring memory impairment following a bilateral lesion limited to field CA1 of the hippocampus. J. Neurosci. 6:2950–2961.

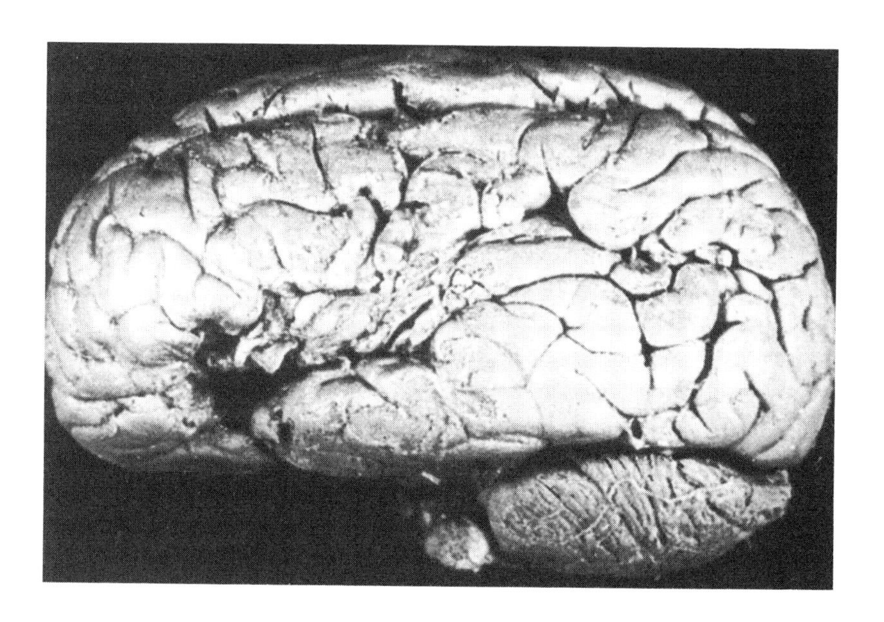

Vignette. **Broca's Speech Area**
Photograph of the brain of Paul Broca's patient called "Tan." (With permission from the Musée Dupuytren; see Finger, S., 1994. *Origins of Neuroscience*.)

22

The Cerebral Cortex and Thalamus

The cerebral cortex is intimately related to most major structures in the brain and spinal cord, and it has reached an enormous degree of development in the human brain. In general, the organization of the cerebral cortex is characterized by a mixture of laminar and vertical (columnar) features. Its intrinsic organization is very complicated, but thanks to the histotechnical revolution which has taken place during the last 20 years, it has now become possible to elucidate some of the local circuit connections in the cerebral cortex.

Sensory impulses impinge on the cerebral cortex, where they are processed and transformed as they become part of the mechanisms responsible for perception and ultimately translated into action, if needed. Indeed, a variety of both corticocortical and corticosubcortical neuron circuits are being exploited in the process of preparing for an appropriate response to the information received. The cerebral cortex is also responsible for cognition and thoughts and it is instrumental in the elaboration of various types of intellectual activities, including learning, memory, and language functions.

The development of the cerebral cortex and the topographic anatomy of the cerebral lobes and their convolutions were discussed and depicted in Chapters 2 and 4, and this chapter will focus attention on the intrinsic organization and extrinsic connections of this complicated structure, and on some of its functional and clinical aspects. The thalamus will also be discussed in the context of the thalamocortical relations.

Almost half of the brain's volume is represented by the cerebral cortex, which makes it by far the largest structure in the brain. The cerebral cortex can be subdivided into *isocortex* and *allocortex*. Isocortex is characterized by six layers, whereas allocortex ("other cortex") exhibits a more variable laminar pattern, usually between three and five layers. Allocortex can be further subdivided into *paleocortex* (olfactory cortex, Chapter 12) and *archicortex* (cortex of the hippocampus formation, Chapter 21). The isocortex, which occupies 95% of the cerebral cortex, is phylogenetically the youngest cortex and is therefore referred to as *neocortex*.

Structure of the Cerebral Cortex

Cortical Layers

The cerebral cortex is a 2- to 4-mm-thick, highly convoluted sheath of gray substance, of which more than half is hidden in the depth of the various sulci and gyri. The nerve cell bodies of the cerebral cortex are disposed in more or less well-defined layers as shown by Nissl stains. Although the arrangement of these layers varies in different parts of the cortex, it is usually possible to distinguish six different layers throughout the neocortex (Fig. 22–1).

1. The *molecular layer* is a fiber-rich superficial layer with few cell bodies. It contains a meshwork of horizontally running axons and the tufts of apical dendrites from pyramidal cells whose cell bodies are located in deeper layers.

2 and 3. The *external granular* and *external pyramidal layers* contain mostly small and medium-sized pyramidal cells. Each pyramidal cell has an apical dendrite, which extends toward the surface of the cortex, and several basal dendrites that arborize at the base of the cell body (Fig. 22–1). The pyramidal cells in these two layers give rise to the association and commissural fibers.

4. The *internal granular layer* is composed of many stellate (star-shaped) cells in addition to pyramidal cells. Its upper part contains a mixture of pyramidal and stellate cells, whereas the deeper part contains mostly stellate cells. The deeper part, which is especially well developed in primary sensory areas (e.g., primary visual cortex, Chapter 13), is the major recipient of the thalamocortical fibers. Layer 4 also contains a large number of horizontally running myelinated fibers, the outer band of Baillarger, referred to as the line or "stripe" of Gennari in the primary visual cortex: thus the term *striate cortex*.

5. The *internal pyramidal layer* contains a mixture of cell types, including medium-sized and large pyramidal cells, which has given the layer its name. This is the layer that contains the giant pyramidal cells of Betz in the motor cortex. The pyramidal cells of this layer give rise to most of the corticofugal fibers to subcortical regions except the corticothalamic fibers, which originate in layer 6.

6. The *multiform layer* contains a variety of stellate and pyramidal cells in addition to a large number of cells with elongated cell bodies, called *fusiform cells*.

Granular and Agranular Cortex

The cortical areas that receive the primary sensory pathways are characterized by a granular type of cortex, in which layers 2 and 4 are especially well developed and contain a large number of small cells, whereas it is difficult to recognize layers 3 and 5 with pyramidal cells. In the motor cortex, which typifies agranular cortex, layers 2 and 4 are poorly developed, whereas the pyramidal layers are well defined. The granular and agranular types of cortex represent the two extremes in a continuum of structural types.

Brodmann's Map

Although all cortical areas contain the same types of cells and the same general organization and intrinsic circuits (see below), different parts of the cerebral cortex show variations on a common cortical design, e.g., in the form of relative thickness or concentrations of different cell types. These variations can usually be correlated with a specific pattern of afferent and efferent connections, which means that different types of data are being processed in the different areas, and the result of this processing, which is basically the same in all parts of the cerebral cortex, is disseminated to different regions of the nervous system.

On the basis of the regional variations in cytoarchitecture, the cerebral cortex can be divided into different areas; the best known of these cytoarchitectural maps is the one published by the German histologist Korbinian Brodmann in 1909 (Fig. 22–2). He identified 52 different regions, which he

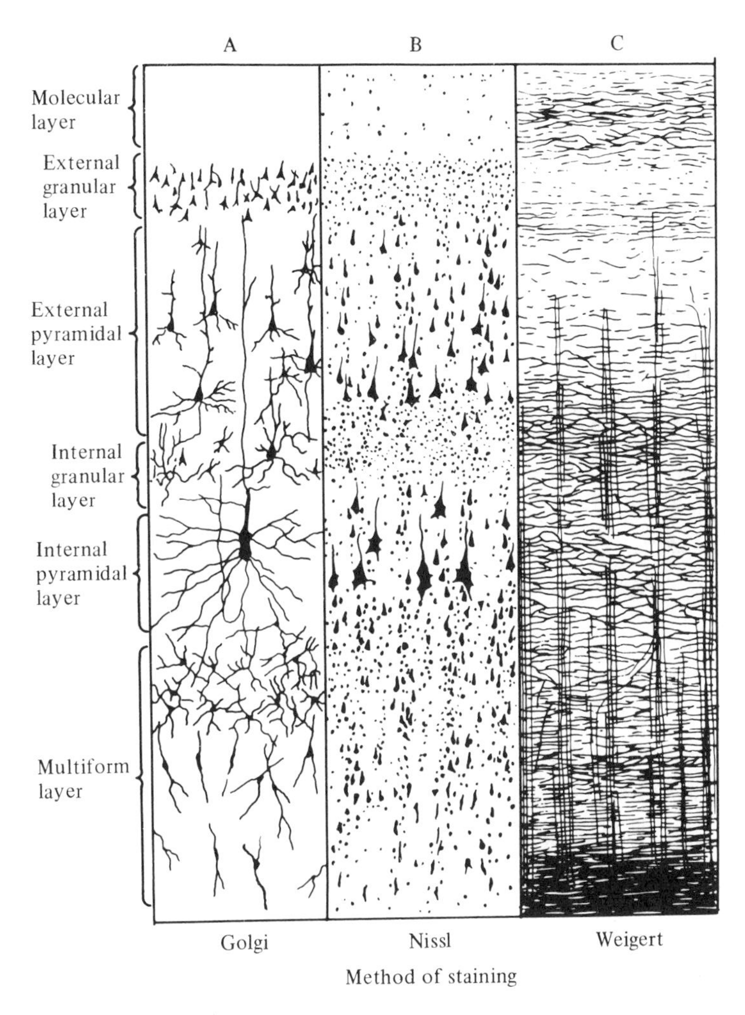

Fig. 22–1. Cortical Layers (after Brodmann)

numbered from 1 to 52 in the order in which he studied them. With a few exceptions, the borders between Brodmann's areas are not sharp. Nonetheless, several of Brodmann's areas have been found to be related to specific functional systems, and Brodmann's numbers are of great interest to clinicians and scientists, who often use the numbers for reference purpose. For instance, areas 3, 1, and 2 represent the somatosensory cortex (Chapter 9); areas 4, 6, and 8 the motor cortices (Chapter 15); areas 17, 18, and 19 the visual cortex (Chapter 13); and areas 41 and 42 the auditory cortex (Chapter 14).

Cortical Columns

Whereas the cortical layers represent a tangential or longitudinal arrangement, other histologic features reflect a vertical (perpendicular to the surface) organization. Neuronal cell bodies often seem to be aligned in a columnar fashion in Nissl-stained sections, and apical dendrites and many myelinated axons, either solitary or in bundles, tend to have a vertical orientation. Investigations of interneuronal connections also suggest the existence of considerable communication in the vertical direction, i.e., between nearby cells in different layers. Physiologic studies, likewise, have indicated the presence of a vertical organization in the form of cell columns, in which the cells in a given column share similar functional attributes. Columnar organization, therefore, seems to be a principle of great functional importance.

Functional columns were first described in the somatosensory cortex, where they are a few hundred µm wide and consist of hundreds to thousands of neurons. These functional units are activated by specific stimuli. For instance, some columns are activated by touch, others by position of a joint, etc.

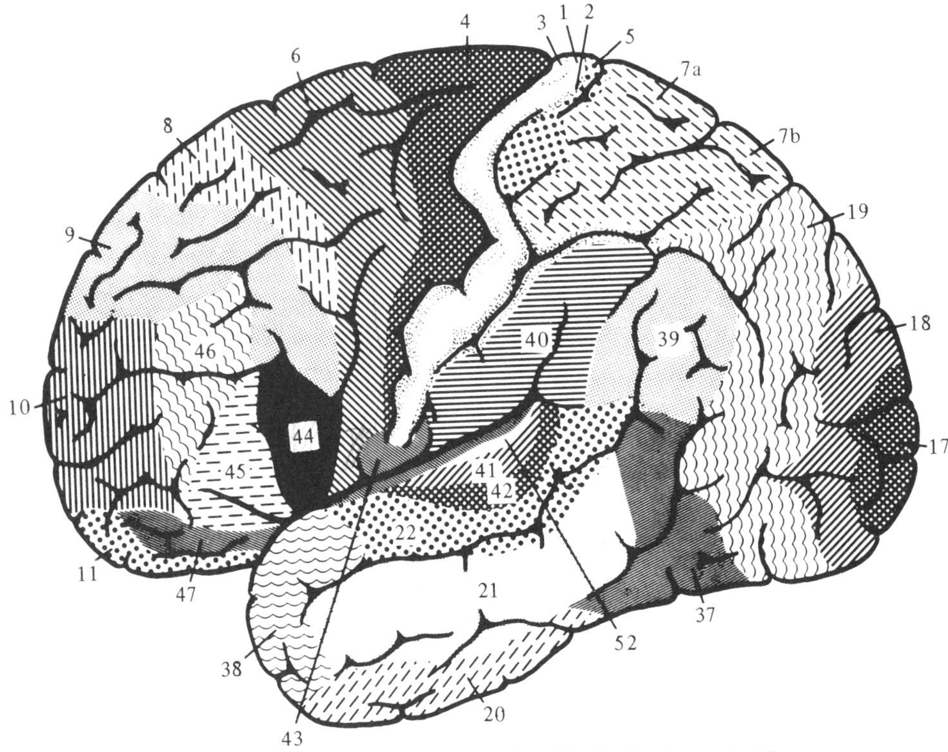

Fig. 22–2. Brodmann's Map
From Bindman, L. and O. Lippold, 1981. The Neurophysiology of the Cerebral Cortex. (With permission from University of Texas Press.)

Since the neurons within a single column share a certain number of functional properties, a vertical column appears to represent an elementary functional unit, which in the case of the somatosensory cortex is both place- and modality-specific. Functional columns in the visual cortex, e.g., ocular dominance and orientation columns, are described in Chapter 13.

The discovery of functional columns, or slabs, was greeted with great enthusiasm, and for good reasons. Functional columns, which have been discovered in both sensory and motor areas and even in a few other areas where some information exists about the receptive field properties, have demonstrated that the cerebral cortex is characterized by a more far-reaching segregation of functions than anyone could have imagined. The attributes of the sensory environment can be understood in terms of emergent properties, such as orientation selectivity, which are manifest at the level of the response specificity of individual cells and at the level of the columnar functional architecture of the cortex. Why cells sharing similar properties are grouped into columns is probably related to the mechanisms by which these properties are generated and the way they are integrated at subsequent stages of processing. It is likely that such grouping allows the cortex to minimize the total length of axons interconnecting cells.

Cortical Neurons and Intrinsic Circuits

The cerebral cortex contains many billions of neurons, and the intrinsic circuitry is extremely complicated. A multitude of interneurons, collateral pathways, and synaptic relationships provide a seemingly unlimited number of possibilities for impulse transmission. Even an optimist may be inclined to wonder if a universal principle of cortical circuitry exists, and if it does, will it ever be possible to elucidate the fundamental plan of cortical circuitry? However, what seemed like an impossible task only a few years ago is now within the bounds of possibility thanks to the histotechnical revolution that has taken place during the last 20 years (Chapter 7). Another reason for optimism is the fact that there are many striking similarities between neocortical regions in spite of the cytoarchitectonic differences. For instance, the number of neurons is surprisingly constant from region to region (except for the primary visual cortex, which contains two times as many neurons as other cortical areas), and the proportion of excitatory and inhibitory neurons is roughly the same in all parts of the neocortex. There is also a striking uniformity in the way the extrinsic afferent and efferent pathways relate to various cortical layers, as described in more detail below.

There are basically two types of neurons in the cerebral cortex, *spiny* and *aspiny* neurons. Those with spiny dendrites are either pyramidal or stellate (starlike) in shape. They are generally considered to be excitatory in nature, using glutamate or aspartate as transmitters. Pyramidal neurons (Fig. 22–3), which are present in all cortical layers except layer 1, represent the main output neuron of the cerebral cortex, but they are also characterized by extensive axon collateral systems, which are part of the intrinsic circuits. The spiny stellate neurons (Fig. 22–4) are present only in layer 4, and their axonal arborizations are usually confined to the cortex.

The neurons with aspiny dendrites are referred to as *smooth neurons*, of which there are many types with different dendritic and axonal branching patterns. All smooth neurons are believed to be inhibitory interneurons, and most of them use γ-aminobutyric acid (GABA) as their transmitter, often in combination with one or several neuropeptides (e.g., somatostatin, substance P, vasoactive intestinal polypeptide or cholecystokinin).

A summary of the most significant intracortical connections is shown in Fig. 22–5. The example is taken from the primary visual cortex, which has been studied in considerable detail by many research groups using various combinations of anterograde and retrograde tracer methods, intracellular labeling, and immunohistochemistry on both the light and the electron microscopic levels. An especially useful method of investigation has been the anatomic study of intracellularly labeled neurons, whose physiologic properties were examined before the injection of the intracellular tracer, e.g., horseradish peroxidase (HRP).

Although input to the cerebral cortex comes from a variety of sources, including the basal forebrain cholinergic system as well as the various

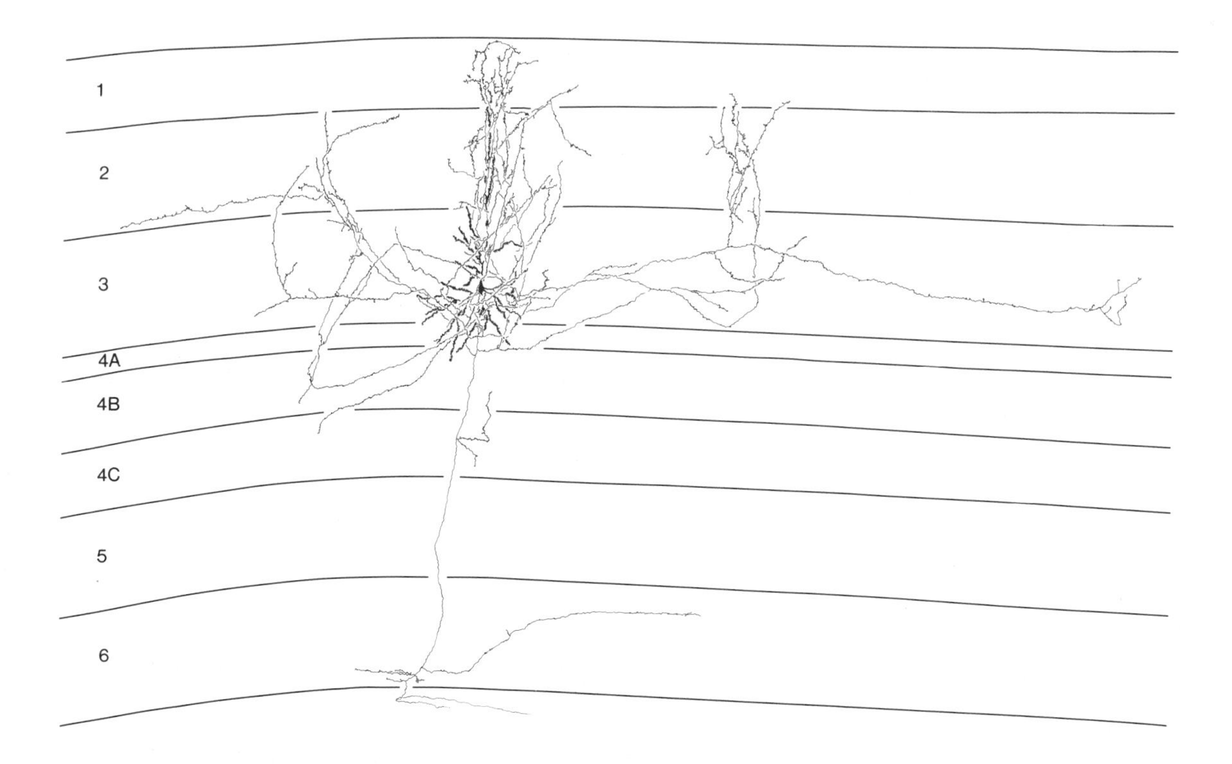

Fig. 22–3. Cortical Pyramidal Cell
Pyramidal cell from monkey striate cortex. Note the horizontal axons and their terminal clusters. (From Gilbert, C. D., 1992. Horizontal integration of cortical dynamics. Neuron 9:1–13. With permission from the author and Cell Press, Cambridge, MASS.)

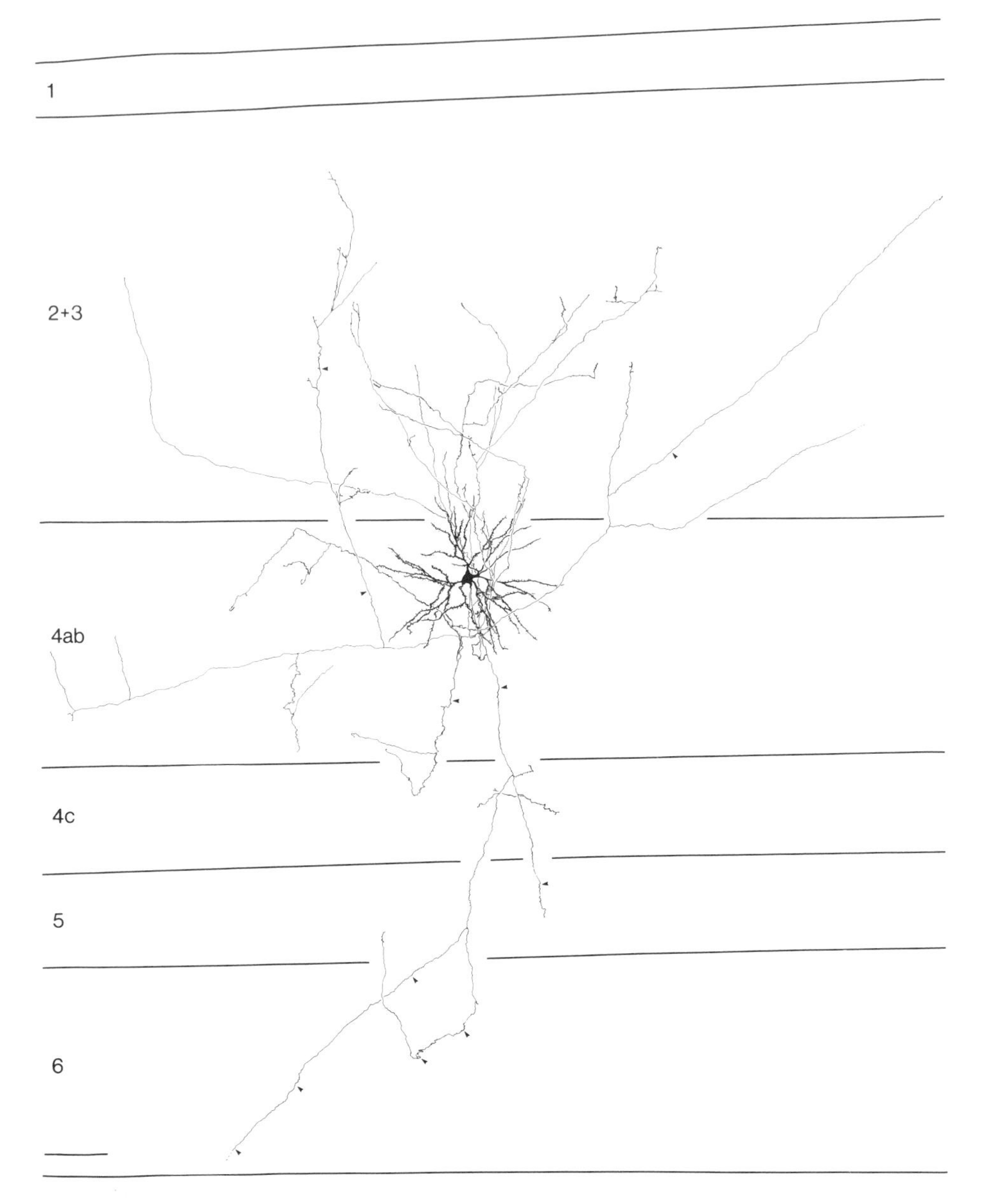

Fig. 22–4. Cortical Spiny Stellate Cell
Spiny stellate cell from cat striate cortex. (From Gilbert, C. D. and T. N. Wiesel, 1979. Morphology and intracortical projections of functionally characterized neurons in the cat visual cortex, Nature, Vol. 280, 12 July. With permission from the authors and Macmillan Journal, LTO, London.)

monoamine systems, the thalamocortical system represents the major extrinsic input to the cerebral cortex. The thalamocortical fibers, which terminate in layer 4 and to some extent in layer 6 and in the superficial layers, establish excitatory synaptic contacts primarily on the spines of the spiny stellate cells and pyramidal cells. They also excite smooth (inhibiting) neurons in layer 4. It is estimated that each thalamic fiber establishes synaptic contact with several thousand neurons.

Cells in layer 4 generally project to layers 2 and 3, whose neurons in turn project to layer 5 cells. Axon collaterals of layer 5 cells reach layer 6 cells, which project back to layer 4, thereby closing an important cortical loop (indicated by the four numbered black arrows in Fig. 22–5). Since at each stage in this circuit some additional processing takes place, the response properties become increasingly complex and specialized. Note that the pyramidal neurons that form part of this intrinsic

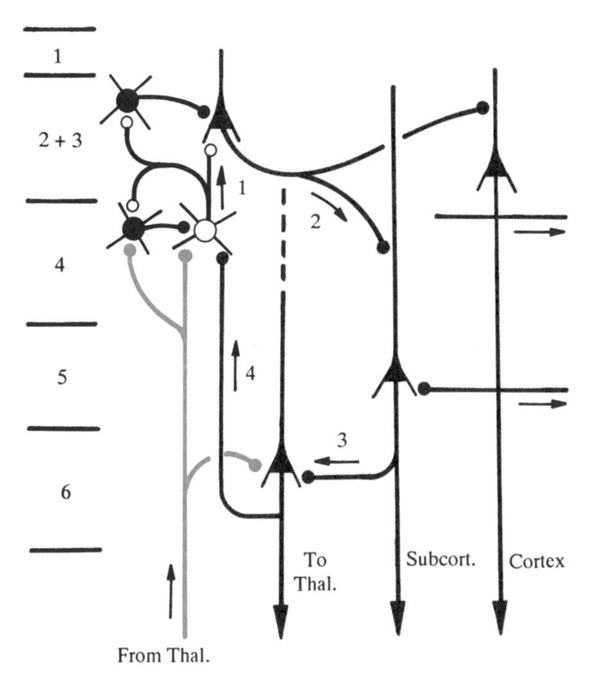

Fig. 22–5. Intrinsic Cortical Circuits
Highly simplified drawing illustrating the main types of cortical neurons and one of the basic intrinsic circuits within a functional column. The excitatory cells are represented by pyramidal neurons (*red*) and spiny stellate neurons (*open circle*). Inhibitory interneurons are indicated by *solid black circles*. The four numbered arrows identify one of many important intrinsic loops, which can be activated by the thalamocortical input to layer 4. (The sketch was prepared by Dr. Charles Gilbert at Rockefeller University.)

loop are also projection neurons in the sense that their main axon projects to other cortical and subcortical regions. For instance, descending fibers from pyramidal cells in layers 2 and 3 reach other cortical areas, either on the same side (association fibers) or in the opposite hemisphere (commissural fibers). Layer 5 pyramidal cells project to various subcortical regions including the basal ganglia, brain stem and spinal cord, whereas the axons from layer 6 pyramidal neurons project to the thalamus.

It is important to recognize that the intrinsic loop emphasized above is only one of several local circuits that can be identified even in the highly simplified diagram shown in Fig. 22–5. In addition to the many connections that provide for communication in a vertical direction, there are also tangential (horizontal) connections in both the superficial and the deep cell layers. Inhibitory interneurons provide for lateral inhibition, which serves to enhance contrast and sharpen boundaries between stimulated and nonstimulated areas in the receptor surface. Excitatory tangential intrinsic fibers, which cover distances of several millimeters, serve to integrate activities in different columns, e.g., by linking distant cell groups with similar orientation preference in the visual cortex. This important fact is indicated by the axon collaterals that emanate from the pyramidal neuron in the upper right corner of Fig. 22–5.

In summarizing this highly simplified account of the intrinsic organization of the cerebral cortex, one might say that the intrinsic cortical circuits include combinations of vertical and tangential connections. The vertical connections, which serve to tie together neurons in different layers into functional columns, appear to take care of the analysis of local attributes, whereas the horizontal connections deal with lateral integration for more global properties.

Extrinsic Connections Including the Thalamocortical Relations

Corticofugal and Corticopetal Connections

The multitude of corticofugal projections is astounding, and it reflects the fact that the cerebral cortex influences practically every functional system in the brain and spinal cord. Most major subcortical structures receive corticofugal projections. Examples of corticofugal pathways include the descending projections to striatum, brainstem and spinal cord, which have already been described in Chapters 15 and 16. All parts of the cortex are also characterized by projections to different parts of the thalamus (see below). Many of the corticofugal projections pass through the internal capsule.

The corticopetal projections are no less impressive. Foremost among these are the various thalamocortical pathways, many of which have been described as part of the sensory systems (Chapters 9, 13, and 14), or as links in basal ganglia or cerebellar projections to the cerebral cortex (Chapters 16 and 17). Important "extrathalamic" corticopetal pathways include the monoaminergic systems and the cholinergic basal forebrain projection system, which were described in some detail in Chapters 10 and 16. Their corticopetal components are briefly summarized below under the heading Extrathalamic Corticopetal Modulatory Systems.

The amygdaloid body, finally, is also related to many parts of the cerebral cortex, including the hippocampal formation, by a widespread system of reciprocal connections (Chapter 20).

Thalamus and Thalamocortical Relations

The *thalamus*, which is located underneath the lateral ventricle, is an egg-shaped nuclear mass (Fig. 22–6C) that has been divided into more than

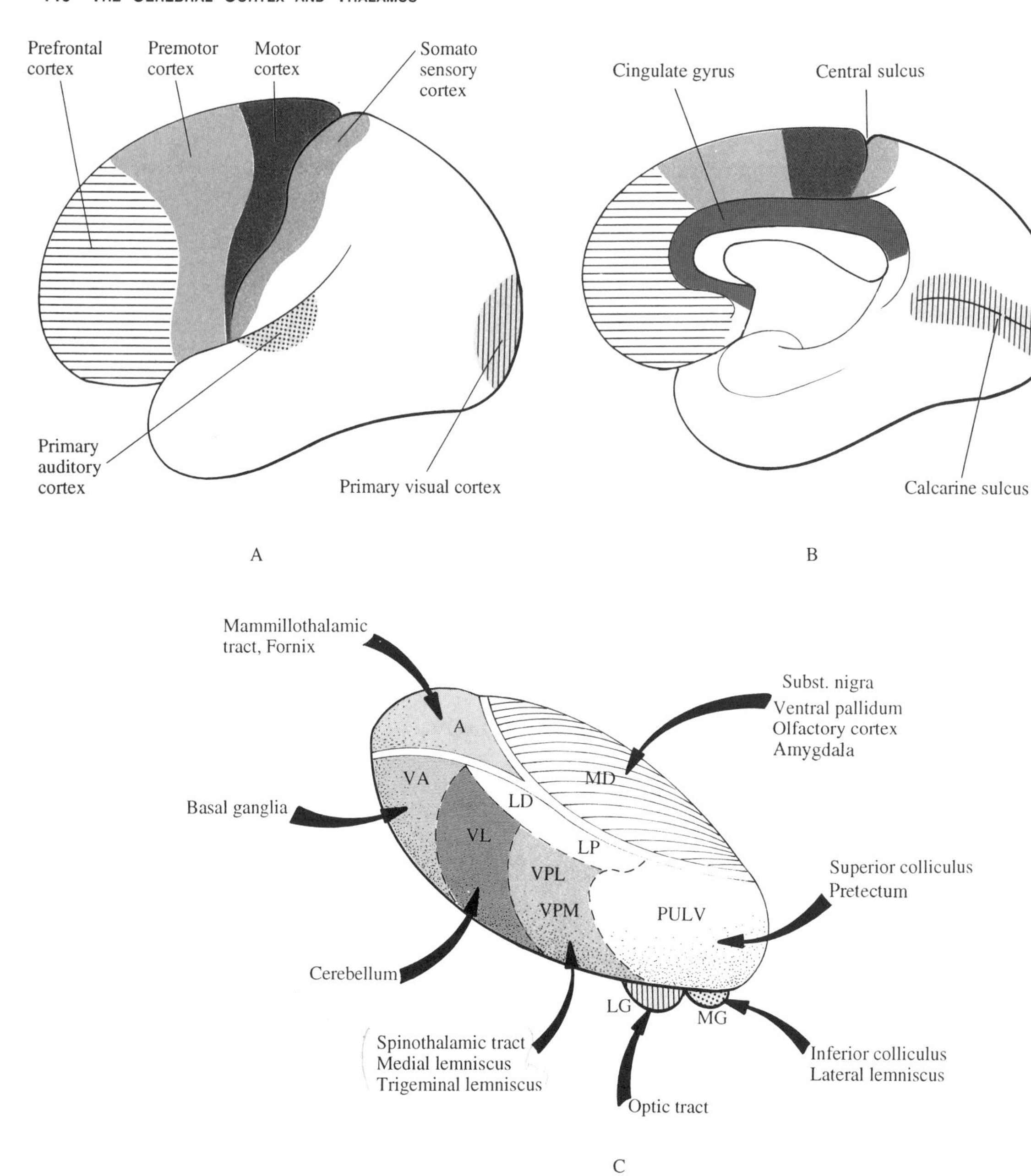

Fig. 22–6. Thalamocortical Relations

A and B. Lateral and medial surfaces of the cerebral hemisphere showing the main projection areas of the various thalamic nuclei.

C. Dorsolateral view of the thalamus demonstrating its main subdivisions and their subcortical input.

50 different nuclei. However, the *internal medullary lamina* serves to simplify the subject in the sense that it separates the principal thalamic nuclei into a *medial* group on one side and a *ventrolateral* group on the other side of the internal medullary lamina. The anterior part of the internal medullary lamina, furthermore, splits into two lamellae, which surround an anterior group of nuclei.

There are also important cell groups within the medullary laminae, referred to as *intralaminar nuclei*, and one nucleus, the *reticular thalamic nucleus*, surrounds the rest of the thalamus like a shield, especially on its lateral side. Although the relations of the thalamic nuclei with other parts of the nervous system are manifold, all thalamic nuclei, except the reticular nucleus (see below), are characterized by projections to the cerebral cortex, and many of them also by projections to the striatum. Whereas the principal or major relay nuclei (depicted in Fig. 22–6C) project to one or a few well-defined cortical fields, many of the intralaminar nuclei project diffusely to many areas of the cortex. Hence, the intralaminar nuclei (and the reticular thalamic nucleus) are sometimes referred to somewhat inappropriately as "diffuse" or "nonspecific" thalamic nuclei.

Principal Thalamic Nuclei

Anterior Nuclear Group (A). This consists of three subdivisions and receives a significant part of its input from the mammillary body. It is connected with the cingulate gyrus on the medial side of the hemisphere, and is part of Papez' circuit (Chapter 21).

Medial Nuclear Group. The main nucleus in this group, the *mediodorsal nucleus* (MD), is related in an orderly fashion to extensive parts of the frontal lobe, i.e., the prefrontal cortex including the orbitofrontal cortex. The MD, which receives input from other parts of the thalamus and from a variety of subcortical structures including the amygdaloid body, parts of the basal ganglia, and the midbrain reticular formation, is important for memory functions (see below).

Ventrolateral Nuclear Group. A ventral group contains three major nuclei: the *ventral posterior nucleus* (VP), which is the thalamic relay for somatosensory input to the cerebral cortex (Chapter 9), and the *ventral lateral* (VL) and *ventral anterior* (VA) nuclei, which were described as relay nuclei in the cerebellar and basal ganglia circuits to motor and premotor cortical areas (Chapters 16 and 17).

Posterior Group. This group includes the lateral posterior nucleus (LP) and the large pulvinar thalami. These nuclei are connected reciprocally with extensive areas in the parietal, occipital, and temporal cortices, and they receive significant input also from the superior colliculus and pretectum. Their role in visual functions has attracted considerable interest (Chapter 13). The *lateral geniculate body* (LG) and the *medial geniculate body* (MG), finally, which serve as relay nuclei for visual and auditory information to the cerebral cortex (Chapters 13 and 14), are usually included in the posterior group of thalamic nuclei.

Summary of the Thalamocortical Relations

Figure 22–6 summarizes in a very schematic fashion how the principal thalamic nuclei are related to different parts of the cerebral cortex. Besides the thalamocortical projections from the principal nuclei in the ventral group and from the medial and lateral geniculate nuclei, there are prominent projections from the mediodorsal thalamic nucleus (which is the main nucleus in the medial group) to the prefrontal cortex, and from the anterior group of nuclei to the cingulate cortex. The pulvinar–lateral posterior complex is remarkable for its large size and widespread reciprocal relationships with the parieto-occipito-temporal association areas. It collaborates closely with these areas in many important functions including visual and language functions.

The scheme of the thalamocortical connections in Fig. 22–6 accentuates the close reciprocal relations that exist between the principal thalamic nuclei and specific regions of the cerebral cortex. However, it should be emphasized that the connections between the thalamus and the cortex are in reality more complicated. First, there are multiple channels in these thalamocortical systems in the sense that the various principal thalamic nuclei contain different sets of relay cells that project to different subdivisions of the cortex (e.g., Chapter 9) or to different cortical laminae. Second, several of the thalamic nuclei contain cells that project more diffusely to much wider areas than those shown in Fig. 22–6. This is especially the case for several nuclei in the internal medullary lamina, the so-called intralaminar nuclei, which are part of the ascending activating system (Chapter 10).

Other prominent projections from the intralaminar and midline nuclei reach the striatum, and since the intralaminar nuclei (and related nuclei in the midline, the midline nuclei) receive input not only from the brain stem but also from the cerebellum and the spinal cord, they emerge as important components of circuits through which many subcortical structures can affect both the cerebral cortex and the basal ganglia.

Finally, an intimate relationship exists between the thalamocortical fiber systems and the reticular thalamic nucleus, which occupies a strategic position as it surrounds the thalamus, especially on its lateral side (see also Figs. 5–10A and 5–11A). This nucleus, which is populated almost exclusively by inhibitory GABA-ergic neurons, is traversed by all fibers in the thalamic radiations interconnecting the thalamus and the cerebral cortex. The reticular thalamic nucleus is the only one of the thalamic nuclei that does not project to the cerebral cortex; instead, it receives axon collaterals from the corticothalamic and thalamocortical fibers penetrating the nucleus, and it is reciprocally related to the other thalamic nuclei and to the mesencephalic reticular formation. It is therefore in an excellent position to monitor the activity in the corticothalamic and thalamocortical channels, and to transmit information back to the thalamus and the midbrain.

Extrathalamic Corticopetal Modulatory Systems

Important components of the monoamine systems (Chapter 10) and the basal forebrain cholinergic system (Chapter 16) project directly to the cerebral cortex without relay in the thalamus. These extrathalamic pathways originate in circumscribed nuclei in the brain stem and caudal hypothalamus (monoamine systems), and the diagonal band nuclei and basal nucleus of Meynert (cholinergic and GABA-ergic systems). Although these neuronal systems apparently modulate the activity of the cerebral cortex, they are not necessarily nonspecific and diffuse in their actions. As a matter of fact, several of these systems are characterized by both a regional and a laminar specificity in their pattern of terminations in the cerebral cortex.

The Serotonergic Corticopetal System. The corticopetal component of the serotonergic system takes its origin in the rostral group of raphe nuclei located in the mesencephalon and rostral pons (Fig. 10–6). In general, it seems that both the regional and the laminar innervation by the serotonergic system is complementary to the noradrenaline innervation (see below). For instance, whereas the densest serotonergic innervation coincides with the laminae that receive the thalamic input, the noradrenaline innervation appears to be more pronounced in the layers giving rise to corticofugal fibers. The brain stem serotonergic neurons are more numerous than any of the other monoaminergic neurons, but their functions are not well understood. Nonetheless, the widely distributed serotonergic pathways have been implicated in a large number of functions, including pain control and sleep (Chapter 10).

The Dopaminergic Corticopetal System. The cerebral cortex receives dopaminergic fibers from a band of cells located in the ventral tegmental area and the substantia nigra, pars compacta (Fig. 10–7). All parts of the cerebral cortex receive dopaminergic fibers, with the densest innervation reaching the motor cortex. The prefrontal and temporal association areas are also densely innervated, whereas the innervation becomes increasingly sparse in the caudal parts of the cerebral cortex.

Recently, catecholamine-containing, presumably dopaminergic cells have been discovered in the cerebral cortex and underlying white matter in the human. These cells are peculiar in the sense that they also contain GABA, but little is known regarding their anatomic affiliations. Although these catecholaminergic cells are rather widely scattered, they may still surpass the mesencephalic dopamine cells in absolute numbers considering the expanse of the cerebral cortex.

The Noradrenergic Corticopetal System. The noradrenergic corticopetal fibers emanate from cells in the locus ceruleus, located underneath the floor of the fourth ventricle at the rostral level of the pons (Figs. 10–8 and 10–9). As mentioned in relation to the serotonergic innervation of the cortex, there is a certain degree of laminar complementarity in the terminal pattern of serotonergic and noradrenergic fibers. This is evident at least in some of the primary sensory areas, where there is a concentration of serotonergic terminations in the fourth cortical layer and of noradrenergic terminal branches in the third layer and the two deep layers. This has been seen as an indication that the serotonin system is involved primarily in the initial stages of cortical processing, whereas the noradrenergic system may be more engaged in higher-order information processing. As indicated in Chapter 10, the activity of the noradrenaline system correlates closely with the state of arousal, in the sense that the noradrenergic neurons are especially active during situations of extreme vigilance and attention.

The Histaminergic Corticopetal System. The histamine-containing cells are located in the tuberomammillary nucleus in the posterolateral hypothalamus (Fig. 10–10). The corticopetal histaminergic fibers pass through the internal capsule and the caudate-putamen to reach all parts of the cerebral cortex in moderate numbers and with no apparent regional or laminar predominance. Besides its effect on autonomic and neuroendocrine activities, the functions of the histaminergic system are little known.

The Cholinergic and GABA-ergic Corticopetal Systems. This extensive collection of large and medium-sized corticopetal cells is concentrated especially in the septum, the diagonal band of Broca, and the basal nucleus of Meynert (Fig. 16–15). The cells in the septal-diagonal band area, many of which are GABA-ergic, project to the hippocampus, whereas the mostly cholinergic cells in the basal nucleus of Meynert and neighboring regions project to the rest of the cerebral cortex. The corticopetal cholinergic system is believed to be the most important system for cortical arousal, i.e., providing the necessary conditions for the cortical neurons to perform efficiently in the processing and transformation of the incoming information (Chapter 16). A similar activating effect on thalamic mechanisms is proposed for the ascending cholinergic reticular pathway, which originates in the pedunculopontine-laterodorsal tegmental complex in the brain stem (Fig. 10–12).

Association Fibers

By far the largest part of the white substance in the cerebral hemisphere is accounted for by the association fibers. Blunt dissection of the major fiber systems reveals a number of these important connections (Chapter 4, Third Dissection), which form the basis for cooperation between different cortical areas within the same hemisphere. For instance, short arcuate fibers, or U-fibers, connect adjacent gyri with each other, whereas areas in different lobes are joined by long fiber bundles such as the cingulum, arcuate fasciculus, superior and inferior occipitofrontal fasciculus, and uncinate fasciculus. Gross anatomic dissections, however, cannot reveal the details of these multifarious connections, and our knowledge of this subject is based primarily on experimental neuroanatomic studies in primates.

As expected, the results obtained indicate that the association pathways are very elaborate, yet precisely organized. Some of the general organizational principles are illustrated in Fig. 22–7, but the actual situation is considerably more complicated than these highly schematic figures indicate. One cortical region usually receives input from a large number of cortical areas, and in turn projects to many other cortical regions. For instance, neuroanatomists have traced in excess of several hundred corticocortical pathways between the various striate and extrastriate visual areas alone, and some prefrontal cortical regions may be related to as many as 15 to 20 different regions in the various lobes.

Short association fibers connect the primary sensory areas in a sequential fashion with neighboring areas, and as indicated by the bidirectional arrows in the highly schematic drawing in Fig. 22–7A, association fibers are in general reciprocal. However, reciprocity does not necessarily mean that only the cells that give rise to a specific projection in turn receive fibers from the region to which they project. The neighboring areas related to the same sensory modality are sometimes referred to as unimodal "association" areas, such as Brodmann's areas 18 and 19 in the occipital lobe. As indicated in Chapter 13, however, areas 18 and 19 contain several visual areas, each associated with a specific aspect of visual function. The term "association" areas as opposed to primary sensory (or motor) areas, therefore, does not adequately describe a very complicated situation, in which segregation of function within the same sensory system is likely to continue beyond the primary receptive cortical areas, but there are also precisely organized "intrinsic" associative connections between the different channels within the system.

Area 39 (angular gyrus) in the inferior parietal lobule is an example of a multimodal cortical area, where impulses from different sensory modalities converge. Other important multimodal cortical areas are located in the prefrontal cortex and in the parahippocampal gyrus on the medial side of the temporal lobe. It is important to realize, however, that even in the multimodal cortical regions, there are sites that are connected predominantly with one sensory modality.

The long association fibers (Fig. 22–7B) are also organized according to certain general principles. For instance, except for some fibers between primary somatosensory areas and primary motor cortex (area 4), the primary sensory areas are not connected to each other by long association pathways. Also, there is a tendency for first-order sensory association areas (e.g., area 5 in the somatosensory cortex) to be connected with the premotor region (area 6), and for higher-order association areas (e.g., area 7) to be connected with multimodal regions in, for instance, the prefrontal cortex (e.g., area 46) or parahippocampal gyrus. The situation is similar in regard to the motor regions, i.e., whereas primary motor cortex (area 4) is related to premotor cortex (area 6) and supplementary motor area (SMA), it is primarily area 6 that communicates with the prefrontal cortex.

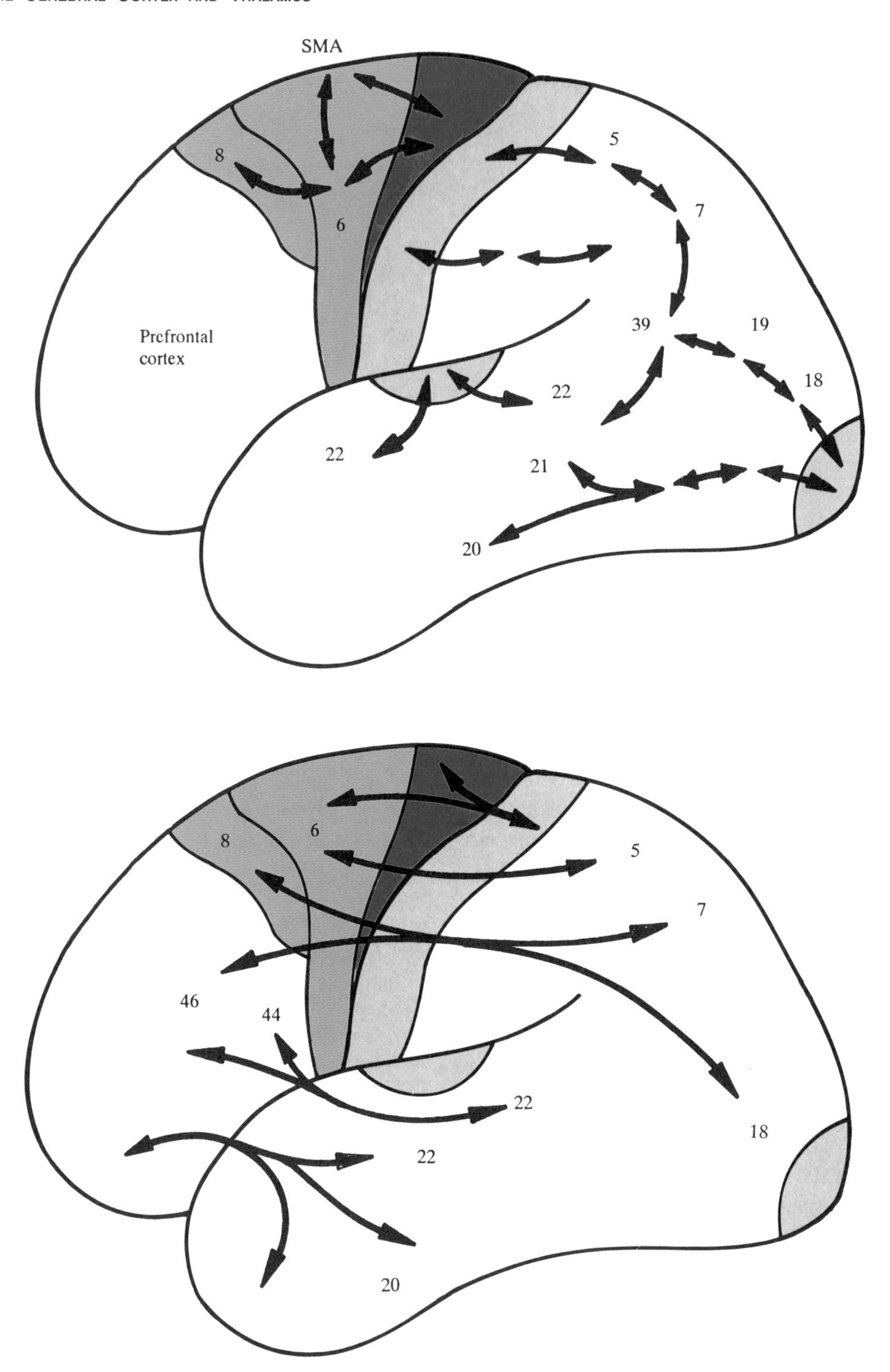

Fig. 22–7. Association Pathways

A. Highly schematic figure showing the sequential processing of information in cerebral cortex (*arrows*). Association fibers are in general reciprocal.

B. Some examples of long association fibers. There is a tendency for first-order sensory association areas to be connected with the premotor cortex, and for higher-order association areas to be connected with multimodal regions.

Commissural Fibers

The overwhelming majority of the over 300 million commissural fibers are contained in a massive fiber bundle, the *corpus callosum*. Commissural fibers between the two temporal lobes pass in both the corpus callosum and the anterior commissure. The commissural connections are not uniformly distributed across the cerebral hemisphere; some areas, e.g., the hand and foot areas in the somatosensory and motor cortices, and part of visual area 17, do not receive commissural fibers. The commissural fibers permit the transfer of information and learning from one hemisphere to the other (see below).

Cortical Functions

Since cortical functions related to the motor system (Chapter 15) and the different sensory systems (Chapters 9, 12, 13, and 14) have already been discussed in considerable detail, this and the following sections will focus primarily on higher cognitive functions, including language and memory, which are critically dependent on the cerebral cortex.

The pioneer American neuroanatomist C. Judson Herrick referred to the cerebral cortex as the "organ of civilization," and there is little doubt that the dramatic expansion of the cerebral cortex and the concomitant development of the neuronal mechanisms for language are in large part responsible for man's superiority over other animals. Although much of our understanding regarding cortical motor and sensory functions has been obtained in animal experiments, especially in the monkey, human cognitive functions including language functions have until recently been based primarily on studies of patients with cortical lesions. In the past, a meaningful correlation between clinical symptoms and structural damage could usually be made only after a postmortem examination of the patient's brain. With the introduction of modern imaging methods during the last two decades, however, the situation has improved considerably. As exemplified by the magnetic resonance images (MRIs) shown in several of the previous chapters, the living brain can be "sliced" in any plane, and the sections provide a clear picture of the intracranial anatomy including a clearly visible difference between gray and white matter. This provides the opportunity to map precisely the location of a focal lesion in the living brain (see also Introduction to Chapter 5).

Electrical stimulation of the cerebral cortex is another method that has yielded valuable information regarding higher cortical functions. The stimulation can be produced with low-voltage electrical current in awake patients whose cortex has been exposed under local anesthesia in preparation for surgery. This procedure was used very successfully in the middle of this century by the Canadian neurosurgeon Wilder Penfield and his colleagues, who made a series of important discoveries including those related to the representation of the various body parts in the cortical motor and somatosensory areas (Fig. 15–8), the localization of speech areas, and the memory functions of the temporal lobe (see also Language Disorders in Clinical Notes, below).

Another profitable approach to the study of human cognition has been the study of so-called *split-brain patients*, i.e., patients whose corpus callosum has been sectioned to prevent the spread of epileptic seizures from one hemisphere to the other.

With the aid of *positron emission tomography* (PET), functional–anatomic mapping of the human brain is rapidly reaching new frontiers, as demonstrated by the scans showing cortical areas of increased blood flow during different kinds of language-related functions (Fig. 22–8). PET relies on the systemic administration of radiolabeled tracers, which can be used to measure different tissue functions such as glucose metabolism and blood flow, or to study neurotransmitter systems by labeling substances known to bind to their receptors. The method for studying glucose metabolism, for instance, relies on the use of deoxyglucose (Chapter 7), which is labeled by positron-emitting isotopes with short half-lives (e.g., carbon 11 or fluorine 18) to limit exposure to radiation. Since the isotopes have short half-lives, an on-site accelerator, such as a cyclotron must be available. The concentration of the radioactive tracer in various brain regions is determined by a ring of detectors that pick up the gamma rays generated by the isotopes as they decay. The PET images can be constructed with computations similar to those used in computed tomography (CT) and MRI.

Many of the techniques related to PET are difficult and the instruments used are expensive. PET scanning, therefore, is conducted only in a few centers for the time being. However, the technique has already provided much valuable information regarding metabolic events and brain functions under normal and pathologic conditions, even if the interpretation of the results obtained is not always as straightforward as it appears from the few examples included in this book. With reduced cost and improved techniques, PET scanning is

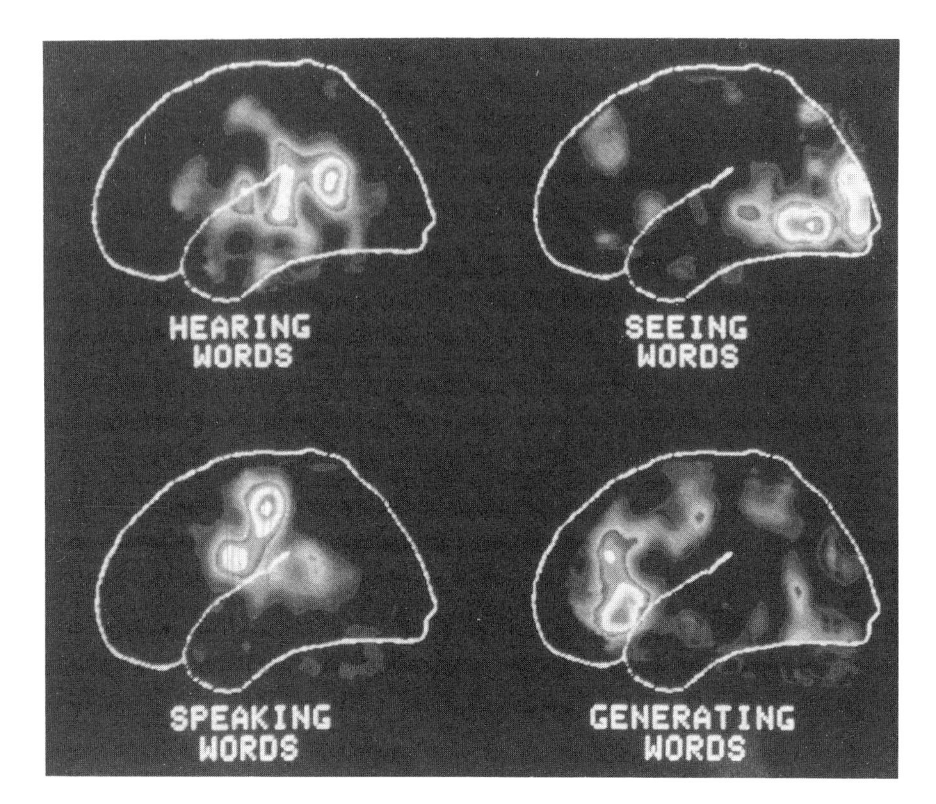

Fig. 22–8. Cerebral Blood Flow During Different Language Tasks
Positron emission tomography (PET) showing increase of blood flow in different parts of the cerebral cortex during different language-related functions. (Images, produced by Dr. Marcus Raichle, were reproduced from *Scientific American*, September 1992, Mind and Brain, Special Issue. With permission of Scientific American, Inc.)

likely to become more widely available and will undoubtedly play an increasing role in the study and diagnosis of many brain disorders.

Functional Specialization and Cerebral Asymmetry

In 1861 the French anthropologist-physician Pierre-Paul Broca provided the first clear evidence of functional specialization of the cerebral cortex when he studied a patient who had lost his ability to speak. Since the only word the patient could utter was "*tan*", the patient became widely known as "Tan." Shortly after he had been examined by Broca, the patient died and Broca discovered during the postmortem examination that the man had a large pathologic softening in the left frontal lobe, which Broca identified as the source of his speech problem. The central focus of the lesion was the posterior part of the inferior frontal gyrus, which is now referred to as *Broca's area* (see Vignette).

Although the idea that different functions were localized in different parts of the brain had been around for a long time, Broca's discovery provided the first clear and authoritative evidence of cerebral localization, and as illustrated in several of the previous chapters, the principle of functional specialization is now firmly established. But as we shall see in The Anatomic Substrate of Cognition (below), functional specialization does not translate directly into a mosaic of centers for different functions.

One of the best-known examples of functional asymmetry or lateralization is related to language, which in more than 95% of the population is lateralized to the left hemisphere. In other words, the left hemisphere is the dominant hemisphere for speech and language functions. It is also dominant for motor planning (praxic) skills. The right hemisphere, on the other hand, is dominant for spatial abilities and for some aspects of music. There is also some evidence that it is more involved in various aspects of emotional behavior than the left hemisphere.

The lateralization of language functions to the left hemisphere was in the past believed to be directly related to a larger temporal plane on the left side of the brain. However, the temporal plane is larger on the left side in only about 65% of the population, and as indicated in Chapter 14, there is currently no evidence to support the theory that this anatomic asymmetry evolved to support language functions in the left hemisphere.

Most people are right-handed, which indicates that the left motor cortex is usually more resourceful than the motor areas on the right side in regard to hand skills. Handedness, however, is not an absolute phenomenon, and the percentage of left-handed people varies between 5% and 20% depending on the task at hand, e.g., writing, hammering, dealing cards, etc. Some persons, furthermore, are ambidextrous, i.e., they can use either hand with the same ease. It should also be emphasized that there is not a strong relationship between handedness and language functions, since in the majority (about 70%) of left-handed people, language is still lateralized to the left hemisphere.

Corpus Callosum and Interhemispheric Transfer

In patients with severe and disabling epileptic seizures, the corpus callosum is sometimes sectioned (commissurotomy) to prevent the spread of epileptic discharge from one hemisphere to the other. Although such "split-brain" patients can function surprisingly well—there are no obvious changes in intellect and behavior—they are unable to perform certain tasks, because information cannot be transferred from one hemisphere to the other. For example, if an object is placed in the patient's right hand (with eyes closed), the patient can name the object, because the sensory information reaches the left hemisphere, which is dominant for speech and language functions. If the object is placed in the left hand, however, the patient will not be able to name the object, because the right hemisphere does not have access to the memory for language in the left hemisphere.

The degree to which various functions are lateralized to one or the other hemisphere can be studied following transection of the callosal fibers. The American neuroscientist Roger Sperry and his collaborators made a series of important observations regarding cortical functions following experimental commissurotomies in cats or therapeutic callosal transections in humans. The results obtained showed that each hemisphere has its own independent mechanisms for the processing of learning, thoughts, and memory, which are not available to the other hemisphere if the fibers in the corpus callosum are interrupted. Sperry was awarded the Nobel Prize in 1981 for these discoveries. Split-brain studies continue to provide valuable information on cortical functions.

Without detracting from any of the acclaim that Sperry and his collaborators deserve, it is nonetheless interesting to note that the German neurologist Hugo Liepmann drew similar conclusions concerning the functions of the corpus callosum from his studies of stroke patients at the turn of the century. Following occlusion of the anterior cerebral artery, which effectively transects the callosal fibers, Liepmann noticed that the patient, when given a command, could execute a movement with the right hand but not with the left. Since Broca had already shown that language functions are lateralized to the left hemisphere, Liepmann's interpretation, in essence correct, was as follows: The verbal instruction was comprehended only by the patient's left hemisphere, and the patient had no problem executing the movement with the right hand, since all neuronal mechanisms took place in the left hemisphere, including the final transfer of the motor program to the left motor cortex for deliberate activation of the right hand. However, the destruction of the corpus callosum prevented the transfer of the appropriate information from the left hemisphere to the right motor cortex, which would have to be activated in order to carry out the movement with the left hand. Inability to carry out a purposeful motor act, despite the absence of motor or sensory deficits, is referred to as *apraxia* (see Clinical Notes).

The Anatomic Substrate of Cognition

The traditional notion held that integration of neuronal activity for perception and recall is achieved by a unidirectional serial processing of information from primary sensory and motor areas, via unimodal association areas, to higher-order multimodal integrative regions. The theory also implied that an increasingly refined representation of different features took place within the various centers along hierarchical pathways towards an area of ultimate integration in more anterior parts of the cerebral cortex, such as the prefrontal or anterior temporal region. However, no such final area for integration has ever been identified, and the notion that perception and awareness are dependent solely on a hierarchical serial processing of information has now been more or less abandoned in favor of theories that invoke interaction between many different structures or highly specialized modules and functional subsystems, or that make reference to the simultaneous activation of large neuronal assemblies or distributed neuronal networks as the basis for mental activity. For instance, it was mentioned that one of the theories for integration in the visual system suggests a multistage process with simultaneous activation of a number of segregated visual channels and specialized areas involving both striate and extrastriate cortical regions (Chapter 13). The contribution of reciprocal or re-entrant corticocortical circuits is crucial in any such theory.

As a further elaboration on the theme of distributed networks, it has been suggested that the brain might bind entities and events together by multiregional activation from so-called *convergence* or *activation zones* (Damasio, 1989). As already mentioned in some of the previous chapters (e.g., Chapters 9 and 13), experimental investigations and clinical experience have confirmed the existence of a striking anatomic segregation of fragmentation of functional processing within the sensory cortices, and it has been suggested that the integration of different submodalities is dependent on a simultaneous activation of multiple sites within the sensory cortices. The notion of convergence zones implies that recall and recognition of entities and categories depend on the simultaneous retroactivation of the fragmentary records within multiple cortical regions through feedback activity from convergence zones, which are located within the sensory association cortices and which contain the "binding codes" for entity binding. The convergence zones would be predetermined by the arrangement of the associative connections, although their anatomic organization would be modifiable by learning. Area 39 (Fig. 22–7A), for instance, might serve as a convergence zone for entities coded by three different modalities. Needless to say, according to this and similar hypotheses invoking the idea of distributed networks, no single area can be identified for integration of sensory activity.

Records of spatial and temporal relationships among entities, which in turn would signify complex events, are thought to be coded in convergence zones, which are presumed to be located in more anterior parts of the cerebral cortex. Recognition and recall of events, in other words, would be dependent on the simultaneous multiregional reactivation of records of entities and features in multiple regions located in the posterior parts of the cerebral cortex.

The study of the human mind, which used to be the domain of philosophers and psychologists including some clinicians, has now become a legitimate research topic in the field of basic neuroscience (*Scientific American*, September 1992); and there has already been significant progress in the analysis and identification of brain systems related to some of the "higher" brain functions. The study of memory is a case in point.

Memory

The ability to learn and form new memories is critically dependent on the hippocampus and neighboring cortical areas in the medial temporal lobe (Chapter 21). Lesions in the hippocampal formation, therefore, produce severe *amnesia*, i.e., inability to recall facts and events. It is important to realize, however, that the hippocampus interacts with the rest of the neocortex, especially the sensory association areas, in memory functions. As a matter of fact, both immediate memory (short-term memory) and the ability to retain old and remote memories are still intact following lesions in the hippocampus, which indicates that various neocortical areas rather than the hippocampus formation are essential both for the initial stages of memory-related functions and for the final storage of the type of memory that requires conscious deliberations, i.e., *declarative (explicit) memory*.

Memory functions are also severely affected by bilateral lesions in the mediodorsal thalamus and its cortical projection area, i.e., the prefrontal cortex. The type of memory that is related to the prefrontal cortical areas has been referred to as *working memory*. A widely accepted notion is that the prefrontal cortex is important for organizing and controlling our behavior, i.e., for ordering and producing sequences in our behavior, and it has been suggested that working memory makes this possible, in the sense that it provides the moment-to-moment awareness and retrieval of stored information for the purpose of making informed decisions and judgments.

Other memory functions, including motor skills and habits and classical conditioning, which do not require conscious participation, are preserved following lesions in the hippocampal formation or the prefrontal cortex. These types of learning or memory functions, which can be referred to as *nondeclarative (implicit) memories*, are dependent on the specific brain systems engaged in the learning experience, e.g., the amygdaloid body for fear conditioning and the basal ganglia for motor skills.

The idea that the basic wiring diagram of the brain can be modified as a result of experience has been around for a long time, and many scientists, including the eminent Spanish neuroanatomist Ramón y Cajal, have suggested that the formation of new synapses may provide the anatomic basis for learning. That learning is related to plastic changes and remodeling of neuronal pathways is now well established. Much interest has focused on changes in synaptic efficacy and the fact that synaptic efficacy can be improved in different ways, e.g., from changes in the strength of existing synapses to the establishment of new connections with concomitant increase in the number of synapses. Indeed, it is now fairly certain that the architecture of our brains is significantly modified as a result of learning and exercise. For instance, it has been shown in primate experiments that increased use or repeated exercise of some of the fingers is likely to lead to increased cortical representation of those fingers. *Long-term potentiation*, i.e., increased synaptic efficacy as a result of brief periods of stimulation, is another phenomenon which has received much attention because of its possible role in memory functions (see "Amino Acids" in Chapter 23).

Clinical Notes

Frontal Lobes

The cortical surface of the frontal lobe is divided into a caudal component related to motor functions and a rostral part, the prefrontal cortex, which has increased dramatically in phylogenetic development; it represents about one-quarter of the entire cerebral cortex in the human brain.

Lesions of the motor-premotor areas (areas 4, 6, and 8) cause spastic paralysis on the opposite side of the body (Chapter 15), whereas lesions in area 44 (part of Broca's area, see below) and neighboring area 4 in the dominant hemisphere produce motor speech defects. Lesions in the rostral parts of the frontal lobes give rise to changes in the person's personality. Clues regarding the functions of the frontal lobes have to a large extent been gained by studying patients with frontal lobe damage, e.g., war injuries, or by observing the effect of psychosurgical operations on the prefrontal cortex.

The New England railroad worker Phineas Gage is the most famous "lobotomized" person in the history of medicine. According to a report published in a medical journal more than a century ago, this unfortunate man was struck by a thick iron bar which was propelled through his forehead by an explosion. Miraculously, the man survived and was able to walk with help to see a doctor. The crowbar had inflicted massive damage to the midsections of his frontal lobes, which resulted in a striking change of his personality. According to a publication of the Massachusetts Medical Society in Boston Harlow, J. M. 1868. Recovery from the passage of an iron bar through the head. Bulletin of the Massachusetts Medical Society, 2:3–20: "Gage lived for twelve years afterwards; but whereas before the injury he had been a most efficient and capable foreman in charge of laborers, afterwards he was unfit to be given such work. He became fitful and irreverent, indulged at times in the grossest profanity, and showed little respect for his fellow man. He was impatient of restraint of advice, at times obstinate, yet capricious and vacillating. A child in his intellectual capacity and manifestations, he had the animal passions of a strong man." In essence, "He was no longer Gage."

As illustrated by the "Boston Crowbar Case," it is clear that frontal lobe damage can lead to dramatic changes in a person's personality and conduct. The term "frontal lobe syndrome" is often used to denote a prefrontal lesion, but it should be recognized that the symptoms can vary considerably from patient to patient. The effects of a prefrontal lesion depend on the size and the location of the lesion, and also to some extent on the patient's premorbid personality. The more dramatic effects are seen after bilateral lesions, whereas the symptoms can be subtle and elusive following unilateral involvement. Nonetheless, it is generally recognized that the changes following pathologic derangement of the frontal lobe (i.e., its prefrontal parts) can be conceived of as a disorder of personality, with lack of drive and spontaneity as the most significant symptoms. Apathy, lack of initiative, and cognitive deficits are more pronounced in lesions of the dorsolateral parts of the frontal lobe, whereas orbitofrontal lesions are more apt to produce changes in mood and affect, including impulsive and inappropriate behavior.

One of the problems in the analysis of frontal lobe functions and in the appreciation of symptoms following its destruction appears to be that so many different functional-anatomic systems are represented in the frontal lobes. There is a multitude of connections between the prefrontal cortex and other parts of the CNS; different parts of the prefrontal cortex are reciprocally related not only to association areas in the other cerebral lobes (Fig. 22–7B), but also to the hippocampal formation and the amygdaloid body, and prominent projections reach the prefrontal cortex from the medial and anterior thalamus. Both clinical and experimental studies of prefrontal functions have been notoriously difficult to interpret, and it has not been easy to put forward a unified hypothesis of frontal lobe function. As mentioned earlier, however, the notion of working memory has been promoted as an important aspect of frontal lobe function. If this function is derailed by a lesion in the prefrontal regions, it will be difficult to plan for the future or to initiate a plan of action.

Parietal Lobes

The anterior part of the parietal lobe contains the somatosensory projection areas (areas 3, 1, and 2) in the postcentral gyrus (Chapter 9). Circumscribed lesions in this part of the parietal lobe cause defects in sensory discriminative functions primarily on the opposite side of the body, whereas the perception of sensory stimuli is still relatively intact.

The posterior parietal cortex, which merges imperceptably with the occipital and the temporal lobes, includes the superior parietal lobule (areas 5 and 7) and the inferior parietal lobule (supramarginal and angular gyri). The inferior parietal lobule is often included as part of Wernicke's speech area, which will be discussed in Language Disorders below.

The posterior parietal cortex, represented by areas 5 and 7, is interposed between the somatosensory cortex in the postcentral gyrus and the visual areas of the occipital lobe. It is therefore easy to appreciate that this region is of special importance for the integration of tactile and visual information. Furthermore, and as indicated in Fig. 22–7B, there are important associative connections between the posterior parietal cortex and the premotor and prefrontal cortices, and these connections are part of the anatomic substrate for sensory and visually guided movements.

Lesions involving the posterior parietal cortex are characterized by complex sensory defects. The most striking symptoms are neglect of sensory stimuli and disturbance of spatial relationships. For instance, a patient with a posterior parietal lesion, especially on the right side, may not be aware of the contralateral half of the body. The absence of awareness of illness, such as hemiplegia, is called *anosognosia*. Neglect of the contralateral side of the body (e.g., in dressing or shaving) is also common, especially following a lesion in the right parietal lobe. Bilateral lesions may result in a general lack of spatial perception and topographic sense; the patient is unable to grasp an object or draw a simple diagram, or even describe how to get from his home to work.

A lesion in the dominant, usually the left, parietal lobe can result in a number of characteristic symptoms, which together form the *Gerstmann syndrome*. The prominent features are inability to recognize and distinguish the different fingers on the two hands (finger agnosia), confusion of the left and right sides of the body, and the inability to calculate and to write.

Agnosia and Apraxia. The lack of understanding of the significance of sensory stimuli is referred to as *agnosia*. *Tactile agnosia* is the inability to recognize objects by touch. It is related to, but different from, *astereognosis*, which is a loss of the ability to judge the form of an object by touch. Agnosias are caused not only by parietal lesions but also by lesions of the occipital or temporal cortices. *Visual agnosia* is a failure to interpret visual images, and *auditory agnosia* is a failure to recognize what is heard.

Disorder of execution of learned movements that cannot be accounted for by weakness, ataxia, or sensory loss are termed *apraxia*. Apraxias involving both sides of the body are most often seen in left parietal lobe lesions that interrupt the association pathways between the parietotemporal junction area and the premotor areas in the frontal lobe. The term apraxia is not used if the failure to carry out a command is due to a severe deficit of comprehension, e.g., in a patient suffering from sensory aphasia (see below).

Temporal Lobes

A number of different functional–anatomic systems are concentrated in the temporal lobes. The cortical sensory areas for hearing (areas 41 and 42) are located in Heschl's gyrus in the depth of the lateral fissure, and the cortical receptive area for olfaction in the uncus of the parahippocampal gyrus. Clinical notes related to the cortical structures of these two sensory systems were discussed in Chapters 14 and 12.

The posterior part of the superior temporal gyrus (area 22) and the angular gyrus (area 39) in the parieto-occipito-temporal junction area in the dominant hemisphere are part of the central language zone, which is important for understanding speech and written language (see below). Large parts of the inferotemporal cortex belong to the extrastriate visual areas, and lesions in these regions produce various types of visual recognition defects (Chapter 13).

Meyer's Loop. In this context, it is worth emphasizing that the ventral fibers in the optic radiation, which are related to the lower part of the retina, make a detour into the temporal lobe before proceeding to the visual cortex. A temporal lobe lesion that interrupts this loop produces a contralateral upper homonymous visual defect (Fig. 13–16).

Other important systems are located in the medial part of the temporal lobe. As mentioned earlier, the temporal part of the olfactory cortex is located in the uncus of the parahippocampal gyrus, and deep to the uncus is the amygdaloid body, which is an important modulator of emotional behavior (Chapter 20). Bordering on the caudal aspect of the amygdaloid body is the hippocampus formation, which is crucially involved in memory functions (Chapter 21).

Occipital Lobes

The occipital lobes are related exclusively to visual functions. Their functional anatomy and clinical correlations were discussed in considerable detail in Chapter 13, and will not be repeated here.

Epilepsy

Epilepsy is a common disorder characterized by paroxysmal attacks of brain dysfunction, the nature of which depends on the anatomic location of the disturbance. Although alteration of consciousness is frequently seen, it is not the rule. There may be motor, sensory, or autonomic disturbances, depending on the portion of the brain involved, but even if the seizures are predominantly focal they may secondarily generalize. Seizures are symptoms of epilepsy, which is the result of an inherited or acquired disorder. Inherited epilepsy is referable to the genome, and several epilepsies have now been genetically identified. Genetic epilepsies are known as idiopathic. The underlying defect appears to be operative at the ion channel level of the cell membrane, resulting in an imbalance between excitatory and inhibitory neurotransmitter function. The former function is primarily glutamatergic and the latter is the result of GABA- or glycine-mediated activities. Epileptic seizures due to acquired disease (in utero, at birth, or any time thereafter) may be the first symptom of infections, malformations, tumors, trauma, or vascular disorder.

There are basically two types of seizures, *convulsive* and *nonconvulsive*. Convulsive seizures may be partial (focal) or generalized. Although seizure activity may start in any part of the brain, cortical seizures have attracted special attention because of the characteristic symptomatology which reflects the signature of the area predominantly involved at the outset. Lesions in the region of the central sulcus, for example, may give rise to motor or sensory symptoms. Occipital seizures are characterized by visual symptoms, and temporal lobe disturbances may be reflected by dysmnesic (characterized by impairment of memory) or autonomic symptomatology. Focal seizures may spread, and spreading excitation will mobilize adjacent areas of cortex and lead to a so-called Jacksonian[1] seizure, which involves structures according to their cortical representation (see the homunculus, Fig. 15–8B). Partial seizures leading to alteration of consciousness are known as *complex partial seizures* and are especially likely to arise from the frontal or temporal cortex. Seizures which are generalized from the beginning include generalized tonic–clonic seizures (grand mal) with loss of consciousness, and tonic contraction followed by clonic or rhythmic jerking which may persist for several minutes. The violent contractions produce rigid distortion of the body, breathing is impaired, and the skin may turn blue from lack of oxygen. Other types of generalized seizures include sudden single or repetitive rapid jerks known as *myoclonus*. Occasionally, paroxysmal seizure activity may be predominantly inhibitory and lead to atonic or drop attacks, or absence staring spells.

1. The famous British neurologist Hughlings Jackson, who lived in the nineteenth century, suggested that epilepsy results from an abnormal discharge of neurons. On the basis of careful clinical examinations, Jackson predicted that the various parts of the body must be represented in a somatotopic fashion in the motor cortex.

Language Disorders

The importance of language, i.e., the comprehension and communication of ideas and feelings, can hardly be overestimated. Language is the very essence of social intercourse, and a severe language disorder, *aphasia*, is devastating to the patient.

Language functions are localized to the left hemisphere in more than 95% of the population, and the areas around the lateral or Sylvian fissure (the perisylvian region) are most closely related to language functions (Fig. 22–9). Traditionally, a distinction has been made between two major cortical language areas. The expressive speech area or Broca's area (area 44 and adjoining part of area 45) is concerned primarily with motor aspects of speech. Another, more extensive, region in the posterior part of the superior temporal gyrus and the parieto-occipito-temporal junction area, the receptive language area, is of great importance for comprehension of spoken and written language. The

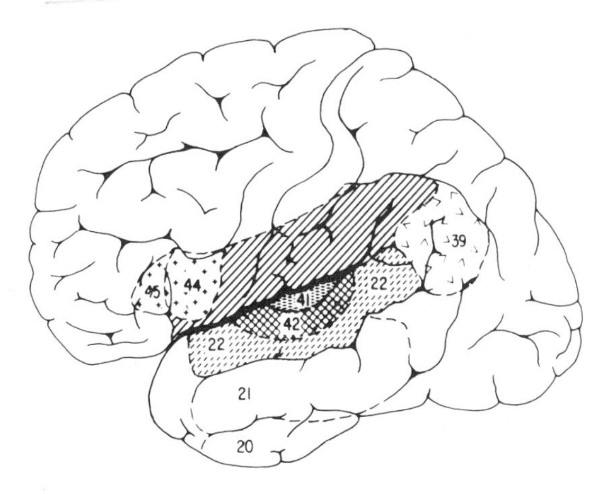

Fig. 22–9. The Perisylvian Language Zone
Diagram of the left hemisphere showing the classic language areas with Brodmann's numbers. The entire perisylvian region appears to form a contiguous "language zone". (From Adams and Victor, 1993, see Suggested Readings. With permission from the authors and McGraw-Hill, Inc., New York.)

posterior region can be further subdivided into two receptive areas: one subserving the perception of spoken language (Heschl's gyrus, areas 41 and 24, and the posterior part of the superior temporal gyrus, area 22), and another related primarily to perception of written language (angular gyrus, area 39).

Although this conventional and highly schematic presentation of the anatomic basis of language functions serves as a first approximation, it is important to recognize it for what it is, i.e., a gross simplification. In all likelihood, language depends on a much larger region indicated by the areas shown in Fig. 22–9. Indeed, the contiguity between the different perisylvian areas believed to subserve language functions can be more easily appreciated by unfolding the cerebral cortex (Gazzaniga, 1989). It may be more appropriate to speak of one continuous perisylvian "language zone." The examples in Fig. 22–8, moreover, show that many areas outside the perisylvian region are involved in language functions. Recent stimulation mapping of the cerebral cortex in patients undergoing surgery for epilepsy, furthermore, has shown that the location of the language areas can vary considerably from patient to patient. For instance, Broca's area, although usually confined to the posterior part of the inferior frontal gyrus, can extend into parts of the middle frontal gyrus or further rostrally in the inferior frontal gyrus. Nonetheless, the traditional subdivision into an expressive (motor) and a receptive (sensory) language area serves as a convenient approximation for clinical use.

A lesion in Broca's area results in *motor* or *nonfluent aphasia* (*Broca's aphasia*). Although the muscles of articulation may not be paralyzed, the patient can speak only with great difficulty, and the speech is nonfluent (telegraphic) with only a few words to a phrase; the speech is agrammatical although the few spoken words may be appropriate. A patient with Broca's aphasia usually has a right-sided hemiplegia as well, because of involvement of the left precentral gyrus.

Another type of language disorder, *sensory aphasia*, appears when the lesion is located in the posterior part of the superior temporal gyrus (area 22) and in the angular and supramarginal gyri (areas 39 and 40). A sensory aphasia is characterized by difficulty in understanding spoken words and sentences (*word deafness*), in reading, (*alexia*), or in writing (*agraphia*). Some call this type of aphasia *fluent aphasia* or *Wernicke's aphasia*, in honor of the German physician Carl Wernicke who made extensive studies on aphasic patients in the nineteenth century and thereby laid the groundwork for our present understanding.

In summary, language functions are very complicated, and clinical-pathologic studies have shown that language disorders are hardly ever purely motor or purely sensory. A patient with Broca's motor aphasia, for instance, also has some difficulty in grammar and in comprehending language. The different areas in the perisylvian language zone are closely interconnected with each other as well as with many other parts of the brain, including nearby areas in the prefrontal cortex and the sensory association areas in the other lobes. Connections with the thalamus and the basal ganglia are also likely to be of importance in this context, although little is known regarding corticosubcortical relations in language functions.

Mathematics or music may be similarly affected with involvement of specific areas of cortex, and syntactic and prosodic (relating to rhythm and pitch of speech) elements, as well as memory, may be individually involved in patients undergoing change in ability to communicate. Finally, it should be noted that the affective-prosodic elements of speech appear to be dependent on the right hemisphere rather than the left.

Clinical Example

Epilepsy

A 47-year-old left-handed man presented with a 15-year history of seizures.

Twenty years earlier, while in military service, he was struck by shrapnel which entered the right frontal region. He was rendered unconscious for about an hour, and when he returned to consciousness he had a left hemiparesis and expressive speech difficulty. His wound was debrided and thereafter he made a good recovery, clearing his hemiparesis and aphasia. He took anticonvulsant medication prophylactically for about two years. After discharge from the service, he became a car dealer.

Five years later, while enjoying a cup of coffee, he suddenly experienced jerking, beginning in the thumb and forefinger of the left hand. He dropped the cup, and as he watched the convulsive movements spread to involve the forearm and shoulder, his head drew to the left, and he was unable to speak though he tried to call out. After about 2 minutes, his seizure ceased and he was left with weakness of the left arm for a further 10 to 15 minutes; he regained the power of speech within the same time.

Following this, he had similar episodes at approximately 2- to 3-week intervals, and some of the attacks progressed to loss of consciousness and urinary incontinence. Each time, however, the attack commenced in the thumb and index finger of the left hand and spread to the face and occasionally to the leg before consciousness was lost. Occasionally, he was able to abort an attack by grasping his forearm just above the wrist.

On examination, he was a cooperative person who showed a mild left lower facial weakness and slight diminution of power in the long finger extensors and wrist extensors of the left hand. Rapid alternating movements were less well performed on the left than on the right, and the stretch reflexes were hyperactive in the left upper extremities as compared with the right. The remainder of the neurologic examination was entirely normal, though tests of speech function revealed an occasional paraphasic speech error and an occasional word substitution on reading.

Skull X-rays revealed a frontal skull defect resulting from the posttraumatic surgical procedure. The EEG (electroencephalogram) showed right frontal electrographic seizure activity during sleep.

1. What type of seizure does the patient suffer from?
2. Why do the seizures often start in the thumb and forefingers?
3. Why did the patient have a mild aphasia?

Discussion

This is a partial seizure of posttraumatic origin. The term Jacksonian seizure is applied to simple partial seizures of either the motor or the sensory variety when in their propagation across the cerebral cortex, they reveal a motor or sensory march (Fig. 15–8B). Because of the organization of the human cerebral cortex, in which the majority of the surface is involved with the function of thumb, forefinger, mouth, and tongue, these areas are predominantly involved in irritative cortical phenomena. At any time, the seizure may become secondarily generalized by propagating to centrencephalic structures (i.e., in the brain stem; see Ascending Activating System in Chapter 10) and thus to both hemispheres simultaneously.

Following such a seizure, a postictal or Todd's paralysis is common, which may be caused by inhibition or by neuronal exhaustion resulting from increased metabolic activity or clearing in the seizure focus. The patient's speech disturbance indicates that his language areas are located in the right hemisphere, which is dominant for speech in about one of three left-handed people. The speech deficit, like the paralysis, is transitory.

Patients can frequently abort the spread of an epileptic march by grasping the extremity above the place where the seizure phenomenon is occurring. The reason for this is not clear, but it may have to do with evocation of inhibitory neurotransmitters.

Patients with posttraumatic epilepsy can usually be treated successfully with anticonvulsant medication, and only rarely do such patients require neurosurgical therapy with excision of the epileptic focus.

Suggested Reading

1. Adams, R. D. and M. Victor, 1993. Principles of Neurology. 7th ed. McGraw-Hill, New York, pp. 378–430.
2. Bentivoglio, M. and R. Spreatico, 1988. Cellular Thalamic Mechanisms. Excerpta Medica: Amsterdam.
3. Gilbert, C. D., 1992. Horizontal integration and cortical dynamics. Neuron 9:1–13.
4. Damasio, A. R., 1989. The brain binds entities and events by multiregional activation from convergence zones. Neural Comput. 1:123–132.
5. Damasio, H., T. Grabowski, R. Frank, A. M. Galaburda, and A. R. Damasio, 1994. The return of Phineas Gage: Clues about the brain from the skull of a famous patient. Science 265:1102.
6. Douglas, R. J. and K. A. C. Martin, 1990. Neocortex. In G. M. Shepherd (ed.): The synaptic Organization of the Brain. Oxford University Press: New York, pp. 389–438.
7. Engel, J. Jr., 1989. Seizures in Epilepsy. F. A. Davis Co., Philadelphia. Fuster, J. M., 1989. The Prefrontal Cortex. Anatomy, Physiology, and Neuropsychology of the Frontal Lobe. 2nd ed. Raven Press, New York.
8. Gazzaniga, M. S., 1989. Organization of the human brain. Science 245:947–952.
9. Jones, E. G., 1985. The Thalamus. Plenum Press: New York.
10. Kolb, B. and I. Q. Whishaw, 1989. Fundamentals of Human Neuropsychology. 3rd ed. W. H. Freeman and Co., New York, pp. 347–522.
11. Mesulam, M.-M., 1990. Large-scale neurocognitive networks and distributed processing for attention, language and memory. Ann. Neurol. 28:597–613.
12. Ojemann, G. A., W. W. Sutherling, R. P. Lesser, et al., 1993. Cortical Stimulation in Surgical Treatment of the Epilepsies. 2nd ed. Raven Press, New York, pp. 399–414.
13. Peters, A. and E. G. Jones, 1984. Cellular Components of the Cerebral Cortex. Cerebral Cortex, Vol. 1. Plenum Press: New York.
14. Rakic, P. and W. Singer, 1988. Neurobiology of Neocortex. John Wiley & Sons: New York.
15. Scientific American, 1992. Mind and Brain, 267 (September), Special Issue.
16. Thierry, A.-M., J. Glowinski, P. S. Goldman-Rakic and J. Christen (eds.), 1993. Motor and cognitive functions of the prefrontal cortex. Spinger Verlag, Berlin.

Catecholamines

The three catecholamines, dopamine, noradrenaline, and adrenaline, are synthesized in a common metabolic pathway which uses the enzymes shown in this vignette. Neurons containing only tyrosine hydroxylase (TH) and dihydroxyphenylalanine (DOPA)-decarboxylase produce dopamine. Neurons which in addition contain dopamine-β-hydroxylase (DBH) produce noradrenaline, whereas those which also contain phenylethanolamine *N*-methyl transferase (PNMT) will produce adrenaline. [From Moore, Robert Y. 1981. Fluorescence histochemical methods. In: L. Heimer and M. J. Robards (eds.): *Neuroanatomical Tract-Tracing Methods*. Plenum Press, New York. With permission.]

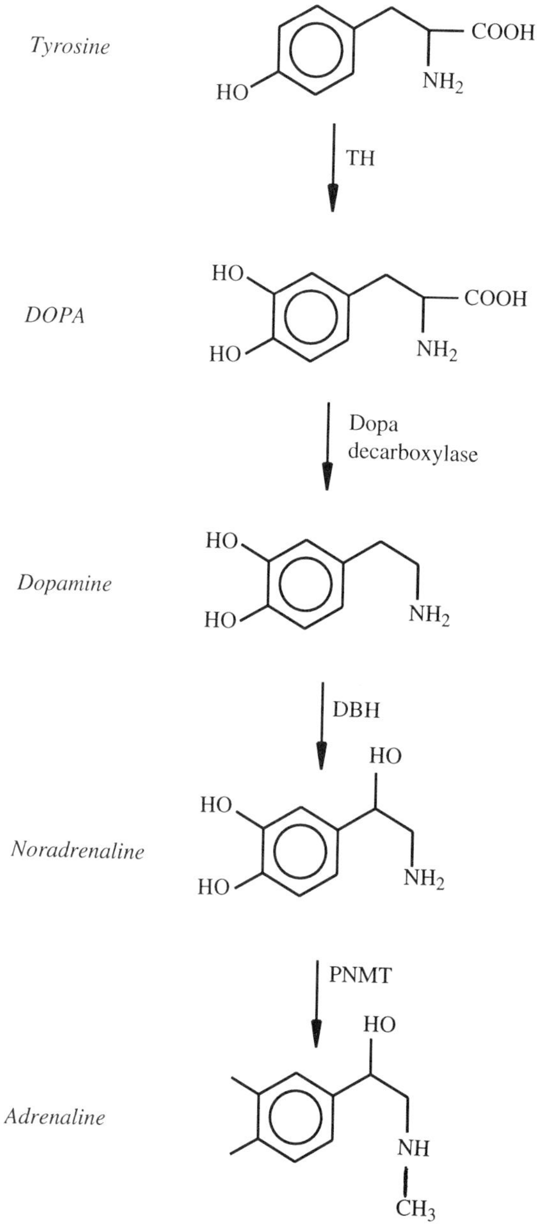

Tyrosine

HO COOH NH_2

TH

DOPA

HO HO COOH NH_2

Dopa decarboxylase

Dopamine

HO HO NH_2

DBH

Noradrenaline

HO HO HO NH_2

PNMT

Adrenaline

HO NH CH_3

23

Neurotransmitters

Although information on the distribution and function of neurotransmitters, and the identification of their receptors, is important in efforts to develop rational drug therapy, such knowledge will be of little use if the chemical anatomy of the specific brain disorder in question remains a mystery.

The discovery, in the late 1950s and early 1960s, therefore that Parkinson's disease is in large part related to reduction of dopamine in the striatum caused by degeneration of the nigrostriatal pathway was hailed as a scientific breakthrough, and it served as an inspiration for the many neuroscientists who are trying to get a handle on the debilitating disorders affecting the human brain. But the study of neuronal dysfunctions and of transmitters and their receptors in postmortem human brains is a difficult undertaking, and progress has been slow in coming. It is now clear, however, that many cases of depression are accompanied by decreased serotonin function, and increased dopamine activity, possibly coupled with derangement of some of the neuropeptides, e.g., neurotensin and cholecystokinin, seems to characterize some forms of schizophrenia. Deficits produced by degeneration of cholinergic and monoaminergic systems can be expected in some forms of dementia.

About 50 different chemical substances are known transmitters or modulators, and more will undoubtedly be discovered. The classical neurotransmitters, including acetylcholine, some amino acids, and the monoamines, are small molecules that usually are fast acting, whereas the large neuropeptide molecules in general have a slower and more prolonged action.

Introduction

The renowned English physiologist Sir Charles Sherrington introduced the term *synapse* and discussed its importance in intercellular communication around the turn of the century. At about the same time, the suggestion was made that an adrenaline-like substance was liberated at the sympathetic neuromuscular junction. Inspired by these notions, a large group of investigators turned their attention to the problems of chemical transmission, but it remained for the German pharmacologist Otto Loewi to finally demonstrate chemical transmission in 1921. He did this in a classic experiment in which he showed that a heart could be excited or inhibited by submerging it in a solution that had previously been in contact with a heart whose sympathetic or parasympathetic (vagus) nerves had been stimulated. A few years later, the "Vagusstoff," i.e., acetylcholine, was identified in the laboratory of the British physiologist Henry Dale.

A momentous event occurred in the early 1960s with the development of the Falck–Hillarp fluorescence histochemical method for the demonstration of biogenic amines. By this method it became possible for the first time to identify and characterize monoamine neurons and their projections on the basis of their transmitter content. The subsequent introduction of the immunocytochemical techniques and in situ hybridization histochemistry has made it possible to characterize all kinds of neurons on the basis of their transmitters and modulators.

Acetylcholine and the monoamines are well characterized as transmitters, and their related cell groups are notable in the sense that they are confined to certain parts of the brain. Their axons, on the other hand, are in general widely distributed, and these neuronal systems, therefore, are in an excellent position to influence or modulate the functions in many different parts of the brain, which may be involved in such complex behaviors as arousal, sleep–wake functions, and affect. However, as exemplified by the cholinergic striatal interneurons or the tuberoinfundibular dopamine pathway (Chapter 18), these transmitters are also involved in more specific neuronal mechanisms, or in pathways with restricted distribution.

Several amino acids have been recognized as neurotransmitters, and quantitatively they are undoubtedly the major transmitters. Based on their physiologic effect, the amino acids can be subdivided into two groups, the inhibitory and the excitatory amino acids. Only the prominent members of these two groups will be reviewed below.

During the last 20 years, a large number of peptides have gained recognition as neurotransmitters or modulators. Usually, two or more neuropeptides coexist within the same neuron with a conventional neurotransmitter; in other instances classical transmitters, e.g., GABA and acetylcholine coexist. The phenomenon of coexistence, which was first described by the Swedish histologist Tomas Hökfelt and his colleagues, has important implications; it provides for great flexibility in interneuronal communication, and it also suggests to the clinician that there may be several ways to interfere pharmacologically with the functions of a specific fiber system.

The effect produced by a synapse, i.e., excitatory or inhibitory, is dependent on the type of transmitter released and the properties of the postsynaptic receptors (Chapter 6). γ-aminobutyric acid (GABA) and glycine are always inhibitory, and glutamate and aspartate are always excitatory. However, other neurotransmitters, such as acetylcholine, serotonin, and dopamine, can be either excitatory or inhibitory, as exemplified by dopamine in the nigrostriatal system (Chapter 16). Receptors, in the form of macromolecules, have been identified for all the major transmitters as well as for many of the neuropeptides, and many of the transmitters have multiple receptors. The study of receptors and their interactions with various ligands is one of the most active fields in neuroscience, largely because the therapeutic effect of neuroactive drugs is usually related to their interaction with receptors.

The monoamine and acetylcholine systems are described in detail in Chapters 10 and 16, but they will be briefly reviewed here for the sake of completeness. Several of the other transmitters and modulators included in this summary have been discussed in the context of the various functional-anatomical systems described in the previous chapters.

Acetylcholine

Acetylcholine (ACh), which was identified more than 60 years ago, is the main neurotransmitter in the peripheral nervous system. The lack of a specific method for localizing ACh in neurons hampered the study of cholinergic pathways in the CNS for many years, but it is now clearly established that ACh is an important transmitter also in the CNS.

The breakdown of ACh is facilitated by acetylcholinesterase (AChE), which can be localized by histochemical methods. AChE, however, has also been identified in some noncholinergic pathways, and is therefore not a perfect marker for cholinergic neurons. With the development of antibodies against choline acetyltransferase (ChAT), the enzyme that catalyzes the synthesis of ACh, it is now possible to identify cholinergic neurons by immunocytochemical methods. The cholinergic pathways, accordingly, have been mapped in considerable detail (Chapter 16).

One of the most prominent cholinergic systems arises in the basal nucleus of Meynert (Fig. 18-1B). The large neurons in this nucleus and in nearby regions of the basal forebrain project to widespread regions of the cerebral cortex and several other forebrain regions. This system appears to be of special significance for cortical arousal and related mechanisms of learning and memory. One hypothesis that has attracted much attention is that the symptoms of Alzheimer's disease are due in part to degeneration of these basal forebrain cholinergic neurons. As indicated in Chapter 21, however, cortical association areas, including the hippocampus formation, are also the focus of pathologic changes in this disease.

Another prominent ascending cholinergic system originates in the pedunculopontine-laterodorsal tegmental complex in the upper brain stem (Figs. 10-12 and 16-15). The brain stem cholinergic system, like the basal nucleus of Meynert, is part of the ascending activating system (Chapter 10).

Some of the interneurons in the striatum use ACh as their transmitter. Although loss of striatal cholinergic interneurons is one of the features of Huntington's disease, the main pathologic event in this disease is a loss of GABA-ergic projection neurons (see Clinical Notes, Chapter 16). Unlike the situation in Alzheimer's disease, the corticopetal basal forebrain cholinergic neurons are usually not involved in Huntington's disease.

Monoamines

The monoamine transmitters, i.e., serotonin, dopamine, noradrenaline, adrenaline, and histamine, are derived from amino acids, and most of the monoaminergic pathways take their origin from cell groups in the brain stem (Chapter 10). The catecholamines (dopamine, noradrenaline, and adrenaline) are derived from tyrosine (see Vignette), serotonin from tryptophan, and histamine from histidine.

Dopamine

The mesencephalic dopamine cells, located primarily in the substantia nigra, pars compacta, and the ventral tegmental area, form a massive projection system, the mesotelencephalic dopaminergic system (Fig. 10-7). This ascending system reaches all parts of the striatum and many other regions in the telencephalon, e.g., the amygdaloid body including the extended amygdala, septal, and olfactory areas as well as the cerebral cortex. Dopamine-containing cell bodies are also located in several forebrain regions including the hypothalamus, olfactory bulb and retina.

Degeneration of the nigrostriatal dopaminergic system is the most characteristic pathologic feature in Parkinson's disease (see Clinical Notes, Chapter 16). Other transmitter systems are usually also affected in Parkinson's disease, especially the noradrenaline system, which may in part explain why restoring dopamine by L-DOPA therapy is not sufficient to alleviate the symptoms of Parkinson's disease. Dopamine also appears to play a significant role in schizophrenia (see Clinical Notes, Chapter 16).

Noradrenaline and Adrenaline

Noradrenaline (norepinephrine) is released from all postganglionic sympathetic fibers except those to the sweat glands, which are innervated by cholinergic fibers. Noradrenaline is also an important transmitter in the CNS, where more than half of the noradrenergic cells are located in the locus ceruleus.

The noradrenergic projections are widespread, as they affect practically every major region in the brain and spinal cord (Fig. 10-9). In general, it appears that the noradrenergic projections enhance the ability of their target neurons to respond to whatever input they have to deal with. Most of the ascending noradrenergic fibers originate in the locus ceruleus, which appears to play an especially important role in the sleep-wake cycle. Ascending adrenergic fibers are much more restricted in their distribution than noradrenergic fibers, in the sense that they reach primarily the paraventricular nucleus, periaqueductal gray, and locus ceruleus. A stimulatory influence on eating behavior can be mentioned as one of the many potential effects that noradrenaline and adrenaline exert on the paraventricular nucleus (Chapter 18).

The intermediate reticular zone in the medulla oblongata and caudal pons (Chapter 10) contains both noradrenergic and adrenergic cells, which are involved in circuits related to visceral, cardiovascular, and respiratory functions.

Serotonin

The cells of origin of the serotonin system are located mainly in the raphe nuclei, which form an extensive, more or less continuous collection of cell groups close to the midline throughout the brain stem (Fig. 10-6). The axons of the cells in the rostral group of raphe nuclei reach all parts of the forebrain, including the cerebral cortex. Some of the raphe nuclei project to the cerebellum, and those located in the medulla oblongata send projections to the spinal cord.

Like the noradrenergic system, the serotonin neurons appear to play an important role in the sleep–wake cycle. The activity of the serotonin system decreases during rapid eye movement (REM) sleep when muscle relaxation is pronounced, which is consistent with a serotonergic excitatory effect on the motor neurons. Food intake, hormone secretion, sexual behavior, and thermoregulation also seem to be influenced by the serotonin system (Chapter 18). Descending serotonin pathways are important components of a central pain control system (Chapter 9), and they are also involved in many other functions including motor activity and cardiovascular regulation.

Histamine

The histaminergic projections, with cells of origin located in the tuberomammillary nucleus in the posterolateral hypothalamus (Fig. 18–3C), reach all major regions of the forebrain and several structures in the brain stem and spinal cord (Fig. 10–10). It has only recently become evident that histamine is a transmitter in the CNS, but its functional significance is not well known. The most compelling evidence relates to its role in autonomic and neuroendocrine functions (Chapter 10).

Amino Acids

γ-Aminobutyric Acid (GABA) and Glycine

Both GABA and glycine are inhibitory transmitters. GABA is derived from glutamic acid, a reaction which is catalyzed by the enzyme glutamic acid decarboxylase (GAD). GABA is often colocalized with other classic transmitters, e.g., acetylcholine, or with different peptides like somatostatin and cholecystokinin, or with enkephalin and substance P, as is the case in the striatopallidal and striatonigral pathways (Chapter 16).

The distribution of GABA-ergic neurons and pathways has been successfully revealed by the aid of immunohistochemical techniques using antibodies against GAD or GABA, and it is now evident that most of the inhibitory interneurons in the CNS are GABA-ergic. Well-known examples are the Renshaw cells in the spinal cord and the basket cells and other inhibitory local circuit neurons in the cerebellar cortex. A large number of GABA-ergic local circuit neurons are also located in other parts of the brain and spinal cord.

GABA is also used by several groups of projection neurons, including the inhibitory pathway from the Purkinje cells to the intracerebellar nuclei, many corticopetal projection neurons in the basal forebrain, and the striatopallidal and striatonigral projections. Indeed, the concentrations of GABA in the globus pallidus and the substantia nigra are among the highest in the brain.

Two major types of GABA receptors have been identified, GABA-A and GABA-B, with GABA-A being the more widely distributed. The GABA-A receptor complex plays a major role in the control of neuronal excitability throughout the brain, and there is considerable evidence that drugs like the benzodiazepines (e.g., Valium and Librium) owe their tranquilizing effect to the fact that they interact with the GABA-A receptor complex.

Whereas GABA is present throughout the CNS, glycine is more restricted in its distribution to primarily the spinal cord and brain stem. Several of the inhibitory interneurons in the spinal cord contain glycine, which is sometimes colocalized with GABA. Glycine, like GABA, is also present in some projection fibers, especially in the auditory system.

Glutamate and Aspartate

These two amino acids are the major excitatory neurotransmitters, but they are also involved in many other metabolic functions. For instance, as indicated earlier, glutamate is a precursor to GABA, and it is also a component of proteins. The presence of glutamate or its synthetic enzyme, therefore, does not necessarily mean that the neuron uses glutamate as its transmitter. Glutamate is more widely distributed and more potent than aspartate, and a number of major pathways are known to use glutamate as their transmitter, including the various corticofugal pathways, e.g., the corticostriatal, corticothalamic, corticopontine, and corticospinal projections. Glutamate is also the main transmitter in cortical associational and commissural connections. Furthermore, it is involved in several of the sensory systems, as well as in perforant path and other pathways through the hippocampus.

There are several different classes of excitatory amino acid receptors, which are defined by the action of specific agonists. One of the receptor types, the NMDA receptors, which are selectively activated by *N*-methyl-D-aspartic acid, have attracted attention because they are believed to play a role in *long-term potentiation* (LTP), a mechanism believed to be related to memory. LTP, which is long-lasting (days to weeks) enhancement of synaptic transmission, can be obtained in certain pathways by a brief period of high-frequency stimulation. This phenomenon was first described in the hippocampus, where it is most easily obtained, and since the hippocampus appears to be of special significance for memory, LTP has been promoted as a potential candidate for the neural mechanisms of learning and memory. There is some evidence that LTP may be related to the formation of new boutons, which would mean an increase in the amount of transmitters available.

It has been shown that too much glutamate can be harmful to neurons, and glutamate, therefore, has also attracted much interest as a neurotoxin and as a potential culprit in neurodegenerative disorders. The idea is that any condition that permits excessive stimulation of glutamate receptors (e.g., excessive release or deficient reuptake of the transmitter) might result in excitotoxic degeneration of neurons. This has been suggested as one of the pathophysiologic mechanisms involved in Huntington's disease (Chapter 16). Excessive release of glutamate, furthermore, has been suggested as a possible mechanism for brain damage in epilepsy (Chapter 20) and in hypoxic ischemia. Destruction of glutamatergic pathways, on the other hand, may play a role in schizophrenia (Chapter 16).

Neuropeptides

Close to 50 peptides, i.e., short chains of amino acids, have been identified in the brain, and most of these appear to play important roles as neurotransmitters, modulators, or hormones. Substance P was the first peptide to be discovered in the brain and the intestines in 1931, but it was not isolated until many years later. As discussed in Chapter 18, other peptides were first identified as pituitary hormones, e.g., adrenocorticotropin (ACTH), oxytocin, and vasopressin, or as hypothalamic releasing or inhibiting hormones, e.g., corticotropin-releasing hormone (CRH), thyrotropin-releasing hormone (TRH), and somatostatin (SST). Several gastrointestinal peptides besides substance P are also present in the brain, e.g., vasoactive intestinal polypeptide (VIP), cholecystokinin (CCK), and neurotensin (NT). The opioid peptides, i.e., enkephalins, dynorphins, and endorphins, represent another family of neuropeptides. The term *endorphin* (endogenous morphine) was introduced in the mid-1970s when an endogenous morphine-like factor was discovered. It was suggested that the endorphins act on specific receptor sites that can also be utilized by the narcotic drug morphine. The opioid peptides have attracted enormous interest, not least because of their apparent role in pain regulation (Chapter 9).

Although there are certainly variations in regard to the distribution of the various peptides, some areas of the CNS are especially rich in many of the peptidergic cell bodies and terminals listed below. These areas include the extended amygdala (the centromedial amygdaloid complex and the bed nucleus of the stria terminalis), striatum, and hypothalamus in the forebrain, the periaqueductal gray, raphe nuclei, and vagal-solitary complex in the brain stem, and the substantia gelatinosa in the spinal cord. Many of the interneurons in the cerebral cortex are also peptidergic, whereas the peptide content is considerably less in the thalamus and all but absent in the cerebellum, which, however, contains some of the peptide receptors.

The amino acid and monoamine transmitters, which are produced directly from dietary sources by the action of a few enzymes, can be produced in any part of the neuron including the synaptic terminals (Chapter 6). The neuropeptides, on the other hand, are synthesized by messenger RNA in relation to ribosomes, which are almost without exception localized in the cell body. The neuropeptides, or rather "neuropeptide precursors" (large polyprotein molecules), which are subsequently cleaved by the action of proteolytic enzymes to form the neuroactive peptides, are therefore packaged into vesicles and transported to the terminals.

Enkephalins, Dynorphins, and Endorphins

The best-known of the opioid peptides are met-enkephalin (met-ENK) and leu-enkephalin (leu-ENK), which are derived from proenkephalin. They are identical pentapeptides except for the terminal amino acid, which is methionine in met-ENK and leucine in leu-ENK. Enkephalinergic cell bodies and terminals are widely distributed in the brain and spinal cord, especially in the areas mentioned above as typical for their rich content of neuropeptides. The external segment of the globus pallidus and the substantia nigra, pars reticulata, show an exceptionally dense staining of enkephalinergic fibers, which is a reflection of the fact that they represent the two main targets for axons originating in enkephalinergic striatal projection neurons (Chapter 16).

The superior ability of morphine to relieve severe pain is hardly a secret to anyone; in fact, it was introduced in medicine as an analgesic at the beginning of the last century. The euphoric effect of the opiates is also legend, but so are the many undesirable side effects on the respiratory, cardiovascular, and gastrointestinal systems. With this in mind, and considering the distribution of the opioid peptides, i.e., in the basal ganglia and in the "central autonomic network" (extended amygdala, hypothalamus, periaqueductal gray, and vagal-solitary complex), as well as in the substantia gelatinosa, it is easy to appreciate that the "brain's own morphine," i.e., the endogenous opioid peptides, are involved in functions ranging from the regulation of movements and of mood and emotions, to control of hypothalamohypohysial and autonomic functions, to modulation of pain transmission.

Cell bodies and terminals containing dynorphins (derived from prodynorphin) are widely distributed in a pattern reminiscent of that of the enkephalins, although the two types of opioid peptides are generally not located in the same neuron. Nonetheless, since the enkephalins and dynorphins have a similar distribution, they are likely to be involved in similar functions.

β-Endorphin is the main opioid peptide derived from proopiomelanocortin. The precursor is confined to cell bodies in the pituitary, the arcuate nucleus of the hypothalamus, and the nucleus solitarius. The fibers from these cell groups reach many regions in the basal forebrain and brain stem, including the periaqueductal gray and the raphe magnus, where β-endorphin appears to play a role in the pain-inhibiting system.

Oxytocin and Vasopressin (ADH)

The involvement of these two peptides in the hypothalamic projection to the neurohypophysis and to autonomic centers in the brain stem and spinal cord was discussed in Chapters 18 and 19. Vasopressin fibers from the paraventricular hypothalamic nucleus also project to the median eminence, where the peptide is secreted into the hypohysial portal system, apparently together with CRH, for the purpose of regulating the secretion of ACTH from the anterior pituitary.

Adrenocorticotropin (ACTH)

ACTH, like β-endorphin, is derived primarily from the proopiomelanocortin-containing cell group in the arcuate nucleus, but the ACTH fibers, like the β-endorphin fibers, reach many areas in both the forebrain and the brain stem.

Corticotropin-Releasing Hormone (CRH)

CRH-containing cells are located in the paraventricular hypothalamic nucleus (Chapter 18) and in many other parts of the basal forebrain and brain stem, including the extended amygdala, the lateral preoptic–hypothalamic continuum, the periaqueductal gray, the parabrachial nucleus, and the nucleus of the solitary tract. Brain stem areas contain ing the noradrenergic cell groups, e.g., the locus ceruleus and the intermediate reticular zone, are also characterized by the presence of CRH-positive cells.

As indicated above, CRH is involved in the regulation of ACTH release from the adenohypophysis, and it is probably no accident that the distribution of CRH-containing cell groups and circuits in the CNS in general matches that of the ACTH fibers. CRH, in conjunction with several other transmitters such as noradrenaline and vasopressin, appears to be especially important for mobilizing the various bodily functions during stress.

Thyrotropin-Releasing Hormone (TRH)

Axons from TRH-containing cell bodies in the periventricular hypothalamic area and paraventricular nucleus project to the median eminence, which has the highest TRH concentration in the brain. TRH, like other hypophysiotropic hormones, are discharged into the perivascular spaces surrounding the portal capillaries. Besides this neuroendocrine part of the TRH system, TRH-containing cell bodies and fibers are present in other parts of the hypothalamus and in the brain stem, especially in the motor cranial nerve nuclei and in the caudal raphe nuclei and adjacent reticular formation, which give rise to descending projections to spinal cord motor regions. TRH fibers, accordingly, are well placed to be involved in the modulation of motor functions.

Luteinizing Hormone-Releasing Hormone (LHRH)

This hormone is also known as gonadotropin-releasing hormone (GnRH), and its neuroendocrine functions are well known (see discussion of reproduction in Chapter 18). Many LHRH-positive neurons are located outside the basomedial hypothalamus, especially in the extended amygdala (medial amygdaloid nucleus and bed nucleus of stria terminalis), medial preoptic area, and septum-diagonal band area.

Somatostatin (SST)

Somatostatinergic neurons are widely distributed in the brain, but their functions are not well understood. Somatostatin-containing interneurons are present in both the cerebral cortex and the striatum. Many structures in the basal forebrain, including the extended amygdala and the hypothalamus, especially the periventricular region and the preoptic–anterior continuum, contain a large number of somatostatinergic neurons. Some of the hypothalamic SST neurons project to the median eminence, and it is known that SST inhibits the release of growth hormone (GH), thyrotropin (TSH), and prolactin (PRL).

Other hypothalamic SST neurons project to nearby regions of the hypothalamus and to the extended amygdala, the septal area, and the habenula. Downstream projections to the brain stem and even the spinal cord have been described, and they are augmented by fibers from somatostatinergic cells in the brain stem. Even the spinal cord, especially the substantia gelatinosa and the spinal ganglia, have SST-containing cells, and it has been suggested that somatostatin is involved in the regulation of pain perception.

Considering its extensive distribution in the brain, somatostatin is bound to have a number of important functions besides modulating pain transmission. In general, it enhances the action of ACh, and it is often colocalized with GABA in axon terminals, but the significance of this is not known.

Substance P (SP)

Substance P (P stands for preparation) is widely distributed throughout the central and peripheral nervous system including the autonomic nervous system and endocrine glands. Its distribution pattern in the CNS is reminiscent of that of the enkephalins, which has promoted the idea that the two types of peptides have a close functional relationship. The dorsal root ganglia contain a large number of SP-positive cells, and SP neurons are also located in the dorsal horn of the spinal cord, especially in the substantia gelatinosa. Substance P, therefore, is believed to play an important role in the transmission of sensory impulses, and its role in pain transmission was specifically addressed in Chapter 9. SP and glutamate may be cotransmitters in some of the sensory fibers coming into the spinal cord, in which case SP is likely to lower the threshold for excitation by the glutamate. Other peptides, e.g., CCK and VIP, may also be involved in sensory mechanisms at the level of the spinal cord.

Brain stem regions noted for their content of SP include the vagal–solitary complex and the raphe nuclei, as well as the periaqueductal gray and neighboring reticular formation, where some of the cholinergic cells in the pedunculopontine–laterodorsal tegmental complex (Chapter 10) contain SP. Many of the GABA-ergic striatal projection neurons contain SP, and the majority of these fibers project to the internal segment of the globus pallidus and the substantia nigra (Chapter 16), which therefore belong to regions with the highest concentration of SP in the brain. The hypothalamus contains only scattered SP neurons, but it is richly innervated by SP-containing axons, some of which come from the amygdala.

Cholecystokinin (CCK)

CCK is another gut peptide, which is widely distributed in the brain and spinal cord. High concentrations are found in the cerebral cortex, where it appears to be located in both excitatory and inhibitory interneurons. The amygdaloid body and many of the hypothalamic nuclei, including the supraoptic and paraventricular nuclei, contain CCK neurons, and CCK has attracted considerable interest because of its inhibitory effect on eating (Chapter 18).

Some of the gut–brain peptides, including CCK and neurotensin (see below), are of special interest in schizophrenia research because of their interplay with dopamine. In short, some of the target areas for the mesotelencephalic dopamine projections, i.e., the ventral striatum, extended amygdala, and frontal cortex, all of which have been implicated in schizophrenia (see "Emerging Concepts of Basal Forebrain Organization" in Chapter 16), are also characterized by high concentrations of CCK, and there is some indication that CCK inhibits dopamine in some of these target areas.

Vasoactive Intestinal Polypeptide (VIP)

Many of the excitatory interneurons in the cerebral cortex contain VIP, which is also well represented in the striatum, extended amygdala, and hypothalamus, where VIP-positive cell bodies are concentrated especially in the suprachiasmatic nucleus. VIP is present in high concentration in the median eminence and the portal blood, and it is reported to have a modulating effect on the release of many hormones. Other areas with high concentrations of VIP include the periaqueductal gray and the nucleus solitarius. VIP is also prominent in the autonomic nervous system and is known as a potent vaso- and bronchodilator.

Neurotensin (NT)

Neurotensin-containing cell bodies in the forebrain are located especially in the extended amygdala and hypothalamus, where the peptide is generally believed to affect several of the functions related to the hypothalamohypophysial system, including the release of some of the pituitary hormones. NT, like the opioid peptides, is known for its pain-inhibiting effects, and it is present in high concentration in both the periaqueductal gray and the substantia gelatinosa. NT is colocalized with dopamine in the tuberoinfundibular pathway (Chapter 10), and it reportedly has a stimulating effect on the release of both GH and PRL.

NT, like CCK (see above), has been of considerable interest to neuropsychiatrists since it was discovered that the NT concentration in the target areas of the ascending mesotelencephalic dopamine pathway could be differentially affected by the administration of antipsychotic drugs.

Angiotensin (ANG)

Angiotensin (derived from the renin–angiotensin system of the kidney) and its role in the regulation of water balance and blood pressure was discussed in the context of the neurohypophysial hormones in Chapter 18. It is now clear that neurons in the CNS can produce its own angiotensin, but the angiotensin-containing cell bodies are more restricted in their location than many of the other peptidergic neurons, in the sense that they are located primarily in the periventricular region of the hypothalamus, including the paraventricular and supraoptic nuclei. As a neurotransmitter, angiotensin stimulates the release of vasopressin and oxytocin from the posterior pituitary, and it also modulates the release of several other pituitary hormones. The regions most affected by the release of angiotensin in the medulla oblongata are the general region in and around the dorsal motor nucleus of the vagus and the nucleus of the solitary tract, both of which are associated with cardiovascular, respiratory, and autonomic functions.

Suggested Reading

1. Björklund, A. and T. Hökfelft (eds.): 1984–1992. Handbook of Chemical Neuroanatomy. Vols. 2–3 (1984), Vol. 4 (1985), Vol. 9 (1990), and Vols. 10–11 (1992). Elsevier, Amsterdam.
2. Cooper, J. R., F. E. Bloom and R. H. Roth, 1991. The Biochemical Basis of Neuropharmacology. 6th ed. Oxford University Press: New York.
3. Hökfelt, T., K. Fuxe and B. Pernow, 1987. Coexistence of neuronal messengers: A new principle in chemical transmission. Progr. Brain Res. 68: 1–411.
4. Hökfelt, T., 1991. Neuropeptides in perspective: The last ten years. Neuron 7:867–879.
5. Nieuwenhuys, R., 1985. Chemoarchitecture of the Brain. Springer-Verlag, Berlin.
6. Ottersen, O. P., O. P. Hielle, K. K. Osen and J. H. Laake, 1994. Amino acid transmitters. In G. Paxinos (ed.): The Rat Nervous System (2nd. ed.) Academic Press: San Diego.
7. Riederer, P., N. Kopp, and J. Pearson, 1990. An Introduction to Neurotransmission in Health and Disease. Oxford University Press: Oxford.

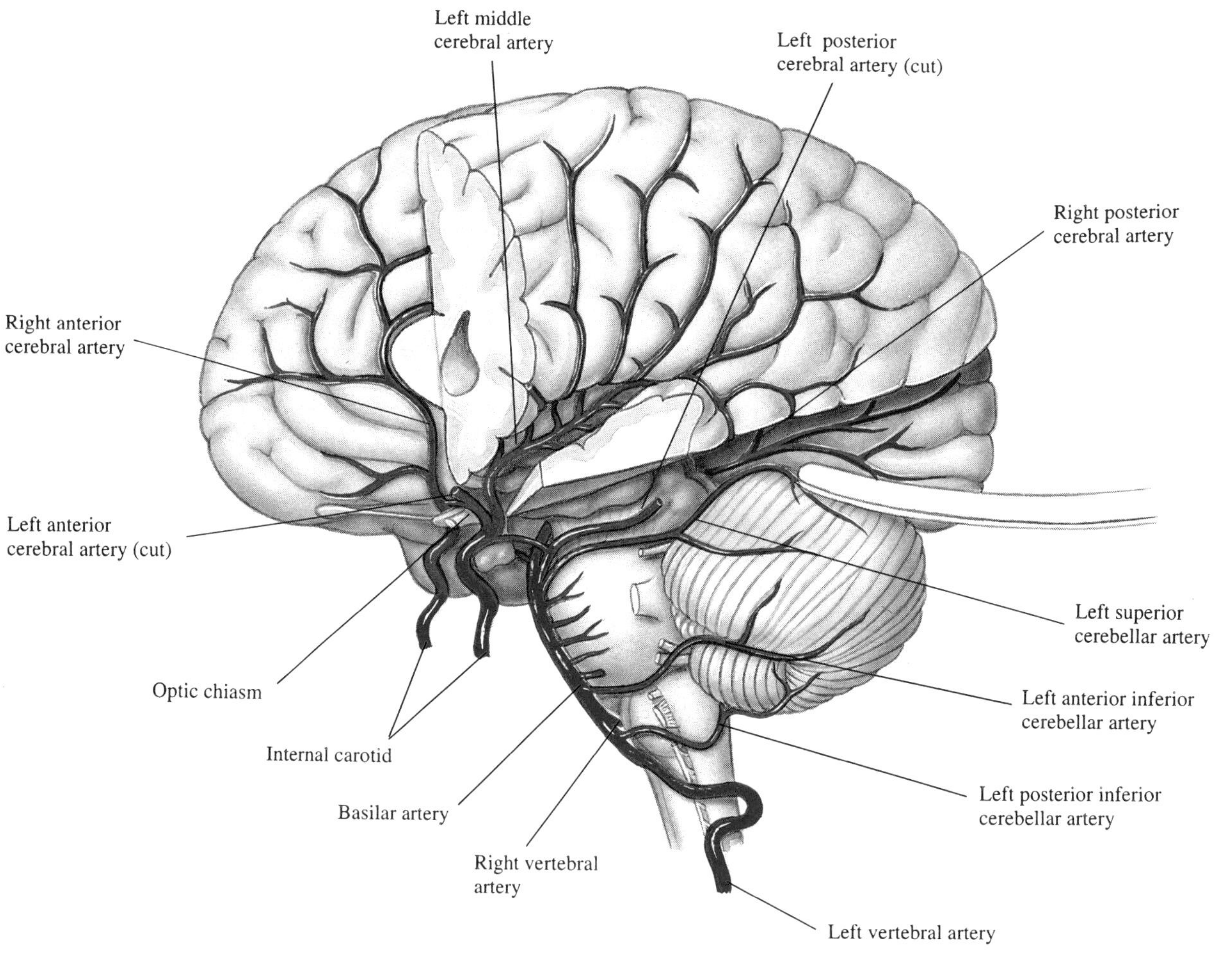

Arterial Supply of the Brain
Semidiagrammatic figure showing the two major arteries, the internal carotid artery and the vertebral artery, carrying blood to the brain. Part of the left hemisphere has been removed in order to show the distribution of the anterior and posterior cerebral arteries on the medial side of the hemisphere. The three main cerebellar arteries are also illustrated. (Art by Medical and Scientific Illustration, Crozet, Virginia.)

24

Cerebrovascular System

Cerebrovascular disease (stroke) is one of the most common of all disorders and a major cause of disability and death. Continuous and vigorous research, however, is gradually improving our abilities to diagnose and prevent strokes, and an increasing number of stroke victims are being successfully treated by a competent medical staff.

As in most diseases of the nervous system, the clinical features of stroke are to a large extent dependent on the location of the pathologic process, and the examination of a stroke patient is likely to become an exercise in futility unless one has a basic understanding of the main arteries and their distribution. Fortunately, it is relatively easy to acquire such knowledge.

The frequency of cerebrovascular disorders makes the cerebrovascular system one of the most important subjects in clinical neurology. Pathologic changes in brain arteries alone are responsible for almost 50% of all neuro[illegible] a general hospital, and patients w[illegible] occlusion or a hemorr[illegible] arteries are frequent[illegible] practice. Since the [illegible] syndrome cannot be [illegible] knowledge of the co[illegible] major arteries, the morphology [illegible] vessels is a subject of the highest priority in any neuroscience course for medical students. Blood vessels and vascular disorders related to the meninges are discussed in Chapter 3.

The cerebrovascular system can be visualized radiographically by injecting a contrast medium into the carotid or vertebral arteries. Roentgenograms obtained in this manner are termed angiograms. To be able to interpret the angiograms and to appreciate various abnormalities of the cerebrovascular system, its normal anatomic appearance must be known. The anatomy of the cerebrovascular system is a matter of genuine concern for most experts on the CNS, regardless of whether they are interested in its clinical–pathologic aspects or more basic physiologic problems of blood flow.

Cerebral Blood Flow

Regulation

Although the brain represents only 2% of the body weight, it receives about 15% of the cardiac output. It cannot effectively store either oxygen or glucose, and is critically dependent on a continuous blood supply.

Although the cerebral vessels are supplied with autonomic nerves, neurogenic control is apparently of minor importance under normal conditions. Instead, the cerebral blood flow appears to be regulated locally by metabolic factors such as H^+, K^+, and adenosine. This local control of blood flow permits a discrete distribution in response to regional variations in activity (Fig. 22–8). In addition, the brain has the capacity to maintain relatively constant blood flow despite large changes in systemic blood pressure. This phenomenon, which is known as *autoregulation*, also appears to be under metabolic control. However, when mean arterial pressure decreases to less than 50–70 mmHg, autoregulation fails and blood flow decreases.

Cerebral blood flow changes with altered levels of oxygen (Pa_{O_2}) and carbon dioxide (Pa_{CO_2}) in the arterial blood. When Pa_{O_2} is decreased to 50 mm Hg (hypoxia), cerebral blood flow is increased twofold. Changes in Pa_{CO_2} have an even more dramatic effect, with hypercarbia causing an increase and hypocarbia a decrease in blood flow.

[illegible]ting and Syncope

[illegible]den reduction in blood flow to the entire [illegible]n inadequate amount of oxygen in the [illegible] result in *faintness*, characterized by a [illegible]ness and dizziness. If the deficiencies [illegible] blood flow are more pronounced, [illegible]ay suffer a more or less complete loss of consc[illegible]sness, i.e., *syncope*. Episodic attacks of faintness and syncope are common, and can be caused by a number of conditions, including cardiac disease, defective vasovagal reflexes, or strong emotions.

Stroke

A cerebrovascular disease that results in a sudden appearance of neurologic deficits is referred to as *stroke*. This term includes both blockade of a cerebral artery (or vein) leading to cerebral infarction, and hemorrhage from an artery (or vein). *Transient ischemic attacks* (*TIAs*) are often forerunners of a major stroke.

Cerebral Thrombosis and Hemorrhage

Arterial strokes are much more common than venous strokes. The immediate cause is usually an occlusion of a vessel by a thrombus (blood clot or cholesterol deposit) or bleeding from a ruptured vessel, which results in destruction of nervous tissue. Atherosclerosis, hypertension, and cardiac disease are the main contributing factors to stroke. Bleeding that spreads through the subarachnoid space rather than within the brain tissue is called a *subarachnoid hemorrhage* (*SAH*). It is often the result of a ruptured aneurysm or trauma. Less frequently, SAH is caused by a rupture of an *arteriovenous malformation* (*AVM*).

Collateral Circulation

Although the location and size of the pathologic process to a large extent determine the symptoms of a stroke, the effect on the brain tissue of a gradually developing occlusion is also dependent on the available anastomotic channels and the possibility for collateral circulation. The clinical importance of the *circle of Willis* (Fig. 4–4) should be emphasized in this context. A thrombosis in the internal carotid artery is likely to cause a massive cerebral infarct resulting in the death of the patient, but it may also go unnoticed if there is an efficient collateral circulation in the circle of Willis.

Anastomotic channels between the external carotid and the internal carotid systems in the region of the ophthalmic artery may be of great functional importance in the case of a carotid artery thrombosis. Collateral circulation through superficial pial anastomoses between the major arteries of the cerebrum (anterior, middle, and posterior cerebral arteries) can also be quite effective. Similar anastomoses exist in the cerebellum between the superior cerebellar, anterior inferior cerebellar, and posterior inferior cerebellar arteries. The perforating arteries to the deep structures of the cerebrum, cerebellum, and brain stem, however, are in effect "end arteries," and occlusions of these arteries usually have severe effects.

Stroke Therapy

Stroke therapy is aimed at restoring circulation and preventing further strokes. For TIAs an effective therapy has proven to be endarterectomy with surgical removal of an obstruction usually in the carotid artery at the level of the bifurcation. Similar operations on the upper intracranial parts of the vertebral artery, where most atherosclerotic plaques in the vertebral–basilar system are located, have traditionally been considered too difficult. However, great progress has recently been made in endovascular procedures which utilize a balloon to dilate the constriction. There is hope that atherosclerotic plaques in the vertebral arteries, and maybe even in some of the other large arteries in the base of the brain, can be successfully treated in the near future. Partial occlusion of internal carotid or vertebral–basilar systems can often be arrested by anticoagulation.

Vascular Anatomy and Summary of Neurovascular Syndromes

The brain is supplied with blood through four large trunks, the paired internal carotid arteries and vertebral arteries, which approach the ventral brain surface where they unite to form the *circle of Willis* (Figs. 4–1 and vignette).

Each internal carotid artery enters the skull through the carotid canal, and at the level of the optic chiasm bifurcates to form the anterior cerebral and the middle cerebral arteries. The *internal carotid system* and its two main branches, the anterior and middle cerebral arteries, supply approximately the rostral two-thirds of the brain including the main parts of the basal ganglia and the internal capsule (Fig. 24–2). Cerebrovascular lesions in the territory of the internal carotid artery are characterized by contralateral signs: hemiplegia, hemianesthesia, and hemianopia. If the deficits are on the right side of the body there is usually aphasia as well, as a result of infarction in the distribution of the left middle cerebral artery (see Clinical Examples).

The two vertebral arteries, which enter the skull through the foramen magnum, unite at the pontomedullary junction to form the basilar artery (Fig. 4–1). The basilar artery divides into the two posterior cerebral arteries at the pontomesencephalic junction. The *vertebral–basilar system* supplies the cerebellum, the brain stem, most of the thalamus, and the posterior parts of the hemispheres. Lesions in this system may result in cerebellar ataxia, various brain stem syndromes, and, in case of involvement of the posterior cerebral artery, a homonymous hemianopia.

Internal Carotid System

Internal Carotid Artery

This artery arises from the common carotid artery in the neck and passes through the carotid canal of the petrous bone. It then turns medially and proceeds through the cavernous sinus. In doing so, it curves in a S-like shape, forming the *carotid siphon* (Fig. 24–5). The intracranial portion of the internal carotid artery bifurcates into the middle and the anterior cerebral artery lateral to the optic chiasm. Before it divides, however, it gives rise to the following branches:

1. Inferior and superior hypophysial arteries
2. Ophthalmic artery
3. Posterior communicating artery
4. Anterior choroidal artery.

Vascular Syndrome. Atheromatous plaques often form at bifurcations, and the region of the carotid bifurcation is a common site of internal carotid occlusion. The clinical picture of carotid artery thrombosis varies according to the compensatory capacity of the circle of Willis. Carotid thrombosis can be completely silent or it can cause massive destruction of those parts of the ipsilateral hemisphere that are supplied by the middle and anterior cerebral arteries (Fig. 24–1). Most often the territory of the middle cerebral artery is affected, giving rise to contralateral weakness and sensory loss, most pronounced in the face and arms. The speech areas are usually involved if the lesion is on the left side. Involvement of the anterior cerebral artery usually gives rise to sensorimotor deficits in the contralateral foot and leg.

Many of the strokes related to the internal carotid artery are preceded by TIAs. These TIAs, like more

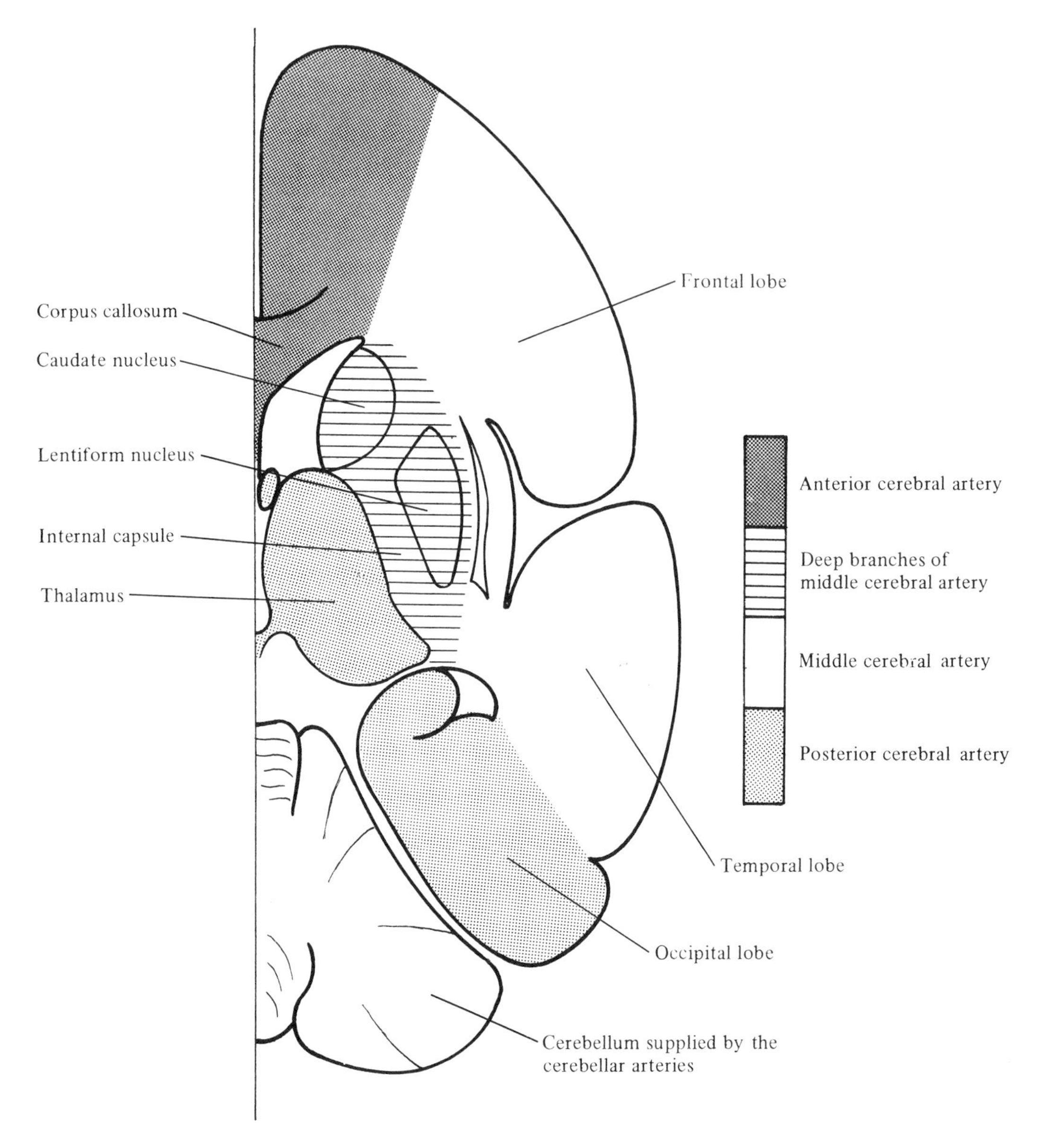

Fig. 24–1. Regional Blood Supply of the Main Cerebral Arteries
Diagram of the cerebral hemisphere showing the territories of the main cerebral vessels in a horizontal section. (Modified from C. M. Fisher, 1975. The Anatomy and Pathology of the Cerebral Vasculature. In: Modern Concepts of Cerebrovascular Disease [Ed. J. S. Meyer]. Spectrum Publications, Inc., New York, Toronto, London, Sidney.)

permanent strokes, manifest themselves in many forms depending on the artery that is temporarily occluded: monocular blindness (ophthalmic artery), hand weakness (middle cerebral artery), and speech and hand disorders (middle cerebral artery of dominant hemisphere). Since sclerotic plagues in the internal carotid artery can often be removed surgically, or thrombosis treated medically, it is important to be aware of these warning attacks before the brain is permanently damaged.

Middle Cerebral Artery

This is the major branch of the internal carotid artery. It proceeds laterally toward the lateral fissure between the frontal and temporal lobes (Fig. 4–4). It lies below the anterior perforated space on

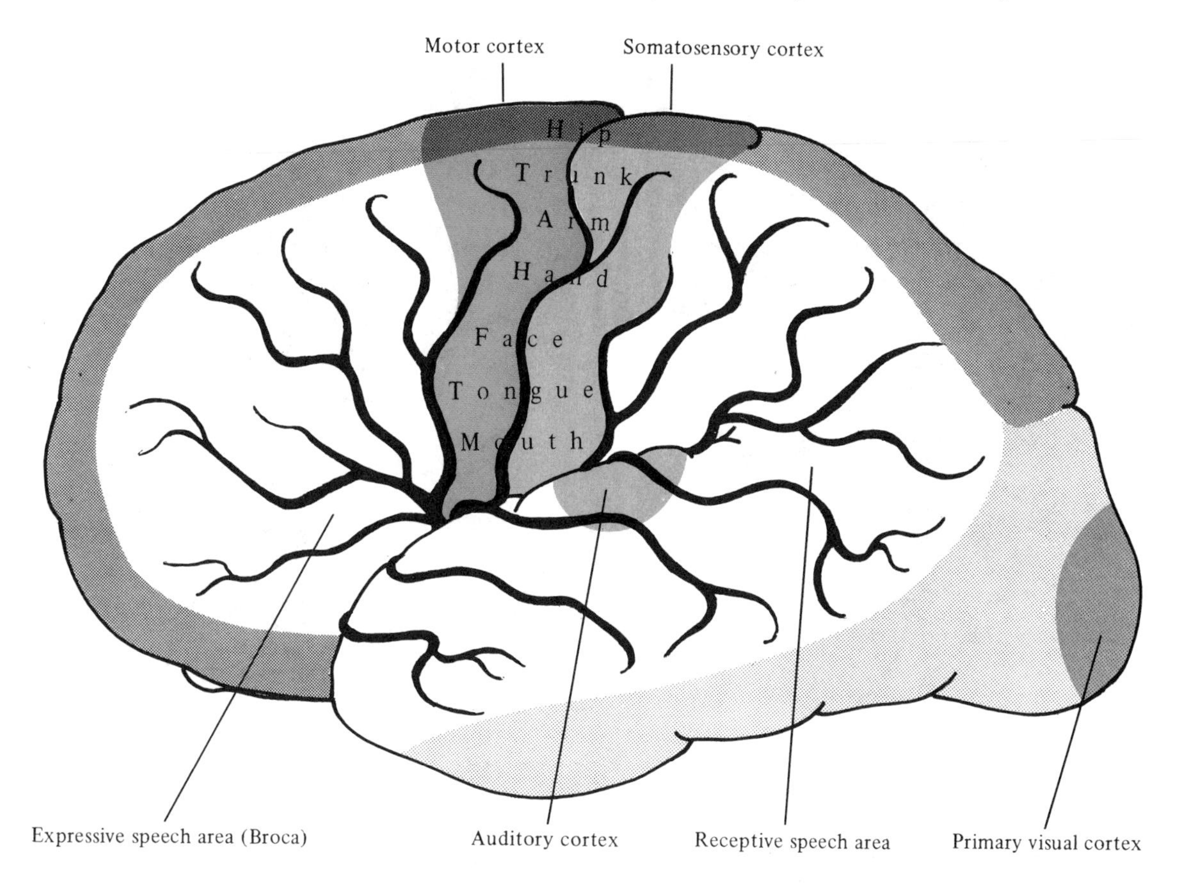

Fig. 24–2. Arterial Blood Supply of the Lateral Brain Surface

Diagram showing the territories of the main cerebral vessels, the branches of the middle cerebral artery, and the main area of cerebral localization on the lateral surface of the left hemisphere. (*Dark gray* = territory of anterior cerebral artery; *light gray* = territory of posterior cerebral artery.)

its way to the lateral fissure. A series of 7–10 branches, the *striate arteries*, arise at right angles from the proximal segment of the middle cerebral artery and perforate this space. The medial group of the striate arteries is joined by the *recurrent artery of Heubner* from the anterior cerebral artery. The striate arteries supply the main part of the striatum, including the fundus striati, the globus pallidus, and the internal capsule.

When the middle cerebral artery reaches the lateral fissure at the level of the limen insulae it divides into several cortical branches on the surface of the insula. Most of these branches exhibit a tortuous course on the surface of the insula before they pass out of the lateral fissure to distribute to most of the cortex and underlying white matter on the lateral side of the hemisphere (Fig. 24–2).

Vascular Syndrome. The middle cerebral artery and its branches supply the main part of the cerebral hemisphere, including not only the basal ganglia and a major part of the internal capsule, but also important motor centers in the frontal lobe and somatosensory centers in the parietal lobe. The speech areas are also supplied by this artery, usually on the left side. A neurovascular syndrome resulting from occlusion of the stem of the middle cerebral artery, therefore, includes contralateral hemiplegia and hemianesthesia, homonymous heminopsia, as well as global aphasia in most patients with left-sided lesions.

Of all cerebral arteries, the middle cerebral artery is the one most often affected by a cerebrovascular accident. The large majority of *cerebral embolisms* (detached blood clot, often from the heart), affect the middle cerebral artery, and the site of occlusion depends on the size of the embolus (see Middle Cerebral Artery Infarction in Clinical Examples). Most of the emboli are small and usually reach far peripherally into some of the cortical branches, thereby producing only part of the full-

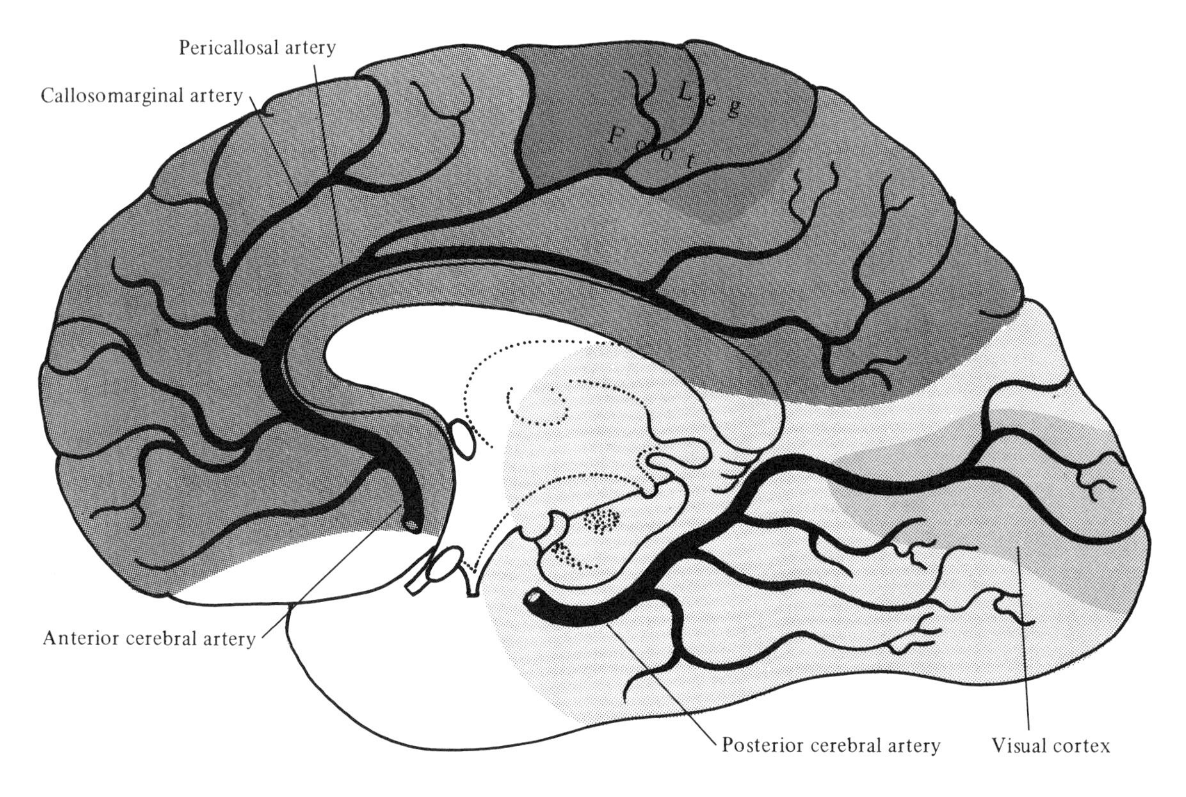

Fig. 24–3. Arterial Blood Supply of the Medial Brain Surface
Diagram of the medial aspect of the brain showing the territories of the main cerebral vessels and the cortical branches of the anterior and posterior cerebral arteries and their relations to the main areas of cerebral localization.

fledged middle cerebral artery syndrome. The embolic stroke is characterized by an abrupt onset.

The striate arteries are a common site of cerebral hemorrhages that usually include the basal ganglia and the internal capsule. A typical hemorrhagic stroke has a sudden onset, develops gradually during several hours, and is often fatal. Better control of hypertension has apparently brought about a gratifying reduction of hemorrhagic strokes in many countries.

Anterior Cerebral Artery

The anterior cerebral artery is smaller than the middle cerebral artery. The proximal trunk gives rise to perforating branches to the striatum and the anterior hypothalamus. It then proceeds to the interhemispheric fissure, where it communicates with the opposite anterior cerebral artery through the *anterior communicating artery* (Fig. 4–4). The distal trunk beyond the anterior communicating artery proceeds to the medial side of the frontal lobe toward the genu of the corpus callosum, where it divides into a *callosomarginal* and a *pericallosal trunk* (Fig. 24–3). Their branches supply most of the corpus callosum as well as the medial surfaces of the frontal and parietal lobes, including the leg area in the paracentral lobule.

Vascular Syndrome. The anterior cerebral artery syndrome is characterized primarily by paralysis of the contralateral leg and foot. Behavioral disturbances, e.g., mental confusion and slowness, may also be apparent.

Vertebral–Basilar System

Vertebral Artery

The two vertebral arteries, which carry about one-third of the blood to the brain, represent the first branches of the subclavian arteries. They ascend through the transverse foramina of the cervical vertebrae before entering the skull through the

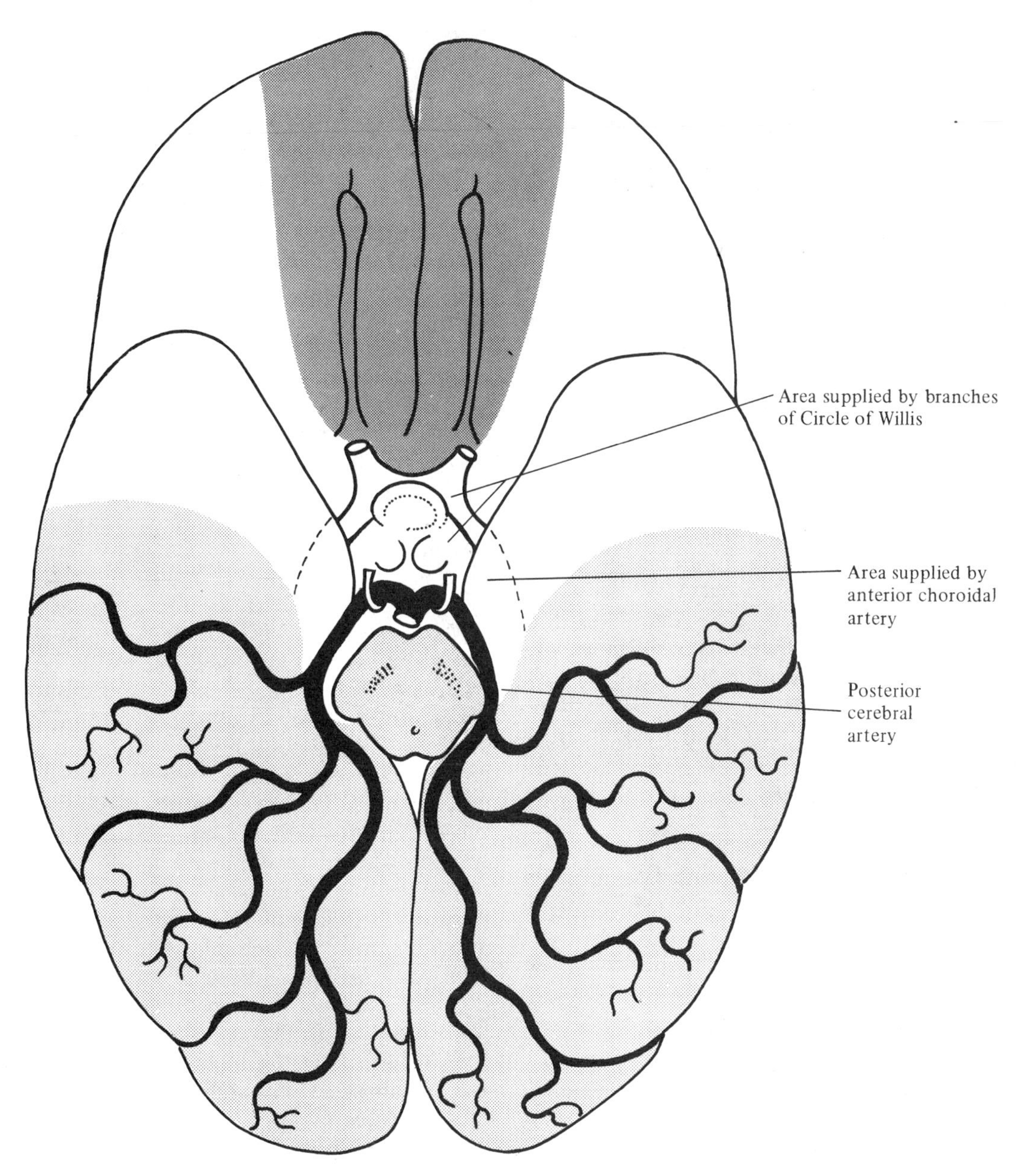

Fig. 24–4. Arterial Blood Supply of the Basal Surface of the Brain
Diagram of the ventral brain surface showing the territories of the main cerebral vessels as well as the branches of the posterior cerebral artery. (*Dark gray* = territory of anterior cerebral artery; *white* = territory of middle cerebral artery.)

foramen magnum. The arteries give off segmental branches both to the intraspinal circulation and to the neck muscles throughout their extracranial course. The muscular branches may be able to serve as anastomotic channels between the external carotid system and the vertebral–basilar system. The vertebral arteries supply the medulla oblongata with small bulbar branches and they give off two larger branches, the *anterior spinal artery* and the *posterior inferior cerebellar artery* (*PICA*) (Fig. 4–1). The PICA is one of the most variable arteries in the vertebral–basilar system. Although it often comes from the vertebral artery and makes a contorted loop on the side of the medulla before it reaches the inferior surface of the cerebellum, it may also have a common origin with the anterior inferior cerebellar artery from the basilar artery.

Vascular Syndromes. The two vertebral arteries are seldom of equal size, and if the larger artery is occluded, the smaller artery may not always be able to compensate for the deficiency. This can be devastating, especially if the collateral circulation is insufficient. The signs following occlusion of the vertebral artery are also characterized by variability. A well-known syndrome, the *lateral medullary syndrome of Wallenberg* (see Clinical Example, Chapter 11) is caused by an infarct in the lateral part of the medulla oblongata (Fig. 11–16). It is characterized by contralateral impairment of pain and temperature in the trunk and extremities (spinothalamic tract), ipsilateral impairment of pain and temperature on the face (descending tract and nucleus of V), ipsilateral Horner's syndrome of miosis, ptosis, and decreased sweating (descending sympathetic fibers), dysphagia and dysarthria (nucleus ambiguus), vertigo, nausea, and nystagmus (vestibular nuclei), and sometimes ipsilateral limb ataxia (inferior cerebellar peduncle). Although the lateral medullary syndrome is most often caused by an occlusion of the vertebral artery, it can also be caused by a lesion in the PICA or the anterior inferior cerebellar artery.

Occlusion of the vertebral artery can sometimes result in a *medial medullary syndrome* with ipsilateral paralysis of the tongue (XII), contralateral paralysis of arm and leg (corticospinal tract), and contralateral impairment of touch and position sense (medial lemniscus).

Basilar Artery

The basilar artery is formed by the union of the two vertebral arteries at the pontomedullary junction. It gives rise to the *anterior inferior cerebellar arteries*, the *internal auditory arteries*, the *pontine arteries*, and the *superior cerebellar arteries* before it bifurcates into the two *posterior cerebral arteries* at the upper border of the pons.

Vascular Syndromes. Since the brain stem contains a large number of structures including the cranial nerve nuclei and long fiber tracts, it is obvious that vascular insufficiency of the basilar artery or one of its many branches may present a variety of different symptoms depending on the site of the occlusion, as well as the efficiency of the collateral circulation.

A total occlusion of the basilar artery may occur suddenly or following TIAs. The full-fledged *basilar artery syndrome* includes disturbance of consciousness, often coma (reticular formation), tetraplegia (corticospinal tracts), impaired sensation (long somatosensory pathways), impaired vision or blindness (visual cortex), disorders of eye movements, facial paralysis and nystagmus (cranial nerve nuclei and their interconnection), and cerebellar ataxia (cerebellar peduncles and cerebellar hemispheres). Death usually follows within a few days. A large pontine hemorrhage is described in Clinical Examples, Chapter 11.

More restricted syndromes occur if the vascular lesion is related to some of the branches from the basilar artery. Occlusion of paramedian arteries involves the corticospinal tract, the medial lemniscus, and the nuclei of cranial nerves III, IV, and VI. Infarcts related to the circumferential arteries usually include parts of the cerebellar peduncles and the cerebellum, the spinothalamic tract, and the nuclei of cranial nerves V, VII, and VIII. The following syndromes may serve as examples:

Medial pontine syndrome. Depending on the dorsal and lateral extent of the lesion, the medial pontine syndrome can include some or all of the following symptoms (compare Fig. 10–2): contralateral paralysis of arm and leg (corticospinal tract), homolateral paralysis of the face (VII), contralateral impairment of touch and position sense (medial lemiscus), and inward deviation of the homolateral eye (VI). There may also be paralysis of conjugate gaze to the side of the lesion (see Eye Movements, Chapter 11). If the lesion is located somewhat more rostral in the pons, the VIth and VIIth cranial nerves may escape involvement.

Lateral pontine syndrome (anterior inferior cerebellar artery). The lateral pontine syndrome often includes impairment of pain and temperature on the contralateral side of the body (spinothalamic tract), impaired facial sensation on the homolateral side (V), vertigo, nystagmus and deafness on the homolateral side (VIII), homolateral facial paralysis (VII), paralysis of conjugate gaze to the side of the lesion (VI), and cerebellar ataxia on the homolateral side (middle cerebellar peduncle and cerebellar hemisphere).

Mediobasal mesencephalic syndrome of Weber (see Clinical Examples, Chapter 11) includes contralateral paralysis of arm and leg (corticospinal tract) and ipsilateral paralysis of the oculomotor nerve, i.e., homolateral ophthalmoplegia with ptosis, dilation of pupil, absent light reflex, and outward deviation of the eye (III). Although Weber's syndrome is often due to an aneurysm in the posterior part of the circle of Willis (see Clinical

Examples, Chapter 11), it can also result from involvement of the paramedian thalamoperforating branches of the basilar or posterior cerebral artery, in which case the red nucleus and its cerebellar connections may be involved, giving rise to crossed tremor and cerebellar ataxia.

Posterior Cerebral Artery

Thalamoperforating branches from the proximal part of this artery enter the posterior perforate substance to reach the medial part of the mesencephalon, the hypothalamus, and the thalamus. Distal to the *posterior communicating* artery, other perforating branches, the *thalamogeniculate arteries*, enter the cerebral peduncle to reach the thalamus and geniculate bodies. The posterior cerebral artery also gives rise to several *posterior choroidal arteries*, which enter the choroid plexus, where they form anastomoses with the anterior choroidal artery (Fig. 4–3). The *cortical branches*, finally, supply the medial sides of the temporal and occipital lobes including the visual cortex (Figs. 24–3 and 24–4).

Vascular Syndrome. An occlusion of the proximal trunk of the posterior cerebral artery damages both the thalamus and the occipital lobe. The involvement of the visual cortex in the occipital lobe results in homonymous hemianopia, whereas the thalamic involvement causes a "*thalamic syndrome*," which consists of hemiparesis, impairment of superficial and deep sensation, spontaneous and agonizing pain, and choreatheoid movements, ataxia, or tremor, all on the side of the lesion.

Normal Angiograms

Cerebral angiography is a relatively safe and very valuable procedure for the diagnosis of various pathologic processes especially related to the cerebrovascular system. It is usually performed by catheterizing the femoral artery. Following cannulation of the femoral artery, a contrast medium can be injected into the appropriate cerebral vessels. Normal routine angiograms are illustrated in Fig. 24–5.

In recent years MRI angiography has developed into a useful screening method for evaluating the intracranial vasculature, although it has not yet replaced conventional angiography.

Clinical Examples

Middle Cerebral Artery Territory Infarct

A 68-year-old left-handed female with coronary artery disease was admitted with crushing chest pain. Electrocardiogram showed an acute anterior myocardial infarction and she was admitted to the Coronary Care Unit.

She improved over the next 3 days but then, suddenly, developed more chest pain and a cardiogram showed extension of her myocardial infarct. Her condition deteriorated quickly with the development of ventricular fibrillation but, fortunately, she was converted back to normal sinus rhythm with electric counter shock.

After recovery from the counter shock, she was able to speak and had no difficulty understanding spoken and written language. It was noted, however, that she had a right homonymous hemianopsia. Her right lower facial muscles were very weak, but she could wrinkle her brow well on the right side. Her tongue deviated slightly to the right when protruded. She had a right hemiparesis with moderate weakness in the arm, but only mild weakness in the leg. She also had impaired stereognosis and graphesthesia in her right upper extremity but could still recognize pain in her right arm, although she had difficulty localizing the site of the stimulus.

An emergency arteriogram was performed. The angiograms from that study (Figs. 24–6A and B) were obtained after selective catheterization of the left internal carotid artery and injection of X-ray dye.

1. What vessel is occluded?
2. Is the lesion demonstrated on the arteriograms appropriate to her neurologic deficits?
3. Why are the weakness and sensory deficits more pronounced in the arm than in the leg?
4. Why is only the lower part of her face on the right side paralyzed?
5. Why was she not aphasic?

Discussion

The patient developed sudden neurologic deficits in the wake of an acute myocardial infarction and cardiac arrhythmia. Since the arteriogram demonstrates occlusion of the left middle cerebral artery (arrow), it is likely that her heart condition resulted in the development of an embolus, which entered the general circulation and lodged itself in the middle cerebral artery. Note the absence of the middle cerebral artery and most of its branches in both the frontal and the lateral projections (Figs. 24–6A and B).

The area of the left hemisphere supplied by the middle cerebral artery includes the motor and sensory areas controlling face, upper extremity, and trunk on the right side of the body (Fig. 24–2).

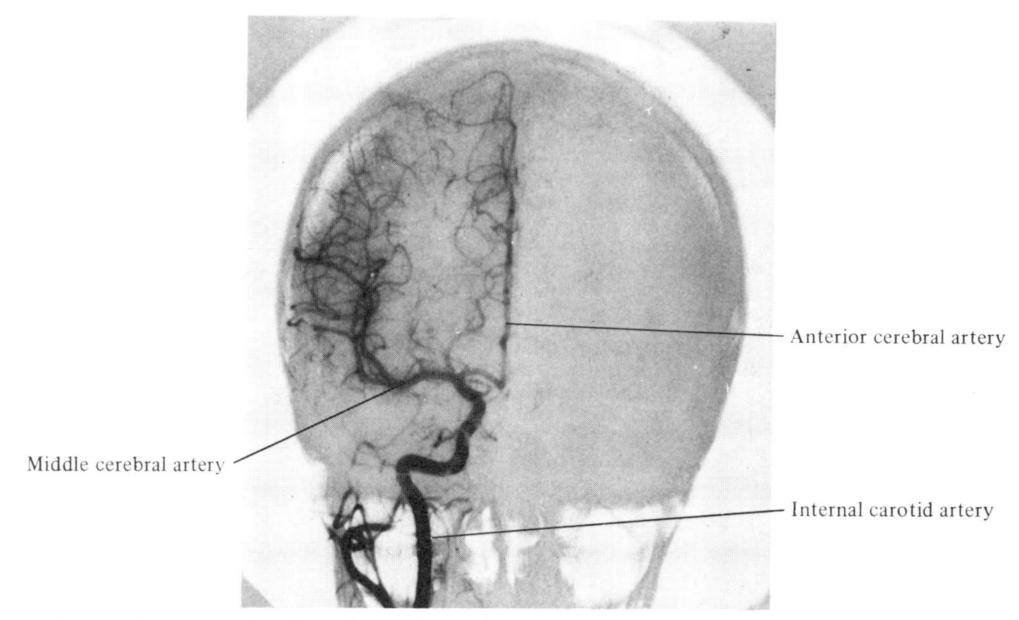

A. Internal carotid artery and its branches in frontal projection

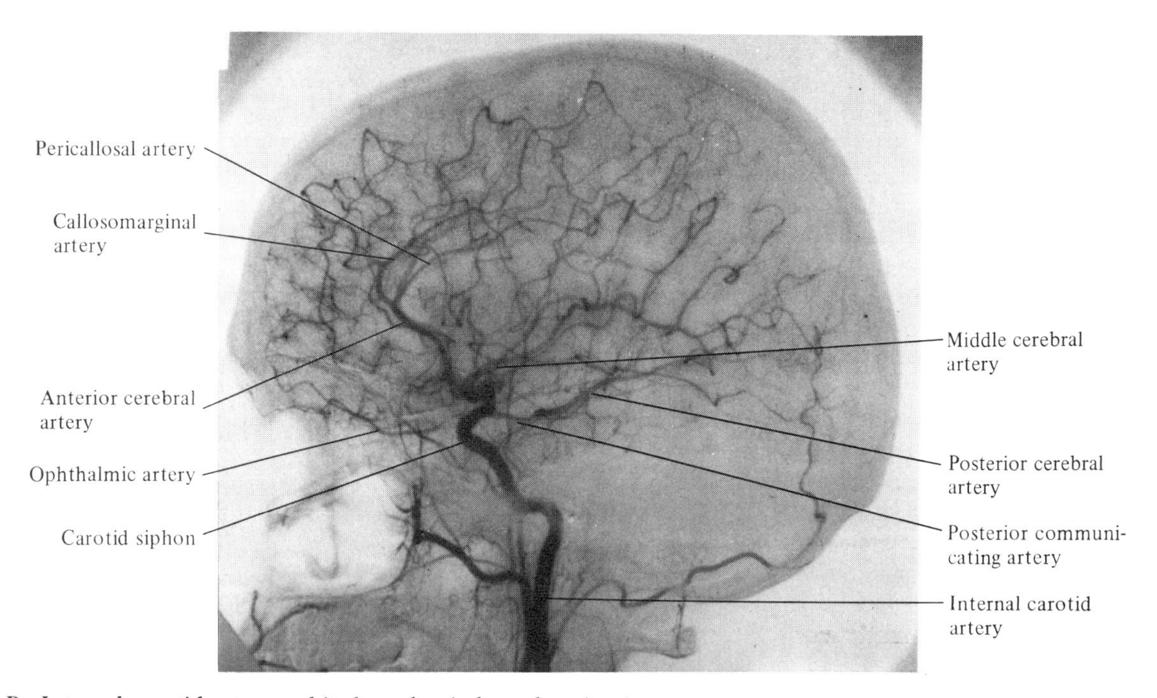

B. Internal carotid artery and its branches in lateral projection

Fig. 24–5. Normal Angiograms

A. Internal carotid artery and its branches in frontal projection.

B. Internal carotid artery and its branches in lateral projection. Note that the posterior communicating artery and the main stem of the posterior cerebral artery have also been filled.

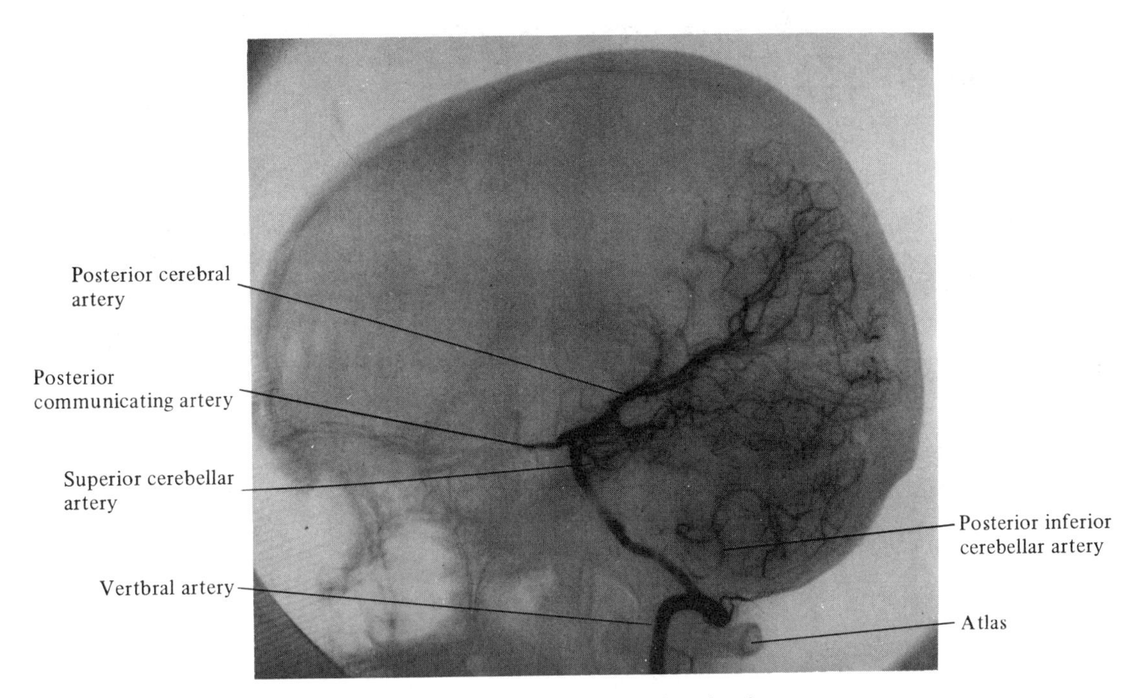

C. Vertebral and basilar arteries and their branches in lateral projection

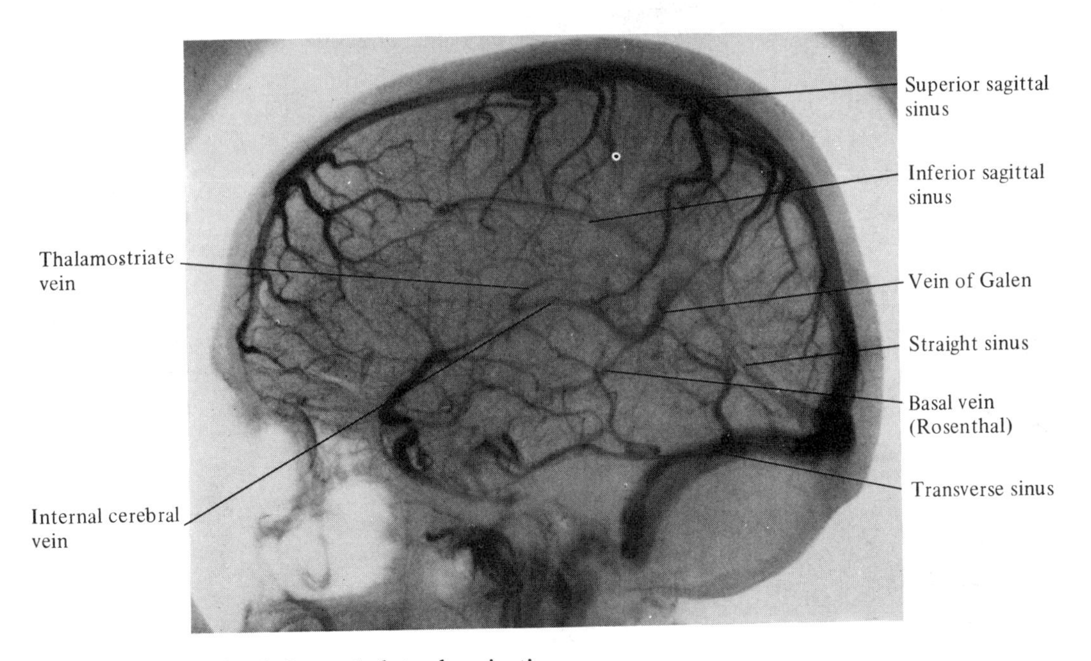

D. Cerebral veins and dural sinuses in lateral projection

C. Vertebral and basilar arteries and their branches in lateral projection.
D. Cerebral veins and dural sinuses in lateral projection.

(**A–D**, Micrograph courtesy of L. Morris. The angiograms were obtained by the subtraction method, which gives an excellent visualization of vessel detail.)

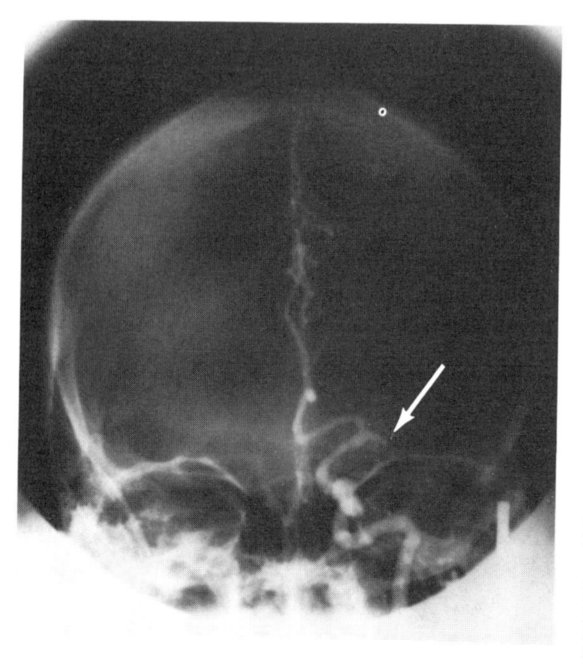

A

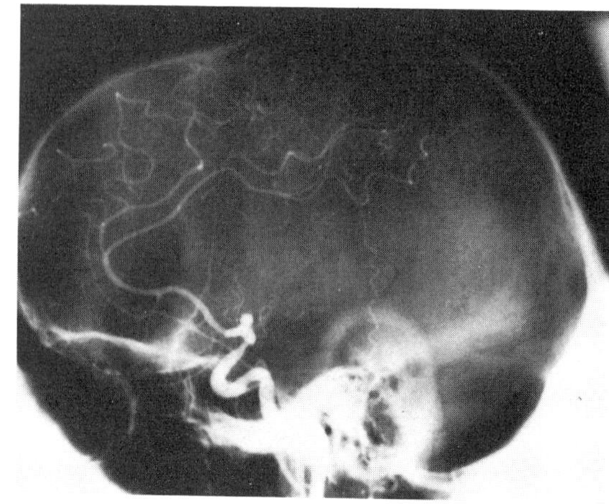

B

Fig. 24–6A and B.
The angiograms were kindly provided by Dr. L. Morris.
A (left). Middle Cerebral Artery Occlusion (Frontal Projection)
B (above). Middle Cerebral Artery Occlusion (Lateral Projection)

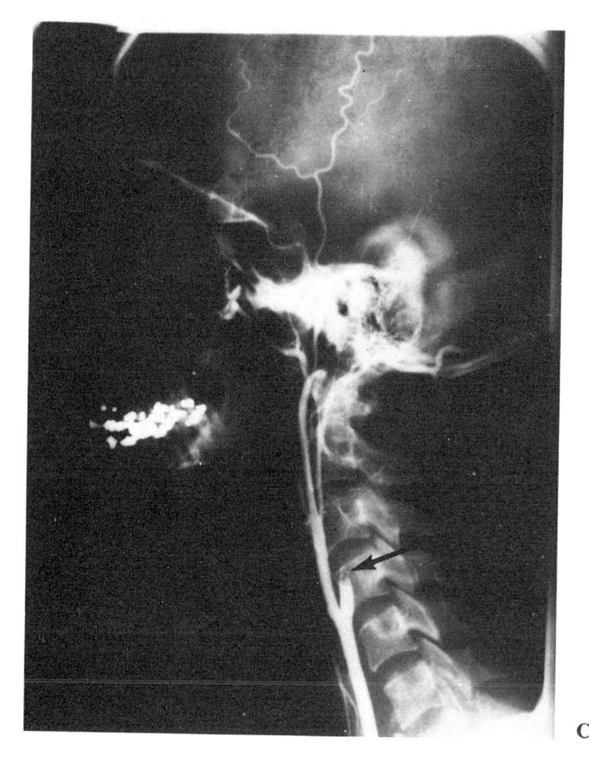

C

Fig. 24–6C. Internal Carotid Artery Occlusion
The angiogram was kindly provided by Dr. L. Morris.

The optic radiation in the left hemisphere is also involved. Motor cortex subserving leg movement is supplied by the anterior cerebral artery (Fig. 24–3) and is preserved in middle cerebral artery occlusions. Furthermore, since the lower limb is less affected than the upper limb, the corticofugal and corticopetal fibers connected with the cortical leg area must be largely intact. In other words, the occlusion of the middle cerebral artery has not deprived the internal capsule of all its blood supply.

The subdivisions of the facial nuclei controlling upper facial muscles receive bilateral corticobulbar innervation in most individuals (Fig. 11–7). Therefore, upper facial muscle function is usually preserved in unilateral cortical infarctions.

The motor and sensory language areas, which are supplied by the middle cerebral artery, are usually located in the left hemisphere (see "Language Disorders", Chapter 22), especially in right-handed people. This patient is left-handed, however, and the right hemisphere is dominant for language in about 30% of left-handed people. Since the patient did not become aphasic, her right hemisphere is apparently dominant in regard to language.

Internal Carotid Artery Occlusion

A middle-aged right-handed male was well until the evening of admission when he suddenly found himself unable to speak. Nor could he walk because of profound right-sided weakness.

On examination, he was alert but could not speak and was unable to understand requests. He had a right homonymous hemianopsia, and he could not look to the right. Lateral gaze to the left, however, was normal. There was weakness of the right lower facial muscles, and a profound right-sided hemiparesis involving arm and leg equally. Muscle tone and stretch reflexes were reduced on the right. Within half an hour he slowly began to get drowsier and he was comatose within 3 hours.

An arteriogram (Fig. 24–6C), which was obtained after injection of dye into the left common carotid artery, shows complete occlusion of the left internal carotid artery 2 cm distal to its origin (arrow).

1. Why was the patient aphasic?
2. Why did he have a right homonymous hemianopsia?
3. Why was the hemiparesis of equal degree in his arm and leg?
4. Why was he unable to look voluntarily to the right?
5. Why did he gradually become comatose?

Discussion

Sudden occlusion of the internal carotid artery usually occurs as a result of extensive atherosclerosis. The occlusion is usually followed by rapid infarction of brain tissue. The size of the infarction is a function of the amount of collateral circulation that can be provided from the left posterior cerebral artery and from flow across the circle of Willis from the vertebral basilar system or the internal carotid artery of the opposite side (see Fig. 4–1).

The aphasia arose because of infarction of cortical language areas in the left hemisphere (Fig. 22–9). The right homonymous hemianopsia most likely appeared because of damage to the left optic radiation, which is normally supplied by the middle cerebral artery. Equal weakness in arm and leg suggests that the infarct included the distribution areas of both the middle cerebral artery (arm and face) and the anterior cerebral artery (leg) (see Figs. 24–2 and 24–3). Infarction of the left frontal eye field, which is supplied by rostral branches of the middle cerebral artery, caused inability to voluntarily gaze to the right.

Symptoms following occlusion of the internal carotid artery are characterized by great variability, which is a product of the availability of collateral circulation. The clinical picture often resembles that of a middle cerebral artery occlusion (see above), especially if the anterior communicating artery is able to provide blood supply to the anterior cerebral artery on the affected side. In this patient there was little sign of collateral circulation, and the occlusion caused a massive infarct involving the anterior two-thirds of the hemisphere. The large size of the infarct and concomitant hemispheral swelling made the patient comatose and he died a few days later.

Anterior Spinal Artery Syndrome

A 70-year-old man with known hypertension and atherosclerosis was awakened in the middle of the night with excruciating upper abdominal pain that radiated into his back. Within minutes his legs became paralyzed and numb.

On examination his blood pressure was 200/120. A loud bruit was heard over his abdominal aorta and his femoral pulses were reduced in intensity. Cranial nerves and upper extremity functions were normal, but there was flaccid paraplegia of the lower extremities with reduced muscle tone and absent stretch reflexes. Pain and temperature sensations were completely lost below the level of T8 bilaterally, whereas vibration and position sense were intact. An emergency abdominal aortogram demonstrated a dissecting aneurysm of his upper abdominal aorta, and the *radicular artery of Adamkiewicz* was apparently occluded.

His blood pressure was carefully lowered and he underwent successful emergency repair of his aneurysm. Examination 10 days later showed some power in his legs, but his right leg was much weaker than the left. Muscle tone and stretch reflexes were now increased in both legs and he had bilateral Babinski signs. Touch, position, and vibratory sensations in his legs remained intact. He could also feel pain and temperature in his right leg, but not in the left leg or lower trunk below the level of the T10 dermatome.

1. Where is the initial lesion in the CNS?
2. Why were touch, position, and vibratory sensations in his legs preserved?
3. Why were his upper extremity motor and sensory functions preserved?
4. As he recovered, his deficits became asymmetric in the sense that his right leg was weaker than his left leg but certain sensory functions were more impaired in his left side compared to his right side. How do you explain this discrepancy?

Discussion

The patient suffered from an acutely dissecting aneurysm of his aorta with occlusion of the artery of Adamkiewicz (radicular artery of the lumbar enlargement), which is the major supplier of blood to the anterior spinal artery in the lower thoracic and lumbosacral part of the spinal cord. This resulted in infarction of the anterior two-thirds of the spinal cord below the midthoracic level. The posterior one-third of the spinal cord, which is supplied from intracranial and multiple extracranial branches, is usually much less affected by sudden occlusion of any single vessel.

Touch, position, and vibratory sensations in the lower extremities were preserved because the posterior columns and dorsolateral fasciculi, which are nourished by the posterior spinal arteries, had not been damaged (see Fig. 9–5). His upper extremity functions were preserved because the part of the anterior spinal artery that supplies the cervical and upper thoracic region receives blood from the vertebral arteries or from branches of the subclavian arteries. Motor paralysis and dissociated sensory loss below the level of the lesion are typical for an anterior spinal artery syndrome.

As spinal cord infarctions resolve, asymmetrical deficits commonly arise. In this case the spinothalamic and corticospinal tracts appear to be more damaged on the right than on the left side.

Suggested Reading

1. Adams, R. D. and M. Victor, 1993. Principles of Neurology, 5th ed. McGraw-Hill: New York, pp. 669–748.
2. Brodal, A., 1973. Self-observations and neuroanatomical considerations after a stroke. Brain 96:675–694.
3. Kirkwood, J. R., 1990. Essentials of Neuroimaging. Churchill Livingstone: New York, pp. 1–133.
4. Phillis, J. W., 1993. The Regulation of Cerebral Blood Flow. CRC Press: Boca Raton.
5. Poirier, J., F. Gray, and R. Escourolle, 1990. Manual of Basic Neuropathy, 3rd ed. W. B. Saunders Co.: Philadelphia, 1990.

Appendix

Peripheral Nerves

Figures A.2–A.16 are reproduced from Medical Research Council, Memorandum No. 45, Aids to the Examination of the Peripheral Nervous System. With permission of Her Majesty's Stationery Office, London.

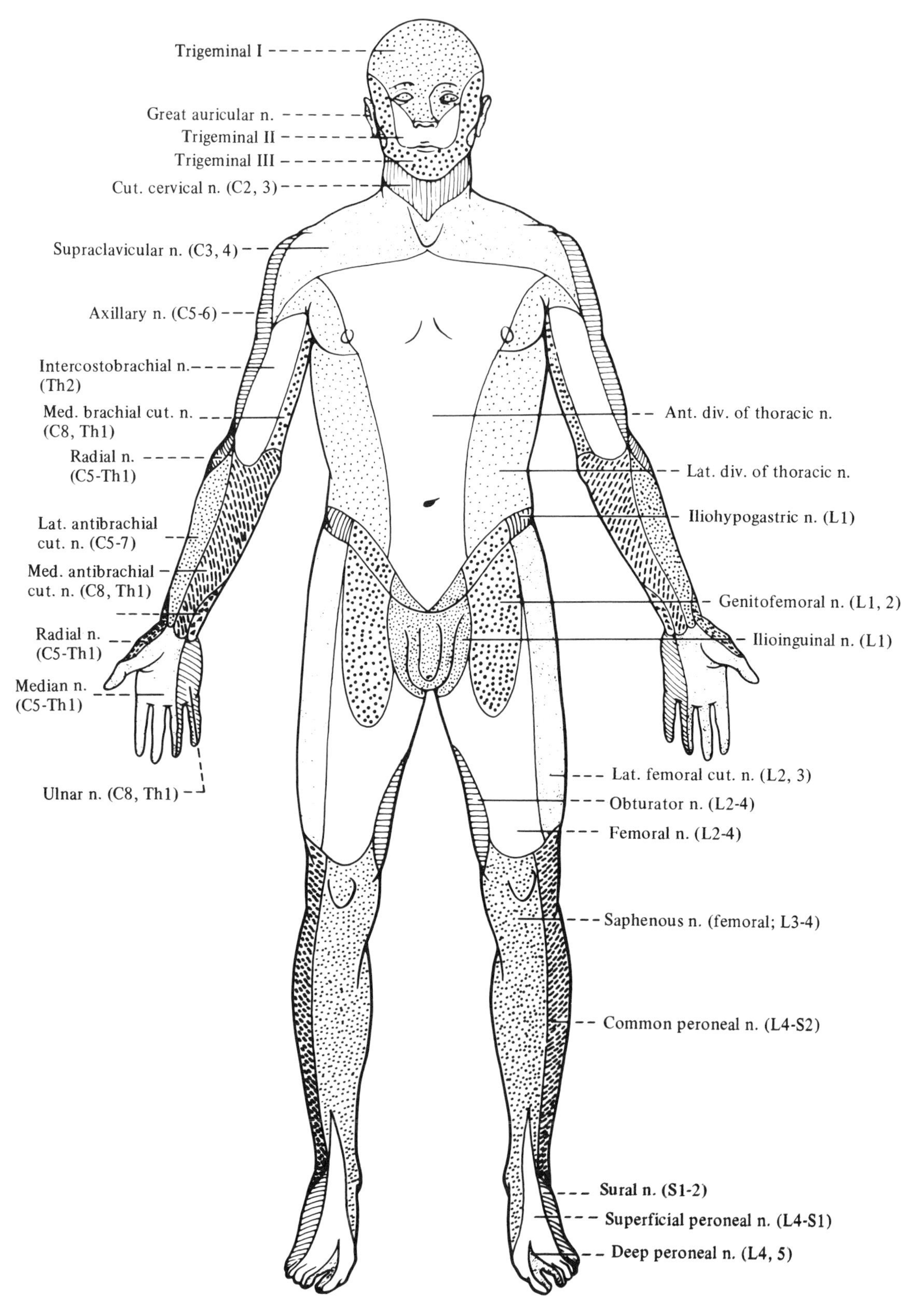

Fig. A–1A. The Cutaneous Distribution of the Peripheral Nerves
A. On the anterior aspect of the body.

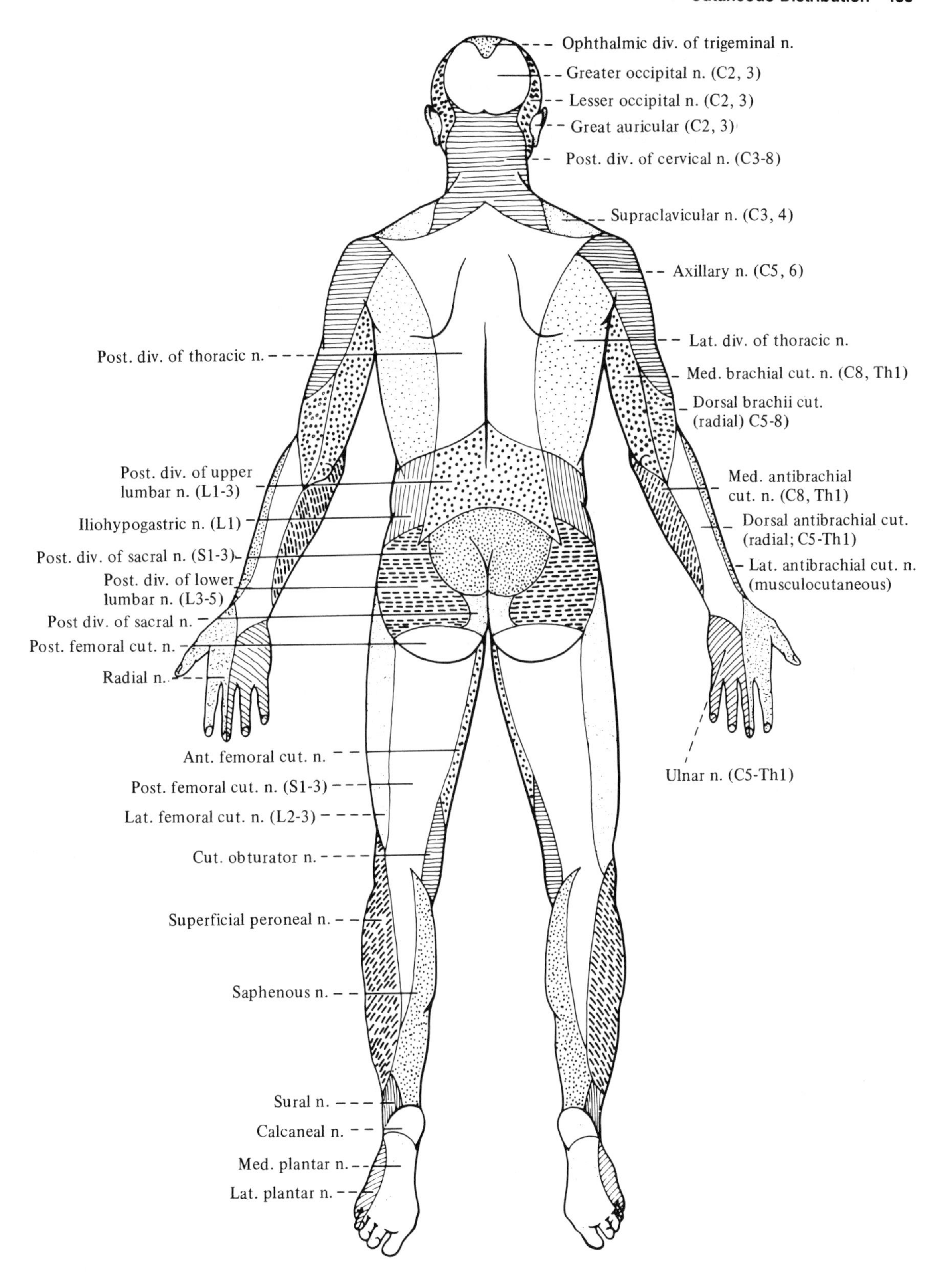

Fig. A–1B.
On the posterior aspect of the body. (A & B from R. de Jong, 1979. The Neurologic Examination. JB Lippincott, Philadelphia. With permission of the publisher.)

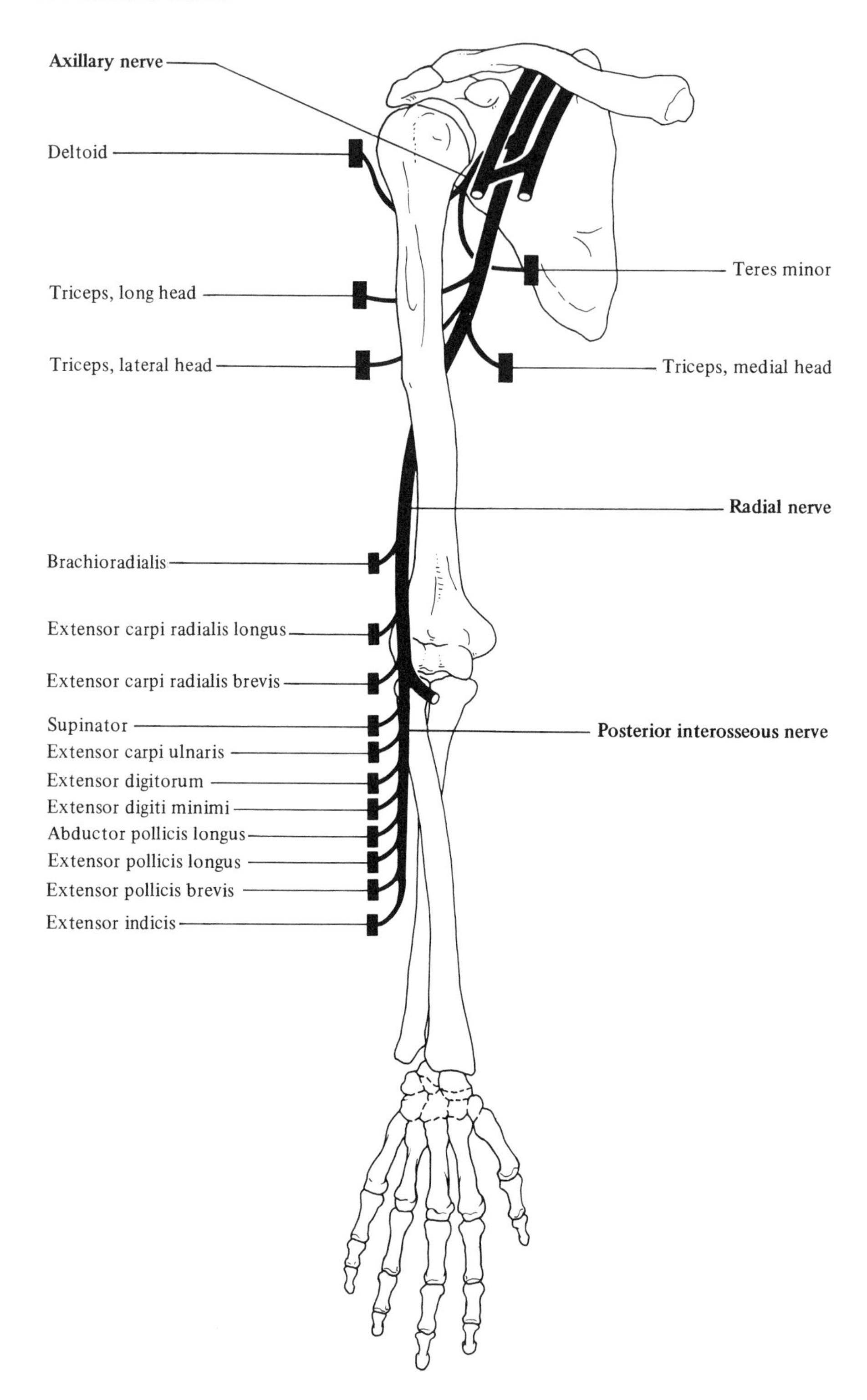

Fig. A–2.
Diagram of the axillary and radial nerves and the muscles which they supply. (Modified from Pitres and Testut.)

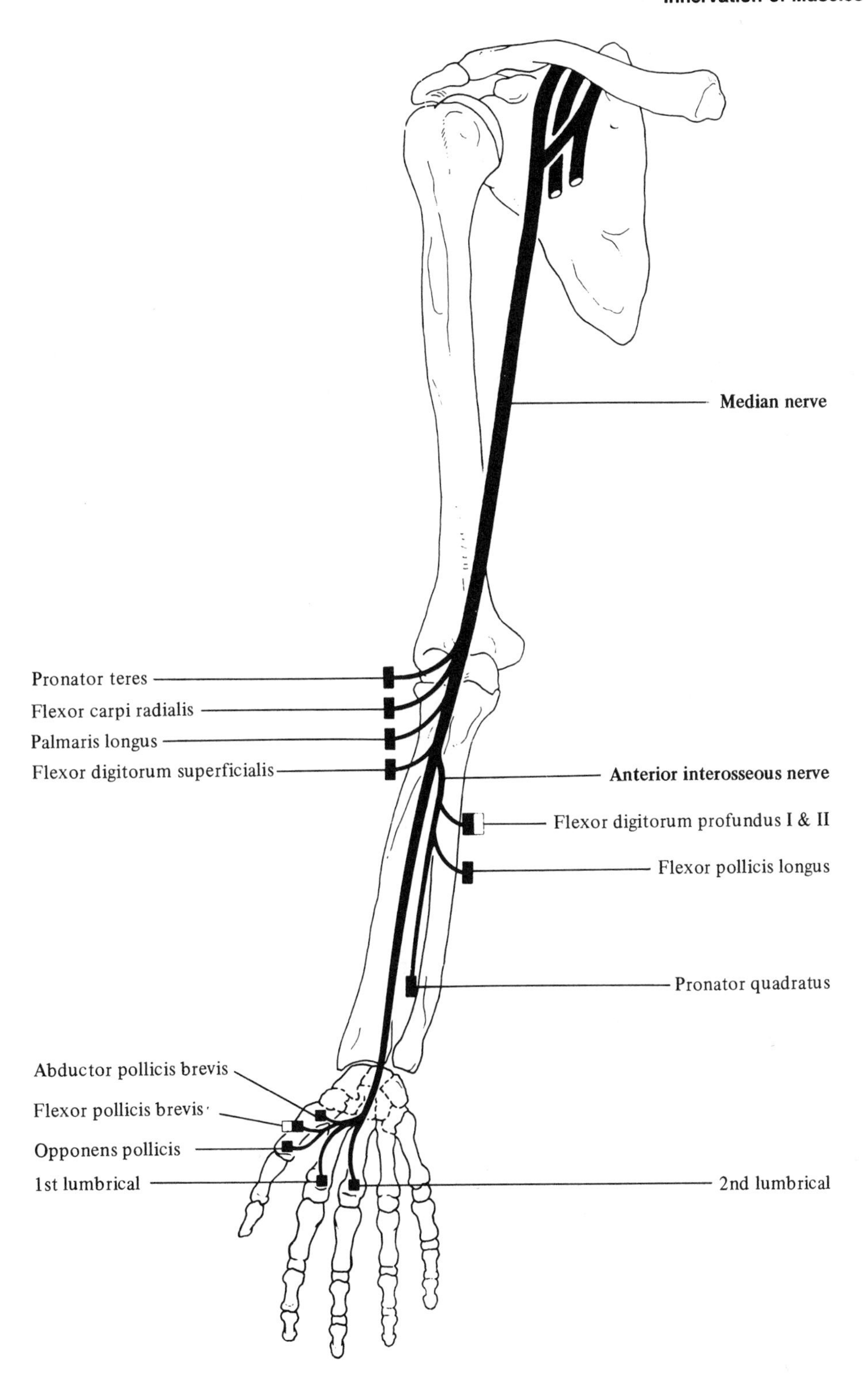

Fig. A–3.
Diagram of the median nerve and the muscles which it supplies. (Modified from Pitres and Testut.) Note: the white rectangle signifies that the muscle indicated receives a part of its nerve supply from another peripheral nerve.

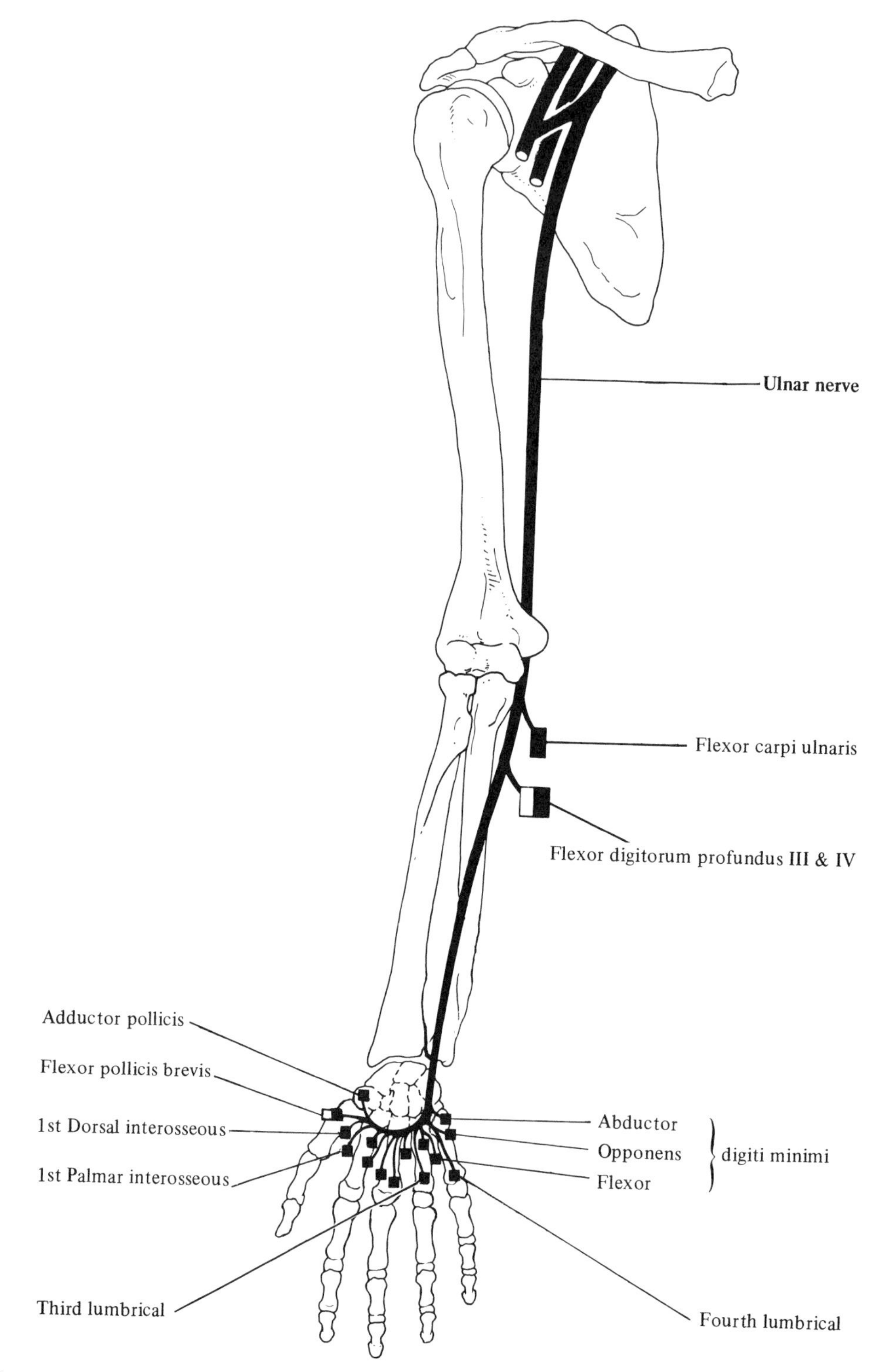

Fig. A–4.
Diagram of the ulnar nerve and the muscles which it supplies. (Modified from Pitres and Testut.)

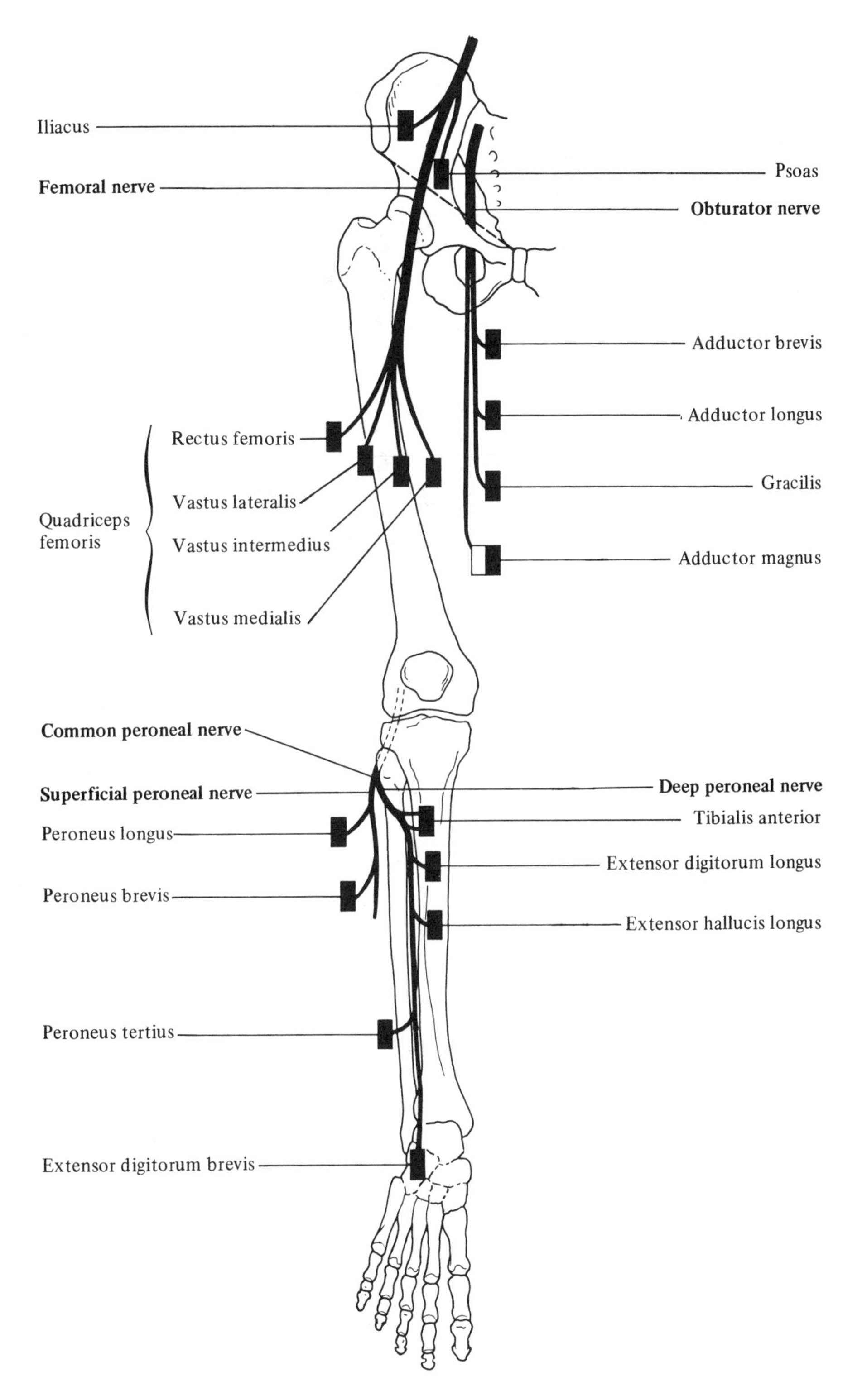

Fig. A–5.
Diagram of the nerves on the anterior aspect of the lower limb, and the muscles which they supply. (Modified from Pitres and Testut.)

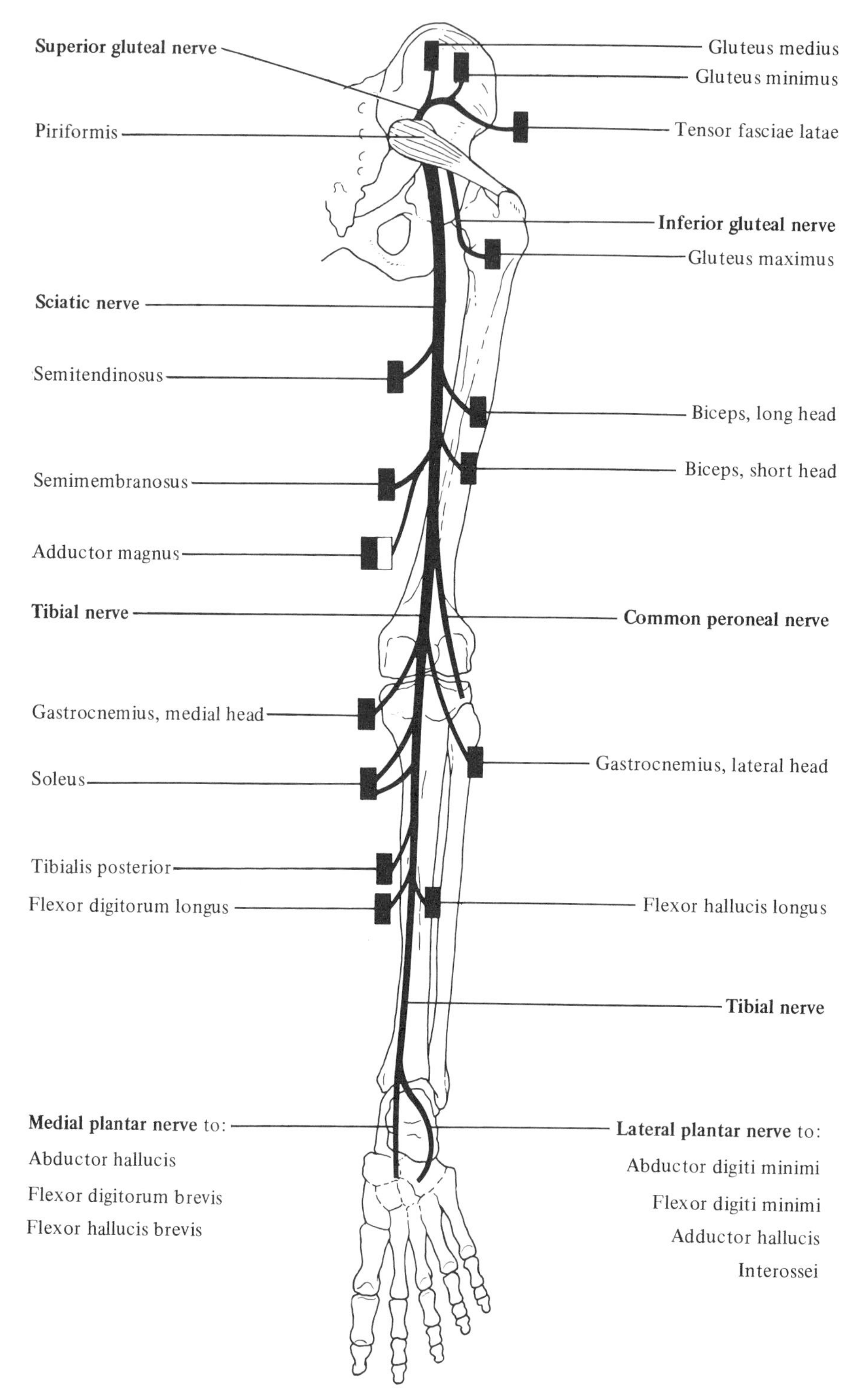

Fig. A–6.
Diagram of the nerves on the posterior aspect of the lower limb, and the muscles which they supply. (Modified from Pitres and Testut.)

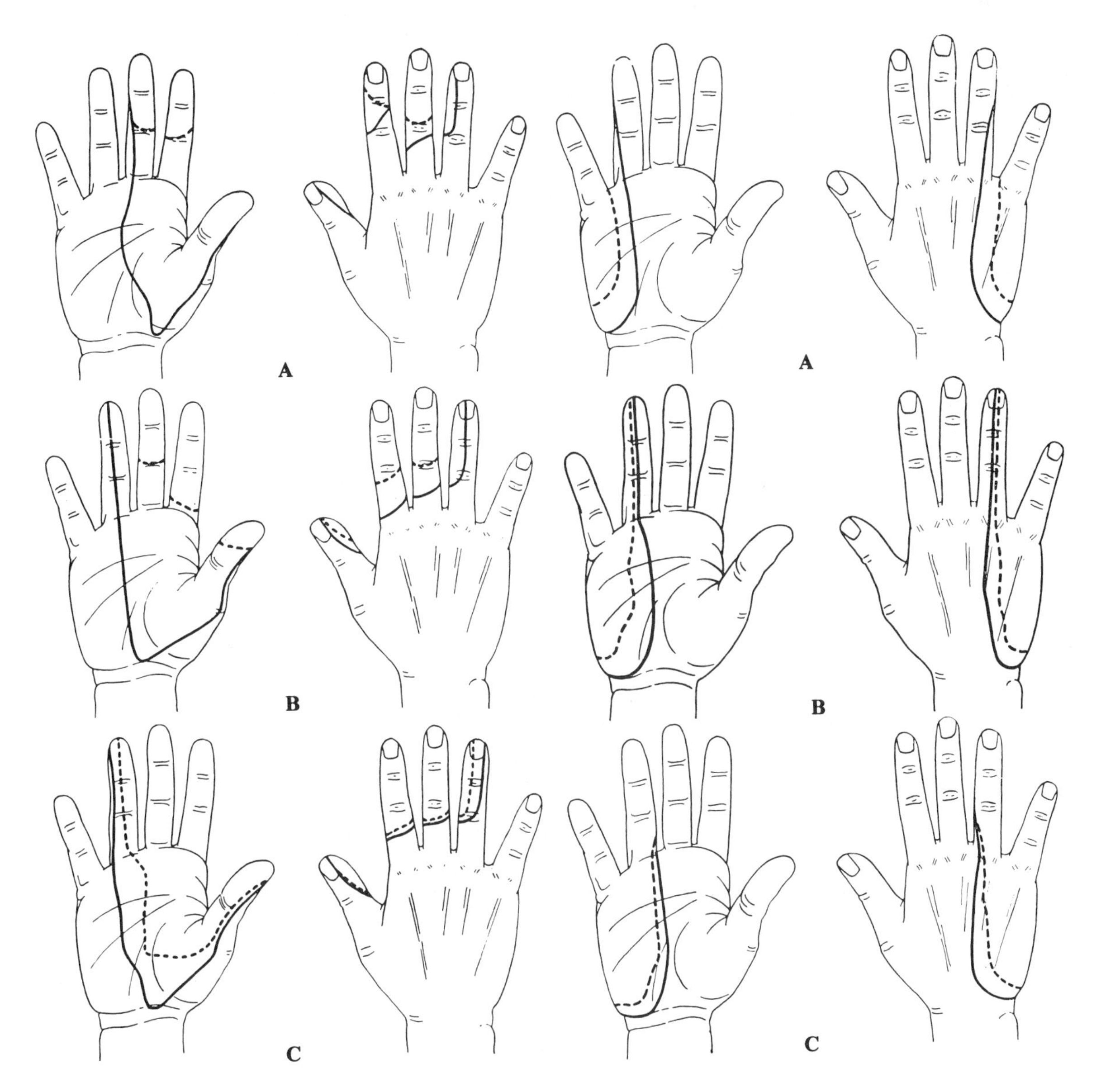

Fig. A–7.
The approximate areas within which sensory changes may be found in lesions of the median nerve. **A** small area, **B** average area, and **C** large area. Light touch, continuous line; pin-prick, dotted line. Immediately after complete division of the median nerve, the insensitive area on the palm of the hand may be somewhat larger than is shown. (Modified from Head and Sherren (1905) *Brain*: 28, 116.)

Fig. A–8.
The approximate areas within which sensory changes may be found in lesions of the ulnar nerve. **A** small area, **B** average area, and **C** large area. Light touch, continuous line; pin-prick, dotted line. Immediately after complete division of the ulnar nerve, the insensitive area on the palm of the hand may be somewhat larger than is shown. (Modified from Head and Sherren.)

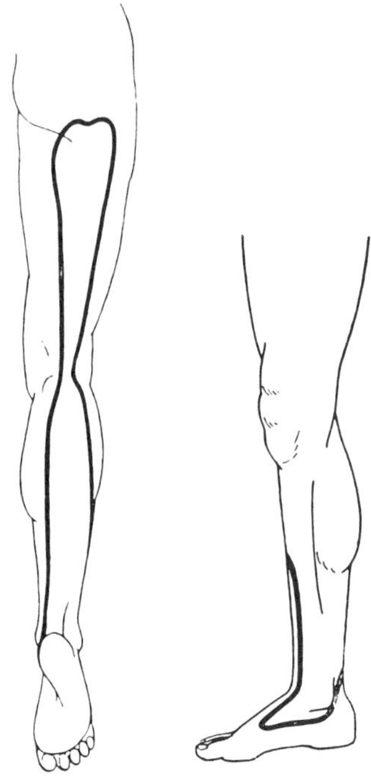

Fig. A-9.
The approximate area within which sensory changes may be found in lesions of both the sciatic and the posterior cutaneous nerve of the thigh.

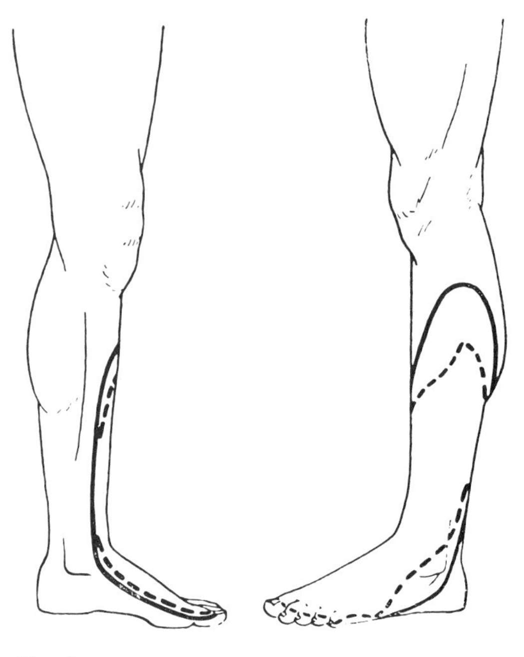

Fig. A-10.
The approximate area within which sensory changes may be found in lesions of the common peroneal nerve above the origin of the superficial peroneal nerve. Light touch, continuous line; pin-prick, dotted line. (Modified from M. R. C. Special Report No. 54, 1920.)

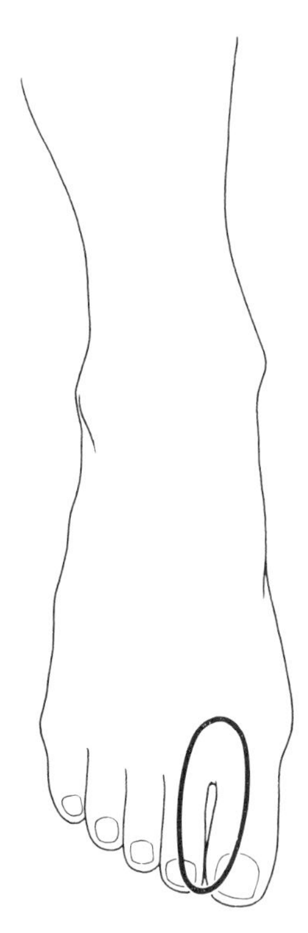

Fig. A-11.
The approximate area within which sensory changes may be found in lesions of the deep peroneal nerve.

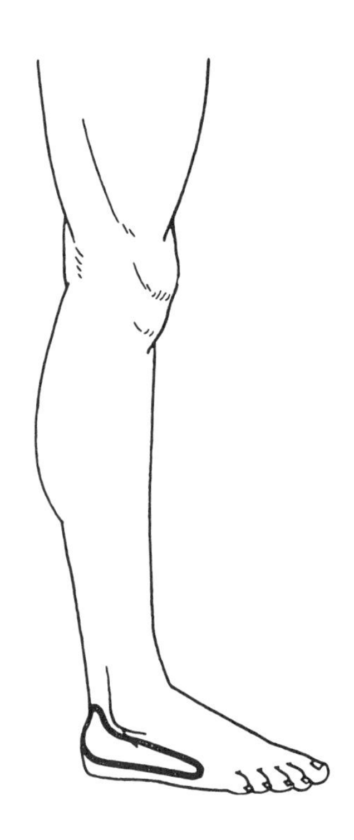

Fig. A-12.
The approximate area within which sensory changes may be found in lesions of the sural nerve.

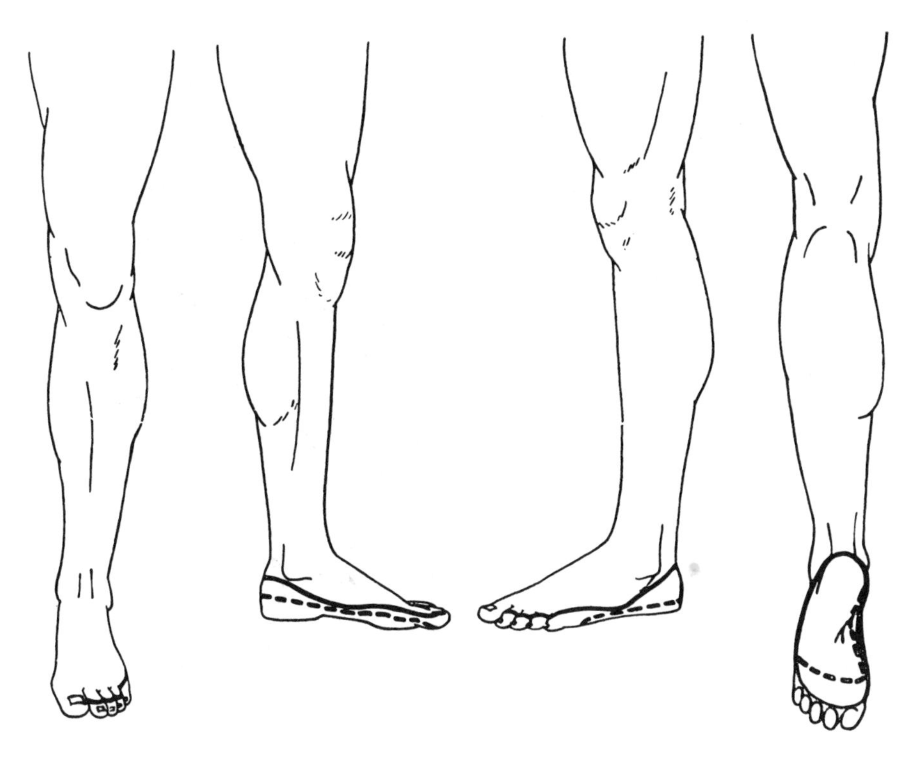

Fig. A–13.
The approximate area within which sensory changes may be found in lesions of the tibial nerve. Light touch, continuous line; pin-prick, dotted line. (Modified from M. R. C. Special Report No. 54, 1920.)

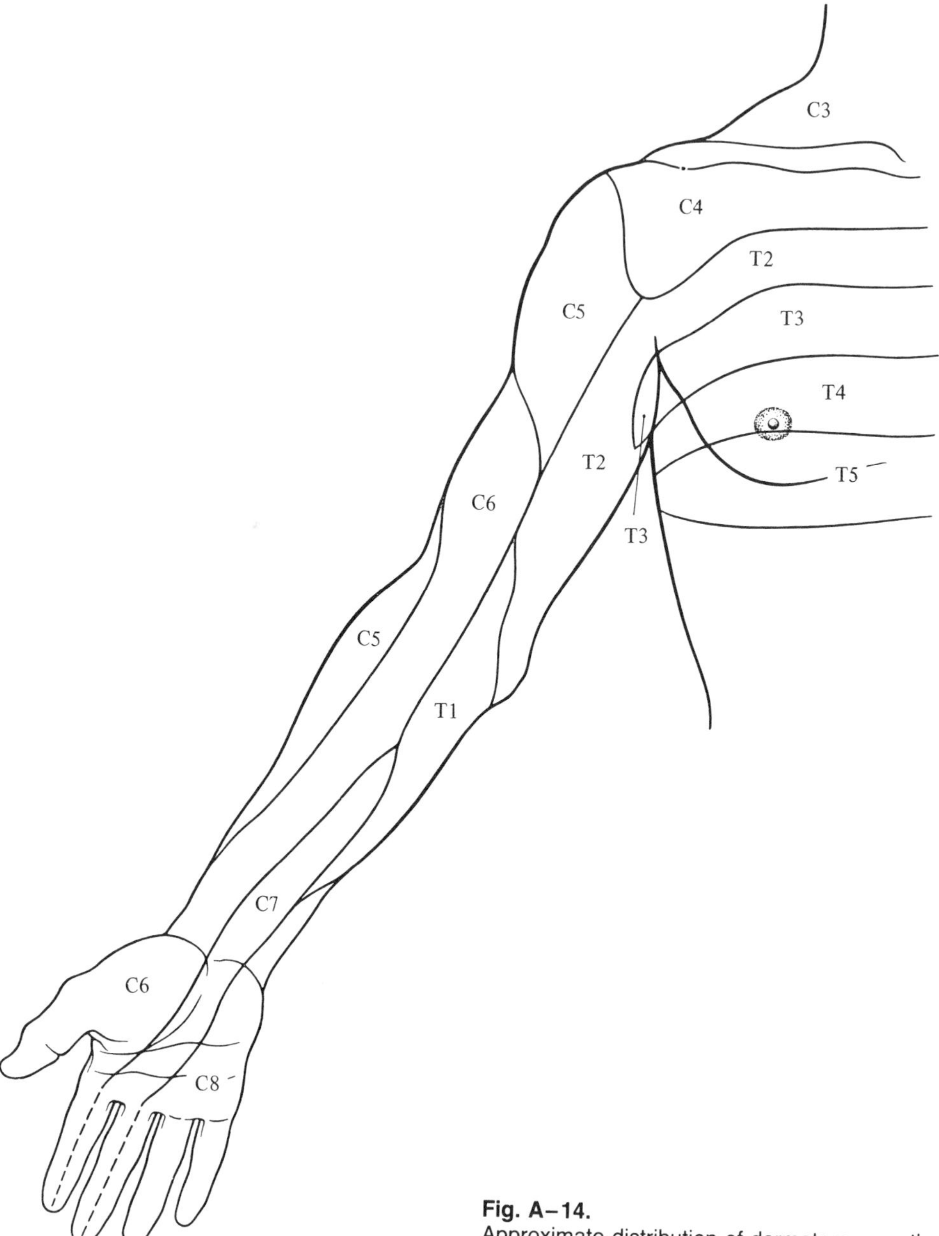

Fig. A–14.
Approximate distribution of dermatomes on the anterior aspect of the upper limb.

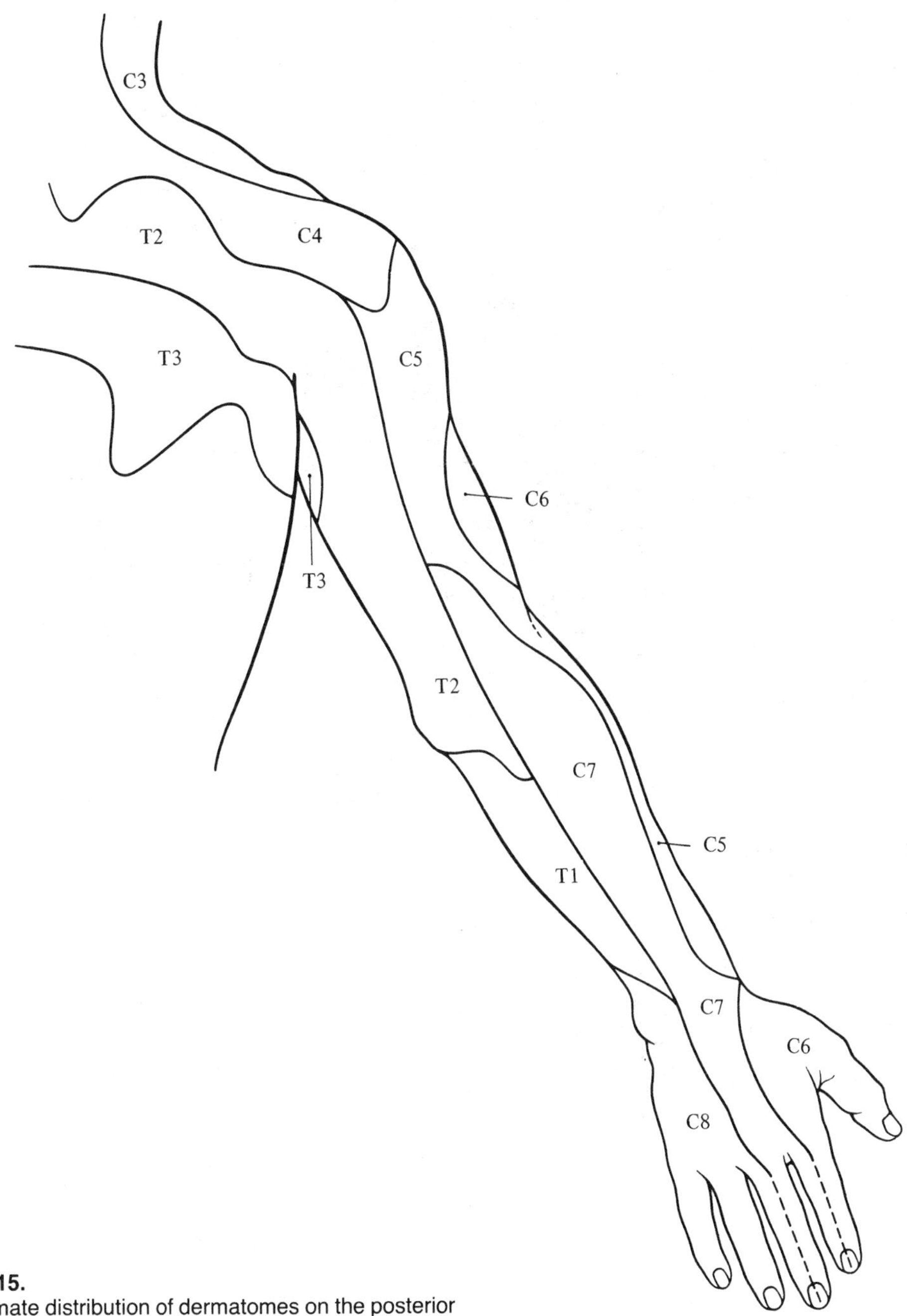

Fig. A–15.
Approximate distribution of dermatomes on the posterior aspect of the upper limb.

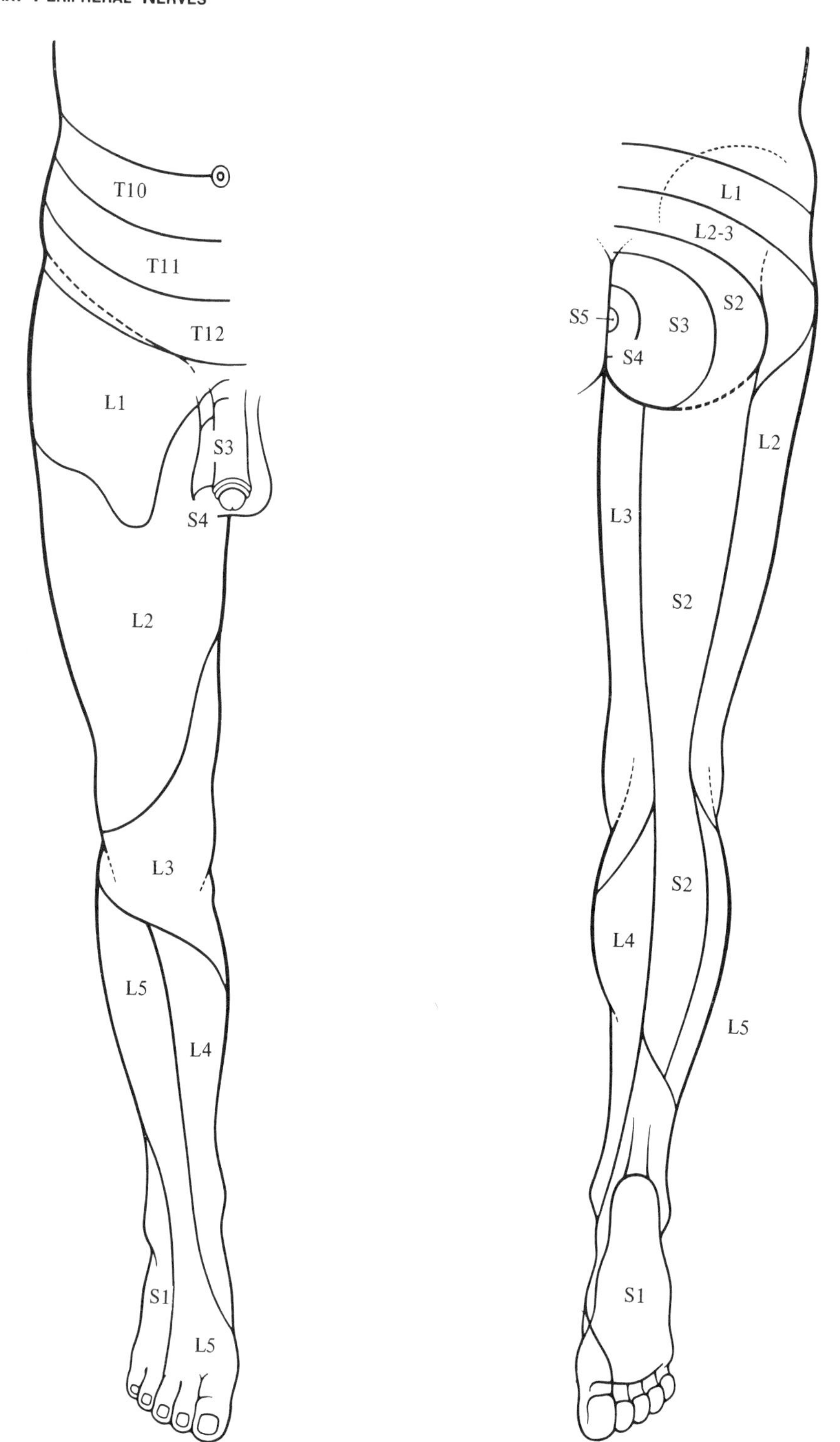

Fig. A–16.
Approximate distribution of dermatomes on the lower limb.

General References in Neuroanatomy and Related Fields

Adams, R. D. and M. Victor, 1993. Principles of Neurology, 5th ed. McGraw-Hill: New York.

Asbury, A. K., G. M. McKhann, and W. I. McDonald, 1992. Diseases of the Nervous System. Clinical Neurology, 2nd ed., Vols. 1 and 2. W. B. Saunders Company: Philadelphia.

Barr, M. L. and J. A. Kiernan, 1988. The Human Nervous System. An Anatomical Viewpoint, 5th ed. J. B. Lippincott Co.: Philadelphia.

Brodal, A., 1981. Neurological Anatomy in Relation to Clinical Medicine. Oxford University Press: New York.

Brodal, P., 1992. The Central Nervous System. Oxford University Press: New York.

Burt, A. M., 1993. Textbook of Neuroanatomy. W. B. Saunders Co.: Philadelphia.

Carpenter, M. B., 1991. Core Text of Neuroanatomy, 4th ed. Williams and Wilkins: Baltimore.

Daube, J. R., 1986. Medical Neurosciences: An Approach to Anatomy, Pathology, and Physiology by Systems and Levels, 2nd ed. Little, Brown, and Company: Boston.

de Groot, J., 1991. Correlative Neuroanatomy, 21st ed. Appleton and Lange: Norwalk, Connecticut.

Duus, P. and R. Lindenberg, 1989. Topical Diagnosis in Neurology, 2nd ed. Thieme: New York.

Finger, S., 1994. Origins of Neuroscience: A History of Explorations into Brain Function. Oxford University Press: New York.

Gilman, S. and S. Winans-Newman, 1992. Manter and Gatz's Essentials of Clinical Neuroanatomy and Neurophysiology, 8th ed. F. A. Davis Co.: Philadelphia.

Groves, P. M. and G. V. Rebec, 1991. Introduction to Biological Psychology, 4th ed. William C. Brown Publisher: Dubugue, Indiana.

Guyton, A. C., 1987. Basic Neuroscience. Anatomy and Physiology. W. B. Saunders Co.: Philadelphia.

Haines, D. E., 1985. Correlative Neuroanatomy. The Anatomical Basis for some Common Neurologic Deficits. Urban & Schwarzenberg: Baltimore.

Haines, D. E., 1991. Neuroanatomy: An Atlas of Structures, Sections and Systems. Urban & Schwarzenberg: Baltimore.

Hall, Z. W., 1991. An Introduction to Molecular Biology. Sinauer Associates: Sunderland, Massachusetts.

Haymaker, W., 1969. Bing's Local Diagnosis in Neurological Diseases, 15th ed. C. V. Mosby Company: St. Louis.

Heilman, K. M. and E. Valenstein, 1993. Clinical Neuropsychology, 3rd ed. Oxford University Press: New York.

Jones, E.G., 1985. The Thalamus. Plenum Press: New York.

Kandel, E. R., J. H. Schwartz, and T. M. Jessell, 1991. Principles of Neural Science, 3rd ed. Elsevier: New York.

Kirkwood, J. R., 1990. Essentials of Neuroimaging. Churchill Livingstone: New York.

Lockard, I., 1991. Desk Reference for Neuroscience, 2nd ed. Springer-Verlag: New York.

Martin, J. H., 1989. Neuroanatomy. Text and Atlas. Elsevier: New York.

Mesulam, M.-M., 1985. Principles of Behavioral Neurology. F. A. Davis Co.: Philadelphia.

Netter, F. H., 1983. The Ciba Collection of Medical Illustrations. Vol. 1: Nervous System. Part 1: Anatomy and Physiology. CIBA Pharmaceutical Co.: West Caldwell, New Jersey.

Nicholls, J. G. and A. R. Martin, 1992. From Neuron to Brain: A Cellular Approach to the Function of the Nervous System, 3rd ed. Sinauer Associates: Sunderland, Massachusetts.

Nieuwenhuys, R., 1985. Chemoarchitecture of the Brain. Springer-Verlag: New York.

Nieuwenhuys, R., J. Voogd, and C. van Huijzen, 1988. The Human Central Nervous System. A Synopsis and Atlas, 3rd ed. Springer-Verlag: New York.

Noback, C. R., N. L. Strominger, and R. J. Demarest, 1991. The Human Nervous System, 4th ed. Lea and Febiger: Philadelphia.

Nolte, J., 1993. The Human Brain: An Introduction to Its Functional Anatomy, 3rd ed. Mosby-Year Book: St. Louis.

Okazaki, H., 1989. Fundamentals of Neuropathology. Morphologic Basis of Neurologic Disorders, 2nd ed. Igaku-Shoin Medical Publishers: New York.

Paxinos, G. (ed.), 1990. The Human Nervous System. Academic Press: San Diego.

Pearlman, A. L. and R. C. Collins (eds.), 1989. Neurobiology of Disease. Oxford University Press: New York.

Peters, A., S. L. Palay, and H. deF. Webster, 1991. The Fine Structure of the Nervous System, 3rd ed. Oxford University Press: New York.

Poirier, J., F. Gray, and R. Escourolle, 1990. Manual of Basic Neuropathy, 3rd ed. W. B. Saunders Co.: Philadelphia.

Roland, P. E., 1993. Brain Activation. Wiley-Liss: New York.

Romero-Sierra, C., 1986. Neuroanatomy: A Conceptual Approach. Churchill Livingstone: New York.

Sartor, K., 1992. MR Imaging of the Skull and Brain. A Correlative Text-Atlas. Springer-Verlag: Berlin, pp. 53–115.

Schmidt, R. F., 1985. Fundamentals of Neurophysiology, 3rd ed. Springer-Verlag: New York.

Shepherd, G. M., 1990. The Synaptic Organization of the Brain, 3rd ed. Oxford University Press: New York.

Shepherd, G. M., 1994. Neurobiology, 3rd ed. Oxford University Press: New York.

Snell, R. S., 1992. Clinical Neuroanatomy for Medical Students, 3rd ed. Little, Brown and Company: Boston.

Index